代数符号

| Notations | Definition | Examples | Explanation |
| --- | --- | --- | --- |
| $\equiv$ | $A \equiv B$,<br><br>indicates that A is defined<br>by B | $f(x) \equiv x^2$ | $f(x)$ is defined by $x^2$. |
| $:=$ | $A := B$,<br><br>indicates that A is defined<br>by B | $f(x) := x^3$ | $f(x)$ is defined by $x^3$. |
| $\infty$ | infinity | $\lim_{n \to \infty} n^2 = \infty$ | When $n$ approaches infinity,<br>$n^2$ also approaches infinity. |
| $\lfloor x \rfloor$ | $\lfloor x \rfloor$ is the greatest integer<br>that is less than or equal to<br>$x$ | $\lfloor 2.3 \rfloor = 2$,<br>$\lfloor -2.3 \rfloor = -3$ | 2 is the greatest integer that is<br>less than 2.3.<br>$-3$ is the greatest integer that is<br>less than $-2.3$. |
| $[x]$ | $[x]$ is the greatest integer<br>that is less than or equal to<br>$x$ | $[2.3] = 2$,<br>$[-2.3] = -3$ | 2 is the greatest integer that is<br>less than 2.3.<br>$-3$ is the greatest integer that is<br>less than $-2.3$. |
| $\sum$ | Summation(Sigma) | $\sum_{n=1}^{10} n = 55$ | $1 + 2 + \cdots + 10 = 55$ |
| $\prod$ | continued multiplication<br>(Pi) | $\prod_{n=1}^{5} n = 120$ | $1 \times 2 \ldots \times 5 = 120$ |
| $e$ | $e := \sum_{n=0}^{\infty} \frac{1}{n!}$,<br><br>(Euler's number) | $e^2 \cdot e^3 = e^5$ | Based on Exponential Law,<br>$e^2 \cdot e^3 = e^5$. |
| $\pi$ | $\pi = 3.1415926\ldots$ (Pi,<br>Ratio of the circumference<br>of a circle to its diameter) | $\sin\frac{\pi}{2} = 1$ | When the angle of $\theta$ equals $\frac{\pi}{2}$,<br>$\sin\theta$ equals $1$ |

排列组合符号

| Notations | Definition | Examples | Explanation |
|---|---|---|---|
| ! | $n! = 1 \times 2 \times .... \times n$ | $3!$ | $3! = 1 \times 2 \times 3 = 6$ |
| $P_k^n$ | $P_k^n = \dfrac{n!}{(n-k)!}$ | $P_2^5 = \dfrac{5!}{3!}$ | $P_2^5 = \dfrac{5!}{3!} = 5 \times 4 = 20$ |
| $C_k^n$ | $C_k^n = \dfrac{n!}{k!\,(n-k)!}$ | $C_2^5 = \dfrac{5!}{3!\,2!}$ | $C_2^5 = \dfrac{5!}{3!\,2!} = 10$ |

集合符号

| Notations | Definition | Examples | Explanation |
|---|---|---|---|
| { } | set | $\{1,2,3,5\}$ | The set consist of these elements 1,2,3,5. |
| $\cup$ | union | $\{1,3\} \cup \{2,5\} = \{1,2,3,5\}$ | Union of $\{1,3\}$ and $\{2,5\}$ is equal to $\{1,2,3,5\}$. |
| $\cap$ | intersect | $\{1,3,5\} \cap \{1,2,5,7\} = \{1,5\}$ | Intersection of $\{1,3,5\}$ and $\{1,2,5,7\}$ is equal to $\{1,5\}$. |
| $\subset$ | subset | $\{1,5\} \subset \{1,3,5\}$ | $\{1,5\}$ is the subset of $\{1,3,5\}$. |
| $\not\subset$ | not a subset | $\{1,3,5\} \not\subset \{1,5\}$ | Because $3 \notin \{1,5\}$, $\{1,3,5\}$ is not the subset of $\{1,5\}$. |
| $\subseteq$ | subset or equal | $\{1,5\} \subseteq \{1,3,5\}$, $\{1,5\} \subseteq \{1,5\}$, $\{1,3,5\} \not\subseteq \{1,5\}$ | $\{1,5\}$ is the subset of $\{1,3,5\}$. Because $\{1,5\} = \{1,5\}$, $\{1,5\} \subseteq \{1,5\}$. $\{1,3,5\}$ is not the subset of $\{1,5\}$. |
| $A^c$ | A's complement set | $\sqrt{3} \in Q^c$ | $\sqrt{3}$ can't be written in a fraction. Hence, it's an irrational number. |
| $A\backslash B$ | A set, excluding $A \cap B$ elements | $\{1,3,5\}\backslash\{5,7\} = \{1,3\}$ | The intersection for the two sets is $\{5\}$, so taking away $\{5\}$ from $\{1,3,5\}$ equals $\{1,3\}$. |
| $\in$ | belong to | $a \in \{a,b,c\}$ | $a$ belongs to the set $\{a,b,c\}$. |
| $\notin$ | does not | $d \notin \{a,b,c\}$ | $d$ doesn't belong to the set |

| | belong to | | $\{a, b, c\}$. |
|---|---|---|---|
| N | positive integers | $N = \{1,2,3,\dots\}$, $3 \in N$ | By definition, the set of positive integers is equal to $\{1,2,3,\dots\}$. 3 belong to the set of positive integers. |
| Z | integers | $Z = \{\dots, -2, -1, 0, 1, 2, \dots\}$, $-2 \in Z$ | By definition, we have $Z = \{\dots, -2, -1, 0, 1, 2, \dots\}$. $-2$ belongs to the set of integers. |
| Q | rational numbers | $Q = \left\{\dfrac{m}{n} : m \in Z, n \in Z \setminus \{0\}\right\}$, $-\dfrac{4}{3} \in Q$ | Rational numbers set includes all numbers with a fraction format. $-\dfrac{4}{3}$ is a rational number. |
| $Q^c$ | irrational numbers | $\sqrt{3} \in Q^c$ | $\sqrt{3}$ can't be written in a fraction. Hence, it's an irrational number. |
| R | real numbers | $R = Q \cup Q^c$, $-\dfrac{4}{3} \in R, \sqrt{3} \in R$ | Real numbers are the union of irrational numbers and rational numbers. $-\dfrac{4}{3}$ is a real number, $\sqrt{3}$ is a real number |

逻辑符号

| Notations | Definition | Examples | Explanation |
|---|---|---|---|
| $\vee$ | or | $x \in A \vee x \in B \Rightarrow x \in A \cup B$ | If $x$ belongs to $A$ or B then $x$ belongs to $A \cup B$. |
| $\wedge$ | and | $x < 1 \wedge x > 0 \Rightarrow 0 < x < 1$ | If $x < 1$ and $x > 0$ then $0 < x < 1$. |
| $\Rightarrow$ | implies | $x < 1 \Rightarrow x < 2$ | $x < 1$ implies $x < 2$. |
| $\Leftrightarrow$ | if and only if | $|x| < 1 \Leftrightarrow -1 < x < 1$ | If $|x| < 1$ then $-1 < x < 1$. Conversely, if $-1 < x < 1$ |

| | | | then $|x| < 1$. |
|---|---|---|---|
| $\because$ | because | $\because x < 1 \quad \therefore x < 2$ | Because $x < 1, x < 2$. |
| $\therefore$ | so | $\because x < 1 \quad \therefore x < 2$ | $x < 1$, so $x < 2$. |
| $\forall$ | for each | $\forall x \in Q, \exists m \in Z, n \in Z\backslash\{0\}$ <br> s.t. $x = \dfrac{m}{n}$ | By definition of ration numbers, for each rational number $x$, there extis $m \in Z, n \in Z\backslash\{0\}$ such that $x = \dfrac{m}{n}$. |
| $\exists$ | There exists | $\forall x \in Q, \exists m \in Z, n \in Z\backslash\{0\}$ <br> s.t. $x = \dfrac{m}{n}$ | By definition of ration numbers, for each rational number $x$, there extis $m \in Z, n \in Z\backslash\{0\}$ such that $x = \dfrac{m}{n}$. |

微积分符号

| Notations | Definition | Examples | Explanation |
|---|---|---|---|
| $\displaystyle\lim_{x\to c} f(x) = L$ | $\forall \varepsilon > 0, \exists \delta > 0$ such that $0 < |x - c| < \delta$ $\Rightarrow |f(x) - L| < \varepsilon$ | $\displaystyle\lim_{x\to 2} x^2 = 4$ | When $x$ approaches 2, the limit of $x^2$ is 4 |
| $\varepsilon$ | epsilon | $\forall \varepsilon > 0$ | for each $\varepsilon > 0$ |
| $\delta$ | delta | $\forall \varepsilon > 0,$ $\exists \delta > 0 \; s.t. ...$ | for each $\varepsilon > 0$, there exist $\delta > 0$ such that .... |
| $e$ | $e := \displaystyle\sum_{n=0}^{\infty} \dfrac{1}{n!},$ <br> (Euler's number) | $\displaystyle\lim_{x\to\infty} \left(1 + \dfrac{1}{x}\right)^x = e$ | It can be proven with L'Hôpital's rule. |
| $f'(x)$ | derivative | $(x^2)' = 2x$ | $x^2$ has derivative $(x^2)' = 2x$. |
| $\dfrac{\partial f(x, y)}{\partial x}$ | partial derivative | $\dfrac{\partial}{\partial x}(x^2 y) = 2xy$ | $x^2 y$ has the partial derivative |

| | | | |
|---|---|---|---|
| | | | $\dfrac{\partial}{\partial x}(x^2 y) = 2xy.$ |
| $\displaystyle\int$ | integral | $\displaystyle\int 2x\,dx = x^2$ | $2x$ has indefinite integral $x^2$. |
| $\displaystyle\iint$ | double integral | Find $\displaystyle\iint_R y e^{xy}\,dA =?$ | Find the double integral of $ye^{xy}$ in the range $R$. |
| $\displaystyle\iiint$ | triple integral | Find $\displaystyle\iiint_V x^2\,dV =?$ | Find the triple integral of $x^2$ in the range V |
| $\displaystyle\oint$ | closed contour / line integral | Find $\displaystyle\oint y^3 dx + x^3\,dy$ $=?,\; C:x^2 + y^2 = r^2$ | Find the closed line integral $\displaystyle\oint y^3 dx + x^3\,dy$ $=?$, where $C:x^2 + y^2 = r^2$. |
| $\displaystyle\oiint$ | closed surface integral | Find $\displaystyle\oiint_S \vec{F}\cdot\vec{n}\,dA =?,$ where $\vec{F} = (x,y,z)$ | Find the closed surface integral $\displaystyle\oiint_S \vec{F}\cdot\vec{n}\,dA =?,$ $\vec{F} = (x,y,z).$ |

# 第七章　　多变量函数的微分与应用

首先将介绍多变量函数极限存在的定义，判断函数于原点的极限是否存在时，令 $y = mx$，接着观察 $\lim\limits_{(x,y) \to (0,0)} f(x, mx)$ 是否与 $m$ 值有关，若有关则找确切两个相异 $m$ 值说明极限不存在；若 $\lim\limits_{(x,y) \to (0,0)} f(x, mx)$ 与 $m$ 值无关，有可能藉由曲线的方式逼近 $(0,0)$ 使得极限不存在，如果藉由曲线逼近 $(0,0)$ 的极限值仍与 $m$ 值无关，则有很大可能 $f(x,y)$ 于 $(0,0)$ 极限值存在，此时则使用 $\varepsilon, \delta$ 说明极限存在；接着介绍多变量函数连续的定义，判断多变量函数是否连续的流程与说明多变量函数极限是否存在的流程是类似的，观察

$$\lim\limits_{(x,y) \to (0,0)} f(x, mx)$$ 是否与 $m$ 值有关，若有关则找确切两个相异 $m$ 值说明极限不存在

使得 $f(x,y)$ 于 $(0,0)$ 不连续，若 $\lim\limits_{(x,y) \to (0,0)} f(x, mx)$ 与 $m$ 值无关，有可能藉由曲线的方式逼近 $(0,0)$ 使得极限不存在，藉此说明 $f(x,y)$ 于 $(0,0)$ 不连续，如果藉由曲线逼近 $(0,0)$ 的极限值仍与 $m$ 值无关且等于 $f(0,0)$，则有很大可能 $f(x,y)$ 于 $(0,0)$ 连续，此时则使用 $\varepsilon, \delta$ 说明 $f(x,y)$ 于 $(0,0)$ 连续。

接着讨论多变量函数的偏导数，多变量函数的可微分性与多变量函数的 Chain Rule，判断多变量函数是否可微分的计算过程会需要计算多变量函数的偏导数，因此有效率且精确算出多变量函数的偏导数是重要的；从双变量函数可微分的定义，能观察出如果 $f(x,y)$ 于 $(0,0)$ 可微分，则藉由 $(h,k)$ 从任意方向逼近 $(0,0)$ 的极限值必须等于 $0$，因此，令 $h = mk$，当 $f(x,y)$ 于 $(0,0)$ 可微分，则

$$\lim\limits_{h \to 0, k \to 0} \left| \frac{f(mk, k) - f(0,0) - \frac{\partial f}{\partial x}(0,0)mk - \frac{\partial f}{\partial y}(0,0)k}{\sqrt{(mk)^2 + k^2}} \right| = 0$$

并且此极限值与 $m$ 值无关，此外，偏导数存在也是双变数函数可微分的必要条件；与 $m$ 值有关时，则试着找出 $m \in R$ 使得此极限值不等于零来说明 $f(x,y)$ 于 $(0,0)$ 不可微分；如果与 $m$ 值无关则令 $\varepsilon > 0$，取出明确 $\delta$ 的值，使得 $h^2 + k^2 < \delta$

$$\Rightarrow \left| \frac{f(h, k) - f(0,0) - \frac{\partial f}{\partial x}(0,0)h - \frac{\partial f}{\partial y}(0,0)k}{\sqrt{h^2 + k^2}} \right| < \varepsilon$$

此外，多变量函数的 Chain Rule 可以看成是多变量合成函数的微分

接着讨论求多变量函数的极值问题，从宏观的角度来看，最重要的两个定理分别为双变量函数求极值的二阶导数检定法与 Lagrange multiplier；无限制条件的双变量函数求极值问题使用二阶导数检定法，其证明过程用到多变量函数的 Taylor's Formula，此外，多变量函数 Taylor's Formula 的证明过程用到了单变量函数 Taylor's Formula；因此，在无限制条件下，用来求双变量函数极值的二阶导数检定法，用到了单变量函数的 Taylor's Formula；在有限制条件求极值问题的 Lagrange multiplier，其证明过程用到多变量函数的隐函数定理(Implicit Function Theorem)，多变量函数的隐函数定理在证明过程中引用了多变量函数的反函数定理而多变量函数的反函数定理证明用到了单变量函数的反函数定理，换句话说，Lagrange multiplier 的证明最终用到了单变量函数的反函数定理

最后说明如何使用上述两个定理求多变量函数的极值，这类问题可分为无限制式条件与有限制式条件；无限制式求双变量函数极值的问题，通常使用求极值的二阶导数检定法求解；有限制式条件通常是等式或不等式，如果是等式的情况则直接使用 Lagrange multiplier 求解；如果是不等式的情况(如封闭区域)则将封闭区域拆解为在封闭区域内与封闭区域边界(有等式)的两个问题，在封闭区域内求极值问题时，则藉由双变量函数求极值的二阶导数检定法找出极值点，再判断极值点是否落在封闭区域内，另一方面，在封闭区域边界求极值问题则用 Lagrange multiplier 求解，最后比较封闭区域内与封闭区域边界的极值点，假设两者于各自限制式皆为相对最大值，则取出较大者当作于此不等式的绝对最大值，绝对最小值也是类似的作法；此外，也可能给数个点求这些点所围区域内的极值点，这时候需多花一道工将封闭区域的方程式求出

## 7.1 求多变量函数的极限

**【定义】** 双变量函数的极限定义

$f(x,y)$ 于 $(x_0,y_0)$ 的极限值存在且等于 $L$ $\Leftrightarrow \lim\limits_{(x,y)\to(x_0,y_0)} f(x,y) = L$

$\Leftrightarrow \forall \varepsilon > 0, \exists \delta > 0,\ \text{such that } \sqrt{(x-x_0)^2+(y-y_0)^2} < \delta \Rightarrow |f(x,y)-L| < \varepsilon$

**【定义】** 多变量函数的极限定义

$f(x_1,\ldots,x_n)$ 于 $(c_1,\ldots,c_n)$ 的极限值存在且等于 $L$ $\Leftrightarrow \lim\limits_{(x_1,\ldots,x_n)\to(c_1,\ldots,c_n)} f(x_1,\ldots,x_n) = L$

$$\Leftrightarrow \forall \varepsilon > 0, \exists \delta > 0, \ \text{such that} \ \sqrt{(x_1 - c_1)^2 + \cdots + (x_n - c_n)^2} < \delta \Rightarrow |f(x_1, \ldots, x_n) - L| < \varepsilon$$

  判断双变数函数 $f(x,y)$ 于 $(x_0, y_0)$ 的极限是否存在的的解法流程：从双变量函数的极限定义，能观察出如果 $f(x,y)$ 于 $(x_0, y_0)$ 的极限值存在，则从任意方向逼近 $(x_0, y_0)$ 的极限值必须相等，一般的手法为，令 $y - y_0 = m(x - x_0)$ 为过 $(x_0, y_0)$ 的直线，当 $f(x,y)$ 于 $(x_0, y_0)$ 极限值存在, 则

$$\lim_{(x,y) \to (x_0, y_0)} f(x,y) = \lim_{(x,y) \to (x_0, y_0)} f(x, m(x - x_0) + y_0)$$

且

$$\lim_{(x,y) \to (x_0, y_0)} f(x, m(x - x_0) + y_0) \quad \text{与 } m \text{ 值无关}$$

值得注意的是 $\lim\limits_{(x,y) \to (x_0, y_0)} f(x, m(x - x_0) + y_0)$ 与 $m$ 值无关是 $f(x,y)$ 于 $(x_0, y_0)$ 极限值存在的必要条件并非充分条件, 底下展示一双变数函数 $f(x,y)$ 于 $(0,0)$ 的极限值与 $m$ 值无关但 $f(x,y)$ 于 $(0,0)$ 的极限值不存在

<u>范例说明:</u>

Determine whether $f(x,y) = \dfrac{xy^2}{x^2 + y^4}$ has a limit as $(x,y) \to (0,0)$.

令 $y = mx$ 则 $f(x,y) = \dfrac{xy^2}{x^2 + y^4} = \dfrac{m^2 x^3}{x^2 + m^4 x^4}$

$\therefore \lim\limits_{(x,y) \to (0,0)} f(x, mx)$ 与 $m$ 值无关 且 $\lim\limits_{(x,y) \to (0,0)} f(x, mx) = 0$

令 $x = y^2$ 则 $f(x,y) = \dfrac{y^4}{y^4 + y^4} = \dfrac{1}{2}$ 且 $\lim\limits_{(x,y) \to (0,0)} f(y^2, y) = \dfrac{1}{2}$

$\because$ 若 $f(x,y)$ 于 $(0,0)$ 的极限值存在，则从任意方向逼近 $(0,0)$ 的极限值必须相等

$\therefore f(x,y)$ 于 $(0,0)$ 极限值不存在

  常见的考试类型为 $(x_0, y_0) = (0,0)$, 假设 $(x_0, y_0) = (0,0)$ 且 $f(x,y)$ 于 $(0,0)$ 极限值存在，

令 $y = mx$, 则 $\lim\limits_{(x,y) \to (0,0)} f(x,y) = \lim\limits_{(x,y) \to (0,0)} f(x, mx)$ 且 $\lim\limits_{(x,y) \to (0,0)} f(x, mx)$ 与 $m$ 值无关,

因此, $\lim\limits_{(x,y) \to (0,0)} f(x, mx)$ 与 $m$ 值无关是 $f(x,y)$ 于 $(0,0)$ 极限值存在的必要条件

此外，假设 $\exists\, m_1 \neq m_2 \in R$ 使得 $\displaystyle\lim_{(x,y)\to(0,0)} f(x, m_1 x) \neq \lim_{(x,y)\to(0,0)} f(x, m_2 x)$

则 $f(x,y)$ 于 $(0,0)$ 极限值不存在, 也就是, 令 $y = mx$, 当 $\displaystyle\lim_{(x,y)\to(0,0)} f(x, mx)$ 与 $m$ 值有关时,

则找出 $m_1 \neq m_2 \in R$ 使得 $\displaystyle\lim_{(x,y)\to(0,0)} f(x, m_1 x) \neq \lim_{(x,y)\to(0,0)} f(x, m_2 x)$ 來证明 $f(x,y)$ 于 $(0,0)$

极限值不存在, 在求多变量函数极限的考试类型当中, 以求双变量函数极限的题型最常出现, 读者若能熟悉双变量题型, 求三变量函数的题型都能够掌握；总结上述几点, 底下介绍双变量函数于原点 $(0,0)$ 的极限是否存在的考试类型与解题流程

考试类型:
Type 1.

求双变数函数 $f(x,y)$ 于 $(0,0)$ 的极限, $\displaystyle\lim_{(x,y)\to(0,0)} f(x, y) =?$, 其中函数 $f(x,y)$ 满足:

$$\text{令}\, y = mx, \quad \text{当} \lim_{(x,y)\to(0,0)} f(x, mx) = L \text{ 与 } m \text{值无关}$$

解题流程:
令 $\varepsilon > 0$, 取 $\delta$ 或明确找出 $\delta$ 的值, 使得 $x^2 + y^2 < \delta \Rightarrow |f(x,y) - L| < \varepsilon$

<u>補充说明:</u>
证明过程常藉由不等式 $x^2 + y^2 \geq 2|xy|$, 推导出 $|f(x,y) - L| < g(x^2 + y^2) < g(\delta) < \varepsilon$, 并且能得知 $g$ 明确的样貌, 最后一个不等式的 $g(\delta) < \varepsilon$ 能帮助明确定义 $\delta$。值得注意的是推

论过程中, 当 $\displaystyle\lim_{(x,y)\to(0,0)} f(x, mx) = L$ 与 $m$ 值无关时, 只能说当下这个函数有可能于 $(0,0)$ 的

极限值存在, 也有可能藉由曲线的方式逼近 $(0,0)$ 使得极限不存在, 如果藉由曲线逼近 $(0,0)$ 的极限值仍为 $L$, 则有很大机率极限值存在, 此时则走上述的流程令 $\varepsilon > 0$, 明确取出 $\delta$ 的值, 使得 $x^2 + y^2 < \delta \Rightarrow |f(x,y) - L| < \varepsilon$

<u>范例说明:</u>

(I) 求 $\displaystyle\lim_{(x,y)\to(0,0)} \frac{3x^2 y}{x^2 + y^2} =?$

令 $y = mx$ 则 $\displaystyle\lim_{(x,y)\to(0,0)} \frac{3x^2(mx)}{x^2 + (mx)^2} = \lim_{(x,y)\to(0,0)} \frac{3mx}{1 + m^2} = 0$ 与 $m$ 值无关

$$\because x^2 + y^2 \geq 2|xy| \quad \therefore \frac{1}{x^2 + y^2} \leq \frac{1}{2|xy|} \Rightarrow \left|\frac{3x^2y}{x^2 + y^2}\right| \leq \left|\frac{3x^2y}{2xy}\right| = \left|\frac{3x}{2}\right|$$

Claim: $\displaystyle\lim_{(x,y)\to(0,0)} \frac{3x^2y}{x^2 + y^2} = 0$

令 $\epsilon > 0$ 取 $\delta < \left(\frac{2\epsilon}{3}\right)^2$

则 $x^2 + y^2 < \delta \Rightarrow \left|\frac{3x^2y}{x^2 + y^2}\right| \leq \left|\frac{3x^2y}{2xy}\right| = \left|\frac{3x}{2}\right| < \left|\frac{3\sqrt{x^2 + y^2}}{2}\right| < \frac{3\sqrt{\delta}}{2} < \frac{3}{2} \cdot \frac{2\epsilon}{3} = \epsilon$

$$\therefore \lim_{(x,y)\to(0,0)} \frac{3x^2y}{x^2 + y^2} = 0$$

(II) 求 $\displaystyle\lim_{(x,y)\to(0,0)} \frac{x^4 + y^4}{x^2 + y^2} = ?$

令 $y = mx$ 则 $\displaystyle\lim_{(x,y)\to(0,0)} \frac{x^4 + y^4}{x^2 + y^2} = \lim_{(x,y)\to(0,0)} \frac{x^2 + m^4x^2}{1 + m^2} = 0$ 与 $m$ 值无关

$\because x^4 + y^4 = (x^2 + y^2)^2 - 2x^2y^2$

$$\therefore \lim_{(x,y)\to(0,0)} \frac{x^4 + y^4}{x^2 + y^2} = \lim_{(x,y)\to(0,0)} \frac{(x^2 + y^2)^2 - 2x^2y^2}{x^2 + y^2}$$

$$= \lim_{(x,y)\to(0,0)} x^2 + y^2 - \lim_{(x,y)\to(0,0)} \frac{2x^2y^2}{x^2 + y^2} = -\lim_{(x,y)\to(0,0)} \frac{2x^2y^2}{x^2 + y^2}$$

Claim: $\displaystyle\lim_{(x,y)\to(0,0)} \frac{2x^2y^2}{x^2 + y^2} = 0$

$\because x^2 + y^2 \geq 2|xy| \quad \therefore \frac{1}{x^2 + y^2} \leq \frac{1}{2|xy|} \Rightarrow \left|\frac{2x^2y^2}{x^2 + y^2}\right| \leq \left|\frac{2x^2y^2}{2xy}\right| = |xy|$

令 $\epsilon > 0$ 取 $\frac{\delta}{2} < \epsilon$

则 $x^2 + y^2 < \delta \Rightarrow \left|\frac{2x^2y^2}{x^2 + y^2}\right| \leq \left|\frac{2x^2y^2}{2xy}\right| = |xy| < \left|\frac{x^2 + y^2}{2}\right| < \frac{\delta}{2} < \epsilon$

$$\therefore \lim_{(x,y)\to(0,0)} \frac{2x^2y^2}{x^2 + y^2} = 0 \Rightarrow \lim_{(x,y)\to(0,0)} \frac{x^4 + y^4}{x^2 + y^2} = 0$$

Type 2.

求双变数函数 $f(x,y)$ 于 $(0,0)$ 的极限，$\lim\limits_{(x,y)\to(0,0)} f(x,y) =?$，其中函数 $f(x,y)$ 满足：

$$\text{令 } y = mx, \quad \text{当 } \lim\limits_{(x,y)\to(0,0)} f(x,mx) \text{ 与 } m \text{ 值有关时}$$

解题流程：

取两个相异 $m$ 值，$m_1 \cdot m_2$，计算两者的极限值，使得

$$\lim\limits_{(x,y)\to(0,0)} f(x,m_1 x) \neq \lim\limits_{(x,y)\to(0,0)} f(x,m_2 x) \quad \therefore \lim\limits_{(x,y)\to(0,0)} f(x,y) \text{ 不存在}$$

范例说明：

(I) 求 $\lim\limits_{(x,y)\to(0,0)} \dfrac{x^2 - 2y^2}{x^2 + y^2} =?$

令 $y = mx$ 则 $\lim\limits_{(x,y)\to(0,0)} \dfrac{x^2 - 2y^2}{x^2 + y^2} = \lim\limits_{(x,y)\to(0,0)} \dfrac{1 - 2m^2}{1 + m^2}$ 与 $m$ 值有关

取 $m = 1$ 则 $\lim\limits_{(x,y)\to(0,0)} \dfrac{x^2 - 2y^2}{x^2 + y^2} = \lim\limits_{(x,y)\to(0,0)} \dfrac{1 - 2m^2}{1 + m^2} = -\dfrac{1}{2}$

取 $m = \sqrt{2}$ 则 $\lim\limits_{(x,y)\to(0,0)} \dfrac{x^2 - 2y^2}{x^2 + y^2} = \lim\limits_{(x,y)\to(0,0)} \dfrac{1 - 2m^2}{1 + m^2} = -1$

$\therefore \lim\limits_{(x,y)\to(0,0)} \dfrac{x^2 - 2y^2}{x^2 + y^2}$ 不存在

(II) 求 $\lim\limits_{(x,y)\to(0,0)} \dfrac{x^3 y}{x^6 + 2y^2} =?$

令 $y = mx$ 则 $\lim\limits_{(x,y)\to(0,0)} \dfrac{x^3 y}{x^6 + 2y^2} = \lim\limits_{(x,y)\to(0,0)} \dfrac{m}{1 + 2m^2}$ 与 $m$ 值有关

取 $m = -1$ 则 $\lim\limits_{(x,y)\to(0,0)} \dfrac{x^3 y}{x^6 + 2y^2} = \lim\limits_{(x,y)\to(0,0)} \dfrac{m}{1 + 2m^2} = -\dfrac{1}{3}$

取 $m = 1$ 则 $\lim\limits_{(x,y)\to(0,0)} \dfrac{x^3 y}{x^6 + 2y^2} = \lim\limits_{(x,y)\to(0,0)} \dfrac{m}{1 + 2m^2} = \dfrac{1}{3}$

$\therefore \lim\limits_{(x,y)\to(0,0)} \dfrac{x^3 y}{x^6 + 2y^2}$ 不存在

Type 3.

求三变数函数 $f(x,y,z)$ 于 $(0,0,0)$ 的极限, $\displaystyle\lim_{(x,y,z)\to(0,0,0)}f(x,y,z)=?$，其中函数 $f(x,y,z)$ 满足:

$$\text{令}\ y=z=mx,\quad \text{当}\ \lim_{(x,y,z)\to(0,0,0)}f(x,mx,mx)=L\ \text{与}\ m\ \text{无关时}$$

解题流程:

令 $\varepsilon>0$，取 $\delta$ 或明确找出 $\delta$ 的值，使得 $x^2+y^2+z^2<\delta \Rightarrow |f(x,y,z)-L|<\varepsilon$

范例说明:

(I) 求 $\displaystyle\lim_{(x,y,z)\to(0,0,0)}\dfrac{x^2y}{x^2+y^4+z^6}=?$

令 $y=z=mx$ 则 $\dfrac{x^2y}{x^2+y^4+z^6}=\dfrac{mx^3}{x^2+m^4x^4+m^6x^6}=\dfrac{mx}{1+m^4x^2+m^6x^4}$

$\therefore \displaystyle\lim_{(x,y,z)\to(0,0,0)}f(x,mx,mx)=0$ 与 $m$ 无关

$\because x^2+y^4+z^6\ge x^2 \qquad \therefore \left|\dfrac{x^2y}{x^2+y^4+z^6}\right|\le\left|\dfrac{x^2y}{x^2}\right|=|y|$

令 $\epsilon>0$ 取 $\delta^{\frac{1}{2}}<\varepsilon$

则 $x^2+y^2+z^2<\delta \Rightarrow \left|\dfrac{x^2y}{x^2+y^4+z^6}\right|<|y|<(x^2+y^2+z^2)^{\frac{1}{2}}<\delta^{\frac{1}{2}}<\epsilon$

$\therefore \displaystyle\lim_{(x,y,z)\to(0,0,0)}\dfrac{x^2y}{x^2+y^4+z^6}=0$

值得注意的是 $\displaystyle\lim_{(x,y,z)\to(0,0,0)}f(x,mx,mx)=0$ 与 $m$ 无关是 $f(x,y,z)$ 于 $(0,0,0)$ 的极限值存在

的必要条件并非充分条件，也就是推论的过程中，如果出现

"因为 $\displaystyle\lim_{(x,y,z)\to(0,0,0)}f(x,mx,mx)=0$ 与 $m$ 无关，所以 $f(x,y,z)$ 于 $(0,0,0)$ 的极限存在"

为错误的论述，底下展示一多变数函数做说明

范例说明:

Determine whether $f(x,y,z)=\dfrac{xyz}{x^2+y^4+z^4}$ has a limit as $(x,y,z)\to(0,0,0)$

令 $y=z=mx$ 则 $f(x,y,z)=\dfrac{xyz}{x^2+y^4+z^4}=\dfrac{m^3x^3}{x^2+m^4x^4+m^4x^4}=\dfrac{m^3x}{1+m^4x^2+m^4x^2}$

$$\therefore \lim_{(x,y,z)\to(0,0,0)} f(x,mx,mx) = 0 \text{ 与} m\text{无关且} \lim_{(x,y,z)\to(0,0,0)} f(x,mx,mx) = 0, \text{ 然而}$$

令 $x = mt^2$, $y = z = t$ 则 $f(x,y,z) = \dfrac{mt^4}{(m^2+2)t^4} = \dfrac{m}{m^2+2}$

取 $m = -1$ 则 $\lim_{(x,y,z)\to(0,0,0)} \dfrac{xyz}{x^2+y^4+z^4} = \lim_{(x,y,z)\to(0,0,0)} \dfrac{m}{m^2+2} = -\dfrac{1}{3}$

取 $m = 1$ 则 $\lim_{(x,y,z)\to(0,0,0)} \dfrac{xyz}{x^2+y^4+z^4} = \lim_{(x,y,z)\to(0,0,0)} \dfrac{m}{m^2+2} = \dfrac{1}{3}$

$$\therefore \lim_{(x,y,z)\to(0,0,0)} \dfrac{xyz}{x^2+y^4+z^4} \text{ 不存在}$$

## Type 4.

求三变数函数 $f(x,y,z)$ 于 $(0,0,0)$ 的极限, $\lim_{(x,y,z)\to(0,0,0)} f(x,y,z) = ?$, 其中函数 $f(x,y,z)$ 满足:

$$\text{令} y = z = mx, \text{ 当} \lim_{(x,y,z)\to(0,0,0)} f(x,mx,mx) \text{ 与} m\text{有关时}$$

解题流程:

取两个相异值 $m_1$、$m_2$ 使得

$$\lim_{(x,y,z)\to(0,0,0)} f(x,m_1x,m_1x) \neq \lim_{(x,y,z)\to(0,0,0)} f(x,m_2x,m_2x) \quad \therefore \lim_{(x,y,z)\to(0,0,0)} f(x,y,z) \text{ 不存在}$$

<u>范例说明:</u>

(I) 求 $\lim_{(x,y,z)\to(0,0,0)} \dfrac{3xy}{x^2+y^2+z^2} = ?$

令 $y = z = mx$ 则 $\dfrac{3xy}{x^2+y^2+z^2} = \dfrac{3mx^2}{x^2+m^2x^2+m^2x^2} = \dfrac{3m}{2m^2+1}$

取 $m = -1$ 则 $\lim_{(x,y,z)\to(0,0,0)} \dfrac{3xy}{x^2+y^2+z^2} = \lim_{(x,y,z)\to(0,0,0)} \dfrac{3m}{2m^2+1} = -1$

取 $m = 1$ 则 $\lim_{(x,y,z)\to(0,0,0)} \dfrac{3xy}{x^2+y^2+z^2} = \lim_{(x,y,z)\to(0,0,0)} \dfrac{3m}{2m^2+1} = 1$

因此 $\lim_{(x,y,z)\to(0,0,0)} \dfrac{3xy}{x^2+y^2+z^2} \text{ 不存在}$

Example 1.

$$求\quad \lim_{(x,y)\to(0,0)} \frac{x^2 - 2y^2}{x^2 + y^2} =?$$

【解】

令 $y = mx$ 则 $\dfrac{x^2 - 2y^2}{x^2 + y^2} = \dfrac{x^2 - 2m^2x^2}{x^2 + m^2x^2} = \dfrac{x^2(1 - 2m^2)}{x^2(1 + m^2)} = \dfrac{1 - 2m^2}{1 + m^2}$

$\therefore \lim_{(x,y)\to(0,0)} \dfrac{x^2 - 2y^2}{x^2 + y^2} = \lim_{(x,y)\to(0,0)} \dfrac{1 - 2m^2}{1 + m^2}$

取 $m = 1$ 则 $\lim_{(x,y)\to(0,0)} \dfrac{x^2 - 2y^2}{x^2 + y^2} = \lim_{(x,y)\to(0,0)} \dfrac{1 - 2m^2}{1 + m^2} = -\dfrac{1}{2}$

取 $m = \sqrt{2}$ 则 $\lim_{(x,y)\to(0,0)} \dfrac{x^2 - 2y^2}{x^2 + y^2} = \lim_{(x,y)\to(0,0)} \dfrac{1 - 2m^2}{1 + m^2} = -1$

$\therefore \lim_{(x,y)\to(0,0)} \dfrac{x^2 - 2y^2}{x^2 + y^2}$ 不存在

Example 2.

$$求\quad \lim_{(x,y)\to(0,0)} \left(\frac{x^2 - y^2}{x^2 + 2y^2}\right)^2 =?$$

【解】

令 $y = mx$ 则 $\left(\dfrac{x^2 - y^2}{x^2 + 2y^2}\right)^2 = \left(\dfrac{x^2 - m^2x^2}{x^2 + 2m^2x^2}\right)^2 = \left(\dfrac{x^2(1 - m^2)}{x^2(1 + 2m^2)}\right)^2 = \dfrac{(1 - m^2)^2}{(1 + 2m^2)^2}$

$\therefore \lim_{(x,y)\to(0,0)} \left(\dfrac{x^2 - y^2}{x^2 + 2y^2}\right)^2 = \lim_{(x,y)\to(0,0)} \dfrac{(1 - m^2)^2}{(1 + 2m^2)^2}$

取 $m = 1$ 则 $\lim_{(x,y)\to(0,0)} \left(\dfrac{x^2 - y^2}{x^2 + 2y^2}\right)^2 = \lim_{(x,y)\to(0,0)} \dfrac{(1 - m^2)^2}{(1 + 2m^2)^2} = 0$

取 $m = \sqrt{2}$ 则 $\lim_{(x,y)\to(0,0)} \left(\dfrac{x^2 - y^2}{x^2 + 2y^2}\right)^2 = \lim_{(x,y)\to(0,0)} \dfrac{(1 - m^2)^2}{(1 + 2m^2)^2} = \dfrac{1}{25}$

$\therefore \lim_{(x,y)\to(0,0)} \left(\dfrac{x^2 - y^2}{x^2 + 2y^2}\right)^2$ 不存在

Example 3.

$$\text{求} \lim_{(x,y)\to(0,0)} \frac{x^2+y^2}{|x|+|y|} = ?$$

【解】

$$\because (|x|+|y|)^2 = |x|^2+|y|^2+2|xy| \geq x^2+y^2 \quad \therefore |x|+|y| \geq \sqrt{x^2+y^2}$$

$$\therefore \frac{x^2+y^2}{|x|+|y|} \leq \frac{x^2+y^2}{\sqrt{x^2+y^2}} = \sqrt{x^2+y^2}$$

Claim: $\displaystyle\lim_{(x,y)\to(0,0)} \frac{x^2+y^2}{|x|+|y|} = 0$

令 $\epsilon > 0$ 取 $\delta^{\frac{1}{2}} < \varepsilon$ 则 $x^2+y^2 < \delta \Rightarrow \dfrac{x^2+y^2}{|x|+|y|} \leq \dfrac{x^2+y^2}{\sqrt{x^2+y^2}} = \sqrt{x^2+y^2} < \delta^{\frac{1}{2}} < \epsilon$

因此 $\displaystyle\lim_{(x,y)\to(0,0)} \frac{x^2+y^2}{|x|+|y|} = 0$

Example 4.

$$\text{求} \lim_{(x,y)\to(0,0)} \frac{xy\cos(x^2+y^2)}{\sqrt{x^2+y^2}} = ?$$

【解】

$$\because x^2+y^2 \geq 2|xy| \quad \therefore \frac{1}{\sqrt{x^2+y^2}} \leq \frac{1}{\sqrt{2|xy|}} \Rightarrow \left|\frac{xy\cos(x^2+y^2)}{\sqrt{x^2+y^2}}\right| \leq \left|\frac{xy}{\sqrt{2|xy|}}\right| < |xy|^{\frac{1}{2}}$$

Claim: $\displaystyle\lim_{(x,y)\to(0,0)} \frac{xy\cos x^2+y^2}{\sqrt{x^2+y^2}} = 0$

令 $\epsilon > 0$ 取 $\delta^{\frac{1}{2}} < \varepsilon$

则 $x^2+y^2 < \delta \Rightarrow \left|\dfrac{xy\cos(x^2+y^2)}{\sqrt{x^2+y^2}}\right| \leq \left|\dfrac{xy}{\sqrt{2|xy|}}\right| < |xy|^{\frac{1}{2}} < (x^2+y^2)^{\frac{1}{2}} < \delta^{\frac{1}{2}} < \epsilon$

因此 $\displaystyle\lim_{(x,y)\to(0,0)} \frac{xy\cos x^2+y^2}{\sqrt{x^2+y^2}} = 0$

Example 5.

$$\text{求} \lim_{(x,y,z)\to(0,0,0)} \frac{\sin(x^2+y^2+z^2)}{\sqrt{x^2+y^2+z^2}} = ?$$

【解】

令 $t = x^2 + y^2 + z^2$ 　则 $\dfrac{\sin(x^2 + y^2 + z^2)}{\sqrt{x^2 + y^2 + z^2}} = \dfrac{\sin t}{\sqrt{t}} = \dfrac{\sqrt{t}\sin t}{t}$

$$\lim_{(x,y,z)\to(0,0,0)} \frac{\sin(x^2 + y^2 + z^2)}{\sqrt{x^2 + y^2 + z^2}} = \lim_{t\to 0} \frac{\sqrt{t}\sin t}{t}$$

藉由罗比达法则　$\displaystyle\lim_{t\to 0}\frac{\sin t}{t} = \lim_{t\to 0}\frac{\cos t}{1} = 1 \Rightarrow \lim_{t\to 0}\frac{\sqrt{t}\sin t}{t} = 0$

$\therefore \displaystyle\lim_{(x,y,z)\to(0,0,0)} \frac{\sin x^2 + y^2 + z^2}{\sqrt{x^2 + y^2 + z^2}} = 0$

Example 6.

$$求 \quad \lim_{(x,y)\to(0,0)} \frac{3x^2 y}{x^2 + y^2} = ?$$

【解】

$\because x^2 + y^2 \geq 2|xy| \quad \therefore \dfrac{1}{x^2 + y^2} \leq \dfrac{1}{2|xy|} \Rightarrow \left|\dfrac{3x^2 y}{x^2 + y^2}\right| \leq \left|\dfrac{3x^2 y}{2xy}\right| = \left|\dfrac{3x}{2}\right|$

Claim: $\displaystyle\lim_{(x,y)\to(0,0)} \frac{3x^2 y}{x^2 + y^2} = 0$

令 $\epsilon > 0$　取 $\delta < \left(\dfrac{2\varepsilon}{3}\right)^2$

则 $x^2 + y^2 < \delta \Rightarrow \left|\dfrac{3x^2 y}{x^2 + y^2}\right| \leq \left|\dfrac{3x^2 y}{2xy}\right| = \left|\dfrac{3x}{2}\right| < \left|\dfrac{3\sqrt{x^2 + y^2}}{2}\right| < \dfrac{3\sqrt{\delta}}{2} < \dfrac{3}{2}\cdot\dfrac{2\varepsilon}{3} = \epsilon$

$\therefore \displaystyle\lim_{(x,y)\to(0,0)} \frac{3x^2 y}{x^2 + y^2} = 0$

Example 7.

$$求 \quad \lim_{(x,y)\to(0,0)} \frac{2xy^2}{x^2 + y^2} = ?$$

【解】

$\because x^2 + y^2 \geq 2|xy| \quad \therefore \dfrac{1}{x^2 + y^2} \leq \dfrac{1}{2|xy|} \Rightarrow \left|\dfrac{2xy^2}{x^2 + y^2}\right| \leq \left|\dfrac{2xy^2}{2xy}\right| = |y|$

Claim: $\displaystyle\lim_{(x,y)\to(0,0)}\frac{2xy^2}{x^2+y^2}=0$

令 $\epsilon>0$ 取 $\delta<\varepsilon$ 则 $x^2+y^2<\delta\Rightarrow\left|\dfrac{2xy^2}{x^2+y^2}\right|\leq\left|\dfrac{2xy^2}{2xy}\right|=|y|<\left|\sqrt{x^2+y^2}\right|<\delta<\epsilon$

$\therefore\displaystyle\lim_{(x,y)\to(0,0)}\frac{2xy^2}{x^2+y^2}=0$

Example 8.

$$求\quad\lim_{(x,y,z)\to(0,0,0)}\frac{xyz}{x^2+y^4+z^4}=?$$

【解】

令 $x=mt^2,\ y=z=t$ 则 $\quad\dfrac{mt^4}{(m^2+2)t^4}=\dfrac{m}{m^2+2}$

取 $m=-1$ 则 $\displaystyle\lim_{(x,y,z)\to(0,0,0)}\frac{xyz}{x^2+y^4+z^4}=\lim_{(x,y,z)\to(0,0,0)}\frac{m}{m^2+2}=-\frac{1}{3}$

取 $m=1$ 则 $\displaystyle\lim_{(x,y,z)\to(0,0,0)}\frac{xyz}{x^2+y^4+z^4}=\lim_{(x,y,z)\to(0,0,0)}\frac{m}{m^2+2}=\frac{1}{3}$

因此 $\displaystyle\lim_{(x,y,z)\to(0,0,0)}\frac{xyz}{x^2+y^4+z^4}$ 不存在

Example 9.

$$求\quad\lim_{(x,y,z)\to(0,0,0)}\frac{x^2y}{x^2+y^4+z^6}=?$$

【解】

$\because x^2+y^4+z^6\geq x^2\qquad\therefore\left|\dfrac{x^2y}{x^2+y^4+z^6}\right|\leq\left|\dfrac{x^2y}{x^2}\right|=|y|$

令 $\epsilon>0$ 取 $\delta^{\frac{1}{2}}<\varepsilon$

则 $x^2+y^2+z^2<\delta\Rightarrow\left|\dfrac{x^2y}{x^2+y^4+z^6}\right|<|y|<(x^2+y^2+z^2)^{\frac{1}{2}}<\delta^{\frac{1}{2}}<\epsilon$

因此 $\displaystyle\lim_{(x,y,z)\to(0,0,0)}\frac{x^2y}{x^2+y^4+z^6}=0$

Example 10.

$$\underset{(x,y)\to(0,0)}{\text{求 }\lim} \frac{x^4 + y^4}{x^2 + y^2} =?$$

【解】

$$\because x^4 + y^4 = (x^2 + y^2)^2 - 2x^2y^2$$

$$\therefore \lim_{(x,y)\to(0,0)} \frac{x^4 + y^4}{x^2 + y^2} = \lim_{(x,y)\to(0,0)} \frac{(x^2 + y^2)^2 - 2x^2y^2}{x^2 + y^2}$$

$$= \lim_{(x,y)\to(0,0)} x^2 + y^2 - \lim_{(x,y)\to(0,0)} \frac{2x^2y^2}{x^2 + y^2} = - \lim_{(x,y)\to(0,0)} \frac{2x^2y^2}{x^2 + y^2}$$

Claim: $\lim_{(x,y)\to(0,0)} \frac{2x^2y^2}{x^2 + y^2} = 0$

$$\because x^2 + y^2 \geq 2|xy| \quad \therefore \frac{1}{x^2 + y^2} \leq \frac{1}{2|xy|} \Rightarrow \left|\frac{2x^2y^2}{x^2 + y^2}\right| \leq \left|\frac{2x^2y^2}{2xy}\right| = |xy|$$

令 $\epsilon > 0$ 取 $\frac{\delta}{2} < \varepsilon$

则 $x^2 + y^2 < \delta \Rightarrow \left|\frac{2x^2y^2}{x^2 + y^2}\right| \leq \left|\frac{2x^2y^2}{2xy}\right| = |xy| < \left|\frac{x^2 + y^2}{2}\right| < \frac{\delta}{2} < \epsilon$

因此 $\lim_{(x,y)\to(0,0)} \frac{2x^2y^2}{x^2 + y^2} = 0 \Rightarrow \lim_{(x,y)\to(0,0)} \frac{x^4 + y^4}{x^2 + y^2} = 0$

Example 11.

$$\underset{(x,y)\to(0,0)}{\text{求 }\lim} \frac{x^3y}{x^6 + 2y^2} =?$$

【解】

令 $y = mx^3$ 则 $\dfrac{x^3y}{x^6 + 2y^2} = \dfrac{x^3(mx^3)}{x^6 + 2m^2x^6} = \dfrac{m}{1 + 2m^2}$

取 $m = -1$ 则 $\lim_{(x,y)\to(0,0)} \dfrac{x^3y}{x^6 + 2y^2} = \lim_{(x,y)\to(0,0)} \dfrac{m}{1 + 2m^2} = -\dfrac{1}{3}$

取 $m = 1$ 则 $\lim_{(x,y)\to(0,0)} \dfrac{x^3y}{x^6 + 2y^2} = \lim_{(x,y)\to(0,0)} \dfrac{m}{1 + 2m^2} = \dfrac{1}{3}$

因此 $\lim_{(x,y)\to(0,0)} \dfrac{x^3y}{x^6 + 2y^2}$ 不存在

Example 12.

$$求 \lim_{(x,y)\to(0,0)} \frac{x - y^2}{x + 2y^2} = ?$$

【解】

令 $y = m\sqrt{x}$ 则 $\dfrac{x - y^2}{x + 2y^2} = \dfrac{x - m^2 x}{x + 2m^2 x} = \dfrac{1 - m^2}{1 + 2m^2}$

取 $m = \sqrt{2}$ 则 $\displaystyle\lim_{(x,y)\to(0,0)} \frac{x - y^2}{x + 2y^2} = \lim_{(x,y)\to(0,0)} \frac{1 - m^2}{1 + 2m^2} = \frac{-1}{5}$

取 $m = 1$ 则 $\displaystyle\lim_{(x,y)\to(0,0)} \frac{x - y^2}{x + 2y^2} = \lim_{(x,y)\to(0,0)} \frac{1 - m^2}{1 + 2m^2} = 0$

因此 $\displaystyle\lim_{(x,y)\to(0,0)} \frac{x - y^2}{x + 2y^2}$ 不存在

Example 13.

$$求 \lim_{(x,y)\to(0,0)} \frac{xy^5}{x^2 + y^{10}} = ?$$

【解】

令 $y = (mx)^{\frac{1}{5}}$ 则 $\dfrac{xy^5}{x^2 + y^{10}} = \dfrac{mx^2}{(1 + m^2)x^2} = \dfrac{m}{1 + m^2}$

取 $m = -1$ 则 $\displaystyle\lim_{(x,y)\to(0,0)} \frac{xy^5}{x^2 + y^{10}} = \lim_{(x,y)\to(0,0)} \frac{m}{1 + m^2} = -\frac{1}{2}$

取 $m = 1$ 则 $\displaystyle\lim_{(x,y)\to(0,0)} \frac{xy^5}{x^2 + y^{10}} = \lim_{(x,y)\to(0,0)} \frac{m}{1 + m^2} = \frac{1}{2}$

因此 $\displaystyle\lim_{(x,y)\to(0,0)} \frac{xy^5}{x^2 + y^{10}}$ 不存在

Example 14.

$$求 \lim_{(x,y)\to(0,0)} \frac{x^2 y^2}{\sqrt{x^2 + y^2}} = ?$$

【解】

$$\because x^2 + y^2 \geq 2|xy| \quad \therefore \frac{1}{\sqrt{x^2+y^2}} \leq \frac{1}{\sqrt{2|xy|}} \Rightarrow \left|\frac{x^2y^2}{\sqrt{x^2+y^2}}\right| \leq \left|\frac{x^2y^2}{\sqrt{2|xy|}}\right| = |xy|^{\frac{3}{2}}$$

Claim: $\displaystyle \lim_{(x,y)\to(0,0)} \frac{x^2y^2}{\sqrt{x^2+y^2}} = 0$

令 $\epsilon > 0$ 取 $\delta^{\frac{3}{2}} < \varepsilon$ 则 $x^2 + y^2 < \delta$

$$\Rightarrow \left|\frac{x^2y^2}{\sqrt{x^2+y^2}}\right| \leq \left|\frac{x^2y^2}{\sqrt{2|xy|}}\right| = |xy|^{\frac{3}{2}} < (x^2+y^2)^{\frac{3}{2}} < \delta^{\frac{3}{2}} < \epsilon$$

因此 $\displaystyle \lim_{(x,y)\to(0,0)} \frac{x^2y^2}{\sqrt{x^2+y^2}} = 0$

Example 15.

$$求 \lim_{(x,y)\to(0,0)} \frac{(xy)^p}{\sqrt{x^2+y^2}} =?, \quad \forall p > \frac{1}{2}$$

【解】

令 $p > \dfrac{1}{2}$

$$\because x^2 + y^2 \geq 2|xy| \quad \therefore \frac{1}{\sqrt{x^2+y^2}} \leq \frac{1}{\sqrt{2|xy|}} \Rightarrow \left|\frac{(xy)^p}{\sqrt{x^2+y^2}}\right| \leq \left|\frac{(xy)^p}{\sqrt{2|xy|}}\right| < |xy|^{p-\frac{1}{2}}$$

Claim: $\displaystyle \lim_{(x,y)\to(0,0)} \frac{(xy)^p}{\sqrt{x^2+y^2}} = 0$

令 $\epsilon > 0$ 取 $\delta^{p-\frac{1}{2}} < \varepsilon$

则 $x^2 + y^2 < \delta \Rightarrow \left|\dfrac{(xy)^p}{\sqrt{x^2+y^2}}\right| \leq \left|\dfrac{(xy)^p}{\sqrt{2|xy|}}\right| < |xy|^{p-\frac{1}{2}} < (x^2+y^2)^{p-\frac{1}{2}} < \delta^{p-\frac{1}{2}} < \epsilon$

因此 $\displaystyle \lim_{(x,y)\to(0,0)} \frac{(xy)^p}{\sqrt{x^2+y^2}} = 0$

Example 16.

$$假设 f(x,y) = (x+y)\sin\frac{1}{x}\sin\frac{1}{y} \quad 则 \lim_{x\to 0}\left(\lim_{y\to 0} f(x,y)\right) =?$$

【解】

Claim: $\lim\limits_{y\to 0} f(x,y)$ 不存在

令 $y_{1,n} = \dfrac{1}{2n\pi + \dfrac{\pi}{2}}$ 且 $y_{2,n} = \dfrac{1}{2n\pi + \dfrac{3\pi}{2}}$ 则 $\lim\limits_{n\to\infty} y_{1,n} = 0$ 且 $\lim\limits_{n\to\infty} y_{2,n} = 0$

则 $\lim\limits_{n\to\infty} f(x,y_{1,n}) = \lim\limits_{n\to\infty}\left( x + \dfrac{1}{2n\pi + \dfrac{\pi}{2}} \right)\sin\dfrac{1}{x}\sin\left(2n\pi + \dfrac{\pi}{2}\right) = \sin\dfrac{1}{x}$

$\lim\limits_{n\to\infty} f(x,y_{2,n}) = \lim\limits_{n\to\infty}\left( x + \dfrac{1}{2n\pi + \dfrac{3\pi}{2}} \right) + \sin\dfrac{1}{x}\sin\left(2n\pi + \dfrac{3\pi}{2}\right) = -\sin\dfrac{1}{x}$

令 $x_m = \dfrac{1}{2m\pi + \dfrac{\pi}{2}}$ 则 $\lim\limits_{m\to\infty} x_m = 0$

$\lim\limits_{m\to\infty}\left( \lim\limits_{n\to\infty} f(x_m,y_{1,n}) \right) = \lim\limits_{m\to\infty}\sin\left(2m\pi + \dfrac{\pi}{2}\right) = 1$

$\lim\limits_{m\to\infty}\left( \lim\limits_{n\to\infty} f(x_m,y_{2,n}) \right) = -\lim\limits_{m\to\infty}\sin\left(2m\pi + \dfrac{\pi}{2}\right) = -1$

因此 $\lim\limits_{x\to 0}\left( \lim\limits_{y\to 0} f(x,y) \right)$ 不存在

Example 17.

$$\text{求}\ \lim_{y\to 0}\left( \lim_{x\to 0}\frac{\cos x \tan y}{x - y} \right) = ?$$

【解】

$$\because \lim_{x\to 0}\frac{\cos x \tan y}{x - y} = \frac{\cos 0 \tan y}{-y} = \frac{\tan y}{-y}$$

$$\therefore \lim_{y\to 0}\left( \lim_{x\to 0}\frac{\cos x \tan y}{x - y} \right) = \lim_{y\to 0}\left( \frac{\tan y}{-y} \right) = -\lim_{y\to 0}\left( \frac{\tan y}{y} \right) = -1$$

Example 18.

$$\text{求}\ \lim_{(x,y,z)\to(0,0,0)}\frac{4xy + xz + z^2}{x^2 + 2y^2} = ?$$

【解】

令 $z = y = mx$ 则 $\dfrac{4mx^2 + mx^2 + m^2x^2}{(1 + 2m^2)x^2} = \dfrac{5m + m^2}{1 + 2m^2}$

取 $m = -1$ 则 $\displaystyle\lim_{(x,y,z)\to(0,0,0)} \dfrac{4xy + xz + z^2}{x^2 + 2y^2} = \lim_{(x,y,z)\to(0,0,0)} \dfrac{5m + m^2}{1 + 2m^2} = -\dfrac{4}{3}$

取 $m = 1$ 则 $\displaystyle\lim_{(x,y,z)\to(0,0,0)} \dfrac{4xy + xz + z^2}{x^2 + 2y^2} = \lim_{(x,y,z)\to(0,0,0)} \dfrac{5m + m^2}{1 + 2m^2} = \dfrac{6}{3} = 2$

因此 $\displaystyle\lim_{(x,y,z)\to(0,0,0)} \dfrac{4xy + xz + z^2}{x^2 + 2y^2}$ 不存在

Example 19.

$$\text{求} \quad \lim_{(x,y,z)\to(0,0,0)} \dfrac{3xy}{x^2 + y^2 + z^2} = ?$$

【解】

令 $y = mx$ 且 $z = mx$ 则 $\dfrac{3xy}{x^2 + y^2 + z^2} = \dfrac{3mx^2}{x^2 + m^2x^2 + m^2x^2} = \dfrac{3m}{2m^2 + 1}$

取 $m = -1$ 则 $\displaystyle\lim_{(x,y,z)\to(0,0,0)} \dfrac{3xy}{x^2 + y^2 + z^2} = \lim_{(x,y,z)\to(0,0,0)} \dfrac{3m}{2m^2 + 1} = -1$

取 $m = 1$ 则 $\displaystyle\lim_{(x,y,z)\to(0,0,0)} \dfrac{3xy}{x^2 + y^2 + z^2} = \lim_{(x,y,z)\to(0,0,0)} \dfrac{3m}{2m^2 + 1} = 1$

因此 $\displaystyle\lim_{(x,y,z)\to(0,0,0)} \dfrac{3xy}{x^2 + y^2 + z^2}$ 不存在

Example 20.

$$\text{求} \quad \lim_{(x,y,z)\to(0,0,0)} \dfrac{x^4 + y^4}{x^2 + y^2 + z^4} = ?$$

【解】

$\because x^2 + y^2 + z^2 \geq x^2 + y^2 \quad \therefore \left|\dfrac{x^4 + y^4}{x^2 + y^2 + z^4}\right| \leq \left|\dfrac{x^4 + y^4}{x^2 + y^2}\right|$

Claim: $\displaystyle\lim_{(x,y)\to(0,0)} \dfrac{x^4 + y^4}{x^2 + y^2} = 0$

$\because x^4 + y^4 = (x^2 + y^2)^2 - 2x^2y^2$

$$\therefore \lim_{(x,y)\to(0,0)} \frac{x^4 + y^4}{x^2 + y^2} = \lim_{(x,y)\to(0,0)} \frac{(x^2 + y^2)^2 - 2x^2y^2}{x^2 + y^2}$$

$$= \lim_{(x,y)\to(0,0)} x^2 + y^2 - \lim_{(x,y)\to(0,0)} \frac{2x^2y^2}{x^2 + y^2} = - \lim_{(x,y)\to(0,0)} \frac{2x^2y^2}{x^2 + y^2}$$

Claim: $\lim_{(x,y)\to(0,0)} \dfrac{2x^2y^2}{x^2 + y^2} = 0$

$$\because x^2 + y^2 \geq 2|xy| \quad \therefore \frac{1}{x^2 + y^2} \leq \frac{1}{2|xy|} \Rightarrow \left| \frac{2x^2y^2}{x^2 + y^2} \right| \leq \left| \frac{2x^2y^2}{2xy} \right| = |xy|$$

$$令 \; \epsilon > 0 \; 取 \; \frac{\delta}{2} < \varepsilon \; 则 \; x^2 + y^2 < \delta \Rightarrow \left| \frac{2x^2y^2}{x^2 + y^2} \right| \leq \left| \frac{2x^2y^2}{2xy} \right| = |xy| < \left| \frac{x^2 + y^2}{2} \right| < \frac{\delta}{2} < \epsilon$$

$$因此 \; \lim_{(x,y)\to(0,0)} \frac{2x^2y^2}{x^2 + y^2} = 0 \Rightarrow \lim_{(x,y)\to(0,0)} \frac{x^4 + y^4}{x^2 + y^2} = 0$$

$$\Rightarrow \lim_{(x,y,z)\to(0,0,0)} \frac{x^4 + y^4}{x^2 + y^2 + z^4} = 0$$

## 7.2  判断多变量函数是否连续

【**定义**】双变量函数连续的定义

$$f(x,y) 于 (x_0, y_0) 连续 \Leftrightarrow \lim_{(x,y)\to(x_0,y_0)} f(x,y) = f(x_0, y_0)$$

$$\Leftrightarrow \forall \varepsilon > 0, \; \exists \delta > 0, \; \text{such that} \; \sqrt{(x - x_0)^2 + (y - y_0)^2} < \delta \Rightarrow |f(x,y) - f(x_0, y_0)| < \varepsilon$$

【**定义**】多变量函数连续的定义

$$f(x_1, \ldots, x_n) 于 (c_1, \ldots, c_n) 连续 \Leftrightarrow \lim_{(x_1,\ldots,x_n)\to(c_1,\ldots,c_n)} f(x_1, \ldots, x_n) = f(c_1, \ldots, c_n)$$

$$\Leftrightarrow \forall \varepsilon > 0, \; \exists \delta > 0, \; \text{such that}$$
$$\sqrt{(x_1 - c_1)^2 + \cdots + (x_n - c_n)^2} < \delta \Rightarrow |f(x_1, \ldots, x_n) - f(c_1, \ldots, c_n)| < \varepsilon$$

从双变量函数的连续定义，能观察出如果 $f(x,y)$ 于 $(x_0, y_0)$ 连续，则从任意方向逼近 $(x_0, y_0)$ 的极限值必须等于 $f(x_0, y_0)$，因此，令 $y - y_0 = m(x - x_0)$ 为过 $(x_0, y_0)$ 的直线，若 $f(x,y)$ 于 $(x_0, y_0)$ 连续，则

$$\lim_{(x,y)\to(x_0,y_0)} f(x, m(x-x_0)+y_0) = f(x_0, y_0)$$

且

$$\lim_{(x,y)\to(x_0,y_0)} f(x, m(x-x_0)+y_0) \text{ 与} m \text{值无关}$$

常见的考试题型为 $(x_0, y_0) = (0,0)$, 假设 $f(x,y)$ 于 $(0,0)$ 连续, 令 $y = mx$ 则

$\lim\limits_{(x,y)\to(0,0)} f(x, mx) = f(0,0)$ 且 $\lim\limits_{(x,y)\to(0,0)} f(x, mx)$ 与 $m$ 值无关, 因此, $\lim\limits_{(x,y)\to(0,0)} f(x, mx)$

与 $m$ 值无关是 $f(x,y)$ 于 $(0,0)$ 连续的必要条件, 值得注意的是, $\lim\limits_{(x,y)\to(x_0,y_0)} f(x, mx)$

与 $m$ 值无关是 $f(x,y)$ 于 $(0,0)$ 连续的必要条件并非充分条件, 底下展示一双变数函数 $f(x,y)$ 作说明

范例说明:

假设 $f(x,y) = \begin{cases} \dfrac{xy^2}{x^2+y^4}, & (x,y) \neq 0 \\ 0, & (x,y) = 0 \end{cases}$ , 判断 $f(x,y)$ 在 $(0,0)$ 是否连续?

令 $y = mx$ 则 $f(x,y) = \dfrac{xy^2}{x^2+y^4} = \dfrac{m^2 x^3}{x^2 + m^4 x^4}$

$\therefore \lim\limits_{(x,y)\to(0,0)} f(x, mx)$ 与 $m$ 值无关且 $\lim\limits_{(x,y)\to(0,0)} f(x, mx) = 0$,

令 $x = y^2$ 则 $f(x,y) = \dfrac{y^4}{y^4 + y^4} = \dfrac{1}{2}$ 且 $\lim\limits_{(x,y)\to(0,0)} f(y^2, y) = \dfrac{1}{2} \neq f(0,0) = 0$

$\therefore f(x,y)$ 于 $(0,0)$ 不连续

此外, 如果 $\exists\, m \in R$ s.t. $\lim\limits_{(x,y)\to(0,0)} f(x, mx) \neq f(0,0)$ 则 $f(x,y)$ 于 $(0,0)$ 不连续, 因此,

令 $y = mx$, 当 $\lim\limits_{(x,y)\to(0,0)} f(x, mx)$ 与 $m$ 值有关, 则取 $m \in R$ 使得 $\lim\limits_{(x,y)\to(0,0)} f(x, mx) \neq f(0,0)$

以说明 $f(x,y)$ 于 $(0,0)$ 不连续, 判断多变量函数是否连续的考试类型当中, 以双变量函数的题型最常出现, 读者若能熟悉双变量题型, 其它类型都能够掌握。总结上述两点, 底下说明双变量函数 $f(x,y)$ 于原点 $(0,0)$ 是否连续的考试类型与解题流程

考试类型:

Type 1.

給定双变数函数$f(x,y)$, 判断$f(x,y)$于$(0,0)$是否连续, 其中$f(x,y)$满足:

$$令 y = mx, \quad 当 \lim_{(x,y)\to(0,0)} f(x,mx) = f(0,0) \ 与 m 值无关时$$

解题流程:

令$\varepsilon > 0$, 取$\delta$或明确找出$\delta$的值, 使得$x^2 + y^2 < \delta \Rightarrow |f(x,y) - f(0,0)| < \varepsilon$

补充说明:

常藉由 $x^2 + y^2 \geq 2|xy|$, 推导出$|f(x,y) - f(0,0)| < g(x^2 + y^2) < g(\delta) < \varepsilon$, 并且能得知$g$明确的样貌, 最后一个不等式的$g(\delta) < \varepsilon$能帮助明确定义$\delta$

范例说明:

令$y = mx$

(I)设$f(x,y) = \begin{cases} \dfrac{3x^2 y}{x^2 + y^2}, & (x,y) \neq 0 \\ 0, & (x,y) = 0 \end{cases}$, 则 $f(x,y)$ 在$(0,0)$ 是否连续?

令 $y = mx$ 则 $\lim_{(x,y)\to(0,0)} \dfrac{3x^2(mx)}{x^2 + (mx)^2} = \lim_{(x,y)\to(0,0)} \dfrac{3mx}{1 + m^2} = 0$ 与$m$值无关

Claim: $\left| \dfrac{3x^2 y}{x^2 + y^2} \right| \leq \left| \dfrac{3\sqrt{x^2 + y^2}}{2} \right|$

$\because x^2 + y^2 \geq 2|xy| \quad \therefore \dfrac{1}{x^2 + y^2} \leq \dfrac{1}{2|xy|} \Rightarrow \left| \dfrac{3x^2 y}{x^2 + y^2} \right| \leq \left| \dfrac{3x^2 y}{2xy} \right| = \left| \dfrac{3x}{2} \right| \leq \left| \dfrac{3\sqrt{x^2 + y^2}}{2} \right|$

令 $\epsilon > 0$ 取 $\delta < (\dfrac{2\varepsilon}{3})^2$ 则 $x^2 + y^2 < \delta \Rightarrow \left| \dfrac{3x^2 y}{x^2 + y^2} \right| \leq \left| \dfrac{3\sqrt{x^2 + y^2}}{2} \right| < \dfrac{3\sqrt{\delta}}{2} < \dfrac{3}{2} \cdot \dfrac{2\varepsilon}{3} = \epsilon$

因此 $f(x,y)$ 在$(0,0)$连续

(II)若$f(x,y) = \begin{cases} \dfrac{x^3 - y^3}{x^2 + y^2}, & (x,y) \neq 0 \\ 0, & (x,y) = 0 \end{cases}$, 则 $f(x,y)$ 在$(0,0)$ 是否连续?

令 $y = mx$ 则 $\lim_{(x,y)\to(0,0)} \dfrac{x^3 - y^3}{x^2 + y^2} = \lim_{(x,y)\to(0,0)} \dfrac{x - m^3 x}{1 + m^2} = 0$ 与$m$值无关

$\because x^3 - y^3 = (x-y)(x^2 + xy + y^2)$ 且 $x^2 + y^2 \geq 2|xy|$

$\therefore |x^3 - y^3| \leq |x-y||x^2 + |xy| + y^2| \leq |x-y|\left|x^2 + \dfrac{x^2+y^2}{2} + y^2\right|$

$\therefore |x^3 - y^3| \leq \dfrac{3}{2}|x-y||x^2+y^2| \Rightarrow \left|\dfrac{x^3-y^3}{x^2+y^2}\right| \leq \dfrac{3}{2}|x-y|$

$\because |x-y| = \sqrt{|x-y|^2} \leq \sqrt{x^2+y^2+|2xy|} \leq \sqrt{2(x^2+y^2)}$

令 $\epsilon > 0$ 取 $3\sqrt{\dfrac{\delta}{2}} < \epsilon$ 则 $x^2+y^2 < \delta \Rightarrow \left|\dfrac{x^3-y^3}{x^2+y^2}\right| \leq \dfrac{3}{2}|x-y| \leq \dfrac{3}{2}\sqrt{2(x^2+y^2)} < 3\sqrt{\dfrac{\delta}{2}} < \epsilon$

因此 $f(x,y)$ 在 $(0,0)$ 连续

Type 2.

給定双变数函数 $f(x,y)$，判断 $f(x,y)$ 于 $(0,0)$ 是否连续，其中 $f(x,y)$ 满足：

$$\text{令 } y = mx, \quad \text{当} \lim_{(x,y)\to(0,0)} f(x,mx) \text{ 与 } m \text{ 值有关时}$$

解题流程：

取 $m$ 值且计算极限值使得 $\lim\limits_{(x,y)\to(0,0)} f(x,mx) \neq f(0,0)$ $\quad \therefore f(x,y)$ 于 $(0,0)$ 不连续

<u>范例说明：</u>

令 $y = mx$

(I) 设 $f(x,y) = \begin{cases} \dfrac{x^2-y^2}{x^2+y^2}, & (x,y) \neq 0 \\ 0, & (x,y) = 0 \end{cases}$ ，则 $f(x,y)$ 在 $(0,0)$ 是否连续？

令 $y = mx$ 则 $\lim\limits_{(x,y)\to(0,0)} \dfrac{x^2-y^2}{x^2+y^2} = \lim\limits_{(x,y)\to(0,0)} \dfrac{1-m^2}{1+m^2}$ 与 $m$ 值有关

取 $m = \sqrt{2}$ 则 $\lim\limits_{(x,y)\to(0,0)} f(x,y) = \lim\limits_{(x,y)\to(0,0)} \dfrac{1-m^2}{1+m^2} = -\dfrac{1}{3} \neq f(0,0) = 0$

因此 $f(x,y)$ 在 $(0,0)$ 不连续

(II) 设 $f(x,y) = \begin{cases} \dfrac{x^2 y}{x^3+y^3}, & (x,y) \neq 0 \\ 0, & (x,y) = 0 \end{cases}$ ，则 $f(x,y)$ 在 $(0,0)$ 是否连续？

令 $y = mx$ 则 $\lim\limits_{(x,y)\to(0,0)} \dfrac{x^2 y}{x^3+y^3} = \lim\limits_{(x,y)\to(0,0)} \dfrac{m}{1+m^3}$ 与 $m$ 值有关

取 $m = 1$ 则 $\lim\limits_{(x,y)\to(0,0)} f(x,y) = \lim\limits_{(x,y)\to(0,0)} \dfrac{m}{1+m^3} = \dfrac{1}{2} \neq f(0,0)$

因此 $f(x,y)$ 在 $(0,0)$ 不连续

Type 3.

給定三变数函数 $f(x,y,z)$，判断 $f(x,y,z)$ 于 $(0,0,0)$ 是否连续，其中 $f(x,y,z)$ 满足：

$$令 y = z = mx,\ 当 \lim\limits_{(x,y,z)\to(0,0,0)} f(x,mx,mx) = L\ 与\ m\ 无关时$$

解题流程：

令 $\varepsilon > 0$，取 $\delta$ 或明确找出 $\delta$ 的值，使得 $x^2 + y^2 + z^2 < \delta \Rightarrow |f(x,y,z) - f(0,0,0)| < \varepsilon$

<u>范例说明：</u>

(I) 设 $f(x,y,z) = \begin{cases} \dfrac{x^2 y}{x^2 + y^4 + z^6}, & (x,y,z) \neq 0 \\ 0, & (x,y,z) = 0 \end{cases}$，则 $f(x,y,z)$ 在 $(0,0,0)$ 是否连续？

令 $y = z = mx$ 则 $\dfrac{x^2 y}{x^2 + y^4 + z^6} = \dfrac{mx^3}{x^2 + m^4 x^4 + m^6 x^6} = \dfrac{mx}{1 + m^4 x^2 + m^6 x^4}$

$\therefore \lim\limits_{(x,y,z)\to(0,0,0)} f(x,mx,mx) = 0$ 与 $m$ 无关

$\because x^2 + y^4 + z^6 \geq x^2 \qquad \therefore \left| \dfrac{x^2 y}{x^2 + y^4 + z^6} \right| \leq \left| \dfrac{x^2 y}{x^2} \right| = |y|$

令 $\epsilon > 0$ 取 $\delta^{\frac{1}{2}} < \varepsilon$

则 $x^2 + y^2 + z^2 < \delta \Rightarrow \left| \dfrac{x^2 y}{x^2 + y^4 + z^6} \right| < |y| < (x^2 + y^2 + z^2)^{\frac{1}{2}} < \delta^{\frac{1}{2}} < \epsilon$

$\therefore f(x,y,z)$ 在 $(0,0,0)$ 是否连续

值得注意的是 $\lim\limits_{(x,y,z)\to(0,0,0)} f(x,mx,mx) = 0$ 与 $m$ 无关是 $f(x,y,z)$ 于 $(0,0,0)$ 连续的必要条件

并非充分条件，也就是推论的过程中，如果出现

$$"因为 \lim\limits_{(x,y,z)\to(0,0,0)} f(x,mx,mx) = 0\ 与\ m\ 无关，所以 f(x,y,z) 于 (0,0,0) 连续"$$

为错误的论述，底下展示一多变数函数做说明

范例说明:

假设 $f(x,y,z) = \begin{cases} \dfrac{xyz}{x^2 + y^4 + z^4}, & (x,y,z) \neq 0 \\ 0, & (x,y,z) = 0 \end{cases}$ ，判断 $f(x,y,z)$ 在 $(0,0,0)$ 是否连续

令 $y = z = mx$ 则 $f(x,y,z) = \dfrac{xyz}{x^2 + y^4 + z^4} = \dfrac{m^3 x^3}{x^2 + m^4 x^4 + m^4 x^4} = \dfrac{m^3 x}{1 + m^4 x^2 + m^4 x^2}$

$\therefore \lim\limits_{(x,y,z)\to(0,0,0)} f(x,mx,mx) = 0$ 与 $m$ 无关且 $\lim\limits_{(x,y,z)\to(0,0,0)} f(x,mx,mx) = 0$，然而

令 $x = mt^2$，$y = z = t$ 则 $f(x,y,z) = \dfrac{mt^4}{(m^2+2)t^4} = \dfrac{m}{m^2+2}$

取 $m = -1$ 则 $\lim\limits_{(x,y,z)\to(0,0,0)} \dfrac{xyz}{x^2+y^4+z^4} = \lim\limits_{(x,y,z)\to(0,0,0)} \dfrac{m}{m^2+2} = -\dfrac{1}{3}$

取 $m = 1$ 则 $\lim\limits_{(x,y,z)\to(0,0,0)} \dfrac{xyz}{x^2+y^4+z^4} = \lim\limits_{(x,y,z)\to(0,0,0)} \dfrac{m}{m^2+2} = \dfrac{1}{3}$

$\therefore \lim\limits_{(x,y,z)\to(0,0,0)} \dfrac{xyz}{x^2+y^4+z^4}$ 不存在　　$\therefore f(x,y,z)$ 在 $(0,0,0)$ 不连续

Type 4.

给定三变数函数 $f(x,y,z)$，判断 $f(x,y,z)$ 于 $(0,0,0)$ 是否连续，其中 $f(x,y,z)$ 满足:

$$\text{令} y = z = mx, \quad \text{当} \lim\limits_{(x,y,z)\to(0,0,0)} f(x,mx,mx) \text{ 与 } m \text{有关时}$$

解题流程:

取两个相异 $m_1$、$m_2$ 使得

$$\lim\limits_{(x,y,z)\to(0,0,0)} f(x,m_1x,m_1x) \neq \lim\limits_{(x,y,z)\to(0,0,0)} f(x,m_2x,m_2x) \quad \therefore \lim\limits_{(x,y,z)\to(0,0,0)} f(x,y,z) \text{ 不连续}$$

范例说明:

(I) 设 $f(x,y,z) = \begin{cases} \dfrac{3xy}{x^2 + y^2 + z^2}, & (x,y,z) \neq 0 \\ 0, & (x,y,z) = 0 \end{cases}$ ，则 $f(x,y,z)$ 在 $(0,0,0)$ 是否连续?

令 $y = z = mx$ 则 $\dfrac{3xy}{x^2+y^2+z^2} = \dfrac{3mx^2}{x^2 + m^2x^2 + m^2x^2} = \dfrac{3m}{2m^2+1}$

取 $m = -1$ 则 $\lim\limits_{(x,y,z)\to(0,0,0)} \dfrac{3xy}{x^2+y^2+z^2} = \lim\limits_{(x,y,z)\to(0,0,0)} \dfrac{3m}{2m^2+1} = -1$

取 $m = 1$ 则 $\displaystyle \lim_{(x,y,z)\to(0,0,0)} \frac{3xy}{x^2+y^2+z^2} = \lim_{(x,y,z)\to(0,0,0)} \frac{3m}{2m^2+1} = 1$

$\therefore \displaystyle \lim_{(x,y,z)\to(0,0,0)} \frac{3xy}{x^2+y^2+z^2}$ 不存在 $\quad \therefore f(x,y,z)$ 在 $(0,0,0)$ 不连续

**Example 1.**

$$设 f(x,y) = \begin{cases} (x+y)\sin\dfrac{1}{x}\sin\dfrac{1}{y}, & (x,y) \neq (0,0) \\ 0, & (x,y) = (0,0) \end{cases}, \quad 则 f(x,y) 在 (0,0)$$

是否连续?

【解】

$\because (x+y)^2 = x^2 + y^2 + 2xy \leq (|x|+|y|)^2 = x^2 + y^2 + 2|xy|$ 且 $x^2 + y^2 \geq 2|xy|$

$\therefore (x+y)^2 \leq 2(x^2+y^2)$

令 $\epsilon > 0$ 取 $\sqrt{2\delta} < \epsilon$

则 $x^2 + y^2 < \delta \Rightarrow \left| (x+y)\sin\dfrac{1}{x}\sin\dfrac{1}{y} \right| \leq |x+y| < \sqrt{2(x^2+y^2)} < \sqrt{2\delta} < \epsilon$

$\therefore \displaystyle \lim_{(x,y)\to(0,0)} f(x,y) = 0 = f(0,0), \quad$ 因此 $f(x,y)$ 在 $(0,0)$ 连续

**Example 2.**

$$设 f(x,y) = \begin{cases} \dfrac{x^2-y^2}{x^2+y^2}, & (x,y) \neq 0 \\ 0, & (x,y) = 0 \end{cases}, \quad 则 f(x,y) 在 (0,0) 是否连续?$$

【解】

令 $y = mx$ 则 $\dfrac{x^2-y^2}{x^2+y^2} = \dfrac{x^2-m^2x^2}{x^2+m^2x^2} = \dfrac{x^2(1-m^2)}{x^2(1+m^2)} = \dfrac{1-m^2}{1+m^2}$

$\therefore \displaystyle \lim_{(x,y)\to(0,0)} \frac{x^2-y^2}{x^2+y^2} = \lim_{(x,y)\to(0,0)} \frac{1-m^2}{1+m^2}$

取 $m = \sqrt{2}$ 则 $\displaystyle \lim_{(x,y)\to(0,0)} f(x,y) = \lim_{(x,y)\to(0,0)} \frac{x^2-y^2}{x^2+y^2} = \lim_{(x,y)\to(0,0)} \frac{1-m^2}{1+m^2} = -\frac{1}{3} \neq f(0,0) = 0$

因此 $f(x,y)$ 在 $(0,0)$ 不连续

Example 3.

$$设 f(x,y) = \begin{cases} \dfrac{x^2 y}{x^4 + y^2}, & (x,y) \neq 0 \\ 0, & (x,y) = 0 \end{cases}, \ 则 f(x,y) 在(0,0) 是否连续?$$

【解】

$$令 \ y = mx^2 \ 则 \ \frac{x^2 y}{x^4 + y^2} = \frac{x^2(m^2 x^2)}{x^4 + m^4 x^4} = \frac{m^2}{1 + m^4} \quad \therefore \lim_{(x,y)\to(0,0)} \frac{x^2 y}{x^4 + y^2} = \lim_{(x,y)\to(0,0)} \frac{m^2}{1 + m^4}$$

$$取 m = 1 \ 则 \lim_{(x,y)\to(0,0)} f(x,y) = \lim_{(x,y)\to(0,0)} \frac{x^2 y}{x^4 + y^2} = \lim_{(x,y)\to(0,0)} \frac{m^2}{1 + m^4} = \frac{1}{2} \neq f(0,0)$$

因此 $f(x,y)$ 在 $(0,0)$ 不连续

Example 4.

$$设 f(x,y) = \begin{cases} \dfrac{x^2 y}{x^3 + y^3}, & (x,y) \neq 0 \\ 0, & (x,y) = 0 \end{cases}, \ 则 f(x,y) 在(0,0) 是否连续?$$

【解】

$$令 \ y = mx \ 则 \ \frac{x^2 y}{x^3 + y^3} = \frac{mx^3}{x^3 + m^3 x^3} = \frac{m}{1 + m^3}$$

$$\therefore \lim_{(x,y)\to(0,0)} \frac{x^2 y}{x^3 + y^3} = \lim_{(x,y)\to(0,0)} \frac{m}{1 + m^3}$$

$$取 m = 1 \ 则 \lim_{(x,y)\to(0,0)} f(x,y) = \lim_{(x,y)\to(0,0)} \frac{x^2 y}{x^3 + y^3} = \lim_{(x,y)\to(0,0)} \frac{m}{1 + m^3} = \frac{1}{2} \neq f(0,0)$$

因此 $f(x,y)$ 在 $(0,0)$ 不连续

Example 5.

$$设 f(x,y) = \begin{cases} \dfrac{x^3 - y^3}{x^2 + y^2}, & (x,y) \neq 0 \\ 0, & (x,y) = 0 \end{cases}, \ 则 f(x,y) 在(0,0) 是否连续?$$

【解】

$$\because x^3 - y^3 = (x - y)(x^2 + xy + y^2) \ 且 \ x^2 + y^2 \geq 2|xy|$$

$$\therefore |x^3 - y^3| \leq |x - y||x^2 + |xy| + y^2| \leq |x - y| \left| x^2 + \frac{x^2 + y^2}{2} + y^2 \right|$$

$$\therefore |x^3 - y^3| \leq \frac{3}{2}|x - y||x^2 + y^2| \Rightarrow \left| \frac{x^3 - y^3}{x^2 + y^2} \right| \leq \frac{3}{2}|x - y|$$

$$\because |x - y| = \sqrt{|x-y|^2} \leq \sqrt{x^2 + y^2 + |2xy|} \leq \sqrt{2(x^2 + y^2)}$$

$$令 \epsilon > 0 \ 取 \ 3\sqrt{\frac{\delta}{2}} < \epsilon \ 则 \ x^2 + y^2 < \delta \Rightarrow \left|\frac{x^3 - y^3}{x^2 + y^2}\right| \leq \frac{3}{2}|x - y| \leq \frac{3}{2}\sqrt{2(x^2 + y^2)} < 3\sqrt{\frac{\delta}{2}} < \epsilon$$

因此 $f(x,y)$ 在 $(0,0)$ 连续

Example 6.

$$设 f(x,y) = \begin{cases} \dfrac{3x^2 y}{x^2 + y^2}, & (x,y) \neq 0 \\ 0, & (x,y) = 0 \end{cases}, \ 则 \ f(x,y) \ 在(0,0) \ 是否连续?$$

【解】

$$\text{Claim:} \ \left|\frac{3x^2 y}{x^2 + y^2}\right| \leq \left|\frac{3\sqrt{x^2 + y^2}}{2}\right|$$

$$\because x^2 + y^2 \geq 2|xy| \quad \therefore \frac{1}{x^2 + y^2} \leq \frac{1}{2|xy|} \Rightarrow \left|\frac{3x^2 y}{x^2 + y^2}\right| \leq \left|\frac{3x^2 y}{2xy}\right| = \left|\frac{3x}{2}\right| \leq \left|\frac{3\sqrt{x^2 + y^2}}{2}\right|$$

$$令 \epsilon > 0 \ 取 \ \delta < \left(\frac{2\epsilon}{3}\right)^2 \ 则 \ x^2 + y^2 < \delta \Rightarrow \left|\frac{3x^2 y}{x^2 + y^2}\right| \leq \left|\frac{3\sqrt{x^2 + y^2}}{2}\right| < \frac{3\sqrt{\delta}}{2} < \frac{3}{2} \cdot \frac{2\epsilon}{3} = \epsilon$$

因此 $f(x,y)$ 在 $(0,0)$ 连续

Example 7.

$$设 f(x,y) = \begin{cases} \dfrac{2xy^2}{x^2 + y^2}, & (x,y) \neq 0 \\ 0, & (x,y) = 0 \end{cases}, \ 则 \ f(x,y) \ 在(0,0) \ 是否连续?$$

【解】

$$\text{Claim:} \ \left|\frac{2xy^2}{x^2 + y^2}\right| \leq \left|\sqrt{x^2 + y^2}\right|$$

$$\because x^2 + y^2 \geq 2|xy| \quad \therefore \frac{1}{x^2 + y^2} \leq \frac{1}{2|xy|} \Rightarrow \left|\frac{2xy^2}{x^2 + y^2}\right| \leq \left|\frac{2xy^2}{2xy}\right| = |y| < \left|\sqrt{x^2 + y^2}\right|$$

$$令 \epsilon > 0 \ 取 \ \sqrt{\delta} < \varepsilon$$

则 $x^2 + y^2 < \delta \Rightarrow \left| \dfrac{2xy^2}{x^2 + y^2} \right| \leq \left| \sqrt{x^2 + y^2} \right| < \sqrt{\delta} < \epsilon \Rightarrow f(x,y)$ 在 $(0,0)$ 连续

Example 8.

$$\text{设} f(x,y) = \begin{cases} \dfrac{x^4 + y^4}{x^2 + y^2}, & (x,y) \neq 0 \\ 0, & (x,y) = 0 \end{cases}, \quad \text{则 } f(x,y) \text{ 在} (0,0) \text{ 是否连续?}$$

【解】

Claim: $\left| \dfrac{x^4 + y^4}{x^2 + y^2} \right| \leq \dfrac{3}{2} |x^2 + y^2|$

$\because x^4 + y^4 = (x^2 + y^2)^2 - 2x^2y^2$  $\therefore \dfrac{x^4 + y^4}{x^2 + y^2} = \dfrac{(x^2 + y^2)^2 - 2x^2y^2}{x^2 + y^2} = x^2 + y^2 - \dfrac{2x^2y^2}{x^2 + y^2}$

$\because x^2 + y^2 \geq 2|xy|$  $\therefore \dfrac{1}{x^2 + y^2} \leq \dfrac{1}{2|xy|} \Rightarrow \left| \dfrac{2x^2y^2}{x^2 + y^2} \right| \leq \left| \dfrac{2x^2y^2}{2xy} \right| = |xy| \leq \left| \dfrac{x^2 + y^2}{2} \right|$

$\therefore \left| \dfrac{x^4 + y^4}{x^2 + y^2} \right| = \left| x^2 + y^2 - \dfrac{2x^2y^2}{x^2 + y^2} \right| \leq |x^2 + y^2| + \left| \dfrac{2x^2y^2}{x^2 + y^2} \right| \leq \dfrac{3}{2} |x^2 + y^2|$

令 $\epsilon > 0$ 取 $\dfrac{3\delta}{2} < \varepsilon$

则 $x^2 + y^2 < \delta \Rightarrow \left| \dfrac{x^4 + y^4}{x^2 + y^2} \right| \leq \dfrac{3}{2} |x^2 + y^2| < \dfrac{3\delta}{2} < \epsilon \Rightarrow f(x,y)$ 在 $(0,0)$ 连续

Example 9.

$$\text{设} f(x,y) = \begin{cases} \dfrac{xy^4}{x^2 + y^8}, & (x,y) \neq 0 \\ 0, & (x,y) = 0 \end{cases}, \quad \text{则 } f(x,y) \text{ 在} (0,0) \text{ 是否连续?}$$

【解】

令 $y = (mx)^{\frac{1}{4}}$ 则 $\dfrac{xy^4}{x^2 + y^8} = \dfrac{mx^2}{(1 + m^2)x^2} = \dfrac{m}{1 + m^2}$

取 $m = -1$ 则 $\displaystyle\lim_{(x,y)\to(0,0)} \dfrac{xy^4}{x^2 + y^8} = \lim_{(x,y)\to(0,0)} \dfrac{m}{1 + m^2} = -\dfrac{1}{2} \neq 0 = f(0,0)$

因此 $f(x,y)$ 在 $(0,0)$ 不连续

Example 10.

设 $f(x,y) = \begin{cases} \dfrac{x^2 y^2}{\sqrt{x^2 + y^2}}, & (x,y) \neq 0 \\ 0, & (x,y) = 0 \end{cases}$ ，则 $f(x,y)$ 在 $(0,0)$ 是否连续?

【解】

$\because x^2 + y^2 \geq 2|xy| \quad \therefore \dfrac{1}{\sqrt{x^2+y^2}} \leq \dfrac{1}{\sqrt{2|xy|}} \Rightarrow \left|\dfrac{x^2 y^2}{\sqrt{x^2+y^2}}\right| \leq \left|\dfrac{x^2 y^2}{\sqrt{2|xy|}}\right| < |xy|^{\frac{3}{2}}$

Claim: $\displaystyle\lim_{(x,y)\to(0,0)} \dfrac{x^2 y^2}{\sqrt{x^2+y^2}} = 0$

令 $\epsilon > 0$ 取 $\delta^{\frac{3}{2}} < \varepsilon$ 则 $x^2 + y^2 < \delta \Rightarrow \left|\dfrac{x^2 y^2}{\sqrt{x^2+y^2}}\right| \leq \left|\dfrac{x^2 y^2}{\sqrt{2|xy|}}\right| < |xy|^{\frac{3}{2}} < (x^2+y^2)^{\frac{3}{2}} < \delta^{\frac{3}{2}} < \epsilon$

$\therefore \displaystyle\lim_{(x,y)\to(0,0)} \dfrac{x^2 y^2}{\sqrt{x^2+y^2}} = 0 \Rightarrow f(x,y)$ 在 $(0,0)$ 连续

Example 11.

设 $f(x,y) = \begin{cases} \dfrac{(xy)^p}{\sqrt{x^2 + y^2}}, & (x,y) \neq 0 \\ 0, & (x,y) = 0 \end{cases}$ ，则 $f(x,y)$ 在 $(0,0)$ 是否连续?, $\forall p > \dfrac{1}{2}$

【解】

令 $p > \dfrac{1}{2}$

$\because x^2 + y^2 \geq 2|xy| \quad \therefore \dfrac{1}{\sqrt{x^2+y^2}} \leq \dfrac{1}{\sqrt{2|xy|}} \Rightarrow \left|\dfrac{(xy)^p}{\sqrt{x^2+y^2}}\right| \leq \left|\dfrac{(xy)^p}{\sqrt{2|xy|}}\right| < |xy|^{p-\frac{1}{2}}$

Claim: $\displaystyle\lim_{(x,y)\to(0,0)} \dfrac{(xy)^p}{\sqrt{x^2+y^2}} = 0$

令 $\epsilon > 0$ 取 $\delta^{p-\frac{1}{2}} < \varepsilon$

则 $x^2 + y^2 < \delta \Rightarrow \left|\dfrac{(xy)^p}{\sqrt{x^2+y^2}}\right| \leq \left|\dfrac{(xy)^p}{\sqrt{2|xy|}}\right| < |xy|^{p-\frac{1}{2}} < (x^2+y^2)^{p-\frac{1}{2}} < \delta^{p-\frac{1}{2}} < \epsilon$

因此 $\displaystyle\lim_{(x,y)\to(0,0)} \dfrac{(xy)^p}{\sqrt{x^2+y^2}} = 0 \Rightarrow f(x,y)$ 在 $(0,0)$ 连续

Example 12.

设 $f(x,y) = \begin{cases} \dfrac{x^2 + y^2}{|x| + |y|}, & (x,y) \neq 0 \\ 0, & (x,y) = 0 \end{cases}$ ，则 $f(x,y)$ 在 $(0,0)$ 是否连续？

【解】

$\because (|x| + |y|)^2 = |x|^2 + |y|^2 + 2|xy| \geq x^2 + y^2 \quad \therefore |x| + |y| \geq \sqrt{x^2 + y^2}$

$\therefore \dfrac{x^2 + y^2}{|x| + |y|} \leq \dfrac{x^2 + y^2}{\sqrt{x^2 + y^2}} = \sqrt{x^2 + y^2}$

Claim: $\displaystyle\lim_{(x,y)\to(0,0)} \dfrac{x^2 + y^2}{|x| + |y|} = 0$

令 $\epsilon > 0$ 取 $\delta^{\frac{1}{2}} < \varepsilon$ 则 $x^2 + y^2 < \delta \Rightarrow \dfrac{x^2 + y^2}{|x| + |y|} \leq \dfrac{x^2 + y^2}{\sqrt{x^2 + y^2}} = \sqrt{x^2 + y^2} < \delta^{\frac{1}{2}} < \epsilon$

$\therefore \displaystyle\lim_{(x,y)\to(0,0)} \dfrac{x^2 + y^2}{|x| + |y|} = 0 \quad \Rightarrow f(x,y)$ 在 $(0,0)$ 连续

Example 13.

设 $f(x,y) = \begin{cases} \dfrac{xy\cos(x^2 + y^2)}{\sqrt{x^2 + y^2}}, & (x,y) \neq 0 \\ 0, & (x,y) = 0 \end{cases}$ ，则 $f(x,y)$ 在 $(0,0)$ 是否连续？

【解】

$\because x^2 + y^2 \geq 2|xy| \quad \therefore \dfrac{1}{\sqrt{x^2 + y^2}} \leq \dfrac{1}{\sqrt{2|xy|}} \Rightarrow \left| \dfrac{xy\cos x^2 + y^2}{\sqrt{x^2 + y^2}} \right| \leq \left| \dfrac{xy}{\sqrt{2|xy|}} \right| < |xy|^{\frac{1}{2}}$

Claim: $\displaystyle\lim_{(x,y)\to(0,0)} \dfrac{xy\cos(x^2 + y^2)}{\sqrt{x^2 + y^2}} = 0$

令 $\epsilon > 0$ 取 $\delta^{\frac{1}{2}} < \varepsilon$

则 $x^2 + y^2 < \delta \Rightarrow \left| \dfrac{xy\cos(x^2 + y^2)}{\sqrt{x^2 + y^2}} \right| \leq \left| \dfrac{xy}{\sqrt{2|xy|}} \right| < |xy|^{\frac{1}{2}} < (x^2 + y^2)^{\frac{1}{2}} < \delta^{\frac{1}{2}} < \epsilon$

$\therefore \displaystyle\lim_{(x,y)\to(0,0)} \dfrac{xy\cos(x^2 + y^2)}{\sqrt{x^2 + y^2}} = 0 \Rightarrow f(x,y)$ 在 $(0,0)$ 连续

Example 14.

设 $f(x,y,z) = \begin{cases} \dfrac{\sin(x^2 + y^2 + z^2)}{\sqrt{x^2 + y^2 + z^2}}, & (x,y,z) \neq 0 \\ 0, & (x,y,z) = 0 \end{cases}$ ，

则 $f(x,y,z)$ 在 $(0,0,0)$ 是否连续?

【解】

令 $t = x^2 + y^2 + z^2$ 则 $\dfrac{\sin(x^2+y^2+z^2)}{\sqrt{x^2+y^2+z^2}} = \dfrac{\sin t}{\sqrt{t}} = \dfrac{\sqrt{t}\sin t}{t}$

$$\lim_{(x,y,z)\to(0,0,0)} \frac{\sin(x^2+y^2+z^2)}{\sqrt{x^2+y^2+z^2}} = \lim_{t\to 0} \frac{\sqrt{t}\sin t}{t}$$

藉由罗比达法则 $\lim\limits_{t\to 0}\dfrac{\sin t}{t} = \lim\limits_{t\to 0}\dfrac{\cos t}{1} = 1 \Rightarrow \lim\limits_{t\to 0}\dfrac{\sqrt{t}\sin t}{t} = 0$

$\therefore \lim\limits_{(x,y,z)\to(0,0,0)} \dfrac{\sin(x^2+y^2+z^2)}{\sqrt{x^2+y^2+z^2}} = 0 \Rightarrow f(x,y,z)$ 在 $(0,0,0)$ 连续

Example 15.

$$设 f(x,y,z) = \begin{cases} \dfrac{xyz}{x^2+y^4+z^4} & (x,y,z) \neq 0 \\ 0, & (x,y,z) = 0 \end{cases}, \quad 则 f(x,y,z) 在 (0,0,0) 是否连续?$$

【解】

令 $x = mt^2,\ y = z = t$ 则 $\dfrac{mt^4}{(m^2+2)t^4} = \dfrac{m}{m^2+2}$

取 $m=1$ 则 $\lim\limits_{(x,y,z)\to(0,0,0)}\dfrac{xyz}{x^2+y^4+z^4} = \lim\limits_{(x,y,z)\to(0,0,0)}\dfrac{m}{m^2+2} = \dfrac{1}{3} \neq 0 = f(0,0,0)$

因此 $f(x,y,z)$ 在 $(0,0,0)$ 不连续

Example 16.

$$设 f(x,y,z) = \begin{cases} \dfrac{x^2 y}{x^2+y^4+z^4} & (x,y,z) \neq 0 \\ 0, & (x,y,z) = 0 \end{cases}, \quad 则 f(x,y,z) 在 (0,0,0) 是否连续?$$

【解】

$\because x^2 + y^4 + z^4 \geq x^2 \quad \therefore \left|\dfrac{x^2 y}{x^2+y^4+z^4}\right| \leq \left|\dfrac{x^2 y}{x^2}\right| = |y|$

令 $\epsilon > 0$ 取 $\delta^{\frac{1}{4}} < \varepsilon$

则 $x^2 + y^4 + z^4 < \delta \Rightarrow \left|\dfrac{x^2 y}{x^2+y^4+z^4}\right| < |y| < (x^2+y^4+z^4)^{\frac{1}{4}} < \delta^{\frac{1}{4}} < \epsilon$

因此 $\displaystyle\lim_{(x,y,z)\to(0,0,0)} \frac{x^2y}{x^2+y^4+z^4} = 0 \Rightarrow f(x,y,z)$ 在 $(0,0,0)$ 连续

Example 17.

$$设 f(x,y,z) = \begin{cases} \dfrac{3xy+xz+z^2}{x^2+2y^2}, & (x,y,z) \neq 0 \\ 0, & (x,y,z) = 0 \end{cases}, \quad 则 f(x,y,z) 在 (0,0,0) 是否连续?$$

【解】

令 $z = y = mx$ 则 $\dfrac{3mx^2+mx^2+m^2x^2}{(1+2m^2)x^2} = \dfrac{4m+m^2}{1+2m^2}$

取 $m = 1$ 则 $\displaystyle\lim_{(x,y,z)\to(0,0,0)} \frac{3xy+xz+z^2}{x^2+2y^2} = \lim_{(x,y,z)\to(0,0,0)} \frac{4m+m^2}{1+2m^2} = \frac{5}{3} \neq f(0,0,0) = 0$

因此 $f(x,y,z)$ 在 $(0,0,0)$ 不连续

## 7.3 求多变量函数的偏导数

此节介绍如何求多变量函数的偏导数，值得注意的是，在下一节判断多变量函数是否可微分的过程中需要求得偏导数的值，因此读者需多加练习偏导数的例题，以利在判断多变量函数是否可微分时能更加迅速

【定义】双变量函数偏导数的定义

$(i) f(x,y)$ 对于 $x$ 在 $(x_0,y_0)$ 可偏微分 $\Leftrightarrow \displaystyle\lim_{h\to 0} \frac{f(h+x_0,y_0)-f(x_0,y_0)}{h}$ 存在

如果上述极限存在时则用 $\dfrac{\partial f}{\partial x}(x_0,y_0)$ 表示或用 $f_x(x_0,y_0)$ 表示

$(ii) f(x,y)$ 对于 $y$ 在 $(x_0,y_0)$ 可偏微分 $\Leftrightarrow \displaystyle\lim_{h\to 0} \frac{f(x_0,y_0+h)-f(x_0,y_0)}{h}$ 存在

如果上述极限存在时则用 $\dfrac{\partial f}{\partial y}(x_0,y_0)$ 表示或用 $f_y(x_0,y_0)$ 表示

【定义】多变量函数偏导数的定义
$f(x_1,...,x_n)$ 对于 $x_j$ 在 $(c_1,...,c_n)$ 可偏微分

$$\Leftrightarrow \lim_{h \to 0} \frac{f(c_1, c_2, \ldots, h + c_j, \ldots, c_n) - f(c_1, c_2, \ldots, c_j, \ldots, c_n)}{h} \text{ 存在}, \quad 1 \le j \le n$$

此外，如果上述极限存在时，则用 $\dfrac{\partial f}{\partial x_j}(c_1, \ldots, c_n)$ 表示或用 $f_{x_j}(c_1, \ldots, c_n)$ 表示

多变量函数偏导数的考试类型当中，以双变量函数的题型最常出现，读者若能熟悉双变量题型，其余类型都能够掌握。

考试类型:

Type 1.

給定双变数函数 $f(x, y)$，求 $\dfrac{\partial f}{\partial x}(x_0, y_0) = ?$，$\dfrac{\partial f}{\partial y}(x_0, y_0) = ?$

解题流程:

Step1.

使用定义求 $\lim\limits_{h \to 0} \dfrac{f(h + x_0, y_0) - f(x_0, y_0)}{h} = ?$ 以及 $\lim\limits_{h \to 0} \dfrac{f(x_0, y_0 + h) - f(x_0, y_0)}{h} = ?$

Step2.

如果 $\lim\limits_{h \to 0} \dfrac{f(h + x_0, y_0) - f(x_0, y_0)}{h}$ 存在则 $\dfrac{\partial f}{\partial x}(x_0, y_0) = \lim\limits_{h \to 0} \dfrac{f(h + x_0, y_0) - f(x_0, y_0)}{h}$

如果 $\lim\limits_{h \to 0} \dfrac{f(x_0, y_0 + h) - f(x_0, y_0)}{h}$ 存在则 $\dfrac{\partial f}{\partial y}(x_0, y_0) = \lim\limits_{h \to 0} \dfrac{f(x_0, y_0 + h) - f(x_0, y_0)}{h}$

如果 $\lim\limits_{h \to 0} \dfrac{f(h + x_0, y_0) - f(x_0, y_0)}{h}$ 不存在则 $\dfrac{\partial f}{\partial x}(x_0, y_0)$ 不存在(偏导数不存在)

如果 $\lim\limits_{h \to 0} \dfrac{f(x_0, y_0 + h) - f(x_0, y_0)}{h}$ 不存在则 $\dfrac{\partial f}{\partial y}(x_0, y_0)$ 不存在(偏导数不存在)

Type 2.

給定双变数函数 $f(x, y)$，求 $f_{xx}(x_0, y_0) = ?$，$f_{xy}(x_0, y_0) = ?$，$f_{yx}(x_0, y_0) = ?$，$f_{yy}(x_0, y_0) = ?$
底下说明求 $f_{yx}(x_0, y_0)$ 的方法，其它三者方法类似

解题流程:

Step1.

$$f_{yx}(0,0) := \lim_{h \to 0} \frac{f_y(h, 0) - f_y(0,0)}{h}$$

Step2.

求 $f_y(h,0) =?$ 且 $f_y(0,0) =?$

$$f_y(h,0) = \lim_{t \to 0} \frac{f(h,0+t) - f(h,0)}{t} \quad 且 \quad f_y(0,0) = \lim_{t \to 0} \frac{f(0,0+t) - f(0,0)}{t}$$

Step3.

把 $f_y(h,0)$、$f_y(0,0)$ 带入 Step1. 求 $\displaystyle\lim_{h \to 0} \frac{f_y(h,0) - f_y(0,0)}{h} =?$

Example 1.

(1) 设 $f(x,y) = 3x - x^2y^2 + 2x^3y + y$，求 $\dfrac{\partial f}{\partial x}(x,y)$，$\dfrac{\partial f}{\partial y}(x,y) =?$

(2) 设 $f(x,y) = (x^2y + xy + x)^2$，求 $\dfrac{\partial f}{\partial x}(x,y) =?$

【解】

(1)

$$\frac{\partial f}{\partial x} = 3 - 2xy^2 + 6x^2y, \quad \frac{\partial f}{\partial y} = -2x^2y + 2x^3 + 1$$

(2)

$$\frac{\partial f}{\partial x} = 2(x^2y + xy + x)(2xy + y + 1)$$

Example 2.

设 $f(x,y) = \tan^{-1}\dfrac{y}{x}$，求 $\dfrac{\partial f}{\partial x}(3,1)$，$\dfrac{\partial f}{\partial y}(2,2) =?$

【解】

$$\frac{\partial f}{\partial x}(3,1) = \lim_{h \to 0} \frac{f(h+3,1) - f(3,1)}{h} = \lim_{h \to 0} \frac{\tan^{-1}\frac{1}{3+h} - \tan^{-1}\frac{1}{3}}{h} = \lim_{h \to 0} \frac{-(3+h)^{-2}}{1 + \left(\frac{1}{3+h}\right)^2} = \frac{-1}{10}$$

$$\frac{\partial f}{\partial y}(2,2) = \lim_{h \to 0} \frac{f(2,2+h) - f(2,2)}{h} = \lim_{h \to 0} \frac{\tan^{-1}\frac{2+h}{2} - \tan^{-1}1}{h}$$

$$= \lim_{h \to 0} \frac{\frac{1}{2}}{1 + \left(\frac{2+h}{2}\right)^2} = \frac{1}{4}$$

Example 3.

$$设 f(x,y) = \int_x^{y^3} \cos t\, dt,\quad 求\ \frac{\partial f}{\partial x}(x,y) - \frac{\partial f}{\partial y}(x,y) =?$$

【解】

$$\frac{\partial f}{\partial x} - \frac{\partial f}{\partial y} = -\cos x - 3y^2 \cos y^3$$

Example 4.

$$设 f(x,y) = \sin(x^4 + x^2 y^2) + \cos(xy + y^3),\quad 求\ f_x =?$$

【解】

$$f_x = \cos(x^3 + x^2 y)\,(4x^3 + 2xy^2) - y\sin(xy + y^3)$$

Example 5.

$$设\ f(x,y,z) = x^{y^z}\ (x>0, y>0, z>0),\quad 求 f_x =?,\ f_y =?,\ f_z =?$$

【解】

$$\because f(x,y,z) = x^{y^z} = e^{y^z \ln x} \quad \therefore f_x(x,y,z) = x^{y^z-1} \cdot y^z$$

$$\therefore f_y(x,y,z) = e^{y^z \ln x}(z\,y^{z-1}\ln x) = x^{y^z}(z\,y^{z-1}\ln x)$$

$$\therefore f_z(x,y,z) = e^{y^z \ln x}(y^z \ln x \ln y)$$

Example 6.

$$设 f(x,y) = x^3 y + e^{xy^2},\quad 求 f_x =?,\ f_y =?,\ f_{xy} =?,\ f_{yx} =?$$

【解】

$$f_x(x,y) = 3x^2 y + y^2 e^{xy^2},\quad f_y(x,y) = x^3 + 2xy e^{xy^2}$$

$$f_{xy}(x,y) = 3x^2 + 2y e^{xy^2} + 2xy^3 e^{xy^2},\quad f_{yx}(x,y) = 3x^2 + 2y e^{xy^2} + 2xy^3 e^{xy^2}$$

Example 7.

$$设 f(x,y) = x^2 \tan^{-1}\frac{y}{x},\quad 求 \frac{\partial^2 f}{\partial x \partial y}(1,0) =?$$

【解】

$$\because \frac{\partial f}{\partial y}(x,y) = \frac{x}{1 + \left(\frac{y}{x}\right)^2} = \frac{x^3}{x^2 + y^2},\quad \frac{\partial^2 f}{\partial x \partial y}(x,y) = \frac{3x^2(x^2 + y^2) - x^3(2x)}{(x^2 + y^2)^2} = \frac{x^4 + 3x^2 y^2}{(x^2 + y^2)^2}$$

$$\therefore \frac{\partial^2 f}{\partial x \partial y}(1,0) = 1$$

Example 8.

$$设 f(x,y) = x^2 - 2x + y^2 + 2y + 3, \quad 求\frac{\partial f}{\partial x}(x,y) =?, \quad \frac{\partial f}{\partial y}(x,y) =?$$

【解】

$$\frac{\partial f}{\partial x}(x,y) = 2x - 2, \quad \frac{\partial f}{\partial y}(x,y) = 2y + 2$$

Example 9.

$$设 f(x,y) = x^2 e^{2y} - \cos\frac{x}{y} + x^2 y^3, \quad 求 f_{xy}(x,y) =?, \quad f_{yy}(x,y) =?$$

【解】

$$\frac{\partial f}{\partial x}(x,y) = 2xe^{2y} + \frac{1}{y}\sin\frac{x}{y} + 2xy^3, \quad \frac{\partial f}{\partial y}(x,y) = 2x^2 e^{2y} + \left(\frac{-x}{y^2}\right)\sin\frac{x}{y} + 3x^2 y^2$$

$$f_{xy}(x,y) = \frac{\partial}{\partial y}\left(\frac{\partial f(x,y)}{\partial x}\right) = \frac{\partial}{\partial y}\left(2xe^{2y} + \frac{1}{y}\sin\frac{x}{y} + 2xy^3\right)$$

$$= 4xe^{2y} - \frac{1}{y^2}\sin\frac{x}{y} - \frac{x}{y^3}\cos\frac{x}{y} + 6xy^2$$

$$f_{yy}(x,y) = \frac{\partial}{\partial y}\left(\frac{\partial f(x,y)}{\partial y}\right) = \frac{\partial}{\partial y}\left(2x^2 e^{2y} + \left(\frac{-x}{y^2}\right)\sin\frac{x}{y} + 3x^2 y^2\right)$$

$$= 4x^2 e^{2y} + (2xy^3)\sin\frac{x}{y} + \left(\frac{x^2}{y^4}\right)\cos\frac{x}{y} + 6x^2 y$$

Example 10.

$$设 f(x,y) = \sin x^2 + y^2, \quad 求 f_{xx}(x,y) =?, \quad f_{yx}(x,y) =?, \quad f_{yy}(x,y) =?$$

【解】

$$f_x(x,y) = 2x\cos(x^2 + y^2), \quad f_y(x,y) = 2y\cos(x^2 + y^2)$$

$$f_{xx}(x,y) = \frac{\partial}{\partial x}\left(\frac{\partial f(x,y)}{\partial x}\right) = \frac{\partial}{\partial x}(2x\cos(x^2 + y^2)) = 2\cos(x^2 + y^2) - 4x^2\sin(x^2 + y^2)$$

$$f_{yx}(x,y) = \frac{\partial}{\partial x}\left(\frac{\partial f(x,y)}{\partial y}\right) = \frac{\partial}{\partial x}(2y\cos(x^2 + y^2)) = -4xy\sin(x^2 + y^2)$$

$$f_{yy}(x,y) = \frac{\partial}{\partial y}\left(\frac{\partial f(x,y)}{\partial y}\right) = \frac{\partial}{\partial y}(2y\cos(x^2+y^2)) = 2\cos(x^2+y^2) - 4y^2\sin(x^2+y^2)$$

Example 11.

设 $f(x,y) = e^{ny}\sin nx$，求 $f_{xx} + f_{yy} =$? 且 试证 $f_{xy} = f_{yx}$

【解】

Claim: $f_{xx} + f_{yy} = 0$

$\because f_x = ne^{ny}\cos nx$ 且 $f_y = ne^{ny}\sin nx$

$\therefore f_{xx} = \dfrac{\partial f_x}{\partial x} = -n^2 e^{ny}\sin nx$ 且 $f_{yy} = \dfrac{\partial f_y}{\partial y} = n^2 e^{ny}\sin nx \Rightarrow f_{xx} + f_{yy} = 0$

Claim: $f_{xy} = f_{yx}$

$\because f_{xy} = \dfrac{\partial f_x}{\partial y} = n^2 e^{ny}\cos nx$ 且 $f_{yx} = \dfrac{\partial f_y}{\partial x} = n^2 e^{ny}\cos nx \qquad \therefore f_{xy} = f_{yx}$

Example 12.

设 $f(x,y) = \dfrac{x}{x^2+y^2}$，求 $\dfrac{\partial^2 f}{\partial x^2}(x,y) + \dfrac{\partial^2 f}{\partial y^2}(x,y) =$?

【解】

$$\frac{\partial f}{\partial x}(x,y) = \frac{\partial}{\partial x}(x(x^2+y^2)^{-1}) = (x^2+y^2)^{-1} - 2x^2(x^2+y^2)^{-2}$$

$$\frac{\partial f}{\partial y}(x,y) = \frac{\partial}{\partial y}(x(x^2+y^2)^{-1}) = -2xy(x^2+y^2)^{-2}$$

$$\frac{\partial^2 f}{\partial x^2}(x,y) = \frac{\partial}{\partial x}\left(\frac{\partial f}{\partial x}(x,y)\right) = -2x(x^2+y^2)^{-2} - 4x(x^2+y^2)^{-2} + 8x^3(x^2+y^2)^{-3}$$

$$\frac{\partial^2 f}{\partial y^2}(x,y) = \frac{\partial}{\partial y}\left(\frac{\partial f}{\partial y}(x,y)\right) = -2x(x^2+y^2)^{-2} + 8xy^2(x^2+y^2)^{-3}$$

$\because 8x^3(x^2+y^2)^{-3} + 8xy^2(x^2+y^2)^{-3} = 8x(x^2+y^2)^{-2} \quad \therefore \dfrac{\partial^2 f(x,y)}{\partial x^2} + \dfrac{\partial^2 f(x,y)}{\partial y^2} = 0$

Example 13.

$$\text{设} f(x,y) = \frac{e^{-\sqrt{x^2+y^2}}}{\sqrt{x^2+y^2}}, \quad \text{求} \frac{\partial f}{\partial x}(x,y) = ?$$

【解】

$$\text{令 } r = \sqrt{x^2+y^2} \text{ 则 } f(r) = \frac{e^{-r}}{r}$$

$$\frac{\partial f}{\partial x}(x,y) = \frac{\partial f(r)}{\partial r} \cdot \frac{\partial r}{\partial x}, \quad \frac{\partial f(r)}{\partial r} = -\frac{e^{-r}}{r} - \frac{e^{-r}}{r^2}, \quad \frac{\partial r}{\partial x} = x(x^2+y^2)^{-\frac{1}{2}} = \frac{x}{r}$$

$$\frac{\partial f}{\partial x}(x,y) = \frac{\partial f(r)}{\partial r} \cdot \frac{\partial r}{\partial x} = -\frac{xe^{-\sqrt{x^2+y^2}}}{x^2+y^2} - \frac{xe^{-\sqrt{x^2+y^2}}}{(x^2+y^2)^{\frac{3}{2}}}$$

Example 14.

$$\text{设} f(x,y,z) = (x^2+y^2+z^2)^{-\frac{1}{2}}, \quad \text{试证 } \frac{\partial^2 f}{\partial x^2} + \frac{\partial^2 f}{\partial y^2} + \frac{\partial^2 f}{\partial z^2} = 0$$

【解】

$$\because \frac{\partial f}{\partial x} = -\frac{2x}{2}(x^2+y^2+z^2)^{-\frac{3}{2}} = -x(x^2+y^2+z^2)^{-\frac{3}{2}}$$

$$\frac{\partial f}{\partial y} = -y(x^2+y^2+z^2)^{-\frac{3}{2}} \text{ 且 } \frac{\partial f}{\partial z} = -z(x^2+y^2+z^2)^{-\frac{3}{2}}$$

$$\therefore \frac{\partial^2 f}{\partial x^2} = -(x^2+y^2+z^2)^{-\frac{3}{2}} + \frac{3x}{2}(x^2+y^2+z^2)^{-\frac{5}{2}}(2x)$$

$$= -(x^2+y^2+z^2)^{-\frac{3}{2}} + 3x^2(x^2+y^2+z^2)^{-\frac{5}{2}}$$

$$\frac{\partial^2 f}{\partial y^2} = -(x^2+y^2+z^2)^{-\frac{3}{2}} + \frac{3y}{2}(x^2+y^2+z^2)^{-\frac{5}{2}}(2y)$$

$$= -(x^2+y^2+z^2)^{-\frac{3}{2}} + 3y^2(x^2+y^2+z^2)^{-\frac{5}{2}}$$

$$\frac{\partial^2 f}{\partial z^2} = -(x^2+y^2+z^2)^{-\frac{3}{2}} + \frac{3z}{2}(x^2+y^2+z^2)^{-\frac{5}{2}}(2z)$$

$$= -(x^2+y^2+z^2)^{-\frac{3}{2}} + 3z^2(x^2+y^2+z^2)^{-\frac{5}{2}}$$

$$\therefore \frac{\partial^2 f}{\partial x^2} + \frac{\partial^2 f}{\partial y^2} + \frac{\partial^2 f}{\partial z^2}$$

$$= -3(x^2 + y^2 + z^2)^{-\frac{3}{2}} + 3x^2(x^2 + y^2 + z^2)^{-\frac{5}{2}} + 3y^2(x^2 + y^2 + z^2)^{-\frac{5}{2}}$$

$$+3z^2(x^2 + y^2 + z^2)^{-\frac{5}{2}}$$

$$= -3(x^2 + y^2 + z^2)^{-\frac{3}{2}} + 3(x^2 + y^2 + z^2)^{-\frac{5}{2}}(x^2 + y^2 + z^2)$$

$$= -3(x^2 + y^2 + z^2)^{-\frac{3}{2}} + 3(x^2 + y^2 + z^2)^{-\frac{3}{2}} = 0$$

Example 15.

(1) 设 $f(x,y) = (x^4 + y^4)^{\frac{1}{4}}$, 求 $\dfrac{\partial f}{\partial x}(0,0)$, $\dfrac{\partial f}{\partial y}(0,0) =?$

(2) 设 $f(x,y) = \begin{cases} \dfrac{\tan(2x + y)}{2x + y}, & (x,y) \neq (0,0) \\ 0, & (x,y) = (0,0) \end{cases}$ , 则 $\dfrac{\partial f}{\partial x}(0,0) =?$ , $\dfrac{\partial f}{\partial y}(0,0) =?$

【解】

(1)

$$\because \frac{\partial f}{\partial x}(0,0) = \lim_{h \to 0} \frac{f(h+0,0) - f(0,0)}{h} = \lim_{h \to 0} \frac{(h^4 + 0^4)^{\frac{1}{4}} - 0}{h} = \lim_{h \to 0} \frac{|h|}{h} = \pm 1$$

$$\frac{\partial f}{\partial y}(0,0) = \lim_{h \to 0} \frac{f(0, h+0) - f(0,0)}{h} = \lim_{h \to 0} \frac{(h^4 + 0^4)^{\frac{1}{4}} - 0}{h} = \lim_{h \to 0} \frac{|h|}{h} = \pm 1$$

$$\therefore \frac{\partial f}{\partial x}(0,0), \ \frac{\partial f}{\partial y}(0,0) \ 两者皆不存在$$

(2)

$$\because \frac{\partial f}{\partial x}(0,0) = \lim_{h \to 0} \frac{f(h+0,0) - f(0,0)}{h} = \lim_{h \to 0} \frac{\frac{\tan 2h}{2h} - 0}{h} = \lim_{h \to 0} \frac{1 - 0}{h} = \pm\infty$$

$$\frac{\partial f}{\partial y}(0,0) = \lim_{h \to 0} \frac{f(0, h+0) - f(0,0)}{h} = \lim_{h \to 0} \frac{\frac{\tan h}{h} - 0}{h} = \lim_{h \to 0} \frac{1 - 0}{h} = \pm\infty$$

$$\therefore \frac{\partial f}{\partial x}(0,0), \ \frac{\partial f}{\partial y}(0,0) \ 两者皆不存在$$

Example 16.

$$\text{设 } \frac{\partial f}{\partial x} = x^2 + y^2, \quad \frac{\partial f}{\partial y} = 2xy \ \text{ 及 } f(0,1) = 5, \quad \text{求 } f(x,y) = ?$$

【解】

$$\because \frac{\partial f}{\partial x} = x^2 + y^2 \quad \therefore f(x,y) = \frac{x^3}{3} + xy^2 + c$$

$$\because \frac{\partial f}{\partial y} = 2xy \quad \therefore f(x,y) = xy^2 + c \quad \therefore f(x,y) = \frac{x^3}{3} + xy^2 + c$$

$$\because f(0,1) = 5 \quad \therefore c = 5 \quad \therefore f(x,y) = \frac{x^3}{3} + xy^2 + 5$$

Example 17.

$$\text{设 } f(x,y) = \begin{cases} \sqrt{xy}, & (x,y) \neq 0 \\ 0, & (x,y) = 0 \end{cases}, \quad \text{则 } \frac{\partial f}{\partial x}(0,0) = ?, \quad \frac{\partial f}{\partial y}(0,0) = ?$$

【解】

$$f_x(0,0) = \lim_{h \to 0} \frac{f(h+0,0) - f(0,0)}{h} = \lim_{h \to 0} \frac{\sqrt{h \cdot 0} - 0}{h} = 0$$

$$f_y(0,0) = \lim_{h \to 0} \frac{f(0,h+0) - f(0,0)}{h} = \lim_{h \to 0} \frac{\sqrt{h \cdot 0} - 0}{h} = 0$$

Example 18.

$$\text{设 } f(x,y) = \begin{cases} \dfrac{x^2 - y^2}{\sqrt{x^2 + y^2}}, & (x,y) \neq 0 \\ 0, & (x,y) = 0 \end{cases}, \quad \text{则 } \frac{\partial f}{\partial x}(0,0) = ?, \quad \frac{\partial f}{\partial y}(0,0) = ?$$

【解】

$$f_x(0,0) = \lim_{h \to 0} \frac{f(h+0,0) - f(0,0)}{h} = \lim_{h \to 0} \frac{\frac{h^2}{h} - 0}{h} = 1$$

$$f_y(0,0) = \lim_{h \to 0} \frac{f(0,h+0) - f(0,0)}{h} = \lim_{h \to 0} \frac{\frac{-h^2}{h} - 0}{h} = -1$$

Example 19.

$$\text{设} f(x,y) = \begin{cases} \dfrac{x^3 - y^3}{x^2 + y^2}, & (x,y) \neq 0 \\ 0, & (x,y) = 0 \end{cases}, \quad \text{则} \ \frac{\partial f}{\partial x}(0,0) = ?, \quad \frac{\partial f}{\partial y}(0,0) = ?$$

【解】

$$f_x(0,0) = \lim_{h \to 0} \frac{f(h+0,0) - f(0,0)}{h} = \lim_{h \to 0} \frac{\dfrac{h^3}{h^2} - 0}{h} = 1$$

$$f_y(0,0) = \lim_{h \to 0} \frac{f(0,h+0) - f(0,0)}{h} = \lim_{h \to 0} \frac{\dfrac{-h^3}{h^2} - 0}{h} = -1$$

Example 20.

$$(1)\text{设} f(x,y) = \begin{cases} \dfrac{2x^2 y}{x^5 + y^2}, & (x,y) \neq 0 \\ 0, & (x,y) = 0 \end{cases}, \quad \text{则} \ \frac{\partial f}{\partial x}(0,0) = ?, \quad \frac{\partial f}{\partial y}(0,0) = ?$$

$$(2)\text{设} f(x,y) = \begin{cases} \dfrac{3x^2 y}{x^4 + y^3}, & (x,y) \neq 0 \\ 0, & (x,y) = 0 \end{cases}, \quad \text{则} \ \frac{\partial f}{\partial x}(0,0) = ?, \quad \frac{\partial f}{\partial y}(0,0) = ?$$

【解】

(1)

$$\because \frac{\partial f}{\partial x}(0,0) = \lim_{h \to 0} \frac{f(h+0,0) - f(0,0)}{h} = \lim_{h \to 0} \frac{\dfrac{2h^2 \cdot 0}{h^5 + 0^2} - 0}{h} = \lim_{h \to 0} \frac{0}{h^6} = 0$$

$$\frac{\partial f}{\partial y}(0,0) = \lim_{h \to 0} \frac{f(0,h+0) - f(0,0)}{h} = \lim_{h \to 0} \frac{\dfrac{2 \cdot 0^2 \cdot h}{0^5 + h^2} - 0}{h} = \lim_{h \to 0} \frac{0}{h^3} = 0$$

$$\therefore \frac{\partial f}{\partial x}(0,0) = 0, \quad \frac{\partial f}{\partial y}(0,0) = 0$$

(2)

$$\because \frac{\partial f}{\partial x}(0,0) = \lim_{h \to 0} \frac{f(h+0,0) - f(0,0)}{h} = \lim_{h \to 0} \frac{\dfrac{3 \cdot h^2 \cdot 0}{h^4 + 0^3} - 0}{h} = \lim_{h \to 0} \frac{0}{h^3} = 0$$

$$\frac{\partial f}{\partial y}(0,0) = \lim_{h \to 0} \frac{f(0, h+0) - f(0,0)}{h} = \lim_{h \to 0} \frac{\dfrac{3 \cdot 0^2 \cdot h}{0^4 + h^3} - 0}{h} = \lim_{h \to 0} \frac{0}{h^4} = 0$$

$$\therefore \frac{\partial f}{\partial x}(0,0) = 0, \quad \frac{\partial f}{\partial y}(0,0) = 0$$

Example 21.

$$(1)\,设 f(x,y) = \begin{cases} \dfrac{x^5 - y^5}{x^5 + y^5}, & (x,y) \neq 0 \\ 0, & (x,y) = 0 \end{cases}, \quad 则 \quad \frac{\partial f}{\partial x}(0,0) =?, \quad \frac{\partial f}{\partial y}(0,0) =?$$

$$(2)\,设 f(x,y) = \begin{cases} \dfrac{5x^4 - 2y^3}{x^2 + 3y^2}, & (x,y) \neq 0 \\ 0, & (x,y) = 0 \end{cases}, \quad 则 \quad \frac{\partial f}{\partial x}(0,0) =?, \quad \frac{\partial f}{\partial y}(0,0) =?$$

【解】

(1)

$$\because \frac{\partial f}{\partial x}(0,0) = \lim_{h \to 0} \frac{f(h+0, 0) - f(0,0)}{h} = \lim_{h \to 0} \frac{\dfrac{h^5 - 0^5}{h^5 + 0^5} - 0}{h} = \lim_{h \to 0} \frac{1}{h} = \pm\infty$$

$$\frac{\partial f}{\partial y}(0,0) = \lim_{h \to 0} \frac{f(0, h+0) - f(0,0)}{h} = \lim_{h \to 0} \frac{\dfrac{0^5 - h^5}{0^5 + h^5} - 0 - 0}{h} = \lim_{h \to 0} \frac{-1}{h} = \pm\infty$$

$$\therefore \frac{\partial f}{\partial x}(0,0), \quad \frac{\partial f}{\partial y}(0,0)\,两者皆不存在$$

(2)

$$\because \frac{\partial f}{\partial x}(0,0) = \lim_{h \to 0} \frac{f(h+0, 0) - f(0,0)}{h} = \lim_{h \to 0} \frac{\dfrac{5h^4 - 2 \cdot 0^3}{h^2 + 3 \cdot 0^2} - 0}{h} = \lim_{h \to 0} \frac{5h^2}{h} = 0$$

$$\frac{\partial f}{\partial y}(0,0) = \lim_{h \to 0} \frac{f(0, h+0) - f(0,0)}{h} = \lim_{h \to 0} \frac{\dfrac{5 \cdot 0^4 - 2 \cdot h^3}{0^2 + 3 \cdot h^2} - 0}{h} = \lim_{h \to 0} \frac{-2h}{3h} = \frac{-2}{3}$$

$$\therefore \frac{\partial f}{\partial x}(0,0) = 0, \quad \frac{\partial f}{\partial y}(0,0) = \frac{-2}{3}$$

Example 22.

$$\text{设} f(x,y) = \begin{cases} xy\left(\dfrac{x^2-y^2}{x^2+y^2}\right), & (x,y) \neq 0 \\ 0, & (x,y) = 0 \end{cases}, \qquad \text{求}$$

(1) $f_x(0,0)$    (2) $f_y(0,0)$    (3) $f_{xx}(0,0)$    (4) $f_{yx}(0,0)$    (5) $f_{xy}(0,0)$    (6) $f_{yy}(0,0)$

【解】

(1) $f_x(0,0) = \lim\limits_{h\to 0} \dfrac{f(h+0,0)-f(0,0)}{h} = \lim\limits_{h\to 0}\dfrac{0-0}{h} = 0$

(2) $f_y(0,0) = \lim\limits_{h\to 0} \dfrac{f(0,h)-f(0,0)}{h} = \lim\limits_{h\to 0}\dfrac{0-0}{h} = 0$

(3) $f_{xx}(0,0) = \lim\limits_{h\to 0} \dfrac{f_x(h,0)-f_x(0,0)}{h}$

$\because f_x(h,0) = \lim\limits_{k\to 0} \dfrac{f(h+k,0)-f(h,0)}{k} = \lim\limits_{h\to 0}\dfrac{0-0}{h} = 0$

$\therefore f_{xx}(0,0) = \lim\limits_{h\to 0} \dfrac{f_x(h,0)-f_x(0,0)}{h} = \lim\limits_{h\to 0}\dfrac{0-0}{h} = 0$

(4) $f_{yx}(0,0) = \lim\limits_{h\to 0} \dfrac{f_y(h,0)-f_y(0,0)}{h}$

$\because f_y(h,0) = \lim\limits_{k\to 0} \dfrac{f(h,k)-f(h,0)}{k} = \lim\limits_{k\to 0}\dfrac{hk\left(\dfrac{h^2-k^2}{h^2+k^2}\right)-0}{k} = h$

$\therefore f_{yx}(0,0) = \lim\limits_{h\to 0} \dfrac{f_y(h,0)-f_y(0,0)}{h} = \lim\limits_{h\to 0}\dfrac{h-0}{h} = 1$

(5) $f_{xy}(0,0) = \lim\limits_{h\to 0} \dfrac{f_x(0,h)-f_x(0,0)}{h}$

$\because f_x(0,h) = \lim\limits_{k\to 0} \dfrac{f(k,h)-f(0,h)}{k} = \lim\limits_{k\to 0}\dfrac{hk\left(\dfrac{k^2-h^2}{h^2+k^2}\right)}{k} = -h$

$\therefore f_{xy}(0,0) = \lim\limits_{h\to 0} \dfrac{f_x(0,h)-f_x(0,0)}{h} = \lim\limits_{h\to 0}\dfrac{-h-0}{h} = -1$

(6) $f_{yy}(0,0) = \lim\limits_{h\to 0} \dfrac{f_y(0,h)-f_y(0,0)}{h}$

$\because f_y(0,h) = \lim\limits_{k\to 0} \dfrac{f(0,h+k)-f(0,h)}{k} = \lim\limits_{h\to 0}\dfrac{0-0}{h} = 0$

$$\therefore f_{yy}(0,0) = \lim_{h \to 0} \frac{f_y(0,h) - f_y(0,0)}{h} = \lim_{h \to 0} \frac{0 - 0}{h} = 0$$

Example 23.

$$设 f(x,y) = \begin{cases} (x^2 + y^2)\sin\left(\dfrac{1}{x^2 + y^2}\right), & (x,y) \neq 0 \\ 0, & (x,y) = 0 \end{cases},$$

则 $\dfrac{\partial f}{\partial x}(0,0) = ?$, $\dfrac{\partial f}{\partial y}(0,0) = ?$

【解】

$$f_x(0,0) = \lim_{h \to 0} \frac{f(h+0,0) - f(0,0)}{h} = \lim_{h \to 0} \frac{h^2 \sin(\frac{1}{h^2}) - 0}{h} = 0$$

$$f_y(0,0) = \lim_{h \to 0} \frac{f(0,0+h) - f(0,0)}{h} = \lim_{h \to 0} \frac{h^2 \sin(\frac{1}{h^2}) - 0}{h} = 0$$

Example 24.

$$设 f(x,y) = \begin{cases} (x^2 + y^2)^p \sin\left(\dfrac{1}{x^2 + y^2}\right), & (x,y) \neq 0 \\ 0, & (x,y) = 0 \end{cases}, \quad \forall p > \frac{1}{2}$$

则 $\dfrac{\partial f}{\partial x}(0,0) = ?$, $\dfrac{\partial f}{\partial y}(0,0) = ?$

【解】

$$令 \, p > \frac{1}{2}, \, f_x(0,0) = \lim_{h \to 0} \frac{f(h+0,0) - f(0,0)}{h} = \lim_{h \to 0} \frac{h^{2p} \sin(\frac{1}{h^2}) - 0}{h} = \lim_{h \to 0} h^{2p-1} \sin(\frac{1}{h^2})$$

$$\because p > \frac{1}{2} \quad \therefore \lim_{h \to 0} h^{2p-1} \sin(\frac{1}{h^2}) = 0 \Rightarrow f_x(0,0) = 0$$

$$f_y(0,0) = \lim_{h \to 0} \frac{f(0,0+h) - f(0,0)}{h} = \lim_{h \to 0} \frac{h^{2p} \sin(\frac{1}{h^2}) - 0}{h} = \lim_{h \to 0} h^{2p-1} \sin(\frac{1}{h^2})$$

$$\because p > \frac{1}{2} \quad \therefore \lim_{h \to 0} h^{2p-1} \sin(\frac{1}{h^2}) = 0 \Rightarrow \frac{\partial f}{\partial y}(0,0) = 0$$

Example 25.

$$\text{设} f(x,y) = \begin{cases} x|y|, & (x,y) \neq 0 \\ 0, & (x,y) = 0 \end{cases}, \quad \text{则} \ \frac{\partial f}{\partial x}(0,0) =?, \ \frac{\partial f}{\partial y}(0,0) =?$$

【解】

$$f_x(0,0) = \lim_{h \to 0} \frac{f(h+0,0) - f(0,0)}{h} = \lim_{h \to 0} \frac{0-0}{h} = 0$$

$$f_y(0,0) = \lim_{h \to 0} \frac{f(0,0+h) - f(0,0)}{h} = \lim_{h \to 0} \frac{0-0}{h} = 0$$

Example 26.

$$\text{设} f(x,y) = \begin{cases} x^p|y|, & (x,y) \neq 0 \\ 0, & (x,y) = 0 \end{cases}, \quad \forall p > 0, \ \text{则} \ \frac{\partial f}{\partial x}(0,0) =?, \ \frac{\partial f}{\partial y}(0,0) =?$$

【解】

$$f_x(0,0) = \lim_{h \to 0} \frac{f(h+0,0) - f(0,0)}{h} = \lim_{h \to 0} \frac{0-0}{h} = 0$$

$$f_y(0,0) = \lim_{h \to 0} \frac{f(0,0+h) - f(0,0)}{h} = \lim_{h \to 0} \frac{0-0}{h} = 0$$

Example 27.

$$\text{设} f(x,y) = \begin{cases} (xy)^p \dfrac{x^2 - y^2}{x^2 + y^2}, & (x,y) \neq 0 \\ 0, & (x,y) = 0 \end{cases}, \quad \forall p > 0 \ \text{则} \ \frac{\partial f}{\partial x}(0,0) =?, \ \frac{\partial f}{\partial y}(0,0) =?$$

【解】

$$\frac{\partial f}{\partial x}(0,0) = \lim_{h \to 0} \frac{f(h+0,0) - f(0,0)}{h} = \lim_{h \to 0} \frac{0-0}{h} = 0$$

$$\frac{\partial f}{\partial y}(0,0) = \lim_{h \to 0} \frac{f(0,h+0) - f(0,0)}{h} = \lim_{h \to 0} \frac{0-0}{h} = 0$$

Example 28.

$$\text{设} f(x,y) = \begin{cases} \dfrac{x^2 y}{x^2 + y^2}, & (x,y) \neq 0 \\ 0, & (x,y) = 0 \end{cases}, \quad \text{则} \ \frac{\partial f}{\partial x}(0,0) =?, \ \frac{\partial f}{\partial y}(0,0) =?$$

【解】

$$\frac{\partial f}{\partial x}(0,0) = \lim_{h \to 0} \frac{f(h+0,0) - f(0,0)}{h} = \lim_{h \to 0} \frac{0-0}{h} = 0$$

$$\frac{\partial f}{\partial y}(0,0) = \lim_{h \to 0} \frac{f(0,h+0) - f(0,0)}{h} = \lim_{h \to 0} \frac{0-0}{h} = 0$$

Example 29.

$$\text{设} f(x,y) = \sin(x - ay), \quad \text{试证} \quad \frac{\partial^2 f}{\partial y^2}(x,y) - a^2 \frac{\partial^2 f}{\partial x^2}(x,y) = 0$$

【解】

$$\frac{\partial f}{\partial x}(x,y) = \cos(x - ay), \quad \frac{\partial f}{\partial y}(x,y) = -a\cos(x - ay)$$

$$\frac{\partial^2 f}{\partial x^2}(x,y) = \frac{\partial}{\partial x}\left(\frac{\partial f}{\partial x}(x,y)\right) = -\sin(x - ay)$$

$$\frac{\partial^2 f}{\partial y^2}(x,y) = \frac{\partial}{\partial y}\left(\frac{\partial f}{\partial y}(x,y)\right) = -a^2 \sin(x - ay)$$

$$\therefore \frac{\partial^2 f}{\partial y^2}(x,y) - a^2 \frac{\partial^2 f}{\partial x^2}(x,y) = 0$$

## 7.4 判断多变量函数是否可微分与 Chain Rule

【定义】双变量函数的可微分定义

$f(x,y)$ 于 $(x_0, y_0)$ 可微分 $\iff$

(i) $\dfrac{\partial f}{\partial x}(x_0, y_0)$ 且 $\dfrac{\partial f}{\partial y}(x_0, y_0)$ 皆存在

(ii) $\displaystyle\lim_{h \to 0, k \to 0} \left| \frac{f(h+x_0, k+y_0) - f(x_0, y_0) - \dfrac{\partial f}{\partial x}(x_0, y_0)h - \dfrac{\partial f}{\partial y}(x_0, y_0)k}{\sqrt{h^2 + k^2}} \right| = 0$

【**定义**】多变量函数的可微分定义

$f(x_1, \ldots, x_n)$ 于 $(c_1, \ldots, c_n)$ 可微分 $\Leftrightarrow$

(i) $\dfrac{\partial f}{\partial x_j}(c_1, \ldots, c_n)$ 皆存在, $\forall 1 \leq j \leq n$

(ii) $\displaystyle\lim_{(h_1,\ldots,h_n)\to(0,\ldots,0)} \left| \frac{f(h_1+c_1, \ldots, h_n+c_n) - f(c_1, \ldots, c_n) - \sum_{j=1}^{n} \frac{\partial f}{\partial x_j}(c_1, \ldots, c_n)h_j}{\sqrt{h_1^2 + \cdots + h_n^2}} \right| = 0$

从双变量函数可微分的定义, 能观察出如果 $f(x,y)$ 于 $(x_0, y_0)$ 可微分, 则藉由 $(h,k)$ 从任意方向逼近 $(x_0, y_0)$ 的极限值必须等于 0, 因此, 令 $h = mk$, 若 $f(x,y)$ 于 $(x_0, y_0)$ 可微分, 则

$$\lim_{h\to 0, k\to 0} \left| \frac{f(mk+x_0, k+y_0) - f(x_0, y_0) - \frac{\partial f}{\partial x}(x_0, y_0)mk - \frac{\partial f}{\partial y}(x_0, y_0)k}{\sqrt{(mk)^2 + k^2}} \right| = 0$$

并且此极限值与 $m$ 值无关。此外, 偏导数存在也是双变数函数可微分的必要条件但不是充分条件, 底下展示一双变数函数 $f(x,y)$ 做说明

<u>范例说明:</u>

(I) 假设 $f(x,y) = \begin{cases} \dfrac{x^2 y}{x^2 + y^2}, & (x,y) \neq (0,0) \\ 0, & (x,y) = (0,0) \end{cases}$, 判断 $f(x,y)$ 于 $(0,0)$ 是否可微分

$\dfrac{\partial f}{\partial x}(0,0) = \lim_{h\to 0} \dfrac{f(h+0,0) - f(0,0)}{h} = \lim_{h\to 0} \dfrac{0-0}{h} = 0$

$\dfrac{\partial f}{\partial y}(0,0) = \lim_{h\to 0} \dfrac{f(0, h+0) - f(0,0)}{h} = \lim_{h\to 0} \dfrac{0-0}{h} = 0$

令 $h = mk$ 则 $\displaystyle\lim_{h\to 0, k\to 0} \left| \frac{f(mk, k) - f(0,0) - \frac{\partial f}{\partial x}(0,0)mk - \frac{\partial f}{\partial y}(0,0)k}{\sqrt{(mk)^2 + k^2}} \right|$

$= \displaystyle\lim_{h\to 0, k\to 0} \left| \frac{\frac{(mk)^2 k}{(mk)^2 + k^2}}{\sqrt{(mk)^2 + k^2}} \right| = \lim_{h\to 0, k\to 0} \left| \frac{\frac{m^2}{m^2+1}}{\sqrt{m^2+1}} \right|$ 与 $m$ 值有关

取 $h = k$, 即 $m = 1$, 则

$$\lim_{h\to0,k\to0}\left|\frac{f(h+0,k+0)-f(0,0)-\frac{\partial f}{\partial x}(0,0)h-\frac{\partial f}{\partial y}(0,0)k}{\sqrt{h^2+k^2}}\right|=\lim_{h\to0,k\to0}\left|\frac{\frac{m^2}{m^2+1}}{\sqrt{m^2+1}}\right|=\frac{1}{2\sqrt{2}}\neq0$$

$\therefore f(x,y)$ 在 $(0,0)$ 的偏导数存在且在 $(0,0)$ 不可微分

假设 $f(x,y)$ 于 $(0,0)$ 可微分，令 $y=mx$，则

$$\lim_{h\to0,k\to0}\left|\frac{f(mk,k)-f(0,0)-\frac{\partial f}{\partial x}(0,0)mk-\frac{\partial f}{\partial y}(0,0)k}{\sqrt{(mk)^2+k^2}}\right|=0$$

且此极限值与 $m$ 值无关，因此，极限值与 $m$ 值无关是 $f(x,y)$ 于 $(0,0)$ 可微分的必要条件
此外，

$$假设\exists\,m\in R\,使得\;\lim_{h\to0,k\to0}\left|\frac{f(mk,k)-f(0,0)-\frac{\partial f}{\partial x}(0,0)mk-\frac{\partial f}{\partial y}(0,0)k}{\sqrt{(mk)^2+k^2}}\right|\neq0$$

则 $f(x,y)$ 于 $(0,0)$ 不可微分，换句話说，令 $y=mx$，

$$当\;\lim_{h\to0,k\to0}\left|\frac{f(mk,k)-f(0,0)-\frac{\partial f}{\partial x}(0,0)mk-\frac{\partial f}{\partial y}(0,0)k}{\sqrt{(mk)^2+k^2}}\right|\;与m值有关时,$$

则试着找出 $m\in R$ 使得此极限值 $\neq0$ 来说明 $f(x,y)$ 于 $(0,0)$ 不可微分

接下来介绍 Chain Rule，可以看成是多变量合成函数的微分

【**定理**】Chain Rule

(i)假设多变数函数 $f$ 于开集合 $A$ 为连续可微

(ii)$\vec{r}=\vec{r}(t)$ 为可微分曲线且曲线包含在开集合 $A$ 之内

$$\Rightarrow f(\vec{r}(t))\;可微分\;且\;\frac{d}{dt}f(\vec{r}(t))=\nabla f(\vec{r}(t))\cdot\vec{r}'(t)$$

<u>Proof:</u>

Claim: $\displaystyle\lim_{h\to0}\frac{f(\vec{r}(t+h))-f(\vec{r}(t))}{h}=\nabla f(\vec{r}(t))\cdot\vec{r}'(t)$

$\because A$ is an open set and $\vec{r}(t)$ is continuous

Choose $h>0$ s.t. the line segment joining $\vec{r}(t)$ to $\vec{r}(t+h)$ lies entirely in $A$

$\because$ By mean-value theorem, $\exists$ a point $\vec{c}(h)$ between $\vec{r}(t)$ and $\vec{r}(t+h)$ such that

$f(\vec{r}(t+h))-f(\vec{r}(t))=\nabla f(\vec{c}(h))\cdot(\vec{r}(t+h)-\vec{r}(t))$

$$\Rightarrow \frac{f(\vec{r}(t+h)) - f(\vec{r}(t))}{h} = \nabla f(\vec{c}(h)) \cdot \left( \frac{\vec{r}(t+h) - \vec{r}(t)}{h} \right)$$

Let $h \to 0$, $\vec{c}(h) \to \vec{r}(t)$, $\because \nabla f$ is continuous $\quad \therefore \nabla f(\vec{c}(h)) \to \nabla f(\vec{r}(t))$

$$\because \frac{\vec{r}(t+h) - \vec{r}(t)}{h} \to \vec{r}'(t) \quad \therefore \lim_{h \to 0} \frac{f(\vec{r}(t+h)) - f(\vec{r}(t))}{h} = \nabla f(\vec{r}(t)) \cdot \vec{r}'(t)$$

考试类型:

Type 1.

給定双变数函数 $f(x,y)$,判断 $f(x,y)$ 于 $(0,0)$ 是否可微分,其中 $f(x,y)$ 满足此条件:

$$\text{令 } h = mk, \quad \text{当} \lim_{h \to 0, k \to 0} \left| \frac{f(mk,k) - f(0,0) - \frac{\partial f}{\partial x}(0,0)mk - \frac{\partial f}{\partial y}(0,0)k}{\sqrt{(mk)^2 + k^2}} \right| \text{与} m \text{值无关时}$$

解题流程:

令 $\varepsilon > 0$,取出明确 $\delta$ 的值,使得 $h^2 + k^2 < \delta$

$$\Rightarrow \left| \frac{f(h,k) - f(0,0) - \frac{\partial f}{\partial x}(0,0)h - \frac{\partial f}{\partial y}(0,0)k}{\sqrt{h^2 + k^2}} \right| < \varepsilon$$

补充说明:

过程中,常藉由此不等式 $x^2 + y^2 \geq 2|xy|$,推导出

$$\left| \frac{f(h,k) - f(0,0) - \frac{\partial f}{\partial x}(0,0)h - \frac{\partial f}{\partial y}(0,0)k}{\sqrt{h^2 + k^2}} \right| < g(h^2 + k^2) < g(\delta) < \varepsilon$$

并且能得知函数 $g$ 明确的样貌,最后的不等式 $g(\delta) < \varepsilon$ 能帮助明确定义 $\delta$

<u>范例说明:</u>

(I)假设 $f(x,y) = \begin{cases} xy\left(\dfrac{x^2 - y^2}{x^2 + y^2}\right), & (x,y) \neq (0,0) \\ 0, & (x,y) = (0,0) \end{cases}$

判断 $f(x,y)$ 于 $(0,0)$ 是否可微分

$$\frac{\partial f}{\partial x}(0,0) = \lim_{h \to 0} \frac{f(h+0,0) - f(0,0)}{h} = \lim_{h \to 0} \frac{0-0}{h} = 0$$

$$\frac{\partial f}{\partial y}(0,0) = \lim_{h \to 0} \frac{f(0, h+0) - f(0,0)}{h} = \lim_{h \to 0} \frac{0-0}{h} = 0$$

令 $h = mk$ 则 $\displaystyle\lim_{h\to 0, k\to 0} \left| \frac{f(mk, k) - f(0,0) - \frac{\partial f}{\partial x}(0,0)mk - \frac{\partial f}{\partial y}(0,0)k}{\sqrt{(mk)^2 + k^2}} \right|$

$$= \lim_{h\to 0, k\to 0} \left| \frac{mk^2 \left( \frac{(mk)^2 - k^2}{(mk)^2 + k^2} \right)}{\sqrt{(mk)^2 + k^2}} \right| = 0 \ \text{与} \ m \text{值无关}$$

Claim: $\displaystyle\lim_{h\to 0, k\to 0} \left| \frac{f(h+0, k+0) - f(0,0) - \frac{\partial f}{\partial x}(0,0)h - \frac{\partial f}{\partial y}(0,0)k}{\sqrt{h^2 + k^2}} \right| = 0$

$$\frac{\partial f}{\partial x}(0,0) = \lim_{h\to 0} \frac{f(h+0,0) - f(0,0)}{h} = \lim_{h\to 0} \frac{0 - 0}{h} = 0$$

$$\frac{\partial f}{\partial y}(0,0) = \lim_{h\to 0} \frac{f(0, h+0) - f(0,0)}{h} = \lim_{h\to 0} \frac{0 - 0}{h} = 0$$

$$\because \left| \frac{f(h+0, k+0) - f(0,0) - \frac{\partial f}{\partial x}(0,0)h - \frac{\partial f}{\partial y}(0,0)k}{\sqrt{h^2 + k^2}} \right| = \left| \frac{hk \left( \frac{h^2 - k^2}{h^2 + k^2} \right)}{\sqrt{h^2 + k^2}} \right|$$

$$\leq \left| \frac{hk \left( \frac{h^2 + k^2}{h^2 + k^2} \right)}{\sqrt{h^2 + k^2}} \right| = \left| \frac{hk}{\sqrt{h^2 + k^2}} \right| \leq \left| \frac{hk}{\sqrt{2hk}} \right| \leq \left| \frac{\sqrt{hk}}{\sqrt{2}} \right| \leq \sqrt{\frac{h^2 + k^2}{4}}$$

且 $\displaystyle\lim_{h\to 0, k\to 0} \sqrt{\frac{h^2 + k^2}{4}} = 0$

$$\therefore \lim_{h\to 0, k\to 0} \left| \frac{f(h+0, k+0) - f(0,0) - \frac{\partial f}{\partial x}(0,0)h - \frac{\partial f}{\partial y}(0,0)k}{\sqrt{h^2 + k^2}} \right| = 0$$

$\therefore f(x,y)$ 在 $(0,0)$ 可微分

Type 2.

给定双变数函数 $f(x, y)$，判断 $f(x, y)$ 于 $(0,0)$ 是否可微分，其中 $f(x, y)$ 满足此条件：

令 $h = mk$, 当 $\lim\limits_{h\to 0, k\to 0} \left| \dfrac{f(mk, k) - f(0,0) - \dfrac{\partial f}{\partial x}(0,0)mk - \dfrac{\partial f}{\partial y}(0,0)k}{\sqrt{(mk)^2 + k^2}} \right|$ 与 $m$ 值有关时

解题流程:

Step1.

取 $m$ 值且使得 $\lim\limits_{h\to 0, k\to 0} \left| \dfrac{f(mk, k) - f(0,0) - \dfrac{\partial f}{\partial x}(0,0)mk - \dfrac{\partial f}{\partial y}(0,0)k}{\sqrt{(mk)^2 + k^2}} \right| \neq 0$

说明 $f(x,y)$ 于 $(0,0)$ 不可微分

<u>范例说明:</u>

(I)假设 $f(x,y) = \begin{cases} \dfrac{x^3 - y^3}{x^2 + y^2}, & (x,y) \neq (0,0) \\ 0, & (x,y) = (0,0) \end{cases}$, 判断 $f(x,y)$ 于 $(0,0)$ 是否可微分

$$f_x(0,0) = \lim\limits_{h\to 0} \frac{f(h + 0, 0) - f(0,0)}{h} = \lim\limits_{h\to 0} \frac{\dfrac{h^3}{h^2} - 0}{h} = 1$$

$$f_y(0,0) = \lim\limits_{h\to 0} \frac{f(0, h + 0) - f(0,0)}{h} = \lim\limits_{h\to 0} \frac{\dfrac{-h^3}{h^2} - 0}{h} = -1$$

令 $h = mk$ 则 $\lim\limits_{h\to 0, k\to 0} \left| \dfrac{f(mk, k) - f(0,0) - \dfrac{\partial f}{\partial x}(0,0)mk - \dfrac{\partial f}{\partial y}(0,0)k}{\sqrt{(mk)^2 + k^2}} \right|$

$$= \lim\limits_{h\to 0, k\to 0} \left| \frac{\dfrac{(mk)^3 - k^3}{(mk)^2 + k^2} - mk + k}{\sqrt{(mk)^2 + k^2}} \right| = \lim\limits_{h\to 0, k\to 0} \left| \frac{\dfrac{m^3 - 1}{m^2 + 1} - m + 1}{\sqrt{m^2 + 1}} \right| \quad \text{与 } m \text{ 值有关}$$

取 $h = 2k$, 即 $m = 2$,

$$\lim\limits_{h\to 0, k\to 0} \left| \frac{f(h + 0, k + 0) - f(0,0) - \dfrac{\partial f}{\partial x}(0,0)h - \dfrac{\partial f}{\partial y}(0,0)k}{\sqrt{h^2 + k^2}} \right| = \left| \frac{\dfrac{7}{5} - 1}{\sqrt{5}} \right| \neq 0$$

$\therefore f(x,y)$ 在 $(0,0)$ 不可微分

(II)假设 $f(x,y) = \begin{cases} \dfrac{x^2 y}{x^2 + y^2}, & (x,y) \neq (0,0) \\ 0, & (x,y) = (0,0) \end{cases}$ ，判断 $f(x,y)$ 于 $(0,0)$ 是否可微分

$$\frac{\partial f}{\partial x}(0,0) = \lim_{h \to 0} \frac{f(h+0,0) - f(0,0)}{h} = \lim_{h \to 0} \frac{0-0}{h} = 0$$

$$\frac{\partial f}{\partial y}(0,0) = \lim_{h \to 0} \frac{f(0,h+0) - f(0,0)}{h} = \lim_{h \to 0} \frac{0-0}{h} = 0$$

令 $h = mk$ 则 $\displaystyle\lim_{h \to 0, k \to 0} \left| \frac{f(mk,k) - f(0,0) - \frac{\partial f}{\partial x}(0,0)mk - \frac{\partial f}{\partial y}(0,0)k}{\sqrt{(mk)^2 + k^2}} \right|$

$$= \lim_{h \to 0, k \to 0} \left| \frac{\frac{(mk)^2 k}{(mk)^2 + k^2}}{\sqrt{(mk)^2 + k^2}} \right| = \lim_{h \to 0, k \to 0} \left| \frac{\frac{m^2}{m^2 + 1}}{\sqrt{m^2 + 1}} \right| \quad \text{与} m \text{值有关}$$

取 $h = k$, 即 $m = 1$,

$$\lim_{h \to 0, k \to 0} \left| \frac{f(h+0, k+0) - f(0,0) - \frac{\partial f}{\partial x}(0,0)h - \frac{\partial f}{\partial y}(0,0)k}{\sqrt{h^2 + k^2}} \right| = \lim_{h \to 0, k \to 0} \left| \frac{\frac{m^2}{m^2 + 1}}{\sqrt{m^2 + 1}} \right|$$

$$= \frac{1}{2\sqrt{2}} \neq 0$$

$\therefore f(x,y)$ 在 $(0,0)$ 不可微分

Type 3.

给定多变数函数 $f(x,y,z)$, $\vec{r}(t) = (x(t), y(t), z(t))$, 求 $\dfrac{df}{dt}(\vec{r}(t)) = ?$

解题流程:

Step1.

$$\nabla f = \left( \frac{\partial f}{\partial x}, \frac{\partial f}{\partial y}, \frac{\partial f}{\partial z} \right) \quad \text{且} \quad \vec{r}'(t) = \left( \frac{dx}{dt}, \frac{dy}{dt}, \frac{dz}{dt} \right)$$

Step2.

$$\therefore \frac{df}{dt}(\vec{r}(t)) = \nabla f(\vec{r}(t)) \cdot \vec{r}'(t) = \frac{\partial f}{\partial x}\left(\frac{dx}{dt}\right) + \frac{\partial f}{\partial y}\left(\frac{dy}{dt}\right) + \frac{\partial f}{\partial z}\left(\frac{dz}{dt}\right)$$

<u>范例说明:</u>

(I) $f(x,y) = \dfrac{x^4 + y^4}{4}$, $r(t) = (a\cos t, b\sin t)$, 求 $\dfrac{df}{dt}(\vec{r}(t)) = ?$

$\because \nabla f = (x^3, y^3)$ $\therefore \nabla f(\vec{r}(t)) = (a^3\cos^3 t, b^3\sin^3 t)$

$\because \vec{r}'(t) = (-a\sin t, b\cos t)$

$\therefore \dfrac{df}{dt}(\vec{r}(t)) = \nabla f(\vec{r}(t)) \cdot \vec{r}'(t) = (a^3\cos^3 t, b^3\sin^3 t) \cdot (-a\sin t, b\cos t)$

$= \sin t \cos t\, (b^4\sin^2 t - a^4\cos^2 t)$

(II) $f(x,y,z) = x^2 y + z\cos x$, $r(t) = (t, t^2, t^3)$, 求 $\dfrac{df}{dt}(\vec{r}(t)) = ?$

$\because \nabla f = (2xy - z\sin x, x^2, \cos x)$ $\therefore \nabla f(\vec{r}(t)) = (2t^3 - t^3\sin t, t^2, \cos t)$

$\because \vec{r}'(t) = (1, 2t, 3t^2)$

$\therefore \dfrac{df}{dt}(\vec{r}(t)) = \nabla f(\vec{r}(t)) \cdot \vec{r}'(t) = (2t^3 - t^3\sin t, t^2, \cos t) \cdot (1, 2t, 3t^2)$

$= 4t^3 - t^3\sin t + 3t^2\cos t$

Type 4.

給定多变数函数 $f(x,y,z)$, $x = x(t,s), y = y(t,s), z = z(t,s)$, 求 $\dfrac{\partial f}{\partial t} = ?$ 且 $\dfrac{\partial f}{\partial s} = ?$

解题流程:

Step1.

$\because \nabla f = \left(\dfrac{\partial f}{\partial x}, \dfrac{\partial f}{\partial y}, \dfrac{\partial f}{\partial z}\right)$, $\dfrac{\partial \vec{r}}{\partial t} = \left(\dfrac{\partial x}{\partial t}, \dfrac{\partial y}{\partial t}, \dfrac{\partial z}{\partial t}\right)$ 且 $\dfrac{\partial \vec{r}}{\partial s} = \left(\dfrac{\partial x}{\partial s}, \dfrac{\partial y}{\partial s}, \dfrac{\partial z}{\partial s}\right)$

Step2.

$\therefore \dfrac{\partial f}{\partial t} = \dfrac{\partial f}{\partial x}\left(\dfrac{\partial x}{\partial t}\right) + \dfrac{\partial f}{\partial y}\left(\dfrac{\partial y}{\partial t}\right) + \dfrac{\partial f}{\partial z}\left(\dfrac{\partial z}{\partial t}\right)$

$\therefore \dfrac{\partial f}{\partial s} = \dfrac{\partial f}{\partial x}\left(\dfrac{\partial x}{\partial s}\right) + \dfrac{\partial f}{\partial y}\left(\dfrac{\partial y}{\partial s}\right) + \dfrac{\partial f}{\partial z}\left(\dfrac{\partial z}{\partial s}\right)$

范例说明:

(I) $f(x,y) = x^2 - 2xy + y^3$, $x = s^2\ln t$, $y = 2st^3$, 求 $\dfrac{\partial f}{\partial t} = ?$ 且 $\dfrac{\partial f}{\partial s} = ?$

$\because \dfrac{\partial f}{\partial x} = 2x - 2y$, $\dfrac{\partial f}{\partial y} = -2x + 6y^2$, $\dfrac{\partial x}{\partial t} = \dfrac{s^2}{t}$, $\dfrac{\partial x}{\partial s} = 2s\ln t$, $\dfrac{\partial y}{\partial t} = 6st^2$, $\dfrac{\partial y}{\partial s} = 2t^3$

$\therefore \dfrac{\partial f}{\partial t} = \dfrac{\partial f}{\partial x}\left(\dfrac{\partial x}{\partial t}\right) + \dfrac{\partial f}{\partial y}\left(\dfrac{\partial y}{\partial t}\right) = (2x - 2y)\dfrac{s^2}{t} + (-2x + 6y^2)6st^2$

$$\therefore \frac{\partial f}{\partial s} = \frac{\partial f}{\partial x}\left(\frac{\partial x}{\partial s}\right) + \frac{\partial f}{\partial y}\left(\frac{\partial y}{\partial s}\right) = (2x - 2y)2s\ln t + (-2x + 6y^2)2t^3$$

(II)$f(x, y, z) = x^2 y^3 e^{xz}$, $x = t^2 + s^2$, $y = 2st$, $z = s\ln t$, 求 $\dfrac{\partial f}{\partial s} = ?$

$$\because \frac{\partial f}{\partial s} = \frac{\partial f}{\partial x}\left(\frac{\partial x}{\partial s}\right) + \frac{\partial f}{\partial y}\left(\frac{\partial y}{\partial s}\right) + \frac{\partial f}{\partial z}\left(\frac{\partial z}{\partial s}\right)$$

$$\because \frac{\partial f}{\partial x} = 2xy^3 e^{xz} + zx^2 y^3 e^{xz}, \quad \frac{\partial f}{\partial y} = 3x^2 y^2 e^{xz}, \quad \frac{\partial f}{\partial z} = x^3 y^3 e^{xz},$$

$$\because \frac{\partial x}{\partial s} = 2s, \quad \frac{\partial y}{\partial s} = 2t, \quad \frac{\partial z}{\partial s} = \ln t$$

$$\therefore \frac{\partial f}{\partial s} = \frac{\partial f}{\partial x}\left(\frac{\partial x}{\partial s}\right) + \frac{\partial f}{\partial y}\left(\frac{\partial y}{\partial s}\right) + \frac{\partial f}{\partial z}\left(\frac{\partial z}{\partial s}\right)$$

$$= (2xy^3 e^{xz} + zx^2 y^3 e^{xz})2s + (3x^2 y^2 e^{xz})2t + (x^3 y^3 e^{xz})\ln t$$

Example 1.

(1)设$f(x, y) = \sqrt[3]{xy}$, 试证 $f_x(0,0)$ 与 $f_y(0,0)$都存在而在$(0,0)$不可微分

(2)设$f(x, y) = \begin{cases} \dfrac{x^4 - y^4}{x^3 + y^3}, & (x, y) \neq (0,0) \\ 0, & (x, y) = (0,0) \end{cases}$, 试证$f(x, y)$在$(0,0)$不可微分

【解】

(1)

Claim: $f_x(0,0)$ 与 $f_y(0,0)$都存在

$$f_x(0,0) = \lim_{h \to 0} \frac{f(h + 0,0) - f(0,0)}{h} = \lim_{h \to 0} \frac{\sqrt[3]{h \cdot 0} - 0}{h} = 0$$

$$f_y(0,0) = \lim_{h \to 0} \frac{f(0, h + 0) - f(0,0)}{h} = \lim_{h \to 0} \frac{\sqrt[3]{0 \cdot h} - 0}{h} = 0$$

$$\text{Claim: } \lim_{h \to 0, k \to 0} \left| \frac{f(h + 0, k + 0) - f(0,0) - \frac{\partial f}{\partial x}(0,0)h - \frac{\partial f}{\partial y}(0,0)k}{\sqrt{h^2 + k^2}} \right| \neq 0$$

$$\because \left| \frac{f(h+0,k+0)-f(0,0)-\frac{\partial f}{\partial x}(0,0)h-\frac{\partial f}{\partial y}(0,0)k}{\sqrt{h^2+k^2}} \right| = \left| \frac{\sqrt[3]{hk}}{\sqrt{h^2+k^2}} \right|$$

取 $h=k^2$ 则 $\displaystyle\lim_{h\to 0, k\to 0} \left| \frac{f(h+0,k+0)-f(0,0)-\frac{\partial f}{\partial x}(0,0)h-\frac{\partial f}{\partial y}(0,0)k}{\sqrt{h^2+k^2}} \right|$

$$= \lim_{h\to 0, k\to 0} \left| \frac{\sqrt[3]{hk}}{\sqrt{h^2+k^2}} \right| = \lim_{h\to 0, k\to 0} \left| \frac{\sqrt[3]{k^3}}{\sqrt{k^4+k^2}} \right| = \lim_{k\to 0} \left| \frac{1}{\sqrt{k^2+1}} \right| = 1 \neq 0$$

$\therefore f(x,y)$ 在 $(0,0)$ 不可微分

(2)

Claim: $f_x(0,0)$ 与 $f_y(0,0)$ 都存在

$$f_x(0,0) = \lim_{h\to 0} \frac{f(h+0,0)-f(0,0)}{h} = \lim_{h\to 0} \frac{\frac{h^4}{h^3}-0}{h} = 1$$

$$f_y(0,0) = \lim_{h\to 0} \frac{f(0,h+0)-f(0,0)}{h} = \lim_{h\to 0} \frac{\frac{-h^4}{h^3}-0}{h} = -1$$

Claim: $\displaystyle\lim_{h\to 0, k\to 0} \left| \frac{f(h+0,k+0)-f(0,0)-\frac{\partial f}{\partial x}(0,0)h-\frac{\partial f}{\partial y}(0,0)k}{\sqrt{h^2+k^2}} \right| \neq 0$

$$\because \left| \frac{f(h+0,k+0)-f(0,0)-\frac{\partial f}{\partial x}(0,0)h-\frac{\partial f}{\partial y}(0,0)k}{\sqrt{h^2+k^2}} \right| = \left| \frac{\frac{h^4-k^4}{h^3+k^3}-h+k}{\sqrt{h^2+k^2}} \right|$$

取 $h=2k$ 则

$$\lim_{h\to 0, k\to 0} \left| \frac{f(h+0,k+0)-f(0,0)-\frac{\partial f}{\partial x}(0,0)h-\frac{\partial f}{\partial y}(0,0)k}{\sqrt{h^2+k^2}} \right| = \lim_{h\to 0, k\to 0} \left| \frac{\frac{2}{3}k}{\sqrt{5}k} \right| = \frac{2}{3\sqrt{5}} \neq 0$$

$\therefore f(x,y)$ 在 $(0,0)$ 不可微分


Example 2.

$$设 f(x,y) = \begin{cases} (x^2+y^2)\sin\left(\dfrac{1}{x^2+y^2}\right), & (x,y) \neq (0,0) \\ 0, & (x,y) = (0,0) \end{cases},$$

(1)试证 $f_x(x,y)$ 与 $f_y(x,y)$ 皆存在但在 $(0,0)$ 不连续　(2) 试证 $f(x,y)$ 在 $(0,0)$ 可微分

【解】

(1)

$$f_x(x,y) = 2x\sin\left(\frac{1}{x^2+y^2}\right) + (x^2+y^2)\cos\left(\frac{1}{x^2+y^2}\right)\frac{-2x}{(x^2+y^2)^{-2}}$$

$$= 2x\sin\left(\frac{1}{x^2+y^2}\right) + \cos\left(\frac{1}{x^2+y^2}\right)\frac{-2x}{x^2+y^2}$$

$$f_y(x,y) = 2y\sin\left(\frac{1}{x^2+y^2}\right) + (x^2+y^2)\cos\left(\frac{1}{x^2+y^2}\right)\frac{-2y}{(x^2+y^2)^{-2}}$$

$$= 2y\sin\left(\frac{1}{x^2+y^2}\right) + \cos\left(\frac{1}{x^2+y^2}\right)\frac{-2y}{x^2+y^2}$$

Claim: $\lim\limits_{x\to 0, y\to 0} f_x(x,y) \neq f_x(0,0)$　且　$\lim\limits_{x\to 0, y\to 0} f_y(x,y) \neq f_y(0,0)$

$$f_x(0,0) = \lim_{h\to 0}\frac{f(h+0,0) - f(0,0)}{h} = \lim_{h\to 0}\frac{h^2\sin(\frac{1}{h^2}) - 0}{h} = 0$$

$$f_y(0,0) = \lim_{h\to 0}\frac{f(0,0+h) - f(0,0)}{h} = \lim_{h\to 0}\frac{h^2\sin(\frac{1}{h^2}) - 0}{h} = 0$$

取 $x_n = y_n = \dfrac{1}{\sqrt{n\pi}}$

$$\lim_{n\to\infty} f_x(x_n, y_n) = \lim_{n\to\infty}\cos\left(\frac{1}{x_n^2+y_n^2}\right)\frac{-2x_n}{(x_n^2+y_n^2)} = \lim_{n\to\infty}\cos(2n\pi)\frac{-4n\pi}{\sqrt{n\pi}} = -\infty$$

$$\lim_{n\to\infty} f_y(x_n, y_n) = \lim_{n\to\infty}\cos\left(\frac{1}{x_n^2+y_n^2}\right)\frac{-2y_n}{(x_n^2+y_n^2)} = \lim_{n\to\infty}\cos(2n\pi)\frac{-4n\pi}{\sqrt{n\pi}} = -\infty$$

因此 $\lim\limits_{x\to 0, y\to 0} f_x(x,y) \neq f_x(0,0)$ 且 $\lim\limits_{x\to 0, y\to 0} f_y(x,y) \neq f_y(0,0)$

$\therefore$ $f_x(x,y)$ 与 $f_y(x,y)$ 在 $(0,0)$ 不连续

(2)

Claim: $\lim\limits_{h\to 0, k\to 0}\left|\dfrac{f(h+0, k+0) - f(0,0) - \dfrac{\partial f}{\partial x}(0,0)h - \dfrac{\partial f}{\partial y}(0,0)k}{\sqrt{h^2+k^2}}\right| = 0$

$$\because \left|\frac{f(h+0,k+0)-f(0,0)-\frac{\partial f}{\partial x}(0,0)h-\frac{\partial f}{\partial y}(0,0)k}{\sqrt{h^2+k^2}}\right| = \left|\frac{(h^2+k^2)\sin\left(\frac{1}{h^2+k^2}\right)}{\sqrt{h^2+k^2}}\right|$$

$$= \left|(\sqrt{h^2+k^2})\sin\left(\frac{1}{h^2+k^2}\right)\right| \le \sqrt{h^2+k^2}$$

$$\because \lim_{h\to 0,k\to 0}|\sqrt{h^2+k^2}| = 0 \quad\therefore \lim_{h\to 0,k\to 0}\left|\frac{f(h+0,k+0)-f(0,0)-\frac{\partial f}{\partial x}(0,0)h-\frac{\partial f}{\partial y}(0,0)k}{\sqrt{h^2+k^2}}\right| = 0$$

$\therefore f(x,y)$ 在 $(0,0)$ 可微分

Example 3.

设 $f(x,y) = |x|y$ 试证 (1) $f_x(x,y)$ 与 $f_y(x,y)$ 皆连续 (2) $f(x,y)$ 在 $(0,0)$ 可微分

【解】

(1)

As $x > 0$, $f_x(x,y) = y$ 且 $f_y(x,y) = x$

As $x < 0$, $f_x(x,y) = -y$ 且 $f_y(x,y) = -x$

Claim: $\lim\limits_{x\to 0,y\to 0} f_x(x,y) = f_x(0,0)$ 且 $\lim\limits_{x\to 0,y\to 0} f_y(x,y) = f_y(0,0)$

$$\because f_x(0,0) = \lim_{h\to 0}\frac{f(h+0,0)-f(0,0)}{h} = \lim_{h\to 0}\frac{0-0}{h} = 0$$

$$f_y(0,0) = \lim_{h\to 0}\frac{f(0,0+h)-f(0,0)}{h} = \lim_{h\to 0}\frac{0-0}{h} = 0$$

$$\lim_{x\to 0,y\to 0} f_x(x,y) = \lim_{x\to 0,y\to 0} y = \lim_{x\to 0,y\to 0} -y = 0$$

$$\lim_{x\to 0,y\to 0} f_y(x,y) = \lim_{x\to 0,y\to 0} x = \lim_{x\to 0,y\to 0} -x = 0$$

$$\therefore \lim_{x\to 0,y\to 0} f_x(x,y) = f_x(0,0) \quad 且 \quad \lim_{x\to 0,y\to 0} f_y(x,y) = f_y(0,0)$$

$\therefore f_x(x,y)$ 与 $f_y(x,y)$ 皆连续

(2)

$$\text{Claim:} \quad \lim_{h\to 0,k\to 0}\left|\frac{f(h+0,k+0)-f(0,0)-\frac{\partial f}{\partial x}(0,0)h-\frac{\partial f}{\partial y}(0,0)k}{\sqrt{h^2+k^2}}\right| = 0$$

$$\because \left| \frac{f(h+0, k+0) - f(0,0) - \frac{\partial f}{\partial x}(0,0)h - \frac{\partial f}{\partial y}(0,0)k}{\sqrt{h^2 + k^2}} \right|$$

$$= \left| \frac{|h|k - f(0,0) - \frac{\partial f}{\partial x}(0,0)h - \frac{\partial f}{\partial y}(0,0)k}{\sqrt{h^2 + k^2}} \right| = \left| \frac{|h|k}{\sqrt{h^2 + k^2}} \right| \leq \left| \frac{hk}{\sqrt{2hk}} \right| = \left| \frac{\sqrt{hk}}{\sqrt{2}} \right|$$

$$\because \lim_{h \to 0, k \to 0} \left| \frac{\sqrt{hk}}{\sqrt{2}} \right| = 0$$

$$\therefore \lim_{h \to 0, k \to 0} \left| \frac{f(h+0, k+0) - f(0,0) - \frac{\partial f}{\partial x}(0,0)h - \frac{\partial f}{\partial y}(0,0)k}{\sqrt{h^2 + k^2}} \right| = 0 \quad \therefore f(x,y) 在 (0,0) 可微分$$

Example 4.

$$设 f(x,y) = \begin{cases} (xy)^p \left( \dfrac{x^2 - y^2}{x^2 + y^2} \right), & (x,y) \neq (0,0) \\ 0, & (x,y) = (0,0) \end{cases}, \quad 试证 f(x,y) 在 (0,0) 可微分, \ p > \frac{1}{2}$$

【解】

$$Claim: \lim_{h \to 0, k \to 0} \left| \frac{f(h+0, k+0) - f(0,0) - \frac{\partial f}{\partial x}(0,0)h - \frac{\partial f}{\partial y}(0,0)k}{\sqrt{h^2 + k^2}} \right| = 0$$

$$\frac{\partial f}{\partial x}(0,0) = \lim_{h \to 0} \frac{f(h+0,0) - f(0,0)}{h} = \lim_{h \to 0} \frac{0 - 0}{h} = 0$$

$$\frac{\partial f}{\partial y}(0,0) = \lim_{h \to 0} \frac{f(0, h+0) - f(0,0)}{h} = \lim_{h \to 0} \frac{0 - 0}{h} = 0$$

$$令 p > \frac{1}{2}$$

$$\because \left| \frac{f(h+0, k+0) - f(0,0) - \frac{\partial f}{\partial x}(0,0)h - \frac{\partial f}{\partial y}(0,0)k}{\sqrt{h^2 + k^2}} \right| = \left| \frac{(hk)^p \left( \frac{h^2 - k^2}{h^2 + k^2} \right)}{\sqrt{h^2 + k^2}} \right|$$

$$\leq \left| \frac{(hk)^p \left( \frac{h^2 + k^2}{h^2 + k^2} \right)}{\sqrt{h^2 + k^2}} \right| = \left| \frac{(hk)^p}{\sqrt{h^2 + k^2}} \right| = \left| \frac{(hk)^p}{\sqrt{2hk}} \right| \leq \left| \frac{(hk)^{p - \frac{1}{2}}}{\sqrt{2}} \right|$$

且 $\lim\limits_{h\to 0, k\to 0}(hk)^{p-\frac{1}{2}}=0,\ \ \forall p>\dfrac{1}{2}$

$$\therefore\ \lim_{h\to 0, k\to 0}\left|\frac{f(h+0,k+0)-f(0,0)-\dfrac{\partial f}{\partial x}(0,0)h-\dfrac{\partial f}{\partial y}(0,0)k}{\sqrt{h^2+k^2}}\right|=0$$

$\therefore f(x,y)$ 在 $(0,0)$ 可微分，$\forall p>\dfrac{1}{2}$

Example 5.

$$设 f(x,y)=\begin{cases}\dfrac{xy}{\sqrt{x^2+y^2}}, & (x,y)\neq(0,0)\\[2mm] 0, & (x,y)=(0,0)\end{cases},\ \ 试证 f(x,y) 在 (0,0) 不可微分$$

【解】

$$Claim:\ \lim_{h\to 0, k\to 0}\left|\frac{f(h+0,k+0)-f(0,0)-\dfrac{\partial f}{\partial x}(0,0)h-\dfrac{\partial f}{\partial y}(0,0)k}{\sqrt{h^2+k^2}}\right|\neq 0$$

$$\frac{\partial f}{\partial x}(0,0)=\lim_{h\to 0}\frac{f(h+0,0)-f(0,0)}{h}=\lim_{h\to 0}\frac{0-0}{h}=0$$

$$\frac{\partial f}{\partial y}(0,0)=\lim_{h\to 0}\frac{f(0,h+0)-f(0,0)}{h}=\lim_{h\to 0}\frac{0-0}{h}=0$$

$$\because\left|\frac{f(h+0,k+0)-f(0,0)-\dfrac{\partial f}{\partial x}(0,0)h-\dfrac{\partial f}{\partial y}(0,0)k}{\sqrt{h^2+k^2}}\right|=\left|\frac{\dfrac{hk}{\sqrt{h^2+k^2}}}{\sqrt{h^2+k^2}}\right|$$

取 $h=k$

$$\lim_{h\to 0, k\to 0}\left|\frac{f(h+0,k+0)-f(0,0)-\dfrac{\partial f}{\partial x}(0,0)h-\dfrac{\partial f}{\partial y}(0,0)k}{\sqrt{h^2+k^2}}\right|$$

$$=\lim_{h\to 0, k\to 0}\left|\frac{\dfrac{hk}{\sqrt{h^2+k^2}}}{\sqrt{h^2+k^2}}\right|=\lim_{h\to 0, k\to 0}\left|\frac{h^2}{2h^2}\right|=\frac{1}{2}\neq 0$$

$\therefore f(x,y)$ 在 $(0,0)$ 不可微分

Example 6.

$$\text{设} f(x,y) = \begin{cases} \dfrac{2x\sqrt{y}}{\sqrt{x^4 + y^2}}, & (x,y) \neq (0,0) \\ 0, & (x,y) = (0,0) \end{cases}, \quad \text{试证} f(x,y) \text{于} (0,0) \text{不可微分}$$

【解】

Claim: $\displaystyle\lim_{h\to 0, k\to 0} \left| \dfrac{f(h+0, k+0) - f(0,0) - \dfrac{\partial f}{\partial x}(0,0)h - \dfrac{\partial f}{\partial y}(0,0)k}{\sqrt{h^2 + k^2}} \right| \neq 0$

$$\frac{\partial f}{\partial x}(0,0) = \lim_{h\to 0} \frac{f(h+0,0) - f(0,0)}{h} = \lim_{h\to 0} \frac{\dfrac{2h\cdot 0}{\sqrt{h^4 + 0^2}} - 0}{h} = \lim_{h\to 0} \frac{0}{h^2} = 0$$

$$\frac{\partial f}{\partial y}(0,0) = \lim_{h\to 0} \frac{f(0, h+0) - f(0,0)}{h} = \lim_{h\to 0} \frac{\dfrac{2\cdot 0 \cdot \sqrt{h}}{\sqrt{0^4 + h^2}} - 0}{h} = \lim_{h\to 0} \frac{0}{h^{\frac{3}{2}}} = 0$$

$$\therefore \frac{\partial f}{\partial x}(0,0) = 0, \quad \frac{\partial f}{\partial y}(0,0) = 0$$

$$\because \left| \frac{f(h+0, k+0) - f(0,0) - \dfrac{\partial f}{\partial x}(0,0)h - \dfrac{\partial f}{\partial y}(0,0)k}{\sqrt{h^2 + k^2}} \right| = \left| \frac{\dfrac{2h\sqrt{k}}{\sqrt{h^4 + k^2}}}{\sqrt{h^2 + k^2}} \right|$$

取 $k = h^2$

$$\lim_{h\to 0, k\to 0} \left| \frac{f(h+0, k+0) - f(0,0) - \dfrac{\partial f}{\partial x}(0,0)h - \dfrac{\partial f}{\partial y}(0,0)k}{\sqrt{h^2 + k^2}} \right|$$

$$= \lim_{h\to 0, k\to 0} \left| \frac{\dfrac{2h\sqrt{k}}{\sqrt{h^4 + k^2}}}{\sqrt{h^2 + k^2}} \right| = \lim_{h\to 0, k\to 0} \left| \frac{\dfrac{2h^2}{\sqrt{2}h^2}}{\sqrt{h^2 + h^4}} \right| = \infty \neq 0$$

$$\therefore f(x,y) \text{于} (0,0) \text{不可微分}$$

Example 7.

$$\text{设} f(x, y) = \begin{cases} xy\left(\dfrac{x^2 - y^2}{x^2 + y^2}\right), & (x, y) \neq (0,0) \\ 0, & (x, y) = (0,0) \end{cases}$$

试证 $(1)\, f(x, y)$ 在 $(x, y) = (0,0)$ 连续 $(2)\, f(x, y)$ 在 $(0,0)$ 可微分

【解】

(1)

$\because x^2 + y^2 \geq 2|xy| \qquad \therefore \dfrac{1}{x^2 + y^2} \leq \dfrac{1}{2|xy|}$

$\therefore \left| xy\left(\dfrac{x^2 - y^2}{x^2 + y^2}\right) \right| \leq \left| xy\left(\dfrac{x^2 - y^2}{2xy}\right) \right| \leq \dfrac{x^2 + y^2}{2}$

Claim: $\displaystyle\lim_{x\to0, y\to0} f(x, y) = f(0,0)$

$\because \displaystyle\lim_{x\to0, y\to0} \dfrac{x^2 + y^2}{2} = 0 \qquad \therefore \displaystyle\lim_{x\to0, y\to0} \left| xy\left(\dfrac{x^2 - y^2}{x^2 + y^2}\right) \right| = 0$

$\therefore f(x, y)$ 在 $(x, y) = (0,0)$ 连续

(2)

Claim: $\displaystyle\lim_{h\to0, k\to0} \left| \dfrac{f(h+0, k+0) - f(0,0) - \dfrac{\partial f}{\partial x}(0,0)h - \dfrac{\partial f}{\partial y}(0,0)k}{\sqrt{h^2 + k^2}} \right| = 0$

$\dfrac{\partial f}{\partial x}(0,0) = \displaystyle\lim_{h\to0} \dfrac{f(h+0, 0) - f(0,0)}{h} = \lim_{h\to0} \dfrac{0-0}{h} = 0$

$\dfrac{\partial f}{\partial y}(0,0) = \displaystyle\lim_{h\to0} \dfrac{f(0, h+0) - f(0,0)}{h} = \lim_{h\to0} \dfrac{0-0}{h} = 0$

$\because \left| \dfrac{f(h+0, k+0) - f(0,0) - \dfrac{\partial f}{\partial x}(0,0)h - \dfrac{\partial f}{\partial y}(0,0)k}{\sqrt{h^2 + k^2}} \right| = \left| \dfrac{hk\left(\dfrac{h^2 - k^2}{h^2 + k^2}\right)}{\sqrt{h^2 + k^2}} \right|$

$\leq \left| \dfrac{hk\left(\dfrac{h^2 + k^2}{h^2 + k^2}\right)}{\sqrt{h^2 + k^2}} \right| = \left| \dfrac{hk}{\sqrt{h^2 + k^2}} \right| \leq \left| \dfrac{hk}{\sqrt{2hk}} \right| \leq \left| \dfrac{\sqrt{hk}}{\sqrt{2}} \right| \leq \sqrt{\dfrac{h^2 + k^2}{4}}$

且 $\displaystyle\lim_{h\to0, k\to0} \sqrt{\dfrac{h^2 + k^2}{4}} = 0$

$$\therefore \lim_{h\to 0, k\to 0}\left|\frac{f(h+0,k+0)-f(0,0)-\dfrac{\partial f}{\partial x}(0,0)h-\dfrac{\partial f}{\partial y}(0,0)k}{\sqrt{h^2+k^2}}\right|=0$$

$\therefore f(x,y)$ 在 $(0,0)$ 可微分

Example 8.

$$设 f(x,y)=\begin{cases}\dfrac{x^2 y}{x^2+y^2}, & (x,y)\neq(0,0)\\[2mm] 0, & (x,y)=(0,0)\end{cases}$$

试证 (1) $f(x,y)$ 在 $(x,y)=(0,0)$ 连续 (2) $f(x,y)$ 在 $(0,0)$ 不可微分

【解】

(1)

$\because x^2+y^2\geq 2|xy| \quad \therefore \dfrac{1}{x^2+y^2}\leq\dfrac{1}{2|xy|}\Rightarrow\left|\dfrac{x^2y}{x^2+y^2}\right|\leq\left|\dfrac{x^2y}{2xy}\right|\leq\left|\dfrac{x}{2}\right|$

Claim: $\lim\limits_{x\to 0, y\to 0} f(x,y)=f(0,0)$

$\because \lim\limits_{x\to 0, y\to 0}\left|\dfrac{x}{2}\right|=0 \quad \therefore \lim\limits_{x\to 0, y\to 0}\left|\dfrac{x^2y}{x^2+y^2}\right|=0 \quad \therefore f(x,y)$ 在 $(x,y)=(0,0)$ 连续

(2)

Claim: $\lim\limits_{h\to 0, k\to 0}\left|\dfrac{f(h+0,k+0)-f(0,0)-\dfrac{\partial f}{\partial x}(0,0)h-\dfrac{\partial f}{\partial y}(0,0)k}{\sqrt{h^2+k^2}}\right|\neq 0$

$\dfrac{\partial f}{\partial x}(0,0)=\lim\limits_{h\to 0}\dfrac{f(h+0,0)-f(0,0)}{h}=\lim\limits_{h\to 0}\dfrac{0-0}{h}=0$

$\dfrac{\partial f}{\partial y}(0,0)=\lim\limits_{h\to 0}\dfrac{f(0,h+0)-f(0,0)}{h}=\lim\limits_{h\to 0}\dfrac{0-0}{h}=0$

$\therefore \left|\dfrac{f(h+0,k+0)-f(0,0)-\dfrac{\partial f}{\partial x}(0,0)h-\dfrac{\partial f}{\partial y}(0,0)k}{\sqrt{h^2+k^2}}\right|=\left|\dfrac{\dfrac{h^2k}{h^2+k^2}}{\sqrt{h^2+k^2}}\right|$

取 $h=k$

$$\lim_{h\to 0, k\to 0}\left|\frac{f(h+0,k+0)-f(0,0)-\dfrac{\partial f}{\partial x}(0,0)h-\dfrac{\partial f}{\partial y}(0,0)k}{\sqrt{h^2+k^2}}\right|$$

$$= \lim_{h\to 0, k\to 0} \left| \frac{\dfrac{h^2 k}{h^2 + k^2}}{\sqrt{h^2 + k^2}} \right| = \lim_{h\to 0, k\to 0} \left| \frac{\dfrac{h^3}{2h^2}}{\sqrt{2h^2}} \right| = \frac{1}{2\sqrt{2}} \neq 0$$

$\therefore f(x,y)$ 在 $(0,0)$ 不可微分

Example 9.

$$\text{设} f(x,y) = \begin{cases} \dfrac{2x^2 y}{x^4 + y^2}, & (x,y) \neq (0,0) \\ 0, & (x,y) = (0,0) \end{cases}, \quad \text{试证 } f(x,y) \text{ 于 } (0,0) \text{不可微分}$$

【解】

Claim: $\displaystyle \lim_{h\to 0, k\to 0} \left| \frac{f(h+0, k+0) - f(0,0) - \dfrac{\partial f}{\partial x}(0,0)h - \dfrac{\partial f}{\partial y}(0,0)k}{\sqrt{h^2 + k^2}} \right| \neq 0$

$$\frac{\partial f}{\partial x}(0,0) = \lim_{h\to 0} \frac{f(h+0,0) - f(0,0)}{h} = \lim_{h\to 0} \frac{\dfrac{2h^2 \cdot 0}{h^4 + 0^2} - 0}{h} = \lim_{h\to 0} \frac{0}{h^5} = 0$$

$$\frac{\partial f}{\partial y}(0,0) = \lim_{h\to 0} \frac{f(0, h+0) - f(0,0)}{h} = \lim_{h\to 0} \frac{\dfrac{2 \cdot 0^2 \cdot h}{0^4 + h^2} - 0}{h} = \lim_{h\to 0} \frac{0}{h^3} = 0$$

$$\therefore \frac{\partial f}{\partial x}(0,0) = 0, \quad \frac{\partial f}{\partial y}(0,0) = 0$$

$$\therefore \left| \frac{f(h+0, k+0) - f(0,0) - \dfrac{\partial f}{\partial x}(0,0)h - \dfrac{\partial f}{\partial y}(0,0)k}{\sqrt{h^2 + k^2}} \right| = \left| \frac{\dfrac{2h^2 k}{h^4 + k^2}}{\sqrt{h^2 + k^2}} \right|$$

取 $k = h^2$

$$\lim_{h\to 0, k\to 0} \left| \frac{f(h+0, k+0) - f(0,0) - \dfrac{\partial f}{\partial x}(0,0)h - \dfrac{\partial f}{\partial y}(0,0)k}{\sqrt{h^2 + k^2}} \right|$$

$$= \lim_{h\to 0, k\to 0} \left| \frac{\dfrac{2h^2 k}{h^4 + k^2}}{\sqrt{h^2 + k^2}} \right| = \lim_{h\to 0, k\to 0} \left| \frac{\dfrac{2h^4}{2h^4}}{\sqrt{h^2 + h^4}} \right| = \infty \neq 0$$

$\therefore f(x,y)$ 于 $(0,0)$ 不可微分

Example 10.

$$\text{设}f(x,y) = \begin{cases} \dfrac{3x^2y}{x^3 + y^3}, & (x,y) \neq (0,0) \\ 0, & (x,y) = (0,0) \end{cases}, \quad \text{试证}f(x,y)\text{于}(0,0)\text{不可微分}$$

【解】

Claim: $\displaystyle \lim_{h\to 0, k\to 0} \left| \dfrac{f(h+0, k+0) - f(0,0) - \dfrac{\partial f}{\partial x}(0,0)h - \dfrac{\partial f}{\partial y}(0,0)k}{\sqrt{h^2 + k^2}} \right| \neq 0$

$$\frac{\partial f}{\partial x}(0,0) = \lim_{h\to 0} \frac{f(h+0, 0) - f(0,0)}{h} = \lim_{h\to 0} \frac{\dfrac{3 \cdot h^2 \cdot 0}{h^3 + 0^3} - 0}{h} = \lim_{h\to 0} \frac{0}{h^4} = 0$$

$$\frac{\partial f}{\partial y}(0,0) = \lim_{h\to 0} \frac{f(0, h+0) - f(0,0)}{h} = \lim_{h\to 0} \frac{\dfrac{3 \cdot 0^2 \cdot h}{0^3 + h^3} - 0}{h} = \lim_{h\to 0} \frac{0}{h^4} = 0$$

因此 $\dfrac{\partial f}{\partial x}(0,0) = 0$, $\dfrac{\partial f}{\partial y}(0,0) = 0$

$$\because \left| \frac{f(h+0, k+0) - f(0,0) - \dfrac{\partial f}{\partial x}(0,0)h - \dfrac{\partial f}{\partial y}(0,0)k}{\sqrt{h^2 + k^2}} \right| = \left| \frac{\dfrac{3h^2 k}{h^3 + k^3}}{\sqrt{h^2 + k^2}} \right|$$

取 $k = h$

$$\lim_{h\to 0, k\to 0} \left| \frac{f(h+0, k+0) - f(0,0) - \dfrac{\partial f}{\partial x}(0,0)h - \dfrac{\partial f}{\partial y}(0,0)k}{\sqrt{h^2 + k^2}} \right|$$

$$= \lim_{h\to 0, k\to 0} \left| \frac{\dfrac{3h^2 k}{h^3 + k^3}}{\sqrt{h^2 + k^2}} \right| = \lim_{h\to 0, k\to 0} \left| \frac{\dfrac{3h^3}{2h^3}}{\sqrt{h^2 + h^2}} \right| = \infty \neq 0$$

$\therefore f(x,y)$ 于 $(0,0)$ 不可微分

Example 11.

$(1)\, f(x,y) = \dfrac{x^2}{x^2 + y^2}$, 且 $x = 3\cos t$, $y = 5\sin t$, 求 $\dfrac{df}{dt} = ?$

$(2)\, \text{设}f(x,y) = \sin^{-1}(5x + 2y)$, 且 $x = r^2 e^s$, $y = \sin(rs)$, 求 $\dfrac{\partial f}{\partial r}\, \dfrac{\partial f}{\partial s} = ?$

【解】

(1)

$$\because \frac{df}{dt} = \frac{\partial f}{\partial x}\frac{\partial x}{\partial t} + \frac{\partial f}{\partial y}\frac{\partial y}{\partial t}$$

$$\frac{\partial f}{\partial x} = \frac{2x(x^2+y^2) - x^2(2x)}{(x^2+y^2)^2} = \frac{2xy^2}{(x^2+y^2)^2}, \quad \frac{\partial f}{\partial y} = \frac{-x^2(2y)}{(x^2+y^2)^2} = \frac{-2yx^2}{(x^2+y^2)^2}$$

$$\frac{\partial x}{\partial t} = -3\sin t, \quad \frac{\partial y}{\partial t} = 5\cos t$$

$$\therefore \frac{df}{dt} = \frac{\partial f}{\partial x}\frac{\partial x}{\partial t} + \frac{\partial f}{\partial y}\frac{\partial y}{\partial t} = \frac{2xy^2}{(x^2+y^2)^2}(-3\sin t) + \frac{-2yx^2}{(x^2+y^2)^2}(5\cos t)$$

(2)

$$\because \frac{\partial f}{\partial r} = \frac{\partial f}{\partial x}\frac{\partial x}{\partial r} + \frac{\partial f}{\partial y}\frac{\partial y}{\partial r} \quad 且 \quad \frac{\partial f}{\partial s} = \frac{\partial f}{\partial x}\frac{\partial x}{\partial s} + \frac{\partial f}{\partial y}\frac{\partial y}{\partial s}$$

$$\frac{\partial f}{\partial x} = \frac{5}{\sqrt{1-(5x+2y)^2}}, \quad \frac{\partial f}{\partial y} = \frac{2}{\sqrt{1-(5x+2y)^2}}$$

$$\frac{\partial x}{\partial r} = 2re^s, \quad \frac{\partial y}{\partial r} = s\cos(rs), \quad \frac{\partial x}{\partial s} = r^2 e^s, \quad \frac{\partial y}{\partial s} = r\cos(rs)$$

$$\therefore \frac{\partial f}{\partial r} = \frac{\partial f}{\partial x}\frac{\partial x}{\partial r} + \frac{\partial f}{\partial y}\frac{\partial y}{\partial r} = \frac{10re^s + 2s\cos(rs)}{\sqrt{1-(5x+2y)^2}}$$

$$且 \quad \frac{\partial f}{\partial s} = \frac{\partial f}{\partial x}\frac{\partial x}{\partial s} + \frac{\partial f}{\partial y}\frac{\partial y}{\partial s} = \frac{5r^2 e^s + 2r\cos(rs)}{\sqrt{1-(5x+2y)^2}}$$

Example 12.

$$设 F(x,y) = x^2 y^2 + x + y + 2, \quad x(t) = t^4 - 1, y(t) = t^2 - t^3$$
$$G(t) = F\big(x(t), y(t)\big), \quad 求 \ G'(t) = ?$$

【解】

$$\because G'(t) = \frac{\partial G}{\partial x}\frac{\partial x}{\partial t} + \frac{\partial G}{\partial y}\frac{\partial y}{\partial t}$$

$$\frac{\partial G}{\partial x} = 2xy^2 + 1, \quad \frac{\partial G}{\partial y} = 2yx^2 + 1, \quad \frac{\partial x}{\partial t} = 4t^3, \quad \frac{\partial y}{\partial t} = 2t - 3t^2$$

$$\therefore G'(t) = 4t^3(2xy^2 + 1) + (2t - 3t^2)(2yx^2 + 1)$$

Example 13.

$$设 z = g(x^2 - y^2), \quad 试证 \ y\frac{\partial z}{\partial x} + x\frac{\partial z}{\partial y} = 0$$

【解】

$$\because \frac{\partial z}{\partial x} = g'(x^2 - y^2)(2x), \quad \frac{\partial z}{\partial y} = g'(x^2 - y^2)(-2y)$$

$$\therefore y\frac{\partial z}{\partial x} + x\frac{\partial z}{\partial y} = yg'(x^2 - y^2)(2x) + xg'(x^2 - y^2)(-2y) = 0$$

Example 14.

$$设 z = x^2 + g(xy), \quad 试证 \ x\frac{\partial z}{\partial x} - y\frac{\partial z}{\partial y} = 2x^2$$

【解】

$$\because \frac{\partial z}{\partial x} = 2x + g'(xy)y, \quad \frac{\partial z}{\partial y} = g'(xy)x$$

$$\therefore x\frac{\partial z}{\partial x} - y\frac{\partial z}{\partial y} = x(2x + g'(xy)y) - y(g'(xy)x) = 2x^2$$

Example 15.

$$设 f(x, y, z) = \frac{1}{\sqrt{x^2 + y^2 + z^2}}, \quad 试证 \ f_{xx} + f_{yy} + f_{zz} = 0$$

【解】

$$\because f_x = -\frac{1}{2}(x^2 + y^2 + z^2)^{-\frac{3}{2}}(2x) = -x(x^2 + y^2 + z^2)^{-\frac{3}{2}}$$

$$f_y = -\frac{1}{2}(x^2 + y^2 + z^2)^{-\frac{3}{2}}(2y) = -y(x^2 + y^2 + z^2)^{-\frac{3}{2}}$$

$$f_z = -\frac{1}{2}(x^2 + y^2 + z^2)^{-\frac{3}{2}}(2z) = -z(x^2 + y^2 + z^2)^{-\frac{3}{2}}$$

$$f_{xx} = -(x^2 + y^2 + z^2)^{-\frac{3}{2}} + 3x^2(x^2 + y^2 + z^2)^{-\frac{5}{2}}$$

$$= (x^2 + y^2 + z^2)^{-\frac{5}{2}}\left(3x^2 - (x^2 + y^2 + z^2)\right) = (x^2 + y^2 + z^2)^{-\frac{5}{2}}(2x^2 - y^2 - z^2)$$

$$f_{yy} = (x^2 + y^2 + z^2)^{-\frac{5}{2}}(2y^2 - x^2 - z^2)$$

$$f_{zz} = (x^2 + y^2 + z^2)^{-\frac{5}{2}}(2z^2 - x^2 - y^2)$$

$$\therefore f_{xx} + f_{yy} + f_{zz} = 0$$

Example 16.

$$设 f(x,y,z) = \left(\frac{x-y+z}{x+y-z}\right)^n, \quad 试证 \ xf_x + yf_y + zf_z = 0$$

【解】

$$f_x(x,y,z) = n\left(\frac{x-y+z}{x+y-z}\right)^{n-1}\frac{(x+y-z)-(x-y+z)}{(x+y-z)^2} = 2n\left(\frac{x-y+z}{x+y-z}\right)^{n-1}\frac{y-z}{(x+y-z)^2}$$

$$f_y(x,y,z) = -2n\left(\frac{x-y+z}{x+y-z}\right)^{n-1}\frac{x-z}{(x+y-z)^2}$$

$$f_z(x,y,z) = 2n\left(\frac{x-y+z}{x+y-z}\right)^{n-1}\frac{x-y}{(x+y-z)^2}$$

$$\therefore xf_x + yf_y + zf_z = 2n\left(\frac{x-y+z}{x+y-z}\right)^{n-1}(x(y-z)-y(x-z)+z(x-y)) = 0$$

Example 17.

$$设 \ g(x,y) = f(xy^2), \quad 试证 \ 4x^2 g_{xx} - y^2 g_{yy} + yg_y = 0$$

【解】

$$\because g_x = f'(xy^2)y^2, \quad g_y = f'(xy^2)2xy$$

$$\therefore g_{xx} = f''(xy^2)y^4, \quad g_{yy} = f''(xy^2)4x^2y^2 + 2xf'(xy^2)$$

$$\therefore 4x^2 g_{xx} - y^2 g_{yy} + yg_y$$

$$= 4x^2 f''(xy^2)y^4 - y^2\left(f''(xy^2)4x^2y^2 + 2xf'(xy^2)\right) + yf'(xy^2)2xy = 0$$

Example 18.

$$设 \ w = f(u), u = \frac{xy}{x^2+y^2}, \quad 试证 \ x\frac{\partial w}{\partial x} + y\frac{\partial w}{\partial y} = 0$$

【解】

$$\frac{\partial w}{\partial x} = f'(u)\frac{y(x^2+y^2)-xy(2x)}{(x^2+y^2)^2} = f'(u)\frac{y^3 - x^2 y}{(x^2+y^2)^2}$$

$$\frac{\partial w}{\partial y} = f'(u)\frac{x^3 - y^2 x}{(x^2+y^2)^2}$$

$$x\frac{\partial w}{\partial x} + y\frac{\partial w}{\partial y} = x\left(f'(u)\frac{y^3 - x^2 y}{(x^2 + y^2)^2}\right) + y\left(f'(u)\frac{x^3 - y^2 x}{(x^2 + y^2)^2}\right) = 0$$

Example 19.

$$\text{设 } w = f(x - y, y - x), \quad \text{试证 } \frac{\partial w}{\partial x} + \frac{\partial w}{\partial y} = 0$$

【解】

令 $u = x - y, \ v = y - x$

$$\because \frac{\partial w}{\partial x} = f_u(u, v)\frac{\partial u}{\partial x} + f_v(u, v)\frac{\partial v}{\partial x} = f_u(u, v) - f_v(u, v)$$

$$\frac{\partial w}{\partial y} = f_u(u, v)\frac{\partial u}{\partial y} + f_v(u, v)\frac{\partial v}{\partial y} = -f_u(u, v) + f_v(u, v)$$

$$\therefore \frac{\partial w}{\partial x} + \frac{\partial w}{\partial y} = 0$$

Example 20.

$$\text{设 } u = f(x, y), x = r\cos\theta, y = r\sin\theta, \quad \text{试证 } \frac{\partial^2 u}{\partial x^2} + \frac{\partial^2 u}{\partial y^2} = \frac{\partial^2 u}{\partial r^2} + \frac{1}{r^2}\frac{\partial^2 u}{\partial \theta^2} + \frac{1}{r}\frac{\partial u}{\partial r}$$

【解】

$$\because \frac{\partial u}{\partial r} = \frac{\partial u}{\partial x}\cdot\frac{\partial x}{\partial r} + \frac{\partial u}{\partial y}\cdot\frac{\partial y}{\partial r} = \frac{\partial u}{\partial x}\cdot\cos\theta + \frac{\partial u}{\partial y}\cdot\sin\theta$$

$$\frac{\partial u}{\partial \theta} = \frac{\partial u}{\partial x}\cdot\frac{\partial x}{\partial \theta} + \frac{\partial u}{\partial y}\cdot\frac{\partial y}{\partial \theta} = -\frac{\partial u}{\partial x}\cdot r\sin\theta + \frac{\partial u}{\partial y}\cdot r\cos\theta$$

$$\frac{\partial^2 u}{\partial r^2} = \frac{\partial}{\partial r}\left(\frac{\partial u}{\partial r}\right) = \frac{\partial}{\partial r}\left(\frac{\partial u}{\partial x}\cdot\cos\theta + \frac{\partial u}{\partial y}\cdot\sin\theta\right) = \frac{\partial^2 u}{\partial x^2}\cdot\cos^2\theta + \frac{\partial^2 u}{\partial y^2}\cdot\sin^2\theta$$

$$\frac{\partial^2 u}{\partial \theta^2} = \frac{\partial}{\partial \theta}\left(\frac{\partial u}{\partial \theta}\right) = \frac{\partial}{\partial \theta}\left(-\frac{\partial u}{\partial x}\cdot r\sin\theta + \frac{\partial u}{\partial y}\cdot r\cos\theta\right)$$

$$= \frac{\partial^2 u}{\partial x^2}r^2\sin^2\theta + \frac{\partial^2 u}{\partial y^2}\cdot r^2\cos^2\theta - \frac{\partial u}{\partial x}\cdot r\cos\theta - \frac{\partial u}{\partial y}\cdot r\sin\theta$$

$$\therefore \frac{\partial^2 u}{\partial r^2} + \frac{1}{r^2}\frac{\partial^2 u}{\partial \theta^2} + \frac{1}{r}\frac{\partial u}{\partial r} = \frac{\partial^2 u}{\partial x^2} + \frac{\partial^2 u}{\partial y^2}$$

Example 21.

设 $f(x,y,z) = g(x-y, y-z, z-x)$，试证 $\dfrac{\partial f}{\partial x} + \dfrac{\partial f}{\partial y} + \dfrac{\partial f}{\partial z} = 0$

【解】

令 $t_1 = x - y$，$t_2 = y - z$，$t_3 = z - x$

$\because \dfrac{\partial f}{\partial x} = \dfrac{\partial g}{\partial t_1}\dfrac{\partial t_1}{\partial x} + \dfrac{\partial g}{\partial t_2}\dfrac{\partial t_2}{\partial x} + \dfrac{\partial g}{\partial t_3}\dfrac{\partial t_3}{\partial x}$，$\dfrac{\partial f}{\partial y} = \dfrac{\partial g}{\partial t_1}\dfrac{\partial t_1}{\partial y} + \dfrac{\partial g}{\partial t_2}\dfrac{\partial t_2}{\partial y} + \dfrac{\partial g}{\partial t_3}\dfrac{\partial t_3}{\partial y}$

$\dfrac{\partial f}{\partial z} = \dfrac{\partial g}{\partial t_1}\dfrac{\partial t_1}{\partial z} + \dfrac{\partial g}{\partial t_2}\dfrac{\partial t_2}{\partial z} + \dfrac{\partial g}{\partial t_3}\dfrac{\partial t_3}{\partial z}$ 且 $\left(\dfrac{\partial t_1}{\partial x}, \dfrac{\partial t_2}{\partial x}, \dfrac{\partial t_3}{\partial x}\right) = (1, 0, -1)$，

$\left(\dfrac{\partial t_1}{\partial y}, \dfrac{\partial t_2}{\partial y}, \dfrac{\partial t_3}{\partial y}\right) = (-1, 1, 0)$，$\left(\dfrac{\partial t_1}{\partial z}, \dfrac{\partial t_2}{\partial z}, \dfrac{\partial t_3}{\partial z}\right) = (0, -1, 1)$

$\therefore \dfrac{\partial f}{\partial x} + \dfrac{\partial f}{\partial y} + \dfrac{\partial f}{\partial z} = \dfrac{\partial g}{\partial t_1} - \dfrac{\partial g}{\partial t_3} - \dfrac{\partial g}{\partial t_1} + \dfrac{\partial g}{\partial t_2} - \dfrac{\partial g}{\partial t_2} + \dfrac{\partial g}{\partial t_3} = 0$

Example 22.

设 $f(x,y) = g(x^2 - y^2, y^2 - x^2)$，试证 $y\dfrac{\partial f}{\partial x} + x\dfrac{\partial f}{\partial y} = 0$

【解】

令 $t_1 = x^2 - y^2$，$t_2 = y^2 - x^2$

$\because \dfrac{\partial f}{\partial x} = \dfrac{\partial g}{\partial t_1}\dfrac{\partial t_1}{\partial x} + \dfrac{\partial g}{\partial t_2}\dfrac{\partial t_2}{\partial x}$，$\dfrac{\partial f}{\partial y} = \dfrac{\partial g}{\partial t_1}\dfrac{\partial t_1}{\partial y} + \dfrac{\partial g}{\partial t_2}\dfrac{\partial t_2}{\partial y}$

且 $\left(\dfrac{\partial t_1}{\partial x}, \dfrac{\partial t_2}{\partial x}\right) = (2x, -2x)$，$\left(\dfrac{\partial t_1}{\partial y}, \dfrac{\partial t_2}{\partial y}\right) = (-2y, 2y)$

$\therefore y\dfrac{\partial f}{\partial x} + x\dfrac{\partial f}{\partial y} = y\left(\dfrac{\partial g}{\partial t_1}(2x) + \dfrac{\partial g}{\partial t_2}(-2x)\right) + x\left(\dfrac{\partial g}{\partial t_1}(-2y) + \dfrac{\partial g}{\partial t_2}(2y)\right) = 0$

Example 23.

设 $f(u,v) = g\big(x(u,v), y(u,v)\big)$ 且 $g(x,y) = \dfrac{1}{(x^2 y^2 - 4x + 3y + 1)^p}$ $(p > 0)$

其中 $\begin{cases} x(u,v) = u^2 - 3uv + 2v^2 \\ y(u,v) = u^4 + 3uv^3 - 4v^4 \end{cases}$，求 $\dfrac{\partial f}{\partial u}(1,1) = ?$

【解】

令 $p > 0$

$$\because \frac{\partial f}{\partial u} = \frac{\partial g}{\partial x}\frac{\partial x}{\partial u} + \frac{\partial g}{\partial y}\frac{\partial y}{\partial u}, \quad \frac{\partial g}{\partial x} = -\frac{p(2xy^2 - 4)}{(x^2y^2 - 4x + 3y + 1)^{2p}}$$

$$\frac{\partial g}{\partial y} = -\frac{p(2yx^2 + 3)}{(x^2y^2 - 4x + 3y + 1)^{2p}}, \quad \frac{\partial x}{\partial u} = 2u - 3v, \quad \frac{\partial y}{\partial u} = 4u^3 + 3v^3$$

$$\because (u,v) = (1,1) \quad \therefore (x,y) = (0,0)$$

$$\therefore \frac{\partial f}{\partial u}(1,1) = \frac{\partial g}{\partial x}(0,0)\frac{\partial x}{\partial u}(1,1) + \frac{\partial g}{\partial y}(0,0)\frac{\partial y}{\partial u}(1,1) = 4p(-1) + (-3p)7 = -25p$$

## Example 24.

$$设 f(x,y) = \frac{x^2 y}{1 + x^2 + y^2}, \quad x, y \in R \ 且 \ g(u,v) = f(uv, u^2 + 2v^3), \quad u, v \in R$$

$$求 \ \frac{\partial g}{\partial v}(\sqrt{2}, -1) = ?$$

【解】

$$令 x = uv, \ y = u^2 + 2v^3 \ 则 \ \frac{\partial g}{\partial v} = \frac{\partial f}{\partial x}\frac{\partial x}{\partial v} + \frac{\partial f}{\partial y}\frac{\partial y}{\partial v}$$

$$\because \frac{\partial f}{\partial x} = \frac{2xy(1 + x^2 + y^2) - x^2y(2x)}{(1 + x^2 + y^2)^2} = \frac{2xy(1 + y^2)}{(1 + x^2 + y^2)^2}$$

$$\frac{\partial f}{\partial y} = \frac{x^2(1 + x^2 + y^2) - x^2y(2y)}{(1 + x^2 + y^2)^2} = \frac{x^2 + x^4 - x^2y^2}{(1 + x^2 + y^2)^2}$$

$$且 \ \frac{\partial x}{\partial v} = u, \quad \frac{\partial y}{\partial v} = 6v^2$$

$$令 \ (u,v) = (\sqrt{2}, -1) \ 则 \ (x,y) = (-\sqrt{2}, 0)$$

$$\therefore \frac{\partial g}{\partial v}(-1,-1) = \frac{\partial f}{\partial x}(-\sqrt{2},0)\frac{\partial x}{\partial v}(\sqrt{2},-1) + \frac{\partial f}{\partial y}(-\sqrt{2},0)\frac{\partial y}{\partial v}(\sqrt{2},-1)$$

$$= (0)(\sqrt{2}) + \frac{6}{9}(6) = 4$$

## Example 25.

$$设 f(x,y) = \begin{cases} \dfrac{x^2 - y^2}{\sqrt{x^2 + y^2}}, & (x,y) \neq (0,0) \\ 0, & (x,y) = (0,0) \end{cases} , \quad 试证 f(x,y) 在 (0,0) 不可微分$$

【解】

$$f_x(0,0) = \lim_{h \to 0} \frac{f(h+0,0) - f(0,0)}{h} = \lim_{h \to 0} \frac{\frac{h^2}{h} - 0}{h} = 1$$

$$f_y(0,0) = \lim_{h \to 0} \frac{f(0, h+0) - f(0,0)}{h} = \lim_{h \to 0} \frac{\frac{-h^2}{h} - 0}{h} = -1$$

Claim: $\displaystyle\lim_{h \to 0, k \to 0} \left| \frac{f(h+0, k+0) - f(0,0) - \frac{\partial f}{\partial x}(0,0)h - \frac{\partial f}{\partial y}(0,0)k}{\sqrt{h^2 + k^2}} \right| \neq 0$

$$\because \left| \frac{f(h+0, k+0) - f(0,0) - \frac{\partial f}{\partial x}(0,0)h - \frac{\partial f}{\partial y}(0,0)k}{\sqrt{h^2 + k^2}} \right| = \left| \frac{\frac{h^2 - k^2}{\sqrt{h^2 + k^2}} - h + k}{\sqrt{h^2 + k^2}} \right|$$

$$= \left| \frac{\frac{(h^2 - k^2) - (h - k)(\sqrt{h^2 + k^2})}{\sqrt{h^2 + k^2}}}{\sqrt{h^2 + k^2}} \right|$$

取 $h = 2k$ 则 $\left| \dfrac{\frac{(h^2 - k^2) - (h - k)(\sqrt{h^2 + k^2})}{\sqrt{h^2 + k^2}}}{\sqrt{h^2 + k^2}} \right| = \left| \dfrac{3k^2 - \sqrt{5}k^2}{5k^2} \right|$

$$\therefore \lim_{h \to 0, k \to 0} \left| \frac{f(h+0, k+0) - f(0,0) - \frac{\partial f}{\partial x}(0,0)h - \frac{\partial f}{\partial y}(0,0)k}{\sqrt{h^2 + k^2}} \right|$$

$$= \lim_{h \to 0, k \to 0} \left| \frac{3k^2 - \sqrt{5}k^2}{5k^2} \right| = \frac{3 - \sqrt{5}}{5} \neq 0$$

$\therefore f(x,y)$ 在 $(0,0)$ 不可微分

Example 26.

$$设 f(x,y) = \begin{cases} (x^2 + y^2)^p \sin\left(\dfrac{1}{x^2 + y^2}\right), & (x,y) \neq (0,0) \\ 0, & (x,y) = (0,0) \end{cases}, \quad p > \frac{1}{2}$$

　　试证 $f(x,y)$ 在 $(0,0)$ 可微分

【解】

令 $p > \dfrac{1}{2}$

$$f_x(0,0) = \lim_{h \to 0} \frac{f(h+0,0) - f(0,0)}{h} = \lim_{h \to 0} \frac{h^{2p} \sin(\frac{1}{h^2}) - 0}{h} = \lim_{h \to 0} h^{2p-1} \sin(\frac{1}{h^2})$$

$\because p > \dfrac{1}{2} \quad \therefore \lim_{h \to 0} h^{2p-1} \sin(\frac{1}{h^2}) = 0 \Rightarrow f_x(0,0) = 0$

$$f_y(0,0) = \lim_{h \to 0} \frac{f(0,0+h) - f(0,0)}{h} = \lim_{h \to 0} \frac{h^{2p} \sin(\frac{1}{h^2}) - 0}{h} = \lim_{h \to 0} h^{2p-1} \sin(\frac{1}{h^2})$$

$\because p > \dfrac{1}{2} \quad \therefore \lim_{h \to 0} h^{2p-1} \sin(\frac{1}{h^2}) = 0 \Rightarrow f_y(0,0) = 0$

$$\text{Claim: } \lim_{h \to 0, k \to 0} \left| \frac{f(h+0,k+0) - f(0,0) - \frac{\partial f}{\partial x}(0,0)h - \frac{\partial f}{\partial y}(0,0)k}{\sqrt{h^2+k^2}} \right| = 0$$

$$\because \left| \frac{f(h+0,k+0) - f(0,0) - \frac{\partial f}{\partial x}(0,0)h - \frac{\partial f}{\partial y}(0,0)k}{\sqrt{h^2+k^2}} \right| = \left| \frac{(h^2+k^2)^p \sin \frac{1}{h^2+k^2}}{\sqrt{h^2+k^2}} \right|$$

$$= \left| (h^2+k^2)^{p-\frac{1}{2}} \sin \frac{1}{h^2+k^2} \right| \leq (h^2+k^2)^{p-\frac{1}{2}}$$

$\because \lim_{h \to 0, k \to 0} (h^2+k^2)^{p-\frac{1}{2}} = 0$

$$\therefore \lim_{h \to 0, k \to 0} \left| \frac{f(h+0,k+0) - f(0,0) - \frac{\partial f}{\partial x}(0,0)h - \frac{\partial f}{\partial y}(0,0)k}{\sqrt{h^2+k^2}} \right| = 0$$

$\therefore f(x,y)$ 在 $(0,0)$ 可微分

Example 27.

　　　　设 $f(x,y) = (x|y|)^p, \ p > \dfrac{1}{2}, \ $ 试证 $f(x,y)$ 在 $(0,0)$ 可微分

【解】

令 $p > \dfrac{1}{2}$

$$f_x(0,0) = \lim_{h \to 0} \frac{f(h+0,0) - f(0,0)}{h} = \lim_{h \to 0} \frac{0-0}{h} = 0$$

$$f_y(0,0) = \lim_{h \to 0} \frac{f(0,0+h) - f(0,0)}{h} = \lim_{h \to 0} \frac{0-0}{h} = 0$$

Claim: $\displaystyle \lim_{h \to 0, k \to 0} \left| \frac{f(h+0, k+0) - f(0,0) - \frac{\partial f}{\partial x}(0,0)h - \frac{\partial f}{\partial y}(0,0)k}{\sqrt{h^2 + k^2}} \right| = 0$

$$\because \left| \frac{f(h+0, k+0) - f(0,0) - \frac{\partial f}{\partial x}(0,0)h - \frac{\partial f}{\partial y}(0,0)k}{\sqrt{h^2 + k^2}} \right|$$

$$= \left| \frac{h|k| - f(0,0) - \frac{\partial f}{\partial x}(0,0)h - \frac{\partial f}{\partial y}(0,0)k}{\sqrt{h^2 + k^2}} \right| = \left| \frac{(h|k|)^p}{\sqrt{h^2 + k^2}} \right| \leq \left| \frac{(h|k|)^p}{\sqrt{2|hk|}} \right| < |hk|^{p - \frac{1}{2}}$$

$$\because \lim_{h \to 0, k \to 0} |hk|^{p - \frac{1}{2}} = 0$$

$$\therefore \lim_{h \to 0, k \to 0} \left| \frac{f(h+0, k+0) - f(0,0) - \frac{\partial f}{\partial x}(0,0)h - \frac{\partial f}{\partial y}(0,0)k}{\sqrt{h^2 + k^2}} \right| = 0$$

$\therefore f(x,y)$ 在 $(0,0)$ 可微分


Example 28.

$$设 f(x,y) = \begin{cases} \dfrac{3x\sqrt{y}}{\sqrt{x^3 + y^3}}, & (x,y) \neq (0,0) \\ 0, & (x,y) = (0,0) \end{cases}, \quad 试证 f(x,y) 于 (0,0) 不可微分$$

【解】

Claim: $\displaystyle \lim_{h \to 0, k \to 0} \left| \frac{f(h+0, k+0) - f(0,0) - \frac{\partial f}{\partial x}(0,0)h - \frac{\partial f}{\partial y}(0,0)k}{\sqrt{h^2 + k^2}} \right| \neq 0$

$$\frac{\partial f}{\partial x}(0,0) = \lim_{h \to 0} \frac{f(h+0,0) - f(0,0)}{h} = \lim_{h \to 0} \frac{\frac{3 \cdot h^2 \cdot 0}{\sqrt{h^3 + 0^3}} - 0}{h} = \lim_{h \to 0} \frac{0}{h^{\frac{1}{2}}} = 0$$

$$\frac{\partial f}{\partial y}(0,0) = \lim_{h \to 0} \frac{f(0, h+0) - f(0,0)}{h} = \lim_{h \to 0} \frac{\frac{3 \cdot 0^2 \cdot \sqrt{h}}{\sqrt{0^3 + h^3}} - 0}{h} = \lim_{h \to 0} \frac{0}{h^2} = 0$$

因此 $\dfrac{\partial f}{\partial x}(0,0) = 0, \quad \dfrac{\partial f}{\partial y}(0,0) = 0$

$$\because \left| \frac{f(h+0, k+0) - f(0,0) - \frac{\partial f}{\partial x}(0,0)h - \frac{\partial f}{\partial y}(0,0)k}{\sqrt{h^2 + k^2}} \right| = \left| \frac{\frac{3h\sqrt{k}}{\sqrt{h^3 + k^3}}}{\sqrt{h^2 + k^2}} \right|$$

取 $k = h$

$$\lim_{h \to 0, k \to 0} \left| \frac{f(h+0, k+0) - f(0,0) - \frac{\partial f}{\partial x}(0,0)h - \frac{\partial f}{\partial y}(0,0)k}{\sqrt{h^2 + k^2}} \right|$$

$$= \lim_{h \to 0, k \to 0} \left| \frac{\frac{3h\sqrt{k}}{\sqrt{h^3 + k^3}}}{\sqrt{h^2 + k^2}} \right| = \lim_{h \to 0, k \to 0} \left| \frac{\frac{3h^{\frac{3}{2}}}{\sqrt{2}h^{\frac{3}{2}}}}{\sqrt{h^2 + h^2}} \right| = \infty \neq 0$$

$\therefore f(x,y)$ 于 $(0,0)$ 不可微分

Example 29.

设 $f: R \to R$ 且为可微函数, $f(0) = 0$, $f'(0) = 2$ 且 $F(x) = \displaystyle\int_0^x t^2 f(x^3 - t^3)\, dt$,

求 $\displaystyle\lim_{x \to 0} \frac{F(x)}{x^6} = ?$

【解】

藉由罗比达法则 $\displaystyle\lim_{x \to 0} \frac{F(x)}{x^6} = \lim_{x \to 0} \frac{F'(x)}{6x^5}$

藉由 Leibniz 微分公式 $F'(x) = 3x^2 \displaystyle\int_0^x t^2 f'(x^3 - t^3)\, dt$

$$\therefore \lim_{x \to 0} \frac{F'(x)}{6x^5} = \lim_{x \to 0} \frac{3x^2 \int_0^x t^2 f'(x^3 - t^3)\, dt}{6x^5} = \lim_{x \to 0} \frac{\int_0^x t^2 f'(x^3 - t^3)\, dt}{2x^3}$$

$$= \lim_{x \to 0} \frac{3x^2 \int_0^x t^2 f''(x^3 - t^3)\, dt + x^2 f'(x^3 - x^3)}{6x^2} = \lim_{x \to 0} \frac{x^2 f'(x^3 - x^3)}{6x^2} = \frac{f'(0)}{6} = \frac{1}{3}$$

## 7.5 求方向导数

**【定义】**双变量函数方向导数的定义

(i)給定双变数函数$f(x,y)$，假设$\vec{u}=(u_1,u_2)$为单位向量，$(x_0,y_0)\in R^2$

(ii)若 $\displaystyle\lim_{h\to0}\frac{f(x_0+hu_1,y_0+hu_2)-f(x_0,y_0)}{h}$ 存在，则

称$f(x,y)$在$(x_0,y_0)$沿$\vec{u}=(u_1,u_2)$的方向导数存在且用$D_{\vec{u}}f(x_0,y_0)$表示

**【定义】**多变量函数方向导数的定义

(i)給定多变数函数$f(x_1,\ldots,x_n)$，假设$\vec{u}=(u_1,\ldots,u_n)$为单位向量，$(c_1,\ldots,c_n)\in R^n$

(ii)若 $\displaystyle\lim_{h\to0}\frac{f(c_1+hu_1,..,c_n+hu_n)-f(c_1,\ldots,c_n)}{h}$ 存在，则

称$f(x_1,\ldots,x_n)$在$(c_1,\ldots,c_n)$沿$\vec{u}=(u_1,\ldots,u_n)$的方向导数存在且用$D_{\vec{u}}f(c_1,\ldots,c_n)$表示

考试类型:

Type 1.

給定双变数函数$f(x,y)$，求在$(x_0,y_0)$沿着向量$\vec{u}=(u_1,u_2)$的方向导数，$D_{\vec{u}}f(x_0,y_0)=?$

解题流程:

使用定义求方向导数

如果 $\displaystyle\lim_{h\to0}\frac{f(x_0+hu_1,y_0+hu_2)-f(x_0,y_0)}{h}$ 存在

则函数$f(x,y)$在$(x_0,y_0)$沿向量$(u_1,u_2)$的方向导数 $=\displaystyle\lim_{h\to0}\frac{f(x_0+hu_1,y_0+hu_2)-f(x_0,y_0)}{h}$，

即

$$D_{\vec{u}}f(x_0,y_0)=\lim_{h\to0}\frac{f(x_0+hu_1,y_0+hu_2)-f(x_0,y_0)}{h}$$

此外，常见的题型为求在$(0,0)$沿向量$(u_1,u_2)$的方向导数，即

$$求\ \lim_{h\to0}\frac{f(hu_1,hu_2)-f(0,0)}{h}=?$$

Type 2.

給定三变数函数 $f(x,y,z)$ 求在 $(x_0,y_0,z_0)$ 沿着向量 $\vec{u}=(u_1,u_2,u_3)$ 的方向导数

解题流程:

使用定义求方向导数

如果 $\lim\limits_{h\to 0}\dfrac{f(x_0+hu_1,y_0+hu_2,z_0+hu_3)-f(x_0,y_0,z_0)}{h}$ 存在

则函数 $f(x,y,z)$ 在 $(x_0,y_0,z_0)$ 沿向量 $(u_1,u_2,u_3)$ 的方向导数

$$=\lim\limits_{h\to 0}\dfrac{f(x_0+hu_1,y_0+hu_2,z_0+hu_3)-f(x_0,y_0,z_0)}{h}$$

即

$$D_{\vec{u}}f(x_0,y_0,z_0)=\lim\limits_{h\to 0}\dfrac{f(x_0+hu_1,y_0+hu_2,z_0+hu_3)-f(x_0,y_0,z_0)}{h}$$

此外，常见的题型为求在 $(0,0,0)$ 沿向量 $\vec{u}=(u_1,u_2,u_3)$ 的方向导数，即

$$求 \ \lim\limits_{h\to 0}\dfrac{f(hu_1,hu_2,hu_3)-f(0,0,0)}{h}=?$$

Example 1.

$$设 f(x,y)=\begin{cases}\dfrac{x^2y}{x^2+y^2}, & (x,y)\neq 0\\[2mm] 0, & (x,y)=0\end{cases}$$

求 $f(x,y)$ 在 $(0,0)$ 沿单位向量 $\vec{u}=(u_1,u_2)$ 的方向导数

【解】

$\because D_{\vec{u}}f(x,y)=\lim\limits_{h\to 0}\dfrac{f(x+hu_1,y+hu_2)-f(x,y)}{h},\ \ \vec{u}=(u_1,u_2)$

$\therefore D_{\vec{u}}f(0,0)=\lim\limits_{h\to 0}\dfrac{f(hu_1,hu_2)-f(0,0)}{h}=\lim\limits_{h\to 0}\dfrac{\dfrac{(hu_1)^2 hu_2}{(hu_1)^2+(hu_2)^2}}{h}=\lim\limits_{h\to 0}\dfrac{u_1{}^2u_2}{u_1{}^2+u_2{}^2}$

$=\dfrac{u_1{}^2u_2}{u_1{}^2+u_2{}^2}$

Example 2.

$$设 f(x,y)=\begin{cases}\dfrac{xy^3}{x^3+y^6}, & (x,y)\neq 0\\[2mm] 0, & (x,y)=0\end{cases}$$

求$f(x, y)$在$(0,0)$沿单位向量$\vec{u} = (s, t)$的方向导数

【解】

$$\because D_{\vec{u}}f(x, y) = \lim_{h \to 0} \frac{f(x + hs, y + ht) - f(x, y)}{h}, \quad \vec{u} = (s, t)$$

$$\therefore D_{\vec{u}}f(0,0) = \lim_{h \to 0} \frac{f(hs, ht) - f(0,0)}{h} = \lim_{h \to 0} \frac{\dfrac{hs(ht)^3}{(hs)^3 + (ht)^6}}{h} = \lim_{h \to 0} \frac{h^4 s t^3}{h^4 s^3 + h^7 t^6}$$

$$= \lim_{h \to 0} \frac{s t^3}{s^3 + h^3 t^6} = \frac{t^3}{s^2}$$

Example 3.

设$f(x, y, z) = x^2 + 3y^2 + 4z^2$，求$f(x, y, z)$在$(2,0,1)$沿向量$(2, -1, 0)$的方向导数

【解】

$$\because D_{\vec{u}}f(x, y, z) = \lim_{h \to 0} \frac{f(x + hu_1, y + hu_2, z + hu_3) - f(x, y, z)}{h},$$

$$\vec{u} = (u_1, u_2, u_3) = \left(\frac{2}{\sqrt{5}}, \frac{-1}{\sqrt{5}}, \frac{0}{\sqrt{5}}\right), \quad \text{藉由罗比达法则}$$

$$D_{\vec{u}}f(2,0,1) = \lim_{h \to 0} \frac{f(2 + hu_1, hu_2, 1 + hu_3) - f(2,0,1)}{h}$$

$$= \lim_{h \to 0} \frac{(2 + hu_1)^2 + 3(hu_2)^2 + 4(1 + hu_3)^2 - 8}{h}$$

$$= \lim_{h \to 0} 2u_1(2 + hu_1) + 8u_3(1 + hu_3) = 4u_1 + 8u_3 = \frac{8}{\sqrt{5}}$$

Example 4.

$$\text{设} f(x, y) = \begin{cases} \sqrt{xy}\left(\dfrac{x^2 - y^2}{x^2 + y^2}\right), & (x, y) \neq 0 \\ 0, & (x, y) = 0 \end{cases}$$

求$f(x, y)$在$(x, y) = (0,0)$沿单位向量$\vec{u} = (s, t)$的方向导数

【解】

$$\because D_{\vec{u}}f(x, y) = \lim_{h \to 0} \frac{f(x + hs, y + ht) - f(x, y)}{h}, \quad \vec{u} = (s, t)$$

$$\therefore D_{\vec{u}}f(0,0) = \lim_{h \to 0} \frac{f(hs, ht) - f(0,0)}{h} = \lim_{h \to 0} \frac{\sqrt{h^2 st}\left(\dfrac{(hs)^2 - (ht)^2}{(hs)^2 + (ht)^2}\right)}{h} = \lim_{h \to 0} \frac{\sqrt{st}(s^2 - t^2)}{s^2 + t^2}$$

$$= \frac{\sqrt{st}(s^2 - t^2)}{s^2 + t^2}$$

Example 5.

$$\text{設} f(x,y) = \begin{cases} \dfrac{x^3 - y^3}{x^2 + y^3}, & (x,y) \neq 0 \\ 0, & (x,y) = 0 \end{cases}$$

求 $f(x,y)$ 在 $(x,y) = (0,0)$ 沿單位向量 $\vec{u} = (s,t)$ 的方向導數

【解】

$$\because D_{\vec{u}}f(x,y) = \lim_{h \to 0} \frac{f(x + hs, y + ht) - f(x,y)}{h}, \quad \vec{u} = (s,t)$$

$$\therefore D_{\vec{u}}f(0,0) = \lim_{h \to 0} \frac{f(hs, ht) - f(0,0)}{h} = \lim_{h \to 0} \frac{\dfrac{(hs)^3 - (ht)^3}{(hs)^2 + (ht)^3}}{h} = \lim_{h \to 0} \frac{s^3 - t^3}{s^2} = \frac{s^3 - t^3}{s^2}$$

Example 6.

$$\text{設} f(x,y) = \begin{cases} \dfrac{x^2 - y^2}{\sqrt{x^2 + y^3}}, & (x,y) \neq 0 \\ 0, & (x,y) = 0 \end{cases}$$

求 $f(x,y)$ 在 $(x,y) = (0,0)$ 沿單位向量 $\vec{u} = (s,t)$ 的方向導數

【解】

$$\because D_{\vec{u}}f(x,y) = \lim_{h \to 0} \frac{f(x + hs, y + ht) - f(x,y)}{h}, \quad \vec{u} = (s,t)$$

$$\therefore D_{\vec{u}}f(0,0) = \lim_{h \to 0} \frac{f(hs, ht) - f(0,0)}{h} = \lim_{h \to 0} \frac{\dfrac{(hs)^2 - (ht)^2}{\sqrt{(hs)^2 + (ht)^3}}}{h} = \lim_{h \to 0} \frac{s^2 - t^2}{s} = \frac{s^2 - t^2}{s}$$

Example 7.

$$\text{設} f(x,y) = \begin{cases} \dfrac{\sin(x^2 - y^2)}{x + y}, & (x,y) \neq 0 \\ 0, & (x,y) = 0 \end{cases}$$

求 $f(x,y)$ 在 $(x,y)=(0,0)$ 沿单位向量 $\vec{u}=(s,t)$ 的方向导数

【解】

$$\because D_{\vec{u}}f(x,y)=\lim_{h\to 0}\frac{f(x+hs,y+ht)-f(x,y)}{h},\quad \vec{u}=(s,t)$$

$$\therefore D_{\vec{u}}f(0,0)=\lim_{h\to 0}\frac{f(hs,ht)-f(0,0)}{h}=\lim_{h\to 0}\frac{\dfrac{\sin((hs)^2-(ht)^2)}{hs+ht}}{h}$$

$$=\lim_{h\to 0}\frac{(hs-ht)\cdot\dfrac{\sin((hs)^2-(ht)^2)}{(hs)^2-(ht)^2}}{h}$$

$$\because \lim_{h\to 0}\frac{\sin((hs)^2-(ht)^2)}{(hs)^2-(ht)^2}=1 \quad \therefore D_{\vec{u}}f(0,0)=\lim_{h\to 0}\frac{(hs-ht)\left(\dfrac{\sin(hs)^2-(ht)^2}{(hs)^2-(ht)^2}\right)}{h}=s-t$$

Example 8.

设 $f(x,y)=xe^y-ye^x$，求 $f(x,y)$ 在 $(1,0)$ 沿单位向量 $\left(\dfrac{3}{5},\dfrac{4}{5}\right)$ 的方向导数

【解】

$$\because D_{\vec{u}}f(x,y)=\lim_{h\to 0}\frac{f(x+hu_1,y+hu_2)-f(x,y)}{h},\quad \vec{u}=\left(\frac{3}{5},\frac{4}{5}\right)=(u_1,u_2)$$

藉由罗比达法则

$$D_{\vec{u}}f(0,0)=\lim_{h\to 0}\frac{f(1+u_1h,u_2h)-f(1,0)}{h}=\lim_{h\to 0}\frac{(1+u_1h)e^{u_2h}-u_2he^{1+u_1h}-1}{h}$$

$$=\lim_{h\to 0}u_1e^{u_2h}+u_2(1+u_1h)e^{u_2h}-u_2e^{1+u_1h}-u_1u_2he^{1+u_1h}=\frac{7}{5}-\frac{4e}{5}$$

Example 9.

设 $f(x,y,z)=xy^2+yz^3$，求 $f(x,y,z)$ 在 $(2,-1,1)$ 沿单位向量 $(u_1,u_2,u_3)$ 的方向导数

【解】

$$\because D_{\vec{u}}f(x,y,z)=\lim_{h\to 0}\frac{f(x+hu_1,y+hu_2,z+hu_3)-f(x,y,z)}{h}$$

$\vec{u}=(u_1,u_2,u_3)$，藉由罗比达法则

$$D_{\vec{u}}f(2,-1,1)=\lim_{h\to 0}\frac{f(2+hu_1,-1+hu_2,1+hu_3)-f(2,-1,1)}{h}$$

$$= \lim_{h \to 0} \frac{(2 + hu_1)(-1 + hu_2)^2 + (-1 + hu_2)(1 + hu_3)^3 - 1}{h}$$

$$= \lim_{h \to 0} u_1(-1 + hu_2)^2 + (2 + hu_1)2u_2(-1 + hu_2) + u_2(1 + hu_3)^3$$

$$+ (-1 + hu_2)3u_3(1 + hu_3)^2$$
$$= u_1 - 4u_2 + u_2 - 3u_3 = u_1 - 3u_2 - 3u_3$$

Example 10.

$$\text{设} f(x, y) = \begin{cases} \sqrt{xy}\sin\dfrac{x^2 - y^2}{x^2 + y^2}, & (x, y) \neq 0 \\ 0, & (x, y) = 0 \end{cases}$$

求 $f(x, y)$ 在 $(x, y) = (0,0)$ 沿单位向量 $\vec{u} = (s, t)$ 的方向导数

【解】

$$\because D_{\vec{u}}f(x, y) = \lim_{h \to 0} \frac{f(x + hs, y + ht) - f(x, y)}{h}, \quad \vec{u} = (s, t)$$

$$\therefore D_{\vec{u}}f(0,0) = \lim_{h \to 0} \frac{f(hs, ht) - f(0,0)}{h} = \lim_{h \to 0} \frac{\sqrt{h^2 st}\,\dfrac{\sin((hs)^2 - (ht)^2)}{(hs)^2 + (ht)^2}}{h}$$

$$= \lim_{h \to 0} \frac{\sqrt{h^2 st}((hs)^2 - (ht)^2)\dfrac{\sin((hs)^2 - (ht)^2)}{((hs)^2 + (ht)^2)((hs)^2 - (ht)^2)}}{h}$$

$$\because \lim_{h \to 0} \frac{\sin((hs)^2 - (ht)^2)}{(hs)^2 - (ht)^2} = 1$$

$$\therefore D_{\vec{u}}f(0,0) = \lim_{h \to 0} \frac{\sqrt{h^2 st}((hs)^2 - (ht)^2)\dfrac{\sin(hs)^2 - (ht)^2}{((hs)^2 + (ht)^2)((hs)^2 - (ht)^2)}}{h}$$

$$= \lim_{h \to 0} \frac{\dfrac{\sqrt{h^2 st}((hs)^2 - (ht)^2)}{((hs)^2 + (ht)^2)}}{h} = \frac{\sqrt{st}(s^2 - t^2)}{s^2 + t^2}$$

Example 11.

设 $f(x, y, z) = xyz$

(1) 求在 $(x, y, z) = (1,3,2)$ 沿单位向量 $(u_1, u_2, u_3)$ 的方向导数

(2) 求在$(x, y, z) = (1,3,2)$的最大方向导数为何?

【解】

(1)

$$\because D_{\vec{u}}f(x, y, z) = \lim_{h \to 0} \frac{f(x + hu_1, y + hu_2, z + hu_3) - f(x, y, z)}{h}, \quad \vec{u} = (u_1, u_2, u_3),$$

藉由罗比达法则

$$D_{\vec{u}}f(1,3,2) = \lim_{h \to 0} \frac{f(1 + hu_1, 3 + hu_2, 2 + hu_3) - f(1,3,2)}{h}$$

$$= \lim_{h \to 0} \frac{(1 + hu_1)(3 + hu_2)(2 + hu_3) - 6}{h}$$

$$= \lim_{h \to 0} u_1(3 + hu_2)(2 + hu_3) + u_2(1 + hu_1)(2 + hu_3) + u_3(1 + hu_1)(3 + hu_2)$$

$$= 6u_1 + 2u_2 + 3u_3$$

(2)

已知 $D_{\vec{u}}f(1,3,2) = 6u_1 + 2u_2 + 3u_3$

藉由 Cauchy inequality, $6u_1 + 2u_2 + 3u_3 \leq \sqrt{(6^2 + 2^2 + 3^2)(u_1^2 + u_2^2 + u_3^2)}$

$\because u_1^2 + u_2^2 + u_3^2 = 1$ 且等号成立于 $\dfrac{u_1}{6} = \dfrac{u_2}{2} = \dfrac{u_3}{3}$

因此沿向量 $\left(\dfrac{6}{7}, \dfrac{2}{7}, \dfrac{3}{7}\right)$ 有最大方向导数 7

Example 12.

设 $f(x, y, z) = xy + yz + xz$

(1)在 $(x, y, z) = (1,1,1)$ 沿单位向量 $(u_1, u_2, u_3)$ 的方向导数

(2)在 $(x, y, z) = (1,1,1)$ 的最大方向导数为何?

【解】

(1)

$$\because D_{\vec{u}}f(x, y, z) = \lim_{h \to 0} \frac{f(x + hu_1, y + hu_2, z + hu_3) - f(x, y, z)}{h}, \vec{u} = (u_1, u_2, u_3),$$

藉由罗比达法则

$$D_{\vec{u}}f(1,1,1) = \lim_{h \to 0} \frac{f(1 + hu_1, 1 + hu_2, 1 + hu_3) - f(1,1,1)}{h}$$

$$= \lim_{h\to 0}\frac{(1+hu_1)(1+hu_2)+(1+hu_2)(1+hu_3)+(1+hu_1)(1+hu_3)-3}{h}$$

$$= \lim_{h\to 0}(u_1+u_2+2hu_1u_2)+(u_2+u_3+2hu_3u_2)+(u_1+u_3+2hu_1u_3)=2(u_1+u_2+u_3)$$

(2)

已知 $D_{\vec{u}}f(1,1,1)=2(u_1+u_2+u_3)$

藉由 Cauchy inequality,  $2u_1+2u_2+2u_3 \le \sqrt{(2^2+2^2+2^2)(u_1^2+u_2^2+u_3^2)}$

$\because u_1^2+u_2^2+u_3^2=1$ 且等号成立于 $\dfrac{u_1}{2}=\dfrac{u_2}{2}=\dfrac{u_3}{2}$

因此沿向量 $\left(\dfrac{1}{\sqrt{3}},\dfrac{1}{\sqrt{3}},\dfrac{1}{\sqrt{3}}\right)$ 有最大方向导数 $\sqrt{12}$

Example 13.

$$设 f(x,y)=\begin{cases}\dfrac{\sin(x^3-y^3)}{x^2+xy+y^2}, & (x,y)\neq 0\\[2mm] 0, & (x,y)=0\end{cases}$$

求 $f(x,y)$ 在 $(x,y)=(0,0)$ 沿单位向量 $\vec{u}=(s,t)$ 的方向导数

【解】

$\because D_{\vec{u}}f(x,y)=\lim_{h\to 0}\dfrac{f(x+hs,y+ht)-f(x,y)}{h},\ \vec{u}=(s,t)$

$\therefore D_{\vec{u}}f(0,0)=\lim_{h\to 0}\dfrac{f(hs,ht)-f(0,0)}{h}=\lim_{h\to 0}\dfrac{\dfrac{\sin((hs)^3-(ht)^3)}{(hs)^2+h^2st+(ht)^2}}{h}$

$$= \lim_{h\to 0}\frac{(hs-ht)\dfrac{\sin((hs)^3-(ht)^3)}{(hs)^3-(ht)^3}}{h}$$

$\because \lim_{h\to 0}\dfrac{\sin((hs)^3-(ht)^3)}{(hs)^3-(ht)^3}=1$    $\therefore D_{\vec{u}}f(0,0)=\lim_{h\to 0}\dfrac{(hs-ht)\dfrac{\sin((hs)^3-(ht)^3)}{(hs)^3-(ht)^3}}{h}=s-t$

## 7.6 求曲面的法线与切线方程式

考试类型:

Type 1.

给定曲面 $f(x, y, z) = 0$，求曲面于 $(x_0, y_0, z_0)$ 的单位法向量、切平面与法线方程式

解题流程:

Step1.

求 $\nabla f(x, y, z) = ?$，假设 $\nabla f(x, y, z) = (f_x(x, y, z), f_y(x, y, z), f_z(x, y, z))$

Step2.

因此于 $(x_0, y_0, z_0)$ 的法向量 $= \nabla f(x_0, y_0, z_0) = (f_x(x_0, y_0, z_0), f_y(x_0, y_0, z_0), f_z(x_0, y_0, z_0))$

Step3.

因此于 $(x_0, y_0, z_0)$ 的单位法向量 $= \left( \dfrac{f_x(x_0, y_0, z_0)}{\sqrt{f_x^2 + f_y^2 + f_z^2}}, \dfrac{f_y(x_0, y_0, z_0)}{\sqrt{f_x^2 + f_y^2 + f_z^2}}, \dfrac{f_z(x_0, y_0, z_0)}{\sqrt{f_x^2 + f_y^2 + f_z^2}} \right)$

Step4.

于 $(x_0, y_0, z_0)$ 的切平面:

$$f_x(x_0, y_0, z_0)(x - x_0) + f_y(x_0, y_0, z_0)(y - y_0) + f_z(x_0, y_0, z_0)(z - z_0) = 0$$

Step5.

于 $(x_0, y_0, z_0)$ 的法线方程式: $\dfrac{x - x_0}{f_x(x_0, y_0, z_0)} = \dfrac{y - y_0}{f_y(x_0, y_0, z_0)} = \dfrac{z - z_0}{f_z(x_0, y_0, z_0)}$

Type 2.

给定两曲面 $f_1(x, y, z) = 0$，$f_2(x, y, z) = 0$，求两曲面相交曲线上于 $(x_0, y_0, z_0)$ 的单位切向量

解题流程:

Step1.

求 $\nabla f_1(x, y, z) = ?$，$\nabla f_2(x, y, z) = ?$，假设 $\nabla f_1(x_0, y_0, z_0) = (a, b, c)$，$\nabla f_2(x_0, y_0, z_0) = (d, e, f)$

Step2.

$\because$ 相交曲线的单位切向量垂直于两曲面的单位法向量

$\therefore$ 切向量 $= (a, b, c) \times (d, e, f) = \begin{vmatrix} \vec{i} & \vec{j} & \vec{k} \\ a & b & c \\ d & e & f \end{vmatrix}$

Step3.

单位切向量 $=$ 上述切向量除以它的长度

Example 1.

　　求曲面 $2x^2 + 4yz - 5z^2 = -1$ 于 $Q(0,1,1)$ 的单位法向量

【解】

令 $f(x,y,z) = 2x^2 + 4yz - 5z^2 + 1$ 则 $\nabla f(x,y,z) = (4x, 4z, 4y - 10z)$

因此于 $(0,1,1)$ 的法向量 $= \nabla f(0,1,1) = (0, 4, -6)$

$\therefore$ 过 $(0,1,1)$ 单位法向量 $= \left(0, \dfrac{2}{\sqrt{13}}, \dfrac{-3}{\sqrt{13}}\right)$

Example 2.

　　求曲面 $z = x^3 y^3 + x + 3$ 于 $Q(1,1,5)$ 的单位法向量

【解】

令 $f(x,y,z) = x^3 y^3 + x + 3 - z$ 则 $\nabla f(x,y,z) = (3x^2 y^3 + 1, 3y^2 x^3, -1)$

因此于 $(1,1,5)$ 的法向量 $= \nabla f(1,1,5) = (4, 3, -1)$

$\therefore$ 过 $(1,1,5)$ 单位法向量 $= \left(\dfrac{4}{\sqrt{26}}, \dfrac{3}{\sqrt{26}}, \dfrac{-1}{\sqrt{26}}\right)$

Example 3.

　　求曲面 $2xz^2 - 3yx - 4x = 5$ 于 $Q(-1,1,1)$ 的切平面与法线方程式

【解】

令 $f(x,y,z) = 2xz^2 - 3yx - 4x$ 则 $\nabla f(x,y,z) = (2z^2 - 3y - 4, -3x, 4xz)$

因此于 $(-1,1,1)$ 的法向量 $= \nabla f(-1,1,1) = (-5, 3, -4)$

$\therefore$ 于 $(-1,1,1)$ 切平面：$-5(x+1) + 3(y-1) - 4(z-1) = 0$

$\therefore$ 于 $(-1,1,1)$ 法线方程式：$\dfrac{x+1}{-5} = \dfrac{y-1}{3} = \dfrac{z-1}{-4}$

Example 4.

　　求曲面 $x^2 + 2y^2 + z^2 = 4$ 于 $Q(1,1,1)$ 的切平面与法线方程式

【解】

令 $f(x,y,z) = x^2 + 2y^2 + z^2$ 则 $\nabla f(x,y,z) = (2x, 4y, 2z)$

因此于 $(1,1,1)$ 的法向量 $= \nabla f(1,1,1) = (2, 4, 2)$

$\therefore$ 于 $(1,1,1)$ 切平面：$1(x-1) + 2(y-1) + 1(z-1) = 0$

$\therefore$ 于 $(1,1,1)$ 法线方程式: $\dfrac{x-1}{1} = \dfrac{y-1}{2} = \dfrac{z-1}{1}$

Example 5.

求曲面 $x^3 + 2y^3 + 3z^3 - 2xyz = 3$ 与 $x^2 + y^2 + z^2 = 2,$ 相交的曲线上于 $(1,1,0)$ 上的单位切向量 $=?$

【解】

令 $f_1(x,y,z) = x^3 + 2y^3 + 3z^3 - 2xyz$ 则 $\nabla f_1(x,y,z) = (3x^2 - 2yz, 6y^2 - 2xz, 9z^2 - 2xy)$

令 $f_2(x,y,z) = x^2 + y^2 + z^2$ 则 $\nabla f_2(x,y,z) = (2x, 2y, 2z)$

$\therefore \nabla f_1(1,1,0) = (3,6,-2),\ \nabla f_2(1,1,0) = (2,2,0)$

$\because$ 相交曲线的单位切向量垂直于两曲面的单位法向量

且 $(3,6,-2) \times (2,2,0) = \begin{vmatrix} i & j & k \\ 3 & 6 & -2 \\ 2 & 2 & 0 \end{vmatrix} = (4,-4,-6)$

$\therefore$ 单位法向量 $= \left( \dfrac{2}{\sqrt{17}}, \dfrac{-2}{\sqrt{17}}, \dfrac{-3}{\sqrt{17}} \right)$

Example 6.

求曲面 $\dfrac{x^2}{9} + y^2 + \dfrac{z^2}{4} = 1$ 与曲面 $x^3 = 9yz + 4$ 相交的曲线上于点 $\left( 2, \dfrac{1}{3}, \dfrac{4}{3} \right)$ 的单位切向量 $=?$

【解】

令 $f_1(x,y,z) = \dfrac{x^2}{9} + y^2 + \dfrac{z^2}{4}$ 则 $\nabla f_1(x,y,z) = \left( \dfrac{2x}{9}, 2y, \dfrac{z}{2} \right)$

令 $f_2(x,y,z) = x^3 - 9yz - 4$ 则 $\nabla f_2(x,y,z) = (3x^2, -9z, -9y)$

$\therefore \nabla f_1\left( 2, \dfrac{1}{3}, \dfrac{4}{3} \right) = \left( \dfrac{4}{9}, \dfrac{2}{3}, \dfrac{2}{3} \right),\ \ \nabla f_2\left( 2, \dfrac{1}{3}, \dfrac{4}{3} \right) = (12, -12, -3)$

$\because$ 相交曲线的单位切向量垂直于两曲面的单位法向量

且 $(2,3,3) \times (4,-4,-1) = \begin{vmatrix} i & j & k \\ 2 & 3 & 3 \\ 4 & -4 & -1 \end{vmatrix} = (9,14,-20)$

$\therefore$ 单位法向量 $= \left( \dfrac{9}{\sqrt{677}}, \dfrac{14}{\sqrt{677}}, \dfrac{-20}{\sqrt{677}} \right)$

Example 7.

假设曲面 $z^2 + xy - 2x - y^2 = 3$, 试求曲面上哪些点的切平面与 $z = 3$ 平行

【解】

令 $f(x, y, z) = z^2 + xy - 2x - y^2 - 3$ 则 $\nabla f(x, y, z) = (y - 2, x - 2y, 2z)$

令切平面过 $(x_0, y_0, z_0)$ 则其法向量为 $(y_0 - 2, x_0 - 2y_0, 2z_0)$

$\because$ 切平面与 $z = 3$ 平行　$\therefore (y_0 - 2, x_0 - 2y_0) = (0,0) \Rightarrow y_0 = 2, x_0 = 4,$

$\because$ 曲面 $z^2 + xy - 2x - y^2 = 3$ 过切点 $(4, 2, z_0)$

$\therefore z_0{}^2 + 8 - 8 - 4 = 3 \Rightarrow z_0 = \pm\sqrt{7}$　$\therefore$ 过 $\left(4, 2, \pm\sqrt{7}\right)$ 的切平面与 $z = 3$ 平行

Example 8.

假设曲面 $x^2 - 4xy - 2y^2 + 12x - 12y - z - 3 = 0$, 试求曲面上哪些点的切平面与
$z = 2$ 平行

【解】

令 $f(x, y, z) = x^2 - 4xy - 2y^2 + 12x - 12y - z - 3$

则 $\nabla f(x, y, z) = (2x - 4y + 12, -4x - 4y - 12, -1)$

令切平面过 $(x_0, y_0, z_0)$ 则其法向量为 $(2x_0 - 4y_0 + 12, -4x_0 - 4y_0 - 12, -1)$

$\because$ 切平面与 $z = 2$ 平行

$\therefore (2x_0 - 4y_0 + 12, -4x_0 - 4y_0 - 12) = (0,0) \Rightarrow y_0 = 1, x_0 = -4,$

$\because$ 曲面 $x^2 - 4xy - 2y^2 + 12x - 12y - z - 3 = 0$ 过切点 $(-4, 1, z_0)$

$\therefore 4^2 - 4(-4) - 2 + 12(-4) - 12 - z_0 - 3 = 0 \Rightarrow z_0 = -33$

$\therefore$ 过 $(-4, 1, -33)$ 的切平面与 $z = 2$ 平行

Example 9.

求曲面 $x^2yz + 3y^2 = 2xz^2 - 8z$ 于 $Q(1, 2, -1)$ 的切平面与法线方程式

【解】

令 $f(x, y, z) = x^2yz + 3y^2 - 2xz^2 + 8z$

则 $\nabla f(x, y, z) = (2xyz - 2z^2, x^2z + 6y, x^2y - 4xz + 8)$

因此于 $(1, 2, -1)$ 的法向量 $= \nabla f(1, 2, -1) = (-6, 11, 14)$

$\therefore$ 于 $(1, 2, -1)$ 切平面: $-6(x - 1) + 11(y - 2) + 14(z + 1) = 0$

$\therefore$ 于 $(1, 2, -1)$ 法线方程式: $\dfrac{x - 1}{-6} = \dfrac{y - 2}{11} = \dfrac{z + 1}{14}$

## 7.7 求多变量函数的极值

求多变量函数极值的考试类型包含:使用算术平均大于等于几何平均求极值、使用 Cauchy Inequality 找极值、无限制条件的多变量函数求极值、限制条件为等式的多变量函数求极值、限制条件为不等式的多变量函数求极值、多变量函数于有界且封闭的定义域求极值。无限制条件的多变量函数求极值的问题主要用到多变量函数求极值的二阶导数检定法，限制条件为等式的多变量函数求极值主要用到 Lagrange multiplier，限制条件为不等式的多变量函数求极值相当于是同时处理前面这两个问题，因此这两个方法皆会用到；多变量函数于有界且封闭的定义域求极值相当于要先写下明确的不等式限制条件，再求极值。

为了方便起见，先介绍底下的符号,如果实数值函数 $f: R^n \to R$ 于 $c \in R^n$ 的所有一阶偏导数皆存在则

$$f^{(1)}(c:s) := \sum_{i=1}^{n} D_i f(c) s_i, \quad \forall s \in R^n$$

如果实数值函数 $f: R^n \to R$ 于 $c \in R^n$ 的所有二阶偏导数皆存在则

$$f^{(2)}(c:s) := \sum_{i=1}^{n} \sum_{j=1}^{n} D_{i,j} f(c) s_i s_j, \quad \forall s \in R^n$$

如果实数值函数 $f: R^n \to R$ 于 $c \in R^n$ 的所有三阶偏导数皆存在则

$$f^{(3)}(c:s) := \sum_{i=1}^{n} \sum_{j=1}^{n} \sum_{k=1}^{n} D_{i,j,k} f(c) s_i s_j s_k, \quad \forall s \in R^n$$

其余依此类推

为了说明多变量函数求极值的二阶导数检定法，首先介绍多变量函数的 Taylor's Formula，而多变量函数的 Taylor's Formula 推论过程则是用到了单变量函数的 Taylor's Formula

【定理】多变量函数的 Taylor's Formula

(i)给定一实数值函数 $f: R^n \to R$,假设小于等于 $m-1$ 阶的偏导数于开集合 $B$ 皆可微分

(ii)若 $x, y \in B$ 且 $L(x, y) \subseteq B$ 则

$$\exists z \in L(x, y) \text{ s.t. } f(y) = f(x) + \sum_{k=1}^{m-1} \frac{f^{(k)}(x: y - x)}{k!} + \frac{f^{(m)}(z: y - x)}{m!}$$

<u>Proof:</u>

Let $x, y \in B$. $\because B$ is open $\therefore \exists \delta > 0$ s.t. $x + s(y - x) \in B$, $\forall s \in (-\delta, 1 + \delta)$

Let $h(t) = f(x + s(y - x))$, $\forall s \in (-\delta, 1 + \delta)$, then $f(y) - f(x) = h(1) - h(0)$

By one-dimension Taylor formula, we have

$$h(1) - h(0) = \sum_{k=1}^{m-1} \frac{h^{(k)}(0)}{k!} + \frac{h^{(m)}(s')}{m!}, \quad \text{where } 0 < s' < 1.$$

Let $q(s) = x + s(y - x)$, then $h(s) = f(q(s))$ and $q_k'(s) = y_k - x_k$.

By using Chain Rule, we have $h'(s) = \sum_{j=1}^{n} D_j f(q(s))(y_j - x_j) = f'(q(s) : y - x)$

Applying Chain Rule to $h'(s)$, we have

$$h''(s) = \sum_{i=1}^{n} \sum_{j=1}^{n} D_{i,j} f(q(s))(y_j - x_j)(y_i - x_i) = f''(q(s) : y - x)$$

$\therefore h^{(m)}(s) = f^{(m)}(q(s) : y - x)$

Let $z = x + s'(y - x) \in L(x, y)$ then

$$f(y) - f(x) = h(1) - h(0) = \sum_{k=0}^{m-1} \frac{h^{(k)}(0)}{k!} + \frac{h^{(m)}(s')}{m!} = \sum_{k=0}^{m-1} \frac{f^{(k)}(x : y - x)}{k!} + \frac{f^{(m)}(z : y - x)}{m!}$$

【定义】平稳点(stationary point)与鞍点(saddle point)

(i)如果函数$f$于$c$可微分且$\nabla f(c) = 0$则点$c$称为函数$f$的平稳点(stationary point)

(ii)如果以平稳点$c$为中心的任意开球$B_\delta(c)$，$\exists x_0, x_1 \in B_\delta(c)$s.t. $f(x_0) > f(c)$

    且 $f(x_1) < f(c)$则$c$称为鞍点(saddle point)

值得讨论的是，上述定理如果考虑$m = 2$以及$c$是平稳点(stationary point)的情形，则

$$f(c + s) - f(c) = \nabla f(c) \cdot s + \frac{1}{2} f''(z : s) \quad \text{其中 } z \in L(c, c + s)$$

$\because c$ 是 stationary point $\therefore \nabla f(c) = 0$，上式可改写为

$$f(c + s) - f(c) = \frac{1}{2} f''(z : s) = \frac{1}{2} f''(c : s) + \left( \frac{1}{2} f''(z : s) - \frac{1}{2} f''(c : s) \right)$$

$$\because \frac{1}{2}\big(f''(\mathbf{z}:\mathbf{s}) - f''(\mathbf{c}:\mathbf{s})\big) \le \frac{1}{2}\sum_{i=1}^{n}\sum_{j=1}^{n}\big|D_{i,j}f(\mathbf{z}) - D_{i,j}f(\mathbf{c})\big|\,\|\mathbf{s}\|^2$$

如果函数$f$的所有二阶偏导数在$\mathbf{c}$连续则

$$\lim_{\mathbf{s}\to\mathbf{0}}\frac{1}{2}\big(f''(\mathbf{z}:\mathbf{s}) - f''(\mathbf{c}:\mathbf{s})\big) = 0$$

接下來藉由这项观察推导多变数函数求极值的方法

**【定理】**多变量函数求极值的二阶导数检定法

給定一实数值函数$f: R^n \to R$，假设$\exists\,\delta > 0$ 使得函数$f$的所有二阶偏导数于$B_\delta(\mathbf{c})$內皆存在且在平稳点$\mathbf{c}$连续，

$$P(\mathbf{s}) := \frac{1}{2}f''(\mathbf{c}:\mathbf{s}) = \frac{1}{2}\sum_{i=1}^{n}\sum_{j=1}^{n}D_{i,j}f(\mathbf{c})s_i s_j.$$

则

(i)若 $P(\mathbf{s}) < 0,\ \forall \mathbf{s} \neq \mathbf{0}$，则 $f$ 于 $\mathbf{c}$ 有相对极大值

(ii)若 $P(\mathbf{s}) > 0,\ \forall \mathbf{s} \neq \mathbf{0}$，则 $f$ 于 $\mathbf{c}$ 有相对极小值

(iii)若 $\exists\,\mathbf{s_1}, \mathbf{s_2} \neq \mathbf{0}$ 使得 $P(\mathbf{s_1}) < 0$ 且 $P(\mathbf{s_2}) > 0$，则 $\mathbf{c}$ 为 $f$ 的鞍点

<u>Proof:</u>

(i)Assume $P(\mathbf{s}) < 0,\ \forall \mathbf{s} \neq \vec{0}$. Let $B = \{\mathbf{s}: \|\mathbf{s}\| = 1\}$, then $P(\mathbf{s}) < 0,\ \forall \mathbf{s} \in B$

$\because P(\mathbf{s})$ is continuous on $B$ and $B$ is a compact set.

$\therefore P(\mathbf{s})$ has a maximum $M$ on $B$ and $M < 0$.

$\because P(\alpha\mathbf{s}) = \alpha^2 P(\mathbf{s}),\ \forall \alpha \in R$ and choose $\alpha = \dfrac{1}{\|\mathbf{s}\|}$, then $\alpha\mathbf{s} \in B$

$\therefore \alpha^2 P(\mathbf{s}) = P(\alpha\mathbf{s}) \le M \Rightarrow P(\mathbf{s}) \le M\|\mathbf{s}\|^2$

$\because f(\mathbf{c} + \mathbf{s}) - f(\mathbf{c}) = P(\mathbf{s}) + R(\mathbf{s}) \le M\|\mathbf{s}\|^2 + |R(\mathbf{s})|$

where $R(\mathbf{s}) = \dfrac{1}{2}f''(\mathbf{z}:\mathbf{s}) - \dfrac{1}{2}f''(\mathbf{c}:\mathbf{s})$

$\because |R(\mathbf{s})| \le \dfrac{1}{2}\sum_{i=1}^{n}\sum_{j=1}^{n}\big|D_{i,j}f(\mathbf{z}) - D_{i,j}f(\mathbf{c})\big|\,\|\mathbf{s}\|^2$ and $f$ is continuous at $\mathbf{c}$

Choose $\delta_1 > 0$ s.t. $|R(\mathbf{s})| < \dfrac{-M\|\mathbf{s}\|^2}{2},\ \forall 0 < \|\mathbf{s}\| < \delta_1$

Then $f(c + s) - f(c) \le M\|s\|^2 + |R(s)| \le \dfrac{M\|s\|^2}{2} < 0, \quad \forall 0 < \|s\| < \delta_1$

$\Rightarrow f$ has a relative maximum at $c$

(ii) To prove (ii), we apply part (i) to $-f$

(iii) $\forall \alpha > 0$, we have $f(c + \alpha s) - f(c) = \alpha^2 P(s) + R(\alpha s)$

$\because \lim\limits_{t \to 0} R(t) = 0$, choose $\delta_2 > 0$ s.t. $0 < |R(\alpha s)| < \dfrac{\alpha^2 |P(s)|}{2}, \quad \forall 0 < \alpha < \delta_2$

$\therefore f(c + \alpha s) - f(c)$ and $P(s)$ has the same sign

$\therefore$ if $\exists s_1, s_2 \ne 0$ s.t. $P(s_1) < 0$ and $P(s_2) > 0$ then $f$ has a saddle point at $c$.


多变量函数求极值的考试题型当中以双变量函数求极值最为常见

【定理】双变量函数求极值的二阶导数检定法

給定实数值函数 $f(x, y): R^2 \to R$，假设 $\exists \delta > 0$ 使得函数 $f$ 的所有二阶偏导数于 $B_\delta(c)$ 內皆存在且在平稳点 $c$ 连续，

$$\Delta(c) := \det \begin{bmatrix} f_{xx}(c) & f_{xy}(c) \\ f_{xy}(c) & f_{yy}(c) \end{bmatrix} = f_{xx}(c)f_{yy}(c) - \left(f_{xy}(c)\right)^2$$

Then

(i) 若 $\Delta(c) > 0$ 且 $f_{xx}(c) > 0$，则 $f$ 于 $c$ 有相对极小值.

(ii) 若 $\Delta(c) > 0$ 且 $f_{xx}(c) < 0$，则 $f$ 于 $c$ 有相对极大值

(iii) 若 $\Delta(c) < 0$，则 $c$ 为 $f$ 的鞍点、

Proof:

$\because n = 2 \qquad \therefore P(x, y) = \dfrac{\alpha x^2 + 2\beta xy + \gamma y^2}{2}$, where $\alpha = f_{xx}(c), \; \beta = f_{xy}(c), \; \gamma = f_{yy}(c)$

Let $\alpha \ne 0$, then

$$P(x, y) = \frac{\alpha x^2 + 2\beta xy + \gamma y^2}{2} = \frac{\alpha^2 x^2 + 2\beta\alpha xy + \alpha\gamma y^2}{2\alpha}$$

$$= \frac{(\alpha x + \beta y)^2 + (2\beta\alpha xy + \alpha\gamma y^2 - 2\alpha x\beta y - \beta^2 y^2)}{2\alpha} = \frac{(\alpha x + \beta y)^2 + \Delta(c) \cdot y^2}{2\alpha}$$

$\therefore$ If $\Delta(c) > 0$ and $\alpha > 0$, then $P(x, y) > 0, \forall (x, y) \ne (0,0) \Rightarrow f$ has a relative minimum at $c$.

$\therefore$ If $\Delta(c) > 0$ and $\alpha < 0$, then $P(x, y) < 0, \forall (x, y) \ne (0,0) \Rightarrow f$ has a relative maximum at $c$.

Assume $\Delta(c) < 0$, then $\exists$ two line s.t. $P(x, y) = 0$

$\because P(x, y)$ is continuous at $c$

$\therefore \forall r > 0, \exists (x_0, y_0), (x_1, y_1) \in B_r(c)$ s.t. $P(x_0, y_0) < 0$ and $P(x_1, y_1) > 0$

$\therefore f$ has a saddle point at $c$

上述介绍的方法主要用于无限制条件求多变量函数极值的问题，底下的 Lagrange multiplier 主要用于有限制条件的情形

【定理】Lagrange multiplier

Assume $f: R^n \to R$, $\boldsymbol{x_0}$ is a local extreme point of $f$ subject to $g_1(\boldsymbol{x}) = \cdots = g_m(\boldsymbol{x}) = 0$ and

$$\begin{vmatrix} \dfrac{\partial g_1(\boldsymbol{x_0})}{\partial x_{r_1}} & \dfrac{\partial g_1(\boldsymbol{x_0})}{\partial x_{r_2}} & \cdots & \dfrac{\partial g_1(\boldsymbol{x_0})}{\partial x_{r_m}} \\ \dfrac{\partial g_2(\boldsymbol{x_0})}{\partial x_{r_1}} & \dfrac{\partial g_2(\boldsymbol{x_0})}{\partial x_{r_2}} & \cdots & \dfrac{\partial g_2(\boldsymbol{x_0})}{\partial x_{r_m}} \\ \vdots & \vdots & \ddots & \vdots \\ \dfrac{\partial g_m(\boldsymbol{x_0})}{\partial x_{r_1}} & \dfrac{\partial g_m(\boldsymbol{x_0})}{\partial x_{r_2}} & \cdots & \dfrac{\partial g_m(\boldsymbol{x_0})}{\partial x_{r_m}} \end{vmatrix} \neq 0, \quad (Eq1)$$

for at leats one choice of $r_1 < r_2 < \cdots < r_m$ in $\{1, 2, \ldots, n\}$ where $m < n$. Then there exist constants $\lambda_1, \lambda_2, \ldots, \lambda_m$ such that

$$\frac{\partial f(\boldsymbol{x_0})}{\partial x_i} - \sum_{j=1}^{m} \lambda_j \frac{\partial g_j(\boldsymbol{x_0})}{\partial x_i} = 0, \quad \forall 1 \leq i \leq n.$$

Proof:

Let $r_p = p, 1 \leq p \leq m$ and $A = \begin{bmatrix} \dfrac{\partial g_1(\boldsymbol{x_0})}{\partial x_1} & \dfrac{\partial g_1(\boldsymbol{x_0})}{\partial x_2} & \cdots & \dfrac{\partial g_1(\boldsymbol{x_0})}{\partial x_m} \\ \dfrac{\partial g_2(\boldsymbol{x_0})}{\partial x_1} & \dfrac{\partial g_2(\boldsymbol{x_0})}{\partial x_2} & \cdots & \dfrac{\partial g_2(\boldsymbol{x_0})}{\partial x_m} \\ \vdots & \vdots & \ddots & \vdots \\ \dfrac{\partial g_m(\boldsymbol{x_0})}{\partial x_1} & \dfrac{\partial g_m(\boldsymbol{x_0})}{\partial x_2} & \cdots & \dfrac{\partial g_m(\boldsymbol{x_0})}{\partial x_m} \end{bmatrix}$

then (Eq1) becomes

$$\det(A) \neq 0 \quad (Eq2).$$

Let $\boldsymbol{x_0} = (\boldsymbol{x_0'}; \boldsymbol{t_0}), \boldsymbol{x_0'} = (x_{0,1}, x_{0,2}, \ldots, x_{0,m})$ and $\boldsymbol{t_0} = (x_{0,m+1}, x_{0,m+2}, \ldots, x_{0,n})$. Apply the Implicit Function Theorem to (Eq2) then there exist continuously differentiable functions $h_p = h_p(\boldsymbol{t}), 1 \leq p \leq m$, defined on a neighborhood $B_\delta(\boldsymbol{t_0})$ s.t.

$(h_1(\boldsymbol{t}), h_2(\boldsymbol{t}), \ldots, h_m(\boldsymbol{t}), \boldsymbol{t}) \in D, \quad \forall \boldsymbol{t} \in B_\delta(\boldsymbol{t_0}), \quad (h_1(\boldsymbol{t_0}), h_2(\boldsymbol{t_0}), \ldots, h_m(\boldsymbol{t_0}), \boldsymbol{t_0}) = \boldsymbol{x_0} \quad (Eq3)$

and

$$g_p(h_1(\boldsymbol{t}), h_2(\boldsymbol{t}), \ldots, h_m(\boldsymbol{t}), \boldsymbol{t}) = 0, \quad \forall \boldsymbol{t} \in B_\delta(\boldsymbol{t_0}), \quad 1 \leq p \leq m \quad (Eq4).$$

Let $A^T \begin{bmatrix} \lambda_1 \\ \lambda_2 \\ \vdots \\ \lambda_m \end{bmatrix} = \begin{bmatrix} f_{x_1}(\boldsymbol{x_0}) \\ f_{x_2}(\boldsymbol{x_0}) \\ \vdots \\ f_{x_m}(\boldsymbol{x_0}) \end{bmatrix}$ with (Eq1), then $\dfrac{\partial f(\boldsymbol{x_0})}{\partial x_i} - \displaystyle\sum_{j=1}^{m} \lambda_j \dfrac{\partial g_j(\boldsymbol{x_0})}{\partial x_i} = 0, \ \forall 1 \le i \le m.$

For $m + 1 \le i \le n,$ differentiating (Eq4) with respect to $x_i$ and using (Eq3) yields

$$\frac{\partial g_p(\boldsymbol{x_0})}{\partial x_i} + \sum_{j=1}^{m} \frac{\partial g_p(\boldsymbol{x_0})}{\partial x_j} \frac{\partial h_j(\boldsymbol{x_0})}{\partial x_j} = 0, \ \forall 1 \le p \le m.$$

$\because \boldsymbol{x_0}$ is a local extreme point of $f$ subject to $g_1(\boldsymbol{x}) = g_1(\boldsymbol{x}) = \cdots = g_m(\boldsymbol{x}) = 0$

$\therefore \boldsymbol{t_0}$ is an unconstrained local extreme point of $f(h_1(\boldsymbol{t}), h_2(\boldsymbol{t}), \ldots, h_m(\boldsymbol{t}), \boldsymbol{t})$

$$\therefore \frac{\partial f(\boldsymbol{x_0})}{\partial x_i} + \sum_{j=1}^{m} \frac{\partial f(\boldsymbol{x_0})}{\partial x_j} \frac{\partial h_j(\boldsymbol{x_0})}{\partial x_i} = 0$$

$\therefore \exists$ nontrivial solution $\left( 1, \dfrac{\partial h_1(\boldsymbol{x_0})}{\partial x_1}, \dfrac{\partial h_2(\boldsymbol{x_0})}{\partial x_2}, \ldots, \dfrac{\partial h_m(\boldsymbol{x_0})}{\partial x_m} \right)$ such that

Let $B = \begin{bmatrix} \dfrac{\partial f(\boldsymbol{x_0})}{\partial x_i} & \dfrac{\partial g_1(\boldsymbol{x_0})}{\partial x_i} & \dfrac{\partial g_2(\boldsymbol{x_0})}{\partial x_i} & \cdots & \dfrac{\partial g_m(\boldsymbol{x_0})}{\partial x_i} \\ \dfrac{\partial f(\boldsymbol{x_0})}{\partial x_1} & \dfrac{\partial g_1(\boldsymbol{x_0})}{\partial x_1} & \dfrac{\partial g_2(\boldsymbol{x_0})}{\partial x_1} & \cdots & \dfrac{\partial g_m(\boldsymbol{x_0})}{\partial x_1} \\ \dfrac{\partial f(\boldsymbol{x_0})}{\partial x_2} & \dfrac{\partial g_1(\boldsymbol{x_0})}{\partial x_2} & \dfrac{\partial g_2(\boldsymbol{x_0})}{\partial x_2} & \cdots & \dfrac{\partial g_m(\boldsymbol{x_0})}{\partial x_2} \\ \vdots & \vdots & \vdots & \ddots & \vdots \\ \dfrac{\partial f(\boldsymbol{x_0})}{\partial x_m} & \dfrac{\partial g_1(\boldsymbol{x_0})}{\partial x_m} & \dfrac{\partial g_2(\boldsymbol{x_0})}{\partial x_m} & \cdots & \dfrac{\partial g_m(\boldsymbol{x_0})}{\partial x_m} \end{bmatrix}$ then $B^T \begin{bmatrix} 1 \\ \dfrac{\partial h_1(\boldsymbol{x_0})}{\partial x_1} \\ \dfrac{\partial h_2(\boldsymbol{x_0})}{\partial x_2} \\ \vdots \\ \dfrac{\partial h_m(\boldsymbol{x_0})}{\partial x_m} \end{bmatrix} = 0$

$\therefore \det(B^T) = 0 \Rightarrow \det(B) = 0$

$\therefore \exists$ not all zero constants $c = (c_0, c_1, c_2, \ldots, c_m)$ such that $Bc = 0$

If $c_0 = 0$ then using (Eq2) implies $c_1 = c_2 = \cdots = c_m = 0,$ a contradition. $\therefore c_0 \ne 0$

$\therefore B \begin{bmatrix} c_0 \\ c_1 \\ c_2 \\ \vdots \\ c_m \end{bmatrix} = 0 \Rightarrow B \begin{bmatrix} 1 \\ c_1' \\ c_2' \\ \vdots \\ c_m' \end{bmatrix} = 0 \Rightarrow A^T \begin{bmatrix} -c_1' \\ -c_2' \\ \vdots \\ -c_m' \end{bmatrix} = \begin{bmatrix} \dfrac{\partial f(\boldsymbol{x_0})}{\partial x_1} \\ \dfrac{\partial f(\boldsymbol{x_0})}{\partial x_2} \\ \vdots \\ \dfrac{\partial f(\boldsymbol{x_0})}{\partial x_m} \end{bmatrix},$ where $c_j' = \dfrac{c_j}{c_0}$

By (Eq2) and $\dfrac{\partial f(\boldsymbol{x_0})}{\partial x_i} - \displaystyle\sum_{j=1}^{m} \lambda_j \dfrac{\partial g_j(\boldsymbol{x_0})}{\partial x_i} = 0,\ \forall 1 \leq i \leq m$, we have $c_j' = -\lambda_j,\ \forall 1 \leq j \leq m$

$$\Rightarrow B \begin{bmatrix} 1 \\ -\lambda_1 \\ -\lambda_2 \\ \vdots \\ -\lambda_m \end{bmatrix} = 0.$$ Computing the top row of vector on the left, we have

$$\frac{\partial f(\boldsymbol{x_0})}{\partial x_i} - \sum_{j=1}^{m} \lambda_j \frac{\partial g_j(\boldsymbol{x_0})}{\partial x_i} = 0,\ \forall m + 1 \leq i \leq n$$

## 7.7.1 使用算术平均大于等于几何平均求极值

考试类型:

Type 1.

假设 $x + y + z = r$, $x > 0$、$y > 0$、$z > 0$, 求 $x^a y^b z^c$ 的最大值, $\forall a、b、c \in N$

解题流程:

Step1.

$$\because x + y + z = \sum_{i=1}^{a} \frac{x}{a} + \sum_{i=1}^{b} \frac{y}{b} + \sum_{i=1}^{c} \frac{z}{c} \quad 且 \quad \frac{\sum_{i=1}^{n} x_i}{n} \geq \sqrt[n]{\prod_{i=1}^{n} x_i}$$

Step2.

$$\therefore \frac{\sum_{i=1}^{a} \frac{x}{a} + \sum_{i=1}^{b} \frac{y}{b} + \sum_{i=1}^{c} \frac{z}{c}}{a + b + c} \geq \sqrt[a+b+c]{\left(\frac{x}{a}\right)^a \left(\frac{y}{b}\right)^b \left(\frac{z}{c}\right)^c}$$

Step3.

$$\therefore \left(\frac{r}{a + b + c}\right)^{a+b+c} \geq \left(\frac{x}{a}\right)^a \left(\frac{y}{b}\right)^b \left(\frac{z}{c}\right)^c \Rightarrow \left(\frac{r}{a + b + c}\right)^{a+b+c} a^a b^b c^c \geq x^a y^b z^c$$

Step4.

当 $\dfrac{x}{a} = \dfrac{y}{b} = \dfrac{z}{c}$ 时, 上述不等式的等号成立, $x^a y^b z^c$ 最大值 $= \left(\dfrac{r}{a + b + c}\right)^{a+b+c} a^a b^b c^c$

Example 1.

假设 $x + y + z = 9$, $x > 0$、$y > 0$、$z > 0$ 求 $xyz$ 的最大值 $=?$

【解】

$$\because \frac{\sum_{i=1}^{n} x_i}{n} \geq \sqrt[n]{\prod_{i=1}^{n} x_i}, \quad \forall n \in N \quad \therefore \frac{x + y + z}{3} \geq \sqrt[3]{xyz} \Rightarrow \frac{9}{3} \geq \sqrt[3]{xyz} \Rightarrow 3^3 \geq xyz$$

$\because$ 等号成立于 $x = y = z$ 且 $x + y + z = 9$ $\quad \therefore x = y = z = 3$ $\quad \therefore xyz$ 的最大值 $= 27$

Example 2.

假设 $x + y + z = 1$, $x > 0$、$y > 0$、$z > 0$, 求 $xy^2z^3$ 的最大值 $=?$

【解】

$$\because \frac{\sum_{i=1}^{n} x_i}{n} \geq \sqrt[n]{\prod_{i=1}^{n} x_i}, \quad \forall n \in N$$

$$\therefore x + y + z = \frac{6x + 3y + 3y + 2z + 2z + 2z}{6} \geq \sqrt[6]{6x(3y)^2(2z)^3}$$

$\because$ 等号成立于 $6x = 3y = 2z$ 且 $x + y + z = 1$ $\quad \therefore x = \frac{1}{6}, y = \frac{1}{3}, z = \frac{1}{2}$

$$\therefore xy^2z^3 \text{ 的最大值} = \frac{1}{432}$$

Example 3.

假设 $x + y + z = 6$, $x > 0$、$y > 0$、$z > 0$, 求 $xy^2z^3$ 的极值 $=?$

【解】

$$\because \frac{\sum_{i=1}^{n} x_i}{n} \geq \sqrt[n]{\prod_{i=1}^{n} x_i}, \quad \forall n \in N$$

$$\therefore x + y + z = \frac{6x + 3y + 3y + 2z + 2z + 2z}{6} \geq \sqrt[6]{6x(3y)^2(2z)^3}$$

$\because$ 等号成立于 $6x = 3y = 2z$ 且 $x + y + z = 6$  $\therefore x = 1, y = 2, z = 3$

$\therefore xy^2z^3$ 的最大值 $= 108$

## 7.7.2　使用 Cauchy Inequality 找极值

考试类型:

Type 1.

假设 $f(x, y, z) = ax + by + cz$  且 $x^2 + y^2 + z^2 = \alpha^2$, $\alpha > 0$, 求$f$的极值

解题流程:

Step1.

藉由 Cauchy Inequality

$(ax + by + cz)^2 \leq (a^2 + b^2 + c^2)(x^2 + y^2 + z^2) = \alpha^2(a^2 + b^2 + c^2)$

Step2.

$f(x, y, z)$ 最大值 $= \alpha\sqrt{(a^2 + b^2 + c^2)}$, $f(x, y)$ 最小值 $= -\alpha\sqrt{(a^2 + b^2 + c^2)}$

Example 1.

假设 $f(x, y, z) = x + 2y + 3z$  且 $x^2 + y^2 + z^2 = 1$, 求$f$的极值

【解】

藉由 Cauchy inequality 得　$(x + 2y + 3z)^2 \leq (1^2 + 2^2 + 3^2)(x^2 + y^2 + z^2)$

$\Rightarrow |x + 2y + 3z| \leq \sqrt{14}$  $\therefore f(x, y)$ 最大值 $= \sqrt{14}$, $f(x, y)$ 最小值 $= -\sqrt{14}$

Example 2.

假设 $f(x, y, z) = 3x + 2y + 1z$ 且 $x^2 + y^2 + z^2 = 2$, 求$f$的极值

【解】

藉由 Cauchy inequality 得　$(3x + 2y + 1z)^2 \leq (3^2 + 2^2 + 1^2)(x^2 + y^2 + z^2)$

$\Rightarrow |3x + 2y + z| \leq \sqrt{28}$

$\therefore f(x, y)$ 最大值 $= \sqrt{28}$, $f(x, y)$ 最小值 $= -\sqrt{28}$

Example 3.

假设 $f(x, y, z) = x + 3y + 2z$ 且 $x^2 + y^2 + z^2 = 3$, 求 $f$ 的极值

【解】

藉由 Cauchy inequality 得 $(1x + 3y + 2z)^2 \leq (1^2 + 3^2 + 2^2)(x^2 + y^2 + z^2)$

$\Rightarrow |x + 3y + 2z| \leq \sqrt{42}$ $\therefore f(x, y)$ 最大值 $= \sqrt{42}$, $f(x, y)$ 最小值 $= -\sqrt{42}$

### 7.7.3 无限制条件的双变量函数求极值

首先, 回顾之前介绍无限制条件的双变量函数求极值定理

【定理】双变量函数求极值的二阶导数检定法

给定实数值函数 $f(x, y): R^2 \to R$, 假设 $\exists \delta > 0$ 使得函数 $f$ 的所有二阶偏导数于 $B_\delta(c)$ 内皆存在且在平稳点 $c$ 连续,

$$\Delta(c) := \det \begin{bmatrix} f_{xx}(c) & f_{xy}(c) \\ f_{xy}(c) & f_{yy}(c) \end{bmatrix} = f_{xx}(c)f_{yy}(c) - \left(f_{xy}(c)\right)^2$$

Then

(i) 若 $\Delta(c) > 0$ 且 $f_{xx}(c) > 0$, 则 $f$ 于 $c$ 有相对极小值.

(ii) 若 $\Delta(c) > 0$ 且 $f_{xx}(c) < 0$, 则 $f$ 于 $c$ 有相对极大值

(iii) 若 $\Delta(c) < 0$, 则 $c$ 为 $f$ 的鞍点

考试类型:

Type 1.

给定双变数函数 $f(x, y)$, 求 $f(x, y)$ 的极小值与极大值

解题流程:

Step1.

令 $\nabla f(x_0, y_0) = (0,0)$, 找 $(x_0, y_0)$ s.t. $\nabla f(x_0, y_0) = (0,0)$

Step2.

求 $\Delta(x_0, y_0) =?$ 且 $f_{xx}(x_0, y_0) =?$

Step3.

如果 $\Delta(x_0, y_0) > 0$ 且 $f_{xx}(x_0, y_0) > 0$ 则 $f$ 于 $(x_0, y_0)$ 有相对极小值

如果 $\Delta(x_0, y_0) > 0$ 且 $f_{xx}(x_0, y_0) < 0$ 则 $f$ 于 $(x_0, y_0)$ 有相对极大值

Example 1.

　　假设 $f(x, y) = 3x^2 - 2xy + y^2 - 8y + 4$，求 $f$ 的相对极值

【解】

$\because \nabla f(x, y) = \left(\dfrac{\partial f}{\partial x}, \dfrac{\partial f}{\partial y}\right) = (6x - 2y, -2x + 2y - 8)$

令 $\nabla f(x, y) = (0,0)$ 则 $(x, y) = (2,6)$

$\because \Delta(x, y) = \begin{vmatrix} f_{xx} & f_{xy} \\ f_{yx} & f_{yy} \end{vmatrix} = \begin{vmatrix} 6 & -2 \\ -2 & 2 \end{vmatrix} = 8 > 0$　且 $f_{xx}(2,6) = 6 > 0$

$\therefore f(x, y)$ 在 $(2,6)$ 有相对极小值 $f(2,6) = -20$

Example 2.

　　假设 $f(x, y) = 4xy - x^4 - y^4 + 1$，求 $f$ 的相对极值与鞍点

【解】

$\because \nabla f(x, y) = \left(\dfrac{\partial f}{\partial x}, \dfrac{\partial f}{\partial y}\right) = (4y - 4x^3, 4x - 4y^3)$

令 $\nabla f(x, y) = (0,0)$ 则 $(x, y) = (0,0), (1,1), (-1,-1)$

$\because \Delta(x, y) = \begin{vmatrix} f_{xx} & f_{xy} \\ f_{yx} & f_{yy} \end{vmatrix} = \begin{vmatrix} -12x^2 & 4 \\ 4 & -12y^2 \end{vmatrix} = 144x^2y^2 - 16$

$\therefore \Delta(0,0) = -16, \ \Delta(1,1) = \Delta(-1,-1) = 128 > 0$

且 $f_{xx}(1,1) = f_{xx}(-1,-1) = -12 < 0$

$\therefore f(x, y)$ 在 $(1,1)(-1,-1)$ 有相对极大值 $f(1,1) = f(-1,-1) = 3$

Example 3.

　　假设 $f(x, y) = x^2 + y^2 + xy - 3x - 3y + 2$，求 $f$ 的相对极值与鞍点

【解】

$\because \nabla f(x, y) = \left(\dfrac{\partial f}{\partial x}, \dfrac{\partial f}{\partial y}\right) = (2x + y - 3, 2y + x - 3)$

令 $\nabla f(x, y) = (0,0)$ 则 $(x, y) = (1,1)$

$\because \Delta(x, y) = \begin{vmatrix} f_{xx} & f_{xy} \\ f_{yx} & f_{yy} \end{vmatrix} = \begin{vmatrix} 2 & 1 \\ 1 & 2 \end{vmatrix} = 3 > 0$　$\therefore \Delta(1,1) = 3 > 0$ 且 $f_{xx}(1,1) = 2 > 0$

$\therefore f(x, y)$ 在 $(1,1)$ 有相对极小值 $f(1,1) = -1$

Example 4.

假设 $f(x, y) = x^3 + y^3 - 3x - 12y + 22$，求 $f$ 的相对极值与鞍点

【解】

$\because \nabla f(x, y) = \left(\dfrac{\partial f}{\partial x}, \dfrac{\partial f}{\partial y}\right) = (3x^2 - 3, 3y^2 - 12)$

令 $\nabla f(x, y) = (0,0)$ 则 $(x, y) = (1,2), (1,-2), (-1,2), (-1,-2)$

$\because \Delta(x, y) = \begin{vmatrix} f_{xx} & f_{xy} \\ f_{yx} & f_{yy} \end{vmatrix} = \begin{vmatrix} 6x & 0 \\ 0 & 6y \end{vmatrix} = 36xy$

$\therefore \Delta(1,2) = \Delta(-1,-2) = 72, \ \Delta(1,-2) = \Delta(-1,2) = -72$

且 $f_{xx}(-1,-2) = -6 < 0, \ f_{xx}(1,2) = 6 > 0$

$\therefore f(x, y)$ 在 $(-1,-2)$ 有相对极大值 $f(-1,-2) = 40, f(x, y)$ 在 $(1,2)$ 有相对极小值 $f(1,2) = 4$

Example 5.

假设 $f(x, y) = x^3 + y^2 + 2xy - 4x - 3y + 6$，求 $f$ 的相对极值与鞍点

【解】

$\because \nabla f(x, y) = \left(\dfrac{\partial f}{\partial x}, \dfrac{\partial f}{\partial y}\right) = (3x^2 + 2y - 4, 2y + 2x - 3)$

令 $\nabla f(x, y) = (0,0)$ 则 $(x, y) = \left(-\dfrac{1}{3}, \dfrac{11}{6}\right), \left(1, \dfrac{1}{2}\right)$

$\because \Delta(x, y) = \begin{vmatrix} f_{xx} & f_{xy} \\ f_{yx} & f_{yy} \end{vmatrix} = \begin{vmatrix} 6x & 2 \\ 2 & 2 \end{vmatrix} = 12x - 4 > 0$

$\therefore \Delta\left(1, \dfrac{1}{2}\right) = 8 > 0, \ \Delta\left(-\dfrac{1}{3}, \dfrac{11}{6}\right) = -8 < 0$ 且 $f_{xx}\left(1, \dfrac{1}{2}\right) = 6 > 0$

$\therefore f(x, y)$ 在 $\left(1, \dfrac{1}{2}\right)$ 有相对极小值 $f\left(1, \dfrac{1}{2}\right) = \dfrac{11}{4}$

Example 6.

假设 $f(x, y) = x^3 + y^3 - 3xy - 1$，求 $f$ 的相对极值与鞍点

【解】

$\because \nabla f(x, y) = \left(\dfrac{\partial f}{\partial x}, \dfrac{\partial f}{\partial y}\right) = (3x^2 - 3y, 3y^2 - 3x)$

令 $\nabla f(x,y) = (0,0)$ 则 $(x,y) = (1,1), (0,0)$

$\because \Delta(x,y) = \begin{vmatrix} f_{xx} & f_{xy} \\ f_{yx} & f_{yy} \end{vmatrix} = \begin{vmatrix} 6x & -3 \\ -3 & 6y \end{vmatrix} = 36xy - 9$

$\therefore \Delta(1,1) = 27 > 0, \ \Delta(0,0) = -9 < 0$ 且 $f_{xx}(1,1) = 6 > 0$

$\therefore f(x,y)$ 在 $(1,1)$ 有相对极小值 $f(1,1) = -2$

**Example 7.**

假设 $f(x,y) = e^{-(x^2+y^2+2x)}$，求 $f$ 的相对极值与鞍点

【解】

$\because \nabla f(x,y) = \left(\dfrac{\partial f}{\partial x}, \dfrac{\partial f}{\partial y}\right) = (e^{-(x^2+y^2+2x)}(-2x-2), e^{-(x^2+y^2+2x)}(-2y))$

令 $\nabla f(x,y) = (0,0)$ 则 $(x,y) = (-1,0)$

$\because \Delta(x,y) = \begin{vmatrix} f_{xx} & f_{xy} \\ f_{yx} & f_{yy} \end{vmatrix}$

$= \begin{vmatrix} e^{-(x^2+y^2+2x)}((-2x-2)^2-2) & (e^{-(x^2+y^2+2x)}(-2x-2)(-2y) \\ e^{-(x^2+y^2+2x)}(-2y)(-2x-2) & e^{-(x^2+y^2+2x)}((-2y)^2-2) \end{vmatrix}$

$\therefore \Delta(-1,0) = 4e > 0$ 且 $f_{xx}(-1,0) = -2e < 0$

$\therefore f(x,y)$ 在 $(-1,0)$ 有相对极大值 $e$

**Example 8.**

假设 $f(x,y) = \sin x + \sin y + e, \ 0 < x,y < \pi,$ 求 $f$ 的相对极值

【解】

$\because \nabla f(x,y) = \left(\dfrac{\partial f}{\partial x}, \dfrac{\partial f}{\partial y}\right) = (\cos x, \cos y)$

令 $\nabla f(x,y) = (0,0)$ 则 $(x,y) = \left(\dfrac{\pi}{2}, \dfrac{\pi}{2}\right)$

$\because \Delta(x,y) = \begin{vmatrix} f_{xx} & f_{xy} \\ f_{yx} & f_{yy} \end{vmatrix} = \begin{vmatrix} -\sin x & 0 \\ 0 & -\sin y \end{vmatrix} = \sin x \sin y$

$\therefore \Delta\left(\dfrac{\pi}{2}, \dfrac{\pi}{2}\right) = 1 > 0$ 且 $f_{xx}\left(\dfrac{\pi}{2}, \dfrac{\pi}{2}\right) = -1 < 0$

$\therefore f(x,y)$ 在 $\left(\dfrac{\pi}{2}, \dfrac{\pi}{2}\right)$ 有相对极大值 $f\left(\dfrac{\pi}{2}, \dfrac{\pi}{2}\right) = 2 + e$

Example 9.

假设 $f(x,y) = \sin x + \sin y + \sin(x + y) + 1, 0 < x, y < \pi$，求 $f$ 的相对极值

【解】

$$\because \nabla f(x,y) = \left(\frac{\partial f}{\partial x}, \frac{\partial f}{\partial y}\right) = (\cos x + \cos(x + y), \cos y + \cos(x + y))$$

$$= \left(2\cos\left(x + \frac{y}{2}\right)\cos\frac{y}{2}, 2\cos\left(y + \frac{x}{2}\right)\cos\frac{x}{2}\right)$$

令 $\nabla f(x,y) = (0,0)$ 则 $(x,y) = \left(\frac{\pi}{3}, \frac{\pi}{3}\right)$

$$\because \Delta(x,y) = \begin{vmatrix} -\sin x - \sin(x+y) & -\sin(x+y) \\ -\sin(x+y) & -\sin y - \sin(x+y) \end{vmatrix}$$

$$= \sin(x+y)(\sin x + \sin y) + \sin x \sin y$$

$$\therefore \Delta\left(\frac{\pi}{3}, \frac{\pi}{3}\right) = \frac{\sqrt{3}}{2} \cdot \sqrt{3} + \frac{3}{4} = \frac{9}{4} > 0 \text{ 且 } f_{xx}\left(\frac{\pi}{3}, \frac{\pi}{3}\right) = -\sqrt{3} < 0$$

$$\therefore f(x,y) \text{ 在 } \left(\frac{\pi}{3}, \frac{\pi}{3}\right) \text{ 有相对极大值} f\left(\frac{\pi}{3}, \frac{\pi}{3}\right) = \frac{3\sqrt{3}}{2} + 1$$

## 7.7.4　限制条件为等式的多变量函数求极值

回顾之前所介绍限制条件为等式的多变量函数求极值的方法

【定理】Lagrange multiplier

Assume $f: R^n \to R$, $\boldsymbol{x_0}$ is a local extreme point of $f$ subject to $g_1(\boldsymbol{x}) = \cdots = g_m(\boldsymbol{x}) = 0$ and

$$\begin{vmatrix} \dfrac{\partial g_1(\boldsymbol{x_0})}{\partial x_{r_1}} & \dfrac{\partial g_1(\boldsymbol{x_0})}{\partial x_{r_2}} & \cdots & \dfrac{\partial g_1(\boldsymbol{x_0})}{\partial x_{r_m}} \\ \dfrac{\partial g_2(\boldsymbol{x_0})}{\partial x_{r_1}} & \dfrac{\partial g_2(\boldsymbol{x_0})}{\partial x_{r_2}} & \cdots & \dfrac{\partial g_2(\boldsymbol{x_0})}{\partial x_{r_m}} \\ \vdots & \vdots & \ddots & \vdots \\ \dfrac{\partial g_m(\boldsymbol{x_0})}{\partial x_{r_1}} & \dfrac{\partial g_m(\boldsymbol{x_0})}{\partial x_{r_2}} & \cdots & \dfrac{\partial g_m(\boldsymbol{x_0})}{\partial x_{r_m}} \end{vmatrix} \neq 0, \ \text{(Eq1)}$$

for at leats one choice of $r_1 < r_2 < \cdots < r_m$ in $\{1, 2, \ldots, n\}$ where $m < n$. Then there exist constants $\lambda_1, \lambda_2, \ldots, \lambda_m$ such that

$$\frac{\partial f(\boldsymbol{x_0})}{\partial x_i} - \sum_{j=1}^{m} \lambda_j \frac{\partial g_j(\boldsymbol{x_0})}{\partial x_i} = 0, \ \forall 1 \leq i \leq n.$$

考试类型:

Type 1.

給定双变数函数 $f(x, y)$，假设 $C_1(x, y) = 0$ 为限制式，使用 Lagrange multiplier 求函数 $f$ 于此限制式的极值

解题流程:

Step1.

令 $L(x, y, \lambda) = f(x, y) + \lambda C_1(x, y)$，若 $L(x, y, \lambda)$ 极值存在 则 $\begin{cases} \dfrac{\partial L}{\partial x}(x, y) = 0 \\ \dfrac{\partial L}{\partial y}(x, y) = 0 \\ C_1(x, y) = 0 \end{cases}$

Step2.

找 $(x_0, y_0)$ 满足 $\begin{cases} \dfrac{\partial L}{\partial x}(x, y) = 0 \\ \dfrac{\partial L}{\partial y}(x, y) = 0 \\ C_1(x, y) = 0 \end{cases}$

Step3.

找到的 $(x_0, y_0)$ 可能不只一个，比较每个 $f(x_0, y_0)$ 的大小求得 $f(x, y)$ 的相对极大值与相对极小值

Type 2.

給定曲线 $C_1(x, y) = 0$，求曲线上离原点最近与最远的点坐标

解题流程:

Step1.

曲线至点的距离 $= d = \sqrt{x^2 + y^2}$，令 $L(x, y, \lambda) = x^2 + y^2 + \lambda C_1(x, y)$

若 $L(x, y, \lambda)$ 极值存在 则 $\begin{cases} \dfrac{\partial L}{\partial x}(x, y) = 0 \\ \dfrac{\partial L}{\partial y}(x, y) = 0 \\ C_1(x, y) = 0 \end{cases}$

Step2.

找 $(x_0, y_0)$ 满足 $\begin{cases} \dfrac{\partial L}{\partial x}(x,y) = 0 \\ \dfrac{\partial L}{\partial y}(x,y) = 0 \\ C_1(x,y) = 0 \end{cases}$

Step3.

找到的 $(x_0, y_0)$ 可能不只一个，使用 $\sqrt{x_0^2 + y_0^2}$ 比较大小求得极值

Type 3.

给定曲面 $C_1(x,y,z) = 0$，求曲面距离平面 $ax + by + cz = 0$ 最近的坐标
解题流程：
Step1.

曲面至平面的距离 $= \dfrac{|ax + by + cz|}{\sqrt{a^2 + b^2 + c^2}}$，令 $L(x,y,z,\lambda) = ax + by + cz + \lambda C_1(x,y,z)$

若 $L(x,y,z,\lambda)$ 极值存在 则 $\begin{cases} \dfrac{\partial L}{\partial x}(x,y,z) = 0 \\ \dfrac{\partial L}{\partial y}(x,y,z) = 0 \\ \dfrac{\partial L}{\partial z}(x,y,z) = 0 \\ C_1(x,y,z) = 0 \end{cases}$

Step2.

找 $(x_0, y_0, z_0)$ 满足 $\begin{cases} \dfrac{\partial L}{\partial x}(x,y,z) = 0 \\ \dfrac{\partial L}{\partial y}(x,y,z) = 0 \\ \dfrac{\partial L}{\partial z}(x,y,z) = 0 \\ C_1(x,y,z) = 0 \end{cases}$

Step3.

找到的 $(x_0, y_0, z_0)$ 可能不只一个，使用 $|ax_0 + by_0 + cz_0|$ 比较大小求得极值

Type 4.

给定两曲面 $C_1(x,y,z) = 0$ 与 $C_2(x,y,z) = 0$，求位于两曲面 $C_1$ 与 $C_2$ 交线上，高度最高与

最低点分别为何?

解题流程:

Step1.

令 $L(x, y, z, \lambda_1, \lambda_2) = z + \lambda_1 C_1(x, y, z) + \lambda_2 C_2(x, y, z)$

若 $L(x, y, z, \lambda_1, \lambda_2)$ 极值存在 则
$$
\begin{cases}
\dfrac{\partial L}{\partial x}(x, y, z, \lambda_1, \lambda_2) = 0 \\[2mm]
\dfrac{\partial L}{\partial y}(x, y, z, \lambda_1, \lambda_2) = 0 \\[2mm]
\dfrac{\partial L}{\partial z}(x, y, z, \lambda_1, \lambda_2) = 0 \\[2mm]
C_1(x, y, z) = 0 \\
C_2(x, y, z) = 0
\end{cases}
$$

Step2.

找 $(x_0, y_0, z_0)$ 满足
$$
\begin{cases}
\dfrac{\partial L}{\partial x}(x, y, z, \lambda_1, \lambda_2) = 0 \\[2mm]
\dfrac{\partial L}{\partial y}(x, y, z, \lambda_1, \lambda_2) = 0 \\[2mm]
\dfrac{\partial L}{\partial z}(x, y, z, \lambda_1, \lambda_2) = 0 \\[2mm]
C_1(x, y, z) = 0 \\
C_2(x, y, z) = 0
\end{cases}
$$

Step3.

找到的 $(x_0, y_0, z_0)$ 可能不只一个, 使用 $z_0$ 比较大小求得极值

Type 5.

給定两曲面 $C_1(x, y, z) = 0$ 与 $C_2(x, y, z) = 0$, 求原点 $(0,0,0)$ 至两曲面交线的最短距离

解题流程:

Step1.

原点至交线的距离 $= d = \sqrt{x^2 + y^2 + z^2}$

令 $L(x, y, z, \lambda_1, \lambda_2) = x^2 + y^2 + z^2 + \lambda_1 C_1(x, y, z) + \lambda_2 C_2(x, y, z)$

$$\text{若 } L(x,y,z,\lambda_1,\lambda_2) \text{ 极值存在 则}\begin{cases}\dfrac{\partial L}{\partial x}(x,y,z,\lambda_1,\lambda_2) = 0 \\[2mm] \dfrac{\partial L}{\partial y}(x,y,z,\lambda_1,\lambda_2) = 0 \\[2mm] \dfrac{\partial L}{\partial z}(x,y,z,\lambda_1,\lambda_2) = 0 \\[2mm] C_1(x,y,z) = 0 \\[1mm] C_2(x,y,z) = 0\end{cases}$$

Step2.

$$\text{找}(x_0,y_0,z_0)\text{满足}\begin{cases}\dfrac{\partial L}{\partial x}(x,y,z,\lambda_1,\lambda_2) = 0 \\[2mm] \dfrac{\partial L}{\partial y}(x,y,z,\lambda_1,\lambda_2) = 0 \\[2mm] \dfrac{\partial L}{\partial z}(x,y,z,\lambda_1,\lambda_2) = 0 \\[2mm] C_1(x,y,z) = 0 \\[1mm] C_2(x,y,z) = 0\end{cases}$$

Step3.

找到的 $(x_0,y_0,z_0)$ 可能不只一个，使用 $\sqrt{x_0^2 + y_0^2 + z_0^2}$ 比较大小求最短距离

Example 1.

假设 $x^2 + y^2 = 1$，求 $xy$ 的极值

【解】

令 $L(x,y,\lambda) = xy + \lambda(x^2 + y^2 - 1)$

$$L(x,y,\lambda)\text{极值存在} \Rightarrow \begin{cases}\dfrac{\partial L}{\partial x} = y + 2\lambda x = 0 \\[2mm] \dfrac{\partial L}{\partial y} = x + 2\lambda y = 0 \\[2mm] x^2 + y^2 = 1\end{cases} \Rightarrow y^2 = x^2 \quad \therefore x^2 + y^2 = 1 \text{ 且 } y^2 = x^2$$

$\therefore (x,y) = \left(\dfrac{1}{\sqrt{2}},\dfrac{1}{\sqrt{2}}\right),\left(\dfrac{-1}{\sqrt{2}},\dfrac{1}{\sqrt{2}}\right),\left(\dfrac{1}{\sqrt{2}},\dfrac{-1}{\sqrt{2}}\right),\left(\dfrac{-1}{\sqrt{2}},\dfrac{-1}{\sqrt{2}}\right) \Rightarrow xy$ 有极大值 $\dfrac{1}{2}$，极小值 $\dfrac{-1}{2}$

Example 2.

假设 $f(x, y, z) = xyz$, 其中 $2xz + 2yz + xy = 12$, 试求 $f$ 的极大值

【解】

令 $L(x, y, \lambda) = xyz + \lambda(2xz + 2yz + xy - 12)$

$L(x, y, z, \lambda)$ 极值存在 $\Rightarrow$
$$\begin{cases} \dfrac{\partial L}{\partial x} = yz + \lambda(2z + y) = 0 & (1) \\[2mm] \dfrac{\partial L}{\partial y} = xz + \lambda(2z + x) = 0 & (2) \\[2mm] \dfrac{\partial L}{\partial z} = xy + \lambda(2x + 2y) = 0 & (3) \\[2mm] 2xz + 2yz + xy = 12 & (4) \end{cases}$$

By (1)(2) 得 $z(x - y) + \lambda(x - y) = 0$

By (2)(3) 得 $x(y - 2z) + \lambda(2y - 4z) = 0$

By (1)(3) 得 $y(x - 2z) + \lambda(2x - 4z) = 0$

$\therefore \lambda = -\dfrac{x}{2} = -\dfrac{y}{2} = -z \Rightarrow x = y = -2\lambda, z = -\lambda$

$\therefore 2xz + 2yz + xy = 12 \Rightarrow 4\lambda^2 + 4\lambda^2 + 4\lambda^2 = 12 \Rightarrow \lambda = \pm 1$

$\therefore (x, y, z) = (-2, -2, -1), (2, 2, 1) \qquad \therefore f$ 的极大值 $= 4$

Example 3.

假设 $f(x, y) = xy^3$, 其中 $x^2 + 3y^2 = 16$, 试求 $f$ 的极大值

【解】

令 $L(x, y, \lambda) = xy^3 + \lambda(x^2 + 3y^2 - 16)$

$L(x, y, \lambda)$ 极值存在 $\Rightarrow$
$$\begin{cases} \dfrac{\partial L}{\partial x} = y^3 + 2x\lambda = 0 & (1) \\[2mm] \dfrac{\partial L}{\partial y} = 3xy^2 + 6y\lambda = 0 & (2) \\[2mm] x^2 + 3y^2 - 16 = 0 & (3) \end{cases}$$

By (1)(2) 得 $\lambda = -\dfrac{y^3}{2x} = -\dfrac{xy}{2} \Rightarrow x^2 = y^2$

$\because x^2 + 3y^2 = 16 \quad \therefore 4x^2 = 16 \Rightarrow (x, y) = (2, 2), (-2, 2), (2, -2), (-2, -2)$

$\therefore f$ 极大值 $= 16$

Example 4.

假设 $f(x, y) = 4xy$, 其中 $\dfrac{x^2}{9} + \dfrac{y^2}{25} = 1$, 试求 $f$ 的极值

【解】

令 $L(x, y, \lambda) = 4xy + \lambda\left(\dfrac{x^2}{9} + \dfrac{y^2}{25} - 1\right)$

$L(x, y, \lambda)$ 极值存在 $\Rightarrow \begin{cases} \dfrac{\partial L}{\partial x} = 4y + \dfrac{2x\lambda}{9} = 0 & (1) \\[2mm] \dfrac{\partial L}{\partial y} = 4x + \dfrac{2y\lambda}{25} = 0 & (2) \\[2mm] \dfrac{x^2}{9} + \dfrac{y^2}{25} - 1 = 0 & (3) \end{cases}$

By (1)(2)得 $\lambda = \dfrac{-18y}{x} = \dfrac{-50x}{y} \Rightarrow 9y^2 = 25x^2$

$\because \dfrac{x^2}{9} + \dfrac{y^2}{25} - 1 = 0 \qquad \therefore \dfrac{25x^2 + 9y^2}{225} = 1$

$\Rightarrow (x, y) = \left(\dfrac{3}{\sqrt{2}}, \dfrac{5}{\sqrt{2}}\right), \left(\dfrac{3}{\sqrt{2}}, -\dfrac{5}{\sqrt{2}}\right), \left(-\dfrac{3}{\sqrt{2}}, \dfrac{5}{\sqrt{2}}\right), \left(-\dfrac{3}{\sqrt{2}}, -\dfrac{5}{\sqrt{2}}\right)$

$\therefore f$ 极大值 $= 30$, 极小值 $= -30$

Example 5.

求点 $(1, 2, 0)$ 至曲面 $z^2 = x^2 + y^2$ 的最短距离

【解】

曲面至点的距离 $= d = \sqrt{(x-1)^2 + (y-2)^2 + z^2}$

令 $L(x, y, \lambda) = (x-1)^2 + (y-2)^2 + z^2 + \lambda(x^2 + y^2 - z^2)$

$L(x, y, z, \lambda)$ 极值存在 $\Rightarrow \begin{cases} \dfrac{\partial L}{\partial x} = 2(x-1) + 2\lambda x = 0 & (1) \\[2mm] \dfrac{\partial L}{\partial y} = 2(y-2) + 2\lambda y = 0 & (2) \\[2mm] \dfrac{\partial L}{\partial z} = 2z - 2\lambda z = 0 & (3) \\[2mm] z^2 - x^2 + y^2 = 0 & (4) \end{cases}$

藉由 (3) 得 $\lambda = 1$ 或 $z = 0$

As $\lambda = 1$, $2(x-1) + 2\lambda x = 0$, $2(y-2) + 2\lambda y = 0$ 且 $z^2 = x^2 + y^2$

$\therefore 4x - 2 = 0$ 且 $4y - 4 = 0 \Rightarrow (x, y, z) = \left(\dfrac{1}{2}, 1, \pm\dfrac{\sqrt{5}}{2}\right)$

As $z = 0$, $2(x-1) + 2\lambda x = 0$, $2(y-2) + 2\lambda y = 0$ 且 $0 = x^2 + y^2$ $\Rightarrow$ $(x,y,z) = (0,0,0)$

当 $(x,y,z) = (0,0,0)$, $d = \sqrt{(x-1)^2 + (y-2)^2 + z^2} = \sqrt{5}$

当 $(x,y,z) = \left(\dfrac{1}{2}, 1, \pm\dfrac{\sqrt{5}}{2}\right)$, $d = \sqrt{(x-1)^2 + (y-2)^2 + z^2} = \dfrac{\sqrt{10}}{2}$ 为最短距离

Example 6.

求原点 $(0,0,0)$ 至曲面 $z^2 + 2 = (x-y)^2$ 的最短距离

【解】

曲面至原点 $(0,0,0)$ 的距离 $= d = \sqrt{x^2 + y^2 + z^2}$

令 $L(x,y,z,\lambda) = x^2 + y^2 + z^2 + \lambda((x-y)^2 - z^2 - 2)$

$L(x,y,z,\lambda)$ 极值存在 $\Rightarrow$
$$\begin{cases} \dfrac{\partial L}{\partial x} = 2x + 2\lambda(x-y) = 0 & (1) \\[2mm] \dfrac{\partial L}{\partial y} = 2y - 2\lambda(x-y) = 0 & (2) \\[2mm] \dfrac{\partial L}{\partial z} = 2z - 2\lambda z = 0 & (3) \\[2mm] z^2 + 2 - (x-y)^2 = 0 & (4) \end{cases}$$

藉由 (3) 得 $\lambda = 1$ 或 $z = 0$

As $\lambda = 1$, $4x - 2y = 0$, $4y - 2x = 0$ 且 $z^2 + 2 = (x-y)^2$ $\Rightarrow$ $(x,y,z)$ 不存在

As $z = 0$, $x + y = 0$ 且 $2 = (x-y)^2$

$\Rightarrow (x,y,z) = \left(\dfrac{1}{\sqrt{2}}, \dfrac{-1}{\sqrt{2}}, 0\right), \left(\dfrac{-1}{\sqrt{2}}, \dfrac{1}{\sqrt{2}}, 0\right)$

$d = \sqrt{x^2 + y^2 + z^2} = \sqrt{\dfrac{1}{2} + \dfrac{1}{2} + z^2} = 1$ 为最短距离

Example 7.

(i) 假设 $f(x,y,z) = xyz$, 其中 $x^2 + \dfrac{y^2}{4} + \dfrac{z^2}{9} = r^2, (r > 0)$ 试求 $f$ 的极值

(ii) 假设 $f(x,y,z) = xyz$, 其中 $x^3 + y^3 + z^3 = 24$, 试求 $f$ 的极值

【解】

(i)

令 $r > 0$, $L(x, y, z, \lambda) = xyz + \lambda\left(x^2 + \dfrac{y^2}{4} + \dfrac{z^2}{9} - r^2\right)$

$L(x, y, z, \lambda)$ 极值存在 $\Rightarrow$
$$\begin{cases} \dfrac{\partial L}{\partial x} = yz + 2\lambda x = 0 & (1) \\[2mm] \dfrac{\partial L}{\partial y} = xz + \lambda\dfrac{y}{2} = 0 & (2) \\[2mm] \dfrac{\partial L}{\partial z} = xy + \lambda\dfrac{2z}{9} = 0 & (3) \\[2mm] x^2 + \dfrac{y^2}{4} + \dfrac{z^2}{9} - r^2 = 0 & (4) \end{cases}$$

By (1)(2)(3) 得 $\lambda = -\dfrac{yz}{2x} = -\dfrac{2xz}{y} = -\dfrac{9xy}{2z}$

带入 (4) 得 $x^2 + \dfrac{4x^2}{4} + \dfrac{9x^2}{9} = r^2$ $\qquad \therefore (x, y, z) = \left(\pm\dfrac{r}{\sqrt{3}}, \pm\dfrac{2r}{\sqrt{3}}, \pm\dfrac{3r}{\sqrt{3}}\right)$

$\because f\left(\dfrac{r}{\sqrt{3}}, \dfrac{2r}{\sqrt{3}}, \dfrac{3r}{\sqrt{3}}\right) = \dfrac{2r^3}{\sqrt{3}}$

$f\left(-\dfrac{r}{\sqrt{3}}, \dfrac{2r}{\sqrt{3}}, \dfrac{3r}{\sqrt{3}}\right) = f\left(\dfrac{r}{\sqrt{3}}, -\dfrac{2r}{\sqrt{3}}, \dfrac{3r}{\sqrt{3}}\right) = f\left(\dfrac{r}{\sqrt{3}}, \dfrac{2r}{\sqrt{3}}, -\dfrac{3r}{\sqrt{3}}\right) = -\dfrac{2r^3}{\sqrt{3}}$

$f\left(-\dfrac{r}{\sqrt{3}}, -\dfrac{2r}{\sqrt{3}}, \dfrac{3r}{\sqrt{3}}\right) = f\left(\dfrac{r}{\sqrt{3}}, -\dfrac{2r}{\sqrt{3}}, -\dfrac{3r}{\sqrt{3}}\right) = f\left(-\dfrac{r}{\sqrt{3}}, \dfrac{2r}{\sqrt{3}}, -\dfrac{3r}{\sqrt{3}}\right) = \dfrac{2r^3}{\sqrt{3}}$

$f\left(-\dfrac{r}{\sqrt{3}}, -\dfrac{2r}{\sqrt{3}}, -\dfrac{3r}{\sqrt{3}}\right) = -\dfrac{2r^3}{\sqrt{3}}$

$\therefore f$ 有极大值 $\dfrac{2r^3}{\sqrt{3}}$, 极小值 $-\dfrac{2r^3}{\sqrt{3}}$

(ii)

令 $L(x, y, z, \lambda) = xyz + \lambda(x^3 + y^3 + z^3 - 24)$

$L(x, y, z, \lambda)$ 极值存在 $\Rightarrow$
$$\begin{cases} \dfrac{\partial L}{\partial x} = yz + 3\lambda x^2 = 0 & (1) \\[2mm] \dfrac{\partial L}{\partial y} = xz + 3\lambda y^2 = 0 & (2) \\[2mm] \dfrac{\partial L}{\partial z} = xy + 3\lambda z^2 = 0 & (3) \\[2mm] x^3 + y^3 + z^3 = 24 & (4) \end{cases}$$

By (1)(2)(3) 得 $\lambda = -\dfrac{yz}{3x^2} = -\dfrac{xz}{3y^2} = -\dfrac{xy}{3z^2}$

带入(4)得 $x^3 + x^3 + x^3 = 24$　　$\therefore (x,y,z) = (2,2,2)$，因此 $f$ 有极大值 8

Example 8.

  (i) 求曲线 $17x^2 + 12xy + 8y^2 = 400$ 上离原点最近与最远的点坐标

  (ii) 求曲线 $7x^2 - 6\sqrt{3}xy + 13y^2 = 64$ 上离远点最近与最远的点坐标

【解】

(i)

曲线至点的距离 $= d = \sqrt{x^2 + y^2}$

令 $L(x,y,\lambda) = x^2 + y^2 + \lambda(17x^2 + 12xy + 8y^2 - 400)$

$L(x,y,\lambda)$ 极值存在 $\Rightarrow \begin{cases} \dfrac{\partial L}{\partial x} = 2x + \lambda(34x + 12y) = 0 & (1) \\[2mm] \dfrac{\partial L}{\partial y} = 2y + \lambda(12x + 16y) = 0 & (2) \\[2mm] 17x^2 + 12xy + 8y^2 - 400 = 0 & (3) \end{cases}$

By (1)(2) 得 $\lambda = \dfrac{x}{17x + 6y} = \dfrac{y}{6x + 8y} \Rightarrow x(6x + 8y) = y(17x + 6y)$

$\therefore (x - 2y)(2x + y) = 0$

(a) As $x - 2y = 0$

将其带入(3)得 $(x,y) = (4,2), (-4,-2)$ 且 $d = \sqrt{4^2 + 2^2} = 2\sqrt{5}$

(b) As $2x + y = 0$

将其带入(3)得 $(x,y) = (4,-8), (-4,8)$ 且 $d = \sqrt{4^2 + 8^2} = \sqrt{80} = 4\sqrt{5}$

因此在 $(4,2), (-4,-2)$ 有最近距离，在 $(4,-8), (-4,8)$ 有最远距离

(ii)

曲线至点的距离 $= d = \sqrt{x^2 + y^2}$

令 $L(x,y,\lambda) = x^2 + y^2 + \lambda(7x^2 - 6\sqrt{3}xy + 13y^2 - 64)$

$L(x,y,\lambda)$ 极值存在 $\Rightarrow \begin{cases} \dfrac{\partial L}{\partial x} = 2x + \lambda(14x - 6\sqrt{3}y) = 0 & (1) \\[2mm] \dfrac{\partial L}{\partial y} = 2y + \lambda(-6\sqrt{3}x + 26y) = 0 & (2) \\[2mm] 7x^2 - 6\sqrt{3}xy + 13y^2 - 64 = 0 & (3) \end{cases}$

By (1)(2) 得 $\lambda = \dfrac{-x}{7x - 3\sqrt{3}y} = \dfrac{-y}{-3\sqrt{3}x + 13y}$

$\Rightarrow x(-3\sqrt{3}x + 13y) = y(7x - 3\sqrt{3}y) \quad \therefore (7x - 3\sqrt{3}y)(-3\sqrt{3}x + 13y) = 0$

(a) As $7x - 3\sqrt{3}y = 0$

$\because 7x^2 - 6\sqrt{3}xy + 13y^2 - 64 = 0$

$\therefore 7\left(\dfrac{3\sqrt{3}}{7}y\right)^2 - 6\sqrt{3}\left(\dfrac{3\sqrt{3}}{7}y\right)y + 13y^2 - 64 = 0 \quad \therefore y = \pm\sqrt{7}$

$\therefore (x,y) = \left(\dfrac{3\sqrt{21}}{7}, \sqrt{7}\right), \left(-\dfrac{3\sqrt{21}}{7}, -\sqrt{7}\right)$ 且 $d = \sqrt{\left(\dfrac{3\sqrt{21}}{7}\right)^2 + (\sqrt{7})^2} = 2\sqrt{19}$

(b) As $-3\sqrt{3}x + 13y = 0$

$\because 7x^2 - 6\sqrt{3}xy + 13y^2 - 64 = 0$

$\therefore 7\left(\dfrac{13}{3\sqrt{3}}y\right)^2 - 6\sqrt{3}\left(\dfrac{13}{3\sqrt{3}}y\right)y + 13y^2 - 64 = 0 \quad \therefore y = \pm\sqrt{\dfrac{27}{13}}$

$\therefore (x,y) = \left(\sqrt{13}, \sqrt{\dfrac{27}{13}}\right), \left(-\sqrt{13}, -\sqrt{\dfrac{27}{13}}\right)$ 且 $d = \sqrt{(\sqrt{13})^2 + \left(\sqrt{\dfrac{27}{13}}\right)^2} = 2\sqrt{6}$

因此在 $\left(\sqrt{13}, \sqrt{\dfrac{27}{13}}\right), \left(-\sqrt{13}, -\sqrt{\dfrac{27}{13}}\right)$ 有最近距离

在 $\left(\dfrac{3\sqrt{21}}{7}, \sqrt{7}\right), \left(-\dfrac{3\sqrt{21}}{7}, -\sqrt{7}\right)$ 最远距离

Example 9.

　　求锥面 $x^2 + y^2 = z^2$ 与平面 $x + 14z = 10$ 交线上的最高点

【解】

令 $L(x, y, z, \lambda_1, \lambda_2) = z + \lambda_1(x^2 + y^2 - z^2) + \lambda_2(x + 14z - 10)$

$$L(x, y, z, \lambda_1, \lambda_2) \text{极值存在} \Rightarrow \begin{cases} \dfrac{\partial L}{\partial x} = 2x\lambda_1 + \lambda_2 = 0 & (1) \\[2mm] \dfrac{\partial L}{\partial y} = 2y\lambda_1 = 0 & (2) \\[2mm] \dfrac{\partial L}{\partial z} = 1 - 2z\lambda_1 + 14\lambda_2 = 0 & (3) \\[2mm] x^2 + y^2 - z^2 = 0 & (4) \\[2mm] x + 14z - 10 = 0 & (5) \end{cases}$$

By (1)(2)得 $\lambda_1 \neq 0, y = 0$

By (4) $x = \pm z$, $\because x + 14z - 10 = 0, (x, z) = \left(\dfrac{2}{3}, \dfrac{2}{3}\right), \left(\dfrac{-10}{13}, \dfrac{10}{13}\right)$

$\Rightarrow (x, y, z) = \left(\dfrac{2}{3}, 0, \dfrac{2}{3}\right), \left(\dfrac{-10}{13}, 0, \dfrac{10}{13}\right)$ $\quad \therefore$ 高度最高点 $= \dfrac{10}{13}$

Example 10.

求曲面 $x^2 + y^2 + 5 = z$ 上，与平面 $x + 2y - z = 0$ 距离最近点坐标

【解】

曲面至平面的距离 $= d = \dfrac{|x + 2y - z|}{\sqrt{6}}$

令 $L(x, y, \lambda) = x + 2y - z + \lambda(x^2 + y^2 + 5 - z)$

$$L(x, y, z, \lambda) \text{极值存在} \Rightarrow \begin{cases} \dfrac{\partial L}{\partial x} = 1 + 2x\lambda = 0 & (1) \\[2mm] \dfrac{\partial L}{\partial y} = 2 + 2y\lambda = 0 & (2) \\[2mm] \dfrac{\partial L}{\partial z} = -1 - \lambda = 0 & (3) \\[2mm] x^2 + y^2 + 5 - z = 0 & (4) \end{cases}$$

By (1)(2)(3)得 $\lambda = -1, x = \dfrac{1}{2}, y = 1$ 带入(4)得 $z = \dfrac{25}{4}$ $\Rightarrow$ 曲面于 $\left(\dfrac{1}{2}, 1, \dfrac{25}{4}\right)$ 离平面最近

Example 11.

求曲面 $x^2 + y^2 + z - 5 = 0$ 上，与平面 $x + 2y + 3z = 0$ 距离最近坐标

【解】

曲面至平面的距离 $= d = \dfrac{|x + 2y + 3z|}{\sqrt{14}}$

令 $L(x,y,\lambda) = x + 2y + 3z + \lambda(x^2 + y^2 + z - 5)$

$L(x,y,z,\lambda)$ 极值存在 $\Rightarrow$
$$\begin{cases} \dfrac{\partial L}{\partial x} = 1 + 2x\lambda = 0 & (1) \\[2mm] \dfrac{\partial L}{\partial y} = 2 + 2y\lambda = 0 & (2) \\[2mm] \dfrac{\partial L}{\partial z} = 3 + \lambda = 0 & (3) \\[2mm] x^2 + y^2 + z - 5 = 0 & (4) \end{cases}$$

By (1)(2)(3)得 $\lambda = -3, x = \dfrac{1}{6}, y = \dfrac{1}{3}$ 带入(4)得 $z = \dfrac{175}{36}$

$\Rightarrow$ 曲面于 $\left(\dfrac{1}{6}, \dfrac{1}{3}, \dfrac{175}{36}\right)$ 离平面最近

Example 12.

求原点 $(0,0,0)$ 至两平面 $x + y + z = 2$ 与 $x - y - z = 2$ 交线的最短距离

【解】

原点至交线的距离 $= d = \sqrt{x^2 + y^2 + z^2}$

令 $L(x,y,z,\lambda_1,\lambda_2) = x^2 + y^2 + z^2 + \lambda_1(x + y + z - 2) + \lambda_2(x - y - z - 2)$

$L(x,y,z,\lambda_1,\lambda_2)$ 极值存在 $\Rightarrow$
$$\begin{cases} \dfrac{\partial L}{\partial x} = 2x + \lambda_1 + \lambda_2 = 0 & (1) \\[2mm] \dfrac{\partial L}{\partial y} = 2y + \lambda_1 - \lambda_2 = 0 & (2) \\[2mm] \dfrac{\partial L}{\partial z} = 2z + \lambda_1 - \lambda_2 = 0 & (3) \\[2mm] x + y + z - 2 = 0 & (4) \\[1mm] x - y - z - 2 = 0 & (5) \end{cases}$$

By (2)(3)得 $y = z$ 带入(4)(5) 得 $(x,y,z) = (2,0,0)$, 因此最短距离 $d = \sqrt{2^2 + 0^2 + 0^2} = 2$

Example 13.

(i)位于锥面 $x^2 + y^2 = z^2$ 与平面 $x + y + 2z = 2$ 的交线上, 高度最高与最低点分别为?

(ii)位于锥面 $x^2 + y^2 = z^2$ 与平面 $4x - 3y - z = 5$ 的交线上, 高度最高与最低点 =?

【解】

(i)

令 $L(x,y,z,\lambda_1,\lambda_2) = z + \lambda_1(x^2 + y^2 - z^2) + \lambda_2(x + y + 2z - 2)$

$$L(x,y,z,\lambda_1,\lambda_2)\text{极值存在} \Rightarrow \begin{cases} \dfrac{\partial L}{\partial x} = 2x\lambda_1 + \lambda_2 = 0 & (1) \\[2mm] \dfrac{\partial L}{\partial y} = 2y\lambda_1 + \lambda_2 = 0 & (2) \\[2mm] \dfrac{\partial L}{\partial z} = 1 - 2z\lambda_1 + 2\lambda_2 = 0 & (3) \\[2mm] x^2 + y^2 - z^2 = 0 & (4) \\[1mm] x + y + 2z - 2 = 0 & (5) \end{cases}$$

By $(1)(2)$ 得 $(\lambda_1, \lambda_2) = (0,0)$ 或 $x = y$

As $(\lambda_1, \lambda_2) = (0,0)$ 与 $(3)$ 矛盾 因此解不存在

As $x = y$ 带入 $(4)(5)$

得 $(x,y,z) = (-1 - \sqrt{2}, -1 - \sqrt{2}, 2 + \sqrt{2}), (\sqrt{2} - 1, \sqrt{2} - 1, 2 - \sqrt{2})$

因此于 $(-1 - \sqrt{2}, -1 - \sqrt{2}, 2 + \sqrt{2})$ 高度最高, 于 $(\sqrt{2} - 1, \sqrt{2} - 1, 2 - \sqrt{2})$ 高度最低

(ii)

令 $L(x,y,z,\lambda_1,\lambda_2) = z + \lambda_1(x^2 + y^2 - z^2) + \lambda_2(4x - 3y - z - 5)$

$$L(x,y,z,\lambda_1,\lambda_2)\text{极值存在} \Rightarrow \begin{cases} \dfrac{\partial L}{\partial x} = 2x\lambda_1 + 4\lambda_2 = 0 & (1) \\[2mm] \dfrac{\partial L}{\partial y} = 2y\lambda_1 - 3\lambda_2 = 0 & (2) \\[2mm] \dfrac{\partial L}{\partial z} = 1 - 2z\lambda_1 - \lambda_2 = 0 & (3) \\[2mm] x^2 + y^2 - z^2 = 0 & (4) \\[1mm] 4x - 3y - z - 5 = 0 & (5) \end{cases}$$

By $(1)(2)$ 得 $(\lambda_1, \lambda_2) = (0,0)$ 或 $y = \dfrac{-3x}{4}$

As $(\lambda_1, \lambda_2) = (0,0)$, 其与 $(3)$ 矛盾 因此解不存在

As $y = \dfrac{-3x}{4}$ 带入 $(4)(5)$ 得 $(x,y,z) = \left(1, -\dfrac{3}{4}, \dfrac{5}{4}\right), \left(\dfrac{2}{3}, -\dfrac{1}{2}, -\dfrac{5}{6}\right)$

因此于 $\left(1, -\dfrac{3}{4}, \dfrac{5}{4}\right)$ 高度最高, 于 $\left(\dfrac{2}{3}, -\dfrac{1}{2}, -\dfrac{5}{6}\right)$ 高度最低

Example 14.

假设 $f(x,y) = x^2 + xy + 2y^2$, 其中 $x^2 + y^2 = 1$, 试求 $f$ 的极大值

【解】

令 $L(x,y,\lambda) = x^2 + xy + 2y^2 + \lambda(x^2 + y^2 - 1)$

$$L(x, y, \lambda)\text{极值存在} \Rightarrow \begin{cases} \dfrac{\partial L}{\partial x} = 2x + y + 2x\lambda = 0 \quad (1) \\[2mm] \dfrac{\partial L}{\partial y} = x + 4y + 2y\lambda = 0 \quad (2) \\[2mm] x^2 + y^2 - 1 = 0 \qquad\qquad (3) \end{cases}$$

By (1)(2)得 $\lambda = -\dfrac{2x + y}{2x} = -\dfrac{x + 4y}{2y} \Rightarrow 1 + \dfrac{y}{2x} = \dfrac{x}{2y} + 2$

令 $t = \dfrac{y}{x} \Rightarrow t^2 - 2t - 1 = 0 \Rightarrow t = 1 \pm \sqrt{2}$

(i) 当 $\dfrac{y}{x} = 1 + \sqrt{2}$

$\because \dfrac{y}{x} = 1 + \sqrt{2} \quad \therefore y = (1 + \sqrt{2})x \Rightarrow xy = (1 + \sqrt{2})x^2$ 且 $y^2 = (1 + \sqrt{2})xy$

$\because x^2 + y^2 - 1 = 0 \quad \therefore x^2 + (3 + 2\sqrt{2})x^2 = 1 \Rightarrow x^2 = \dfrac{2 - \sqrt{2}}{4}, y^2 = \dfrac{2 + \sqrt{2}}{4}$

$\because y^2 = (1 + \sqrt{2})xy \quad \therefore xy = \dfrac{\frac{2 + \sqrt{2}}{4}}{1 + \sqrt{2}} = \dfrac{\sqrt{2}}{4}$

$\therefore f(x, y) = x^2 + xy + 2y^2 = 1 + y^2 + xy = \dfrac{3 + \sqrt{2}}{2}$

(ii) 当 $\dfrac{y}{x} = 1 - \sqrt{2}$

$\because \dfrac{y}{x} = 1 - \sqrt{2} \quad \therefore y = (1 - \sqrt{2})x \Rightarrow xy = (1 - \sqrt{2})x^2$ 且 $y^2 = (1 - \sqrt{2})xy$

$\because x^2 + y^2 - 1 = 0 \quad \therefore x^2 + (3 - 2\sqrt{2})x^2 = 1 \Rightarrow x^2 = \dfrac{2 + \sqrt{2}}{4}, y^2 = \dfrac{2 - \sqrt{2}}{4}$

$\because y^2 = (1 - \sqrt{2})xy \quad \therefore xy = \dfrac{\frac{2 - \sqrt{2}}{4}}{1 - \sqrt{2}} = -\dfrac{\sqrt{2}}{4}$

$\therefore f(x, y) = x^2 + xy + 2y^2 = 1 + y^2 + xy = 1 + \dfrac{2 - \sqrt{2}}{4} - \dfrac{\sqrt{2}}{4} = \dfrac{3 - \sqrt{2}}{2}$

By(1)(2) $f$的极大值 $= \dfrac{3 + \sqrt{2}}{2}$

## 7.7.5　限制条件为不等式的多变量函数求极值

如果求极值的限制条件为不等式的时候（即 $C_1(x,y) \leq 0$），则将其拆解为 $C_1(x,y)0$ 与等式限制式 $C_1(x,y) = 0$ 求极值；接着使用双变量函数求极值的二阶导数检定法在 $C_1(x,y) < 0$ 找极值以及使用 Lagrang Multiplier 于等式限制式 $C_1(x,y) = 0$ 求函数的极值

考试类型：

Type 1.

給定双变数函数 $f(x,y)$，假设限制条件为不等式 $C_1(x,y) \leq 0$，求 $f(x,y)$ 于此限制式的极值

解题流程：

Step1.

将其拆解为 $C_1(x,y) < 0$ 与等式限制式 $C_1(x,y) = 0$，分别于两个限制式求极值

Step2.

在 $C_1(x,y) < 0$ 求极值与无条件限制式求极值类似，只是算出极值点 $(x_0, y_0)$ 之后得确认其是否落在 $C_1(x,y) < 0$ 范围内：

令 $\nabla f(x_0, y_0) = (0,0)$ 找 $(x_0, y_0)$，求 $\Delta(x_0, y_0) = ?$ 且 $f_{xx}(x_0, y_0) = ?$

如果 $\Delta(x_0, y_0) > 0$ 且 $f_{xx}(x_0, y_0) > 0$ 则 $f$ 于 $(x_0, y_0)$ 有相对极小值

如果 $\Delta(x_0, y_0) > 0$ 且 $f_{xx}(x_0, y_0) < 0$ 则 $f$ 于 $(x_0, y_0)$ 有相对极大值

确认是否 $C_1(x_0, y_0) < 0$

Step3.

接着使用 Lagrang Multiplier 于等式限制式 $C_1(x,y) = 0$ 找极值 $(x_0, y_0)$：

$$\text{令} L(x,y,\lambda) = f(x,y) + \lambda C_1(x,y), \ \text{若} L(x,y,\lambda) \text{极值存在则} \begin{cases} \dfrac{\partial L}{\partial x}(x,y) = 0 \\ \dfrac{\partial L}{\partial y}(x,y) = 0 \\ C_1(x,y) = 0 \end{cases}$$

$$找 (x_0, y_0) 满足 \begin{cases} \dfrac{\partial L}{\partial x}(x,y) = 0 \\ \dfrac{\partial L}{\partial y}(x,y) = 0 \\ C_1(x,y) = 0 \end{cases}，\; 找到的 (x_0, y_0) 可能不只一个, 使用 f(x_0, y_0) 比较大小, 求得$$

$f(x,y)$ 的相对极大值与相对极小值

Step4.

最后比较 Step2 与 Step3 的极值点, 假设两者于各自限制式皆为相对极大值, 则取出较大者当作 $C_1(x,y) \le 0$ 范围内的绝对极大值, 绝对极小值也是类似的作法

Type 2.

给定双变数函数 $f(x,y)$, 假设限制条件为不等式 $C_1(x,y) \le 0$、$C_2(x,y) \le 0$, 求 $f(x,y)$ 于限制条件下的极值

解题流程:

Step1.

将其拆解为在 $C_1(x,y) < 0$、$C_2(x,y) < 0$ 与等式限制式 $C_1(x,y) = 0$、$C_2(x,y) = 0$, 求极值

Step2.

在 $C_1(x,y) < 0$、$C_2(x,y) < 0$ 求极值与无条件限制式求极值类似, 只是算出极值点 $(x_0, y_0)$ 之后得确认其是否落在 $C_1(x,y) < 0, C_2(x,y) < 0$ 范围内:

令 $\nabla f(x_0, y_0) = (0,0)$ 找 $(x_0, y_0)$, 求 $\Delta(x_0, y_0) = ?$ 且 $f_{xx}(x_0, y_0) = ?$

如果 $\Delta(x_0, y_0) > 0$ 且 $f_{xx}(x_0, y_0) > 0$ 则 $f$ 于 $(x_0, y_0)$ 有相对极小值

如果 $\Delta(x_0, y_0) > 0$ 且 $f_{xx}(x_0, y_0) < 0$ 则 $f$ 于 $(x_0, y_0)$ 有相对极大值

确認是否 $C_1(x_0, y_0) < 0$、$C_2(x_0, y_0) < 0$

Step3.

接着使用 Lagrang Multiplier 于等式限制式 $C_1(x,y) = 0$、$C_2(x,y) = 0$ 找极值 $(x_0, y_0)$:

令 $L(x, y, \lambda_1, \lambda_2) = f(x,y) + \lambda C_1(x,y) + \lambda C_2(x,y)$

$$若 L(x, y, \lambda_1, \lambda_2) 极值存在 则 \begin{cases} \dfrac{\partial L}{\partial x}(x, y, \lambda_1, \lambda_2) = 0 \\ \dfrac{\partial L}{\partial y}(x, y, \lambda_1, \lambda_2) = 0 \\ C_1(x,y) = 0 \\ C_2(x,y) = 0 \end{cases}$$

$$找\,(x_0, y_0)\,满足 \begin{cases} \dfrac{\partial L}{\partial x}(x, y, \lambda_1, \lambda_2) = 0 \\[2mm] \dfrac{\partial L}{\partial y}(x, y, \lambda_1, \lambda_2) = 0 \\[2mm] \quad C_1(x, y) = 0 \\ \quad C_2(x, y) = 0 \end{cases}$$

找到的$(x_0, y_0)$可能不只一个,使用$f(x_0, y_0)$比较大小求得 $f(x, y)$的相对极大值与相对极小值

Step4.

最后比较 Step2 与 Step3 的极值点,假设两者于各自限制式皆为相对极大值,则取出较大者当作于$C_1(x, y) \leq 0$、$C_2(x, y) \leq 0$范围内的绝对极大值,绝对极小值也是类似的作法


Example 1.

假设 $f(x, y) = x^2 + y^2 - 3xy + 1$ 且 $x^2 + y^2 - 6 \leq 0$, 试求$f$的极值

【解】

(1)

As $x^2 + y^2 - 6 < 0$

令$f_x = f_y = 0$ 则 $\begin{cases} f_x = 2x - 3y = 0 \\ f_y = 2y - 3x = 0 \end{cases} \Rightarrow (x, y) = (0, 0)$为临界点且$f(0, 0) = 1$

$\because \Delta(0, 0) = \begin{vmatrix} f_{xx}(0,0) & f_{xy}(0,0) \\ f_{yx}(0,0) & f_{yy}(0,0) \end{vmatrix} = \begin{vmatrix} 2 & -3 \\ -3 & 2 \end{vmatrix} < 0 \quad \therefore (0, 0)$非极值点

(2)

As $x^2 + y^2 - 6 = 0$

令 $L(x, y, \lambda) = x^2 + y^2 - 3xy + 1 + \lambda(x^2 + y^2 - 6)$

$$L(x, y, \lambda)极值存在 \Rightarrow \begin{cases} \dfrac{\partial L}{\partial x} = 2x - 3y + 2\lambda x = 0 \\[2mm] \dfrac{\partial L}{\partial y} = 2y - 3x + 2\lambda y = 0 \\[2mm] \quad x^2 + y^2 - 6 = 0 \end{cases} \Rightarrow x^2 - y^2 = 0$$

$\because x^2 - y^2 = 0$ 且 $x^2 + y^2 - 6 = 0$

$\therefore (x, y) = (\sqrt{3}, \sqrt{3}), (-\sqrt{3}, \sqrt{3}), (\sqrt{3}, -\sqrt{3}), (-\sqrt{3}, -\sqrt{3})$

$\Rightarrow f(\sqrt{3}, \sqrt{3}) = -2, f(-\sqrt{3}, -\sqrt{3}) = -2, f(-\sqrt{3}, \sqrt{3}) = 16, f(\sqrt{3}, -\sqrt{3}) = 16$

$\therefore f$有最小值$-2$、有最大值 16

Example 2.

假设 $f(x,y) = x^2 + 3y^2 + y + 1$, $x^2 + y^2 \leq 1$, 试求 $f$ 的极值

【解】

(1)

As $x^2 + y^2 - 1 < 0$

令 $f_x = f_y = 0$ 则 $\begin{cases} f_x = 2x = 0 \\ f_y = 6y + 1 = 0 \end{cases} \Rightarrow (x,y) = \left(0, -\frac{1}{6}\right)$ 为临界点 且 $f\left(0, -\frac{1}{6}\right) = \frac{11}{12}$

$\because \Delta\left(0, -\frac{1}{6}\right) = \begin{vmatrix} f_{xx}\left(0, -\frac{1}{6}\right) & f_{xy}\left(0, -\frac{1}{6}\right) \\ f_{yx}\left(0, -\frac{1}{6}\right) & f_{yy}\left(0, -\frac{1}{6}\right) \end{vmatrix} = \begin{vmatrix} 2 & 0 \\ 0 & 6 \end{vmatrix} = 12 > 0$ 且 $f_{xx} = 2 > 0$

$\therefore f$ 于 $\left(0, -\frac{1}{6}\right)$ 有相对极小值 $\frac{11}{12}$

(2)

As $x^2 + y^2 - 1 = 0$, 令 $L(x,y,\lambda) = x^2 + 3y^2 + y + 1 + \lambda(x^2 + y^2 - 1)$

$L(x,y,\lambda)$ 极值存在 $\Rightarrow \begin{cases} \dfrac{\partial L}{\partial x} = 2x + 2\lambda x = 0 \\ \dfrac{\partial L}{\partial y} = 6y + 1 + 2\lambda y = 0 \\ x^2 + y^2 - 1 = 0 \end{cases} \Rightarrow 4xy + x = 0$

$\because 4xy + x = 0$ 且 $x^2 + y^2 - 1 = 0$ $\therefore (x,y) = \left(\pm\frac{\sqrt{15}}{4}, -\frac{1}{4}\right), (0,1), (0,-1)$

$\Rightarrow f\left(\pm\frac{\sqrt{15}}{4}, -\frac{1}{4}\right) = \frac{15}{8}$, $f(0,1) = 5$, $f(0,-1) = 3$

因此 $f$ 于 $(0,1)$ 有相对极大值 5

Example 3.

假设 $f(x,y) = x^2 - y^2 + 3$, $x^2 + \frac{y^2}{4} \leq 1$ 试求 $f$ 的极值

【解】

(1)

As $x^2 + \frac{y^2}{4} - 1 < 0$

令 $f_x = f_y = 0$ 则 $\begin{cases} f_x = 2x = 0 \\ f_y = -2y = 0 \end{cases} \Rightarrow (x,y) = (0,0)$ 为临界点，且 $f(0,0) = 3$

$\because \Delta(0,0) = \begin{vmatrix} f_{xx}(0,0) & f_{xy}(0,0) \\ f_{yx}(0,0) & f_{yy}(0,0) \end{vmatrix} = \begin{vmatrix} 2 & 0 \\ 0 & -2 \end{vmatrix} = -4 < 0 \quad \therefore (0,0)$ 非极值点

(2)

As $x^2 + \dfrac{y^2}{4} - 1 = 0$, 令 $L(x,y,\lambda) = x^2 - y^2 + 3 + \lambda(x^2 + \dfrac{y^2}{4} - 1)$

$L(x,y,\lambda)$ 极值存在 $\Rightarrow \begin{cases} \dfrac{\partial L}{\partial x} = 2x + 2\lambda x = 0 \\ \dfrac{\partial L}{\partial y} = -2y + \dfrac{\lambda y}{2} = 0 \Rightarrow xy = 0 \\ x^2 + \dfrac{y^2}{4} - 1 = 0 \end{cases}$

$\because xy = 0$ 且 $x^2 + \dfrac{y^2}{4} - 1 = 0 \quad \therefore (x,y) = (0,\pm2), (\pm1,0)$

$\Rightarrow f(0,\pm2) = -1, f(\pm1,0) = 4$

因此 $f$ 于 $(\pm1,0)$ 有相对极大值 $4$, $f$ 于 $(0,\pm2)$ 有相对极小值 $-1$

Example 4.

$\qquad$ 假设 $f(x,y) = x^2 + 2y^2 - x + 1$ 且 $x^2 + y^2 - 4 \leq 0$, 试求 $f$ 的极值

【解】

(1)

As $x^2 + y^2 - 4 < 0$

令 $f_x = f_y = 0$ 则 $\begin{cases} f_x = 2x - 1 = 0 \\ f_y = 4y = 0 \end{cases} \Rightarrow (x,y) = \left(\dfrac{1}{2}, 0\right)$ 为临界点且 $f\left(\dfrac{1}{2}, 0\right) = \dfrac{3}{4}$

$\therefore \Delta\left(\dfrac{1}{2}, 0\right) = \begin{vmatrix} f_{xx}\left(\dfrac{1}{2}, 0\right) & f_{xy}\left(\dfrac{1}{2}, 0\right) \\ f_{yx}\left(\dfrac{1}{2}, 0\right) & f_{yy}\left(\dfrac{1}{2}, 0\right) \end{vmatrix} = \begin{vmatrix} 2 & 0 \\ 0 & 4 \end{vmatrix} > 0$ 且 $f_{xx}\left(\dfrac{1}{2}, 0\right) = 2 > 0$

$\therefore$ As $x^2 + y^2 - 4 < 0$, $f$ 于 $\left(\dfrac{1}{2}, 0\right)$ 有相对极小值 $\dfrac{3}{4}$

(2)

As $x^2 + y^2 - 4 = 0$, 令 $L(x,y,\lambda) = x^2 + 2y^2 - x + 1 + \lambda(x^2 + y^2 - 4)$

$$L(x,y,\lambda) \text{极值存在} \Rightarrow \begin{cases} \dfrac{\partial L}{\partial x} = 2x - 1 + 2\lambda x = 0 \\[2mm] \dfrac{\partial L}{\partial y} = 4y + 2\lambda y = 0 \\[2mm] x^2 + y^2 - 4 = 0 \end{cases} \Rightarrow 2xy + y = 0$$

$$\because 2xy + y = 0 \quad \text{且} \quad x^2 + y^2 - 4 = 0 \quad \therefore (x,y) = \left(-\frac{1}{2}, \pm\frac{\sqrt{15}}{2}\right), (\pm 2, 0)$$

$$\Rightarrow f\left(-\frac{1}{2}, \pm\frac{\sqrt{15}}{2}\right) = \frac{37}{4}, \ f(2,0) = 3, \ f(-2,0) = 7$$

$$\therefore f \text{于} \left(\frac{1}{2}, 0\right) \text{有相对极小值} \frac{3}{4} \text{、于} \left(-\frac{1}{2}, \pm\frac{\sqrt{15}}{2}\right) \text{有相对极大值} \frac{37}{4}$$

Example 5.

假设 $f(x,y) = 4x^2 + 4x - 2y^2 + 6$ 且 $x^2 + y^2 - 3 \le 0$, 试求 $f$ 的极值

【解】

(1)

As $x^2 + y^2 - 3 < 0$, 令 $f_x = f_y = 0$ 则 $\begin{cases} f_x = 8x + 4 = 0 \\ f_y = -4y = 0 \end{cases}$

$$\Rightarrow (x,y) = \left(-\frac{1}{2}, 0\right) \text{为临界点且} f\left(-\frac{1}{2}, 0\right) = 5$$

$$\because \Delta\left(-\frac{1}{2}, 0\right) = \begin{vmatrix} f_{xx}\left(-\frac{1}{2}, 0\right) & f_{xy}\left(-\frac{1}{2}, 0\right) \\ f_{yx}\left(-\frac{1}{2}, 0\right) & f_{yy}\left(-\frac{1}{2}, 0\right) \end{vmatrix} = \begin{vmatrix} 8 & 0 \\ 0 & -4 \end{vmatrix} < 0 \quad \therefore \left(-\frac{1}{2}, 0\right) \text{非极值点}$$

(2)

As $x^2 + y^2 - 3 = 0$, 令 $L(x,y,\lambda) = 4x^2 + 4x - 2y^2 + 6 + \lambda(x^2 + y^2 - 3)$

$$L(x,y,\lambda) \text{极值存在} \Rightarrow \begin{cases} \dfrac{\partial L}{\partial x} = 8x + 4 + 2\lambda x = 0 \\[2mm] \dfrac{\partial L}{\partial y} = -4y + 2\lambda y = 0 \\[2mm] x^2 + y^2 - 3 = 0 \end{cases} \Rightarrow 3xy + y = 0$$

$$\because 3xy + y = 0 \quad \text{且} \quad x^2 + y^2 - 3 = 0 \quad \therefore (x,y) = \left(-\frac{1}{3}, \pm\frac{\sqrt{26}}{3}\right), (\pm\sqrt{3}, 0)$$

$$\Rightarrow f\left(-\frac{1}{3}, \pm\frac{\sqrt{26}}{3}\right) = -\frac{2}{3}, \quad f(\sqrt{3},0) = 18 + 4\sqrt{3}, \quad f(-\sqrt{3},0) = 18 - 4\sqrt{3}$$

因此 $f$ 有相对极小值 $-\dfrac{2}{3}$ 、有相对极大值 $18 + 4\sqrt{3}$

Example 6.

假设 $f(x,y) = x^2 + y^2 - 2x - 2y + 2$,且 $2x + y - 4 \geq 0$, $x + 2y - 4 \geq 0$, 试求 $f$ 的极小值

【解】

(1)

As $2x + y - 4 > 0$ 且 $x + 2y - 4 > 0$

令 $f_x = f_y = 0$ 则 $\begin{cases} f_x = 2x - 2 = 0 \\ f_y = 2y - 2 = 0 \end{cases} \Rightarrow (x,y) = (1,1)$ 为临界点且 $f(1,1) = 0$

$\therefore (1,1)$ 不属于 $2x + y - 4 > 0$ 且 $x + 2y - 4 > 0$

(2)

As $2x + y - 4 = 0$ 且 $x + 2y - 4 = 0$

令 $L(x,y,\lambda_1,\lambda_2) = x^2 + y^2 - 2x - 2y + 2 + \lambda_1(2x + y - 4) + \lambda_2(x + 2y - 4)$

$L(x,y,\lambda_1,\lambda_2)$ 极值存在 $\Rightarrow \begin{cases} \dfrac{\partial L}{\partial x} = 2x - 2 + 2\lambda_1 + \lambda_2 = 0 \\ \dfrac{\partial L}{\partial y} = 2y - 2 + \lambda_1 + 2\lambda_2 = 0 \\ \quad 2x + y - 4 = 0 \\ \quad x + 2y - 4 = 0 \end{cases}$

$\because 2x + y - 4 = 0$ 且 $x + 2y - 4 = 0$ $\therefore (x,y) = \left(\dfrac{4}{3}, \dfrac{4}{3}\right)$

$\Rightarrow 2\lambda_1 + \lambda_2 = \dfrac{2}{3} \quad \lambda_1 + 2\lambda_2 = \dfrac{2}{3} \quad \therefore (\lambda_1, \lambda_2) = \left(\dfrac{2}{9}, \dfrac{2}{9}\right) \quad \therefore f\left(\dfrac{4}{3}, \dfrac{4}{3}\right) = \dfrac{2}{9}$ 为相对极小值

Example 7.

假设 $f(x,y) = 2x^2 + y^2 - 2y + 3$ 且 $x^2 + y^2 \leq 4$,试求 $f$ 的极值

【解】

(1)

As $x^2 + y^2 - 4 < 0$

令 $f_x = f_y = 0$ 则 $\begin{cases} f_x = 4x = 0 \\ f_y = 2y - 2 = 0 \end{cases} \Rightarrow (x, y) = (0,1)$ 为临界点，且 $f(0,1) = 0$

$\because \Delta(0,1) = \begin{vmatrix} f_{xx}(0,1) & f_{xy}(0,1) \\ f_{yx}(0,1) & f_{yy}(0,1) \end{vmatrix} = \begin{vmatrix} 4 & 0 \\ 0 & 2 \end{vmatrix} = 8 > 0$ 且 $f_{xx}(0,1) = 4 > 0$

$\therefore f$ 于 $(0,1)$ 有相对极小值 2

(2)

As $x^2 + y^2 - 4 = 0$, 令 $L(x, y, \lambda) = 2x^2 + y^2 - 2y + 3 + \lambda(x^2 + y^2 - 4)$

$L(x, y, \lambda)$ 极值存在 $\Rightarrow \begin{cases} \dfrac{\partial L}{\partial x} = 4x + 2\lambda x = 0 \\ \dfrac{\partial L}{\partial y} = 2y - 2 + 2\lambda y = 0 \\ x^2 + y^2 - 4 = 0 \end{cases} \Rightarrow 2xy + 2x = 0$

$\because 2xy + 2x = 0$ 且 $x^2 + y^2 - 4 = 0$ $\quad \therefore (x, y) = \left(\pm\sqrt{3}, -1\right) \Rightarrow f\left(\pm\sqrt{3}, -1\right) = 12$

因此 $f$ 于 $\left(\pm\sqrt{3}, -1\right)$ 有相对极大值 12

Example 8.

$\qquad$ 假设 $f(x, y) = e^{x^2 y}$, $x^2 + y^2 \leq 1$, 试求 $f$ 的极值

【解】

(1)

As $x^2 + y^2 - 1 < 0$

令 $f_x = f_y = 0$ 则 $\begin{cases} f_x = e^{x^2 y} 2xy = 0 \\ f_y = e^{x^2 y} x^2 = 0 \end{cases} \Rightarrow (x, y) = (0, y)$ 为临界点且 $f(0, y) = 1$

$\because \Delta(x, y) = \begin{vmatrix} f_{xx} & f_{xy} \\ f_{yx} & f_{yy} \end{vmatrix} = e^{2x^2 y} \begin{vmatrix} (2xy)^2 + 2y & 2x^3 y + 2x \\ 2x^3 y + 2x & x^4 \end{vmatrix} = 12 > 0$

$\because \Delta(0, y) = 0 \qquad \therefore (0, y)$ 非极值点

(2)

As $x^2 + y^2 - 1 = 0$, 令 $L(x, y, \lambda) = e^{x^2 y} + \lambda(x^2 + y^2 - 1)$

$L(x, y, \lambda)$ 极值存在 $\Rightarrow \begin{cases} \dfrac{\partial L}{\partial x} = e^{x^2 y} 2xy + 2\lambda x = 0 \\ \dfrac{\partial L}{\partial y} = e^{x^2 y} x^2 + 2\lambda y = 0 \\ x^2 + y^2 - 1 = 0 \end{cases} \Rightarrow x(2y^2 - x^2) = 0$

$$\because x(2y^2 - x^2) \text{ 且 } x^2 + y^2 - 1 = 0 \quad \therefore (x,y) = \left(\pm\sqrt{\frac{2}{3}}, \pm\sqrt{\frac{1}{3}}\right), (0, \pm 1)$$

$$\Rightarrow f\left(\pm\sqrt{\frac{2}{3}}, \sqrt{\frac{1}{3}}\right) = e^{\frac{2}{3\sqrt{3}}}, \ f\left(\pm\sqrt{\frac{2}{3}}, -\sqrt{\frac{1}{3}}\right) = e^{\frac{-2}{3\sqrt{3}}}, \ f(0, \pm 1) = e^0 = 1$$

因此 $f$ 于 $\left(\pm\sqrt{\frac{2}{3}}, \sqrt{\frac{1}{3}}\right)$ 有相对极大值 $e^{\frac{2}{3\sqrt{3}}}$ $\therefore f$ 于 $\left(\pm\sqrt{\frac{2}{3}}, -\sqrt{\frac{1}{3}}\right)$ 有相对极小值 $e^{\frac{-2}{3\sqrt{3}}}$

Example 9.

假设 $f(x,y) = 6x^2 - 8x + 2y^2 - 4, \ x^2 + y^2 \le 1$ 试求 $f$ 的极值

【解】

(1)

As $x^2 + y^2 - 1 < 0$,  令 $f_x = f_y = 0$ 则 $\begin{cases} f_x = 12x - 8 = 0 \\ f_y = 4y = 0 \end{cases}$

$$\Rightarrow (x,y) = \left(\frac{2}{3}, 0\right) \text{ 为临界点 且 } f\left(\frac{2}{3}, 0\right) = -\frac{20}{3}$$

$$\because \Delta\left(\frac{2}{3}, 0\right) = \begin{vmatrix} f_{xx}\left(\frac{2}{3}, 0\right) & f_{xy}\left(\frac{2}{3}, 0\right) \\ f_{yx}\left(\frac{2}{3}, 0\right) & f_{yy}\left(\frac{2}{3}, 0\right) \end{vmatrix} = \begin{vmatrix} 12 & 0 \\ 0 & 4 \end{vmatrix} = 48 > 0 \text{ 且 } f_{xx}\left(\frac{2}{3}, 0\right) = 12$$

$$\therefore f \text{ 于 } \left(\frac{2}{3}, 0\right) \text{ 有相对极小值 } -\frac{20}{3}$$

(2)

As $x^2 + y^2 - 1 = 0$,  令 $L(x,y,\lambda) = 6x^2 - 8x + 2y^2 - 4 + \lambda(x^2 + y^2 - 1)$

$$L(x,y,\lambda) \text{极值存在} \Rightarrow \begin{cases} \dfrac{\partial L}{\partial x} = 12x - 8 + 2\lambda x = 0 \\ \dfrac{\partial L}{\partial y} = 4y + 2\lambda y = 0 \end{cases} \Rightarrow 8xy - 8y = 0$$

$$\because xy - y = 0 \text{ 且 } x^2 + y^2 - 1 = 0 \quad \therefore (x,y) = (\pm 1, 0)$$

$$\Rightarrow f(-1, 0) = 10, f(1, 0) = -6, \ \text{ 因此 } f \text{ 于 } (-1, 0) \text{ 有相对极大值 } 10$$

### 7.7.6　多变量函数于有界且封闭的定义域求极值

如果求极值的限制条件为封闭区域时, 需先求出明确的不等式限制条件(即$C_1(x,y) \leq 0$),
接着将其拆解为$C_1(x,y) < 0$与等式限制式$C_1(x,y) = 0$求极值

考试类型:

Type 1.

給定双变数函数$f(x,y)$

(1)求$f(x,y)$在三顶点$(0,0)$、$(a,0)$、$(0,b)$所围封闭区域中的极值

(2)求$f(x,y)$在$\{(x,y): 0 \leq x \leq a, 0 \leq y \leq b\}$所围封闭区域中的极值

(3)求$f(x,y)$在$\{(x,y): 0 \leq x \leq a, 0 \leq y \leq b, y + mx \leq c\}$所围封闭区域中的极值

解题流程:

找$C_1(x,y)$使得原先问题转换成限制条件为不等式$C_1(x,y) \leq 0$的多变量函数求极值问题

Example 1.

假设$f(x,y) = 3xy - 6x - 3y + 8$,　在三顶点为$(0,0),(3,0),(0,5)$的封闭区域中,
试求$f$的极值

【解】

三顶点围成的封闭区域 $= \{(x,y): x \geq 0, y \geq 0, y \leq -\frac{5}{3}(x-3)\}$

(1)

As $x > 0, y > 0$ 且 $y < -\frac{5}{3}(x-3)$

令$f_x = f_y = 0$ 则 $\begin{cases} f_x = 3y - 6 = 0 \\ f_y = 3x - 3 = 0 \end{cases} \Rightarrow (x,y) = (1,2)$为临界点,

且 $(1,2) \in \left\{(x,y): x > 0, y > 0, y < -\frac{5}{3}(x-3)\right\}$,　$f(1,2) = 2$

$\because \Delta(1,2) = \begin{vmatrix} f_{xx}(1,2) & f_{xy}(1,2) \\ f_{yx}(1,2) & f_{yy}(1,2) \end{vmatrix} = \begin{vmatrix} 0 & 3 \\ 3 & 0 \end{vmatrix} < 0 \Rightarrow (1,2)$非极值点

(2)

As $(x,y) \in \{(x,0): 0 \leq x \leq 3\} \cup \{(0,y): 0 \leq y \leq 5\}$

$$\cup \left\{(x,y): y = -\frac{5(x-3)}{3}, 0 \leq y \leq 5, 0 \leq x \leq 3\right\}$$

(a)

若 $(x,y) \in \{(x,0): 0 \le x \le 3\}$ 则 $f(x,y) = -6x + 8,\ f(0,0) = 8,\ f(3,0) = -10$

(b)

若 $(x,y) \in \{(0,y): 0 \le y \le 5\}$ 则 $f(x,y) = -3y + 8,\ f(0,0) = 8,\ f(0,5) = -7$

(c)

若 $(x,y) \in \left\{(x,y): y = -\dfrac{5}{3}(x-3), 0 \le y \le 5, 0 \le x \le 3\right\}$

则 $f(x,y) = 3x\left(-\dfrac{5}{3}(x-3)\right) - 6x - 3\left(-\dfrac{5}{3}(x-3)\right) + 8 = -5x^2 + 14x - 7$

$$= -5\left(x - \frac{7}{5}\right)^2 + \frac{14}{5}$$

$\therefore f(x,y)$ 在 $\left(\dfrac{7}{5}, \dfrac{8}{3}\right)$ 有相对极大值 $\dfrac{14}{5}$ $\Rightarrow f(x,y)$ 在 $(0,0)$ 有最大值 $8$，在 $(3,0)$ 有最小值 $-10$

Example 2.

    假设 $f(x,y) = x^2 - 2xy + 2y$，求在 $\{(x,y): 0 \le x \le 3, 0 \le y \le 2\}$ 的极值

【解】

(1)

As $0 < x < 3,\ 0 < y < 2$

令 $f_x = f_y = 0$ 则 $\begin{cases} f_x = 2x - 2y = 0 \\ f_y = -2x + 2 = 0 \end{cases} \Rightarrow (x,y) = (1,1)$

$\because \Delta(1,1) = \begin{vmatrix} f_{xx}(1,1) & f_{xy}(1,1) \\ f_{yx}(1,1) & f_{yy}(1,1) \end{vmatrix} = \begin{vmatrix} 2 & -2 \\ -2 & 0 \end{vmatrix} = -4 < 0 \quad \therefore (1,1)$ 为鞍点

(2)

As $(x,y) \in \{(x,y): 0 \le x \le 3, y = 0 \lor 2\} \cup \{(x,y): x = 0 \lor 3, 0 \le y \le 2\}$

(a)

若 $(x,y) \in \{(x,0): 0 \le x \le 3\}$ 则 $f(x,0) = x^2$

$\therefore f(0,0) = 0$ 有相对极小值，$f(3,0) = 9$ 有相对极大值

(b)

若 $(x,y) \in \{(x,2): 0 \le x \le 3\}$ 则 $f(x,2) = x^2 - 4x + 4$

令 $\dfrac{\partial f(x,2)}{\partial x} = 0$ 则 $2x - 4 = 0 \Rightarrow x = 2$

$\therefore f(2,2) = 0$ 有相对极小值, $f(0,2) = 4$ 有相对极大值

(c)

若 $(x,y) \in \{(0,y):,0 \le y \le 2\}$ 则 $f(0,y) = 2y$

$\therefore f(0,0) = 0$ 有相对极小值, $f(0,2) = 4$ 有相对极大值

(d)

若 $(x,y) \in \{(3,y):,0 \le y \le 2\}$ 则 $f(3,y) = 9 - 4y$

$\therefore f(3,2) = 1$ 有相对极小值, $f(3,0) = 9$ 有相对极大值

因此 $f(3,0) = 9$ 有绝对极大值


Example 3.

假设 $f(x,y) = x + y^2 - xy$, 求在 $\{(x,y): 0 \le x \le 1, 0 \le y \le 2, 2x + y \le 2\}$ 的极值

【解】

(1)

As $(x,y) \in \{(x,y): 0 < x < 1, 0 < y < 2, 2x + y < 2\}$

令 $f_x = f_y = 0$ 则 $\begin{cases} f_x = 1 - y = 0 \\ f_y = 2y - x = 0 \end{cases} \Rightarrow (x,y) = (2,1)$

$\because \Delta(2,1) = \begin{vmatrix} f_{xx}(2,1) & f_{xy}(2,1) \\ f_{yx}(2,1) & f_{yy}(2,1) \end{vmatrix} = \begin{vmatrix} 0 & -1 \\ -1 & 2 \end{vmatrix} = -1 < 0 \Rightarrow (2,1)$ 为鞍点

(2)

As $(x,y) \in \{(x,0): 0 \le x \le 1\} \cup \{(0,y): 0 \le y \le 2\} \cup \{(x,y): 2x + y = 2\}$

(a)

若 $(x,y) \in \{(x,0): 0 \le x \le 1\}$ 则 $f(x,0) = x$

$\therefore f(0,0) = 0$ 有相对极小值, $f(1,0) = 1$ 有相对极大值

(b)

若 $(x,y) \in \{(0,y): 0 \le y \le 2\}$ 则 $f(0,y) = y^2$

$\therefore f(0,0) = 0$ 有相对极小值, $f(0,2) = 4$ 有相对极大值

(c)

若 $(x,y) \in \{(x,y): 2x + y = 2, 0 \le x \le 1, 0 \le y \le 2\}$

则 $f(x, 2 - 2x) = x + (2 - 2x)^2 - x(2 - 2x)$

令 $\dfrac{\partial f(x,y)}{\partial x} = 0$ 则 $1 - 2(4 - 4x) - 2 + 4x = 0 \Rightarrow x = \dfrac{3}{4}, y = \dfrac{1}{2}$

$\therefore f\left(\dfrac{3}{4}, \dfrac{1}{2}\right) = \dfrac{5}{8}$ 有相对极小值, $f(0,2) = 4$ 有相对极大值

因此 $f(0,2) = 4$ 有绝对极大值, $f(0,0) = 0$ 有绝对极小值

Example 4.

假设 $f(x,y) = 4xy^2 - x^2y^2 - xy^3$, 在三顶点为 $(0,0),(0,6),(6,0)$ 的封闭区域中, 试求 $f$ 的极值

【解】

三顶点围成的封闭区域 $= \{(x,y): x \geq 0, y \geq 0, y \leq -(x-6)\}$

(1)

As $x > 0, y > 0$ 且 $y < -(x-6)$

令 $f_x = f_y = 0$ 则 $\begin{cases} f_x = 4y^2 - 2xy^2 - y^3 = 0 \\ f_y = 8xy - 2x^2y - 3xy^2 = 0 \end{cases}$

$$\Rightarrow \begin{cases} f_x = y^2(4 - 2x - y) = 0 \\ f_y = xy(8 - 2x - 3y) = 0 \end{cases} \Rightarrow \begin{cases} f_x = 4 - 2x - y = 0 \\ f_y = 8 - 2x - 3y = 0 \end{cases} \Rightarrow (x,y) = (1,2)$$

$$\because \Delta(x,y) = \begin{vmatrix} f_{xx} & f_{xy} \\ f_{yx} & f_{yy} \end{vmatrix} = \begin{vmatrix} -2y^2 & 8y - 4xy - 3y^2 \\ 8x - 4xy - 3y^2 & 8x - 2x^2 - 6xy \end{vmatrix}$$

且 $\Delta(1,2) > 0$ 且 $f_{xx}(1,2) < 0$

$\therefore f(1,2) = 4$ 有相对极大值

(2)

As $(x,y) \in \{(x,0): 0 \leq x \leq 6\} \cup \{(0,y): 0 \leq y \leq 6\}$

$$\cup \{(x,y): y = -(x-6), 0 \leq y \leq 6, 0 \leq x \leq 6\}$$

(a)

若 $(x,y) \in \{(x,0): 0 \leq x \leq 6\}$ 则 $f(x,y) = 0$

(b)

若 $(x,y) \in \{(0,y): 0 \leq y \leq 6\}$ 则 $f(x,y) = 0$

(c)

若 $(x,y) \in \{(x,y): y = -(x-6), 0 \leq y \leq 6, 0 \leq x \leq 6\}$

则 $f(x,y) = 4x(6-x)^2 - x^2(6-x)^2 - x(6-x)^3 = -2x^3 + 24x^2 - 72x$

令 $\dfrac{\partial f(x,y)}{\partial x} = 0$ 则 $-6x^2 + 48x - 72 = 0 \Rightarrow x = 2$ 或 $6 \Rightarrow (x,y) = (2,4)(6,0)$

$\because f(2,4) = -64$ 且 $f(6,0) = 0$

因此 $f(1,2) = 4$ 有绝对极大值, $f(2,4) = -64$ 有绝对极小值

Example 5.

假设 $f(x,y) = x^4 + y^4 - 4xy + 3$, 求在 $\{(x,y): 0 \le x \le 3, 0 \le y \le 2\}$ 的极值

【解】

(1)

As $0 < x < 3, 0 < y < 2$, 令 $f_x = f_y = 0$ 则 $\begin{cases} f_x = 4x^3 - 4y = 0 \\ f_y = 4y^3 - 4x = 0 \end{cases}$

$\Rightarrow (x,y) = (1,1), (-1,-1), (0,0)$ 且 $(-1,-1) \notin \{(x,y): 0 < x < 3, 0 < y < 2\}$

$\because \Delta(x,y) = \begin{vmatrix} f_{xx} & f_{xy} \\ f_{yx} & f_{yy} \end{vmatrix} = \begin{vmatrix} 12x^2 & -4 \\ -4 & 12y^2 \end{vmatrix} = 144x^2y^2 - 16$

且 $\Delta(1,1) = 128 > 0$, $f_{xx}(1,1) = 12 > 0$

$\therefore f(1,1) = 1$ 有相对极小值

(2)

As $(x,y) \in \{(x,y): 0 \le x \le 3, y = 0 \vee 2\} \cup \{(x,y): x = 0 \vee 3, 0 \le y \le 2\}$

(a)

若 $(x,y) \in \{(x,0): 0 \le x \le 3\}$:

则 $f(x,0) = x^4 + 3 \Rightarrow f(0,0) = 3$ 有相对极小值, $f(3,0) = 84$ 有相对极大值

(b)

若 $(x,y) \in \{(x,2): 0 \le x \le 3\}$ 则 $f(x,2) = x^4 - 8x + 19$

令 $\dfrac{\partial f(x,2)}{\partial x} = 0$ 则 $4x^3 - 8 = 0 \Rightarrow x = 2^{\frac{1}{3}}$

$\therefore f\left(2^{\frac{1}{3}}, 2\right) = 2^{\frac{4}{3}} + 19 - 8 \cdot 2^{\frac{1}{3}}$ 有相对极小值, $f(3,2) = 76$ 有相对极大值

(c)

若 $(x,y) \in \{(0,y):, 0 \le y \le 2\}$ 则 $f(0,y) = y^4 + 3$

$\Rightarrow f(0,0) = 3$ 有相对极小值, $f(0,2) = 19$ 有相对极大值

(d)

若 $(x,y) \in \{(3,y):, 0 \le y \le 2\}$ 则 $f(3,y) = y^4 - 12y + 84$

令 $\dfrac{\partial f(x,y)}{\partial y} = 0$ 则 $4y^3 - 12 = 0 \Rightarrow y = 3^{\frac{1}{3}}$

$\therefore f\left(3, 3^{\frac{1}{3}}\right) = 84 + 3^{\frac{4}{3}} - 12 \cdot 3^{\frac{1}{3}}$ 有相对极小值, $f(3,2) = 76$ 有相对极大值

因此 $f(3,0) = 84$ 有绝对极大值, $f(1,1) = 1$ 有绝对极小值

Example 6.

假设 $f(x,y) = 4xy - y^2 - x^2 - 6x + 1$，在三顶点为 $(0,0),(2,0),(2,6)$ 的封闭区域中，试求 $f$ 的极值

**【解】**

三顶点围成的封闭区域 $= \{(x,y): 0 \leq x \leq 2, 0 \leq y \leq 6, y \geq 3x\}$

$(1)(x,y) \in \{(x,y): 0 < x < 2, 0 < y < 6, y > 3x\}$

令 $f_x = f_y = 0$ 则 $\begin{cases} f_x = 4y - 2x - 6 = 0 \\ f_y = 4x - 2y = 0 \end{cases} \Rightarrow (x,y) = (1,2)$

$\because \Delta(1,2) = \begin{vmatrix} f_{xx}(1,2) & f_{xy}(1,2) \\ f_{yx}(1,2) & f_{yy}(1,2) \end{vmatrix} = \begin{vmatrix} -2 & 4 \\ 4 & -2 \end{vmatrix} = -20 < 0 \Rightarrow (1,2)$ 为鞍点

$(2)\ (x,y) \in \{(x,0): 0 \leq x \leq 2\} \cup \{(2,y): 0 \leq y \leq 6\} \cup \{(x,y): y = 3x, 0 \leq y \leq 6, 0 \leq x \leq 2\}$

(a)

若 $(x,y) \in \{(x,0): 0 \leq x \leq 2\}$ 则 $f(x,0) = -x^2 - 6x + 1$

令 $\dfrac{\partial f(x,0)}{\partial x} = 0$ 则 $-2x - 6 = 0 \quad \therefore x = -3 \notin (0,2)$

$\therefore f(0,0) = 1$ 有相对极大值，$f(2,0) = -15$ 有相对极小值

(b)

若 $(x,y) \in \{(2,y): 0 \leq y \leq 6\}$ 则 $f(2,y) = 8y - y^2 - 15$

令 $\dfrac{\partial f(2,y)}{\partial y} = 0$ 则 $8 - 2y = 0 \quad \therefore y = 4$

$\therefore f(2,0) = -15$ 有相对极小值，$f(2,4) = 1$ 有相对极大值

(c)

若 $(x,y) \in \{(x,y): y = 3x, 0 \leq y \leq 6, 0 \leq x \leq 2\}$

则 $f(x,y) = 4x \cdot 3x - 9x^2 - x^2 - 6x + 1 = 2x^2 - 6x + 1$

令 $\dfrac{\partial f(x,y)}{\partial x} = 0$ 则 $4x - 6 = 0 \Rightarrow x = \dfrac{3}{2}, y = \dfrac{9}{2}$

$\therefore f\left(\dfrac{3}{2}, \dfrac{9}{2}\right) = -\dfrac{7}{2}$ 有相对极小值，$f(0,0) = 1$ 有相对极大值

因此 $f(2,0) = -15$ 有绝对极小值，$f(0,0) = 1$ 有绝对极大值

# 第八章　　重积分

　　本章依序介绍迭代积分、多重积分定义、Fubini's Theorem 以及多重积分的考试类型；求迭代积分可以看成是求数次的单变量定积分，在多重积分的部分，首先说明多变量的实值函数为黎曼可积分的定义，其定义与单变量实值函数为黎曼可积分类似，接着说明黎曼可积分的充要条件：$f(x)$于$I$黎曼可积分 $\Leftrightarrow$ 函数 $f(x)$于$I$满足黎曼条件 $\Leftrightarrow$ $f(x)$于$I$的上黎曼积分 $= f(x)$于下的上黎曼积分，然而，这些充要条件在使用上并不太容易，事实上，任意定义于紧致集(compact set)的多变量实值连续函数皆为黎曼可积分函数，因此，在第七章所介绍的判断多变量函数是否连续则变得相当重要，简单来说，于 7.2 节当中的多变量实数值函数只要定义在紧致集为连续函数则必为黎曼可积分函数；此外，求多变量实值函数的重积分时，主要是藉由 Fubini's Theorem，将问题转为计算数次的单变量定积分，值得注意的是，使用 Fubini's Theorem 的先决条件是多变量实值函数需为黎曼可积分函数

　　常考的积分类型为二重积分与三重积分，无论是二重积分或三重积分皆是以 Fubini's Theorem 当作基础，将计算重积分的问题转成计算多次单变量定积分的问题，因此，求解过程中可能会运用求定积分的方法，例如：变数变换法、分部积分法..等

　　双重积分的考试题型分为：不需调换积分顺序即可求值、需调换积分顺序才能求值、需藉由极坐标转换才能求值、需藉由广义坐标转换才能求值；需熟悉各种考试题型，当推估无法直接求重积分需藉由调换积分顺序时，是否有把握找出调换积分顺序后的积分上下界；此外，使用极坐标转换或广义坐标转换之后，是否也能迅速并精确找出新变量的积分上下界，如果可以做到这些并且有把握算出各种定积分题型，则处理双重积分的各种题型便能游刃有余。三重积分的考试题型分为：不需调换顺序可直接转换成双重积分再转成迭代积分、转成双重积分后出现$x^2 + y^2$，再使用极坐标转换、使用球坐标转换求三重积分

　　接下来讨论重积分与向量微积分的关联性，向量微积分重点章节包含：线积分、曲面积分、Green's Theorem、Stokes' Theorem 与 The Divergence Theorem；线积分可分为纯量函数于曲线 $C$ 的线积分以及向量函数于曲线 $C$ 的线积分，两者线积分最终皆是转换为单变量定积分的类型作计算；接着介绍 Green's Theorem，当被积分函数的一阶偏导数存在且连续时，Green's Theorem 是计算二维封闭曲线积分与双重积分的重要工具，此定理将一个沿着简单封闭平面曲线 $C$ 的线积分与其所包围平面区域 R 的双重积分链接起来，可以帮助将线积分与双重积分做双向的转换，如果线积分的计算是困难时可藉由此定理转成双重

积分，反之，当计算双重积分是困难时，藉由此定理可转成线积分做计算

　　曲面积分可分为给空间曲面求封闭区域的表面积，给纯量函数求曲面 S 的质量以及给向量函数求曲面的通量，这三者最终皆是转换为双重积分的类型作计算，因此，熟悉双重积分的计算是相当重要；Stokes' Theorem 不仅是微积分基本定理在更高维度的推广，同时也是 Green's Theorem 的高维推广，Green's Theorem 是将平面区域的双重积分与平面上二维封闭曲线之线积分做转换，Stokes' Theorem 则是将三维空间封闭曲线的线积分与非封闭曲面 S 上之曲面积分作转换，当藉由 Stokes' Theorem 将三维空间封闭曲线的线积分问题转换为非封闭曲面 S 的曲面积分的问题时，又得藉由投影的方法转换为双重积分做计算，再一次说明求双重积分的能力是重要的；此外，The Divergence Theorem 可以将向量场 $\vec{F}$ 穿越封闭曲面 S 的通量（曲面积分）转换为 $\vec{F}$ 的散度在封闭体积 V 的三重积分，因此求三重积分的能力也是重要的

　　整体来说，未来会藉由上述这些定理将线积分与曲面积分的问题分别转为二重积分与三重积分的问题，因此需培养扎实求解重积分的能力

## 8.1 迭代积分

考试类型：

Type 1.

給定 $f(x,y), g_1(x), g_2(x)$, 求 $\int_a^b \int_{g_1(x)}^{g_2(x)} f(x,y)dydx =?$

解题流程：

Step1.

把 $f(x,y)$ 当中的 $x$ 当成常数，求 $\int_{g_1(x)}^{g_2(x)} f(x,y)dy =?$

Step2.

求 $\int_a^b h(x)dx =?$, 其中 $h(x) = \int_{g_1(x)}^{g_2(x)} f(x,y)dy$

Type 2.

給定 $f(x,y), g_1(y), g_2(y)$, 求 $\int_a^b \int_{g_1(y)}^{g_2(y)} f(x,y)dxdy =?$

解题流程：

Step1.

把 $f(x,y)$ 当中的 $y$ 当成常数，求 $\displaystyle\int_{g_1(y)}^{g_2(y)} f(x,y)dx =?$

Step2.

求 $\displaystyle\int_a^b h(y)dy =?$，其中 $h(y) = \displaystyle\int_{g_1(y)}^{g_2(y)} f(x,y)dx$

Example 1.

$$求 \int_0^1 \int_0^x e^{x^2} dydx =?$$

【解】

$$\int_0^1 \int_0^x e^{x^2} dydx = \int_0^1 xe^{x^2}\, dx = \frac{1}{2}e^{x^2}\Big|_0^1 = \frac{e-1}{2}$$

Example 2.

$$(1)求 \int_0^{\frac{\pi}{2}} \int_0^x \frac{\sin x}{x} dydx =? \quad (2)求 \int_0^{\frac{\pi}{4}} \int_0^x \frac{\sec^2 x}{x} dydx =?$$

【解】

(1)

$$\int_0^{\frac{\pi}{2}} \int_0^x \frac{\sin x}{x} dydx = \int_0^{\frac{\pi}{2}} \left(\frac{\sin x}{x}\right) y\Big|_{y=0}^{y=x} dx = \int_0^{\frac{\pi}{2}} \left(\frac{\sin x}{x}\right) xdx = \int_0^{\frac{\pi}{2}} \sin x\, dx = -\cos x\Big|_{x=0}^{x=\frac{\pi}{2}} = 1$$

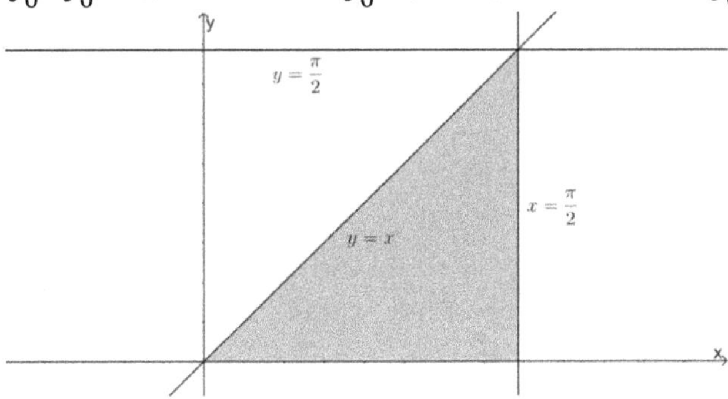


(2)

$$\int_0^{\frac{\pi}{4}} \int_0^x \frac{\sec^2 x}{x} dydx = \int_0^{\frac{\pi}{4}} \left(\frac{\sec^2 x}{x}\right) y\Big|_{y=0}^{y=x} dx = \int_0^{\frac{\pi}{4}} \left(\frac{\sec^2 x}{x}\right) xdx = \int_0^{\frac{\pi}{4}} \sec^2 x\, dx = \tan x\Big|_{x=0}^{x=\frac{\pi}{4}}$$

$$= 1$$

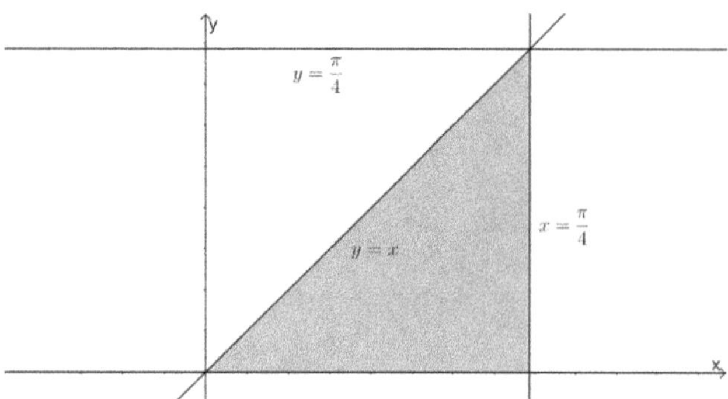


Example 3.

$$求 \int_0^1 \int_0^x \frac{1}{1+x^4}\, dy\, dx =?$$

【解】

$$\int_0^1 \int_0^x \frac{1}{1+x^4}\, dy\, dx = \int_0^1 \frac{x}{1+x^4}\, dx = \left.\frac{\tan^{-1} x^2}{2}\right|_0^1 = \frac{\pi}{8}$$

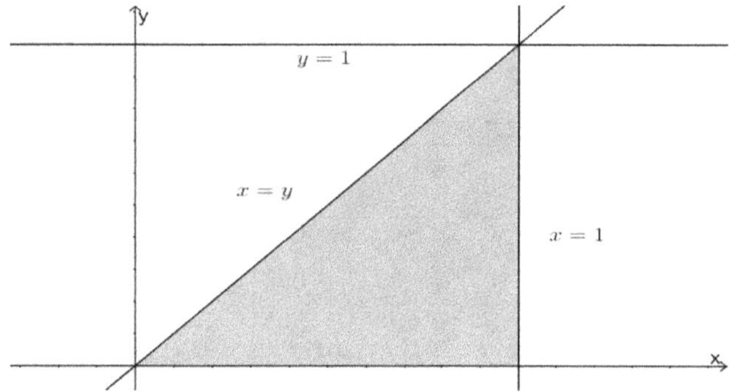


Example 4.

$$求 \int_0^2 \int_0^{2y} e^{y^2}\, dx\, dy =?$$

【解】

$$\int_0^2 \int_0^{2y} e^{y^2}\, dx\, dy = \int_0^2 2y e^{y^2}\, dy = \left. e^{y^2} \right|_0^2 = e^4 - 1$$

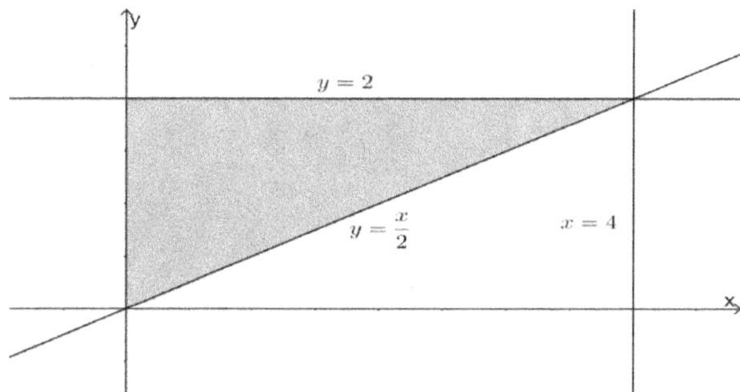

## Example 5.

$$求 \int_0^1 \int_0^x e^{\frac{y}{x}} dy\, dx = ?$$

【解】

$$\int_0^1 \int_0^x e^{\frac{y}{x}} dy\, dx = \int_0^1 x e^{\frac{y}{x}} \Big|_{y=0}^{y=x} dx = (e-1)\int_0^1 x\, dx = \frac{e-1}{2}$$

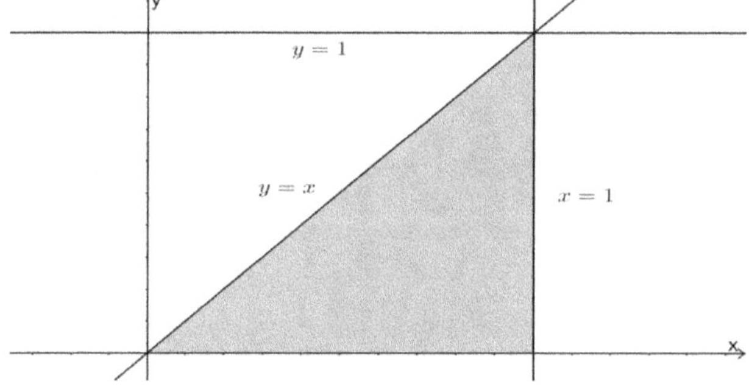

## Example 6.

$$求 \int_0^2 \int_0^x x\sqrt{x^3+1}\, dy\, dx = ?$$

【解】

$$\int_0^2 \int_0^x x\sqrt{x^3+1}\, dy\, dx = \int_0^2 x^2\sqrt{x^3+1}\, dx = \frac{2(x^3+1)^{\frac{3}{2}}}{9}\Bigg|_{x=0}^{x=2} = \frac{2(9)^{\frac{3}{2}}}{9} - \frac{2}{9} = \frac{52}{9}$$

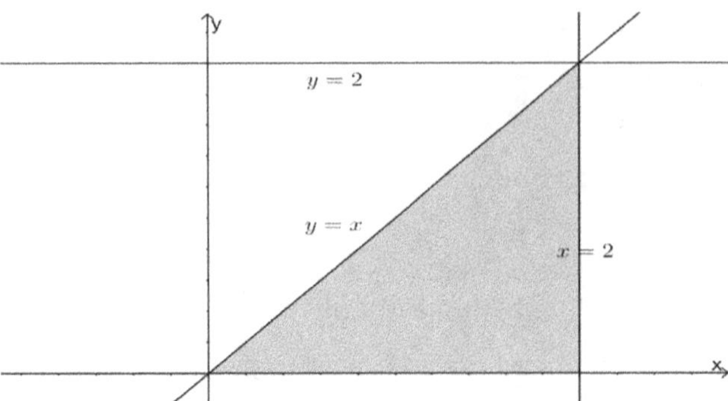

## Example 7.

$$求 \int_0^\pi \int_0^y \frac{\sin y}{y} \cos\frac{x}{y} \, dxdy =?$$

【解】

$$\int_0^\pi \int_0^y \frac{\sin y}{y} \cos\frac{x}{y} \, dxdy = \int_0^\pi \left(\sin y \sin\frac{x}{y}\right)\Big|_{x=0}^{x=y} dy = \int_0^\pi \sin y \,(\sin 1)\, dy = \sin 1\,(-\cos y)|_0^\pi$$

$$= 2\sin 1$$

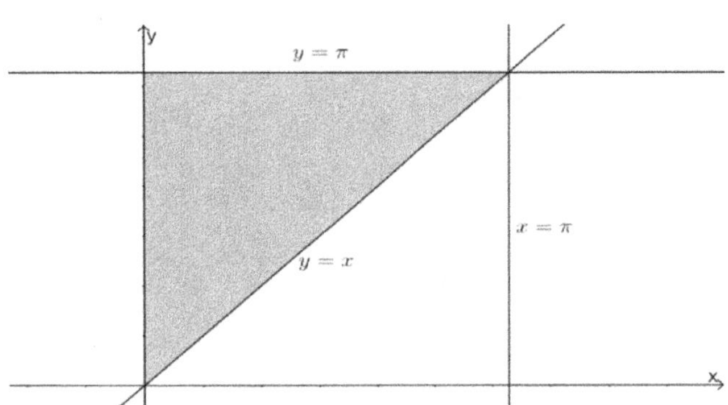

## Example 8.

$$求 \int_0^{\frac{\pi}{2}} \int_0^y \frac{\cos y}{y} \sin\frac{x}{y} \, dxdy =?$$

【解】

$$\int_0^{\frac{\pi}{2}} \int_0^y \frac{\cos y}{y} \sin\frac{x}{y} \, dxdy = \int_0^{\frac{\pi}{2}} \left(-\cos y \cos\frac{x}{y}\right)\Big|_{x=0}^{x=y} dy = \int_0^{\frac{\pi}{2}} \cos y \,(1-\cos 1)\, dy$$

$$= (1-\cos 1)\sin y|_0^{\frac{\pi}{2}} = 1-\cos 1$$

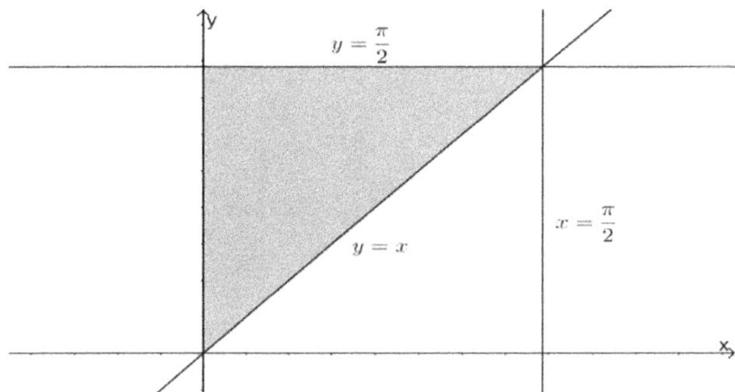

**Example 9.**

$$求 \int_0^2 \int_0^y \sin y^2 \, dxdy = ?$$

【解】

$$\int_0^2 \int_0^y \sin y^2 \, dxdy = \int_0^2 y \sin y^2 \, dy = -\left.\frac{\cos y^2}{2}\right|_0^2 = \frac{1}{2}(1 - \cos 4)$$

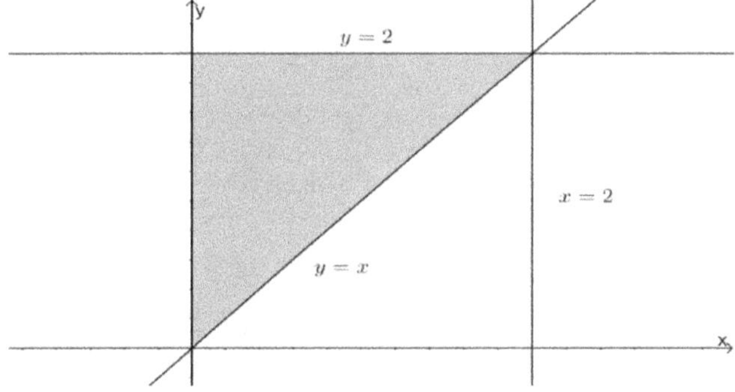

**Example 10.**

$$求 \int_0^{\sqrt{3}} \int_0^x \cos \frac{\pi x^2}{2} \, dydx = ?$$

【解】

$$\int_0^{\sqrt{3}} \int_0^x \cos \frac{\pi x^2}{2} \, dydx = \int_0^{\sqrt{3}} x \cos \frac{\pi x^2}{2} \, dx = \frac{1}{\pi} \left.\sin \frac{\pi x^2}{2}\right|_0^{\sqrt{3}} = -\frac{1}{\pi}$$

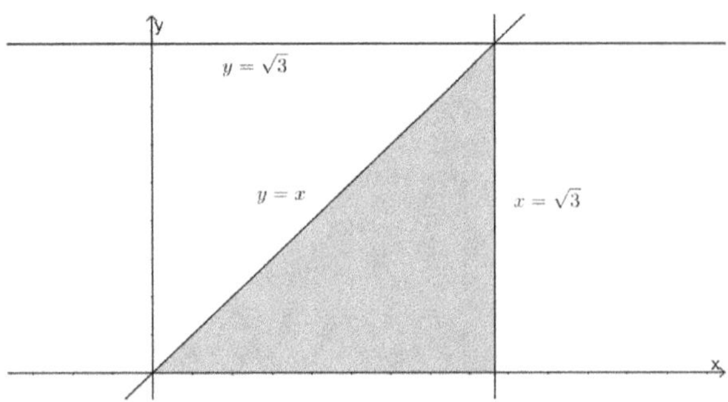

## Example 11.

$$求 \int_0^{\sqrt{3}} \int_0^x \sin\frac{\pi x^2}{2}\, dy dx = ?$$

【解】

$$\int_0^{\sqrt{3}} \int_0^x \sin\frac{\pi x^2}{2}\, dy dx = \int_0^{\sqrt{3}} x\sin\frac{\pi x^2}{2}\, dx = \frac{-1}{\pi}\cos\frac{\pi x^2}{2}\Big|_0^{\sqrt{3}} = \frac{1}{\pi}$$

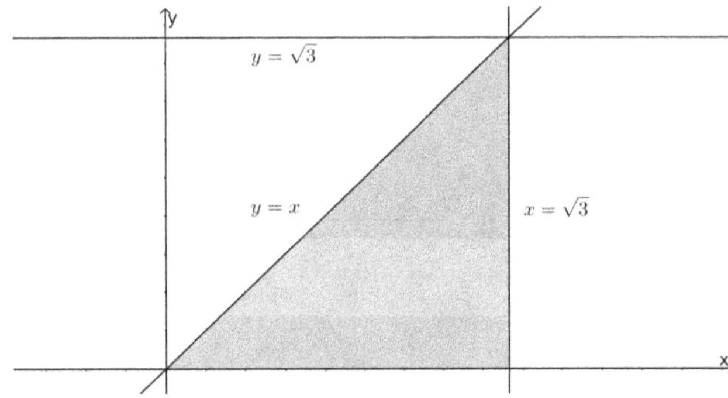

## Example 12.

$$求 \int_0^2 \int_0^y \cos y^2\, dx dy = ?$$

【解】

$$\int_0^2 \int_0^y \cos y^2\, dx dy = \int_0^2 y\cos y^2\, dy = \frac{\sin y^2}{2}\Big|_0^2 = \frac{\sin 4}{2}$$

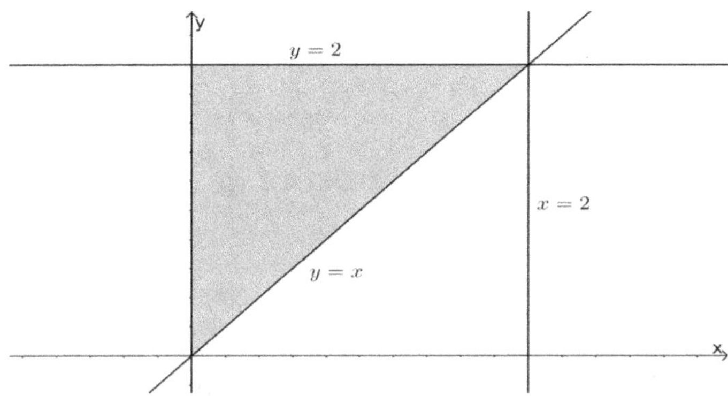

Example 13.

$$求 \int_0^1 \int_0^x e^{-x^2} dy dx = ?$$

【解】

$$\int_0^1 \int_0^x e^{-x^2} dy dx = \int_0^1 x e^{-x^2} dx = -\frac{e^{-x^2}}{2}\bigg|_0^1 = \frac{1}{2}(1 - e^{-1})$$

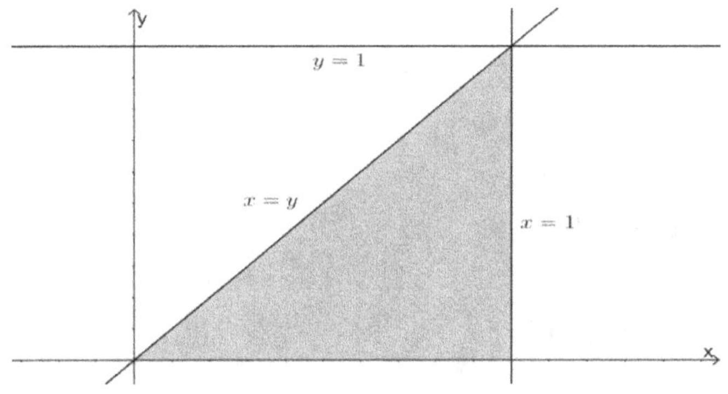

## 8.2 多重积分与 Fubini's Theorem

【定义】$n$ 维区间

假设 $I_j$ 为一维空间的 interval, $\forall j = 1, .., n$ 则

$$I = I_1 \times \cdots \times I_n = \left\{(x_1, \ldots, x_n): x_j \in I_j, \forall j = 1, .., n\right\} 称为 n 维区间,$$

其中 $I_j$ 并没有限定是否为开集合或闭集合或是否有界

**【定义】** $n$维 measure

假设 $I = I_1 \times \cdots \times I_n$ 且 $I_j$ 有界, $\forall j = 1,..,n$ 则

$$\mu(I) := \mu(I_1) \cdot \mu(I_2) \cdots \mu(I_n) \text{ 称为区间} I \text{的} n \text{维 measure}$$

其中 $\mu(I_j)$ 为一维区间 $I_j$ 的长度, $\forall j = 1,...,n$

**【定义】** 在 $\mathbb{R}^n$ 空间的紧致集分割

(i) 假设 $I = I_1 \times \cdots \times I_n$ 为 $\mathbb{R}^n$ 的 compact set, 如果 $P_k$ 是 $I_k$ 的分割, 则 cartesian product $P = P_1 \times \cdots \times P_n$ 则是 $I$ 的分割, 其中如果 $P_k$ 将 $I_k$ 分割为 $n_k$ 个一维的子区间则 $P$ 将 $I$ 分割为 $n_1 \cdots n_k$ 个 $n$ 维区间的聯集

(ii) 如果 $P \subseteq P'$ 则称 $I$ 集合的分割 $P'$ 比分割 $P$ 更精細

**【定义】** 黎曼和

假设 $f$ 定义在紧致区间(compact interval)$I \subseteq \mathbb{R}^n$, 分割 $P$ 将 $I$ 分割为 $k$ 个子区间 $I_1, \ldots, I_k$, 则

$$R(P,f) = \sum_{j=1}^{k} f(s_j)\mu(I_j) \text{ 称为黎曼和(Riemann sum)}$$

其中 $s_j \in I_j$

**【定义】** 黎曼可积分

(i) 函数 $f(x)$ 于 $I \subseteq \mathbb{R}^n$ 黎曼可积分(Riemann Integrable)

$\Leftrightarrow \exists A \in R$ 满足下列条件: $\forall \varepsilon > 0, \exists$ Partition $P_\varepsilon$ of $I$ s.t. $\forall$ Partition $P$ finer than $P_\varepsilon$, we have $|R(P,f) - A| < \varepsilon$, for all Riemann sums $R(P,f)$.

(ii) 此外, 当 $A$ 存在时, 则用 $\displaystyle\int_I f(\boldsymbol{x})d\boldsymbol{x}$ 表示黎曼积分的值, 即 $\displaystyle\int_I f(\boldsymbol{x})d\boldsymbol{x} = A$

**【定义】** 上黎曼和

假设 $P$ 为集合 $I \subseteq \mathbb{R}^n$ 的分割, 函数 $f(\boldsymbol{x})$ 的上黎曼和定义为 $\displaystyle U(P,f) = \sum_{i=1}^{n} \mu(I_k)\sup_{\boldsymbol{x} \in I_k} f(\boldsymbol{x})$

**【定义】** 下黎曼和

假设 $P$ 为集合 $I \subseteq \mathbb{R}^n$ 的分割, 函数 $f(\boldsymbol{x})$ 的下黎曼和定义为 $\displaystyle L(P,f) = \sum_{i=1}^{n} \mu(I_k)\inf_{\boldsymbol{x} \in I_k} f(\boldsymbol{x})$

为了说明黎曼可积分的充要条件，先介绍黎曼条件、上黎曼积分以及下黎曼积分

【**定义**】黎曼条件

函数 $f(\boldsymbol{x})$ 于 $I$ 满足黎曼条件

$\Leftrightarrow \forall \varepsilon > 0, \exists$ partition $P_\varepsilon$ of $I$ such that if $P$ is finer than $\mathrm{P}_\varepsilon$ then $0 \leq U(P, f) - L(P, f) < \varepsilon$

【**定义**】上黎曼积分

$$(U) \int_I f(\boldsymbol{x}) d\boldsymbol{x} = \inf\{U(P, f): \forall \text{ partition } P \in P(I)\}$$

【**定义**】下黎曼积分

$$(L) \int_I f(\boldsymbol{x}) d\boldsymbol{x} = \sup\{L(P, f): \forall \text{ partition } P \in P(I)\}$$

函数于 $R^n$ 空间黎曼可积分之充要条件的证明与 $R^1$ 空间类似，读者请自行练习

【**定理**】黎曼可积分的充要条件

假设 $f: I \subseteq R^n \to R$ 则

$$f(\boldsymbol{x}) \text{ 于 } I \text{ 黎曼可积分} \Leftrightarrow \text{函数 } f(\boldsymbol{x}) \text{ 于 } I \text{ 满足黎曼条件} \Leftrightarrow (U) \int_I f(\boldsymbol{x}) d\boldsymbol{x} = (L) \int_I f(\boldsymbol{x}) d\boldsymbol{x}$$

目前为止，主要说明函数 $f: R^n \to R$ 满足什麼条件时为黎曼可积分函数，接下来把目光聚焦在如何计算二维空间的重积分

【**定理**】Fubini 定理

假设 $f$ 为定义在 $K = [a, b] \times [c, d]$ 上的黎曼可积分函数，$\int_a^b f(x, y) dx$ 存在，$\forall y \in [c, d]$ 且 $\int_c^d f(x, y) dy$ 存在，$\forall x \in [a, b]$ 则

(i) $\int_a^b \int_c^d f(x, y) dy dx$ 且 $\int_c^d \int_a^b f(x, y) dx dy$ 两者皆存在

(ii) $\iint_K f(x, y) dx dy = \int_a^b \int_c^d f(x, y) dy dx = \int_c^d \int_a^b f(x, y) dx dy$

<u>Proof:</u>

Let $\varphi(x) = \int_c^d f(x, y) dy$,

Claim: $\varphi(x)$ is integrable on $[a,b]$ and $\iint_K f(x,y)dxdy = \int_a^b \int_c^d f(x,y)dydx$

Let $P_x, P_y$ be the two partition of $[a,b]$ and $[c,d]$ where $P_x = \{a = x_0 < x_1 < \cdots x_n = b\}$, $P_y = \{c = y_0 < y_1 < \cdots y_n = d\}$. Then $P = P_x \times P_y$ is a partition of K.

Let $m_{ij} = \inf_{K_{ij}} f(x,y)$ and $M_{ij} = \sup_{K_{ij}} f(x,y)$.

Let $m_j(x) = \inf_{y_{j-1}<y<y_j} f(x,y)$ and $M_j(x) = \sup_{y_{j-1}<y<y_j} f(x,y), \quad \forall x \in [x_{i-1}, x_i]$.

Then $m_{ij} \leq m_j(x) \leq M_j(x) \leq M_{ij}$.

$$\therefore \sum_j^n m_{ij}\Delta y_j \leq \sum_j^n m_j(x)\Delta y_j \leq \sum_j^n M_j(x)\Delta y_j \leq \sum_j^n M_{ij}\Delta y_j$$

$$\because \sum_j^n m_j(x)\Delta y_j \leq \varphi(x) = \int_c^d f(x,y)dy \leq \sum_j^n M_j(x)\Delta y_j$$

$$\therefore \sum_j^n m_{ij}(x)\Delta y_j \leq \varphi(x) \leq \sum_j^n M_{ij}(x)\Delta y_j, \quad \forall x \in [a,b]$$

$$\therefore \sum_j^n m_{ij}(x)\Delta y_j \leq \inf_{x\in[x_{i-1},x_i]} \varphi(x) \leq \sup_{x\in[x_{i-1},x_i]} \varphi(x) \leq \sum_j^n M_{ij}(x)\Delta y_j, \forall x \in [x_{i-1},x_i],$$

$i = 1,2.., n$

$$\therefore \sum_i^n \left(\sum_j^n m_{ij}(x)\Delta y_j\right)\Delta x_i \leq \sum_i^n \left(\inf_{x\in[x_{i-1},x_i]} \varphi(x)\right)\Delta x_i \leq \sum_i^n \left(\sup_{x\in[x_{i-1},x_i]} \varphi(x)\right)\Delta x_i$$

$$\leq \sum_i^n \left(\sum_j^n M_{ij}(x)\Delta y_j\right)\Delta x_i$$

$$\therefore L(P,f) \leq L(P_x, f) \leq U(P_x, f) \leq U(P,f)$$

$$\because L(P_x, f) \leq (L)\int_a^b \varphi(x)dx \leq (U)\int_a^b \varphi(x)dx \leq U(P_x, f)$$

$$\therefore L(P,f) \leq (L)\int_a^b \varphi(x)dx \leq (U)\int_a^b \varphi(x)dx \leq U(P,f), \quad \forall \text{ partition } P$$

$$\therefore (L) \iint_K f(x,y)dxdy \leq (L) \int_a^b \varphi(x)dx \leq (U) \int_a^b \varphi(x)dx \leq (U) \iint_K f(x,y)dxdy$$

$\because f$ 为在 $K = [a,b] \times [c,d]$ 上的黎曼可积分函数

$$\therefore (L) \iint_K f(x,y)dxdy = (U) \iint_K f(x,y)dxdy = \iint_K f(x,y)dxdy$$

$$\therefore (L) \int_a^b \varphi(x)dx = (U) \int_a^b \varphi(x)dx = \iint_K f(x,y)dxdy \Rightarrow \int_a^b \varphi(x)dx = \iint_K f(x,y)dxdy$$

By the same way, $\displaystyle \iint_K f(x,y)dxdy = \int_c^d \int_a^b f(x,y)dxdy$

对于 Fubini 定理而言，$f$ 在 $K = [a,b] \times [c,d]$ 上为黎曼可积分函数是重要不可缺少的条件，接下来展示一个双变数函数 $f(x,y)$ 满足条件:

$$\int_a^b f(x,y)dx \text{ 存在, } \forall y \in [c,d] \text{ 且 } \int_c^d f(x,y)dy \text{ 存在, } \forall x \in [a,b],$$

然而

$$\iint_K f(x,y)dxdy \neq \int_a^b \int_c^d f(x,y)dydx \neq \int_c^d \int_a^b f(x,y)dxdy$$

<u>范例说明</u>:

$$f(x,y) = \begin{cases} -\dfrac{1}{x^2}, & 0 < y \leq x \leq 1 \\[2mm] \dfrac{1}{y^2}, & 0 < x \leq y \leq 1 \end{cases} \quad \text{则 } f \text{ 在 } K = [0,1] \times [0,1] \text{ 上为黎曼不可积分函数}$$

$$\because \int_0^1 f(x,y)dx = \int_0^y \frac{1}{y^2}dx + \int_y^1 -\frac{1}{x^2}dx = 1 \text{ 且}$$

$$\int_0^1 f(x,y)dx = \int_0^x -\frac{1}{x^2}dy + \int_x^1 \frac{1}{y^2}dy = -1$$

如果 $f$ 为在 $K = [0,1] \times [0,1]$ 上的黎曼可积分函数 则

$$\iint_K f(x,y)dxdy = \int_a^b \int_c^d f(x,y)dydx = \int_c^d \int_a^b f(x,y)dxdy, \text{ a contradiction}$$

$\therefore f$ 为定义在 $K = [0,1] \times [0,1]$ 上的黎曼不可积分函数

目前为止，Fubini's Theorem 可以帮助将计算双重积分的问题转为迭代积分的问题，其中 $f$ 为在 $K = [a,b] \times [c,d]$ 上的黎曼可积分函数为重要不可缺少的条件，紧接着证明如果

$f(x,y)$于$K = [a,b] \times [c,d]$ 为连续函数则$f$在$K = [a,b] \times [c,d]$上为黎曼可积分函数且

$$\iint_K f(x,y)dA = \int_a^b \int_c^d f(x,y)dydx = \int_c^d \int_a^b f(x,y)dxdy$$

## 【定理】

假设双变数函数 $f(x,y)$于$R = [a,b] \times [c,d]$ 为连续函数

则 $\iint_R f(x,y)dA = \int_a^b \int_c^d f(x,y)dydx = \int_c^d \int_a^b f(x,y)dxdy$

Proof:

Let P be the partition of $[a,b] \times [c,d]$

Let $R_{ij} = [x_{i-1},x_i] \times [y_{j-1},y_j]$, $\forall 0 \leq i \leq m, 0 \leq j \leq n$

Then $[a,b] \times [c,d] = \bigcup_{i=0}^{m}\bigcup_{j=0}^{n}[x_{i-1},x_i] \times [y_{j-1},y_j] = \bigcup_{i=0}^{m}\bigcup_{j=0}^{n}R_{ij}$

Let $m_{ij} = \inf_{R_{ij}} f(x,y)$ and $M_{ij} = \sup_{R_{ij}} f(x,y)$

$\because m_{ij} \leq f(x,y) \leq M_{ij}, \forall (x,y) \in R_{ij}$, $\forall 0 \leq i \leq m, 0 \leq j \leq n$

$\therefore m_{ij}\Delta y_j \leq \int_{y_{j-1}}^{y_j} f(x,y)dy \leq M_{ij}\Delta y_j$, $\forall x \in [x_{i-1},x_i], \forall 0 \leq i \leq m, 0 \leq j \leq n$

$\therefore \sum_{j=0}^{n} m_{ij}\Delta y_j \leq \int_c^d f(x,y)dy \leq \sum_{j=0}^{n} M_{ij}\Delta y_j$, $\forall x \in [x_{i-1},x_i], \forall 0 \leq i \leq m, 0 \leq j \leq n$

$\therefore \sum_{j=0}^{n} m_{ij}\Delta x_i \Delta y_j \leq \int_{x_{i-1}}^{x_i}\int_c^d f(x,y)dydx \leq \sum_{j=0}^{n} M_{ij}\Delta x_i \Delta y_j$

$\therefore \sum_{i=0}^{m}\sum_{j=0}^{n} m_{ij}\Delta x_i \Delta y_j \leq \int_a^b\int_c^d f(x,y)dydx \leq \sum_{i=0}^{m}\sum_{j=0}^{n} M_{ij}\Delta x_i \Delta y_j$

$\Rightarrow L(f,P) \leq \int_a^b\int_c^d f(x,y)dydx \leq U(f,P)$, $\forall$ partition $P$

$\because L(f,P) \leq \iint_R f(x,y)dA \leq U(f,P)$, $\forall$ partition $P$

$$\therefore \iint_R f(x,y)dA = \int_a^b \int_c^d f(x,y)dydx = \int_c^d \int_a^b f(x,y)dxdy$$

**【性质】** 重积分的性质

假设 $f,g: I \subseteq R^n \to R$ 且两者为黎曼可积分

(i) $\displaystyle \int_I \alpha f(\boldsymbol{x}) \pm \beta f(\boldsymbol{x})dx = \alpha \int_I f(\boldsymbol{x})\,dx \pm \beta \int_I f(\boldsymbol{x})\,dx, \ \forall\, \alpha, \beta \in R$

(ii) $\displaystyle \int_I f(\boldsymbol{x})\,dx \geq 0 \text{ if } f(\boldsymbol{x}) \geq 0, \forall \boldsymbol{x} \in R^n$

(iii) $\displaystyle \int_I f(\boldsymbol{x})\,dx \geq \int_I g(\boldsymbol{x})\,dx \text{ if } f(\boldsymbol{x}) \geq g(\boldsymbol{x}), \ \forall \boldsymbol{x} \in R^n$

## 8.3 双重积分的考试类型

双重积分的考试类型包含把双重积分转成迭代积分再计算迭代积分的值、无法直接转定积分，需藉由调换积分顺序、使用极坐标转换求双重积分的值、使用广义坐标转换求双重积分的值

### 8.3.1　把双重积分转成迭代积分再计算迭代积分的值

考试类型：

Type 1.

藉由 Fubinis Theorem 把双重积分 $\displaystyle\iint_R f(x,y)dA$ 转成迭代积分之后可以直接求值，其中

$R = \{(x,y): a \leq x \leq b, c \leq y \leq d\}$

解题流程：

Step1.

藉由 Fubinis Theorem, $\displaystyle\iint_R f(x,y)dA = \int_a^b \int_c^d f(x,y)dydx = \int_c^d \int_a^b f(x,y)dxdy$

Step2.

求 $\displaystyle\int_a^b \int_c^d f(x,y)dydx = ?$ 或 $\displaystyle\int_c^d \int_a^b f(x,y)dxdy = ?$

Type 2.

藉由 Fubinis Theorem 把双重积分转成迭代积分之后，再搭配分部积分求积分值，其中
$R = \{(x, y): a \le x \le b, g_1(x) \le y \le g_2(x)\}$

解题流程：

Step1.

藉由 Fubinis Theorem, $\displaystyle\iint_R f(x,y)dA = \int_a^b \int_{g_1(x)}^{g_2(x)} f(x,y)dydx$

Step2.

假设 $\displaystyle\int_a^b \int_{g_1(x)}^{g_2(x)} f(x,y)dydx = \int_a^b u(x)\,v'(x)dx$, 藉由 integration by parts,

$$\int_a^b u(x)v'(x)dx = u(x)v(x)\big|_{x=a}^{x=b} - \int_a^b u'(x)v(x)dx.$$

Step3.

$$求\ u(x)v(x)\big|_{x=a}^{x=b} - \int_a^b u'(x)v(x)dx = ?$$

<u>范例说明:</u>

(I) 求 $\displaystyle\int_0^2 \int_0^{x^2} e^{\frac{y}{x}}\, dydx = ?$

$$\int_0^2 \int_0^{x^2} e^{\frac{y}{x}}\, dydx = \int_0^2 xe^{\frac{y}{x}}\Big|_{y=0}^{y=x^2} dx = \int_0^2 x(e^x - 1)dx = \int_0^2 xe^x dx - \int_0^2 xdx$$

令 $u = x,\ dv = e^x dx$ 则 $du = dx,\ v = e^x$, 藉由 integration by parts,

$$\int_0^2 xe^x dx = e^2 + 1 \ 且 \int_0^2 xdx = 2$$

(II) 求 $\displaystyle\int_0^1 \int_x^1 \frac{1}{1 + y^2}\, dydx = ?$

$$\int_0^1 \int_x^1 \frac{1}{1 + y^2}\, dydx = \int_0^1 \tan^{-1} y\big|_x^1\, dx = \int_0^1 \frac{\pi}{4} - \tan^{-1} x\ dx$$

令 $u = \tan^{-1} x,\ dv = dx$ 则 $du = \dfrac{dx}{1 + x^2},\ v = x$, 藉由 integration by parts,

$$\int_0^1 \tan^{-1} x\ dx = \frac{\pi}{4} - \frac{\ln 2}{2}$$

Example 1.

$$\text{求} \iint_R 4 - x - y\, dA = ?, \quad R = \{(x, y) : 0 \le x \le 2, 0 \le y \le 2\}$$

【解】

$\because f(x, y) = 4 - x - y$ 在 $R$ 区域为连续函数，藉由 Fubinis Theorem

$$\iint_R 4 - x - y\, dA = \int_0^2 \int_0^2 4 - x - y\, dy\, dx = \int_0^2 4y - xy - \left.\frac{y^2}{2}\right|_{y=0}^{y=2} dx = \int_0^2 8 - 2x - 2\, dx$$

$$= (6x - x^2)\big|_{x=0}^{x=2} = 12 - 4 = 8$$

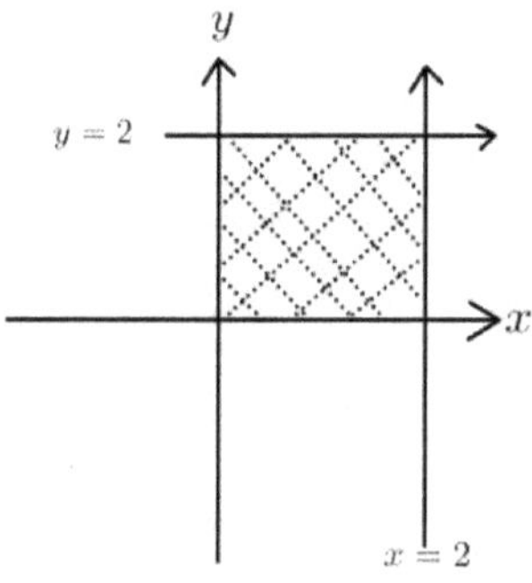

Example 2.

$$\text{求} \iint_R (1 + 8xy)\, dA = ?, \quad R = \{(x, y) : 1 \le x \le 2, 0 \le y \le 2\}$$

【解】

$\because f(x, y) = 1 + 8xy$ 在 $R$ 区域为连续函数，藉由 Fubinis Theorem

$$\iint_R (1 + 8xy)\, dA = \int_0^2 \int_1^2 (1 + 8xy)\, dx\, dy = \int_0^2 (x + 4x^2 y)\big|_{x=1}^{x=2}\, dy = \int_0^2 1 + 12y\, dy$$

$$= (y + 6y^2)\big|_{y=0}^{y=2} = 2 + 6 \cdot 4 = 26$$

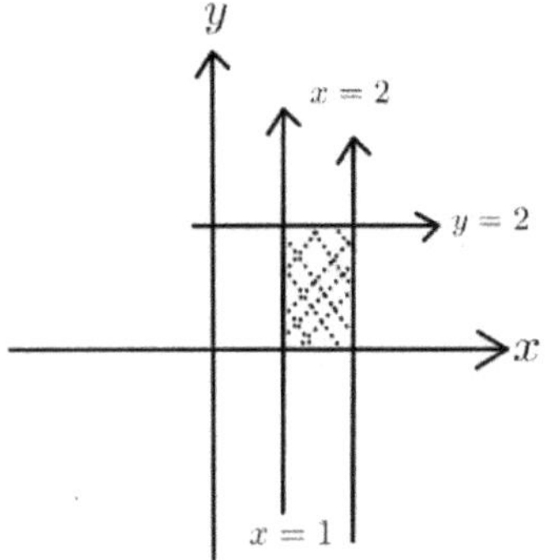

Example 3.

$$\text{求} \iint_R \frac{1}{(x+y)^2}\,dxdy =?, \quad R = \{(x,y): 1 \le x \le 3, 0 \le y \le 2\}$$

【解】

$$\because f(x,y) = \frac{1}{(x+y)^2} \text{ 在 } R \text{ 区域为连续函数, 藉由 Fubinis Theorem}$$

$$\text{则} \iint_R \frac{1}{(x+y)^2}\,dxdy = \int_1^3 \int_0^2 \frac{1}{(x+y)^2}\,dydx = \int_1^3 \frac{-1}{(x+y)}\Big|_{y=0}^{y=2}\,dx = -\int_1^3 \frac{1}{x+2} - \frac{1}{x}\,dx$$

$$= -(\ln(x+2) - \ln x)\big|_{x=1}^{x=3} = -(\ln 5 - \ln 3 - \ln 3) = 2\ln 3 - \ln 5$$

Example 4.

$$\text{求} \iint_R xy\,dA =?, \quad R = \{(x,y): 0 \le x \le 2, 2x - 4 \le y \le 0\}$$

【解】

$$\because f(x,y) = xy \text{ 在 } R \text{ 区域为连续函数, 藉由 Fubinis Theorem}$$

$$\iint_R xy\,dA = \int_0^2 \int_{2x-4}^0 xy\,dydx = \int_0^2 \frac{xy^2}{2}\Big|_{y=2x-4}^{y=0}\,dx = -\int_0^2 \frac{x(2x-4)^2}{2}\,dx$$

$$= -2\int_0^2 x(x-2)^2\,dx = -2\int_0^2 x^3 - 4x^2 + 4x\,dx = -2\left(\frac{x^4}{4} - \frac{4x^3}{3} + 2x^2\right)\Big|_{x=0}^{x=2}$$

$$= -2\left(4 - \frac{32}{3} + 8\right) = \frac{-8}{3}$$

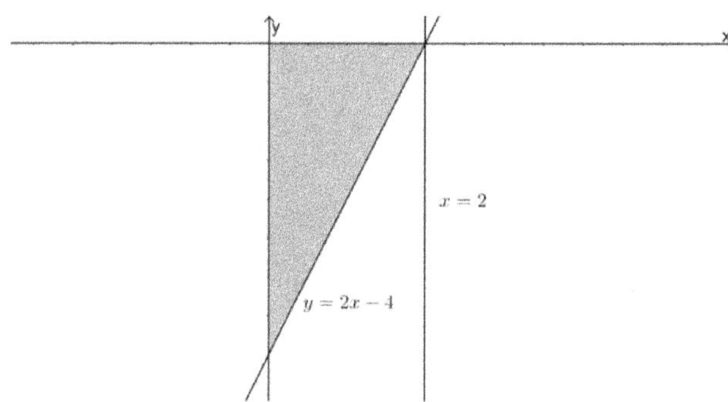

## Example 5.

$$\text{求}\iint_R e^{\frac{y}{x}}dA=?,\quad R=\{(x,y):0\le x\le 2,0\le y\le x^2\}$$

【解】

$\because f(x,y)=e^{\frac{y}{x}}$ 在 $R$ 区域为连续函数，藉由 Fubinis Theorem

$$\iint_R e^{\frac{y}{x}}dA=?\int_0^2\int_0^{x^2}e^{\frac{y}{x}}\,dydx=\int_0^2 xe^{\frac{y}{x}}\Big|_{y=0}^{y=x^2}\,dx=\int_0^2 x(e^x-1)dx=\int_0^2 xe^x dx-\int_0^2 xdx$$

令 $u=x,\ dv=e^x dx$ 则 $du=dx,\ v=e^x$，藉由 Integration By Parts

则 $\displaystyle\int_0^2 xe^x dx=xe^x\big|_{x=0}^{x=2}-\int_0^2 e^x dx=2e^2-(e^2-1)=e^2+1$

$\because\displaystyle\int_0^2 xdx=2\quad\therefore\int_0^2\int_0^{x^2}e^{\frac{y}{x}}\,dydx=\int_0^2 xe^x dx-\int_0^2 xdx=e^2+1-2=e^2-1$

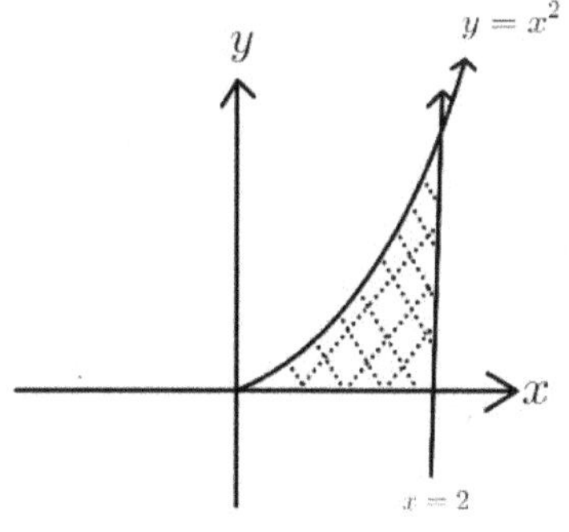

## Example 6.

$$\text{求}\iint_R e^{y^2}dA=?,\quad R=\{(x,y):x-y\le 0,y\le 3,x+3y\ge 0\}$$

【解】

$\because f(x,y) = e^{y^2}$ 在 $R$ 区域为连续函数，藉由 Fubinis Theorem

$$\iint_R e^{y^2} dA = \int_0^3 \int_{-3y}^{y} e^{y^2}\, dxdy = \int_0^3 4ye^{y^2} dy = 2e^{y^2}\Big|_{y=0}^{y=3} = 2(e^9 - 1)$$

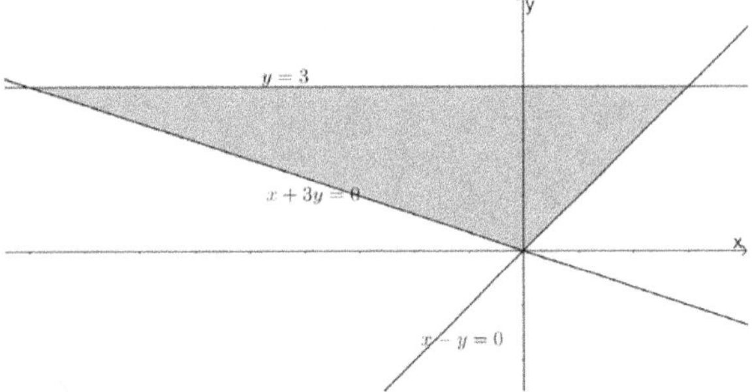

## Example 7.

$$求 \iint_R ye^{xy} dA =?, \quad R = \{(x,y): 0 \le x \le 1, 0 \le y \le 3\}$$

【解】

$\because f(x,y) = ye^{xy}$ 在 $R$ 区域为连续函数，藉由 Fubinis Theorem

$$则 \iint_R ye^{xy} dA = \int_0^3 \int_0^1 ye^{xy}\, dxdy = \int_0^3 e^{xy}\big|_{x=0}^{x=1} dy = \int_0^3 e^y - 1\, dy = (e^3 - 1) - 3$$

$$= e^3 - 4$$

## Example 8.

$$求 \iint_R \frac{1}{1+y^2} dA =?, \quad R = \{(x,y): 0 \le x \le 1, x \le y \le 1\}$$

【解】

$\because f(x,y) = \dfrac{1}{1+y^2}$ 在 $R$ 区域为连续函数，藉由 Fubinis Theorem

$$\iint_R \frac{1}{1+y^2} dA = \int_0^1 \int_x^1 \frac{1}{1+y^2}\, dydx = \int_0^1 \tan^{-1} y\big|_x^1\, dx = \int_0^1 \frac{\pi}{4} - \tan^{-1} x\, dx$$

$$令 u = \tan^{-1} x, \quad dv = dx \text{ 则 } du = \frac{dx}{1+x^2}, \quad v = x, \text{ 藉由 integration by parts}$$

$$\int_0^1 \tan^{-1} x\, dx = x\tan^{-1} x\big|_0^1 - \int_0^1 \frac{xdx}{1+x^2} = \frac{\pi}{4} - \frac{1}{2}\ln(1+x^2)\Big|_0^1 = \frac{\pi}{4} - \frac{\ln 2}{2}$$

$$\therefore \int_0^1 \frac{\pi}{4} - \tan^{-1} x\, dx = \frac{\pi}{4} - \left(\frac{\pi}{4} - \frac{\ln 2}{2}\right) = \frac{\ln 2}{2}$$

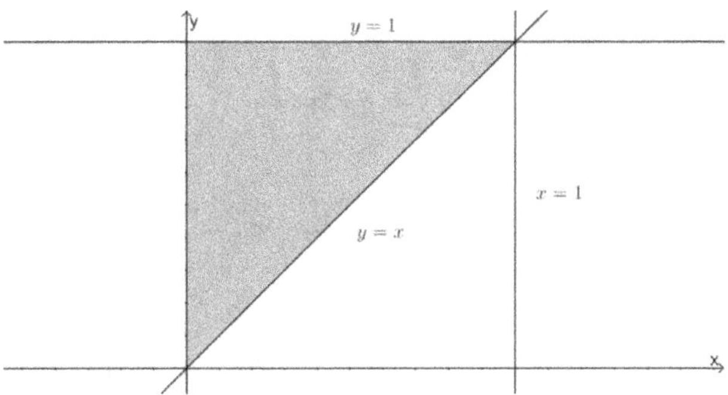

Example 9.

$$\text{求} \iint_R 2x - y^2 \, dA =?, \quad R = \{(x,y): x - y + 1 \leq 0, y \leq 2, x + y - 1 \geq 0\}$$

【解】

$\because f(x,y) = 2x - y^2$ 在 $R$ 区域为连续函数，藉由 Fubinis Theorem

则 $\iint_R 2x - y^2 \, dA = \int_1^2 \int_{1-y}^{y-1} 2x - y^2 \, dx \, dy = \int_1^2 x^2 - y^2 x \big|_{x=1-y}^{x=y-1} \, dy$

$= \int_1^2 (y-1)^2 - y^2(y-1) - ((1-y)^2 - y^2(1-y)) \, dy = \int_1^2 -2y^3 + 2y^2 \, dy$

$= \dfrac{-y^4}{2} \bigg|_{y=1}^{y=2} + \dfrac{2y^3}{3} \bigg|_{y=1}^{y=2} = \dfrac{-17}{6}$

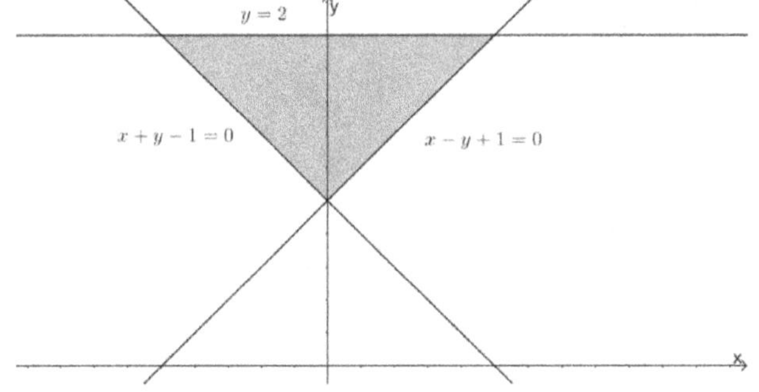

## 8.3.2　无法直接求迭代积分, 需先调换迭代积分顺序

考试类型:

Type 1.

假设 $R = \{(x,y): a \leq x \leq b, f_1(x) \leq y \leq f_2(x)\}$, 无法直接计算 $\displaystyle\int_a^b \int_{f_1(x)}^{f_2(x)} f(x,y)\,dy\,dx =?$

解题流程:

Step1.

找 $c, d, f_3, f_4$ 使得 $R = \{(x,y): f_3(y) \leq x \leq f_4(y), c \leq y \leq d\}$

Step2.

因此 $\displaystyle\int_a^b \int_{f_1(x)}^{f_2(x)} f(x,y)\,dy\,dx = \int_c^d \int_{f_3(y)}^{f_4(y)} f(x,y)\,dx\,dy$

Step3.

计算此积分 $\displaystyle\int_c^d \int_{f_3(y)}^{f_4(y)} f(x,y)\,dx\,dy =?$

Step4.

如果无法计算 $\displaystyle\int_c^d \int_{f_3(y)}^{f_4(y)} f(x,y)\,dx\,dy =?$, 通常会搭配使用变数变换、分部积分求积分值


Type 2.

假设 $R = \{(x,y): f_1(y) \leq x \leq f_2(y), c \leq y \leq d\}$, 无法直接计算 $\displaystyle\int_c^d \int_{f_1(y)}^{f_2(y)} f(x,y)\,dx\,dy$ 时

解题流程:

Step1.

找 $a, b, f_3, f_4$ 使得 $R = \{(x,y): a \leq x \leq b, f_3(x) \leq y \leq f_4(x)\}$

因此 $\displaystyle\int_c^d \int_{f_1(y)}^{f_2(y)} f(x,y)\,dx\,dy = \int_a^b \int_{f_3(x)}^{f_4(x)} f(x,y)\,dy\,dx$

Step2.

计算此积分 $\displaystyle\int_a^b \int_{f_3(x)}^{f_4(x)} f(x,y)\,dy\,dx =?$

Step3.

如果无法计算 $\displaystyle\int_a^b \int_{f_3(x)}^{f_4(x)} f(x,y)\,dy\,dx =?$, 通常会搭配使用变数变换 、分部积分求积分值


Example 1.

求 $\displaystyle\int_0^1 \int_y^1 e^{x^2}\,dx\,dy =?$

【解】

$$\because \{(x,y): y \le x \le 1, 0 \le y \le 1\} = \{(x,y): 0 \le x \le 1, 0 \le y \le x\}$$

$$\therefore \int_0^1 \int_y^1 e^{x^2}\, dxdy = \int_0^1 \int_0^x e^{x^2} dydx = \int_0^1 x e^{x^2}\, dx = \frac{1}{2} e^{x^2}\Big|_0^1 = \frac{e-1}{2}$$

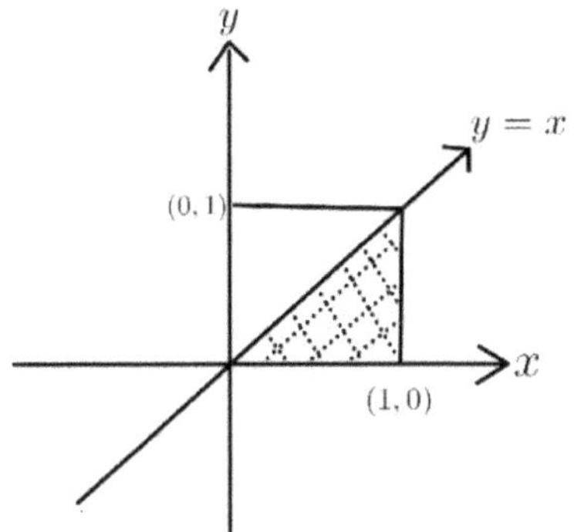


Example 2.

$$求 \int_0^2 \int_{\sqrt{x}}^2 e^{y^3} dydx = ?$$

【解】

$$\because \{(x,y): 0 \le x \le 2, \sqrt{x} \le y \le 2\} = \{(x,y): 0 \le x \le y^2, 0 \le y \le 2\}$$

$$\therefore \int_0^2 \int_{\sqrt{x}}^2 e^{y^3} dydx = \int_0^2 \int_0^{y^2} e^{y^3} dxdy = \int_0^2 y^2 e^{y^3} dy = \frac{e^{y^3}}{3}\Big|_{y=0}^{y=2} = \frac{1}{3}(e^8 - 1)$$

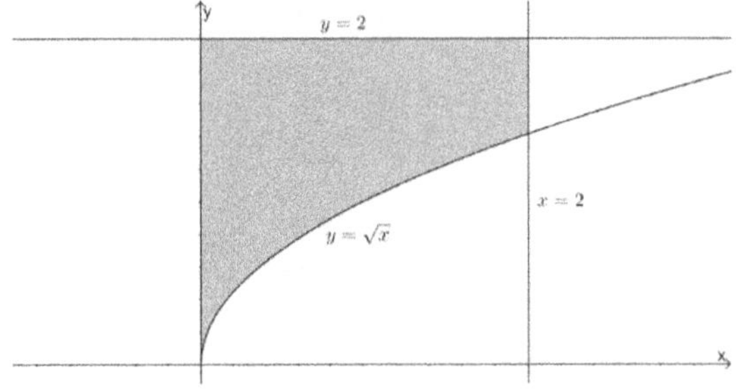


Example 3.

$$求 \int_0^4 \int_{\sqrt{y}}^2 \sqrt{x^3 + 1}\, dxdy = ?$$

【解】

$$\because \{(x,y): \sqrt{y} \le x \le 2, 0 \le y \le 4\} = \{(x,y): 0 \le x \le 2, 0 \le y \le x^2\}$$

$$\therefore \int_0^4 \int_{\sqrt{y}}^2 \sqrt{x^3+1}\,dxdy = \int_0^2 \int_0^{x^2} \sqrt{x^3+1}\,dydx = \int_0^2 x^2\sqrt{x^3+1}\,dx = \frac{2(x^3+1)^{\frac{3}{2}}}{9}\Bigg|_0^2 = \frac{54-2}{9}$$

$$= \frac{52}{9}$$

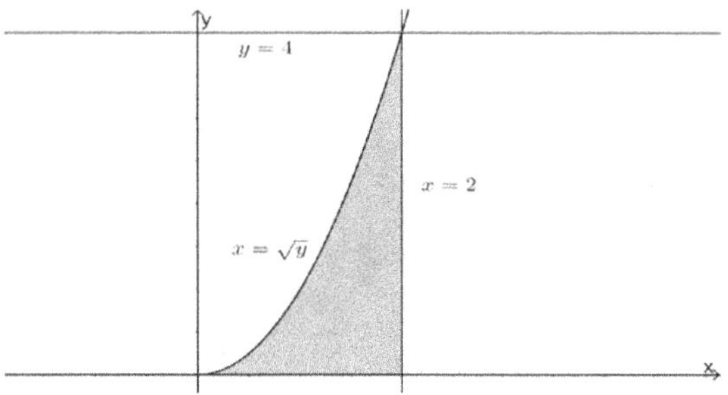

## Example 4.

$$求 \int_0^4 \int_{\sqrt{y}}^2 \frac{y\sqrt{x^2+1}}{x^3}\,dxdy = ?$$

【解】

$$\because \{(x,y): \sqrt{y} \le x \le 2, 0 \le y \le 4\} = \{(x,y): 0 \le x \le 2, 0 \le y \le x^2\}$$

$$\therefore \int_0^4 \int_{\sqrt{y}}^2 \frac{y\sqrt{x^2+1}}{x^3}\,dxdy = \int_0^2 \int_0^{x^2} \frac{y\sqrt{x^2+1}}{x^3}\,dydx = \int_0^2 \frac{x^4\sqrt{x^2+1}}{2x^3}\,dx = \int_0^2 \frac{x\sqrt{x^2+1}}{2}\,dx$$

$$= \frac{(x^2+1)^{\frac{3}{2}}}{6}\Bigg|_0^2 = \frac{5\sqrt{5}-1}{6}$$

## Example 5.

(1)求 $\displaystyle\int_0^{\frac{\pi}{2}}\int_y^{\frac{\pi}{2}}\frac{\sin x}{x}\,dxdy=?$    (2)求 $\displaystyle\int_0^{\frac{\pi}{4}}\int_y^{\frac{\pi}{4}}\frac{\sec^2 x}{x}\,dxdy=?$

【解】

(1)

$\because \{(x,y): y\le x\le\frac{\pi}{2}, 0\le y\le\frac{\pi}{2}\}=\{(x,y): 0\le x\le\frac{\pi}{2}, 0\le y\le x\}$

$\therefore \displaystyle\int_0^{\frac{\pi}{2}}\int_y^{\frac{\pi}{2}}\frac{\sin x}{x}\,dxdy = \int_0^{\frac{\pi}{2}}\int_0^{x}\frac{\sin x}{x}\,dydx = \int_0^{\frac{\pi}{2}}\left(\frac{\sin x}{x}\right)y\Big|_{y=0}^{y=x}dx = \int_0^{\frac{\pi}{2}}\left(\frac{\sin x}{x}\right)x\,dx$

$= \displaystyle\int_0^{\frac{\pi}{2}}\sin x\,dx = -\cos x\Big|_{x=0}^{x=\frac{\pi}{2}} = 1$

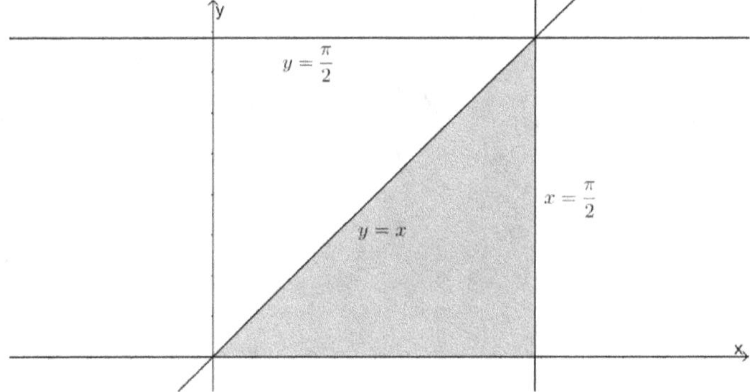


(2)

$\because \{(x,y): y\le x\le\frac{\pi}{4}, 0\le y\le\frac{\pi}{4}\}=\{(x,y): 0\le x\le\frac{\pi}{4}, 0\le y\le x\}$

$\therefore \displaystyle\int_0^{\frac{\pi}{4}}\int_y^{\frac{\pi}{4}}\frac{\sec^2 x}{x}\,dxdy = \int_0^{\frac{\pi}{4}}\int_0^{x}\frac{\sec^2 x}{x}\,dydx = \int_0^{\frac{\pi}{4}}\left(\frac{\sec^2 x}{x}\right)y\Big|_{y=0}^{y=x}dx = \int_0^{\frac{\pi}{4}}\left(\frac{\sec^2 x}{x}\right)x\,dx$

$= \displaystyle\int_0^{\frac{\pi}{4}}\sec^2 x\,dx = \tan x\Big|_{x=0}^{x=\frac{\pi}{4}} = 1$

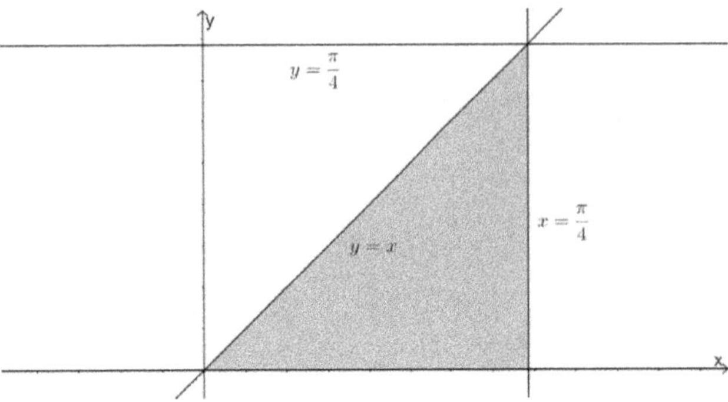


## Example 6.

$$求 \int_0^9 \int_{\sqrt{y}}^3 y \cos x^5 \, dx dy = ?$$

【解】

$$\because \{(x,y): \sqrt{y} \leq x \leq 3, 0 \leq y \leq 9\} = \{(x,y): 0 \leq x \leq 3, 0 \leq y \leq x^2\}$$

$$\therefore \int_0^9 \int_{\sqrt{y}}^3 y \cos x^5 \, dx dy = \int_0^3 \int_0^{x^2} y \cos x^5 \, dy dx = \int_0^3 (\cos x^5) \frac{y^2}{2} \Big|_0^{x^2} dx = \frac{1}{2} \int_0^3 (\cos x^5) x^4 \, dx$$

$$= \frac{1}{10} \sin x^5 \Big|_0^3 = \frac{\sin 243}{10}$$

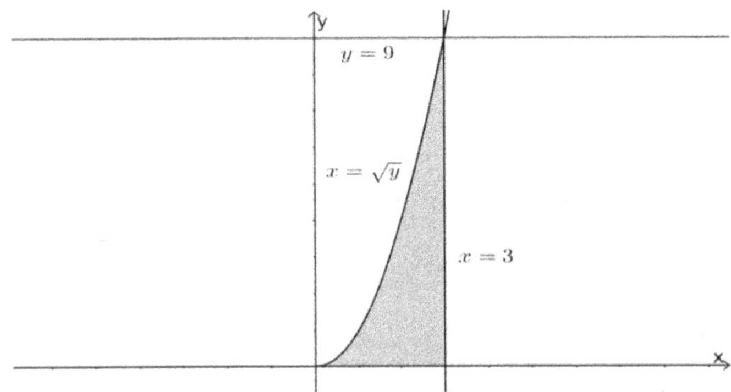


## Example 7.

$$求 \int_0^1 \int_y^1 \frac{1}{1+x^4} \, dx dy = ?$$

【解】

$$\because \{(x,y): y \leq x \leq 1, 0 \leq y \leq 1\} = \{(x,y): 0 \leq x \leq 1, 0 \leq y \leq x\}$$

$$\therefore \int_0^1 \int_y^1 \frac{1}{1+x^4}\,dxdy = \int_0^1 \int_0^x \frac{1}{1+x^4}\,dydx = \int_0^1 \frac{x}{1+x^4}\,dx = \left.\frac{\tan^{-1} x^2}{2}\right|_0^1 = \frac{\pi}{8}$$

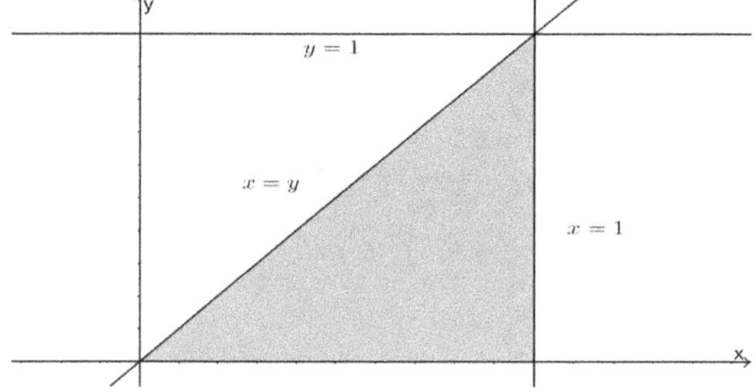


Example 8.

$$求 \int_0^4 \int_{\frac{x}{2}}^2 e^{y^2}\,dydx = ?$$

【解】

$$\because \{(x,y): 0 \le x \le 4, \frac{x}{2} \le y \le 2\} = \{(x,y): 0 \le x \le 2y, 0 \le y \le 2\}$$

$$\therefore \int_0^4 \int_{\frac{x}{2}}^2 e^{y^2}\,dydx = \int_0^2 \int_0^{2y} e^{y^2}\,dxdy = \int_0^2 2y e^{y^2}\,dy = \left.e^{y^2}\right|_0^2 = e^4 - 1$$

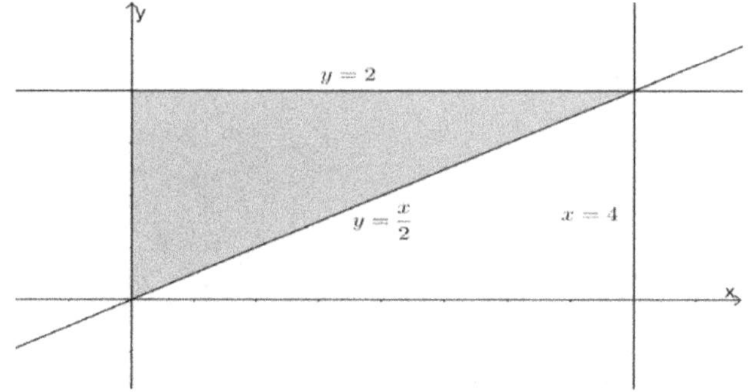


Example 9.

$$求 \int_0^1 \int_y^1 e^{\frac{y}{x}}\,dxdy = ?$$

【解】

$$\because \{(x,y): y \le x \le 1, 0 \le y \le 1\} = \{(x,y): 0 \le x \le 1, 0 \le y \le x\}$$

$$\therefore \int_0^1 \int_y^1 e^{\frac{y}{x}}dxdy = \int_0^1 \int_0^x e^{\frac{y}{x}}dydx = \int_0^1 xe^{\frac{y}{x}}\Big|_{y=0}^{y=x} dx = (e-1)\int_0^1 x\,dx = \frac{e-1}{2}$$

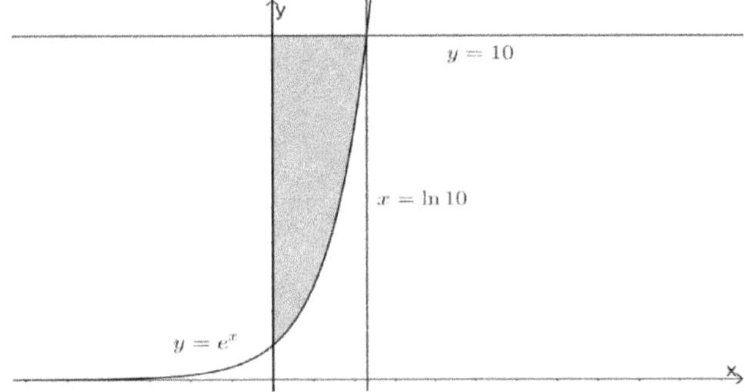


Example 10.

$$求 \int_0^{\ln 10} \int_{e^x}^{10} \frac{1}{\ln y}\,dydx = ?$$

【解】

$$\because \{(x,y): 0 \le x \le \ln 10, e^x \le y \le 10\} = \{(x,y): 0 \le x \le \ln y, 1 \le y \le 10\}$$

$$\therefore \int_0^{\ln 10} \int_{e^x}^{10} \frac{1}{\ln y}\,dydx = \int_1^{10} \int_0^{\ln y} \frac{1}{\ln y}\,dxdy = \int_1^{10} dy = 9$$



Example 11.

$$求 \int_0^2 \int_y^2 x\sqrt{x^3 + 1}\,dxdy = ?$$

【解】

$$\because \{(x,y): y \le x \le 2, 0 \le y \le 2\} = \{(x,y): 0 \le x \le 2, 0 \le y \le x\}$$

$$\therefore \int_0^2 \int_y^2 x\sqrt{x^3 + 1}\,dxdy = \int_0^2 \int_0^x x\sqrt{x^3 + 1}\,dydx = \int_0^2 x^2\sqrt{x^3 + 1}\,dx = \frac{2(x^3 + 1)^{\frac{3}{2}}}{9}\Bigg|_{x=0}^{x=2}$$

$$= \frac{2(9)^{\frac{3}{2}}}{9} - \frac{2}{9} = \frac{52}{9}$$

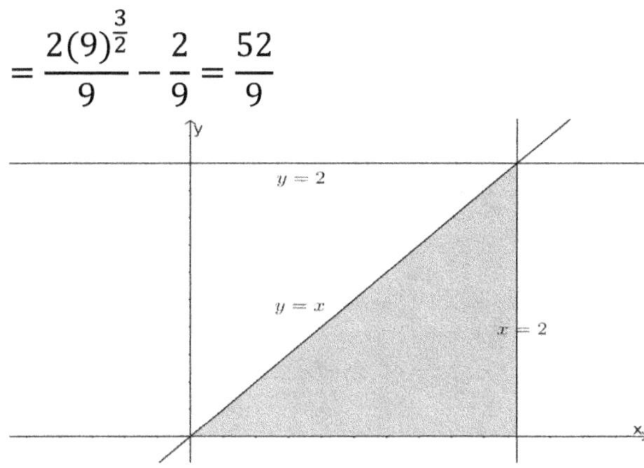

## Example 12.

$$(1) 求 \int_0^2 \int_y^2 e^{\max(x^2, y^2)} dx dy = ? \qquad (2) 求 \int_0^2 \int_0^2 e^{\max(x^2, y^2)} dx dy = ?$$

【解】

(1)

$$\because \{(x,y): y \le x \le 2, 0 \le y \le 2\} = \{(x,y): 0 \le x \le 2, 0 \le y \le x\}$$

$$\therefore \int_0^2 \int_y^2 e^{\max(x^2, y^2)} dx dy = \int_0^2 \int_0^x e^{\max(x^2, y^2)} dy dx = \int_0^2 \int_0^x e^{x^2} dy dx = \int_0^2 e^{x^2} x \, dx$$

$$= \frac{e^{x^2}}{2} \bigg|_{x=0}^{x=2} = \frac{e^4 - 1}{2}$$

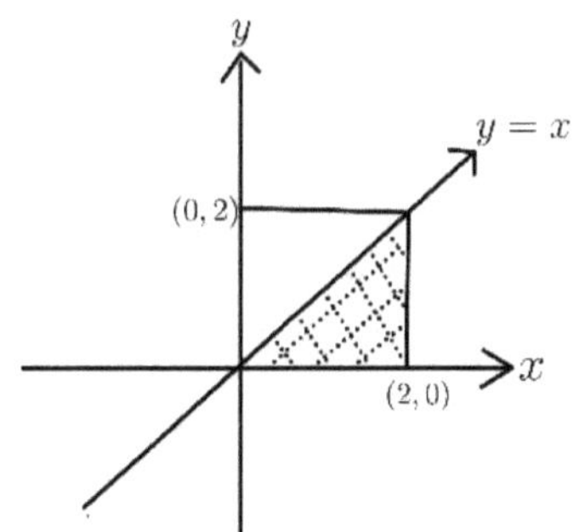

(2)

$$\because \int_0^2 e^{\max(x^2, y^2)} dx = \int_0^y e^{\max(x^2, y^2)} dx + \int_y^2 e^{\max(x^2, y^2)} dx, \ \forall 0 < y < 2$$

$$\therefore \int_0^2 \int_0^2 e^{\max(x^2,y^2)} dxdy = \int_0^2 \left( \int_0^y e^{\max(x^2,y^2)} dx + \int_y^2 e^{\max(x^2,y^2)} dx \right) dy$$

$$\because \int_0^2 \int_0^y e^{\max(x^2,y^2)} dx\, dy = \int_0^2 \left( \int_0^y e^{y^2} dx \right) dy = \int_0^2 y e^{y^2} dy = \frac{e^{y^2}}{2}\bigg|_0^2 = \frac{e^4-1}{2}$$

$$\because \{(x,y): y \le x \le 2, 0 \le y \le 2\} = \{(x,y): 0 \le x \le 2, 0 \le y \le x\}$$

$$\therefore \int_0^2 \left( \int_y^2 e^{\max(x^2,y^2)} dx \right) dy = \int_0^2 \int_0^x e^{\max(x^2,y^2)} dydx = \int_0^2 \int_0^x e^{x^2} dydx = \int_0^2 e^{x^2} x\, dx$$

$$= \frac{e^{x^2}}{2}\bigg|_{x=0}^{x=2} = \frac{e^4-1}{2}$$

$$\therefore \int_0^1 \int_0^1 e^{\max(x^2,y^2)} dxdy = e^4 - 1$$

Example 13.

$$(1)\text{求} \int_0^1 \int_0^1 |x-y| dxdy = ? \quad (2)\text{求} \int_0^2 \int_0^1 \sqrt{|y-x^2|}\, dxdy = ?$$

【解】

(1)

$$\because \int_0^1 |x-y| dx = \int_0^y |x-y| dx + \int_y^1 |x-y| dx = \int_0^y y-x\, dx + \int_y^1 x-ydx, \; \forall 0 < y < 1$$

$$\therefore \int_0^1 \int_0^1 |x-y| dxdy = \int_0^1 \int_0^y y-x\, dxdy + \int_0^1 \int_y^1 x-ydxdy$$

$$\because \int_0^1 \int_0^y y-x\, dxdy = \int_0^1 \left( yx - \frac{x^2}{2} \right)\bigg|_{x=0}^{x=y} dy = \int_0^1 \frac{y^2}{2} dy = \frac{y^3}{6}\bigg|_{y=0}^{y=1} = \frac{1}{6}$$

$$\because \{(x,y): y \le x \le 1, 0 \le y \le 1\} = \{(x,y): 0 \le x \le 1, 0 \le y \le x\}$$

$$\therefore \int_0^1 \int_y^1 x-ydxdy = \int_0^1 \int_0^x x-ydydx = \int_0^1 \left( yx - \frac{y^2}{2} \right)\bigg|_{y=0}^{y=x} dx = \int_0^1 \frac{x^2}{2} dx = \frac{x^3}{6}\bigg|_{x=0}^{x=1} = \frac{1}{6}$$

$$\text{因此} \int_0^1 \int_0^1 |x-y| dxdy = \frac{1}{3}$$

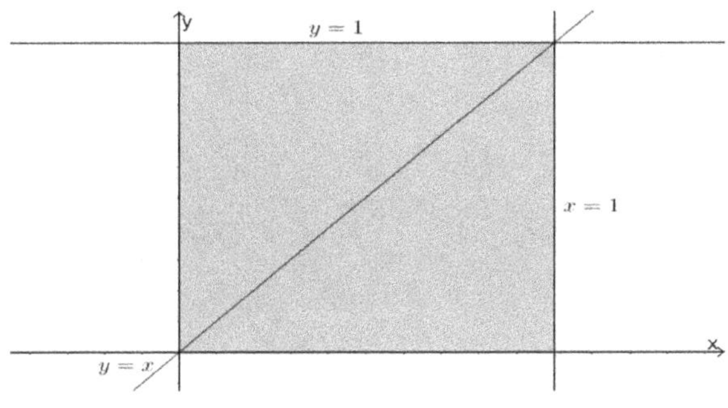


(2)

$\because f(x,y) = \sqrt{|y - x^2|}$ 在 $[0,1] \times [0,2]$ 为连续函数，藉由 Fubinis Theorem

则 $\displaystyle\int_0^2 \int_0^1 \sqrt{|y - x^2|}\, dxdy = \int_0^1 \int_0^2 \sqrt{|y - x^2|}\, dydx$

$\therefore \displaystyle\int_0^1 \int_0^2 \sqrt{|y - x^2|}\, dydx = \int_0^1 \int_0^{x^2} \sqrt{|y - x^2|}\, dydx + \int_0^1 \int_{x^2}^2 \sqrt{|y - x^2|}\, dydx$

$\therefore \displaystyle\int_0^1 \int_0^{x^2} \sqrt{|y - x^2|}\, dydx = \int_0^1 \int_0^{x^2} \sqrt{x^2 - y}\, dydx = \int_0^1 \left.\frac{(-2)(x^2 - y)^{\frac{3}{2}}}{3}\right|_{y=0}^{y=x^2} dx = \int_0^1 \frac{2x^3}{3}\, dx$

$= \left.\dfrac{2x^4}{12}\right|_{x=0}^{x=1} = \dfrac{1}{6}$

$\therefore \displaystyle\int_0^1 \int_{x^2}^2 \sqrt{|y - x^2|}\, dydx = \int_0^1 \int_{x^2}^2 \sqrt{y - x^2}\, dydx = \int_0^1 \left.\frac{2(y - x^2)^{\frac{3}{2}}}{3}\right|_{y=x^2}^{y=2} dx$

$= \displaystyle\int_0^1 \frac{2(2 - x^2)^{\frac{3}{2}}}{3}\, dx$

令 $x = \sqrt{2}\sin\theta$ 则 $dx = \sqrt{2}\cos\theta d\theta$

$\therefore \displaystyle\int_0^1 \frac{2(2 - x^2)^{\frac{3}{2}}}{3}\, dx = \int_0^{\frac{\pi}{4}} \frac{2(2 - 2\sin^2\theta)^{\frac{3}{2}}\sqrt{2}\cos\theta}{3}\, d\theta = \int_0^{\frac{\pi}{4}} \frac{2 \cdot 2\sqrt{2}(1 - \sin^2\theta)^{\frac{3}{2}}\sqrt{2}\cos\theta}{3}\, d\theta$

$= \displaystyle\int_0^{\frac{\pi}{4}} \frac{2 \cdot 2\sqrt{2}(\cos^2\theta)^{\frac{3}{2}}\sqrt{2}\cos\theta}{3}\, d\theta = \int_0^{\frac{\pi}{4}} \frac{8\cos^4\theta}{3}\, d\theta = \frac{8}{3}\int_0^{\frac{\pi}{4}} \left(\frac{1 + \cos 2\theta}{2}\right)^2 d\theta$

$$= \frac{8}{3}\int_0^{\frac{\pi}{4}} \frac{1 + 2\cos 2\theta + \cos^2 2\theta}{4}\, d\theta = \frac{8}{3}\int_0^{\frac{\pi}{4}} \frac{1 + 2\cos 2\theta}{4} + \frac{1 + \cos 4\theta}{8}\, d\theta = \frac{\pi}{4} + \frac{2}{3}$$

$$\therefore \int_0^1 \int_0^2 \sqrt{|y - x^2|}\, dy\, dx = \frac{1}{6} + \frac{\pi}{4} + \frac{2}{3} = \frac{\pi}{4} + \frac{5}{6}$$

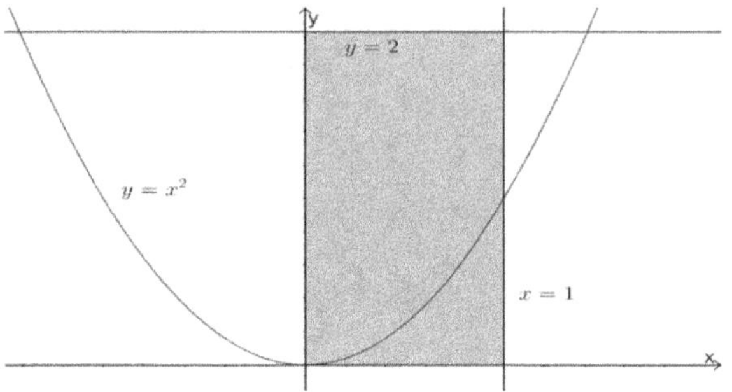


## Example 14.

$$\text{求} \quad \int_0^{\pi} \int_x^{\pi} \frac{\sin y}{y} \cos \frac{x}{y}\, dy\, dx = ?$$

【解】

$$\because \{(x, y): 0 \le x \le \pi, x \le y \le \pi\} = \{(x, y): 0 \le x \le y, 0 \le y \le \pi\}$$

$$\therefore \int_0^{\pi} \int_x^{\pi} \frac{\sin y}{y} \cos \frac{x}{y}\, dy\, dx = \int_0^{\pi} \int_0^{y} \frac{\sin y}{y} \cos \frac{x}{y}\, dx\, dy = \int_0^{\pi} \left(\sin y \sin \frac{x}{y}\right)\Big|_{x=0}^{x=y}\, dy$$

$$= \int_0^{\pi} \sin y\, (\sin 1)\, dy = \sin 1\, (-\cos y)|_0^{\pi} = 2 \sin 1$$

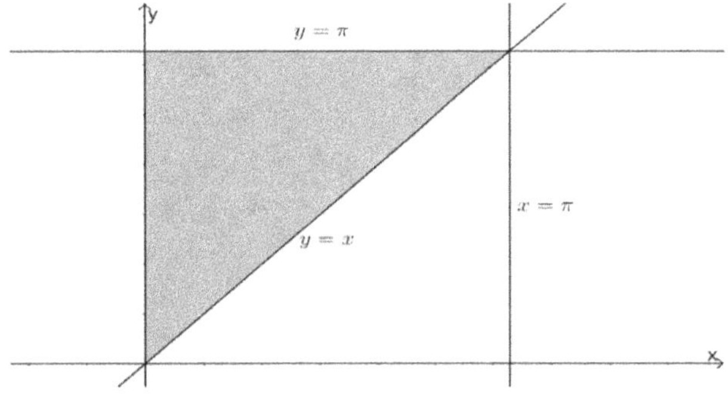


## Example 15.

$$\text{求} \quad \int_0^{\frac{\pi}{2}} \int_x^{\frac{\pi}{2}} \frac{\cos y}{y} \sin \frac{x}{y}\, dy\, dx = ?$$

【解】

$$\because \left\{(x,y): 0 \le x \le \frac{\pi}{2}, x \le y \le \frac{\pi}{2}\right\} = \left\{(x,y): 0 \le x \le y, 0 \le y \le \frac{\pi}{2}\right\}$$

$$\therefore \int_0^{\frac{\pi}{2}} \int_x^{\frac{\pi}{2}} \frac{\cos y}{y} \sin\frac{x}{y}\, dy\, dx = \int_0^{\frac{\pi}{2}} \int_0^y \frac{\cos y}{y} \sin\frac{x}{y}\, dx\, dy = \int_0^{\frac{\pi}{2}} \left(-\cos y \cos\frac{x}{y}\right)\Big|_{x=0}^{x=y}\, dy$$

$$= \int_0^{\frac{\pi}{2}} \cos y\,(1 - \cos 1)\, dy = (1 - \cos 1)\sin y\Big|_0^{\frac{\pi}{2}} = 1 - \cos 1$$

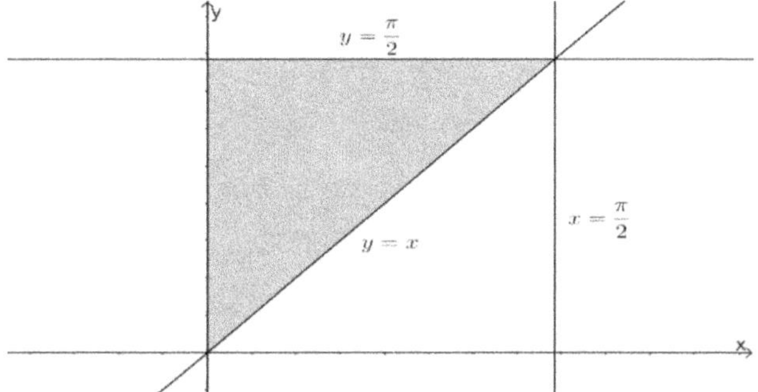


Example 16.

$$求 \int_0^2 \int_x^2 \sin y^2\, dy\, dx = ?$$

【解】

$$\because \{(x,y): 0 \le x \le 2, x \le y \le 2\} = \{(x,y): 0 \le x \le y, 0 \le y \le 2\}$$

$$\therefore \int_0^2 \int_x^2 \sin y^2\, dy\, dx = \int_0^2 \int_0^y \sin y^2\, dx\, dy = \int_0^2 y \sin y^2\, dy = -\frac{\cos y^2}{2}\Big|_0^2 = \frac{1}{2}(1 - \cos 4)$$

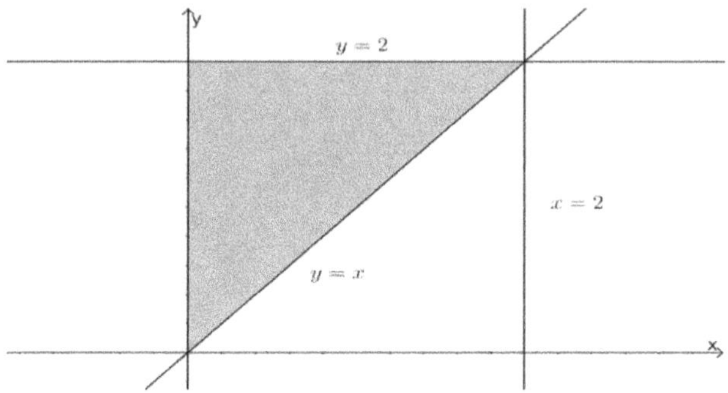


Example 17.

$$求 \iint_R 1\, dA = ?, \quad R = \{(x,y): -3x - y + 6 \le 0, 4x - x^2 - y \ge 0, y \ge 0\}$$

【解】

$$\iint_R 1\,dA = \int_1^2 4x - x^2 - (6 - 3x)\,dx + \int_2^4 4x - x^2\,dx$$

$$= \left(\frac{7x^2}{2} - \frac{x^3}{3} - 6x\right)\Big|_1^2 + \left(2x^2 - \frac{x^3}{3}\right)\Big|_2^4 = \frac{21}{2} - \frac{7}{3} - 6 + 2\cdot 12 - \frac{56}{3} = \frac{15}{2}$$

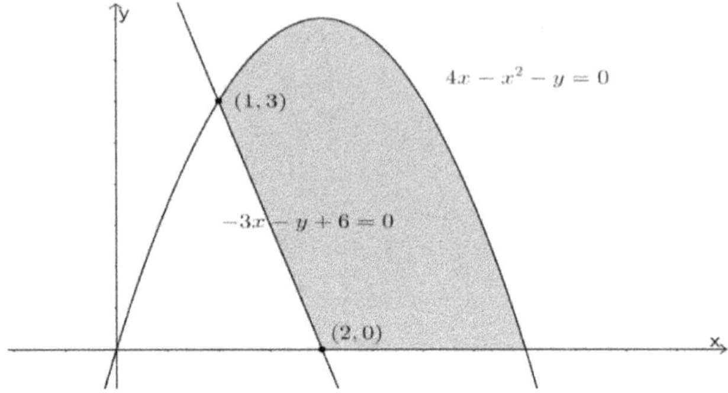


Example 18.

$$求 \iint_R \sqrt{x} - y^2 \,dxdy = ?, \quad R = \{(x,y): y \geq x^2, y \leq x^{\frac{1}{4}}\}$$

【解】

令 $y = x^2 = x^{\frac{1}{4}}$ 则 $x = 0$ 或 $1 \Rightarrow 0 \leq x \leq 1 \Rightarrow R = \{(x,y): 0 \leq x \leq 1, y \geq x^2, y \leq x^{\frac{1}{4}}\}$

$\because f(x,y) = \sqrt{x} - y^2$ 在 $R$ 区域为连续函数, 藉由 Fubinis Theorem

$$则 \iint_R \sqrt{x} - y^2 \,dxdy = \int_0^1 \int_{x^2}^{x^{\frac{1}{4}}} \sqrt{x} - y^2 \,dydx = \int_0^1 \sqrt{x}\,y - \frac{y^3}{3}\Big|_{y=x^2}^{y=x^{\frac{1}{4}}} dx$$

$$= \int_0^1 \sqrt{x}(x^{\frac{1}{4}}) - \frac{x^{\frac{3}{4}}}{3} - \left(\sqrt{x}(x^2) - \frac{x^6}{3}\right) dx = \int_0^1 x^{\frac{3}{4}} - \frac{x^{\frac{3}{4}}}{3} - x^{\frac{5}{2}} + \frac{x^6}{3}\,dx$$

$$= \left(\frac{4x^{\frac{7}{4}}}{7} - \frac{4x^{\frac{7}{4}}}{21} - \frac{2x^{\frac{7}{2}}}{7} + \frac{x^7}{21}\right)\Big|_{x=0}^{x=1} = \frac{2}{7} - \frac{3}{21} = \frac{1}{7}$$

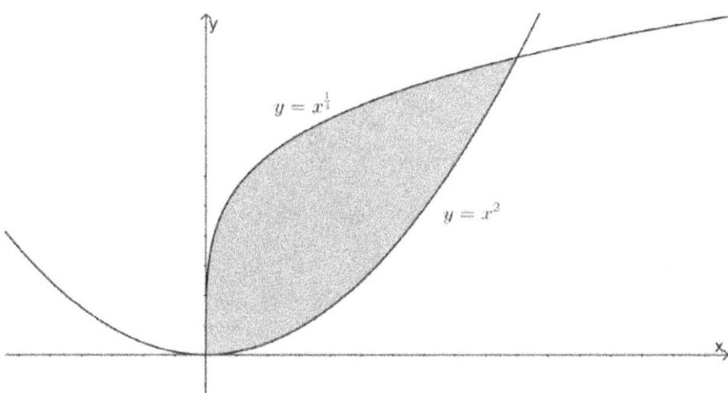

Example 19.

$$求 \int_0^{\sqrt{3}} \int_y^{\sqrt{3}} \cos\frac{\pi x^2}{2}\, dxdy = ?$$

【解】

$$\because \{(x,y): y \le x \le \sqrt{3}, 0 \le y \le \sqrt{3}\} = \{(x,y): 0 \le x \le \sqrt{3}, 0 \le y \le x\}$$

$$\therefore \int_0^{\sqrt{3}} \int_y^{\sqrt{3}} \cos\frac{\pi x^2}{2}\, dxdy = \int_0^{\sqrt{3}} \int_0^{x} \cos\frac{\pi x^2}{2}\, dydx = \int_0^{\sqrt{3}} x\cos\frac{\pi x^2}{2}\, dx = \frac{1}{\pi}\sin\frac{\pi x^2}{2}\Big|_0^{\sqrt{3}} = -\frac{1}{\pi}$$

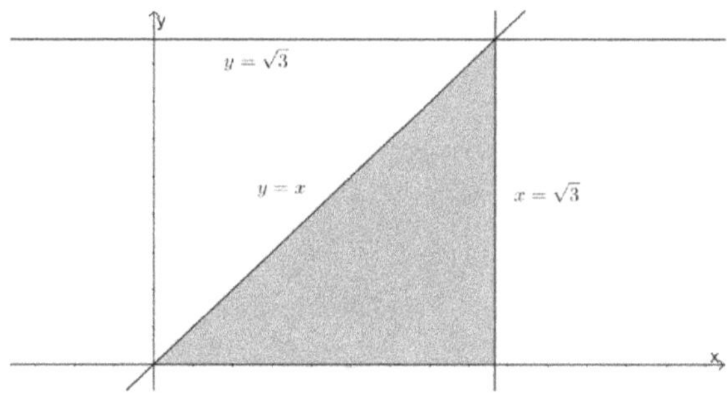

Example 20.

$$求 \int_0^{\sqrt{3}} \int_y^{\sqrt{3}} \sin\frac{\pi x^2}{2}\, dxdy = ?$$

【解】

$$\because \{(x,y): y \le x \le \sqrt{3}, 0 \le y \le \sqrt{3}\} = \{(x,y): 0 \le x \le \sqrt{3}, 0 \le y \le x\}$$

$$\therefore \int_0^{\sqrt{3}} \int_y^{\sqrt{3}} \sin\frac{\pi x^2}{2}\, dxdy = \int_0^{\sqrt{3}} \int_0^{x} \sin\frac{\pi x^2}{2}\, dydx = \int_0^{\sqrt{3}} x\sin\frac{\pi x^2}{2}\, dx = \frac{-1}{\pi}\cos\frac{\pi x^2}{2}\Big|_0^{\sqrt{3}} = \frac{1}{\pi}$$

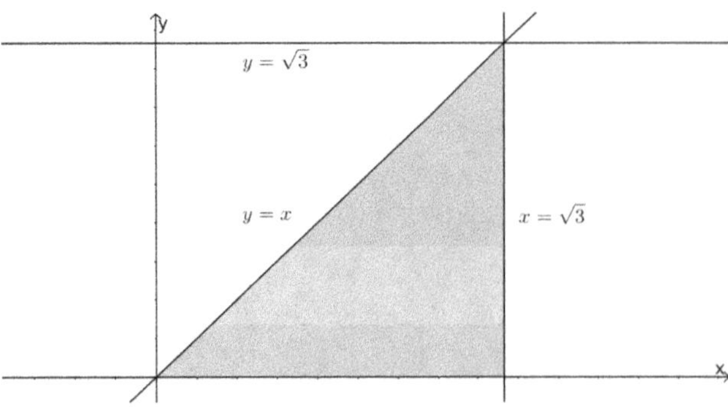

## Example 21.

$$求 \int_0^2 \int_x^2 \cos y^2 \, dydx =?$$

【解】

$$\because \{(x,y): 0 \le x \le 2, x \le y \le 2\} = \{(x,y): 0 \le x \le y, 0 \le y \le 2\}$$

$$\therefore \int_0^2 \int_x^2 \cos y^2 \, dydx = \int_0^2 \int_0^y \cos y^2 \, dxdy = \int_0^2 y \cos y^2 \, dy = \frac{\sin y^2}{2}\Big|_0^2 = \frac{\sin 4}{2}$$

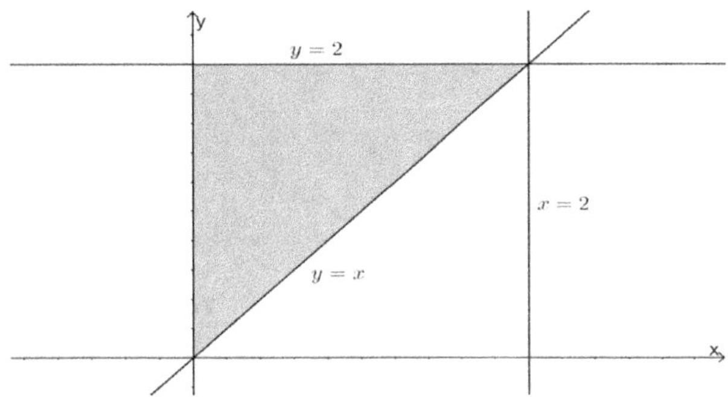

## Example 22.

$$求 \int_0^1 \int_{\sqrt{x}}^1 \sqrt{1+y^3} \, dydx =?$$

【解】

$$\because \{(x,y): 0 \le x \le 1, \sqrt{x} \le y \le 1\} = \{(x,y): 0 \le x \le y^2, 0 \le y \le 1\}$$

$$\therefore \int_0^1 \int_{\sqrt{x}}^1 \sqrt{1+y^3} \, dydx = \int_0^1 \int_0^{y^2} \sqrt{1+y^3} \, dxdy = \int_0^1 y^2 \sqrt{1+y^3} \, dy = \frac{2}{9}(1+y^3)^{\frac{3}{2}}\Big|_0^1$$

$$= \frac{2}{9}(2\sqrt{2}-1)$$

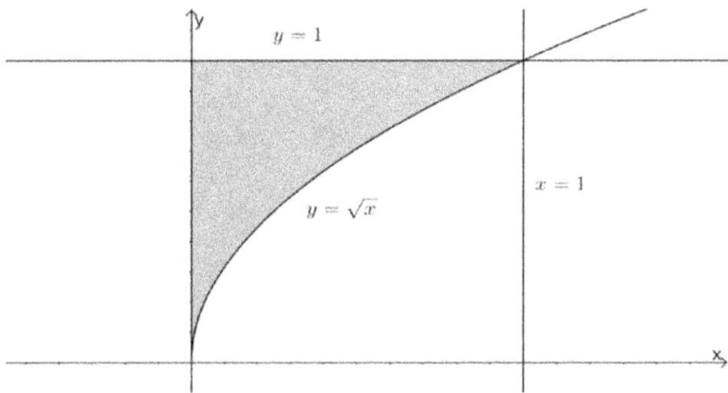

Example 23.

$$\qquad \int_0^1 \int_{2y}^2 \cos x^2 \, dx dy = ?$$

【解】

$\because \{(x,y): 2y \le x \le 2, 0 \le y \le 1\} = \{(x,y): 0 \le x \le 2, 0 \le y \le \dfrac{x}{2}\}$

$$\therefore \int_0^1 \int_{2y}^2 \cos x^2 \, dx dy = \int_0^2 \int_0^{\frac{x}{2}} \cos x^2 \, dy dx = \int_0^2 \frac{x}{2} \cos x^2 \, dx = \left. \frac{\sin x^2}{4} \right|_0^2 = \frac{\sin 4}{4}$$

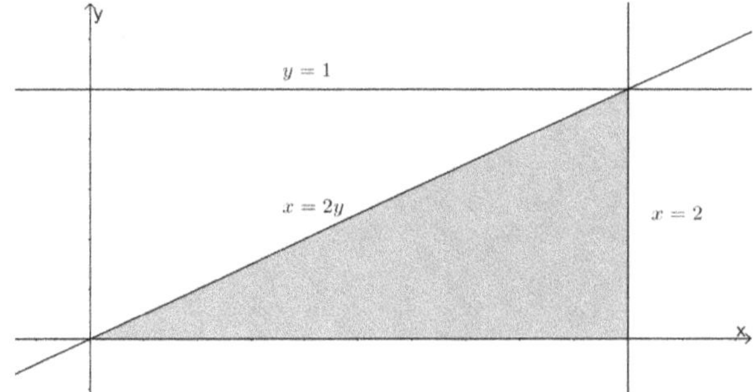

Example 24.

$$\qquad \int_0^1 \int_{2y}^1 \sec^2 x^2 \, dx dy = ?$$

【解】

$\because \{(x,y): 2y \le x \le 1, 0 \le y \le 1\} = \{(x,y): 0 \le x \le 1, 0 \le y \le \dfrac{x}{2}\}$

$$\therefore \int_0^1 \int_{2y}^1 \sec^2 x^2 \, dx dy = \int_0^1 \int_0^{\frac{x}{2}} \sec^2 x^2 \, dy dx = \int_0^1 \frac{x}{2} \sec^2 x^2 \, dx = \left. \frac{\tan x^2}{4} \right|_0^1 = \frac{\tan 1}{4}$$

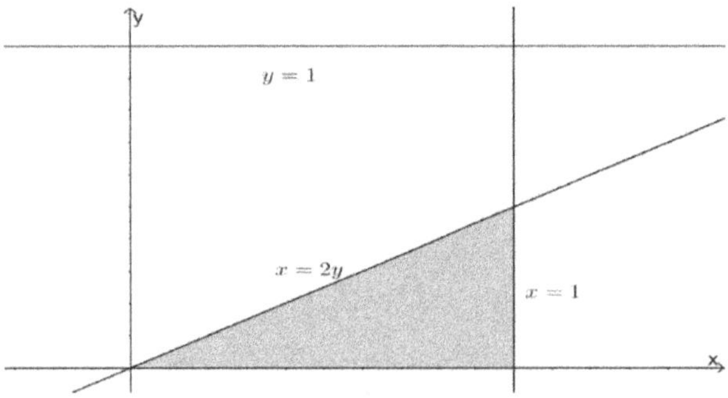


Example 25.

$$求 \int_0^1 \int_y^1 e^{-x^2} dx dy =?$$

【解】

$$\because \{(x,y): y \leq x \leq 1, 0 \leq y \leq 1\} = \{(x,y): 0 \leq x \leq 1, 0 \leq y \leq x\}$$

$$\therefore \int_0^1 \int_y^1 e^{-x^2} dx dy = \int_0^1 \int_0^x e^{-x^2} dy dx = \int_0^1 x e^{-x^2} dx = -\frac{e^{-x^2}}{2}\Big|_0^1 = \frac{1}{2}(1 - e^{-1})$$

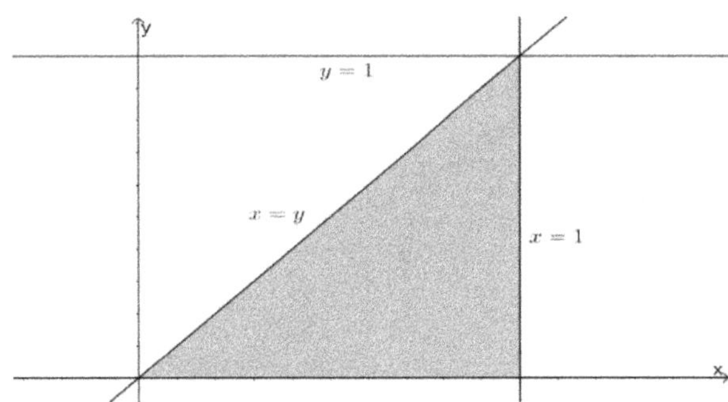


Example 26.

$$求 \int_0^1 \int_{x^2}^1 x^3 \sin y^3 \, dy dx =?$$

【解】

$$\because \{(x,y): 0 \leq x \leq 1, x^2 \leq y \leq 1\} = \{(x,y): 0 \leq x \leq \sqrt{y}, 0 \leq y \leq 1\}$$

$$\therefore \int_0^1 \int_{x^2}^1 x^3 \sin y^3 \, dy dx = \int_0^1 \int_0^{\sqrt{y}} x^3 \sin y^3 \, dx dy = \int_0^1 \frac{y^2 \sin y^3}{4} \, dy = -\frac{\cos y^3}{12}\Big|_0^1 = \frac{1 - \cos 1}{12}$$

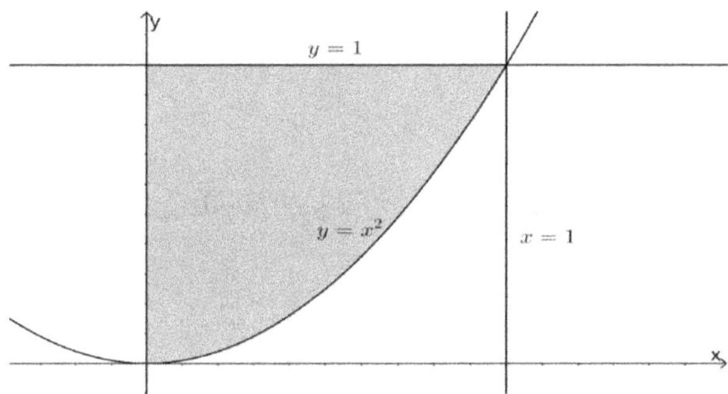

Example 27.

$$求 \int_0^1 \int_{\sin^{-1} y}^{\frac{\pi}{2}} \cos x \sqrt{\cos^2 x + 1}\, dx\, dy = ?$$

【解】

$$\because \left\{ (x,y): \sin^{-1} y \leq x \leq \frac{\pi}{2}, 0 \leq y \leq 1 \right\} = \left\{ (x,y): 0 \leq x \leq \frac{\pi}{2}, 0 \leq y \leq \sin x \right\}$$

$$\therefore \int_0^1 \int_{\sin^{-1} y}^{\frac{\pi}{2}} \cos x \sqrt{\cos^2 x + 1}\, dx\, dy = \int_0^{\frac{\pi}{2}} \int_0^{\sin x} \cos x \sqrt{\cos^2 x + 1}\, dy\, dx$$

$$= \int_0^{\frac{\pi}{2}} \sin x \cos x \sqrt{\cos^2 x + 1}\, dx = \int_0^{\frac{\pi}{2}} \frac{\sin 2x \sqrt{\dfrac{\cos 2x + 3}{2}}}{2}\, dx = -\left. \frac{\left(\dfrac{\cos 2x + 3}{2}\right)^{\frac{3}{2}}}{3} \right|_0^{\frac{\pi}{2}}$$

$$= \frac{2\sqrt{2} - 1}{3}$$

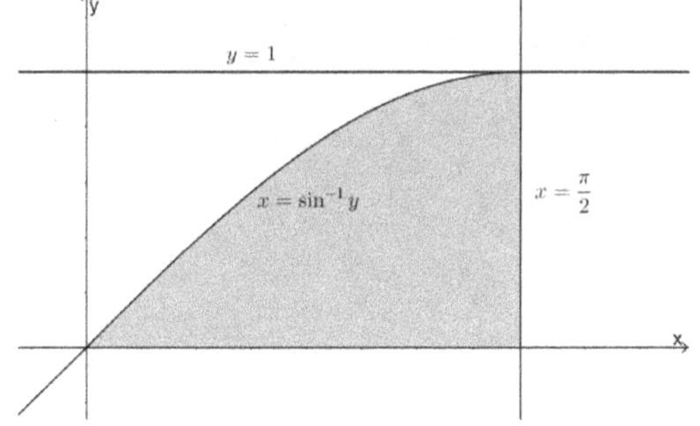

### 8.3.3　使用极坐标转换求双重积分的值

当双重积分的积分区域出现 $x^2 + y^2$ 或者被积分函数出现 $x^2 + y^2$,可视为使用极坐标转换的时机

考试类型:

Type 1.

假设 $R = \{(x,y): 0 \le x^2 + y^2 \le a^2\},\ a > 0,\ $ 求 $\iint_R f(x,y)dA = ?$

解题流程:

Step1.

令 $x = r\cos\theta, y = r\sin\theta$

则 $R = \{(x,y): 0 \le x^2 + y^2 \le a^2\} = \{(r,\theta): 0 \le r \le a, 0 \le \theta \le 2\pi\}$

$$dxdy = \left\| \begin{vmatrix} \dfrac{\partial x}{\partial r} & \dfrac{\partial x}{\partial \theta} \\ \dfrac{\partial y}{\partial r} & \dfrac{\partial y}{\partial \theta} \end{vmatrix} \right\| drd\theta = \left\| \begin{vmatrix} \cos\theta & -r\sin\theta \\ \sin\theta & r\cos\theta \end{vmatrix} \right\| drd\theta = rdrd\theta$$

Step2.

$$\therefore \iint_R f(x,y)dA = \int_0^{2\pi} \int_0^a f(r\cos\theta, r\sin\theta)rdrd\theta$$

Step3.

计算此积分 $\displaystyle\int_0^{2\pi} \int_0^a f(r\cos\theta, r\sin\theta)rdrd\theta = ?$

Step4.

如果无法计算 $\displaystyle\int_0^{2\pi} \int_0^a f(r\cos\theta, r\sin\theta)rdrd\theta = ?$,通常会再搭配使用变数变换、分部积分

<u>范例说明:</u>

假设 $R = \{(x,y): 0 \le x^2 + y^2 \le a^2\},\ a > 0,\ $ 求 $\iint_R f(x,y)dA = ?$

(I)若 $f(x,y) = e^{x^2+y^2}$ 则 $\displaystyle\iint_R f(x,y)dA = \int_0^{2\pi} \int_0^a e^{r^2} rdrd\theta$

(II)若 $f(x,y) = \ln(x^2 + y^2)$ 则 $\iint_R f(x,y)\,dA = \int_0^{2\pi}\int_0^a \ln(r^2)\,r\,dr\,d\theta$

(III)若 $f(x,y) = \sin(x^2 + y^2)$ 则 $\iint_R f(x,y)\,dA = \int_0^{2\pi}\int_0^a \sin(r^2)\,r\,dr\,d\theta$

(IV)若 $f(x,y) = \dfrac{1}{\sqrt{x^2 + y^2}}$ 则 $\iint_R f(x,y)\,dA = \int_0^{2\pi}\int_0^a dr\,d\theta$

(V)若 $f(x,y) = \dfrac{1}{\sqrt{1 + x^2 + y^2}}$ 则 $\iint_R f(x,y)\,dA = \int_0^{2\pi}\int_0^a \dfrac{r}{\sqrt{1 + r^2}}\,dr\,d\theta$

(VI)若 $f(x,y) = e^{-(x^2+y^2)}\cos(x^2 + y^2)$ 则 $\iint_R f(x,y)\,dA = \int_0^{2\pi}\int_0^a e^{-r^2}\cos(r^2)\,r\,dr\,d\theta$

(VII)若 $f(x,y) = \sin^{-1}\dfrac{y}{\sqrt{x^2 + y^2}}$ 则 $\iint_R f(x,y)\,dA = \int_0^{2\pi}\int_0^a \sin^{-1}\left(\dfrac{r\sin\theta}{r}\right)r\,dr\,d\theta$

(VIII)若 $f(x,y) = \sec^2(x^2 + y^2)$ 则 $\iint_R f(x,y)\,dA = \int_0^{2\pi}\int_0^a \sec^2(r^2)\,r\,dr\,d\theta$

Example 1.

求 $\displaystyle\int_0^2\int_0^{\sqrt{4-y^2}} e^{x^2+y^2}\,dx\,dy = ?$

【解】

令 $x = r\cos\theta,\ y = r\sin\theta$

则 $\left\{(x,y): 0 \le x \le \sqrt{4 - y^2}, 0 \le y \le 2\right\} = \left\{(r,\theta): 0 \le r \le 2, 0 \le \theta \le \dfrac{\pi}{2}\right\}$

且 $dx\,dy = \left\|\begin{vmatrix} \dfrac{\partial x}{\partial r} & \dfrac{\partial x}{\partial \theta} \\ \dfrac{\partial y}{\partial r} & \dfrac{\partial y}{\partial \theta} \end{vmatrix}\right\| dr\,d\theta = \left\|\begin{vmatrix} \cos\theta & -r\sin\theta \\ \sin\theta & r\cos\theta \end{vmatrix}\right\| dr\,d\theta = r\,dr\,d\theta$

$\therefore \displaystyle\int_0^2\int_0^{\sqrt{4-y^2}} e^{x^2+y^2}\,dx\,dy = \int_0^{\frac{\pi}{2}}\int_0^2 e^{r^2}\,r\,dr\,d\theta = \int_0^{\frac{\pi}{2}} \dfrac{e^{r^2}}{2}\bigg|_{r=0}^{r=2} d\theta = \int_0^{\frac{\pi}{2}} \dfrac{e^4 - 1}{2}\,d\theta$

$$= \left(\frac{e^4 - 1}{4}\right)\pi$$

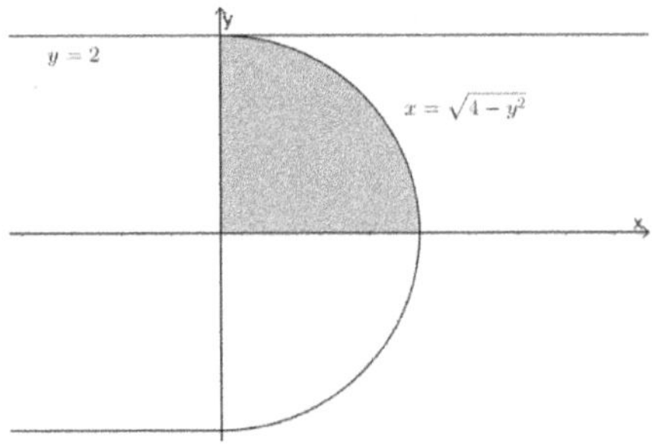


## Example 2.

$$求 \iint_R \ln(x^2 + y^2)\, dA = ?, \ R = \{(x,y): x \geq 0, y \geq 0, 0 \leq x^2 + y^2 \leq 16\}$$

【解】

令$x = r\cos\theta, \ y = r\sin\theta$

则$\{(x,y): x \geq 0, y \geq 0, 0 \leq x^2 + y^2 \leq 16\} = \{(r,\theta): 0 \leq r \leq 4, 0 \leq \theta \leq \frac{\pi}{2}\}$

$$且\ dxdy = \left\|\begin{vmatrix} \dfrac{\partial x}{\partial r} & \dfrac{\partial x}{\partial \theta} \\ \dfrac{\partial y}{\partial r} & \dfrac{\partial y}{\partial \theta} \end{vmatrix}\right\| drd\theta = \left\|\begin{vmatrix} \cos\theta & -r\sin\theta \\ \sin\theta & r\cos\theta \end{vmatrix}\right\| drd\theta = rdrd\theta$$

$$\therefore \iint_R \ln(x^2 + y^2)\, dA = \int_0^{\frac{\pi}{2}} \int_0^4 (\ln r^2) r\, dr\, d\theta = 2\int_0^{\frac{\pi}{2}} \int_0^4 (\ln r) r\, dr\, d\theta$$

藉由 integration by parts

$$则 \int_0^4 (\ln r) r\, dr = \ln r \left(\frac{r^2}{2}\right)\Big|_{r=0}^{r=4} - \int_0^4 \frac{r}{2} dr = \ln r\left(\frac{r^2}{2}\right)\Big|_{r=0}^{r=4} - \frac{r^2}{4}\Big|_{r=0}^{r=4} = 16\ln 2 - 4$$

$$\because 2\int_0^{\frac{\pi}{2}} d\theta = \pi \quad \therefore \iint_R \ln(x^2 + y^2)\, dA = \pi(16\ln 2 - 4)$$

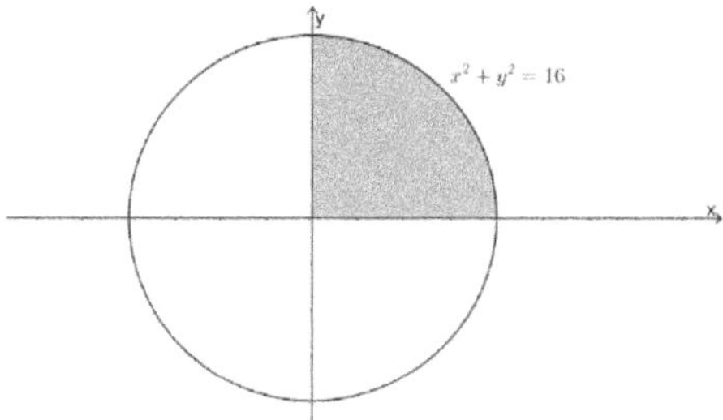

## Example 3.

$$求 \iint_R x^2 y \, dxdy = ?, \quad R = \{(x,y): y \geq 0, \ x^2 + y^2 \leq 4\}$$

【解】

令 $x = r\cos\theta, \ y = r\sin\theta$

则 $\{(x,y): y \geq 0, x^2 + y^2 \leq 4\} = \{(r,\theta): 0 \leq r \leq 2, 0 \leq \theta \leq \pi\}$

且 $dxdy = \left\| \begin{matrix} \dfrac{\partial x}{\partial r} & \dfrac{\partial x}{\partial \theta} \\ \dfrac{\partial y}{\partial r} & \dfrac{\partial y}{\partial \theta} \end{matrix} \right\| drd\theta = \left\| \begin{matrix} \cos\theta & -r\sin\theta \\ \sin\theta & r\cos\theta \end{matrix} \right\| drd\theta = rdrd\theta$

$$\therefore \iint_R x^2 y \, dxdy = \int_0^\pi \int_0^2 (r\cos\theta)^2 (r\sin\theta) r dr d\theta = \int_0^\pi \sin\theta \cos^2\theta \, d\theta \int_0^2 r^4 dr$$

$$= \frac{(-1)\cos^3\theta}{3} \bigg|_0^\pi \cdot \frac{r^5}{5} \bigg|_0^2 = \frac{2}{3} \cdot \frac{32}{5} = \frac{64}{15}$$

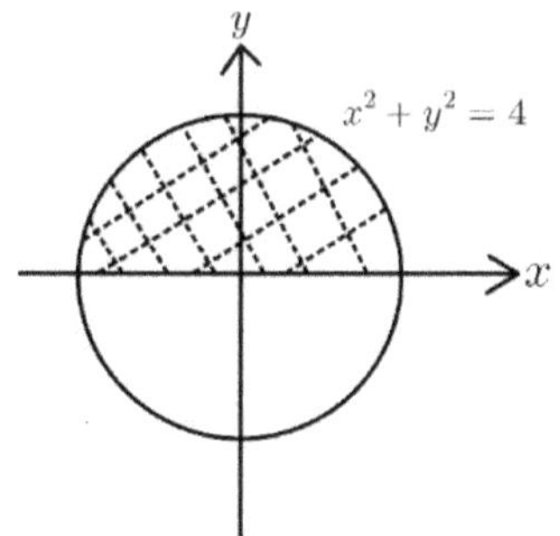

## Example 4.

$$求 \iint_R \sqrt{x^2 + y^2} \, dA = ?, \quad R = \{(x,y): x^2 + y^2 \leq 3x\}$$

【解】

令 $x = r\cos\theta,\ y = r\sin\theta$

则 $\{(x,y): x^2 + y^2 \le 3x\} = \{(r,\theta): 0 \le r \le 3\cos\theta,\ -\dfrac{\pi}{2} \le \theta \le \dfrac{\pi}{2}\}$

且 $dxdy = \left\|\begin{array}{cc} \dfrac{\partial x}{\partial r} & \dfrac{\partial x}{\partial \theta} \\ \dfrac{\partial y}{\partial r} & \dfrac{\partial y}{\partial \theta} \end{array}\right\| drd\theta = \left\|\begin{array}{cc} \cos\theta & -r\sin\theta \\ \sin\theta & r\cos\theta \end{array}\right\| drd\theta = rdrd\theta$

$\therefore \iint_R \sqrt{x^2 + y^2}\,dA = \int_{-\frac{\pi}{2}}^{\frac{\pi}{2}} \int_0^{3\cos\theta} r^2\,drd\theta = \int_{-\frac{\pi}{2}}^{\frac{\pi}{2}} \left. \frac{r^3}{3}\right|_0^{3\cos\theta} d\theta = \int_{-\frac{\pi}{2}}^{\frac{\pi}{2}} 9\cos^3\theta\,d\theta$

$= 9\int_{-\frac{\pi}{2}}^{\frac{\pi}{2}} \cos\theta(1 - \sin^2\theta)\,d\theta = 9\left.\left(\sin\theta - \frac{\sin^3\theta}{3}\right)\right|_{-\frac{\pi}{2}}^{\frac{\pi}{2}} = 12$

Example 5.

$$\text{求}\ \iint_R \frac{1}{\sqrt{x^2 + y^2}}\,dA = ?,\ R = \{(x,y): y \ge 0, 0 \le x^2 + y^2 \le 1\}$$

【解】

令 $x = r\cos\theta,\ y = r\sin\theta$

则 $\{(x,y): y \ge 0, 0 \le x^2 + y^2 \le 1\} = \{(r,\theta): 0 \le r \le 1, 0 \le \theta \le \pi\}$

且 $dxdy = \left\|\begin{array}{cc} \dfrac{\partial x}{\partial r} & \dfrac{\partial x}{\partial \theta} \\ \dfrac{\partial y}{\partial r} & \dfrac{\partial y}{\partial \theta} \end{array}\right\| drd\theta = \left\|\begin{array}{cc} \cos\theta & -r\sin\theta \\ \sin\theta & r\cos\theta \end{array}\right\| drd\theta = rdrd\theta$

$\therefore \iint_R \frac{1}{\sqrt{x^2 + y^2}}\,dA = \int_0^{\pi} \int_0^1 \frac{r}{\sqrt{r^2}}\,drd\theta = \pi$

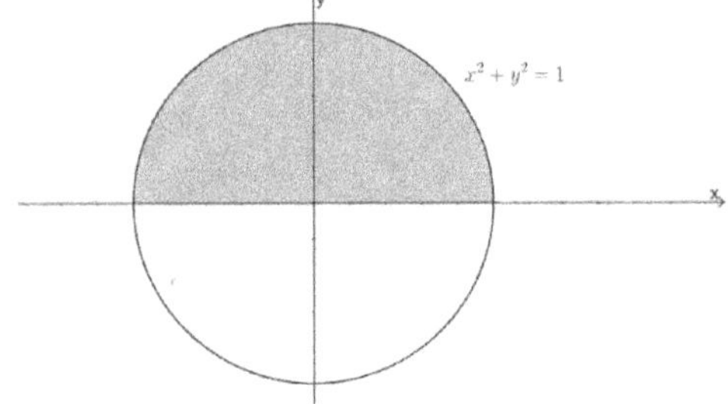

Example 6.

$$\text{求} \int_0^\infty e^{-x^2} dx = ?$$

【解】

$$\because \int_0^\infty e^{-x^2} dx \int_0^\infty e^{-y^2} dy = \int_0^\infty \int_0^\infty e^{-(x^2+y^2)} dy\, dx$$

令 $x = r\cos\theta, \ y = r\sin\theta$

则 $\{(x,y): y \geq 0, 0 \leq x^2 + y^2 \leq \infty\} = \{(r,\theta): 0 \leq r \leq \infty, 0 \leq \theta \leq \dfrac{\pi}{2}\}$

且 $dxdy = \left\| \begin{vmatrix} \dfrac{\partial x}{\partial r} & \dfrac{\partial x}{\partial \theta} \\ \dfrac{\partial y}{\partial r} & \dfrac{\partial y}{\partial \theta} \end{vmatrix} \right\| drd\theta = \left\| \begin{vmatrix} \cos\theta & -r\sin\theta \\ \sin\theta & r\cos\theta \end{vmatrix} \right\| drd\theta = rdrd\theta$

$$\therefore \int_0^\infty \int_0^\infty e^{-(x^2+y^2)} dy\, dx = \int_0^{\frac{\pi}{2}} \int_0^\infty e^{-r^2} rdrd\theta = \int_0^{\frac{\pi}{2}} d\theta \int_0^\infty e^{-r^2} rdr$$

$$= \frac{\pi}{2} \cdot \left. \frac{-e^{-r^2}}{2} \right|_{r=0}^{r=\infty} = \frac{\pi}{4} \Rightarrow \int_0^\infty e^{-x^2} dx = \sqrt{\frac{\pi}{4}}$$

Example 7.

$$\text{求} \int_0^\infty \int_0^\infty e^{-(x^2+y^2)} \cos(x^2 + y^2) dxdy = ?$$

【解】

令 $x = r\cos\theta, \ y = r\sin\theta$

则 $\{(x,y): y \geq 0, 0 \leq x^2 + y^2 \leq \infty\} = \{(r,\theta): 0 \leq r \leq \infty, 0 \leq \theta \leq \dfrac{\pi}{2}\}$

且 $dxdy = \left\| \begin{vmatrix} \dfrac{\partial x}{\partial r} & \dfrac{\partial x}{\partial \theta} \\ \dfrac{\partial y}{\partial r} & \dfrac{\partial y}{\partial \theta} \end{vmatrix} \right\| drd\theta = \left\| \begin{vmatrix} \cos\theta & -r\sin\theta \\ \sin\theta & r\cos\theta \end{vmatrix} \right\| drd\theta = rdrd\theta$

$$\therefore \int_0^\infty \int_0^\infty e^{-(x^2+y^2)} \cos(x^2+y^2)dxdy = \int_0^{\frac{\pi}{2}} \int_0^\infty e^{-r^2}\cos(r^2)rdrd\theta$$

$$= \int_0^{\frac{\pi}{2}} d\theta \int_0^\infty e^{-r^2}\cos(r^2)rdr = \frac{\pi}{2}\int_0^\infty e^{-r^2}\cos(r^2)rdr$$

$$令 u = r^2 \text{ 则 } du = 2rdr \quad \therefore \int_0^\infty e^{-r^2}\cos(r^2)rdr = \frac{1}{2}\int_0^\infty e^{-u}\cos u\,du$$

藉由 integration by parts

$$则 \int_0^\infty e^{-u}\cos u\,du = e^{-u}\sin u|_{u=0}^{u=\infty} + \int_0^\infty e^{-u}\sin u\,du = \int_0^\infty e^{-u}\sin u\,du$$

$$= -e^{-u}\cos u|_{u=0}^{u=\infty} - \int_0^\infty e^{-u}\cos u\,du$$

$$\therefore \int_0^\infty e^{-u}\cos u\,du = \frac{1}{2} \Rightarrow \int_0^\infty e^{-r^2}\cos(r^2)rdr = \frac{1}{4}$$

$$\Rightarrow \int_0^\infty \int_0^\infty e^{-(x^2+y^2)}\cos(x^2+y^2)dxdy = \frac{\pi}{8}$$


Example 8.

$$求 \iint_R \tan^{-1}\frac{y}{x}dA = ?, \ R = \{(x,y): x \geq 0, y \geq 0, 0 \leq x^2+y^2 \leq a^2\}$$

【解】

$$令 x = r\cos\theta, \ y = r\sin\theta$$

$$则 \{(x,y): x \geq 0, y \geq 0, 0 \leq x^2+y^2 \leq a^2\} = \{(r,\theta): 0 \leq r \leq a, 0 \leq \theta \leq \frac{\pi}{2}\}$$

$$且\ dxdy = \left\| \begin{matrix} \dfrac{\partial x}{\partial r} & \dfrac{\partial x}{\partial \theta} \\ \dfrac{\partial y}{\partial r} & \dfrac{\partial y}{\partial \theta} \end{matrix} \right\| drd\theta = \left\| \begin{matrix} \cos\theta & -r\sin\theta \\ \sin\theta & r\cos\theta \end{matrix} \right\| drd\theta = rdrd\theta$$

$$\therefore \iint_R \tan^{-1}\frac{y}{x}dA = \int_0^{\frac{\pi}{2}} \int_0^a \tan^{-1}\left(\frac{r\sin\theta}{r\cos\theta}\right) rdrd\theta = \int_0^{\frac{\pi}{2}} \int_0^a \tan^{-1}(\tan\theta)\,rdrd\theta$$

$$= \frac{\theta^2}{2}\Big|_0^{\frac{\pi}{2}} \cdot \frac{r^2}{2}\Big|_0^a = \frac{(\pi a)^2}{16}$$

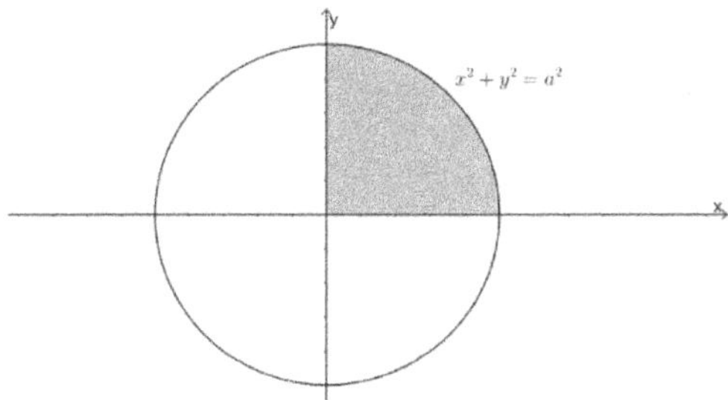

Example 9.

$$\text{求} \iint_R \sin^{-1}\frac{y}{\sqrt{x^2+y^2}}\, dA =?,\ R = \{(x,y): x \geq 0, y \geq 0, 0 \leq x^2 + y^2 \leq a^2\}$$

【解】

令 $x = r\cos\theta,\ y = r\sin\theta$

则 $\{(x,y): x \geq 0, y \geq 0, 0 \leq x^2 + y^2 \leq a^2\} = \{(r,\theta): 0 \leq r \leq a, 0 \leq \theta \leq \frac{\pi}{2}\}$

且 $dxdy = \left\|\begin{vmatrix}\dfrac{\partial x}{\partial r} & \dfrac{\partial x}{\partial \theta}\\[2mm] \dfrac{\partial y}{\partial r} & \dfrac{\partial y}{\partial \theta}\end{vmatrix}\right\| drd\theta = \left\|\begin{vmatrix}\cos\theta & -r\sin\theta\\ \sin\theta & r\cos\theta\end{vmatrix}\right\| drd\theta = rdrd\theta$

$$\therefore \iint_R \sin^{-1}\left(\frac{y}{\sqrt{x^2+y^2}}\right) dA = \int_0^{\frac{\pi}{2}}\int_0^a \sin^{-1}\left(\frac{r\sin\theta}{r}\right) rdrd\theta = \int_0^{\frac{\pi}{2}}\int_0^a \sin^{-1}(\sin\theta)\, rdrd\theta$$

$$= \left.\frac{\theta^2}{2}\right|_0^{\frac{\pi}{2}} \cdot \left.\frac{r^2}{2}\right|_0^a = \frac{(\pi a)^2}{16}$$

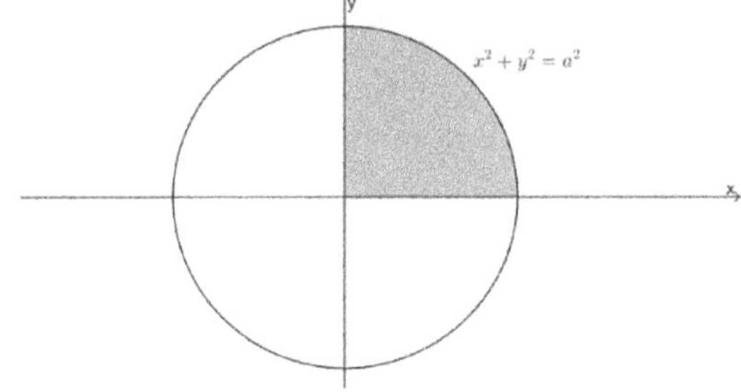

Example 10.

$$\text{求} \iint_R \cos^{-1} \frac{x}{\sqrt{x^2+y^2}}\, dA = ?,\quad R = \{(x,y): x \geq 0, y \geq 0, 0 \leq x^2+y^2 \leq a^2\}$$

【解】

令 $x = r\cos\theta,\ y = r\sin\theta$

则 $\{(x,y): x \geq 0, y \geq 0, 0 \leq x^2+y^2 \leq a^2\} = \{(r,\theta): 0 \leq r \leq a, 0 \leq \theta \leq \frac{\pi}{2}\}$

$$\text{且 } dxdy = \left\| \begin{vmatrix} \dfrac{\partial x}{\partial r} & \dfrac{\partial x}{\partial \theta} \\ \dfrac{\partial y}{\partial r} & \dfrac{\partial y}{\partial \theta} \end{vmatrix} \right\| drd\theta = \left\| \begin{matrix} \cos\theta & -r\sin\theta \\ \sin\theta & r\cos\theta \end{matrix} \right\| drd\theta = rdrd\theta$$

$$\therefore \iint_R \cos^{-1} \frac{x}{\sqrt{x^2+y^2}}\, dA = \int_0^{\frac{\pi}{2}} \int_0^a \cos^{-1}\left(\frac{r\cos\theta}{r}\right) rdrd\theta = \int_0^{\frac{\pi}{2}} \int_0^a \cos^{-1}(\cos\theta)\, rdrd\theta$$

$$= \frac{\theta^2}{2}\bigg|_0^{\frac{\pi}{2}} \cdot \frac{r^2}{2}\bigg|_0^a = \frac{(\pi a)^2}{16}$$

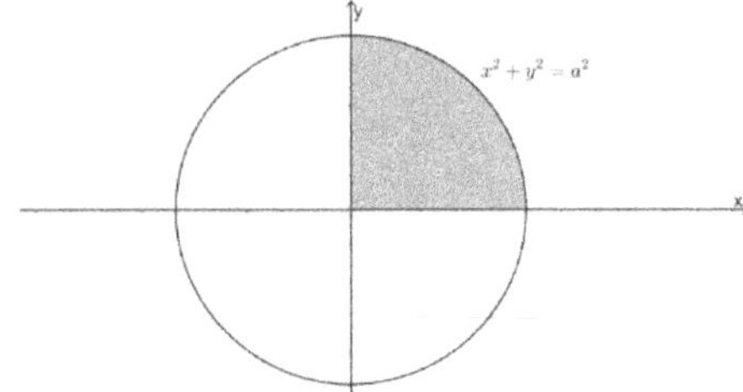


Example 11.

$$\text{求} \int_0^{\frac{1}{2}} \int_{\sqrt{3}x}^{\sqrt{1-x^2}} e^{-x^2-y^2}\, dydx = ?$$

【解】

令 $x = r\cos\theta,\ y = r\sin\theta$ 则

$$\{(x,y): 0 \leq x \leq \frac{1}{2}, \sqrt{3}x \leq y \leq \sqrt{1-x^2}\} = \{(r,\theta): 0 \leq r \leq 1, \frac{\pi}{3} \leq \theta \leq \frac{\pi}{2}\}$$

且 $dxdy = \left\|\begin{vmatrix} \dfrac{\partial x}{\partial r} & \dfrac{\partial x}{\partial \theta} \\ \dfrac{\partial y}{\partial r} & \dfrac{\partial y}{\partial \theta} \end{vmatrix}\right\| drd\theta = \left\|\begin{vmatrix} \cos\theta & -r\sin\theta \\ \sin\theta & r\cos\theta \end{vmatrix}\right\| drd\theta = rdrd\theta$

$\therefore \int_0^{\frac{1}{2}} \int_{\sqrt{3}x}^{\sqrt{1-x^2}} e^{-x^2-y^2}\, dydx = \int_{\frac{\pi}{3}}^{\frac{\pi}{2}} \int_0^1 e^{-r^2} rdrd\theta = \int_0^1 e^{-r^2} rdr \int_{\frac{\pi}{3}}^{\frac{\pi}{2}} d\theta = -\left.\frac{e^{-r^2}}{2}\right|_0^1 \cdot \frac{\pi}{6}$

$= \dfrac{\pi}{12}(1 - e^{-1})$

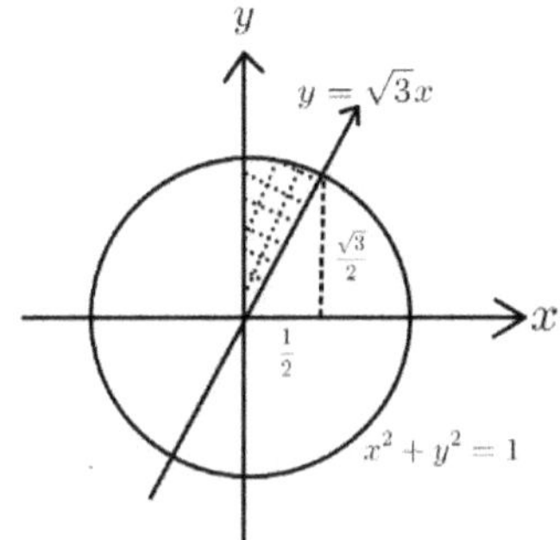


Example 12.

求 $\displaystyle\int_1^2 \int_0^{\sqrt{2x-x^2}} (x^2+y^2)^{-\frac{1}{2}}\, dydx =?$

【解】

令 $x = r\cos\theta,\ y = r\sin\theta$ 则

$\left\{(x,y): 1 \le x \le 2, 0 \le y \le \sqrt{2x-x^2}\right\} = \left\{(r,\theta): \sec\theta \le r \le 2\cos\theta, 0 \le \theta \le \frac{\pi}{4}\right\}$

且 $dxdy = \left\|\begin{vmatrix} \dfrac{\partial x}{\partial r} & \dfrac{\partial x}{\partial \theta} \\ \dfrac{\partial y}{\partial r} & \dfrac{\partial y}{\partial \theta} \end{vmatrix}\right\| drd\theta = \left\|\begin{vmatrix} \cos\theta & -r\sin\theta \\ \sin\theta & r\cos\theta \end{vmatrix}\right\| drd\theta = rdrd\theta$

$\therefore \int_1^2 \int_0^{\sqrt{2x-x^2}} (x^2+y^2)^{-\frac{1}{2}}\, dydx = \int_0^{\frac{\pi}{4}} \int_{\sec\theta}^{2\cos\theta} (r^2)^{-\frac{1}{2}} rdrd\theta = \int_0^{\frac{\pi}{4}} \int_{\sec\theta}^{2\cos\theta} drd\theta$

$= \int_0^{\frac{\pi}{4}} 2\cos\theta - \sec\theta\, d\theta = (2\sin\theta - \ln|\sec\theta + \tan\theta|)\big|_0^{\frac{\pi}{4}} = \sqrt{2} - \ln(1+\sqrt{2})$

Example 13.

$$\text{求} \iint_R \frac{x^2}{(x^2+y^2)^2}\,dxdy =?, \quad R = \{(x,y): a^2 \le x^2 + y^2 \le b^2, 0 < a < b\}$$

【解】

令 $x = r\cos\theta,\ y = r\sin\theta$ 则

$$R = \{(x,y): a^2 \le x^2 + y^2 \le b^2, 0 < a < b\} = \{(r,\theta): a \le r \le b, 0 \le \theta \le 2\pi\}$$

且 $dxdy = \left\| \begin{vmatrix} \dfrac{\partial x}{\partial r} & \dfrac{\partial x}{\partial \theta} \\ \dfrac{\partial y}{\partial r} & \dfrac{\partial y}{\partial \theta} \end{vmatrix} \right\| drd\theta = \left\| \begin{vmatrix} \cos\theta & -r\sin\theta \\ \sin\theta & r\cos\theta \end{vmatrix} \right\| drd\theta = r\,drd\theta$

$$\therefore \iint_R \frac{x^2}{(x^2+y^2)^2}\,dxdy = \int_0^{2\pi}\int_a^b \left(\frac{r^2\cos^2\theta}{r^4}\right) r\,drd\theta = \int_0^{2\pi}\int_a^b \frac{\cos^2\theta}{r}\,drd\theta$$

$$= \int_0^{2\pi}\cos^2\theta\,d\theta \int_a^b \frac{1}{r}\,dr = \int_0^{2\pi}\frac{1+\cos 2\theta}{2}\,d\theta \int_a^b \frac{1}{r}\,dr = \pi \ln\frac{b}{a}$$

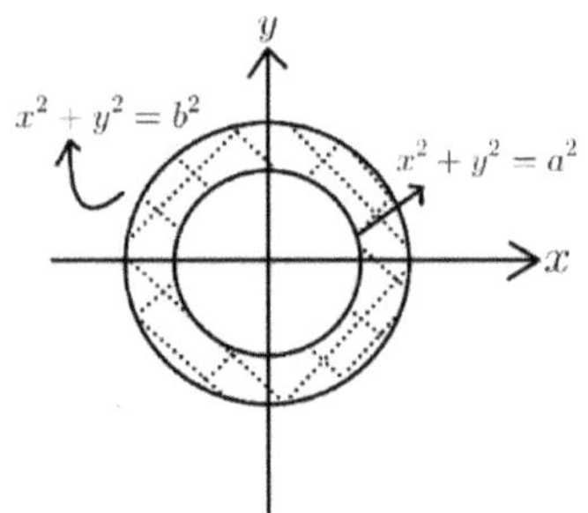


Example 14.

$$\text{求} \int_0^1 \int_{x^2}^x (x^2+y^2)^{-\frac{1}{2}}\,dydx =?$$

【解】

令 $x = r\cos\theta,\ y = r\sin\theta$ 则

$$R = \{(x,y): 0 \le x \le 1, x^2 < y < x\} = \left\{(r,\theta): 0 \le r \le \frac{\sin\theta}{\cos^2\theta}, 0 \le \theta \le \frac{\pi}{4}\right\}$$

且 $dxdy = \left\| \begin{vmatrix} \dfrac{\partial x}{\partial r} & \dfrac{\partial x}{\partial \theta} \\ \dfrac{\partial y}{\partial r} & \dfrac{\partial y}{\partial \theta} \end{vmatrix} \right\| drd\theta = \left\| \begin{vmatrix} \cos\theta & -r\sin\theta \\ \sin\theta & r\cos\theta \end{vmatrix} \right\| drd\theta = r\,drd\theta$

$$\therefore \int_0^1 \int_{x^2}^{x} (x^2 + y^2)^{-\frac{1}{2}} \, dy\,dx = \int_0^{\frac{\pi}{4}} \int_0^{\frac{\sin\theta}{\cos^2\theta}} \frac{1}{r} \cdot r\,dr\,d\theta = \int_0^{\frac{\pi}{4}} \frac{\sin\theta}{\cos^2\theta}\,d\theta = \frac{1}{\cos\theta}\Big|_0^{\frac{\pi}{4}} = \sqrt{2} - 1$$

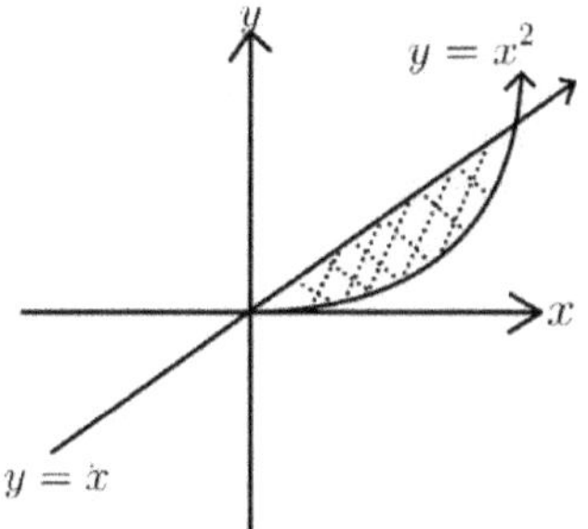


Example 15.

$$求 \int_{-3}^{3} \int_{-\sqrt{9-x^2}}^{\sqrt{9-x^2}} (9 - x^2 - y^2)^{\frac{1}{2}} \, dy\,dx =?$$

【解】

令 $x = r\cos\theta, \ y = r\sin\theta$ 则

$$\left\{(x,y): -3 \le x \le 3, -\sqrt{9-x^2} < y < \sqrt{9-x^2}\right\} = \{(r,\theta): 0 \le r \le 3, 0 \le \theta \le 2\pi\}$$

$$且 \ dxdy = \left\|\begin{vmatrix} \dfrac{\partial x}{\partial r} & \dfrac{\partial x}{\partial \theta} \\ \dfrac{\partial y}{\partial r} & \dfrac{\partial y}{\partial \theta} \end{vmatrix}\right\| dr\,d\theta = \left\|\begin{vmatrix} \cos\theta & -r\sin\theta \\ \sin\theta & r\cos\theta \end{vmatrix}\right\| dr\,d\theta = r\,dr\,d\theta$$

$$\therefore \int_{-3}^{3} \int_{-\sqrt{9-x^2}}^{\sqrt{9-x^2}} (9 - x^2 - y^2)^{\frac{1}{2}} \, dy\,dx = \int_0^{2\pi} \int_0^{3} (9 - r^2)^{\frac{1}{2}} \cdot r\,dr\,d\theta = 2\pi \cdot \frac{-(9-r^2)^{\frac{3}{2}}}{3}\Big|_0^3 = 18\pi$$

Example 16.

$$求 \iint_R \frac{1}{\sqrt{1 + x^2 + y^2}} \, dA =?, \ R = \{(x,y): x^2 + y^2 \le 4\}$$

【解】

令 $x = r\cos\theta, \ y = r\sin\theta$ 则 $\{(x,y): x^2 + y^2 \le 4\} = \{(r,\theta): 0 \le r \le 2, 0 \le \theta \le 2\pi\}$

且 $dxdy = \left\| \begin{vmatrix} \dfrac{\partial x}{\partial r} & \dfrac{\partial x}{\partial \theta} \\[2mm] \dfrac{\partial y}{\partial r} & \dfrac{\partial y}{\partial \theta} \end{vmatrix} \right\| drd\theta = \left\| \begin{vmatrix} \cos\theta & -r\sin\theta \\ \sin\theta & r\cos\theta \end{vmatrix} \right\| drd\theta = rdrd\theta$

$$\therefore \iint_R \frac{1}{\sqrt{1+x^2+y^2}}\, dA = \int_0^{2\pi}\int_0^2 \frac{1}{\sqrt{1+r^2}}\, rdrd\theta = \int_0^{2\pi} d\theta \int_0^2 \frac{rdr}{\sqrt{1+r^2}} = 2\pi \cdot (1+r^2)^{\frac{1}{2}}\Big|_{r=0}^{r=2}$$

$$= 2\pi(\sqrt{5}-1)$$

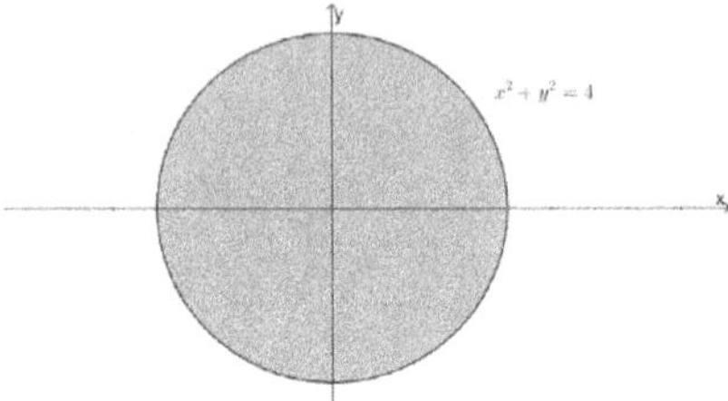


Example 17.

$$求 \iint_R \frac{x^2}{\sqrt{1+(x^2+y^2)^2}}\, dA = ?,\ R = \{(x,y): x^2+y^2 \leq 1, y > 0\}$$

【解】

令 $x = r\cos\theta,\ y = r\sin\theta$ 则 $\{(x,y): x^2+y^2 \leq 1\} = \{(r,\theta): 0 \leq r \leq 1, 0 \leq \theta \leq \pi\}$

且 $dxdy = \left\| \begin{vmatrix} \dfrac{\partial x}{\partial r} & \dfrac{\partial x}{\partial \theta} \\[2mm] \dfrac{\partial y}{\partial r} & \dfrac{\partial y}{\partial \theta} \end{vmatrix} \right\| drd\theta = \left\| \begin{vmatrix} \cos\theta & -r\sin\theta \\ \sin\theta & r\cos\theta \end{vmatrix} \right\| drd\theta = rdrd\theta$

$$\therefore \iint_R \frac{x^2}{\sqrt{1+(x^2+y^2)^2}}\, dA = \int_0^{\pi}\int_0^1 \frac{r^2\sin^2\theta}{\sqrt{1+r^4}}\, rdrd\theta = \int_0^{\pi} \sin^2\theta\, d\theta \int_0^1 \frac{r^3 dr}{\sqrt{1+r^4}}$$

$$= \int_0^{\pi} \frac{1-\cos 2\theta}{2}\, d\theta \int_0^1 \frac{r^3 dr}{\sqrt{1+r^4}} = \frac{\pi}{2} \cdot \frac{1}{2}(1+r^4)^{\frac{1}{2}}\Big|_{r=0}^{r=1} = \frac{\pi}{4}(\sqrt{2}-1)$$

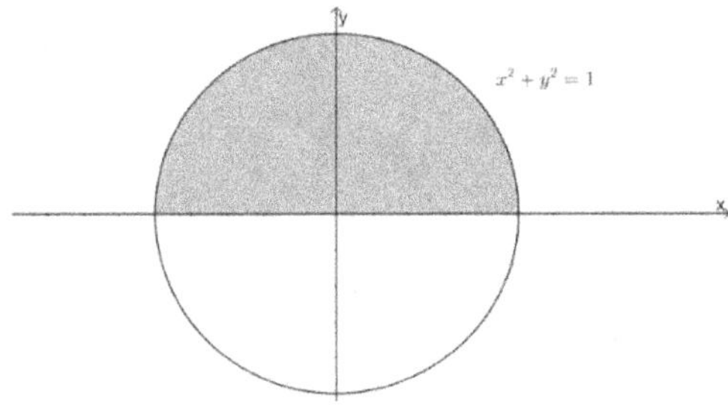

## Example 18.

$$求 \iint_R e^{x^2+y^2}\,dxdy =?\,,\ R = \{(x,y)\colon 0 \le y \le x, x^2 + y^2 \le 4\}$$

【解】

令 $x = r\cos\theta,\ y = r\sin\theta$

则 $\{(x,y)\colon 0 \le y \le x, x^2 + y^2 \le 4\} = \{(r,\theta)\colon 0 \le r \le 2, 0 \le \theta \le \dfrac{\pi}{4}\}$

且 $dxdy = \left\|\begin{matrix} \dfrac{\partial x}{\partial r} & \dfrac{\partial x}{\partial \theta} \\[2mm] \dfrac{\partial y}{\partial r} & \dfrac{\partial y}{\partial \theta} \end{matrix}\right\| drd\theta = \left\|\begin{matrix} \cos\theta & -r\sin\theta \\ \sin\theta & r\cos\theta \end{matrix}\right\| drd\theta = r\,drd\theta$

$\therefore \displaystyle\iint_R e^{x^2+y^2}\,dxdy = \int_0^{\frac{\pi}{4}}\int_0^2 e^{r^2}\,r\,drd\theta = \int_0^{\frac{\pi}{4}} \left.\frac{e^{r^2}}{2}\right|_{r=0}^{r=2} d\theta = \int_0^{\frac{\pi}{4}} \frac{e^4-1}{2}\,d\theta = \left(\frac{e^4-1}{8}\right)\pi$

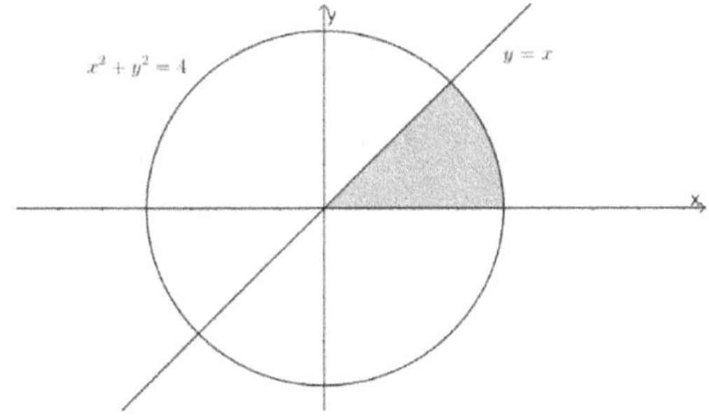

## Example 19.

$$求 \iint_R (x^2 + y^2)e^{\left(x^2+y^2\right)^2}\,dxdy =?\,,\ R = \{(x,y)\colon 0 \le y \le x, x^2 + y^2 \le 4\}$$

【解】

令 $x = r\cos\theta,\ y = r\sin\theta$

则 $\{(x,y): 0 \le y \le x, x^2 + y^2 \le 4\} = \{(r,\theta): 0 \le r \le 2, 0 \le \theta \le \dfrac{\pi}{4}\}$

且 $dxdy = \left\|\begin{vmatrix} \dfrac{\partial x}{\partial r} & \dfrac{\partial x}{\partial \theta} \\ \dfrac{\partial y}{\partial r} & \dfrac{\partial y}{\partial \theta} \end{vmatrix}\right\| drd\theta = \left\|\begin{vmatrix} \cos\theta & -r\sin\theta \\ \sin\theta & r\cos\theta \end{vmatrix}\right\| drd\theta = rdrd\theta$

$$\therefore \iint_R (x^2 + y^2)e^{(x^2+y^2)^2} dxdy = \int_0^{\frac{\pi}{4}} \int_0^2 e^{r^4} r^3 drd\theta = \int_0^{\frac{\pi}{4}} \frac{e^{r^4}}{4}\bigg|_{r=0}^{r=2} d\theta = \int_0^{\frac{\pi}{4}} \frac{e^{16}-1}{4} d\theta$$

$$= \left(\frac{e^{16}-1}{16}\right)\pi$$

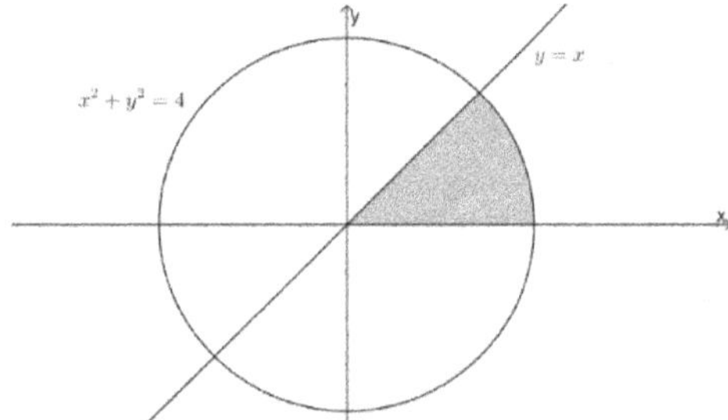

Example 20.

$$求 \iint_R \sqrt{a^2 - x^2 - y^2}\, dA = ?,\ R = \{(x,y): x^2 + y^2 \le a^2\}$$

【解】

令 $x = r\cos\theta,\ y = r\sin\theta$ 则 $\{(x,y): x^2 + y^2 \le a^2\} = \{(r,\theta): 0 \le r \le a, 0 \le \theta \le 2\pi\}$

且 $dxdy = \left\|\begin{vmatrix} \dfrac{\partial x}{\partial r} & \dfrac{\partial x}{\partial \theta} \\ \dfrac{\partial y}{\partial r} & \dfrac{\partial y}{\partial \theta} \end{vmatrix}\right\| drd\theta = \left\|\begin{vmatrix} \cos\theta & -r\sin\theta \\ \sin\theta & r\cos\theta \end{vmatrix}\right\| drd\theta = rdrd\theta$

$$\therefore \iint_R \sqrt{a^2 - x^2 - y^2}\, dA = \int_0^{2\pi} \int_0^a \sqrt{a^2 - r^2}\, r drd\theta = \int_0^{2\pi} d\theta \int_0^a r\sqrt{a^2 - r^2}\, dr$$

$$= 2\pi \cdot \frac{(-1)(a^2 - r^2)^{\frac{3}{2}}}{3}\bigg|_{r=0}^{r=a} = \frac{2\pi a^3}{3}$$

Example 21.

$$\text{求} \iint_R e^{\sqrt{x^2+y^2}}dxdy = ?, \quad R = \{(x,y): x^2+y^2 \leq 4, 0 \leq y \leq x\}$$

【解】

令 $x = r\cos\theta, \ y = r\sin\theta$ 则 $\{(x,y): x^2+y^2 \leq 4\} = \{(r,\theta): 0 \leq r \leq 2, 0 \leq \theta \leq \dfrac{\pi}{4}\}$

且 $dxdy = \left\| \begin{vmatrix} \dfrac{\partial x}{\partial r} & \dfrac{\partial x}{\partial \theta} \\ \dfrac{\partial y}{\partial r} & \dfrac{\partial y}{\partial \theta} \end{vmatrix} \right\| drd\theta = \left\| \begin{vmatrix} \cos\theta & -r\sin\theta \\ \sin\theta & r\cos\theta \end{vmatrix} \right\| drd\theta = rdrd\theta$

$$\therefore \iint_R e^{\sqrt{x^2+y^2}}dxdy = \int_0^{\frac{\pi}{4}} \int_0^2 e^r r \, drd\theta = \int_0^{\frac{\pi}{4}} d\theta \int_0^2 e^r r \, dr$$

藉由 integration by parts 则 $\displaystyle\int_0^2 e^r r \, dr = 2e^2 - (e^2 - 1) = e^2 + 1$

$$\therefore \iint_R e^{\sqrt{x^2+y^2}}dxdy = \frac{\pi}{4}(e^2 + 1)$$

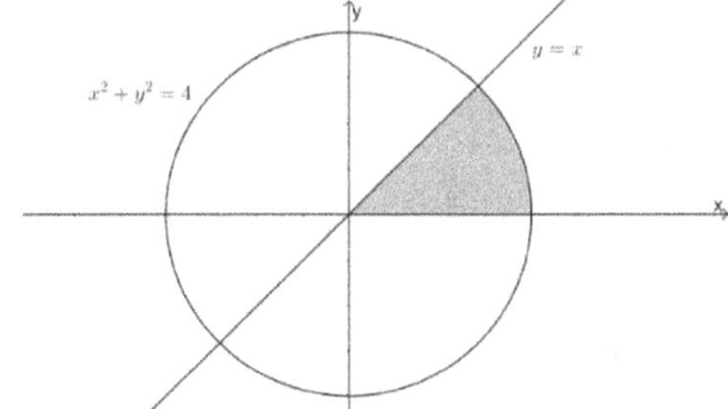


Example 22.

$$(1)\text{求} \iint_R \sin(x^2+y^2)dA = ?, R = \{(x,y): \pi^2 \leq x^2+y^2 \leq 9\pi^2\}$$

$$(2)\text{求} \iint_R \sec^2(x^2+y^2)dA = ?, R = \{(x,y): 0 \leq x^2+y^2 \leq \frac{\pi}{4}\}$$

【解】

(1)

令 $x = r\cos\theta, \ y = r\sin\theta$

则 $\{(x,y): \pi^2 \leq x^2+y^2 \leq 9\pi^2\} = \{(r,\theta): \pi \leq r \leq 3\pi, 0 \leq \theta \leq 2\pi\}$

且 $dxdy = \begin{Vmatrix} \dfrac{\partial x}{\partial r} & \dfrac{\partial x}{\partial \theta} \\ \dfrac{\partial y}{\partial r} & \dfrac{\partial y}{\partial \theta} \end{Vmatrix} drd\theta = \begin{Vmatrix} \cos\theta & -r\sin\theta \\ \sin\theta & r\cos\theta \end{Vmatrix} drd\theta = rdrd\theta$

$$\therefore \iint_R \sin(x^2 + y^2)\, dA = \int_0^{2\pi} \int_\pi^{3\pi} \sin(r^2)\, rdrd\theta = \int_0^{2\pi} d\theta \int_\pi^{3\pi} \sin(r^2)\, rdr$$

$$= 2\pi \cdot \frac{(-1)\cos r^2}{2}\Bigg|_\pi^{3\pi} = (-\pi)(\cos 9\pi^2 - \cos \pi^2)$$

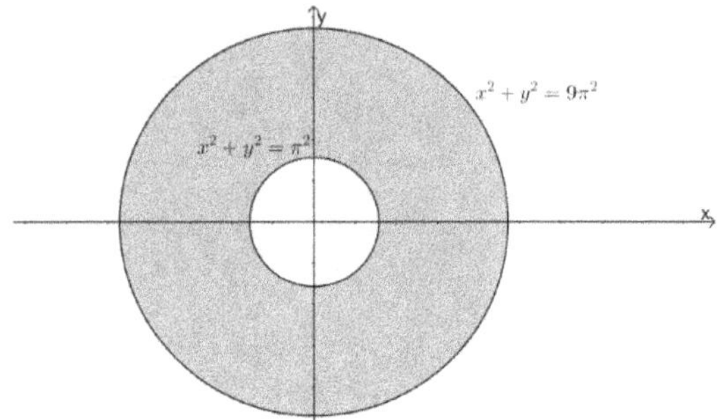


(2)

令 $x = r\cos\theta,\ y = r\sin\theta$

则 $\left\{(x,y): 0 \leq x^2 + y^2 \leq \dfrac{\pi}{4}\right\} = \left\{(r,\theta): 0 \leq r \leq \dfrac{\sqrt{\pi}}{2}, 0 \leq \theta \leq 2\pi\right\}$

且 $dxdy = \begin{Vmatrix} \dfrac{\partial x}{\partial r} & \dfrac{\partial x}{\partial \theta} \\ \dfrac{\partial y}{\partial r} & \dfrac{\partial y}{\partial \theta} \end{Vmatrix} drd\theta = \begin{Vmatrix} \cos\theta & -r\sin\theta \\ \sin\theta & r\cos\theta \end{Vmatrix} drd\theta = rdrd\theta$

$$\therefore \iint_R \sec^2(x^2 + y^2)\, dA = \int_0^{2\pi} \int_0^{\frac{\sqrt{\pi}}{2}} \sec^2(r^2)\, rdrd\theta = \int_0^{2\pi} d\theta \int_0^{\frac{\sqrt{\pi}}{2}} \sec^2(r^2)\, rdr$$

$$= 2\pi \cdot \frac{\tan r^2}{2}\Bigg|_0^{\frac{\sqrt{\pi}}{2}} = \pi$$

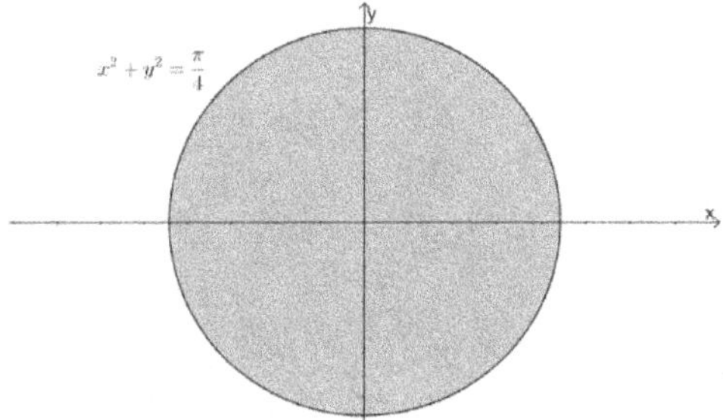


Example 23.

$$求 \int_0^a \int_0^{\sqrt{a^2-x^2}} \frac{1}{(1+x^2+y^2)^{\frac{3}{2}}} \, dydx =?$$

【解】

令 $x = r\cos\theta, \ y = r\sin\theta$ 则

$$\left\{(x,y): 0 \le x \le a, 0 < y < \sqrt{a^2-x^2}\right\} = \{(r,\theta): 0 \le r \le a, 0 \le \theta \le \frac{\pi}{2}\}$$

且 $dxdy = \left\|\begin{vmatrix} \dfrac{\partial x}{\partial r} & \dfrac{\partial x}{\partial \theta} \\ \dfrac{\partial y}{\partial r} & \dfrac{\partial y}{\partial \theta} \end{vmatrix}\right\| drd\theta = \left\|\begin{vmatrix} \cos\theta & -r\sin\theta \\ \sin\theta & r\cos\theta \end{vmatrix}\right\| drd\theta = rdrd\theta$

$$\therefore \int_0^a \int_0^{\sqrt{a^2-x^2}} \frac{1}{(1+x^2+y^2)^{\frac{3}{2}}} \, dydx = \int_0^{\frac{\pi}{2}} \int_0^a \frac{1}{(1+r^2)^{\frac{3}{2}}} \cdot rdrd\theta = -(1+r^2)^{-\frac{1}{2}}\Big|_0^a \cdot \frac{\pi}{2}$$

$$= \frac{\pi}{2}\left(1 - (1+a^2)^{-\frac{1}{2}}\right)$$

Example 24.

$$(1)求 \int_0^{4a} \int_0^{\sqrt{4ax-x^2}} x^2 + y^2 \, dydx =? \quad (2)求 \int_0^{2a} \int_{-\sqrt{2ay-y^2}}^{\sqrt{2ay-y^2}} \sqrt{x^2+y^2} \, dxdy =?$$

【解】

(1)

令 $x = r\cos\theta, \ y = r\sin\theta$

则 $\{(x,y): 0 \le x \le 4a, 0 \le y \le \sqrt{4ax-x^2}\} = \{(r,\theta): 0 \le r \le 4a\cos\theta, 0 \le \theta \le \frac{\pi}{2}\}$

$$\text{且 } dxdy = \left\| \begin{vmatrix} \dfrac{\partial x}{\partial r} & \dfrac{\partial x}{\partial \theta} \\ \dfrac{\partial y}{\partial r} & \dfrac{\partial y}{\partial \theta} \end{vmatrix} \right\| drd\theta = \left\| \begin{vmatrix} \cos\theta & -r\sin\theta \\ \sin\theta & r\cos\theta \end{vmatrix} \right\| drd\theta = rdrd\theta$$

$$\therefore \int_0^{4a} \int_0^{\sqrt{4ax-x^2}} x^2 + y^2 \, dydx = \int_0^{\frac{\pi}{2}} \int_0^{4a\cos\theta} r^3 drd\theta = \int_0^{\frac{\pi}{2}} \frac{r^4}{4} \Big|_0^{4a\cos\theta} d\theta = 64a^4 \int_0^{\frac{\pi}{2}} \cos^4\theta \, d\theta$$

$$\because \cos^4\theta = \left(\frac{1+\cos 2\theta}{2}\right)^2 = \frac{1+2\cos 2\theta + \cos^2 2\theta}{4} = \frac{1+2\cos 2\theta}{4} + \frac{1+\cos 4\theta}{8}$$

$$\therefore \int_0^{\frac{\pi}{2}} \cos^4\theta \, d\theta = \int_0^{\frac{\pi}{2}} \frac{3}{8} + \frac{\cos 2\theta}{2} + \frac{\cos 4\theta}{8} \, d\theta = \frac{3\pi}{16}$$

$$\therefore \int_0^{4a} \int_0^{\sqrt{4ax-x^2}} x^2 + y^2 \, dydx = 12\pi a^4$$

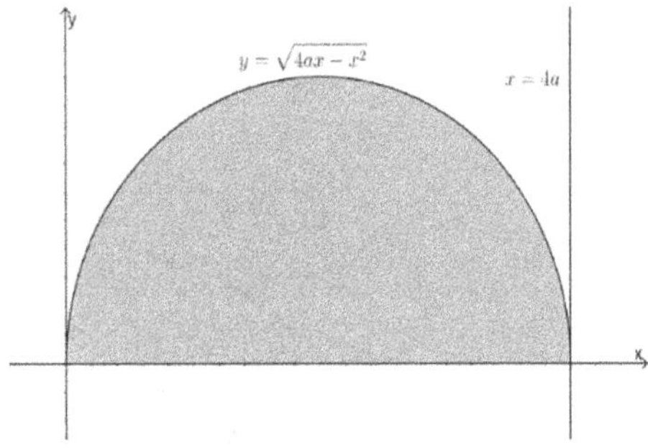


(2)

令$x = r\cos\theta, \ y = r\sin\theta$

则 $\{(x,y): -\sqrt{2ay-y^2} \leq x \leq \sqrt{2ay-y^2}, 0 \leq y \leq 2a\}$

$= \{(r,\theta): 0 \leq r \leq 2a\sin\theta, 0 \leq \theta \leq \pi\}$

$$\text{且 } dxdy = \left\| \begin{vmatrix} \dfrac{\partial x}{\partial r} & \dfrac{\partial x}{\partial \theta} \\ \dfrac{\partial y}{\partial r} & \dfrac{\partial y}{\partial \theta} \end{vmatrix} \right\| drd\theta = \left\| \begin{vmatrix} \cos\theta & -r\sin\theta \\ \sin\theta & r\cos\theta \end{vmatrix} \right\| drd\theta = rdrd\theta$$

$$\therefore \int_0^{2a} \int_{-\sqrt{2ay-y^2}}^{\sqrt{2ay-y^2}} \sqrt{x^2+y^2} \, dxdy = \int_0^{\pi} \int_0^{2a\sin\theta} r^2 drd\theta = \int_0^{\pi} \frac{r^3}{3} \Big|_0^{2a\sin\theta} d\theta$$

$$= \frac{8a^3}{3} \int_0^\pi \sin^3 \theta \, d\theta = \frac{8a^3}{3} \int_0^\pi \sin \theta (1 - \cos^2 \theta) \, d\theta = \frac{8a^3}{3} \left( -\cos \theta + \frac{\cos^3 \theta}{3} \right) \Big|_0^\pi = \frac{32a^3}{9}$$



## 8.3.4 使用广义坐标转换求双重积分的值

当积分区域$R$为下列这几种情形时,可视为使用广义坐标转换的时机

(i)积分区域$R$出现 $ax + by$, $cx + dy$ 且 $ad - bc \neq 0$

(ii)积分区域$R$或被积分函数出现 $ax^2 + bxy + cy^2$, 其中 $4ac - b^2 > 0, a > 0, b > 0$

(iii)积分区域$R$为 $y = ax$, $y = bx$, $xy = c$, $xy = d$ 所围区域, 其中 $abcd \neq 0$

考试类型:

Type 1.

假设$R = \{(x, y) : 0 \leq ax + by \leq s, 0 \leq cx + dy \leq t\}$, $a, b, c, d \in R, s, t > 0$

其中 $ad - bc \neq 0$

求 $\iint_R f(ax + by, cx + dy) dA =$?

解题流程:

Step1.

令 $u = ax + by$, $v = cx + dy$ 则 $x = \dfrac{du - bv}{ad - bc}$ 且 $y = \dfrac{av - cu}{ad - bc}$

$\because dxdy = \left\| \begin{array}{cc} \dfrac{\partial x}{\partial u} & \dfrac{\partial x}{\partial v} \\ \dfrac{\partial y}{\partial u} & \dfrac{\partial y}{\partial v} \end{array} \right\| dudv = \dfrac{1}{|ad - bc|^2} \left\| \begin{array}{cc} d & -b \\ -c & a \end{array} \right\| dudv = \dfrac{1}{|ad - bc|} dudv$

Step2.

$\therefore \iint_R f(ax + by, cx + dy) dA = \dfrac{1}{|ad - bc|} \int_0^t \int_0^s f(u, v) dudv$

Step3.

计算此积分 $\dfrac{1}{|ad - bc|} \displaystyle\int_0^t \int_0^s f(u, v)\,dudv =?$

<u>范例说明:</u>

求 $\displaystyle\iint_R f(ax + by, cx + dy)\,dA =?$, $R = \{(x, y): 0 \le ax + by \le s, 0 \le cx + dy \le t\}$, $a, b, c, d \in R, s, t > 0$

令 $u = ax + by, \ v = cx + dy$ 则 $x = \dfrac{du - bv}{ad - bc}$ 且 $y = \dfrac{av - cu}{ad - bc}$

$\because dxdy = \left\|\begin{vmatrix} \dfrac{\partial x}{\partial u} & \dfrac{\partial x}{\partial v} \\ \dfrac{\partial y}{\partial u} & \dfrac{\partial y}{\partial v} \end{vmatrix}\right\| dudv = \dfrac{1}{|ad - bc|^2} \left\|\begin{vmatrix} d & -b \\ -c & a \end{vmatrix}\right\| dudv = \dfrac{1}{|ad - bc|} dudv$

$\therefore \displaystyle\iint_R f(ax + by, cx + dy)\,dA = \dfrac{1}{|ad - bc|} \int_0^t \int_0^s f(u, v)\,dudv$

(I) 若 $f(ax + by, cx + dy) = (ax + by)(cx + dy)$

则 $\displaystyle\iint_R f(ax + by, cx + dy)\,dA = \dfrac{1}{|ad - bc|} \int_0^t \int_0^s uv\,dudv =?$

(II) 若 $f(ax + by, cx + dy) = e^{\frac{cx+dy}{ax+by}}$

则 $\displaystyle\iint_R f(ax + by, cx + dy)\,dA = \dfrac{1}{|ad - bc|} \int_0^t \int_0^s e^{\frac{v}{u}}\,dudv =?$

(III) 若 $f(ax + by, cx + dy) = (ax + by)^2 + (cx + dy)^2$

则 $\displaystyle\iint_R f(ax + by, cx + dy)\,dA = \dfrac{1}{|ad - bc|} \int_0^t \int_0^s u^2 + v^2\,dudv =?$

Type 2.

假设 $R = \{(x, y): ax^2 + bxy + cy^2 \le d^2\}, \ 4ac - b^2 > 0, a > 0, b > 0$

求 $\displaystyle\iint_R f(ax^2 + bxy + cy^2)\,dA =?$

解题流程:

Step1.

$$\because ax^2 + bxy + cy^2 = a(x + \frac{by}{2a})^2 + \frac{4ac - b^2}{4a}y^2$$

$$令 \sqrt{a}\left(x + \frac{by}{2a}\right) = u, \quad \sqrt{\frac{4ac - b^2}{4a}}\,y = v$$

$$\therefore dxdy = \left\|\begin{vmatrix} \dfrac{\partial x}{\partial u} & \dfrac{\partial x}{\partial v} \\ \dfrac{\partial y}{\partial u} & \dfrac{\partial y}{\partial v} \end{vmatrix}\right\| dudv = \left\|\begin{vmatrix} \dfrac{1}{\sqrt{a}} & 0 \\ \dfrac{2\sqrt{a}}{b} & \sqrt{\dfrac{4a}{4ac - b^2}} \end{vmatrix}\right\| dudv = \frac{2}{\sqrt{4ac - b^2}}\,dudv$$

Step2.

$$\therefore \iint_R f(ax^2 + bxy + cy^2)dA = \iint_{u^2 + v^2 \leq d^2} f(u^2 + v^2)\frac{2}{\sqrt{4ac - b^2}}\,dudv$$

Step3.

$$令 u = r\cos\theta, v = r\sin\theta \text{ 则} \{(u,v): u^2 + v^2 \leq d^2\} = \{(r,\theta): 0 \leq r \leq d, 0 \leq \theta \leq 2\pi\}$$

Step4.

$$\because dudv = \left\|\begin{vmatrix} \dfrac{\partial u}{\partial r} & \dfrac{\partial u}{\partial \theta} \\ \dfrac{\partial v}{\partial r} & \dfrac{\partial v}{\partial \theta} \end{vmatrix}\right\| drd\theta = \left\|\begin{vmatrix} \cos\theta & -r\sin\theta \\ \sin\theta & r\cos\theta \end{vmatrix}\right\| drd\theta = rdrd\theta$$

$$\therefore \iint_{u^2 + v^2 \leq d^2} f(u^2 + v^2)\frac{2}{\sqrt{4ac - b^2}}\,dudv = \int_0^{2\pi}\int_0^d \frac{2f(r^2)}{\sqrt{4ac - b^2}}\,rdrd\theta$$

$$= \frac{2}{\sqrt{4ac - b^2}}\int_0^{2\pi}d\theta\int_0^d f(r^2)rdr$$

Step5.

$$求 \frac{2}{\sqrt{4ac - b^2}}\int_0^{2\pi}d\theta\int_0^d f(r^2)rdr = ?$$

<u>范例说明:</u>

$$求 \iint_R f(ax^2 + bxy + cy^2)dA = ?, R = \{(x,y): ax^2 + bxy + cy^2 \leq d^2\}, 其中$$

$$4ac - b^2 > 0, a > 0, b > 0$$

$$\iint_R f(ax^2 + bxy + cy^2)dA = \iint_{u^2+v^2\leq d^2} f(u^2 + v^2)\frac{2}{\sqrt{4ac - b^2}}\,dudv$$

$$= \frac{2}{\sqrt{4ac - b^2}}\int_0^{2\pi} d\theta \int_0^d f(r^2)r\,dr$$

(I)若 $f(t) = t$ 则 $\iint_R f(ax^2 + bxy + cy^2)dA = \dfrac{2}{\sqrt{4ac - b^2}}\displaystyle\int_0^{2\pi} d\theta \int_0^d r^3\,dr$

(II)若 $f(t) = e^{-t}$ 则 $\iint_R f(ax^2 + bxy + cy^2)dA = \dfrac{2}{\sqrt{4ac - b^2}}\displaystyle\int_0^{2\pi} d\theta \int_0^d e^{-r^2}r\,dr$

(III)若 $f(t) = e^t$ 则 $\iint_R f(ax^2 + bxy + cy^2)dA = \dfrac{2}{\sqrt{4ac - b^2}}\displaystyle\int_0^{2\pi} d\theta \int_0^d e^{r^2}r\,dr$

(IV)若 $f(t) = \ln(t)$ 则 $\iint_R f(ax^2 + bxy + cy^2)dA = \dfrac{2}{\sqrt{4ac - b^2}}\displaystyle\int_0^{2\pi} d\theta \int_0^d \ln(r^2)\,r\,dr$

(V)若 $f(t) = \sin(t)$ 则 $\iint_R f(ax^2 + bxy + cy^2)dA = \dfrac{2}{\sqrt{4ac - b^2}}\displaystyle\int_0^{2\pi} d\theta \int_0^d \sin(r^2)\,r\,dr$

Type 3.

假设 $R$ 为 $y = x^\alpha$, $y = \beta x^\alpha$, $x = y^\alpha$, $x = \gamma y^\alpha$ 所围面积,其中 $\alpha > 1, \beta > 1, \gamma > 1$

求 $\iint_R f\left(\dfrac{y}{x^\alpha}, \dfrac{x}{y^\alpha}\right)dA =?$

解题流程:

Step1.

令 $u = \dfrac{y}{x^\alpha}$, $v = \dfrac{x}{y^\alpha}$ 则 $x = u^{\frac{\alpha}{1-\alpha^2}}v^{\frac{1}{1-\alpha^2}}$, $y = u^{\frac{1}{1-\alpha^2}}v^{\frac{\alpha}{1-\alpha^2}}$

且 $dxdy = \left\|\begin{vmatrix}\dfrac{\partial x}{\partial u} & \dfrac{\partial x}{\partial v}\\[2mm] \dfrac{\partial y}{\partial u} & \dfrac{\partial y}{\partial v}\end{vmatrix}\right\| dudv = \left\|\begin{vmatrix}\dfrac{\alpha u^{\left(\frac{\alpha}{1-\alpha^2}-1\right)}v^{\frac{1}{1-\alpha^2}}}{1-\alpha^2} & \dfrac{u^{\frac{\alpha}{1-\alpha^2}}v^{\left(\frac{1}{1-\alpha^2}-1\right)}}{1-\alpha^2}\\[4mm] \dfrac{u^{\left(\frac{1}{1-\alpha^2}-1\right)}v^{\frac{\alpha}{1-\alpha^2}}}{1-\alpha^2} & \dfrac{\alpha u^{\frac{1}{1-\alpha^2}}v^{\left(\frac{\alpha}{1-\alpha^2}-1\right)}}{1-\alpha^2}\end{vmatrix}\right\| dudv$

$$= \frac{\alpha^2 u^{\frac{\alpha-1+\alpha^2+1}{1-\alpha^2}}v^{\frac{1+\alpha-1+\alpha^2}{1-\alpha^2}} - u^{\frac{\alpha-1+\alpha^2+1}{1-\alpha^2}}v^{\frac{1+\alpha-1+\alpha^2}{1-\alpha^2}}}{(1-\alpha^2)^2}\,dudv = \frac{\alpha^2 u^{\frac{\alpha}{1-\alpha}}v^{\frac{\alpha}{1-\alpha}} - u^{\frac{\alpha}{1-\alpha}}v^{\frac{\alpha}{1-\alpha}}}{(1-\alpha^2)^2}\,dudv$$

$$= \frac{u^{\frac{\alpha}{1-\alpha}}v^{\frac{\alpha}{1-\alpha}}}{\alpha^2 - 1}\,dudv$$

Step2.

令 $R = \left\{ (x, y) : 1 \le \dfrac{y}{x^\alpha} \le \beta, 1 \le \dfrac{x}{y^\alpha} \le \gamma \right\}$ 则 $R = \{(u, v) : 1 \le u \le \beta, 1 \le v \le \gamma\}$

Step3.

$$\therefore \iint_R f\left(\frac{y}{x^\alpha}, \frac{x}{y^\alpha}\right) dA = \int_1^\beta \int_1^\gamma f(u, v) \frac{u^{\frac{\alpha}{1-\alpha}} v^{\frac{\alpha}{1-\alpha}}}{\alpha^2 - 1} \, dv \, du$$

Type 4.

求 $\iint_R f\left(\dfrac{y}{x}, xy\right) dx dy = ?$，其中 $R$ 为 $y = ax,\ y = bx,\ xy = c,\ xy = d$ 所围区域，
$a < b, c < d$

解题流程:

Step1.

令 $\dfrac{y}{x} = u,\ xy = v$ 则 $x = \sqrt{\dfrac{v}{u}},\ y = \sqrt{uv}$

且 $dxdy = \left\| \begin{vmatrix} \dfrac{\partial x}{\partial u} & \dfrac{\partial x}{\partial v} \\ \dfrac{\partial y}{\partial u} & \dfrac{\partial y}{\partial v} \end{vmatrix} \right\| dudv = \left\| \begin{vmatrix} \dfrac{1}{2}\left(\dfrac{v}{u}\right)^{-\frac{1}{2}}\left(-\dfrac{v}{u^2}\right) & \dfrac{1}{2}\left(\dfrac{v}{u}\right)^{-\frac{1}{2}}\left(\dfrac{1}{u}\right) \\ \dfrac{1}{2}(uv)^{-\frac{1}{2}}v & \dfrac{1}{2}(uv)^{-\frac{1}{2}}u \end{vmatrix} \right\| dudv = \dfrac{1}{2u} dudv$

Step2.

令 $R = \left\{ (x, y) : a \le \dfrac{y}{x} \le b, c \le xy \le d \right\}$ 则 $R = \{(u, v) : a \le u \le b, c \le v \le d\}$

Step3.

求 $\iint_R f\left(\dfrac{y}{x}, xy\right) dx dy = \int_a^b \int_c^d \dfrac{f(u, v)}{2u} \, dv \, du$

<u>范例说明:</u>

求 $\iint_R f\left(\dfrac{y}{x}, xy\right) dx dy = ?$，其中 $R$ 为 $y = ax,\ y = bx,\ xy = c,\ xy = d$ 所围区域，$a < b$,
$c < d$

(I) 如果 $f\left(\dfrac{y}{x}, xy\right) = e^{-\frac{y}{x} \cdot xy}$

则 $\displaystyle\iint_R f\left(\frac{y}{x}, xy\right) dxdy = \int_a^b \int_c^d \frac{f(u,v)}{2u} dvdu = \int_a^b \int_c^d \frac{e^{-uv}}{2u} dvdu$

(II) 如果 $f\left(\dfrac{y}{x}, xy\right) = \left(\dfrac{y}{x}\right)^2 + (xy)^2$

则 $\displaystyle\iint_R f\left(\frac{y}{x}, xy\right) dxdy = \int_a^b \int_c^d \frac{f(u,v)}{2u} dvdu = \int_a^b \int_c^d \frac{u^2 + v^2}{2u} dvdu$

Example 1.

$$\text{求} \iint_R x^2 + y^2 dA =?,\ R = \left\{(x,y): \frac{x^2}{a^2} + \frac{y^2}{b^2} \le 4\right\}$$

【解】

令 $x = ar\cos\theta,\ y = br\sin\theta$ 则 $\left\{(x,y): \dfrac{x^2}{a^2} + \dfrac{y^2}{b^2} \le 4\right\} = \{(r,\theta): 0 \le r \le 2, 0 \le \theta \le 2\pi\}$

且 $dxdy = \left\|\begin{matrix} \dfrac{\partial x}{\partial r} & \dfrac{\partial x}{\partial \theta} \\ \dfrac{\partial y}{\partial r} & \dfrac{\partial y}{\partial \theta} \end{matrix}\right\| drd\theta = \left\|\begin{matrix} a\cos\theta & -ra\sin\theta \\ b\sin\theta & rb\cos\theta \end{matrix}\right\| drd\theta = abr\,drd\theta$

$\therefore \displaystyle\iint_R x^2 + y^2 dA = \int_0^{2\pi} \int_0^2 ((ar)^2 \cos^2\theta + (br)^2 \sin^2\theta) abr\, drd\theta$

$= a^3 b \displaystyle\int_0^{2\pi} \cos^2\theta\, d\theta \int_0^2 r^3 dr + ab^3 \int_0^{2\pi} \sin^2\theta\, d\theta \int_0^2 r^3 dr$

$\because \displaystyle\int_0^{2\pi} \cos^2\theta\, d\theta = \int_0^{2\pi} \frac{1 + \cos 2\theta}{2} d\theta = \pi,$

$\displaystyle\int_0^{2\pi} \sin^2\theta\, d\theta = \int_0^{2\pi} \frac{1 - \cos 2\theta}{2} d\theta = \pi$ 且 $\displaystyle\int_0^2 r^3 dr = \frac{16}{4} = 4$

$\therefore \displaystyle\iint_R x^2 + y^2 dA = a^3 b(4\pi) + ab^3(4\pi) = 4ab\pi(a^2 + b^2)$

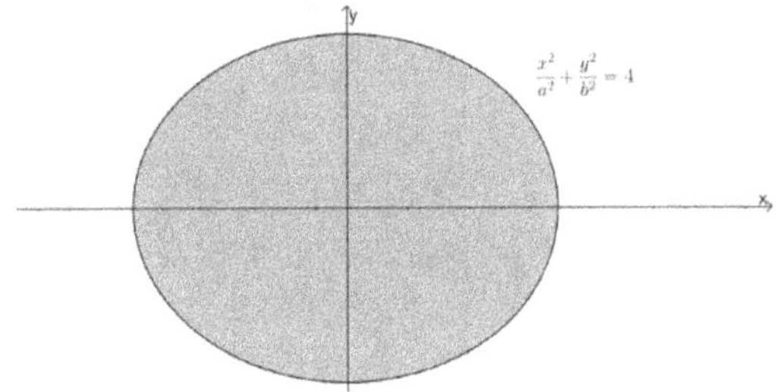

Example 2.

$$\text{求} \iint_R e^{x+y}\, dA =?, \quad R = \{(x,y): |x| + |y| \le 2\}$$

【解】

令 $x + y = u, \ x - y = v$ 则 $\{(x,y): |x| + |y| \le 2\} = \{(u,v): -2 \le u \le 2, -2 \le v \le 2\}$

$$\text{且 } x = \frac{u+v}{2}, \ y = \frac{u-v}{2}, \ dxdy = \left\| \begin{matrix} \dfrac{\partial x}{\partial u} & \dfrac{\partial x}{\partial v} \\ \dfrac{\partial y}{\partial u} & \dfrac{\partial y}{\partial v} \end{matrix} \right\| dudv = \left\| \begin{matrix} \dfrac{1}{2} & \dfrac{1}{2} \\ \dfrac{1}{2} & \dfrac{-1}{2} \end{matrix} \right\| dudv = \left| \frac{-1}{2} \right| dudv$$

$$\therefore \iint_R e^{x+y}\, dA = \int_{-2}^{2} \int_{-2}^{2} \frac{e^u}{2}\, dudv = \frac{1}{2} \int_{-2}^{2} dv \int_{-2}^{2} e^u\, du = 2(e^2 - e^{-2})$$

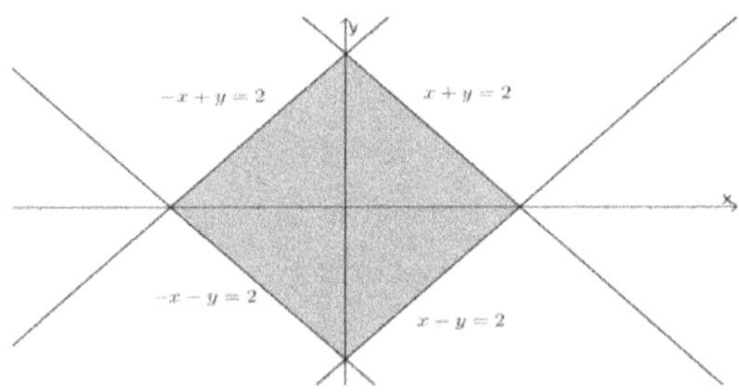


Example 3.

$$\text{求} \iint_R x^2 y\, dA =?, \quad R = \{(x,y): y \ge 0, \ (x-1)^2 + y^2 \le 1\}$$

【解】

令 $x = r\cos\theta + 1, \ y = r\sin\theta$

则 $\{(x,y): (x-1)^2 + y^2 \le 1, y \ge 0\} = \{(r,\theta): 0 \le r \le 1, 0 \le \theta \le \pi\}$

$$\text{且 } dxdy = \left\| \begin{matrix} \dfrac{\partial x}{\partial r} & \dfrac{\partial x}{\partial \theta} \\ \dfrac{\partial y}{\partial r} & \dfrac{\partial y}{\partial \theta} \end{matrix} \right\| drd\theta = \left\| \begin{matrix} \cos\theta & -r\sin\theta \\ \sin\theta & r\cos\theta \end{matrix} \right\| drd\theta = r\,drd\theta$$

$$\therefore \iint_R x^2 y\, dA = \int_0^{\pi} \int_0^1 (r\cos\theta + 1)^2 r\sin\theta\, r\,drd\theta$$

$$= \int_0^{\pi} \int_0^1 (r^2\cos^2\theta + 2r\cos\theta + 1)r\sin\theta\, r\,drd\theta$$

$$= \int_0^\pi \cos^2\theta \sin\theta \, d\theta \int_0^1 r^4 dr + \int_0^\pi 2\cos\theta \sin\theta \, d\theta \int_0^1 r^3 dr + \int_0^\pi \sin\theta \, d\theta \int_0^1 r^2 dr$$

$$= (-1)\frac{\cos^3\theta}{3}\Big|_0^\pi \cdot \frac{r^5}{5}\Big|_0^1 - \frac{\cos 2\theta}{2}\Big|_0^\pi \cdot \frac{r^4}{4}\Big|_0^1 - \cos\theta|_0^\pi \cdot \frac{r^3}{3}\Big|_0^1 = \frac{4}{5}$$

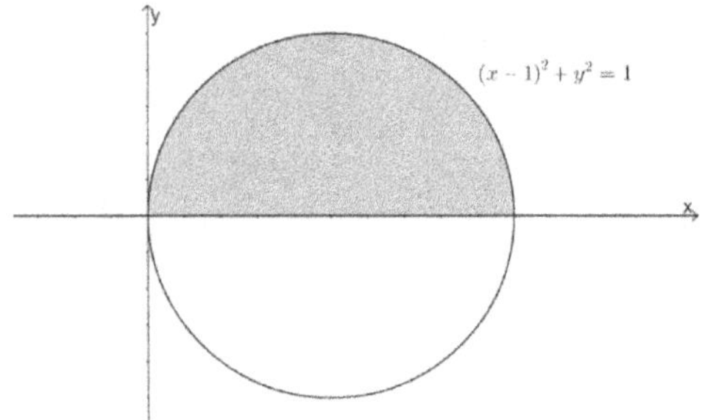


Example 4.

$$求 \iint_R e^{\frac{y-x}{x+y}} dxdy =?, \ R = \{(x,y): x \geq 0, y \geq 0, x+y \leq 3\}$$

【解】

令 $x + y = u, \ y - x = v$

则 $\{(x,y): x \geq 0, y \geq 0, x+y \leq 3\} = \{(u,v): 0 \leq u \leq 3, -u \leq v \leq u\}$

且 $x = \dfrac{u-v}{2}, \ y = \dfrac{u+v}{2}, \ dxdy = \left\|\begin{matrix} \dfrac{\partial x}{\partial u} & \dfrac{\partial x}{\partial v} \\ \dfrac{\partial y}{\partial u} & \dfrac{\partial y}{\partial v} \end{matrix}\right\| dudv = \left\|\begin{matrix} \dfrac{1}{2} & \dfrac{-1}{2} \\ \dfrac{1}{2} & \dfrac{1}{2} \end{matrix}\right\| dudv = |\dfrac{-1}{2}| dudv$

$$\therefore \iint_R e^{\frac{y-x}{x+y}} dxdy = \int_0^3 \int_{-u}^u e^{\frac{v}{u}} \cdot \frac{1}{2} dvdu = \frac{1}{2}\int_0^3 u\, e^{\frac{v}{u}}\Big|_{v=-u}^{v=u} du = \frac{1}{2}\int_0^3 u(e - e^{-1})du$$

$$= \frac{1}{4}(e - e^{-1})u^2\Big|_{u=0}^{u=3} = \frac{9}{4}(e - e^{-1})$$

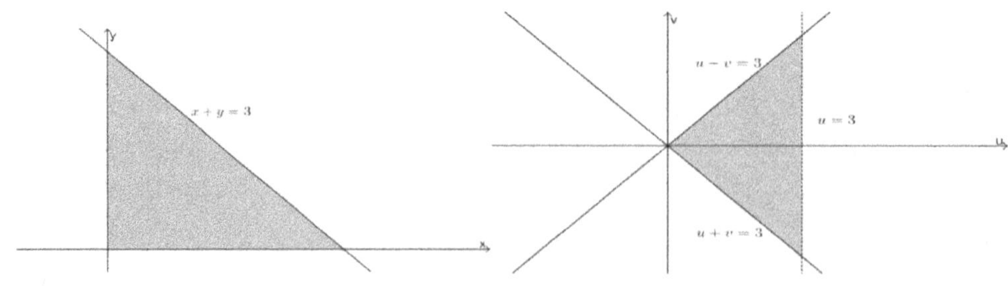

Example 5.

$$\text{试证} \int_{-\infty}^{\infty} \int_{-\infty}^{\infty} e^{-(x^2+2xy+10y^2)} dxdy = \frac{\pi}{3}$$

【解】

$$\because x^2 + 2xy + 10y^2 = (x+y)^2 + 9y^2, \quad 令(x+y) = u, \quad 3y = v$$

$$\text{则} dxdy = \left\| \begin{matrix} \dfrac{\partial x}{\partial u} & \dfrac{\partial x}{\partial v} \\ \dfrac{\partial y}{\partial u} & \dfrac{\partial y}{\partial v} \end{matrix} \right\| dudv = \left\| \begin{matrix} 1 & 0 \\ 1 & \dfrac{1}{3} \end{matrix} \right\| dudv = \frac{1}{3} dudv$$

$$\therefore \int_{-\infty}^{\infty} \int_{-\infty}^{\infty} e^{-(x^2+2xy+10y^2)} dxdy = \int_{-\infty}^{\infty} \int_{-\infty}^{\infty} e^{-(u^2+v^2)} \frac{1}{3} dudv$$

$$令 u = r\cos\theta, \quad v = r\sin\theta$$

$$\text{则} \{(u,v): -\infty \le u \le \infty, -\infty \le v \le \infty\} = \{(r,\theta): 0 \le r \le \infty, 0 \le \theta \le 2\pi\}$$

$$\text{且 } dudv = \left\| \begin{matrix} \dfrac{\partial u}{\partial r} & \dfrac{\partial u}{\partial \theta} \\ \dfrac{\partial v}{\partial r} & \dfrac{\partial v}{\partial \theta} \end{matrix} \right\| drd\theta = \left\| \begin{matrix} \cos\theta & -r\sin\theta \\ \sin\theta & r\cos\theta \end{matrix} \right\| drd\theta = rdrd\theta$$

$$\therefore \int_{-\infty}^{\infty} \int_{-\infty}^{\infty} e^{-(u^2+v^2)} du\,dv = \int_0^{2\pi} \int_0^{\infty} e^{-r^2} rdrd\theta = \int_0^{2\pi} d\theta \int_0^{\infty} e^{-r^2} rdr = 2\pi \cdot \left. \frac{-e^{-r^2}}{2} \right|_{r=0}^{r=\infty} = \pi$$

$$\therefore \int_{-\infty}^{\infty} \int_{-\infty}^{\infty} e^{-(x^2+2xy+10y^2)} dxdy = \frac{\pi}{3}$$

Example 6.

$$(1) 假设 a > 0, \quad b > 0 \text{ 且 } b^2 < 4ac, \quad 试证 \int_{-\infty}^{\infty} \int_{-\infty}^{\infty} e^{-(ax^2+bxy+cy^2)} dxdy = \frac{2\pi}{\sqrt{4ac - b^2}}$$

$$(2) 假设 R = \{(x,y): x^2 + 2xy + 3y^2 \le t^2\}, \quad I(t) = \iint_R e^{-(x^2+2xy+3y^2)} dA,$$

$$(a) \ 求 \ I(1) =? \quad (b) \ \lim_{t \to \infty} I(t) =?$$

【解】

(1)

$$\because ax^2 + bxy + cy^2 = a\left(x + \frac{by}{2a}\right)^2 + \left(\frac{4ac - b^2}{4a}\right) y^2$$

令 $\sqrt{a}\left(x + \dfrac{by}{2a}\right) = u,\ \sqrt{\dfrac{4ac - b^2}{4a}}\, y = v$

则 $dxdy = \left\|\begin{matrix} \dfrac{\partial x}{\partial u} & \dfrac{\partial x}{\partial v} \\[2mm] \dfrac{\partial y}{\partial u} & \dfrac{\partial y}{\partial v} \end{matrix}\right\| dudv = \left\|\begin{matrix} \dfrac{1}{\sqrt{a}} & 0 \\[3mm] \dfrac{2\sqrt{a}}{b} & \sqrt{\dfrac{4a}{4ac - b^2}} \end{matrix}\right\| dudv = \dfrac{2}{\sqrt{4ac - b^2}}\, dudv$

$\therefore \displaystyle\int_{-\infty}^{\infty}\int_{-\infty}^{\infty} e^{-(ax^2+bxy+cy^2)}\,dxdy = \int_{-\infty}^{\infty}\int_{-\infty}^{\infty} e^{-(u^2+v^2)}\dfrac{2}{\sqrt{4ac - b^2}}\,dudv$

令 $u = r\cos\theta,\ v = r\sin\theta$

则 $\{(u,v): -\infty \le u \le \infty, -\infty \le v \le \infty\} = \{(r,\theta): 0 \le r \le \infty, 0 \le \theta \le 2\pi\}$

且 $dudv = \left\|\begin{matrix} \dfrac{\partial u}{\partial r} & \dfrac{\partial u}{\partial \theta} \\[2mm] \dfrac{\partial v}{\partial r} & \dfrac{\partial v}{\partial \theta} \end{matrix}\right\| drd\theta = \left\|\begin{matrix} \cos\theta & -r\sin\theta \\ \sin\theta & r\cos\theta \end{matrix}\right\| drd\theta = r\,drd\theta$

$\therefore \displaystyle\int_{-\infty}^{\infty}\int_{-\infty}^{\infty} e^{-(u^2+v^2)}\,du\,dv = \int_{0}^{2\pi}\int_{0}^{\infty} e^{-r^2} r\,drd\theta = \int_{0}^{2\pi} d\theta \int_{0}^{\infty} e^{-r^2} r\,dr = 2\pi \cdot \dfrac{-e^{-r^2}}{2}\Bigg|_{r=0}^{r=\infty} = \pi$

$\therefore \displaystyle\int_{-\infty}^{\infty}\int_{-\infty}^{\infty} e^{-(u^2+v^2)}\dfrac{2}{\sqrt{4ac - b^2}}\,dudv = \dfrac{2\pi}{\sqrt{4ac - b^2}}$

(2)

$(a) \because x^2 + 2xy + 3y^2 = (x+y)^2 + 2y^2$

令 $(x+y) = u,\ \sqrt{2}\,y = v$ 则 $x = u - \dfrac{v}{\sqrt{2}},\ y = \dfrac{v}{\sqrt{2}}$

且 $dxdy = \left\|\begin{matrix} \dfrac{\partial x}{\partial u} & \dfrac{\partial x}{\partial v} \\[2mm] \dfrac{\partial y}{\partial u} & \dfrac{\partial y}{\partial v} \end{matrix}\right\| dudv = \left\|\begin{matrix} 1 & \dfrac{-1}{\sqrt{2}} \\[3mm] 0 & \sqrt{\dfrac{1}{2}} \end{matrix}\right\| dudv = \sqrt{\dfrac{1}{2}}\, dudv$

$\therefore \displaystyle\iint_{R} e^{-(x^2+2xy+3y^2)}\,dxdy = \iint_{u^2+v^2 \le 1} e^{-(u^2+v^2)}\sqrt{\dfrac{1}{2}}\,dudv$

令 $u = r\cos\theta$, $v = r\sin\theta$ 则 $\{(u,v): u^2 + v^2 \le 1\} = \{(r,\theta): 0 \le r \le 1, 0 \le \theta \le 2\pi\}$

且 $dudv = \left\| \begin{vmatrix} \dfrac{\partial u}{\partial r} & \dfrac{\partial u}{\partial \theta} \\ \dfrac{\partial v}{\partial r} & \dfrac{\partial v}{\partial \theta} \end{vmatrix} \right\| drd\theta = \left\| \begin{vmatrix} \cos\theta & -r\sin\theta \\ \sin\theta & r\cos\theta \end{vmatrix} \right\| drd\theta = rdrd\theta$

$$\therefore \iint_{u^2+v^2\le 1} e^{-(u^2+v^2)} \sqrt{\frac{1}{2}} \, dudv = \sqrt{\frac{1}{2}} \int_0^{2\pi} \int_0^1 e^{-r^2} r\,drd\theta = \sqrt{\frac{1}{2}} \int_0^{2\pi} d\theta \int_0^1 e^{-r^2} r\,dr$$

$$= \sqrt{\frac{1}{2}} \cdot 2\pi \cdot \left. \frac{-e^{-r^2}}{2} \right|_{r=0}^{r=1} = \frac{\pi}{\sqrt{2}}(1 - e^{-1})$$

$(b)$

$$\because I(t) = \sqrt{\frac{1}{2}} \int_0^{2\pi} d\theta \int_0^t e^{-r^2} r\,dr = \sqrt{\frac{1}{2}} \cdot 2\pi \cdot \left. \frac{-e^{-r^2}}{2} \right|_{r=0}^{r=t} = \frac{\pi}{\sqrt{2}}(1 - e^{-t})$$

$$\therefore \lim_{t\to\infty} I(t) = \lim_{t\to\infty} \frac{\pi}{\sqrt{2}}(1 - e^{-t}) = \frac{\pi}{\sqrt{2}}$$

Example 7.

$$求 \iint_R e^{-(x^2+xy+y^2)} dxdy =?, \ R = \{(x,y): x^2 + xy + y^2 \le 4\}$$

【解】

$$\because x^2 + xy + y^2 = \left(x + \frac{y}{2}\right)^2 + \frac{3}{4}y^2, \ \ 令 x + \frac{y}{2} = u, \ \sqrt{\frac{3}{4}}y = v$$

$$则 \ dxdy = \left\| \begin{vmatrix} \dfrac{\partial x}{\partial u} & \dfrac{\partial x}{\partial v} \\ \dfrac{\partial y}{\partial u} & \dfrac{\partial y}{\partial v} \end{vmatrix} \right\| dudv = \left\| \begin{vmatrix} 1 & 0 \\ 2 & \sqrt{\dfrac{4}{3}} \end{vmatrix} \right\| dudv = \frac{2}{\sqrt{3}} dudv$$

$$\therefore \iint_R e^{-(x^2+xy+y^2)} dxdy = \iint_{u^2+v^2\le 4} e^{-(u^2+v^2)} \frac{2}{\sqrt{3}} dudv$$

令 $u = r\cos\theta$, $v = r\sin\theta$

则 $\{(u,v): u^2 + v^2 \le 4\} = \{(r,\theta): 0 \le r \le 2, 0 \le \theta \le 2\pi\}$

$$\text{且 } dudv = \left\|\begin{matrix} \dfrac{\partial u}{\partial r} & \dfrac{\partial u}{\partial \theta} \\ \dfrac{\partial v}{\partial r} & \dfrac{\partial v}{\partial \theta} \end{matrix}\right\| drd\theta = \left\|\begin{matrix} \cos\theta & -r\sin\theta \\ \sin\theta & r\cos\theta \end{matrix}\right\| drd\theta = rdrd\theta$$

$$\therefore \iint_{u^2+v^2\le 4} e^{-(u^2+v^2)}dudv = \int_0^{2\pi}\int_0^2 e^{-r^2}rdrd\theta = \int_0^{2\pi} d\theta \int_0^2 e^{-r^2}rdr = 2\pi \cdot \left.\frac{-e^{-r^2}}{2}\right|_{r=0}^{r=2}$$

$$= \pi(1-e^{-4})$$

$$\therefore \iint_R e^{-(x^2+xy+y^2)}dxdy = \frac{2\pi(1-e^{-4})}{\sqrt{3}}$$

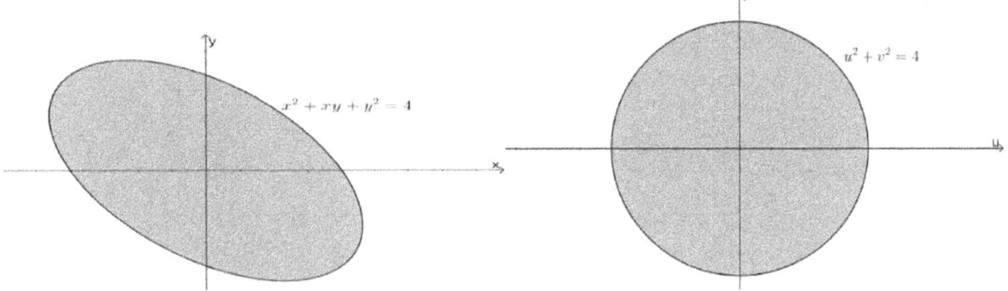


Example 8.

$$\text{求 } \iint_R dA = ?, \ R = \{(x,y): ax^2 + bxy + cy^2 \le \alpha^2\}$$

【解】

$$\because ax^2 + bxy + cy^2 = a\left(x + \frac{by}{2a}\right)^2 + \frac{4ac-b^2}{4a}y^2$$

$$\text{令 } \sqrt{a}\left(x + \frac{by}{2a}\right) = u, \ \sqrt{\frac{4ac-b^2}{4a}}\,y = v$$

$$\text{则 } dxdy = \left\|\begin{matrix} \dfrac{\partial x}{\partial u} & \dfrac{\partial x}{\partial v} \\ \dfrac{\partial y}{\partial u} & \dfrac{\partial y}{\partial v} \end{matrix}\right\| dudv = \left\|\begin{matrix} \dfrac{1}{\sqrt{a}} & 0 \\ \dfrac{2\sqrt{a}}{b} & \sqrt{\dfrac{4a}{4ac-b^2}} \end{matrix}\right\| dudv = \frac{2}{\sqrt{4ac-b^2}}dudv$$

$$\therefore \iint_R dA = \iint_{u^2+v^2\le\alpha^2} \frac{2}{\sqrt{4ac-b^2}}dudv$$

令 $u = r\cos\theta, v = r\sin\theta$ 则 $\{(u,v): u^2 + v^2 \leq \alpha^2\} = \{(r,\theta): 0 \leq r \leq \alpha, 0 \leq \theta \leq 2\pi\}$

且 $dudv = \left\|\begin{vmatrix} \dfrac{\partial u}{\partial r} & \dfrac{\partial u}{\partial \theta} \\ \dfrac{\partial v}{\partial r} & \dfrac{\partial v}{\partial \theta} \end{vmatrix}\right\| drd\theta = \left\|\begin{vmatrix} \cos\theta & -r\sin\theta \\ \sin\theta & r\cos\theta \end{vmatrix}\right\| drd\theta = rdrd\theta$

$$\therefore \iint_{u^2+v^2\leq\alpha^2} \frac{2}{\sqrt{4ac-b^2}} dudv = \int_0^{2\pi}\int_0^{\alpha} \frac{2}{\sqrt{4ac-b^2}} rdrd\theta = \frac{2}{\sqrt{4ac-b^2}} \int_0^{2\pi} d\theta \int_0^{\alpha} rdr$$

$$= \frac{2\pi\alpha^2}{\sqrt{4ac-b^2}}$$

Example 9.

$$求 \iint_R \frac{x^2 - xy + y^2}{2} dA = ?, \quad R = \{(x,y): x^2 - xy + y^2 \leq 2\}$$

【解】

$$\because \frac{x^2 - xy + y^2}{2} = \frac{1}{2}\left(x - \frac{y}{2}\right)^2 + \frac{3}{8}y^2, \quad 令 \sqrt{\frac{1}{2}}\left(x - \frac{y}{2}\right) = u, \quad \sqrt{\frac{3}{8}}y = v$$

则 $dxdy = \left\|\begin{vmatrix} \dfrac{\partial x}{\partial u} & \dfrac{\partial x}{\partial v} \\ \dfrac{\partial y}{\partial u} & \dfrac{\partial y}{\partial v} \end{vmatrix}\right\| dudv = \left\|\begin{vmatrix} \sqrt{2} & 0 \\ -2\sqrt{2} & \sqrt{\dfrac{8}{3}} \end{vmatrix}\right\| dudv = \frac{4}{\sqrt{3}} dudv$

$$\therefore \iint_R \frac{x^2 - xy + y^2}{2} dA = \iint_{u^2+v^2\leq 1} (u^2 + v^2) \frac{4}{\sqrt{3}} dudv$$

令 $u = r\cos\theta, \ v = r\sin\theta$ 则 $\{(u,v): u^2 + v^2 \leq 1\} = \{(r,\theta): 0 \leq r \leq 1, 0 \leq \theta \leq 2\pi\}$

且 $dudv = \left\|\begin{vmatrix} \dfrac{\partial u}{\partial r} & \dfrac{\partial u}{\partial \theta} \\ \dfrac{\partial v}{\partial r} & \dfrac{\partial v}{\partial \theta} \end{vmatrix}\right\| drd\theta = \left\|\begin{vmatrix} \cos\theta & -r\sin\theta \\ \sin\theta & r\cos\theta \end{vmatrix}\right\| drd\theta = rdrd\theta$

$$\therefore \iint_{u^2+v^2\leq 1} (u^2 + v^2)\frac{4}{\sqrt{3}} dudv = \frac{4}{\sqrt{3}}\int_0^{2\pi}\int_0^1 r^2 \cdot rdrd\theta = \frac{4}{\sqrt{3}}\int_0^{2\pi} d\theta \int_0^1 r^3 dr$$

$$= \frac{4}{\sqrt{3}} \cdot 2\pi \cdot \frac{r^4}{4}\bigg|_{r=0}^{r=1} = \frac{2\pi}{\sqrt{3}}$$

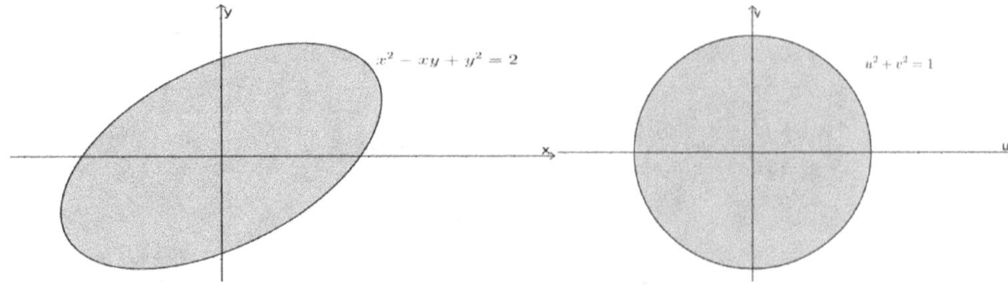

Example 10.

(1) 求由 $x - 2y = -4$, $x - 2y = 1$, $2x - y = 0, 2x - y = 3$ 所围面积

(2) 求由 $x - my = 0$, $x - my = 1$, $nx - y = 0$, $nx - y = 1$ 所围面积, $(m > n > 1)$

(3) 求 $\displaystyle\iint_R \frac{x - 2y}{3x - y} dxdy =?$, $R$ 为四条直线包围区域: $x - 2y = 0, x - 2y = 4$,

$3x - y = 1, 3x - y = 7$

【解】

(1)

令 $x - 2y = u$, $2x - y = v$ 则 $x = -\dfrac{u}{3} + \dfrac{2v}{3}$, $y = -\dfrac{2u}{3} + \dfrac{v}{3}$

且 $dxdy = \left\| \begin{vmatrix} \dfrac{\partial x}{\partial u} & \dfrac{\partial x}{\partial v} \\ \dfrac{\partial y}{\partial u} & \dfrac{\partial y}{\partial v} \end{vmatrix} \right\| dudv = \left\| \begin{vmatrix} \dfrac{-1}{3} & \dfrac{2}{3} \\ \dfrac{-2}{3} & \dfrac{1}{3} \end{vmatrix} \right\| dudv = \dfrac{1}{3} dudv$

令 $R = \{(x, y): -4 \le x - 2y \le 1, 0 \le 2x - y \le 3 \}$ 则 $R = \{(u, v): -4 \le u \le 1, 0 \le v \le 3\}$

$\therefore$ 面积 $= \displaystyle\iint_R 1 dA = \int_{-4}^{1} \int_{0}^{3} \frac{1}{3} dvdu = 5$

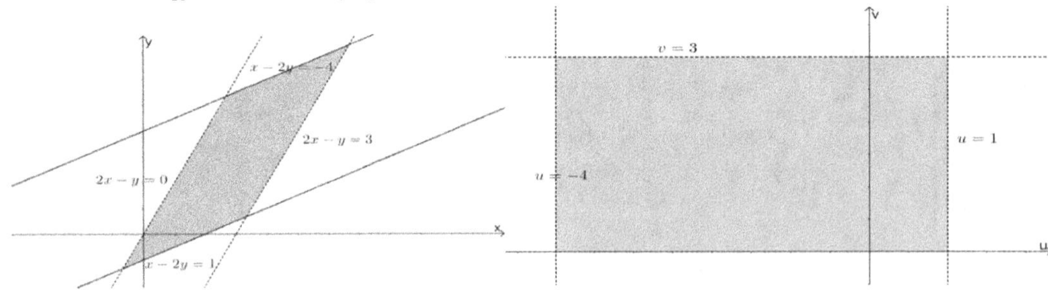

(2)

令 $x - my = u$, $nx - y = v$ 则 $x = \dfrac{mv - u}{mn - 1}$, $y = \dfrac{v - nu}{mn - 1}$

且 $dxdy = \left\|\begin{vmatrix} \dfrac{\partial x}{\partial u} & \dfrac{\partial x}{\partial v} \\ \dfrac{\partial y}{\partial u} & \dfrac{\partial y}{\partial v} \end{vmatrix}\right\| dudv = \left\|\begin{vmatrix} \dfrac{-1}{mn-1} & \dfrac{m}{mn-1} \\ \dfrac{-n}{mn-1} & \dfrac{1}{mn-1} \end{vmatrix}\right\| dudv = \dfrac{1}{mn-1} dudv$

令 $R = \{(x,y): 0 \le x - my \le 1, 0 \le nx - y \le 1\}$ 则 $R = \{(u,v): 0 \le u \le 1, 0 \le v \le 1\}$

$\therefore$ 面积 $= \iint_R 1\, dA = \int_0^1 \int_0^1 \dfrac{1}{mn-1} dudv = \dfrac{1}{mn-1}$

(3)

令 $x - 2y = u$, $\ 3x - y = v$ 则 $x = \dfrac{-u+2v}{5}$, $\ y = \dfrac{-3u+v}{5}$

且 $dxdy = \left\|\begin{vmatrix} \dfrac{\partial x}{\partial u} & \dfrac{\partial x}{\partial v} \\ \dfrac{\partial y}{\partial u} & \dfrac{\partial y}{\partial v} \end{vmatrix}\right\| dudv = \left\|\begin{vmatrix} \dfrac{-1}{5} & \dfrac{2}{5} \\ \dfrac{-3}{5} & \dfrac{1}{5} \end{vmatrix}\right\| dudv = \dfrac{1}{5} dudv$

令 $R = \{(x,y): 0 \le x - 2y \le 4, 1 \le 3x - y \le 7\}$ 则 $R = \{(u,v): 0 \le u \le 4, 1 \le v \le 7\}$

$\therefore \iint_R \dfrac{x-2y}{3x-y} dxdy = \dfrac{1}{5} \int_1^7 \int_0^4 \dfrac{u}{v} dudv = \dfrac{8}{5} \int_1^7 \dfrac{1}{v} dv = \dfrac{8\ln 7}{5}$

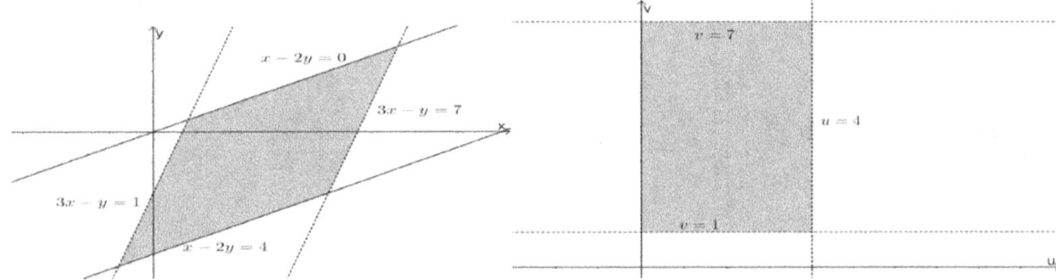


Example 11.

求由 $y = x^2$, $\ y = 3x^2$, $\ x = y^2$, $\ x = 4y^2$ 所围面积

【解】

令 $u = \dfrac{y}{x^2}$, $\ v = \dfrac{x}{y^2}$ 则 $x = u^{-\frac{2}{3}} v^{-\frac{1}{3}}$, $\ y = u^{-\frac{1}{3}} v^{-\frac{2}{3}}$

且 $dxdy = \left\|\begin{vmatrix} \dfrac{\partial x}{\partial u} & \dfrac{\partial x}{\partial v} \\ \dfrac{\partial y}{\partial u} & \dfrac{\partial y}{\partial v} \end{vmatrix}\right\| dudv = \left\|\begin{vmatrix} \dfrac{-2u^{-\frac{5}{3}} v^{-\frac{1}{3}}}{3} & \dfrac{-u^{-\frac{2}{3}} v^{-\frac{4}{3}}}{3} \\ \dfrac{-u^{-\frac{4}{3}} v^{-\frac{2}{3}}}{3} & \dfrac{-2u^{-\frac{1}{3}} v^{-\frac{5}{3}}}{3} \end{vmatrix}\right\| dudv = \dfrac{u^{-2} v^{-2}}{3} dudv$

令 $R = \left\{ (x,y): 1 \leq \dfrac{y}{x^2} \leq 3, 1 \leq \dfrac{x}{y^2} \leq 4 \right\}$ 则 $R = \{(u,v): 1 \leq u \leq 3, 1 \leq v \leq 4\}$

$\therefore$ 面积 $= \iint_R 1\,dA = \int_1^3 \int_1^4 \dfrac{u^{-2}v^{-2}}{3}\,dv\,du = \dfrac{1}{6}$

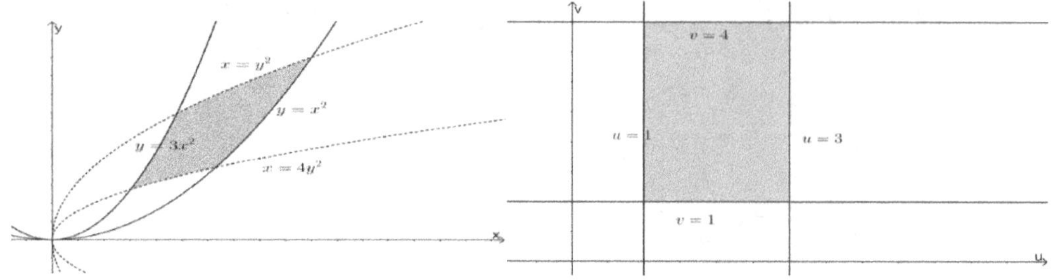

## Example 12.

$$求由 x^2 - 4xy + 4y^2 - 2x - y - 1 = 0 \text{ 与 } y = \dfrac{2}{5} \text{ 所围面积}$$

【解】

$\because x^2 - 4xy + 4y^2 - 2x - y - 1 = (x - 2y)^2 - (2x + y) - 1 = 0$

令 $u = x - 2y$ 且 $v = 2x + y$ 则 $x = \dfrac{u}{5} + \dfrac{2v}{5}, \quad y = -\dfrac{2u}{5} + \dfrac{v}{5}$

且 $dxdy = \left\| \begin{vmatrix} \dfrac{\partial x}{\partial u} & \dfrac{\partial x}{\partial v} \\ \dfrac{\partial y}{\partial u} & \dfrac{\partial y}{\partial v} \end{vmatrix} \right\| dudv = \left\| \begin{vmatrix} \dfrac{1}{5} & \dfrac{2}{5} \\ \dfrac{-2}{5} & \dfrac{1}{5} \end{vmatrix} \right\| dudv = \dfrac{1}{5}\,dudv$

令 $R = \{(u,v): -1 \leq u \leq 3, u^2 - 1 \leq v \leq 2u + 2\}$

$\therefore$ 面积 $= \iint_R 1\,dA = \int_{-1}^3 \int_{u^2-1}^{2u+2} \dfrac{1}{5}\,dv\,du = \dfrac{32}{15}$

## Example 13.

$$求 \iint_R (x - y)^2 \sin^2(x + y)\,dxdy = ?, \quad R \text{为} (\pi, 0),\ (2\pi, \pi),\ (\pi, 2\pi),\ (0, \pi) \text{所围区域}$$

【解】

令 $x + y = u, \quad x - y = v$ 则 $x = \dfrac{u + v}{2}, \quad y = \dfrac{u - v}{2}$

且 $dxdy = \left\| \begin{vmatrix} \dfrac{\partial x}{\partial u} & \dfrac{\partial x}{\partial v} \\[2mm] \dfrac{\partial y}{\partial u} & \dfrac{\partial y}{\partial v} \end{vmatrix} \right\| dudv = \left\| \begin{vmatrix} \dfrac{1}{2} & \dfrac{1}{2} \\[2mm] \dfrac{1}{2} & \dfrac{-1}{2} \end{vmatrix} \right\| dudv = \dfrac{1}{2} dudv$

$\because R = \{(x,y): \pi \le x+y \le 3\pi, -\pi \le x-y \le \pi\}$ $\therefore R = \{(u,v): \pi \le u \le 3\pi, -\pi \le v \le \pi\}$

$\therefore \iint_R (x-y)^2 \sin^2(x+y)\, dxdy = \dfrac{1}{2} \int_{-\pi}^{\pi} \int_{\pi}^{3\pi} v^2 \sin^2 u \, dudv = \dfrac{1}{2} \int_{\pi}^{3\pi} \sin^2 u\, du \int_{-\pi}^{\pi} v^2\, dv$

$= \dfrac{1}{2} \int_{\pi}^{3\pi} \dfrac{1-\cos 2u}{2}\, du \int_{-\pi}^{\pi} v^2\, dv = \dfrac{1}{2}\left(\dfrac{u}{2} - \dfrac{\sin 2u}{4}\right)\Big|_{\pi}^{3\pi} \cdot \dfrac{v^3}{3}\Big|_{-\pi}^{\pi} = \dfrac{\pi^4}{3}$

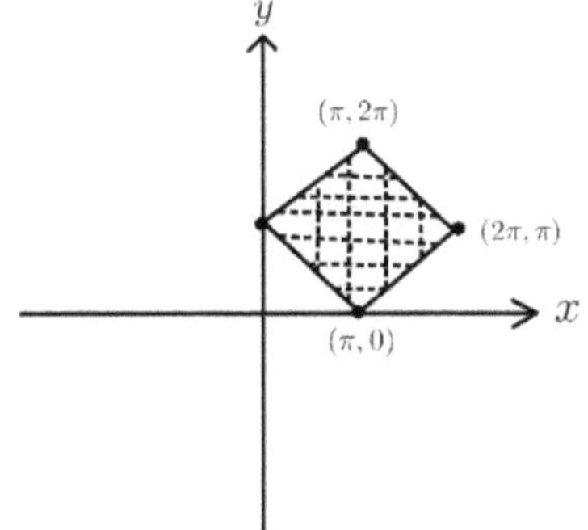
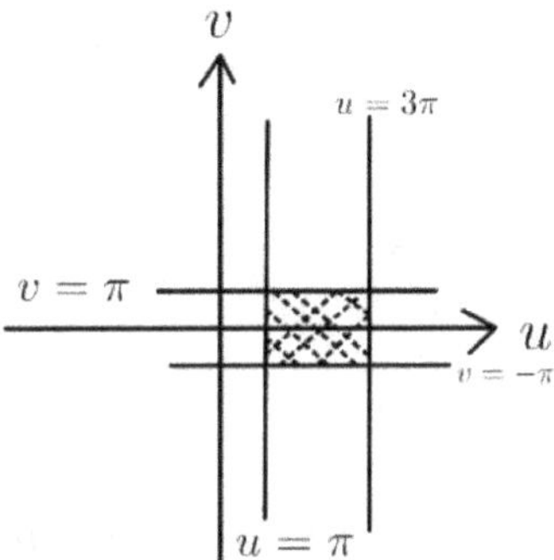

Example 14.

$$\text{求} \iint_R \sqrt{\dfrac{y-x}{x+y}}\, dxdy =?, \quad \text{其中} R \text{为} x=0,\ y=x,\ x+y=1,\ x+y=2 \text{ 所围区域}$$

【解】

令 $y-x=u,\ x+y=v$ 则 $x = \dfrac{v-u}{2},\ y = \dfrac{u+v}{2}$

且 $dxdy = \left\| \begin{vmatrix} \dfrac{\partial x}{\partial u} & \dfrac{\partial x}{\partial v} \\[2mm] \dfrac{\partial y}{\partial u} & \dfrac{\partial y}{\partial v} \end{vmatrix} \right\| dudv = \left\| \begin{vmatrix} \dfrac{-1}{2} & \dfrac{1}{2} \\[2mm] \dfrac{1}{2} & \dfrac{1}{2} \end{vmatrix} \right\| dudv = \dfrac{1}{2} dudv$

$\because R = \{(x,y): y \ge x, x \ge 0, 1 \le x+y \le 2\}$ $\therefore R = \{(x,y): 0 \le u \le v, 1 \le v \le 2\}$

$\therefore \iint_R \sqrt{\dfrac{y-x}{x+y}}\, dxdy = \dfrac{1}{2} \int_1^2 \int_0^v \sqrt{\dfrac{u}{v}}\, dudv = \dfrac{1}{2} \int_1^2 \sqrt{\dfrac{1}{v}} \cdot \dfrac{2u^{\frac{3}{2}}}{3}\Big|_0^v\, dv = \dfrac{1}{3} \int_1^2 v\, dv = \dfrac{1}{2}$

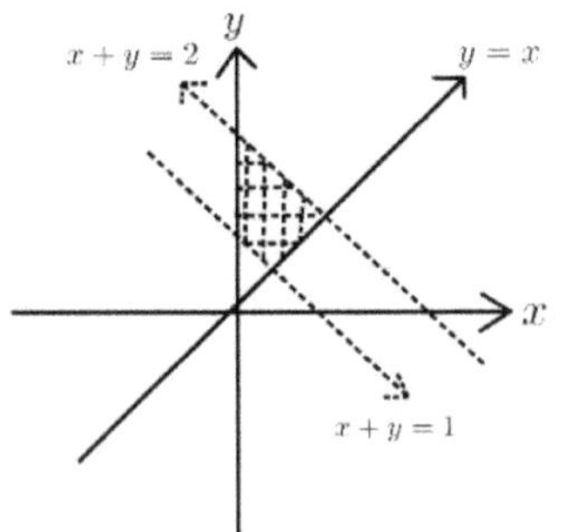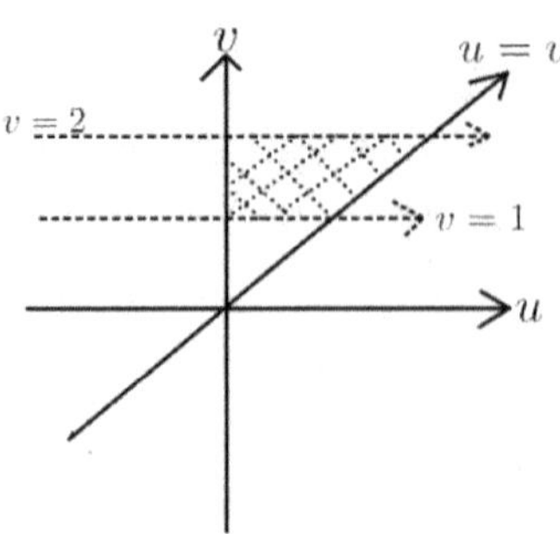

## Example 15.

$$求 \iint_R (x+y)^4 \, dxdy =?, \quad R为 (1,0),(1,3),(2,2),(0,1)所围区域$$

【解】

令 $x+y=u$, $2x-y=v$ 则 $x=\dfrac{u+v}{3}$, $y=\dfrac{2u-v}{3}$

且 $dxdy = \left\| \begin{vmatrix} \dfrac{\partial x}{\partial u} & \dfrac{\partial x}{\partial v} \\ \dfrac{\partial y}{\partial u} & \dfrac{\partial y}{\partial v} \end{vmatrix} \right\| dudv = \left\| \begin{vmatrix} \dfrac{1}{3} & \dfrac{1}{3} \\ \dfrac{2}{3} & \dfrac{-1}{3} \end{vmatrix} \right\| dudv = \dfrac{1}{3} dudv$

$\because R = \{(x,y): 1 \le x+y \le 4, -1 \le 2x-y \le 2 \}$  $\therefore R = \{(u,v): 1 \le u \le 4, -1 \le v \le 2 \}$

$$\therefore \iint_R (x+y)^4 \, dxdy = \frac{1}{3} \int_{-1}^{2} \int_{1}^{4} u^4 \, dudv = \left. \frac{u^5}{5} \right|_1^4 = \frac{1023}{5}$$

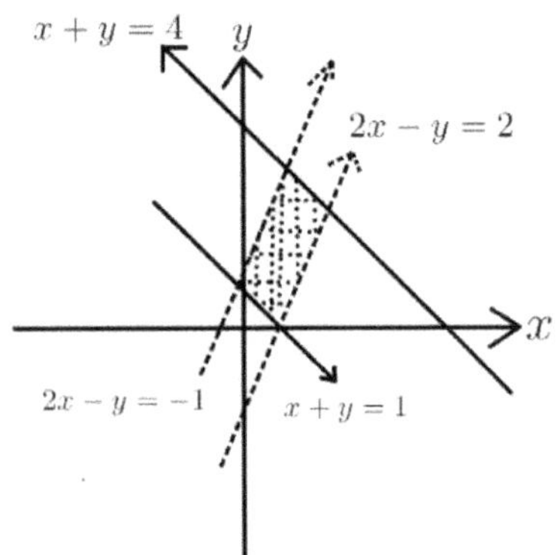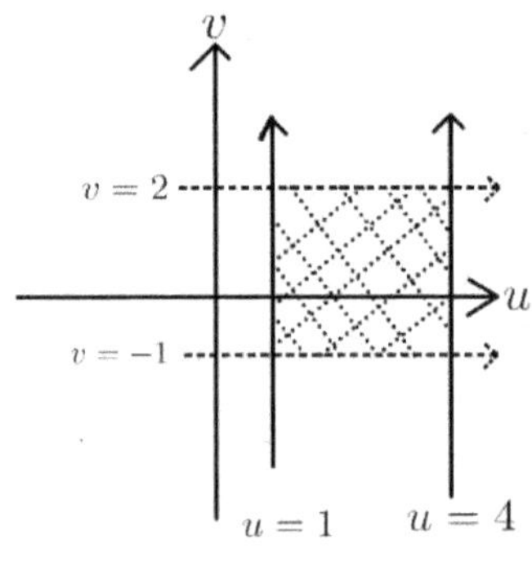

## Example 16.

求 $\iint_R \left(\sqrt{x} + \sqrt{y}\right) dxdy =?$，$R$ 为 $\sqrt{x} + \sqrt{y} = 1$，$x = 0$，$y = 0$ 所围区域

【解】

令 $x = r\cos^4\theta$，$y = r\sin^4\theta$

则 $dxdy = \left\|\begin{matrix} \dfrac{\partial x}{\partial r} & \dfrac{\partial x}{\partial \theta} \\ \dfrac{\partial y}{\partial r} & \dfrac{\partial y}{\partial \theta} \end{matrix}\right\| drd\theta = \left\|\begin{matrix} \cos^4\theta & -4r\cos^3\theta\sin\theta \\ \sin^4\theta & 4r\sin^3\theta\cos\theta \end{matrix}\right\| drd\theta = 4r\sin^3\theta\cos^3\theta\, drd\theta$

$\because R = \{(x,y): x \geq 0,\ y \geq 0, \sqrt{x} + \sqrt{y} \leq 1\} = \{(r,\theta): 0 \leq r \leq 1, 0 \leq \theta \leq \dfrac{\pi}{2}\}$

$\therefore \iint_R \left(\sqrt{x} + \sqrt{y}\right) dxdy = \int_0^{\frac{\pi}{2}} \int_0^1 \sqrt{r}(4r\sin^3\theta\cos^3\theta)\, drd\theta = \int_0^1 4r^{\frac{3}{2}}\, dr \int_0^{\frac{\pi}{2}} \sin^3\theta\cos^3\theta\, d\theta$

$= \int_0^1 4r^{\frac{3}{2}}\, dr \int_0^{\frac{\pi}{2}} \left(\dfrac{\sin 2\theta}{2}\right)^3 d\theta = \dfrac{2}{15}$

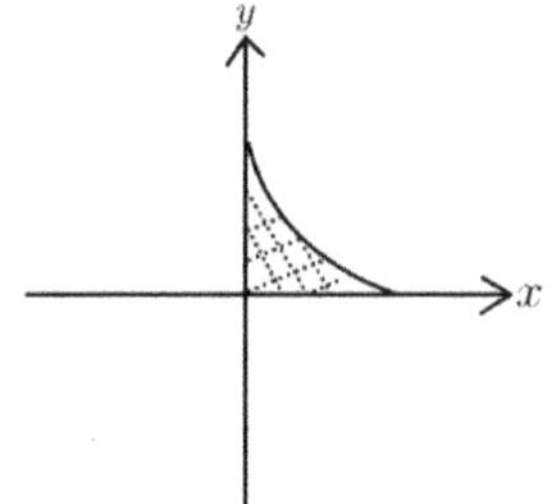

Example 17.

求 $\iint_R x^2 + y^2\, dxdy =?$，其中 $R$ 为 $x^2 - y^2 = 1$，$x^2 - y^2 = 9$，$xy = 2$，$xy = 4$
所围区域

【解】

令 $x^2 - y^2 = u$，$2xy = v$

则 $dudv = \left\|\begin{matrix} \dfrac{\partial u}{\partial x} & \dfrac{\partial u}{\partial y} \\ \dfrac{\partial v}{\partial x} & \dfrac{\partial v}{\partial y} \end{matrix}\right\| dxdy = \left\|\begin{matrix} 2x & -2y \\ 2y & 2x \end{matrix}\right\| dxdy = 4(x^2 + y^2)dxdy$

$\because (x^2 + y^2)^2 = (x^2 - y^2)^2 + (2xy)^2 = u^2 + v^2 \quad \therefore x^2 + y^2 = (u^2 + v^2)^{\frac{1}{2}}$

$$\therefore dxdy = \frac{dudv}{4(u^2 + v^2)^{\frac{1}{2}}}$$

$\because R = \{(x,y): 1 \le x^2 - y^2 \le 9, 4 \le 2xy \le 8\}$   $\therefore R = \{(u,v): 1 \le u \le 9, 4 \le v \le 8\}$

$$\therefore \iint_R x^2 + y^2 \, dxdy = \int_4^8 \int_1^9 \frac{(u^2 + v^2)^{\frac{1}{2}}}{4(u^2 + v^2)^{\frac{1}{2}}} \, dudv = 8$$

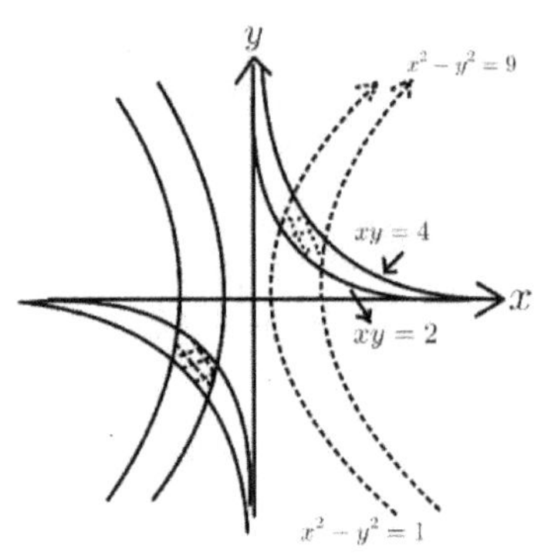


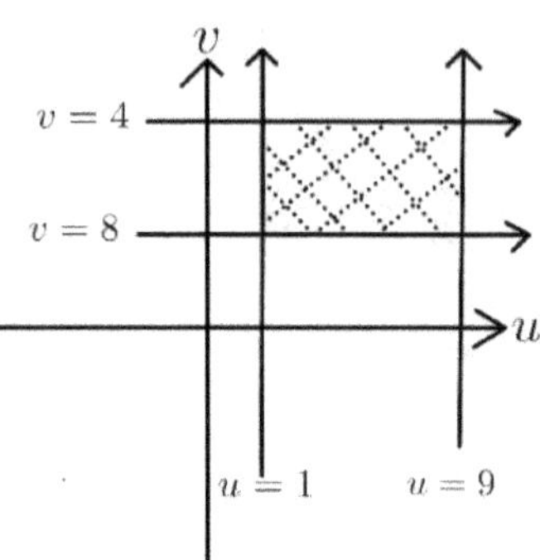


Example 18.

(1) 求 $\iint_R x^2 + y^2 dA = ?$, $R$ 为四条直线包围区域: $-x = y,\ y = -x + 3,$

　　$x - 3 = y,\ y = x$

(2) 求 $\iint_R xy dA = ?$, $R$ 为四条曲线包围于第一象限区域: $x^2 + y^2 = a,$

　　$x^2 + y^2 = b,\ x^2 - y^2 = c,\ x^2 - y^2 = d, (b > a > 0, d > c > 0)$

【解】

(1)

令 $x + y = u,\ x - y = v$ 则 $x = \dfrac{u + v}{2},\ y = \dfrac{u - v}{2}$

且 $dxdy = \left\| \begin{vmatrix} \dfrac{\partial x}{\partial u} & \dfrac{\partial x}{\partial v} \\ \dfrac{\partial y}{\partial u} & \dfrac{\partial y}{\partial v} \end{vmatrix} \right\| dudv = \left\| \begin{vmatrix} \dfrac{1}{2} & \dfrac{1}{2} \\ \dfrac{1}{2} & \dfrac{-1}{2} \end{vmatrix} \right\| dudv = \dfrac{1}{2} dudv$

$\because R = \{(x,y): 0 \le x + y \le 3, 0 \le x - y \le 3\}$   $\therefore R = \{(u,v): 0 \le u \le 3, 0 \le v \le 3\}$

$$\therefore \iint_R x^2 + y^2 dA = \int_0^3 \int_0^3 \left( \left(\frac{u+v}{2}\right)^2 + \left(\frac{u-v}{2}\right)^2 \right) \frac{1}{2} dudv = \frac{1}{4} \int_0^3 \int_0^3 u^2 + v^2 dudv = \frac{27}{2}$$

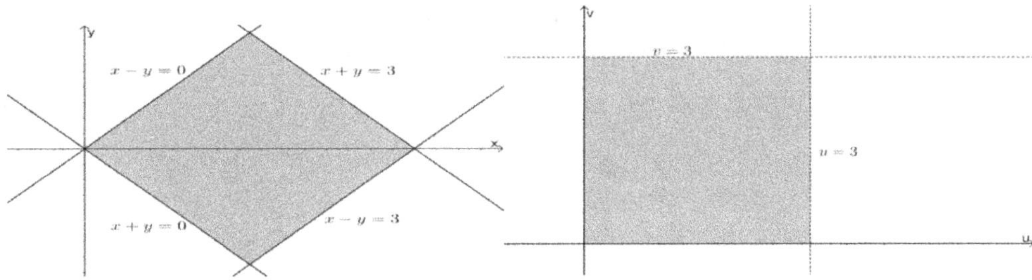

(2)

令 $x^2 + y^2 = u$, $x^2 - y^2 = v$ 则 $x = \dfrac{(u+v)^{\frac{1}{2}}}{\sqrt{2}}$, $y = \dfrac{(u-v)^{\frac{1}{2}}}{\sqrt{2}}$

且 $dxdy = \left\| \begin{vmatrix} \dfrac{\partial x}{\partial u} & \dfrac{\partial x}{\partial v} \\ \dfrac{\partial y}{\partial u} & \dfrac{\partial y}{\partial v} \end{vmatrix} \right\| dudv = \left\| \begin{vmatrix} \dfrac{(u+v)^{-\frac{1}{2}}}{2\sqrt{2}} & \dfrac{(u+v)^{-\frac{1}{2}}}{2\sqrt{2}} \\ \dfrac{(u-v)^{-\frac{1}{2}}}{2\sqrt{2}} & \dfrac{(u-v)^{-\frac{1}{2}}}{2\sqrt{2}} \end{vmatrix} \right\| dudv = \dfrac{(u^2 - v^2)^{-\frac{1}{2}}}{4} dudv$

$\because R = \{(x,y): a \leq x^2 + y^2 \leq b, c \leq x^2 - y^2 \leq d\}$  $\therefore R = \{(u,v): a \leq u \leq b, c \leq v \leq d\}$

$\therefore \iint_R xy dA = \int_c^d \int_a^b \dfrac{(u^2 - v^2)^{\frac{1}{2}}}{2} \cdot \dfrac{(u^2 - v^2)^{-\frac{1}{2}}}{4} dudv = \dfrac{(b-a)(c-d)}{8}$

Example 19.

$$求 \int_0^{\frac{1}{2}} \int_0^{1-2y} e^{\frac{x}{x+2y}} dxdy = ?$$

【解】

令 $x = u$, $x + 2y = v$ 则 $x = u$, $y = \dfrac{-u+v}{2}$

$dxdy = \left\| \begin{vmatrix} \dfrac{\partial x}{\partial u} & \dfrac{\partial x}{\partial v} \\ \dfrac{\partial y}{\partial u} & \dfrac{\partial y}{\partial v} \end{vmatrix} \right\| dudv = \left\| \begin{vmatrix} 1 & 0 \\ \dfrac{-1}{2} & \dfrac{1}{2} \end{vmatrix} \right\| dudv = \dfrac{1}{2} dudv$

令 $R = \left\{(x,y): 0 \leq x \leq 1-2y, 0 \leq y \leq \dfrac{1}{2}\right\}$ 则 $R = \{(u,v): 0 \leq u \leq v, 0 \leq v \leq 1\}$

$$\therefore \int_0^{\frac{1}{2}} \int_0^{1-2y} e^{\frac{x}{x+2y}}\, dxdy = \frac{1}{2}\int_0^1 \int_0^v e^{\frac{u}{v}}\, dudv = \frac{1}{2}\int_0^1 v e^{\frac{u}{v}}\Big|_{u=0}^{u=v}\, dv = \frac{1}{2}\int_0^1 v(e-1)\, dv = \frac{e-1}{4}$$

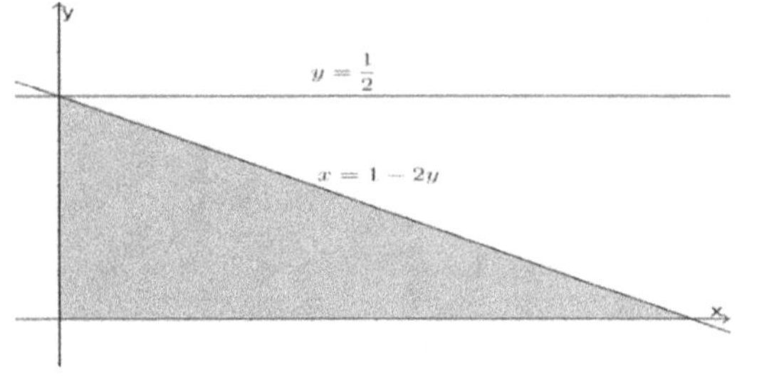


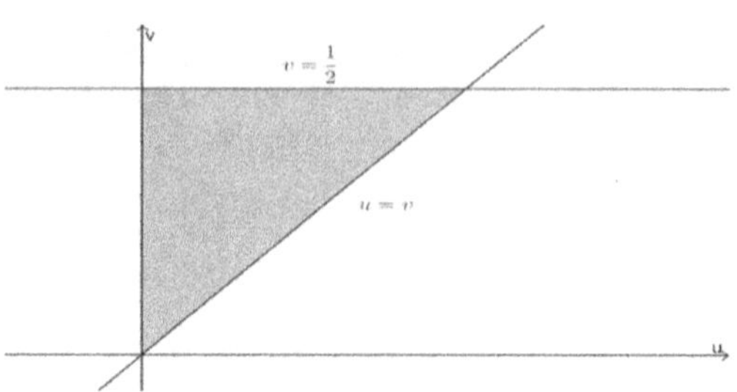


Example 20.

求 $\iint_R 3xy\,dA = ?$, $R$为四条直线包围区域: $x - 2y = 0$, $x - 2y = -4$, $x + y = 4$, $x + y = 1$

【解】

令 $x - 2y = u,\ x + y = v$ 则 $x = \dfrac{u + 2v}{3}$, $y = \dfrac{-u + v}{3}$

且 $dxdy = \left\|\begin{matrix} \dfrac{\partial x}{\partial u} & \dfrac{\partial x}{\partial v} \\ \dfrac{\partial y}{\partial u} & \dfrac{\partial y}{\partial v} \end{matrix}\right\| dudv = \left\|\begin{matrix} \dfrac{1}{3} & \dfrac{2}{3} \\ \dfrac{-1}{3} & \dfrac{1}{3} \end{matrix}\right\| dudv = \dfrac{1}{3} dudv$

$\because R = \{(x, y): -4 \leq x - 2y \leq 0, 1 \leq x + y \leq 4\}$ $\quad \therefore R = \{(u, v): -4 \leq u \leq 0, 1 \leq v \leq 4\}$

$\therefore \iint_R 3xy\,dxdy = \int_1^4 \int_{-4}^0 \left(\dfrac{u + 2v}{3}\right)\left(\dfrac{-u + v}{3}\right) dudv = \dfrac{164}{9}$

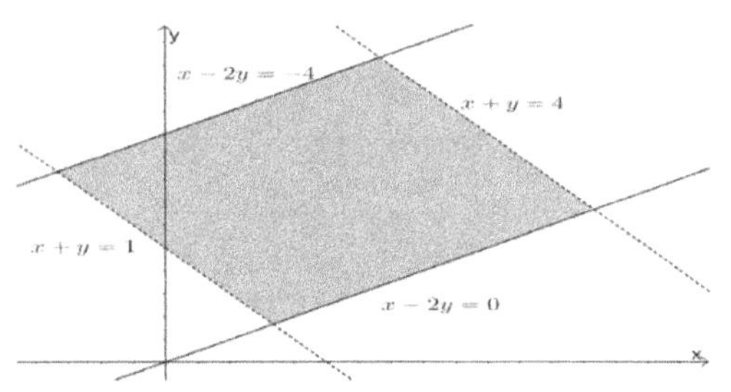


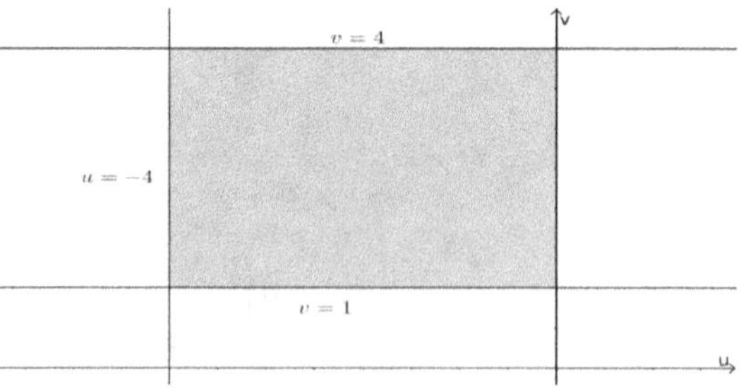


Example 21.

求 $\iint_R (2x+y)dxdy = ?$，$R$ 为曲线：$x^2 - 2xy + y^2 + x + y = 0$ 与 $x + y + 4 = 0$ 包围区域

【解】

$\because x^2 - 2xy + y^2 + x + y = 0 \Leftrightarrow (x-y)^2 = -(x+y)$

令 $x + y = u$，$x - y = v$ 则 $x = \dfrac{u+v}{2}$，$y = \dfrac{u-v}{2}$

且 $dxdy = \left\| \begin{vmatrix} \dfrac{\partial x}{\partial u} & \dfrac{\partial x}{\partial v} \\ \dfrac{\partial y}{\partial u} & \dfrac{\partial y}{\partial v} \end{vmatrix} \right\| dudv = \left\| \begin{vmatrix} \dfrac{1}{2} & \dfrac{1}{2} \\ \dfrac{1}{2} & \dfrac{1}{2} \end{vmatrix} \right\| dudv = \dfrac{1}{2} dudv$

$\because R = \{(u,v): -v^2 \le u \le 0, -2 \le v \le 2\}$

$\therefore \iint_R (2x+y)dxdy = \int_{-2}^{2} \int_{-v^2}^{0} 3u + v \, dudv = -\dfrac{96}{5}$

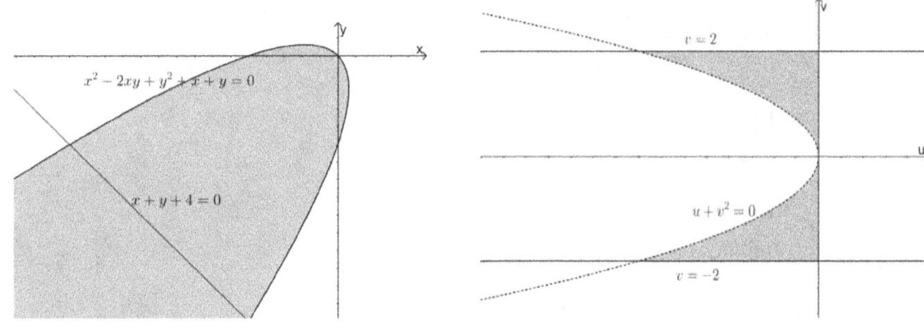

Example 22.

(1) 求 $\iint_R \cos\dfrac{y-x}{x+y} dxdy = ?$，$R$ 为 $(1,0),(3,0),(0,1),(0,3)$ 所围区域：

(2) 求 $\iint_R \sec^2\dfrac{y-x}{x+y} dxdy = ?$，$R$ 为 $(1,0),(3,0),(0,1),(0,3)$ 所围区域

【解】

(1)

令 $y - x = u$，$x + y = v$ 则 $x = \dfrac{-u+v}{2}$，$y = \dfrac{u+v}{2}$

$$\text{且 } dxdy = \left\|\begin{vmatrix}\dfrac{\partial x}{\partial u} & \dfrac{\partial x}{\partial v}\\[2mm] \dfrac{\partial y}{\partial u} & \dfrac{\partial y}{\partial v}\end{vmatrix}\right\| dudv = \left\|\begin{vmatrix}-\dfrac{1}{2} & \dfrac{1}{2}\\[2mm] \dfrac{1}{2} & \dfrac{1}{2}\end{vmatrix}\right\| dudv = \dfrac{1}{2}dudv$$

$\because R = \{(x,y): x \geq 0, y \geq 0, 1 \leq x + y \leq 3\}$  $\therefore R = \{(u,v): -v \leq u \leq v, 1 \leq v \leq 3\}$

$$\therefore \iint_R \cos\frac{y-x}{x+y}dxdy = \frac{1}{2}\int_1^3\int_{-v}^v \cos\frac{u}{v}dudv = \frac{1}{2}\int_1^3 v\sin\frac{u}{v}\Big|_{u=-v}^{u=v} dv = \sin 1 \int_1^3 v\, dv$$

$= 4\sin 1$

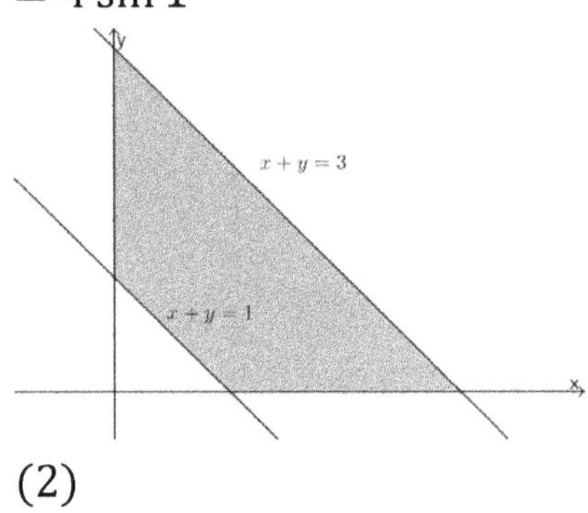
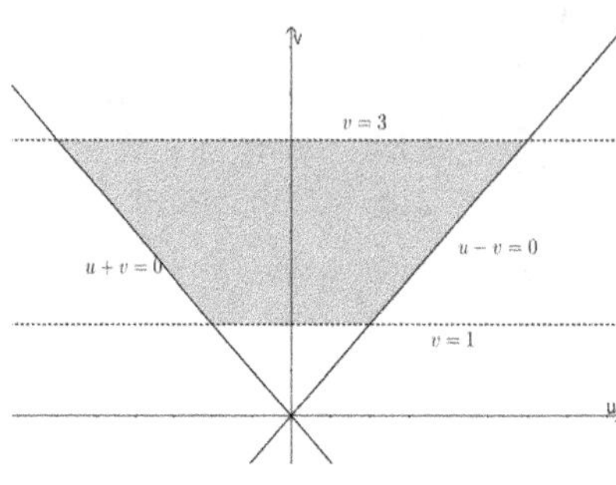



(2)

令 $y - x = u,\ x + y = v$ 则 $x = \dfrac{-u+v}{2},\ y = \dfrac{u+v}{2}$

$$\text{且 } dxdy = \left\|\begin{vmatrix}\dfrac{\partial x}{\partial u} & \dfrac{\partial x}{\partial v}\\[2mm] \dfrac{\partial y}{\partial u} & \dfrac{\partial y}{\partial v}\end{vmatrix}\right\| dudv = \left\|\begin{vmatrix}-\dfrac{1}{2} & \dfrac{1}{2}\\[2mm] \dfrac{1}{2} & \dfrac{1}{2}\end{vmatrix}\right\| dudv = \dfrac{1}{2}dudv$$

$\because R = \{(x,y): x \geq 0, y \geq 0, 1 \leq x + y \leq 3\}$  $\therefore R = \{(u,v): -v \leq u \leq v, 1 \leq v \leq 3\}$

$$\therefore \iint_R \sec^2\frac{y-x}{x+y}dxdy = \frac{1}{2}\int_1^3\int_{-v}^v \sec^2\frac{u}{v}dudv = \frac{1}{2}\int_1^3 v\tan\frac{u}{v}\Big|_{u=-v}^{u=v} dv = \tan 1 \int_1^3 v\, dv$$

$= 4\tan 1$

Example 23.

$$(1)\text{求} \iint_R \left(1 - \frac{x^2}{25} - \frac{y^2}{4}\right)^{\frac{3}{2}} dxdy =?,\ R = \left\{(x,y): \frac{x^2}{25} + \frac{y^2}{4} \leq 1, y \geq 0\right\}$$

$$(2)\text{求} \iint_R \cos(25x^2 + 4y^2)\, dxdy =?,\ R = \left\{(x,y): \frac{x^2}{4} + \frac{y^2}{25} \leq 1\right\}$$

【解】

(1)

令$x = 5r\cos\theta$, $y = 2r\sin\theta$ 则 $R = \{(r,\theta): 0 \le r \le 1, 0 \le \theta \le \pi\}$

且 $dxdy = \left\|\begin{vmatrix}\dfrac{\partial x}{\partial r} & \dfrac{\partial x}{\partial \theta}\\[2mm] \dfrac{\partial y}{\partial r} & \dfrac{\partial y}{\partial \theta}\end{vmatrix}\right\| drd\theta = \left\|\begin{vmatrix}5\cos\theta & -5r\sin\theta\\ 2\sin\theta & 2r\cos\theta\end{vmatrix}\right\| drd\theta = 10rdrd\theta$

$$\therefore \iint_R \left(1 - \frac{x^2}{25} - \frac{y^2}{4}\right)^{\frac{3}{2}} dxdy = \int_0^\pi \int_0^1 (1-r^2)^{\frac{3}{2}} 10rdrd\theta = 10\int_0^\pi d\theta \int_0^1 (1-r^2)^{\frac{3}{2}} rdr$$

$$= 10\pi \cdot \left.\frac{-(1-r^2)^{\frac{5}{2}}}{5}\right|_0^1 = 2\pi$$

(2)

令$x = 2r\cos\theta$, $y = 5r\sin\theta$ 则 $R = \{(r,\theta): 0 \le r \le 1, 0 \le \theta \le 2\pi\}$

且 $dxdy = \left\|\begin{vmatrix}\dfrac{\partial x}{\partial r} & \dfrac{\partial x}{\partial \theta}\\[2mm] \dfrac{\partial y}{\partial r} & \dfrac{\partial y}{\partial \theta}\end{vmatrix}\right\| drd\theta = \left\|\begin{vmatrix}2\cos\theta & -2r\sin\theta\\ 5\sin\theta & 5r\cos\theta\end{vmatrix}\right\| drd\theta = 10rdrd\theta$

$$\therefore \iint_R \cos(25x^2 + 4y^2)\, dxdy = 10\int_0^{2\pi} \int_0^1 \cos(100r^2)\, rdrd\theta$$

$$= 10\int_0^{2\pi} d\theta \int_0^1 \cos(100r^2)\, rdr = 10(2\pi) \cdot \left.\frac{\sin(100r^2)}{200}\right|_0^1 = \frac{\pi \sin 100}{10}$$

Example 24.

$$求 \quad \int_1^2 \int_0^x \frac{1}{\sqrt{x^2 + y^2}}\, dydx =?$$

【解】

令$x = r\cos\theta$, $y = r\sin\theta$

则$\{(x,y): 1 \le x \le 2, 0 \le y \le x\} = \{(r,\theta): \dfrac{1}{\cos\theta} \le r \le \dfrac{2}{\cos\theta}, 0 \le \theta \le \dfrac{\pi}{4}\}$

且 $dxdy = \left\|\begin{vmatrix}\dfrac{\partial x}{\partial r} & \dfrac{\partial x}{\partial \theta}\\[2mm] \dfrac{\partial y}{\partial r} & \dfrac{\partial y}{\partial \theta}\end{vmatrix}\right\| drd\theta = \left\|\begin{vmatrix}\cos\theta & -r\sin\theta\\ \sin\theta & r\cos\theta\end{vmatrix}\right\| drd\theta = rdrd\theta$

$$\therefore \int_1^2 \int_0^x \frac{1}{\sqrt{x^2+y^2}}\,dydx = \int_0^{\frac{\pi}{4}} \int_{\frac{1}{\cos\theta}}^{\frac{2}{\cos\theta}} \frac{r}{\sqrt{r^2}}\,drd\theta = \int_0^{\frac{\pi}{4}} \frac{1}{\cos\theta}\,d\theta = \ln(\sec\theta + \tan\theta)\big|_0^{\frac{\pi}{4}}$$

$$= \ln(\sqrt{2}+1)$$

Example 25.

$$求 \iint_R e^{-xy}\,dxdy = ?, \; R为: y=x, \; y=2x, \; xy=1, \; xy=4 \text{ 所围区域}$$

【解】

$$令 \frac{y}{x} = u, \; xy = v \text{ 则 } x = \sqrt{\frac{v}{u}}, \; y = \sqrt{uv}$$

$$且 \; dxdy = \left\| \begin{vmatrix} \dfrac{\partial x}{\partial u} & \dfrac{\partial x}{\partial v} \\[2mm] \dfrac{\partial y}{\partial u} & \dfrac{\partial y}{\partial v} \end{vmatrix} \right\| dudv = \left\| \begin{vmatrix} \dfrac{1}{2}\left(\dfrac{v}{u}\right)^{-\frac{1}{2}}\left(-\dfrac{v}{u^2}\right) & \dfrac{1}{2}\left(\dfrac{v}{u}\right)^{-\frac{1}{2}}\left(\dfrac{1}{u}\right) \\[2mm] \dfrac{1}{2}(uv)^{-\frac{1}{2}}v & \dfrac{1}{2}(uv)^{-\frac{1}{2}}u \end{vmatrix} \right\| dudv = \frac{1}{2u}\,dudv$$

$$\because R = \left\{ (x,y): 1 \le \frac{y}{x} \le 2, 1 \le xy \le 4 \right\} \quad \therefore R = \{(u,v): 1 \le u \le 2, 1 \le v \le 4\}$$

$$\therefore \iint_R e^{-xy}\,dxdy = 2\int_1^4 \int_1^2 \frac{e^{-v}}{2u}\,dudv = \int_1^4 e^{-v}\,dv \int_1^2 \frac{1}{u}\,du = \left(\frac{1}{e} - \frac{1}{e^4}\right)\ln 2$$

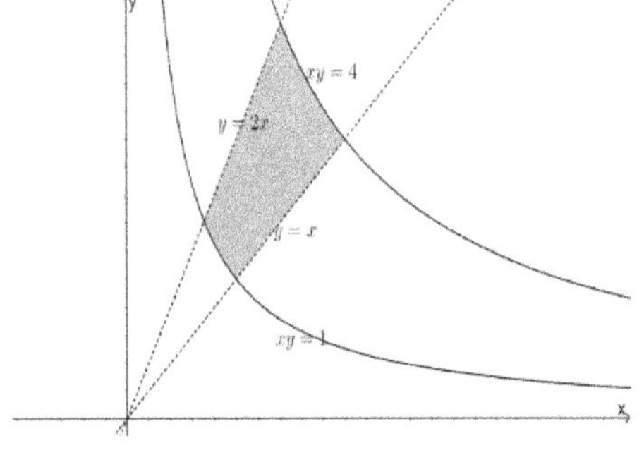


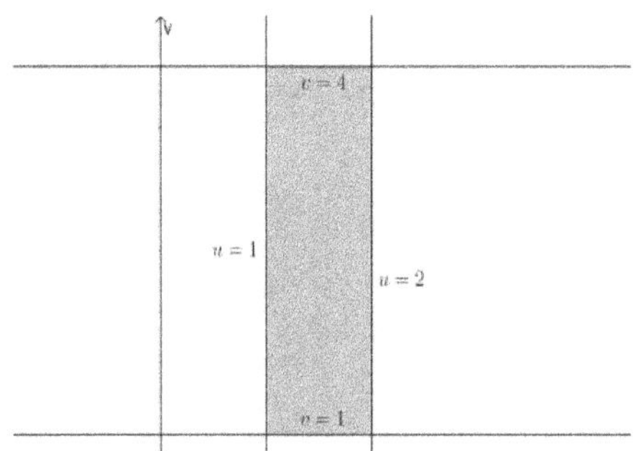


Example 26.

$$求 \iint_R x^2 + y^2\,dxdy = ?, \; R = \left\{ (x,y), 1 \le xy \le 4, 1 \le \frac{y}{x} \le 4 \right\}$$

【解】

$$令 \frac{y}{x} = u, \; xy = v \text{ 则 } x = \sqrt{\frac{v}{u}}, \; y = \sqrt{uv}$$

$$\text{且 } dxdy = \left\| \begin{matrix} \dfrac{\partial x}{\partial u} & \dfrac{\partial x}{\partial v} \\[2mm] \dfrac{\partial y}{\partial u} & \dfrac{\partial y}{\partial v} \end{matrix} \right\| dudv = \left\| \begin{matrix} \dfrac{1}{2}\left(\dfrac{v}{u}\right)^{-\frac{1}{2}}\left(-\dfrac{v}{u^2}\right) & \dfrac{1}{2}\left(\dfrac{v}{u}\right)^{-\frac{1}{2}}\left(\dfrac{1}{u}\right) \\[2mm] \dfrac{1}{2}(uv)^{-\frac{1}{2}}v & \dfrac{1}{2}(uv)^{-\frac{1}{2}}u \end{matrix} \right\| dudv = \dfrac{1}{2u} dudv$$

$$\because R = \left\{(x,y): 1 \le \dfrac{y}{x} \le 4, 1 \le xy \le 4\right\} \qquad \therefore R = \{(u,v): 1 \le u \le 4, 1 \le v \le 4\}$$

$$\iint_R x^2 + y^2 \, dxdy = 2\int_1^4\int_1^4 \left(\dfrac{v}{u^2}+v\right)\dfrac{1}{2}\,dudv = \int_1^4\int_1^4\left(\dfrac{v}{u^2}+v\right)dudv = \int_1^4 \dfrac{15v}{4}\,dv = \dfrac{225}{8}$$

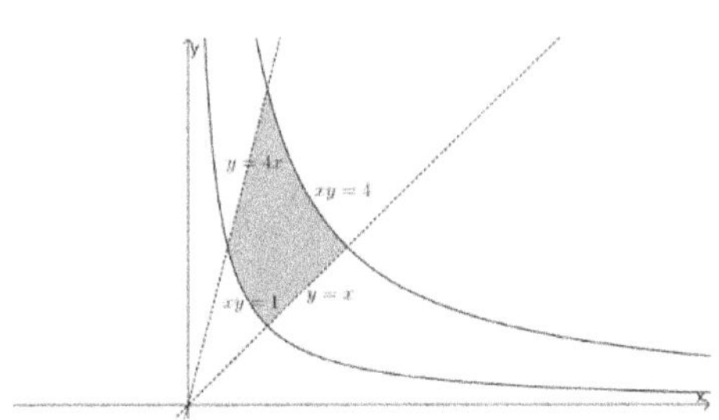 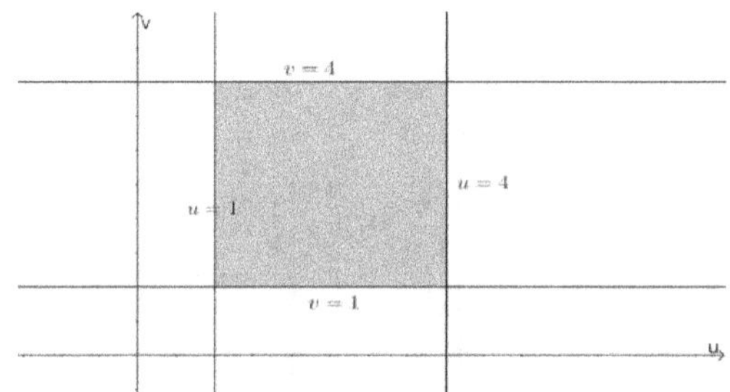

Example 27.

$$\text{求 } \int_{\frac{1}{2}}^{1}\int_{1-x}^{x} \dfrac{1}{\sqrt{x^2+y^2}}\,dydx = ?$$

【解】

令 $x = r\cos\theta, \ y = r\sin\theta$

则 $\left\{(x,y): 1-x \le y \le x, \dfrac{1}{2} \le x \le 1\right\} = \left\{(r,\theta): \dfrac{1}{\cos\theta+\sin\theta} \le r \le \dfrac{1}{\cos\theta}, 0 \le \theta \le \dfrac{\pi}{4}\right\}$

$$\text{且 } dxdy = \left\| \begin{matrix} \dfrac{\partial x}{\partial r} & \dfrac{\partial x}{\partial \theta} \\[2mm] \dfrac{\partial y}{\partial r} & \dfrac{\partial y}{\partial \theta} \end{matrix} \right\| drd\theta = \left| \begin{matrix} \cos\theta & -r\sin\theta \\ \sin\theta & r\cos\theta \end{matrix} \right| drd\theta = rdrd\theta$$

$$\therefore \int_{\frac{1}{2}}^{1}\int_{1-x}^{x} \dfrac{1}{\sqrt{x^2+y^2}}\,dydx = \int_0^{\frac{\pi}{4}}\int_{\frac{1}{\cos\theta+\sin\theta}}^{\frac{1}{\cos\theta}} \dfrac{r}{\sqrt{r^2}}\,drd\theta = \int_0^{\frac{\pi}{4}} \dfrac{1}{\cos\theta} - \dfrac{1}{\cos\theta+\sin\theta}\,d\theta$$

$$= \int_0^{\frac{\pi}{4}} \dfrac{1}{\cos\theta} - \dfrac{1}{\sqrt{2}\sin\left(\theta+\dfrac{\pi}{4}\right)}\,d\theta$$

$$= \ln(\sec\theta + \tan\theta)\Big|_0^{\frac{\pi}{4}} + \frac{1}{\sqrt{2}}\ln(|\csc\left(\theta + \frac{\pi}{4}\right) + \cot\left(\theta + \frac{\pi}{4}\right)|)\Big|_0^{\frac{\pi}{4}} = (1 - \frac{1}{\sqrt{2}})\ln(1 + \sqrt{2})$$

Example 28.

求 $\displaystyle\iint_R (x + y)^2 \, dxdy =?$,

其中 $R$ 为 $x + y = 0$, $x + y = 1$, $2x - y = 0$, $2x - y = 3$ 所围区域

【解】

令 $x + y = u$, $2x - y = v$ 则 $x = \dfrac{u + v}{3}$, $y = \dfrac{2u - v}{3}$

且 $dxdy = \left\|\begin{vmatrix} \dfrac{\partial x}{\partial u} & \dfrac{\partial x}{\partial v} \\ \dfrac{\partial y}{\partial u} & \dfrac{\partial y}{\partial v} \end{vmatrix}\right\| dudv = \left\|\begin{vmatrix} \dfrac{1}{3} & \dfrac{1}{3} \\ \dfrac{2}{3} & \dfrac{-1}{3} \end{vmatrix}\right\| dudv = \dfrac{1}{3} dudv$

$\because R = \{(x,y): 0 \le x + y \le 1, 0 \le 2x - y \le 3\}$ $\therefore R = \{(u,v): 0 \le u \le 1, 0 \le v \le 3\}$

$\therefore \displaystyle\iint_R (x + y)^2 \, dxdy = \frac{1}{3}\int_0^3 \int_0^1 u^2 \, dudv = \frac{u^3}{3}\Big|_0^1 = \frac{1}{3}$

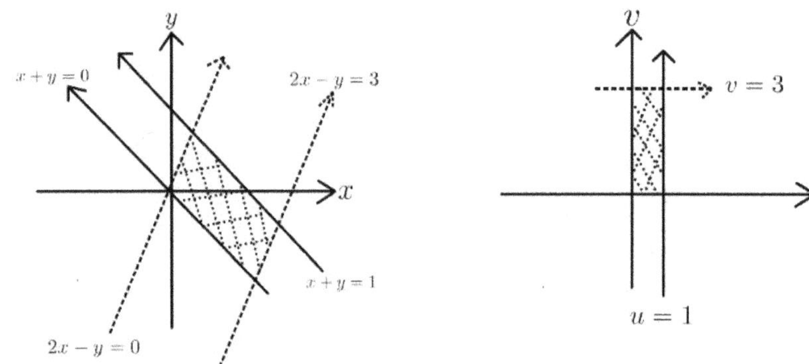


## 8.4 三重积分的考试类型

三重积分的考试类型包含不需调换顺序可直接转换成双重积分再转成定积分、转成双重积分后出现 $x^2 + y^2$,再使用极坐标转换、使用球坐标转换求三重积分

## 8.4.1　不需调换顺序可直接求迭代积分

考试类型：

Type 1.

给定 $f_1(x), f_2(x)$ 以及 $g_1(x,y), g_2(x,y)$，假设 $R = \{(x,y): a \leq x \leq b, f_1(x) \leq y \leq f_2(x)\}$

且 $g_1(x,y) \leq z \leq g_2(x,y)$，求 $\iint_R \int_{g_1(x,y)}^{g_2(x,y)} f(x,y)\,dz\,dA =?$

解题流程：

Step1.

将三重积分换成双重积分，$\iint_R \int_{g_1(x,y)}^{g_2(x,y)} f(x,y)\,dz\,dA = \iint_R f(x,y)\big(g_2(x,y) - g_1(x,y)\big)\,dA$

Step2.

使用 Fubini's Theorem 则

$$\iint_R f(x,y)\big(g_2(x,y) - g_1(x,y)\big)\,dA = \int_a^b \int_{f_1(x)}^{f_2(x)} f(x,y)\big(g_2(x,y) - g_1(x,y)\big)\,dy\,dx$$

Step3.

求 $\displaystyle\int_{f_1(x)}^{f_2(x)} f(x,y)\big(g_2(x,y) - g_1(x,y)\big)\,dy = ?$

假设计算后 $h(x) = \displaystyle\int_{f_1(x)}^{f_2(x)} f(x,y)\big(g_2(x,y) - g_1(x,y)\big)\,dy$

则 $\displaystyle\int_a^b \int_{f_1(x)}^{f_2(x)} f(x,y)\big(g_2(x,y) - g_1(x,y)\big)\,dy\,dx = \int_a^b h(x)\,dx$, 求 $\displaystyle\int_a^b h(x)\,dx = ?$

Type 2.

给定 $f_1(x), f_2(x)$，以及 $g_1(x,y), g_2(x,y)$，假设 $R = \{(x,y): c \leq y \leq d, f_1(y) \leq x \leq f_2(y)\}$

且 $g_1(x,y) \leq z \leq g_2(x,y)$，求 $\iint_R \int_{g_1(x,y)}^{g_2(x,y)} f(x,y,z)\,dz\,dA =?$

解题流程：

Step1.

将三重积分换成双重积分

则 $\iint_R \int_{g_1(x,y)}^{g_2(x,y)} f(x,y,z)\,dz\,dA = \iint_R f(x,y)\big(g_2(x,y) - g_1(x,y)\big)\,dA$

Step2.

使用 Fubini's Theorem 则

$$\iint_R f(x,y)\big(g_2(x,y) - g_1(x,y)\big)dA = \int_c^d \int_{f_1(y)}^{f_2(y)} f(x,y)\big(g_2(x,y) - g_1(x,y)\big)dxdy$$

Step3.

$$求 \int_{f_1(y)}^{f_2(y)} f(x,y)\big(g_2(x,y) - g_1(x,y)\big)dx =?$$

假设计算后 $h(y) = \int_{f_1(y)}^{f_2(y)} f(x,y)\big(g_2(x,y) - g_1(x,y)\big)dx$

则 $\int_c^d \int_{f_1(y)}^{f_2(y)} f(x,y)\big(g_2(x,y) - g_1(x,y)\big)dxdy = \int_c^d h(y)\,dy,\quad 求 \int_c^d h(y)\,dy =?$

Example 1.

(1) 求在 $xy$ 平面上至曲面 $z = x^3 + 2y$ 之间且位于 $y = 2x,\ y = x^2$ 内的区域体积

(2) 假设 V 为 $z > 0,\ z = x^3 + 2y,\ y = 2x,\ y = x^2$ 所围区域, 求 $\iiint_V x\,dV =?$

【解】

(1)

令 $R = \{(x,y): 0 \le x \le 2, x^2 \le y \le 2x\}$

则 体积 $= \iint_R \int_0^{x^3+2y} dzdA = \int_0^2 \int_{x^2}^{2x} x^3 + 2y\,dydx = \int_0^2 x^3 y + y^2\big|_{y=x^2}^{y=2x} dx$

$$= \int_0^2 2x^4 - x^5 + 4x^2 - x^4 dx = \frac{32}{5}$$

(2)

令 $R = \{(x,y): 0 \le x \le 2, x^2 \le y \le 2x\}$

则 $\iiint_V x\,dV = \iint_R \int_0^{x^3+2y} x\,dzdA = \int_0^2 \int_{x^2}^{2x} x(x^3 + 2y)\,dydx = \int_0^2 x\left(x^3 y + y^2\big|_{y=x^2}^{y=2x}\right) dx$

$$= \int_0^2 2x^5 - x^6 + 4x^3 - x^5 dx = \frac{x^6}{6} - \frac{x^7}{7} + x^4\Big|_0^2 = \frac{224 - 384 + 336}{21} = \frac{176}{21}$$

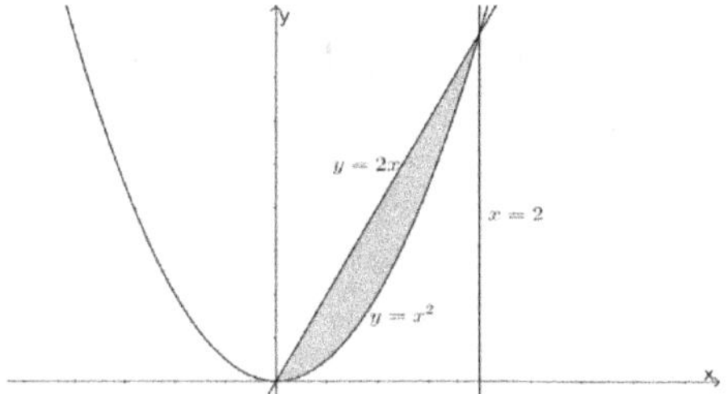

Example 2.

$$求 \int_0^{\frac{\pi}{2}} \int_{\sin 2z}^0 \int_0^{2yz} \sin\frac{x}{y} \, dx\,dy\,dz =?$$

【解】

$$\int_0^{\frac{\pi}{2}} \int_{\sin 2z}^0 \int_0^{2yz} \sin\frac{x}{y} \, dx\,dy\,dz = \int_0^{\frac{\pi}{2}} \int_{\sin 2z}^0 -y\cos\frac{x}{y}\Big|_{x=0}^{x=2yz} \, dy\,dz = \int_0^{\frac{\pi}{2}} \int_{\sin 2z}^0 y - y\cos 2z \, dy\,dz$$

$$= \int_0^{\frac{\pi}{2}} \left(\frac{y^2}{2} - \frac{y^2\cos 2z}{2}\right)\Big|_{y=\sin 2z}^{y=0} \, dz = \int_0^{\frac{\pi}{2}} -\frac{\sin^2 2z}{2} + \frac{\sin^2 2z\cos 2z}{2} \, dz$$

$$= \int_0^{\frac{\pi}{2}} -\frac{1-\cos 4z}{4} + \frac{\sin^2 2z\cos 2z}{2} \, dz = -\frac{\pi}{8}$$

Example 3.

$$求 \int_0^1 \int_0^{\sqrt{\ln 2}} \int_0^1 \frac{xyze^{x^2}}{1+y^2} \, dy\,dx\,dz =?$$

【解】

$$\int_0^1 \int_0^{\sqrt{\ln 2}} \int_0^1 \frac{xyze^{x^2}}{1+y^2} \, dy\,dx\,dz = \int_0^1 \frac{y}{1+y^2} \, dy \int_0^{\sqrt{\ln 2}} xe^{x^2} \, dx \int_0^1 z\,dz$$

$$= \frac{\ln(1+y^2)}{2}\Big|_0^1 \cdot \frac{e^{x^2}}{2}\Big|_0^{\sqrt{\ln 2}} \cdot \frac{z^2}{2}\Big|_0^1 = \frac{\ln 2}{8}$$

Example 4.

　　(1)求在$xy$平面上至平面$x+2y+z=2$之间且位于$2y=x,\ x=0$内的区域体积

(2)假设 V 为 $z > 0,\ x + 2y + z = 2,\ 2y = x,\ x = 0$ 所围区域,求 $\iiint_V x\,dV =?$

【解】

(1)

$$令 R = \left\{(x,y): 0 \le x \le 1, \frac{x}{2} \le y \le 1 - \frac{x}{2}\right\}$$

$$则体积 = \iint_R \int_0^{2-x-2y} dz\,dA = \int_0^1 \int_{\frac{x}{2}}^{1-\frac{x}{2}} 2 - x - 2y\,dy\,dx = \int_0^1 2y - xy - y^2 \Big|_{y=\frac{x}{2}}^{y=1-\frac{x}{2}} dx$$

$$= \int_0^1 1 - 2x + x^2\,dx = \frac{1}{3}$$

(2)

$$令 R = \left\{(x,y): 0 \le x \le 1, \frac{x}{2} \le y \le 1 - \frac{x}{2}\right\}$$

$$则 \iiint_V x\,dV = \iint_R \int_0^{2-x-2y} x\,dz\,dA = \int_0^1 \int_{\frac{x}{2}}^{1-\frac{x}{2}} x(2 - x - 2y)\,dy\,dx$$

$$= \int_0^1 x\left(2y - xy - y^2\Big|_{y=\frac{x}{2}}^{y=1-\frac{x}{2}}\right)dx = \int_0^1 x - 2x^2 + x^3\,dx = \frac{x^2}{2} - \frac{2x^3}{3} + \frac{x^4}{4}\Big|_0^1 = \frac{1}{12}$$

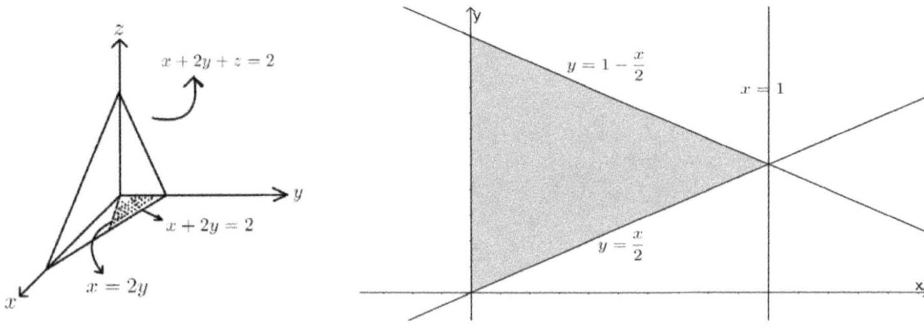


Example 5.

　　求 $y = x, y = 0, z = 0, 6x + 2y + 3z = 6$ 所围的体积

【解】

$$令 R = \{(x,y): 0 \le x \le 1, 0 \le y \le x\}$$

则 $V = \iint_R \dfrac{6 - 6x - 2y}{3} dydx = \int_0^1 \int_0^x 2 - 2x - \dfrac{2y}{3} dydx = \int_0^1 (2 - 2x)y - \dfrac{y^2}{3}\Big|_{y=0}^{y=x} dx$

$= \int_0^1 2x - \dfrac{7x^2}{3} dx = x^2 - \dfrac{7x^3}{9}\Big|_0^1 = \dfrac{2}{9}$

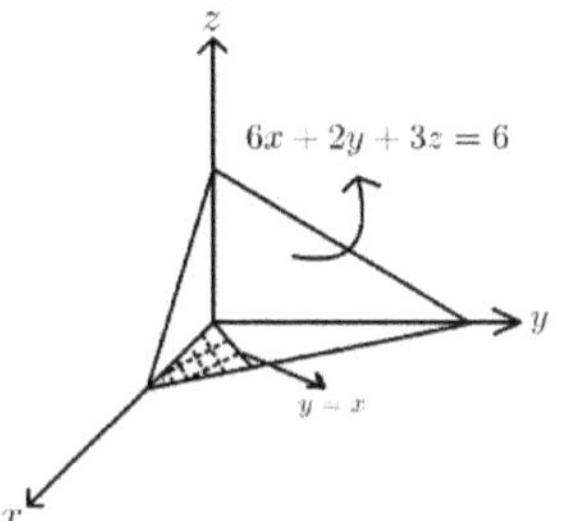


Example 6.

求 $\displaystyle\int_0^{\frac{\pi}{2}} \int_0^z \int_0^y \sin(x + y + z)\, dxdydz =?$

【解】

$\displaystyle\int_0^{\frac{\pi}{2}} \int_0^z \int_0^y \sin(x + y + z)\, dxdydz = \int_0^{\frac{\pi}{2}} \int_0^z -\cos(x + y + z)\big|_0^y\, dydz$

$= \displaystyle\int_0^{\frac{\pi}{2}} \int_0^z \cos(y + z) - \cos(2y + z)\, dydz = \int_0^{\frac{\pi}{2}} \left( \sin(y + z) - \dfrac{\sin(2y + z)}{2} \right)\Big|_{y=0}^{y=z} dz$

$= \displaystyle\int_0^{\frac{\pi}{2}} \sin(2z) - \dfrac{\sin z}{2} - \dfrac{\sin(3z)}{2}\, dz = -\dfrac{\cos 2z}{2} + \dfrac{\cos z}{2} + \dfrac{\cos(3z)}{6}\Big|_0^{\frac{\pi}{2}} = \dfrac{1}{3}$

Example 7.

求 $\displaystyle\int_0^5 \int_{-2}^4 \int_1^2 6xy^2z^3\, dxdydz =?$

【解】

$\displaystyle\int_0^5 \int_{-2}^4 \int_1^2 6xy^2z^3\, dxdydz = \int_0^5 \int_{-2}^4 3x^2y^2z^3\big|_{x=1}^{x=2}\, dydz = \int_0^5 \int_{-2}^4 9y^2z^3\, dydz$

$= \displaystyle\int_0^5 3y^3z^3\big|_{y=-2}^{y=4}\, dz = 216 \int_0^5 z^3\, dz = 33750$

Example 8.

  (1)求由 $x = 0, z = 0, y = x^2$ 与 $y + z = 9$ 所围的体积

  (2)假设 V 为 $x = 0, z = 0, \ y = x^2$ 与 $y + z = 9$ 所围的区域, 求 $\iiint_V x\,dV =?$

【解】

(1)

令$R = \{(x, y): 0 \le x \le 3, x^2 \le y \le 9\}$

则 体积 $= \iint_R \int_0^{9-y} dz\,dA = \int_0^3 \int_{x^2}^9 9 - y\,dy\,dx = \int_0^3 9y - \left.\dfrac{y^2}{2}\right|_{y=x^2}^{y=9} dx$

$$= \int_0^3 9(9 - x^2) - \dfrac{81 - x^4}{2}\,dx = \int_0^3 \dfrac{81}{2} - 9x^2 + \dfrac{x^4}{2}\,dx = \left.\left(\dfrac{81x}{2} - 3x^3 + \dfrac{x^5}{10}\right)\right|_0^3 = \dfrac{324}{5}$$

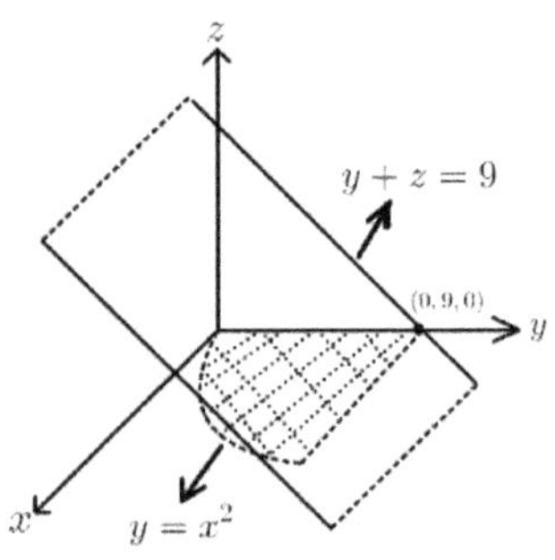


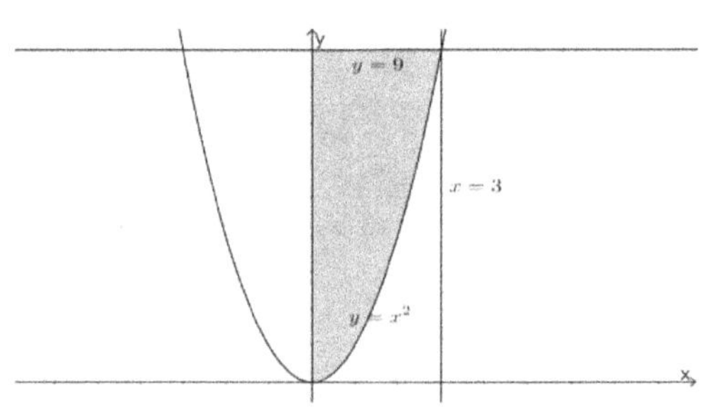


(2)

令$R = \{(x, y): 0 \le x \le 3, x^2 \le y \le 9\}$

则 $\iiint_V x\,dV = \iint_R \int_0^{9-y} x\,dz\,dA = \int_0^3 \int_{x^2}^9 x(9 - y)\,dy\,dx = \int_0^3 x\left(9y - \left.\dfrac{y^2}{2}\right|_{y=x^2}^{y=9}\right) dx$

$$= \int_0^3 x\left(9(9 - x^2) - \dfrac{81 - x^4}{2}\right) dx = \int_0^3 \dfrac{81x}{2} - 9x^3 + \dfrac{x^5}{2}\,dx = \left.\left(\dfrac{81x^2}{4} - \dfrac{9x^4}{4} + \dfrac{x^6}{12}\right)\right|_0^3$$

$$= \dfrac{81}{4}\cdot 9 - \dfrac{9}{4}\cdot 81 + \dfrac{9^3}{12} = \dfrac{243}{4}$$

Example 9.

(1)求由 $3x + 2y + z = 6$, $x = 0$, $y = 0$ 与 $z = 0$ 所围的体积

(2)假设 V 为 $3x + 2y + z = 6$, $x = 0$, $y = 0$ 与 $z = 0$ 所围的区域，求 $\iiint_V x\,dV = ?$

【解】

(1)

令 $R = \{(x, y): 0 \le x \le 2, 0 \le y \le 3 - \dfrac{3x}{2}\}$

$$所围体积 = \iint_R \int_0^{6-3x-2y} dz\,dA = \int_0^2 \int_0^{3-\frac{3x}{2}} 6 - 3x - 2y\,dy\,dx$$

$$= \int_0^2 (6y - 3xy - y^2)\Big|_{y=0}^{y=3-\frac{3x}{2}}\,dx = \int_0^2 9 - 9x + \frac{9x^2}{4}\,dx = \left(9x - \frac{9}{2}x^2 + \frac{3x^3}{4}\right)\Big|_0^2 = 6$$

(2)

令 $R = \{(x, y): 0 \le x \le 2, 0 \le y \le 3 - \dfrac{3x}{2}\}$

$$则 \iiint_V x\,dV = \iint_R \int_0^{6-3x-2y} x\,dz\,dA = \int_0^2 \int_0^{3-\frac{3x}{2}} x(6 - 3x - 2y)\,dy\,dx$$

$$= \int_0^2 x\left((6y - 3xy - y^2)\Big|_{y=0}^{y=3-\frac{3x}{2}}\right)dx = \int_0^2 9x - 9x^2 + \frac{9x^3}{4}\,dx = \left(\frac{9x^2}{2} - 3x^3 + \frac{9x^4}{16}\right)\Big|_0^2 = 3$$

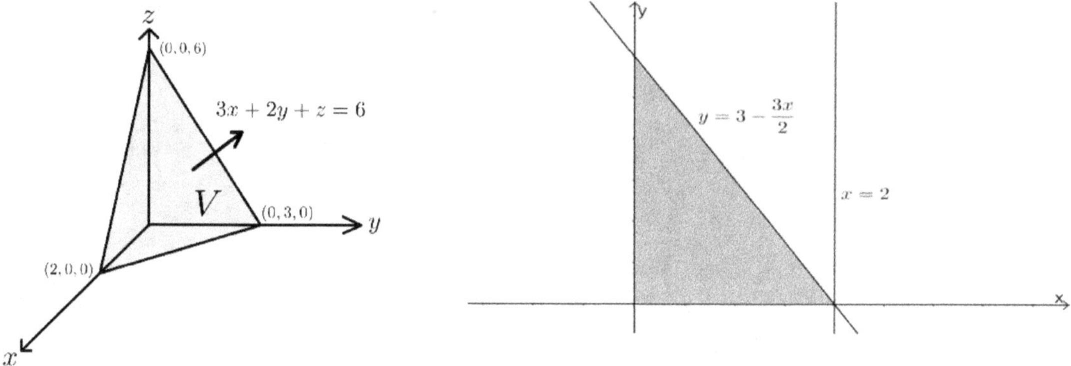


Example 10.

假设 V 为 $x + y + z = 1$, $x = 0$, $y = 0$ 与 $z = 0$ 所围的四面体，求

$$\iiint_V \frac{24}{(1+x+y+z)^4}\,dV = ?$$

【解】

令 $R = \{(x,y): 0 \le x \le 1, 0 \le y \le 1-x\}$

则 $\displaystyle\iiint_V \frac{24}{(1+x+y+z)^4}\,dV = \iint_R \int_0^{1-x-y} \frac{24}{(1+x+y+z)^4}\,dz\,dA$

$\displaystyle = \int_0^1 \int_0^{1-x} -8(1+x+y+z)^{-3}\big|_{z=0}^{z=1-x-y}\,dy\,dx = \int_0^1 \int_0^{1-x} -1 + 8(1+x+y)^{-3}\,dy\,dx$

$\displaystyle = \int_0^1 (-y - 4(1+x+y)^{-2})\big|_{y=0}^{y=1-x}\,dx = \int_0^1 4(1+x)^{-2} + x - 2\,dx$

$\displaystyle = (-4)(1+x)^{-1} + \frac{x^2}{2} - 2x\,\bigg|_0^1 = \frac{1}{2}$

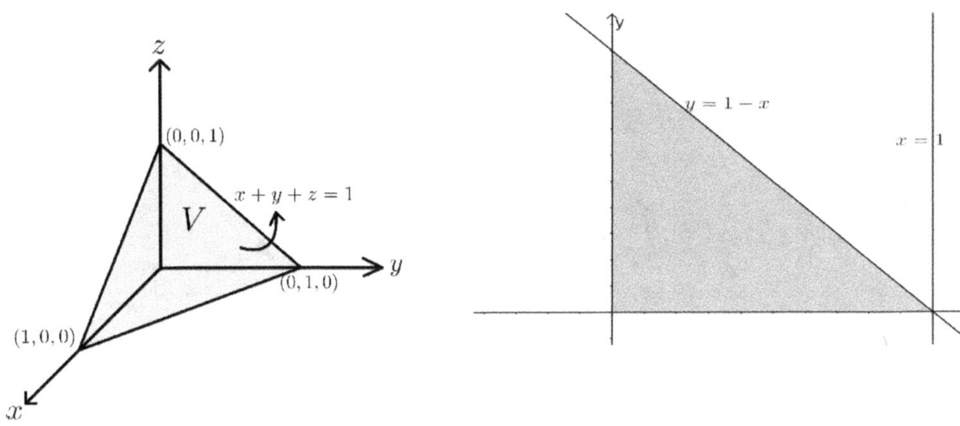


Example 11.

假设 V 为 $4x + 4y + z = 4$, $x = 0$, $y = 0$ 与 $z = 0$ 所围的四面体, 求 $\displaystyle\iiint_V 2x\,dV = ?$

【解】

令 $R = \{(x,y): 0 \le x \le 1, 0 \le y \le 1-x\}$

则 $\displaystyle\iiint_V 2x\,dV = \iint_R \int_0^{4-4x-4y} 2x\,dz\,dA = \int_0^1 \int_0^{1-x} 2xz\big|_{z=0}^{z=4-4x-4y}\,dy\,dx$

$\displaystyle = \int_0^1 \int_0^{1-x} 8x - 8x^2 - 8xy\,dy\,dx = \int_0^1 (8xy - 8x^2y - 4xy^2)\big|_{y=0}^{y=1-x}\,dx$

$$= \int_0^1 8x(1-x) - 8x^2(1-x) - 4x(1-x)^2 \, dx = 2\left(x^2 - \frac{4x^3}{3} + \frac{x^4}{2}\right)\Big|_0^1 = \frac{1}{3}$$

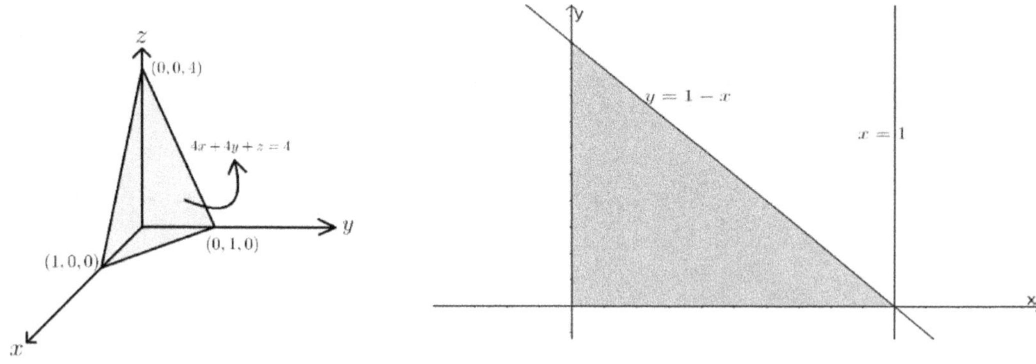


Example 12.

(1)求由$z = 0$, $z = 1 - x - y$, $x = 0$ 与 $y = 0$ 所围的体积

(2)假设 V 为$z = 0$, $z = 1 - x - y$, $x = 0$ 与 $y = 0$ 所围的体积, 求 $\iiint_V x\,dV =$?

【解】

(1)

令$R = \{(x, y) : 0 \le x \le \frac{1}{2}, 0 \le y \le \frac{1}{2} - x\}$

则体积 $= \iint_R \int_0^{1-x-y} dz\,dA = \int_0^{\frac{1}{2}} \int_0^{\frac{1}{2}-x} 1 - x - y\,dy\,dx = \int_0^{\frac{1}{2}} \left(y - xy - \frac{y^2}{2}\right)\Big|_0^{\frac{1}{2}-x} dx$

$$= \int_0^{\frac{1}{2}} \left(\frac{1}{2} - x\right) - x\left(\frac{1}{2} - x\right) - \frac{(\frac{1}{2} - x)^2}{2}\,dx = \int_0^{\frac{1}{2}} \frac{x^2}{2} - x + \frac{3}{8}\,dx = \left(\frac{3x}{8} - \frac{x^2}{2} + \frac{x^3}{6}\right)\Big|_0^{\frac{1}{2}} = \frac{1}{12}$$

(2)

令$R = \{(x, y) : 0 \le x \le \frac{1}{2}, 0 \le y \le \frac{1}{2} - x\}$

则 $\iiint_V x\,dV = \iint_R \int_0^{1-x-y} x\,dz\,dA = \int_0^{\frac{1}{2}} \int_0^{\frac{1}{2}-x} x(1 - x - y)\,dy\,dx$

$$= \int_0^{\frac{1}{2}} x \left( (y - xy - \frac{y^2}{2}) \Big|_0^{\frac{1}{2}-x} \right) dx = \int_0^{\frac{1}{2}} x \left( \left(\frac{1}{2} - x\right) - x\left(\frac{1}{2} - x\right) - \frac{(\frac{1}{2} - x)^2}{2} \right) dx$$

$$= \int_0^{\frac{1}{2}} \frac{x^3}{2} - x^2 + \frac{3x}{8} \, dx = \left( \frac{x^4}{8} - \frac{x^3}{3} + \frac{3x^2}{16} \right) \Big|_0^{\frac{1}{2}} = \frac{1}{8}\left( \frac{1}{16} - \frac{1}{3} + \frac{3}{8} \right) = \frac{5}{384}$$

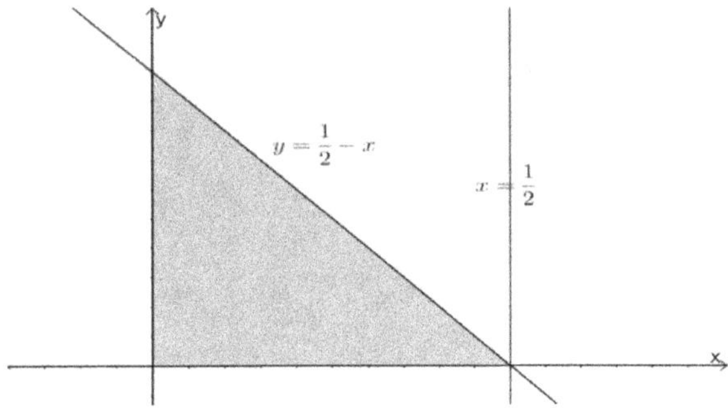


Example 13.

(1)求由 $y + 2z = 2$, $z = x^2$ 与 $y = 0$ 所围的体积

(2)假设 V 为 $y + 2z = 2$, $z = x^2$ 与 $y = 0$ 所围的区域, 求 $\iiint_V x^2 dV =?$

【解】

(1)

令 $R = \{(x, z): -1 \le x \le 1, x^2 \le z \le 1\}$

则 体积 $= \iint_R \int_0^{2-2z} dy dA = 2 \int_{-1}^1 \int_{x^2}^1 1 - z \, dz dx = 2 \int_{-1}^1 (z - \frac{z^2}{2}) \Big|_{x^2}^1 dx$

$$= 2 \int_{-1}^1 \left( \frac{1}{2} - x^2 + \frac{x^4}{2} \right) dx = \frac{16}{15}$$

(2)

令 $R = \{(x, z): -1 \le x \le 1, x^2 \le z \le 1\}$

则 $\iiint_V x^2 dV = \iint_R \int_0^{2-2z} x^2 dy dA = 2 \int_{-1}^1 \int_{x^2}^1 x^2(1 - z) \, dz dx = 2 \int_{-1}^1 x^2 \left( \left(z - \frac{z^2}{2}\right) \Big|_{x^2}^1 \right) dx$

$$= 2 \int_{-1}^1 \left( \frac{x^2}{2} - x^4 + \frac{x^6}{2} \right) dx = 2 \left( \frac{x^3}{6} - \frac{x^5}{5} + \frac{x^7}{14} \right) \Big|_{-1}^1 = \frac{16}{105}$$

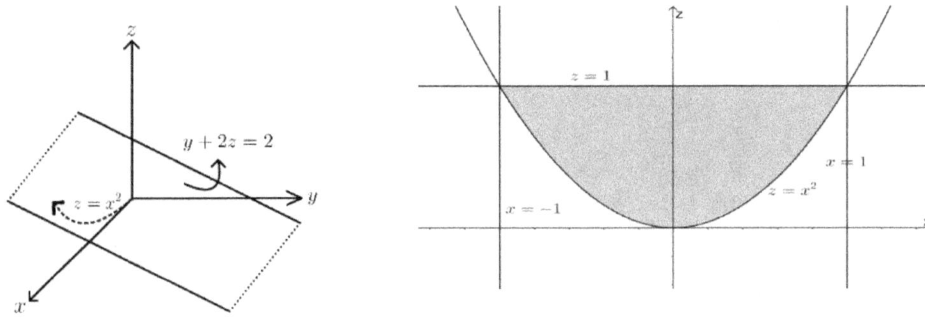

## Example 14.

假设 V 为 $x + y + z = d(d > 0)$, $x = 0$, $y = 0$ 与 $z = 0$ 所围的四面体,

$$求 \iiint_V \frac{24}{(d + x + y + z)^4} dV =?$$

【解】

令 $R = \{(x, y): 0 \le x \le d, 0 \le y \le d - x\}$

$$则 \iiint_V \frac{24}{(d + x + y + z)^4} dV = \iint_R \int_0^{d-x-y} \frac{24}{(d + x + y + z)^4} dz dA$$

$$= \int_0^d \int_0^{d-x} -8(d + x + y + z)^{-3}\Big|_{z=0}^{z=d-x-y} dy dx = \int_0^d \int_0^{d-x} -d^{-3} + 8(d + x + y)^{-3} dy dx$$

$$= \int_0^d (-d^{-3}y - 4(d + x + y)^{-2})\Big|_{y=0}^{y=d-x} dx = \int_0^d 4(d + x)^{-2} + d^{-3}x - 2d^{-2} dx$$

$$= -4(d + x)^{-1} + d^{-3} \cdot \frac{x^2}{2} - 2d^{-2}x \Big|_0^d = \frac{1}{2d}$$

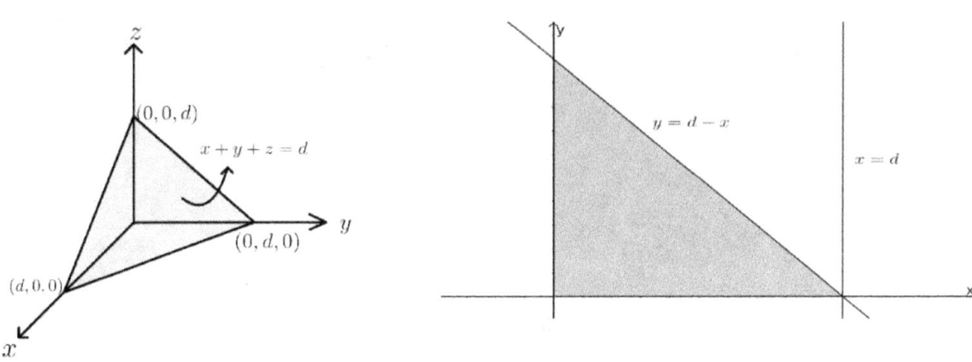

## Example 15.

假设 V 为 $dx + dy + z = d(d > 0)$, $x = 0$, $y = 0$ 与 $z = 0$ 所围的四面体,

求 $\iiint_V 2x\,dV =$?

【解】

令 $R = \{(x, y): 0 \le x \le 1, 0 \le y \le 1 - x\}$

则 $\iiint_V 2x\,dV = \iint_R \int_0^{d-dx-dy} 2x\,dz\,dA = \int_0^1 \int_0^{1-x} 2xz|_{z=0}^{z=d-dx-dy}\,dy\,dx$

$= \int_0^1 \int_0^{1-x} 2d(x - x^2 - xy)\,dy\,dx = 2d \int_0^1 \left(xy - x^2 y - \frac{xy^2}{2}\right)\Big|_{y=0}^{y=1-x}\,dx$

$= 2d \int_0^1 x(1-x) - x^2(1-x) - \frac{x(1-x)^2}{2}\,dx = d\left(\frac{x^2}{2} - \frac{2x^3}{3} + \frac{x^4}{4}\right)\Big|_0^1 = \frac{d}{12}$

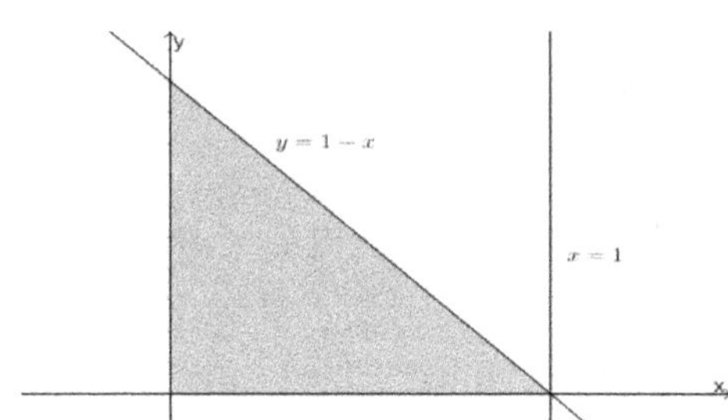


Example 16.

    (1)求由 $z = 3x^2$, $z = 4 - x^2$, $y + 2z = 12$ 与 $y = 0$ 所围的体积

    (2)假设 $V$ 为 $z = 3x^2$, $z = 4 - x^2$, $y + 2z = 12$ 与 $y = 0$ 所围的区域, 求 $\iiint_V x^2\,dV =$?

【解】

(1)

令 $R = \{(x, z): -1 \le x \le 1, 3x^2 \le z \le 4 - x^2\}$

则体积 $= \iint_R \int_0^{12-2z} dy\,dA = 2 \int_{-1}^1 \int_{3x^2}^{4-x^2} 6 - z\,dz\,dx = 2 \int_{-1}^1 \left(6z - \frac{z^2}{2}\right)\Big|_{3x^2}^{4-x^2}\,dx$

$= 8 \int_{-1}^1 x^4 - 5x^2 + 4\,dx = \frac{608}{15}$

(2)

令$R = \{(x,z): -1 \leq x \leq 1, 3x^2 \leq z \leq 4 - x^2\}$

则$\iiint_V x^2 dV = \iint_R \int_0^{12-2z} x^2 dy dA = 2\int_{-1}^1 \int_{3x^2}^{4-x^2} x^2(6-z)dzdx$

$$= 2\int_{-1}^1 x^2 \left( (6z - \frac{z^2}{2}) \Big|_{3x^2}^{4-x^2} \right) dx = 8\int_{-1}^1 x^6 - 5x^4 + 4x^2 dx = 8\left( \frac{x^7}{7} - x^5 + \frac{4x^3}{3} \right) \Big|_{-1}^1$$

$$= 8\left( \frac{2}{7} - 2 + \frac{8}{3} \right) = \frac{160}{21}$$

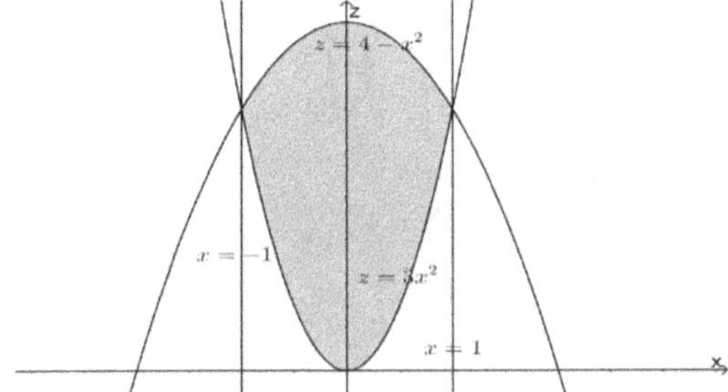


**Example 17.**

(1)求由 $z = x^2$, $z = 1 - y$, $x = 0$ 与 $y = 0$ 所围的体积

(2)假设 V 为 $z = x^2$, $z = 1 - y$, $x = 0$ 与 $y = 0$ 所围的区域, 求 $\iiint_V x dV =$?

【解】

(1)

令$R = \{(x,y): 0 \leq x \leq 1, 0 \leq y \leq 1 - x^2\}$

则体积 $= \iint_R \int_{x^2}^{1-y} dz dA = \int_0^1 \int_0^{1-x^2} 1 - y - x^2 dydx = \int_0^1 \left( y - \frac{y^2}{2} - x^2 y \right) \Big|_0^{1-x^2} dx$

$$= \int_0^1 \frac{x^4}{2} - x^2 + \frac{1}{2} dx = \frac{4}{15}$$

(2)

令$R = \{(x,y): 0 \leq x \leq 1, 0 \leq y \leq 1 - x^2\}$

则$\iiint_V x dV = \iint_R \int_{x^2}^{1-y} x dz dA = \int_0^1 \int_0^{1-x^2} x(1 - y - x^2)dydx$

$$= \int_0^1 x \left( (y - \frac{y^2}{2} - x^2 y) \Big|_0^{1-x^2} \right) dx = \int_0^1 \frac{x^5}{2} - x^3 + \frac{x}{2} dx = \frac{x^6}{12} - \frac{x^4}{4} + \frac{x^2}{4} \Big|_0^1 = \frac{1}{12}$$

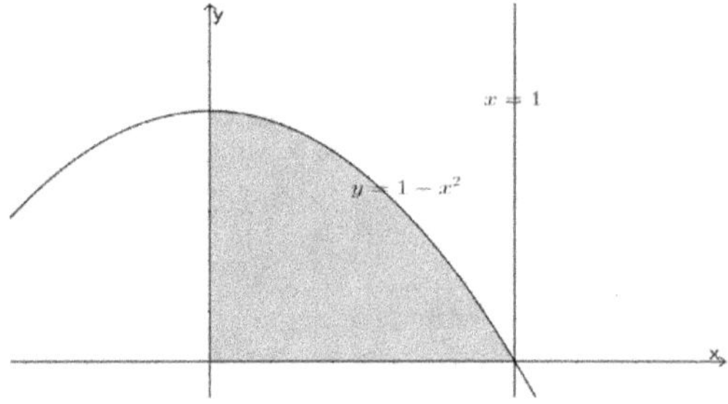

## Example 18.

假设 V 为 $2x + 3y + z = 6, x = 0, y = 0$ 与 $z = 0$ 所围四面体，求 $\iiint_V y^2 dxdydz =?$

【解】

令 $R = \left\{(x, y): 0 \le x \le 3, 0 \le y \le 2 - \dfrac{2x}{3}\right\}$

则 $\iiint_V y^2 dxdydz = \iint_R \int_0^{6-2x-3y} y^2 dzdA = \int_0^3 \int_0^{2-\frac{2x}{3}} y^2 z\big|_{z=0}^{z=6-2x-3y} dydx$

$= \int_0^3 \int_0^{2-\frac{2x}{3}} y^2(6 - 2x - 3y)dydx = \int_0^3 \left(\dfrac{y^3(6-2x)}{3} - \dfrac{3y^4}{4}\right)\Bigg|_{y=0}^{y=2-\frac{2x}{3}} dx$

$= \dfrac{1}{4}\int_0^3 \left(2 - \dfrac{2x}{3}\right)^4 dx = \dfrac{1}{20}\left(-\dfrac{3}{2}\right)\left(2 - \dfrac{2x}{3}\right)^5\Bigg|_0^3 = \dfrac{12}{5}$

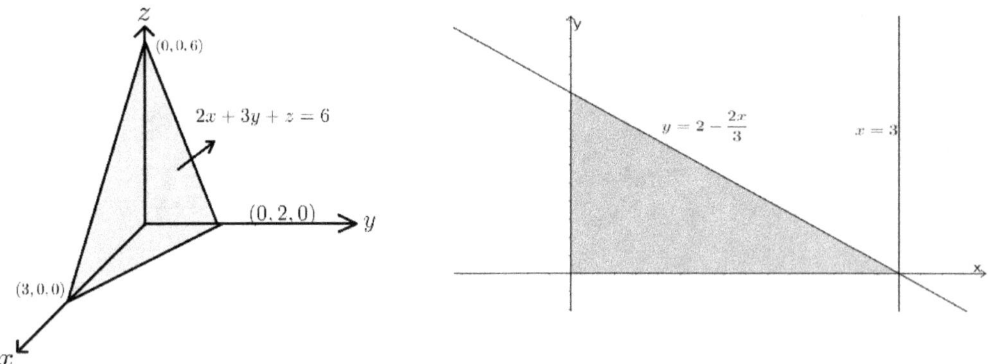

## Example 19.

假设 V 为 $x + y + z = 1$, $x = 0$, $y = 0$ 与 $z = 0$ 所围的四面体，求 $\iiint_V zdV = ?$

【解】

令 $R = \{(x, y): 0 \le x \le 1, 0 \le y \le 1 - x\}$

则 $\iiint_V zdV = \iint_R \int_0^{1-x-y} zdzdA = \int_0^1 \int_0^{1-x} \frac{z^2}{2}\Big|_{z=0}^{z=1-x-y} dydx$

$= \int_0^1 \int_0^{1-x} \frac{(1-x-y)^2}{2} dydx = \frac{-1}{2} \int_0^1 \frac{(1-x-y)^3}{3}\Big|_{y=0}^{y=1-x} dx = \frac{1}{6} \int_0^1 (1-x)^3 \, dx$

$= \frac{-1}{6} \left( \frac{(1-x)^4}{4} \right)\Big|_0^1 = \frac{1}{24}$

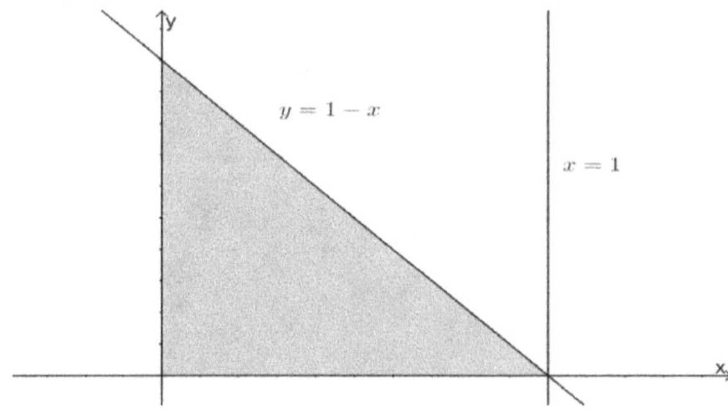


Example 20.

假设 V 为 $x + y + z = d (d > 0)$, $x = 0$, $y = 0$ 与 $z = 0$ 所围四面体，求 $\iiint_V zdV = ?$

【解】

令 $R = \{(x, y): 0 \le x \le d, 0 \le y \le d - x\}$

则 $\iiint_V zdV = \iint_R \int_0^{d-x-y} zdzdA = \int_0^d \int_0^{d-x} \frac{z^2}{2}\Big|_{z=0}^{z=d-x-y} dydx$

$= \int_0^d \int_0^{d-x} \frac{(d-x-y)^2}{2} dydx = \frac{-1}{2} \int_0^d \frac{(d-x-y)^3}{3}\Big|_{y=0}^{y=d-x} dx = \frac{1}{6} \int_0^d (d-x)^3 \, dx$

$$= \frac{-1}{6}\left(\frac{(d-x)^4}{4}\right)\Bigg|_0^d = \frac{d^4}{24}$$

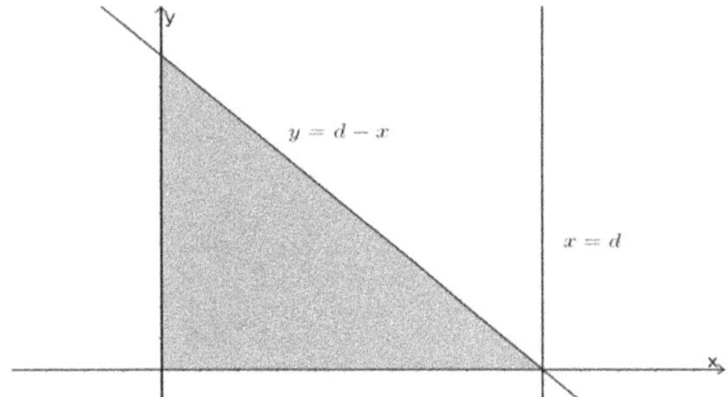


## 8.4.2　转成双重积分后出现 $x^{\wedge}2 + y^{\wedge}2$，再使用极坐标转换

考试类型:

Type 1.

給定 $f_1(x,y)$, $f_2(x,y)$, 假设 $R = \{(x,y): x^2 + y^2 \leq a\}$, $a > 0$ 且 $f_1(x,y) \leq z \leq f_2(x,y)$

求 $V = \iint_R \int_{f_2(x,y)}^{f_1(x,y)} f(x,y)dzdA = ?$

解题流程:

Step1.

$$V = \iint_R \int_{f_2(x,y)}^{f_1(x,y)} f(x,y)dzdA = \iint_R f(x,y)\big(f_1(x,y) - f_2(x,y)\big)dxdy$$

Step2.

令 $x = r\cos\theta, y = r\sin\theta$ 则 $R = \{(x,y): x^2 + y^2 \leq a\} = \{(r,\theta): 0 \leq r \leq a, 0 \leq \theta \leq 2\pi\}$

且 $dxdy = \left\|\begin{vmatrix} \dfrac{\partial x}{\partial r} & \dfrac{\partial x}{\partial \theta} \\ \dfrac{\partial y}{\partial r} & \dfrac{\partial y}{\partial \theta} \end{vmatrix}\right\| drd\theta = \left\|\begin{matrix} \cos\theta & -r\sin\theta \\ \sin\theta & r\cos\theta \end{matrix}\right\| drd\theta = rdrd\theta$

Step3.

$$V = \iint_R f(x,y)\big(f_1(x,y) - f_2(x,y)\big)dxdy$$

$$= \int_0^{2\pi} \int_0^a f(r\cos\theta, r\sin\theta)\big(f_1(r\cos\theta, r\sin\theta) - f_2(r\cos\theta, r\sin\theta)\big)rdrd\theta$$

Step4.

$$求 \int_0^{2\pi} \int_0^a f(r\cos\theta, r\sin\theta)\big(f_1(r\cos\theta, r\sin\theta) - f_2(r\cos\theta, r\sin\theta)\big) r\, dr\, d\theta$$

可能会再搭配使用变量变换、分部积分求积分值


Type 2.

空间中許多的二次曲面，两两交集所围封闭区域于 $xy$ 平面的投影常出现 $\dfrac{x^2}{a^2} + \dfrac{y^2}{b^2} \leq 1$，

得藉由上述方法求值，底下整理有哪些曲面能围出 $\dfrac{x^2}{a^2} + \dfrac{y^2}{b^2} \leq 1$，

首先，复习常见的二次曲面：

圆柱: $x^2 + y^2 = a^2$, $f_1(x,y) \leq z \leq f_2(x,y)$

椭球面: $\dfrac{x^2}{a^2} + \dfrac{y^2}{b^2} + \dfrac{z^2}{c^2} = 1$

椭圆抛物面: $\dfrac{x^2}{a^2} + \dfrac{y^2}{b^2} = \dfrac{z}{c}$

椭圆椎: $\dfrac{x^2}{a^2} + \dfrac{y^2}{b^2} = \dfrac{z^2}{c^2}$

双曲面: $\dfrac{x^2}{a^2} + \dfrac{y^2}{b^2} - \dfrac{z^2}{c^2} = 1$

考试类型：

求圆柱: $x^2 + y^2 = a^2$, $f_1(x,y) \leq z \leq f_2(x,y)$ 所围封闭区域

求椭圆抛物面: $\dfrac{x^2}{a^2} + \dfrac{y^2}{b^2} = \dfrac{z}{c}$ 与 $z = f_1(x,y)$ 所围封闭区域

求两个椭圆抛物面所围封闭区域

求椭圆抛物面与椭圆椎所围封闭区域

求椭球面: $\dfrac{x^2}{a^2} + \dfrac{y^2}{b^2} + \dfrac{z^2}{c^2} = 1$ 与平面 $z = f_1(x,y)$ 所围封闭区域

求椭球面: $\dfrac{x^2}{a^2} + \dfrac{y^2}{b^2} + \dfrac{z^2}{c^2} = 1$ 与椭圆抛物面所围封闭区域

求椭球面: $\dfrac{x^2}{a^2} + \dfrac{y^2}{b^2} + \dfrac{z^2}{c^2} = 1$ 与圆柱 $x^2 + y^2 = r^2$ 所围封闭区域

解题流程：

Step1.

找 $\alpha$, $f_1(x,y)$, $f_2(x,y)$ 使得 $V = \iint_R \int_{f_2(x,y)}^{f_1(x,y)} dzdA = \iint_R f_1(x,y) - f_2(x,y)dxdy$

其中 $R = \{(x,y): x^2 + y^2 \leq \alpha^2\}$, $\alpha > 0$

Step2.

使用上述极坐标转换的解法求 $\iint_R f_1(x,y) - f_2(x,y)dxdy =?$

令 $x = r\cos\theta$, $y = r\sin\theta$ 则 $R = \{(x,y): x^2 + y^2 \leq \alpha\} = \{(r,\theta): 0 \leq r \leq \alpha, 0 \leq \theta \leq 2\pi\}$

且 $dxdy = \left\| \begin{matrix} \dfrac{\partial x}{\partial r} & \dfrac{\partial x}{\partial \theta} \\ \dfrac{\partial y}{\partial r} & \dfrac{\partial y}{\partial \theta} \end{matrix} \right\| drd\theta = \left\| \begin{matrix} \cos\theta & -r\sin\theta \\ \sin\theta & r\cos\theta \end{matrix} \right\| drd\theta = rdrd\theta$

$$\iint_R f_1(x,y) - f_2(x,y)dxdy = \int_0^{2\pi} \int_0^{\alpha} \left( f_1(r\cos\theta, r\sin\theta) - f_2(r\cos\theta, r\sin\theta) \right) rdrd\theta$$

Example 1.

    (1)求在$xy$平面上至抛物面$z = x^2 + y^2$ 之间且位于圆柱 $x^2 + y^2 = a^2$ 区域内的体积

    (2)求圆柱 $x^2 + y^2 = 9$ 在平面 $2x + z = 8$ 与 $z = 0$ 的区域内体积

【解】

(1)

令$R = \{(x,y): x^2 + y^2 \leq a^2\}$ 则 $V = \iint_R \int_0^{x^2+y^2} dzdA = \iint_{x^2+y^2 \leq a^2} x^2 + y^2 dxdy$

令$x = r\cos\theta$, $y = r\sin\theta$ 则 $\{(x,y): x^2 + y^2 \leq a^2\} = \{(r,\theta): 0 \leq r \leq a, 0 \leq \theta \leq 2\pi\}$

且 $dxdy = \left\| \begin{matrix} \dfrac{\partial x}{\partial r} & \dfrac{\partial x}{\partial \theta} \\ \dfrac{\partial y}{\partial r} & \dfrac{\partial y}{\partial \theta} \end{matrix} \right\| drd\theta = \left\| \begin{matrix} \cos\theta & -r\sin\theta \\ \sin\theta & r\cos\theta \end{matrix} \right\| drd\theta = rdrd\theta$

$\therefore \iint_{x^2+y^2 \leq a^2} x^2 + y^2 dxdy = \int_0^{2\pi} \int_0^{a} r^3 drd\theta = \int_0^{2\pi} d\theta \int_0^{a} r^3 dr = \dfrac{\pi a^4}{2}$

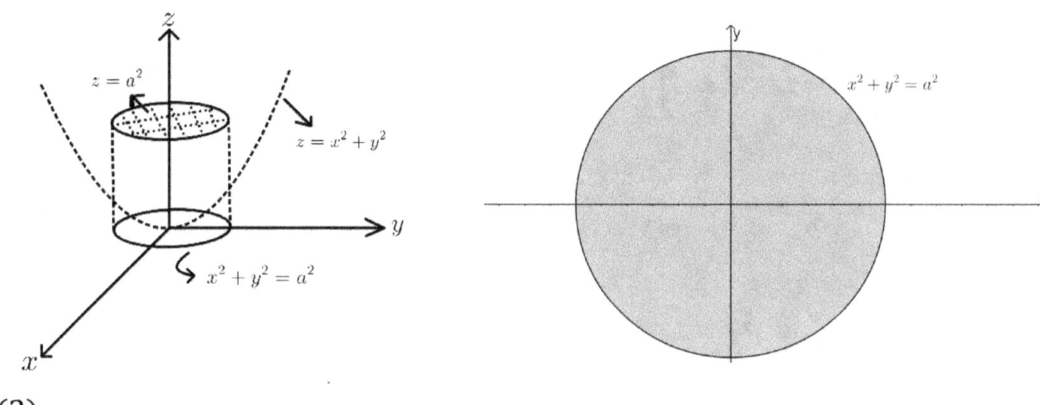

(2)

$$\text{令} R = \{(x,y): x^2 + y^2 \leq 9\} \text{ 则 } V = \iint_R \int_0^{8-2x} dz dA = 2 \iint_{x^2+y^2 \leq 9} 4 - x dx dy$$

$$\text{令} x = r\cos\theta, \ y = r\sin\theta \ \text{则} \ \{(x,y): x^2 + y^2 \leq 9\} = \{(r,\theta): 0 \leq r \leq 3, 0 \leq \theta \leq 2\pi\}$$

$$\text{且} \ dx dy = \left\| \begin{vmatrix} \dfrac{\partial x}{\partial r} & \dfrac{\partial x}{\partial \theta} \\ \dfrac{\partial y}{\partial r} & \dfrac{\partial y}{\partial \theta} \end{vmatrix} \right\| dr d\theta = \left\| \begin{vmatrix} \cos\theta & -r\sin\theta \\ \sin\theta & r\cos\theta \end{vmatrix} \right\| dr d\theta = r dr d\theta$$

$$\therefore 2 \iint_{x^2+y^2 \leq 9} 4 - x dx dy = 2 \int_0^{2\pi} \int_0^3 (4 - r\cos\theta) r \, dr d\theta = 2 \int_0^{2\pi} 2r^2 \big|_0^3 - \frac{r^3 \cos\theta}{3} \Big|_0^3 d\theta$$

$$= 72\pi$$

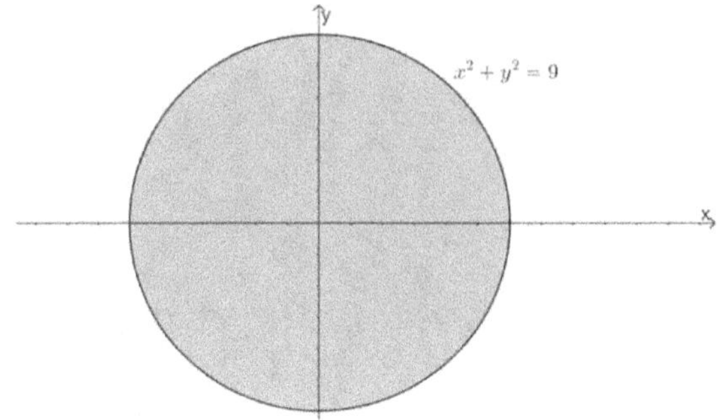

Example 2.

假设 V 为曲面 $x^2 + y^2 = 4z$, $x^2 + y^2 = 8y$, $z = 0$ 所围区域, 求 $\iiint_V dx dy dz =$ ?

【解】

$$\text{令} R = \{(x,y): x^2 + (y-4)^2 \leq 16\}$$

则 $\displaystyle\iiint_V dxdydz = \iint_R \int_0^{\frac{x^2+y^2}{4}} dzdA = \iint_{x^2+(y-4)^2\leq 16} \frac{x^2+y^2}{4}dxdy$

令 $x = 4r\cos\theta$ , $y = 4r\sin\theta$

则 $\{(x,y): x^2+(y-4)^2 \leq 16, y \geq 0\} = \{(r,\theta): 0 \leq r \leq 2\sin\theta, 0 \leq \theta \leq \pi\}$

且 $dxdy = \left\|\begin{vmatrix} \dfrac{\partial x}{\partial r} & \dfrac{\partial x}{\partial \theta} \\[2mm] \dfrac{\partial y}{\partial r} & \dfrac{\partial y}{\partial \theta} \end{vmatrix}\right\| drd\theta = \left\|\begin{vmatrix} 4\cos\theta & -4r\sin\theta \\ 4\sin\theta & 4r\cos\theta \end{vmatrix}\right\| drd\theta = 16rdrd\theta$

$\therefore \displaystyle\iint_{x^2+(y-4)^2\leq 16} \frac{x^2+y^2}{4}dxdy = \int_0^\pi \int_0^{2\sin\theta} 4r^2 \cdot 16rdrd\theta = 64\int_0^\pi \left.\frac{r^4}{4}\right|_0^{2\sin\theta} d\theta$

$= 256\int_0^\pi \sin^4\theta\, d\theta = 256 \cdot \frac{3}{4}\int_0^\pi \sin^2\theta\, d\theta = 256 \cdot \frac{3}{4} \cdot \frac{\pi}{2} = 96\pi$

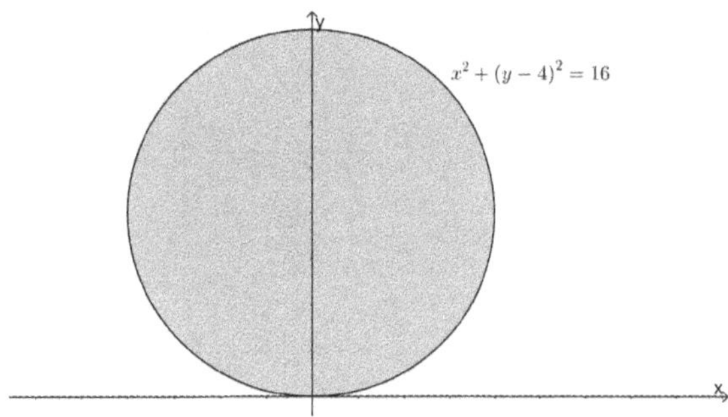


Example 3.

假设 V 为曲面 $x^2 + y^2 = 4$, $2(x^2+y^2) = z$, $xy$平面所围区域, 求 $\displaystyle\iiint_V dxdydz$ =?

【解】

令 $R = \{(x,y): x^2 + y^2 \leq 4\}$

则 $\displaystyle\iiint_V dxdydz = \iint_R \int_0^{2(x^2+y^2)} dzdA = \iint_{x^2+y^2\leq 4} 2(x^2+y^2)dxdy$

令 $x = r\cos\theta$ , $y = r\sin\theta$ 则 $\{(x,y): x^2+y^2 \leq 4\} = \{(r,\theta): 0 \leq r \leq 2, 0 \leq \theta \leq 2\pi\}$

$$\text{且 } dxdy = \left\| \begin{vmatrix} \dfrac{\partial x}{\partial r} & \dfrac{\partial x}{\partial \theta} \\ \dfrac{\partial y}{\partial r} & \dfrac{\partial y}{\partial \theta} \end{vmatrix} \right\| drd\theta = \left\| \begin{vmatrix} \cos\theta & -r\sin\theta \\ \sin\theta & r\cos\theta \end{vmatrix} \right\| drd\theta = rdrd\theta$$

$$\therefore \iint_{x^2+y^2\leq 4} 2(x^2+y^2)dxdy = \int_0^{2\pi}\int_0^2 2r^3 drd\theta = 16\pi$$

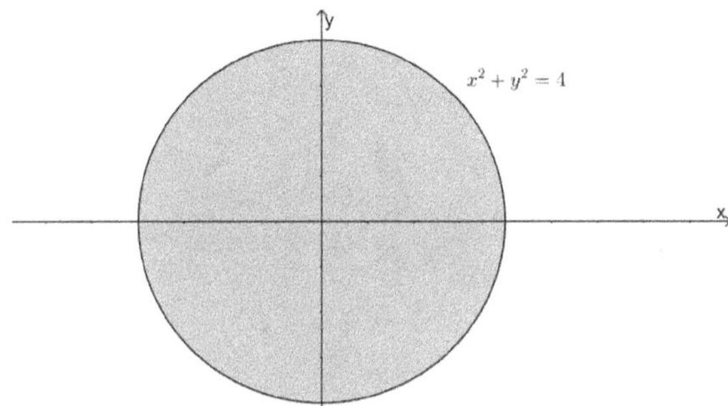


Example 4.

    (1)求抛物体 $z = 2x^2 + 2y^2$ 在 $0 \leq z \leq 8$ 的区域内体积

    (2)假设 V 为 $z = 2x^2 + 2y^2, x > 0, y > 0, z = 8$ 所围区域,求 $\iiint_V xdxdydz =?$

【解】

(1)

$$\text{令}R = \{(x,y): x^2+y^2 \leq 4\} \text{ 则 } V = \iint_R \int_{2x^2+2y^2}^8 dzdA = 2\iint_{x^2+y^2\leq 4} 4-(x^2+y^2)dxdy$$

$$\text{令} x = r\cos\theta, \ y = r\sin\theta \text{ 则 } \{(x,y): x^2+y^2 \leq 4\} = \{(r,\theta): 0 \leq r \leq 2, 0 \leq \theta \leq 2\pi\}$$

$$\text{且 } dxdy = \left\| \begin{vmatrix} \dfrac{\partial x}{\partial r} & \dfrac{\partial x}{\partial \theta} \\ \dfrac{\partial y}{\partial r} & \dfrac{\partial y}{\partial \theta} \end{vmatrix} \right\| drd\theta = \left\| \begin{vmatrix} \cos\theta & -r\sin\theta \\ \sin\theta & r\cos\theta \end{vmatrix} \right\| drd\theta = rdrd\theta$$

$$\therefore 2\iint_{x^2+y^2\leq 4} 4-(x^2+y^2)dxdy = 2\int_0^{2\pi}\int_0^2 4r-r^3 drd\theta = 2\int_0^{2\pi} d\theta \int_0^2 4r - r^3 dr$$

$$= 4\pi\left(2r^2 - \frac{r^4}{4}\right)\Bigg|_0^2 = 16\pi$$

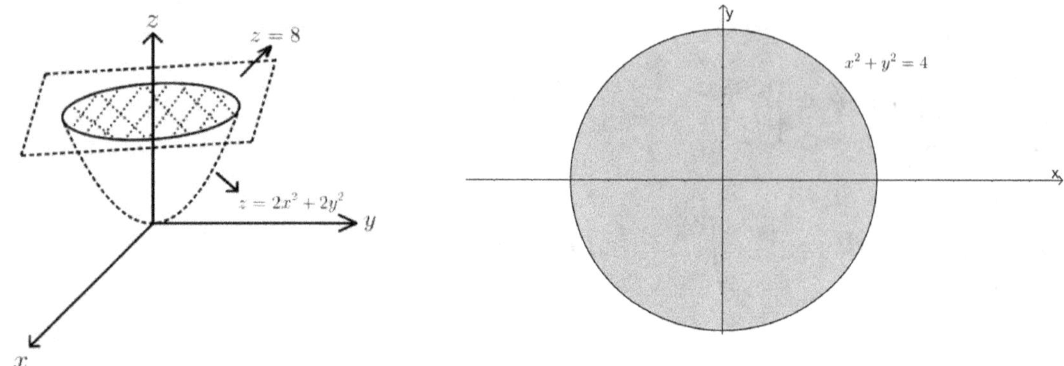

(2)

令 $R = \{(x,y): x^2 + y^2 \leq 4, x > 0, y > 0\}$

则 $\displaystyle\iiint_V x\,dx\,dy\,dz = \iint_R \int_{2x^2+2y^2}^{8} x\,dz\,dA = 2\iint_{x^2+y^2\leq 4} x(4-(x^2+y^2))\,dx\,dy$

令 $x = r\cos\theta, \ y = r\sin\theta$

则 $\{(x,y): x^2 + y^2 \leq 4, x > 0, y > 0\} = \{(r,\theta): 0 \leq r \leq 2, 0 \leq \theta \leq \dfrac{\pi}{2}\}$

且 $dx\,dy = \left\|\begin{matrix} \dfrac{\partial x}{\partial r} & \dfrac{\partial x}{\partial \theta} \\[2mm] \dfrac{\partial y}{\partial r} & \dfrac{\partial y}{\partial \theta} \end{matrix}\right\| dr\,d\theta = \left\|\begin{matrix} \cos\theta & -r\sin\theta \\ \sin\theta & r\cos\theta \end{matrix}\right\| dr\,d\theta = r\,dr\,d\theta$

$\therefore 2\displaystyle\iint_{x^2+y^2\leq 4} x(4-(x^2+y^2))\,dx\,dy = 2\int_0^{\frac{\pi}{2}} \cos\theta\,d\theta \int_0^2 4r^2 - r^4\,dr = 2\left(\dfrac{4r^3}{3} - \dfrac{r^5}{5}\right)\Big|_0^2 = \dfrac{128}{15}$

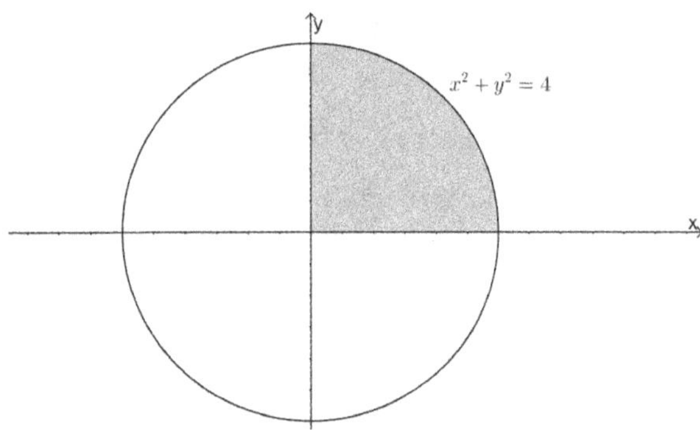

Example 5.

求由曲面 $z = 1 - x^2 - y^2$ 下方与 $z = 1 - y$ 上方所围区域的体积

【解】

令 $R = \{(x,y) : x^2 + \left(y - \dfrac{1}{2}\right)^2 \le \dfrac{1}{4}\}$

则 $V = \iint_R \int_{1-y}^{1-x^2-y^2} dz\,dA = \iint_{x^2+\left(y-\frac{1}{2}\right)^2 \le \frac{1}{4}} \dfrac{1}{4} - x^2 - \left(y - \dfrac{1}{2}\right)^2 dx\,dy$

令 $x = \dfrac{1}{2} r \cos\theta$, $y = \dfrac{1}{2} + \dfrac{1}{2} r \sin\theta$

则 $\{(x,y) : x^2 + \left(y - \dfrac{1}{2}\right)^2 \le \dfrac{1}{4}\} = \{(r, \theta) : 0 \le r \le 1, 0 \le \theta \le 2\pi\}$

且 $dx\,dy = \left\| \begin{matrix} \dfrac{\partial x}{\partial r} & \dfrac{\partial x}{\partial \theta} \\ \dfrac{\partial y}{\partial r} & \dfrac{\partial y}{\partial \theta} \end{matrix} \right\| dr\,d\theta = \left\| \begin{matrix} \dfrac{1}{2}\cos\theta & \dfrac{1}{2} r \sin\theta \\ \dfrac{1}{2}\sin\theta & \dfrac{1}{2} r \cos\theta \end{matrix} \right\| dr\,d\theta = \dfrac{1}{4} r\, dr\,d\theta$

$\therefore \iint_{x^2+(y-\frac{1}{2})^2 \le \frac{1}{4}} \dfrac{1}{4} - x^2 - \left(y - \dfrac{1}{2}\right)^2 dx\,dy = \int_0^{2\pi} \int_0^1 \left(\dfrac{1}{4} - \dfrac{1}{4} r^2\right) \dfrac{1}{4} r\, dr\,d\theta$

$= \dfrac{1}{16} \int_0^{2\pi} d\theta \int_0^1 r - r^3 dr = \dfrac{\pi}{32}$

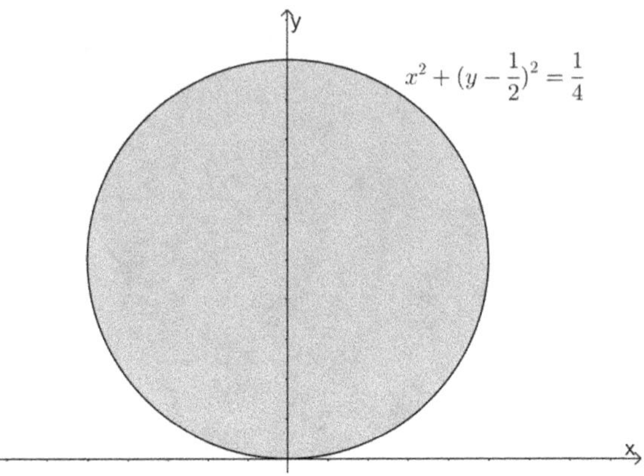


Example 6.

    假设 V 为曲面 $z \le 1 - x^2 - y^2$, $z \ge 1 - y$, $x > 0, y > 0, z > 0$ 所围区域

求 $\iiint_V x\,dx\,dy\,dz =?$

【解】

令 $R = \{(x,y): x^2 + \left(y - \dfrac{1}{2}\right)^2 \le \dfrac{1}{4}, x > 0, y > 0\}$

则 $\iiint_V x\,dx\,dy\,dz = \iint_R \int_{1-y}^{1-x^2-y^2} dz\,dA = \iint_R x\left(\dfrac{1}{4} - x^2 - \left(y - \dfrac{1}{2}\right)^2\right)dx\,dy$

令 $x = \dfrac{1}{2}r\cos\theta,\ y = \dfrac{1}{2} + \dfrac{1}{2}r\sin\theta$

则 $\{(x,y): x^2 + \left(y - \dfrac{1}{2}\right)^2 \le \dfrac{1}{4}, x > 0, y > 0\} = \{(r,\theta): 0 \le r \le 1, 0 \le \theta \le \dfrac{\pi}{2}\}$

且 $dx\,dy = \left\|\begin{matrix} \dfrac{\partial x}{\partial r} & \dfrac{\partial x}{\partial \theta} \\ \dfrac{\partial y}{\partial r} & \dfrac{\partial y}{\partial \theta} \end{matrix}\right\| dr\,d\theta = \left\|\begin{matrix} \dfrac{1}{2}\cos\theta & \dfrac{1}{2}r\sin\theta \\ \dfrac{1}{2}\sin\theta & \dfrac{1}{2}r\cos\theta \end{matrix}\right\| dr\,d\theta = \dfrac{1}{4}r\,dr\,d\theta$

$\therefore \iint_R x\left(\dfrac{1}{4} - x^2 - \left(y - \dfrac{1}{2}\right)^2\right)dx\,dy = \int_0^{\frac{\pi}{2}} \dfrac{1}{2}r\cos\theta \int_0^1 \left(\dfrac{1}{4} - \dfrac{1}{4}r^2\right)\dfrac{1}{4}r\,dr\,d\theta$

$= \dfrac{1}{32}\int_0^{\frac{\pi}{2}} \cos\theta\,d\theta \int_0^1 r^2 - r^4\,dr = \dfrac{1}{240}$

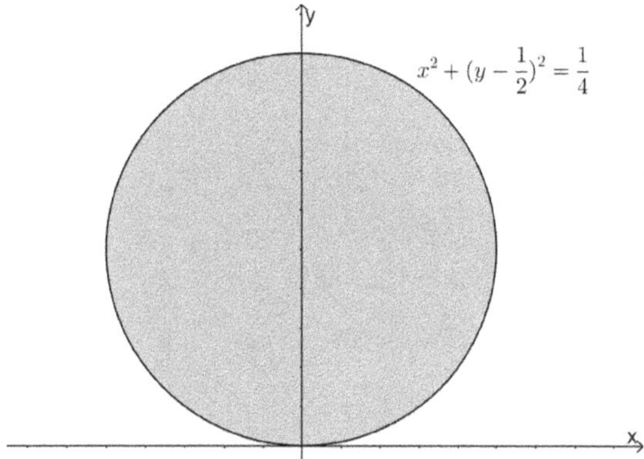


Example 7.

(1)求抛物体 $6x^2 + 6y^2 + z = 20$ 与 $z = 8$ 所围区域的体积

(2)假设 V 为 $6x^2 + 6y^2 + z = 20,\ x > 0,\ y > 0,\ z > 0,\ z = 8$ 所围区域

求 $\displaystyle\iiint_V xdxdydz =?$

【解】

(1)

令$R = \{(x,y): x^2 + y^2 \le 2 \}$ 则 $\displaystyle V = \iint_R \int_8^{20-6x^2-6y^2} dzdA = 2\iint_{x^2+y^2\le2} 6 - 3x^2 - 3y^2 dxdy$

令$x = r\cos\theta,\ y = r\sin\theta$ 则 $\{(x,y): x^2 + y^2 \le 2\} = \{(r,\theta): 0 \le r \le \sqrt{2}, 0 \le \theta \le 2\pi\}$

且 $dxdy = \left\| \begin{matrix} \dfrac{\partial x}{\partial r} & \dfrac{\partial x}{\partial \theta} \\ \dfrac{\partial y}{\partial r} & \dfrac{\partial y}{\partial \theta} \end{matrix} \right\| drd\theta = \left\| \begin{matrix} \cos\theta & -r\sin\theta \\ \sin\theta & r\cos\theta \end{matrix} \right\| drd\theta = rdrd\theta$

$\therefore 2\displaystyle\iint_{x^2+y^2\le2} 6 - 3x^2 - 3y^2 dxdy = 2\int_0^{2\pi}\int_0^{\sqrt{2}} 6r - 3r^3 drd\theta = 2\int_0^{2\pi} d\theta \int_0^{\sqrt{2}} 6r - 3r^3 dr$

$= 4\pi\left( 3r^2 - \dfrac{3r^4}{4} \right)\Big|_0^{\sqrt{2}} = 12\pi$

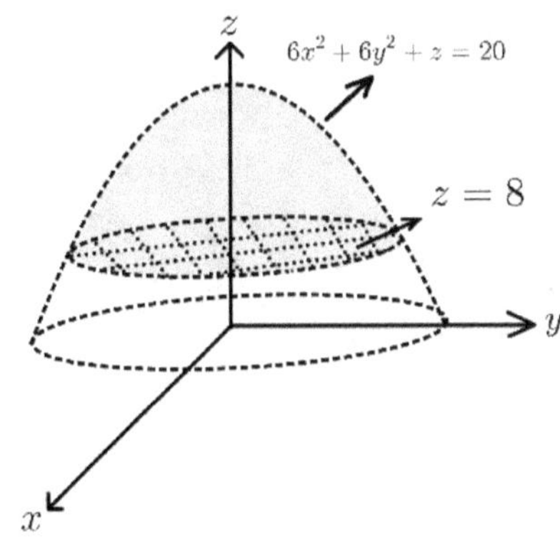


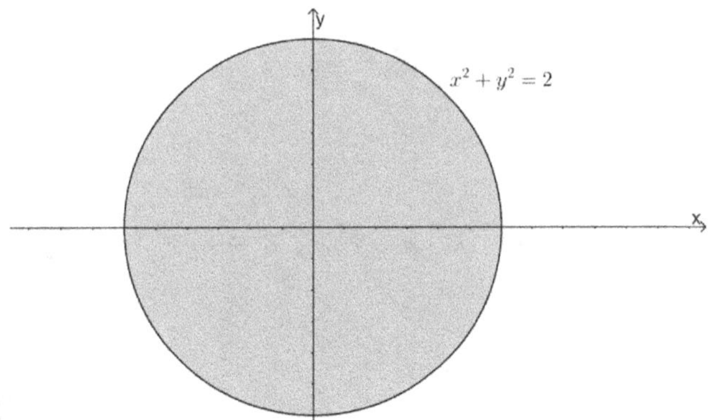


(2)

令$R = \{(x,y): x^2 + y^2 \le 2, x > 0, y > 0 \}$

则 $\displaystyle V = \iint_R \int_8^{20-6x^2-6y^2} xdzdA = 2\iint_{x^2+y^2\le2, x>0, y>0} x(6 - 3x^2 - 3y^2)dxdy$

令$x = r\cos\theta,\ y = r\sin\theta$

则 $\{(x,y): x^2 + y^2 \le 2, x > 0, y > 0\} = \{(r,\theta): 0 \le r \le \sqrt{2}, 0 \le \theta \le \dfrac{\pi}{2}\}$

且 $dxdy = \left\| \begin{vmatrix} \dfrac{\partial x}{\partial r} & \dfrac{\partial x}{\partial \theta} \\ \dfrac{\partial y}{\partial r} & \dfrac{\partial y}{\partial \theta} \end{vmatrix} \right\| drd\theta = \left\| \begin{vmatrix} \cos\theta & -r\sin\theta \\ \sin\theta & r\cos\theta \end{vmatrix} \right\| drd\theta = rdrd\theta$

$\therefore \iiint_V xdxdydz = 2 \iint_{x^2+y^2 \leq 2, x>0, y>0} x(6 - 3x^2 - 3y^2)dxdy$

$= 2 \int_0^{\frac{\pi}{2}} \cos\theta \int_0^{\sqrt{2}} 6r^2 - 3r^4 drd\theta = 2 \int_0^{\frac{\pi}{2}} \cos\theta \, d\theta \int_0^{\sqrt{2}} 6r^2 - 3r^4 dr = 2\left(2r^3 - \dfrac{3r^5}{5}\right)\Big|_0^{\sqrt{2}}$

$= \dfrac{16\sqrt{2}}{5}$

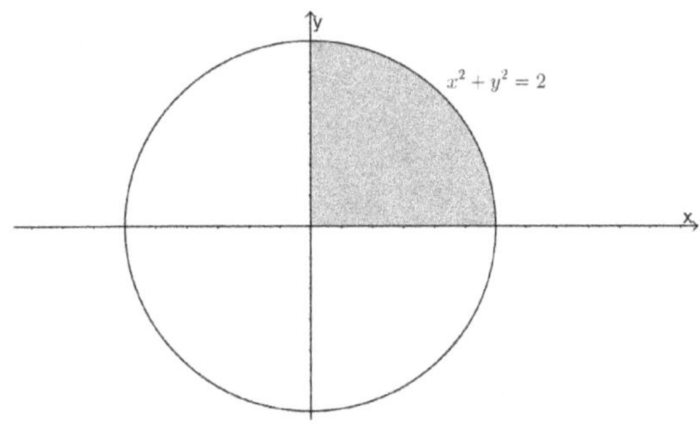


Example 8.

求两抛物面 $z = 5x^2 + 5y^2$ 与 $z = 6 - 7x^2 - y^2$ 所围体积

【解】

令 $R = \{(x, y): 2x^2 + y^2 \leq 1\}$

则 $V = \iint_R \int_{5x^2+5y^2}^{6-7x^2-y^2} dzdA = \iint_R z\big|_{5x^2+5y^2}^{6-7x^2-y^2} dA = \iint_R 6 - 12x^2 - 6y^2 dA$

令 $x = \dfrac{u}{\sqrt{2}}, \ y = v$ 则 $dxdy = \left\| \begin{vmatrix} \dfrac{\partial x}{\partial u} & \dfrac{\partial x}{\partial v} \\ \dfrac{\partial y}{\partial u} & \dfrac{\partial y}{\partial v} \end{vmatrix} \right\| dudv = \left\| \begin{vmatrix} \dfrac{1}{\sqrt{2}} & -0 \\ 0 & 1 \end{vmatrix} \right\| dudv = \dfrac{1}{\sqrt{2}} dudv$

且 $R = \{(u, v): u^2 + v^2 \leq 1\}$

$\therefore \iint_R 6 - 12x^2 - 6y^2 dA = \dfrac{6}{\sqrt{2}} \iint_{u^2+v^2 \leq 1} 1 - u^2 - v^2 dudv$

令 $u = r\cos\theta$ , $v = r\sin\theta$ 则 $\{(u,v): u^2 + v^2 \leq 1\} = \{(r,\theta): 0 \leq r \leq 1, 0 \leq \theta \leq 2\pi\}$

且 $dudv = \left\| \begin{vmatrix} \dfrac{\partial u}{\partial r} & \dfrac{\partial u}{\partial \theta} \\ \dfrac{\partial v}{\partial r} & \dfrac{\partial v}{\partial \theta} \end{vmatrix} \right\| drd\theta = \left\| \begin{vmatrix} \cos\theta & -r\sin\theta \\ \sin\theta & r\cos\theta \end{vmatrix} \right\| drd\theta = r\,drd\theta$

$\therefore V = \dfrac{6}{\sqrt{2}} \iint_{u^2+v^2\leq 1} 1 - u^2 - v^2\, dudv = \dfrac{6}{\sqrt{2}} \int_0^1 \int_0^{2\pi} (1-r^2) r\, d\theta dr = \dfrac{3\sqrt{2}}{2}$

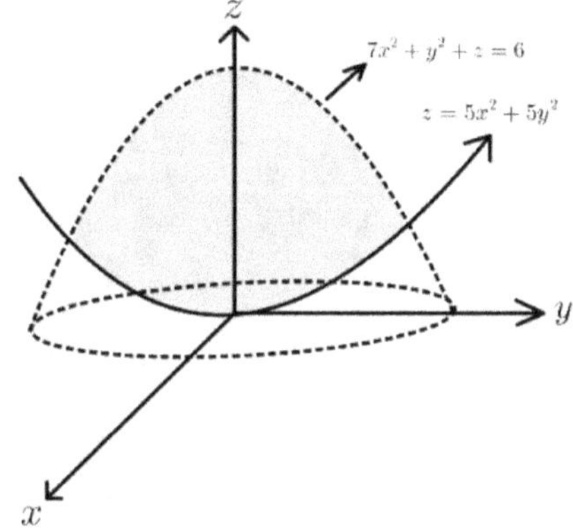


Example 9.

计算球体 $x^2 + y^2 + z^2 \leq 6$ 与 $z \geq x^2 + y^2$ 交集区域体积

【解】

令 $R = \{(x,y): x^2 + y^2 \leq 2\}$

则 $V = \iint_R \int_{x^2+y^2}^{\sqrt{6-x^2-y^2}} dzdA = \iint_{x^2+y^2\leq 2} \sqrt{6-x^2-y^2} - x^2 - y^2\, dxdy$

令 $x = r\cos\theta$, $y = r\sin\theta$ 则 $\{(x,y): x^2 + y^2 \leq 2\} = \{(r,\theta): 0 \leq r \leq \sqrt{2}, 0 \leq \theta \leq 2\pi\}$

且 $dxdy = \left\| \begin{vmatrix} \dfrac{\partial x}{\partial r} & \dfrac{\partial x}{\partial \theta} \\ \dfrac{\partial y}{\partial r} & \dfrac{\partial y}{\partial \theta} \end{vmatrix} \right\| drd\theta = \left\| \begin{vmatrix} \cos\theta & -r\sin\theta \\ \sin\theta & r\cos\theta \end{vmatrix} \right\| drd\theta = r\,drd\theta$

$\therefore \iint_{x^2+y^2\leq 2} \sqrt{6-x^2-y^2} - x^2 - y^2\, dxdy = \int_0^{2\pi} \int_0^{\sqrt{2}} (\sqrt{6-r^2} - r^2) r\, drd\theta$

$$= (2\pi) \cdot \left( \frac{(-1)(6-r^2)^{\frac{3}{2}}}{3} - \frac{r^4}{4} \right)\Bigg|_{r=0}^{r=\sqrt{2}} = (2\pi) \cdot \left( \frac{-8+6\sqrt{6}}{3} - 1 \right) = \frac{2\pi}{3}(6\sqrt{6} - 11)$$

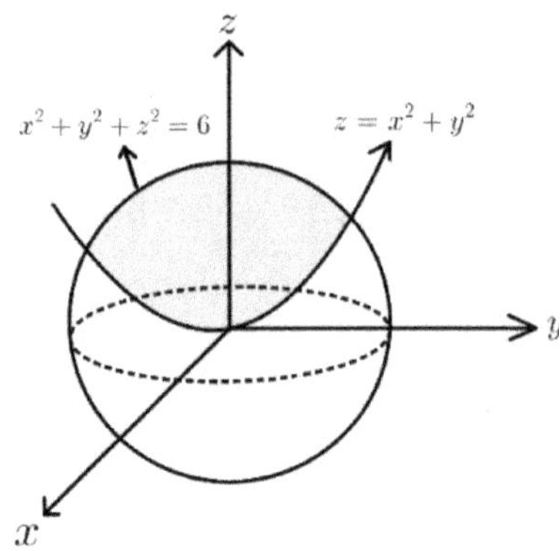


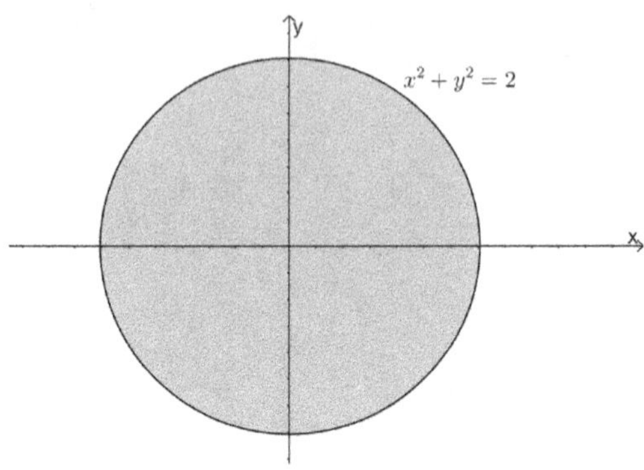


Example 10.

计算球体$x^2 + y^2 + z^2 = 4a^2$ 与圆柱: $x^2 + (y-a)^2 = a^2$所围区域的体积

【解】

令$R = \{(x,y): x^2 + (y-a)^2 \leq a^2\}$

则 $V = \iint_R \int_{-\sqrt{4a^2-x^2-y^2}}^{\sqrt{4a^2-x^2-y^2}} dzdA = \iint_{x^2+(y-a)^2\leq a^2} 2\sqrt{4a^2 - x^2 - y^2}\,dxdy$

令$x = ar\cos\theta$ , $y = ar\sin\theta$

则 $\{(x,y): x^2 + (y-a)^2 \leq a^2\} = \{(r,\theta): 0 \leq r \leq 2\sin\theta , 0 \leq \theta \leq \pi\}$

且 $dxdy = \left\| \begin{vmatrix} \dfrac{\partial x}{\partial r} & \dfrac{\partial x}{\partial \theta} \\[6pt] \dfrac{\partial y}{\partial r} & \dfrac{\partial y}{\partial \theta} \end{vmatrix} \right\| drd\theta = \left\| \begin{matrix} a\cos\theta & ar\sin\theta \\ a\sin\theta & ar\cos\theta \end{matrix} \right\| drd\theta = a^2 r\, drd\theta$

$\therefore \iint_{x^2+(y-a)^2\leq a^2} 2\sqrt{4a^2 - x^2 - y^2}\,dxdy = 2\int_0^\pi \int_0^{2\sin\theta} (\sqrt{4a^2 - a^2 r^2})a^2 r\, drd\theta$

$= 4a^3 \int_0^{\frac{\pi}{2}} \int_0^{2\sin\theta} (\sqrt{4 - r^2})r\, drd\theta = 4a^3 \int_0^{\frac{\pi}{2}} \left( \frac{(-1)(4-r^2)^{\frac{3}{2}}}{3} \right)\Bigg|_0^{2\sin\theta} d\theta$

$= 32a^3 \int_0^{\frac{\pi}{2}} \frac{1 - \cos^3\theta}{3} d\theta = 32a^3 \left( \frac{\pi}{6} - \frac{2}{9} \right)$

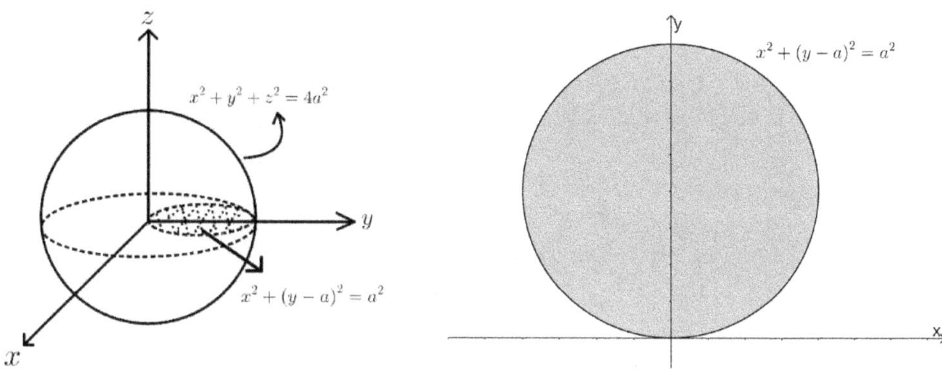

Example 11.

求由 $z^2 = x^2 + y^2$ 与 $nz = x^2 + y^2$ 所围区域的体积, $n \in N$

【解】

令 $R = \{(x,y): x^2 + y^2 \leq n^2\}$

则 $V = \iint_R \int_{\frac{x^2+y^2}{n}}^{\sqrt{x^2+y^2}} dz dA = \iint_{x^2+y^2 \leq n^2} \sqrt{x^2 + y^2} - \left(\frac{x^2 + y^2}{n}\right) dxdy$

令 $x = r\cos\theta, y = r\sin\theta$ 则 $\{(x,y): x^2 + y^2 \leq n^2\} = \{(r,\theta): 0 \leq r \leq n, 0 \leq \theta \leq 2\pi\}$

且 $dxdy = \left\|\begin{matrix} \dfrac{\partial x}{\partial r} & \dfrac{\partial x}{\partial \theta} \\ \dfrac{\partial y}{\partial r} & \dfrac{\partial y}{\partial \theta} \end{matrix}\right\| drd\theta = \left\|\begin{matrix} \cos\theta & -r\sin\theta \\ \sin\theta & r\cos\theta \end{matrix}\right\| drd\theta = rdrd\theta$

$\therefore \iint_{x^2+y^2 \leq n^2} \sqrt{x^2 + y^2} - \left(\frac{x^2 + y^2}{n}\right) dxdy = \int_0^{2\pi} \int_0^n r^2 - \frac{r^3}{n} drd\theta = 2\pi\left(\frac{n^3}{3} - \frac{n^3}{4}\right) = \frac{\pi n^3}{6}$

Example 12.

求 $\int_0^3 \int_0^{\sqrt{9-x^2}} \int_0^2 \sqrt{x^2 + y^2}\, dz dy dx =?$

【解】

令 $x = r\cos\theta, y = r\sin\theta$

则 $\{(x,y): 0 \leq x \leq 3, 0 \leq y \leq \sqrt{9 - x^2}\} = \{(r,\theta): 0 \leq r \leq 3, 0 \leq \theta \leq \frac{\pi}{2}\}$

$$\text{且 } dxdy = \left\| \begin{vmatrix} \dfrac{\partial x}{\partial r} & \dfrac{\partial x}{\partial \theta} \\ \dfrac{\partial y}{\partial r} & \dfrac{\partial y}{\partial \theta} \end{vmatrix} \right\| drd\theta = \left\| \begin{vmatrix} \cos\theta & -r\sin\theta \\ \sin\theta & r\cos\theta \end{vmatrix} \right\| drd\theta = rdrd\theta$$

$$\therefore \int_0^3 \int_0^{\sqrt{9-x^2}} \int_0^2 \sqrt{x^2+y^2}\, dzdydx = \int_0^3 \int_0^{\frac{\pi}{2}} 2r^2 d\theta dr = \left.\frac{2r^3}{3}\right|_0^3 \frac{\pi}{2} = 9\pi$$

Example 13.

    (1)求由 $z = x^2 + 3y^2 + 1$ 与 $z - 9 + x^2 + y^2 = 0$ 所围区域的体积

    (2)假设 V 为 $z = x^2 + 3y^2 + 1, z - 9 + x^2 + y^2 = 0,\ x > 0,\ y > 0, z > 0,$

        所围区域 求 $\iiint_V xdxdydz =?$

【解】

(1)

令$R = \{(x,y): x^2 + 2y^2 \leq 4\,\}$ 则 $V = \iint_R \int_{x^2+3y^2+1}^{9-x^2-y^2} dzdA = \iint_{x^2+2y^2\leq4} 8 - 2x^2 - 4y^2 dxdy$

令$x = 2r\cos\theta,\ y = \sqrt{2}r\sin\theta$ 则 $\{(x,y): x^2 + 2y^2 \leq 4\} = \{(r,\theta): 0 \leq r \leq 1, 0 \leq \theta \leq 2\pi\}$

$$\text{且 } dxdy = \left\| \begin{vmatrix} \dfrac{\partial x}{\partial r} & \dfrac{\partial x}{\partial \theta} \\ \dfrac{\partial y}{\partial r} & \dfrac{\partial y}{\partial \theta} \end{vmatrix} \right\| drd\theta = \left\| \begin{vmatrix} 2\cos\theta & -2r\sin\theta \\ \sqrt{2}\sin\theta & \sqrt{2}r\cos\theta \end{vmatrix} \right\| drd\theta = 2\sqrt{2}rdrd\theta$$

$$\therefore \iint_{x^2+2y^2\leq4} 8 - 2x^2 - 4y^2 dxdy = 2\sqrt{2} \int_0^{2\pi} \int_0^1 8r - 8r^3 drd\theta = 16\sqrt{2} \int_0^{2\pi} d\theta \int_0^1 r - r^3 dr$$
$$= 8\sqrt{2}\pi$$

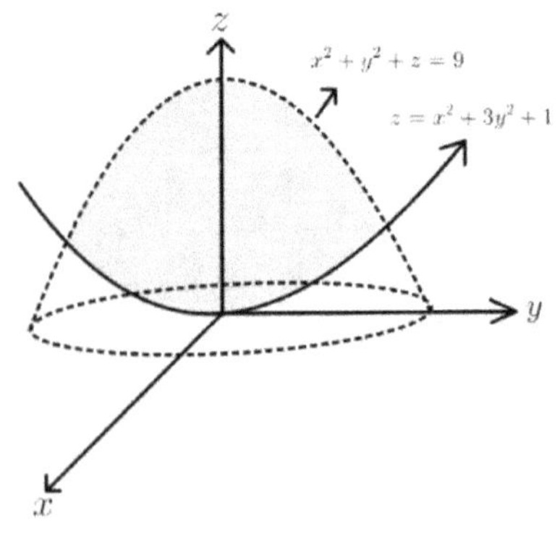


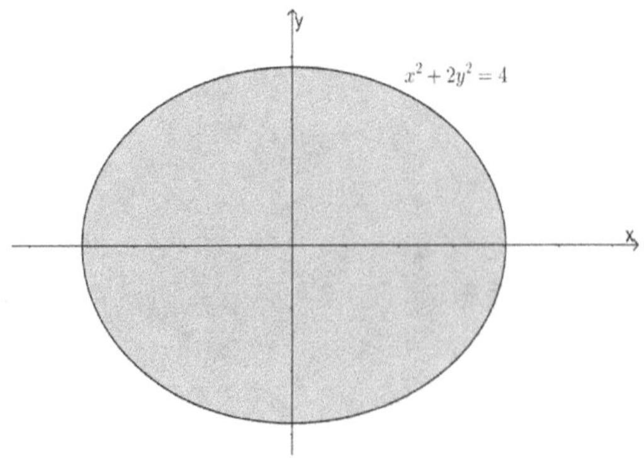


(2)

令 $R = \{(x, y): x^2 + 2y^2 \leq 4, x > 0, y > 0\}$

则 $\iiint_V xdxdydz = \iint_R \int_{x^2+3y^2+1}^{9-x^2-y^2} xdzdA = \iint_{x^2+2y^2\leq 4, x>0, y>0} x(8 - 2x^2 - 4y^2)dxdy$

令 $x = 2r\cos\theta$, $y = \sqrt{2}r\sin\theta$

则 $\{(x, y): x^2 + 2y^2 \leq 4\} = \{(r, \theta): 0 \leq r \leq 1, 0 \leq \theta \leq \frac{\pi}{2}\}$

且 $dxdy = \left\| \begin{vmatrix} \dfrac{\partial x}{\partial r} & \dfrac{\partial x}{\partial \theta} \\ \dfrac{\partial y}{\partial r} & \dfrac{\partial y}{\partial \theta} \end{vmatrix} \right\| drd\theta = \left\| \begin{vmatrix} 2\cos\theta & -2r\sin\theta \\ \sqrt{2}\sin\theta & \sqrt{2}r\cos\theta \end{vmatrix} \right\| drd\theta = 2\sqrt{2}rdrd\theta$

$\therefore \iint_{x^2+2y^2\leq 4, x>0, y>0} x(8 - 2x^2 - 4y^2)dxdy = 2\sqrt{2} \int_0^{\frac{\pi}{2}} 2r\cos\theta \int_0^1 8r^2 - 8r^4 drd\theta$

$= 32\sqrt{2} \int_0^{\frac{\pi}{2}} \cos\theta \, d\theta \int_0^1 r^2 - r^4 dr = \dfrac{64\sqrt{2}}{15}$

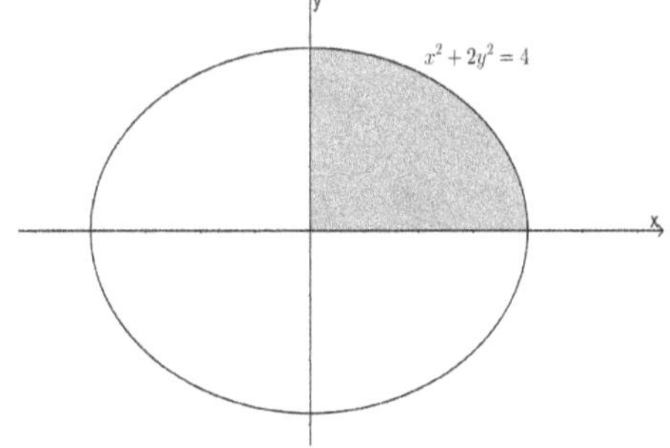


Example 14.

假设 V 为 $z = x^2 + y^2$, $x = 0$, $y = 0$ 与 $z = 4$ 所围的区域, 求 $\iiint_V 3xdxdydz = ?$

【解】

令 $R = \{(x, y): x \geq 0, y \geq 0, x^2 + y^2 \leq 4\}$

则 $V = \iint_R \int_{x^2+y^2}^4 3xdzdA = \iint_{x^2+y^2\leq 4} 3x(4 - x^2 - y^2)dxdy$

令 $x = r\cos\theta, y = r\sin\theta$ 则 $\{(x, y): x^2 + y^2 \leq 4\} = \{(r, \theta): 0 \leq r \leq 2, 0 \leq \theta \leq \frac{\pi}{2}\}$

$$\text{且 } dxdy = \left\| \begin{vmatrix} \dfrac{\partial x}{\partial r} & \dfrac{\partial x}{\partial \theta} \\ \dfrac{\partial y}{\partial r} & \dfrac{\partial y}{\partial \theta} \end{vmatrix} \right\| drd\theta = \left\| \begin{vmatrix} \cos\theta & -r\sin\theta \\ \sin\theta & r\cos\theta \end{vmatrix} \right\| drd\theta = rdrd\theta$$

$$\therefore \iint_{x^2+y^2 \leq 4} 3x(4 - x^2 - y^2)dxdy = \int_0^{\frac{\pi}{2}} \int_0^2 3r\cos\theta\,(4 - r^2)rdrd\theta$$

$$= \int_0^{\frac{\pi}{2}} \cos\theta d\theta \int_0^2 (12r^2 - 3r^4)dr = \frac{64}{5}$$

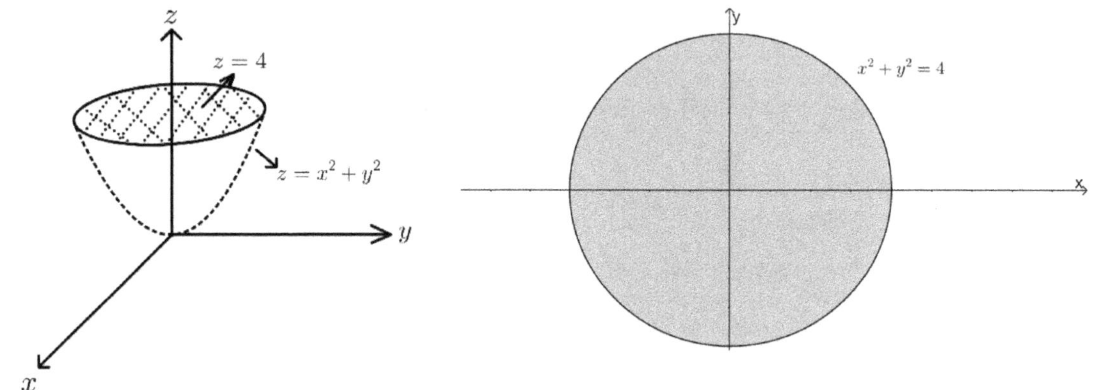


Example 15.

$$\text{求 } \int_0^4 \int_0^{\sqrt{16-x^2}} \int_0^1 \ln\sqrt{x^2 + y^2}\, dzdydx = ?$$

【解】

令 $x = r\cos\theta$, $y = r\sin\theta$

则 $\{(x, y): 0 \leq x \leq 4, 0 \leq y \leq \sqrt{16 - x^2}\} = \{(r, \theta): 0 \leq r \leq 4, 0 \leq \theta \leq \dfrac{\pi}{2}\}$

$$\text{且 } dxdy = \left\| \begin{vmatrix} \dfrac{\partial x}{\partial r} & \dfrac{\partial x}{\partial \theta} \\ \dfrac{\partial y}{\partial r} & \dfrac{\partial y}{\partial \theta} \end{vmatrix} \right\| drd\theta = \left\| \begin{vmatrix} \cos\theta & -r\sin\theta \\ \sin\theta & r\cos\theta \end{vmatrix} \right\| drd\theta = rdrd\theta$$

$$\therefore \int_0^4 \int_0^{\sqrt{16-x^2}} \int_0^1 \ln\sqrt{x^2 + y^2}\, dzdydx = \int_0^4 \int_0^{\frac{\pi}{2}} r\ln r\, d\theta dr = \int_0^{\frac{\pi}{2}} d\theta \int_0^4 r\ln r\, dr$$

$$= \frac{\pi}{2} \cdot \left( \frac{r^2 \ln r}{2} - \frac{r^2}{4} \right) \Bigg|_0^4 = \pi(8 \ln 2 - 2)$$

Example 16.

(1)求由$xy$平面,$z = x^2 - 2$ 与 $x^2 + y^2 = 1$ 所围区域的体积

(2)求由 $z = 1 - 2x^2 - 2y^2$ 与 $z = (x^2 + y^2)^2 - 2$ 所围区域的体积

【解】

(1)

令$R = \{(x,y): x^2 + y^2 \le 1 \}$则 $V = \iint_R \int_{x^2-2}^0 dz dA = \iint_{x^2+y^2 \le 1} 2 - x^2 dx dy$

令$x = r \cos\theta, y = r \sin\theta$ 则 $\{(x,y): x^2 + y^2 \le 1\} = \{(r,\theta): 0 \le r \le 1, 0 \le \theta \le 2\pi\}$

且 $dxdy = \begin{Vmatrix} \dfrac{\partial x}{\partial r} & \dfrac{\partial x}{\partial \theta} \\ \dfrac{\partial y}{\partial r} & \dfrac{\partial y}{\partial \theta} \end{Vmatrix} drd\theta = \begin{Vmatrix} \cos\theta & -r\sin\theta \\ \sin\theta & r\cos\theta \end{Vmatrix} drd\theta = rdrd\theta$

$\therefore \iint_{x^2+y^2\le 1} 2 - x^2 dx dy = \int_0^{2\pi} \int_0^1 (2 - r^2\cos^2\theta) r dr d\theta = \int_0^{2\pi} 1 - \frac{\cos^2\theta}{4} d\theta$

$= \int_0^{2\pi} 1 - \frac{1}{4}\left( \frac{1 + \cos 2\theta}{2} \right) d\theta = \frac{7\pi}{4}$

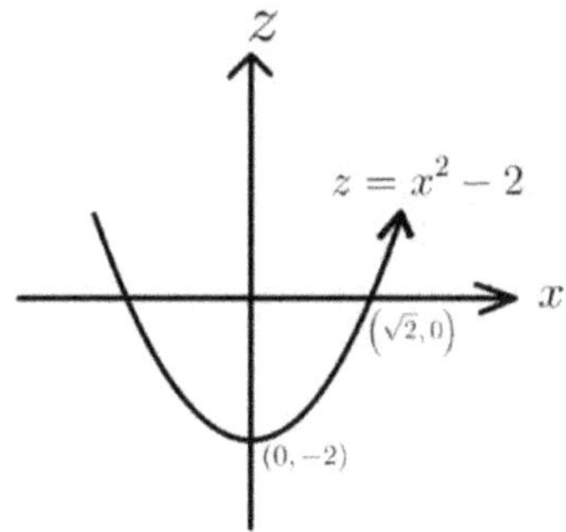


(2)

令$R = \{(x,y): x^2 + y^2 \le 1 \}$

则 $V = \iint_R \int_{(x^2+y^2)^2-2}^{1-2x^2-2y^2} dz dA = \iint_{x^2+y^2\le 1} 1 - 2x^2 - 2y^2 - ((x^2 + y^2)^2 - 2) dx dy$

令$x = r \cos\theta, y = r \sin\theta$ 则 $\{(x,y): x^2 + y^2 \le 1\} = \{(r,\theta): 0 \le r \le 1, 0 \le \theta \le 2\pi\}$

且 $dxdy = \left\| \begin{vmatrix} \dfrac{\partial x}{\partial r} & \dfrac{\partial x}{\partial \theta} \\ \dfrac{\partial y}{\partial r} & \dfrac{\partial y}{\partial \theta} \end{vmatrix} \right\| drd\theta = \left\| \begin{vmatrix} \cos\theta & -r\sin\theta \\ \sin\theta & r\cos\theta \end{vmatrix} \right\| drd\theta = rdrd\theta$

$\therefore \iint_{x^2+y^2\leq 1} 1 - 2x^2 - 2y^2 - ((x^2+y^2)^2 - 2)dxdy = \int_0^{2\pi}\int_0^1 (-r^4 - 2r^2 + 3)rdrd\theta$

$= \int_0^{2\pi} d\theta \int_0^1 (-r^5 - 2r^3 + 3r)dr = 2\pi\left(-\dfrac{1}{6} - \dfrac{2}{4} + \dfrac{3}{2}\right) = \dfrac{5\pi}{3}$

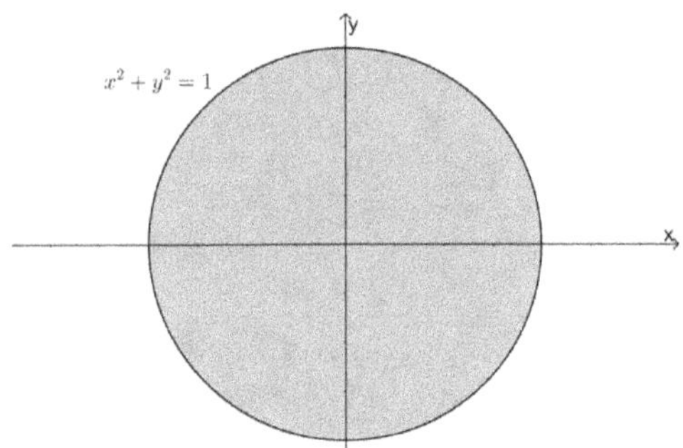


Example 17.

$$求 \int_0^1 \int_0^{\sqrt{1-x^2}} \int_0^1 e^{\sqrt{x^2+y^2}} dzdydx = ?$$

【解】

令 $x = r\cos\theta , y = r\sin\theta$

则 $\{(x,y): 0 \leq x \leq 1, 0 \leq y \leq \sqrt{1-x^2}\} = \{(r,\theta): 0 \leq r \leq 1, 0 \leq \theta \leq \dfrac{\pi}{2}\}$

且 $dxdy = \left\| \begin{vmatrix} \dfrac{\partial x}{\partial r} & \dfrac{\partial x}{\partial \theta} \\ \dfrac{\partial y}{\partial r} & \dfrac{\partial y}{\partial \theta} \end{vmatrix} \right\| drd\theta = \left\| \begin{vmatrix} \cos\theta & -r\sin\theta \\ \sin\theta & r\cos\theta \end{vmatrix} \right\| drd\theta = rdrd\theta$

$\therefore \int_0^1 \int_0^{\sqrt{1-x^2}} \int_0^1 e^{\sqrt{x^2+y^2}} dzdydx = \int_0^1 \int_0^{\frac{\pi}{2}} re^r d\theta dr = \int_0^{\frac{\pi}{2}} d\theta \int_0^1 e^r rdr = \dfrac{\pi}{2}\cdot(re^r - e^r)|_0^1 = \dfrac{\pi}{2}$

Example 18.

(1)计算球体 $x^2 + y^2 + z^2 = 25$ 被圆柱 $x^2 + y^2 = 16$ 所切除体积

(2)计算球体$x^2 + y^2 + z^2 = 25$ 被圆柱$x^2 + y^2 = 16$ 切除后剩余体积

**【解】**

(1)

令$R = \{(x,y): x^2 + y^2 \leq 16\}$

则 $V = \iint_R \int_{-\sqrt{25-x^2-y^2}}^{\sqrt{25-x^2-y^2}} dz\,dA = \iint_{x^2+y^2\leq 16} 2\sqrt{25 - x^2 - y^2}\,dx\,dy$

令$x = r\cos\theta, y = r\sin\theta$ 则 $\{(x,y): x^2 + y^2 \leq 16\} = \{(r,\theta): 0 \leq r \leq 4, 0 \leq \theta \leq 2\pi\}$

且 $dx\,dy = \left\| \begin{vmatrix} \dfrac{\partial x}{\partial r} & \dfrac{\partial x}{\partial \theta} \\ \dfrac{\partial y}{\partial r} & \dfrac{\partial y}{\partial \theta} \end{vmatrix} \right\| dr\,d\theta = \left\| \begin{matrix} \cos\theta & -r\sin\theta \\ \sin\theta & r\cos\theta \end{matrix} \right\| dr\,d\theta = r\,dr\,d\theta$

$\therefore \iint_{x^2+y^2\leq 16} 2\sqrt{25 - x^2 - y^2}\,dx\,dy = 2\int_0^{2\pi} d\theta \int_0^4 \sqrt{25 - r^2}\,r\,dr$

$= 2(2\pi) \cdot \dfrac{(-1)(25-r^2)^{\frac{3}{2}}}{3}\bigg|_0^4 = \dfrac{392\pi}{3}$

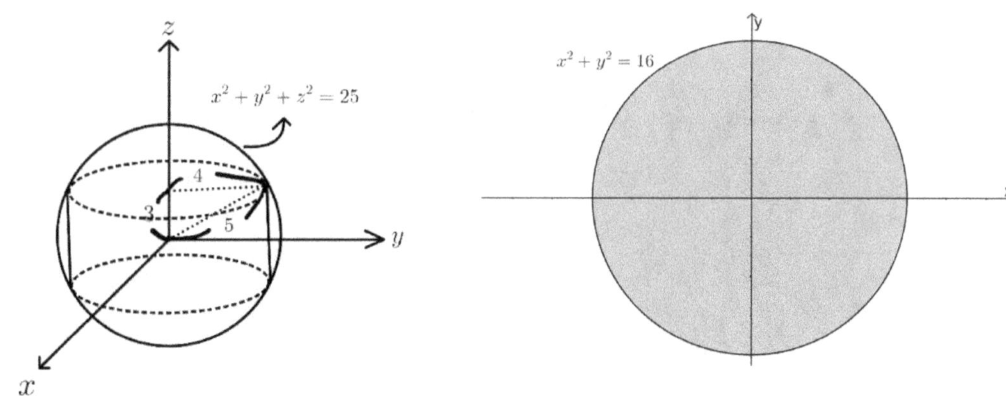


(2)

令$R = \{(x,y): 16 \leq x^2 + y^2 \leq 25\}$

则 $V = \iint_R \int_{-\sqrt{25-x^2-y^2}}^{\sqrt{25-x^2-y^2}} dz\,dA = \iint_{16\leq x^2+y^2\leq 25} 2\sqrt{25 - x^2 - y^2}\,dx\,dy$

令$x = r\cos\theta, y = r\sin\theta$ 则 $\{(x,y): x^2 + y^2 \leq 1\} = \{(r,\theta): 4 \leq r \leq 5, 0 \leq \theta \leq 2\pi\}$

$$\text{且 } dxdy = \left\|\begin{matrix} \dfrac{\partial x}{\partial r} & \dfrac{\partial x}{\partial \theta} \\ \dfrac{\partial y}{\partial r} & \dfrac{\partial y}{\partial \theta} \end{matrix}\right\| drd\theta = \left\|\begin{matrix} \cos\theta & -r\sin\theta \\ \sin\theta & r\cos\theta \end{matrix}\right\| drd\theta = rdrd\theta$$

$$\therefore \iint_{16\leq x^2+y^2\leq 25} 2\sqrt{25-x^2-y^2}\,dxdy = 2\int_0^{2\pi} d\theta \int_4^5 \sqrt{25-r^2}\,rdr$$

$$= 2(2\pi)\cdot \frac{(-1)(25-r^2)^{\frac{3}{2}}}{3}\Bigg|_4^5 = 36\pi$$

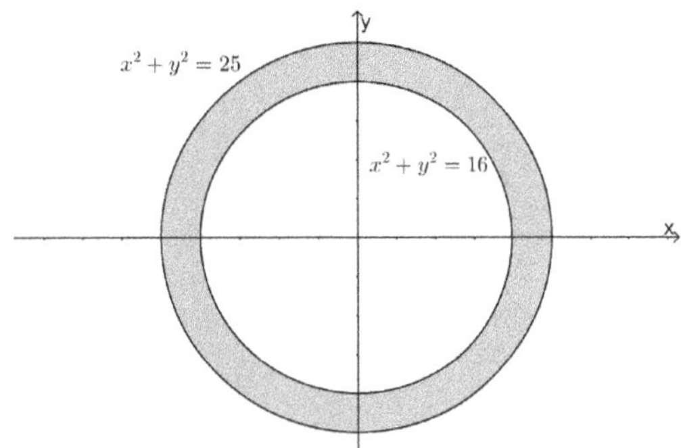


Example 19.

　　计算球体 $x^2+y^2+z^2=4$ 被圆柱 $x^2+y^2=2x$ 所切除体积

【解】

令 $R = \{(x,y): (x-1)^2+y^2 \leq 1\}$

则 $V = \iint_R \int_{-\sqrt{4-x^2-y^2}}^{\sqrt{4-x^2-y^2}} dzdA = \iint_{(x-1)^2+y^2\leq 1} 2\sqrt{4-x^2-y^2}\,dxdy$

令 $x = r\cos\theta,\ y = r\sin\theta$

则 $\{(x,y): (x-1)^2+y^2 \leq 1\} = \{(r,\theta): 0 \leq r \leq 2\cos\theta, 0 \leq \theta \leq \pi\}$

$$\text{且 } dxdy = \left\|\begin{matrix} \dfrac{\partial x}{\partial r} & \dfrac{\partial x}{\partial \theta} \\ \dfrac{\partial y}{\partial r} & \dfrac{\partial y}{\partial \theta} \end{matrix}\right\| drd\theta = \left\|\begin{matrix} a\cos\theta & ar\sin\theta \\ a\sin\theta & ar\cos\theta \end{matrix}\right\| drd\theta = rdrd\theta$$

$$\therefore \iint_{(x-1)^2+y^2\leq 1} 2\sqrt{4-x^2-y^2}\,dxdy = 2\int_0^{\pi} \int_0^{2\cos\theta} (\sqrt{4-r^2})rdrd\theta$$

$$= 4 \int_0^{\frac{\pi}{2}} \int_0^{2\cos\theta} (\sqrt{4 - r^2}) r \, dr \, d\theta = 4 \int_0^{\frac{\pi}{2}} \left( \frac{(-1)(4 - r^2)^{\frac{3}{2}}}{3} \right) \Bigg|_0^{2\cos\theta} d\theta = 32 \int_0^{\frac{\pi}{2}} \frac{1 - \sin^3\theta}{3} d\theta$$

$$= 32 \left( \frac{\pi}{6} - \frac{2}{9} \right)$$

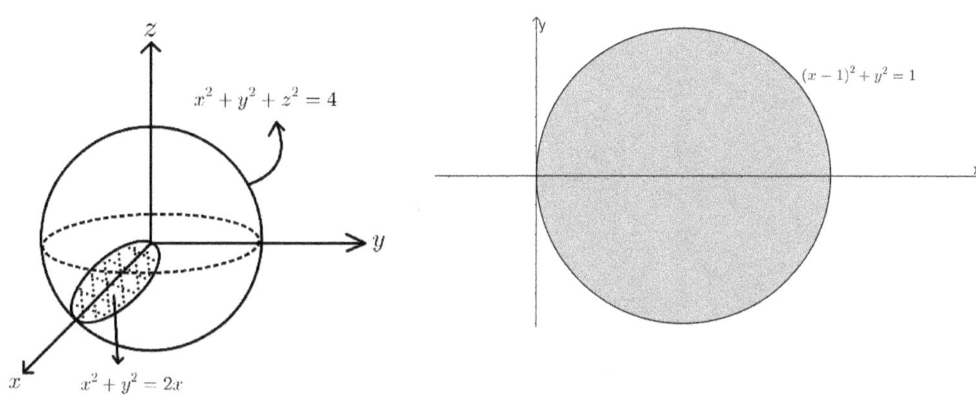


Example 20.

  (1)计算球体$x^2 + y^2 + z^2 = 16$ 被圆柱$r = 4\cos\theta$ 所切除的体积

  (2)计算球体$x^2 + y^2 + z^2 = 4$ 被圆柱$r = 2\sin\theta$ 所切除的体积

【解】

(1)

令$R = \{(x, y) : (x - 2)^2 + y^2 \leq 4, x \geq 0\}$

则 $V = \iint_R \int_{-\sqrt{16 - x^2 - y^2}}^{\sqrt{16 - x^2 - y^2}} dz \, dA = \iint_{(x-2)^2 + y^2 \leq 4} 2\sqrt{16 - x^2 - y^2} \, dx \, dy$

令$x = r\cos\theta, y = r\sin\theta$

则 $\{(x, y) : (x - 2)^2 + y^2 \leq 4, x \geq 0\} = \{(r, \theta) : 0 \leq r \leq 4\cos\theta, \frac{-\pi}{2} \leq \theta \leq \frac{\pi}{2}\}$

且 $dx \, dy = \left\| \begin{vmatrix} \dfrac{\partial x}{\partial r} & \dfrac{\partial x}{\partial \theta} \\ \dfrac{\partial y}{\partial r} & \dfrac{\partial y}{\partial \theta} \end{vmatrix} \right\| dr \, d\theta = \left\| \begin{vmatrix} \cos\theta & -r\sin\theta \\ \sin\theta & r\cos\theta \end{vmatrix} \right\| dr \, d\theta = r \, dr \, d\theta$

$\therefore \iint_{(x-2)^2 + y^2 \leq 4} 2\sqrt{16 - x^2 - y^2} \, dx \, dy = 2 \int_{\frac{-\pi}{2}}^{\frac{\pi}{2}} \int_0^{4\cos\theta} \sqrt{16 - r^2} \, r \, dr \, d\theta$

$$= 2 \int_{\frac{-\pi}{2}}^{\frac{\pi}{2}} \left.\frac{(-1)(16-r^2)^{\frac{3}{2}}}{3}\right|_{r=0}^{r=4\cos\theta} d\theta = \frac{256}{3} \int_0^{\frac{\pi}{2}} 1-(1-\cos^2\theta)^{\frac{3}{2}}d\theta = \frac{256}{3} \int_0^{\frac{\pi}{2}} 1-\sin^3\theta \, d\theta$$

$$= \left(\frac{\pi}{2} - \frac{2}{3}\right)\frac{256}{3}$$

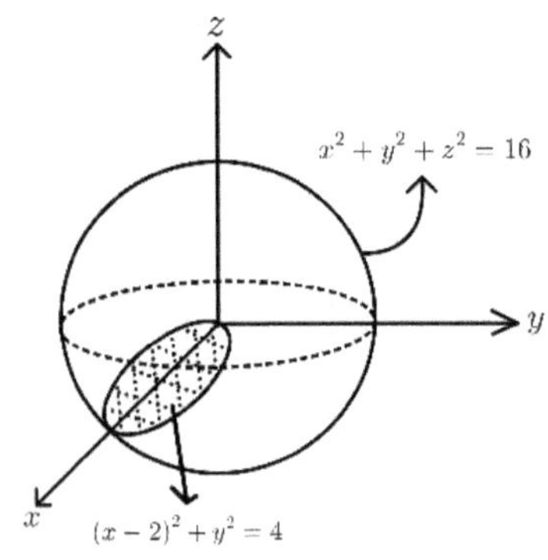


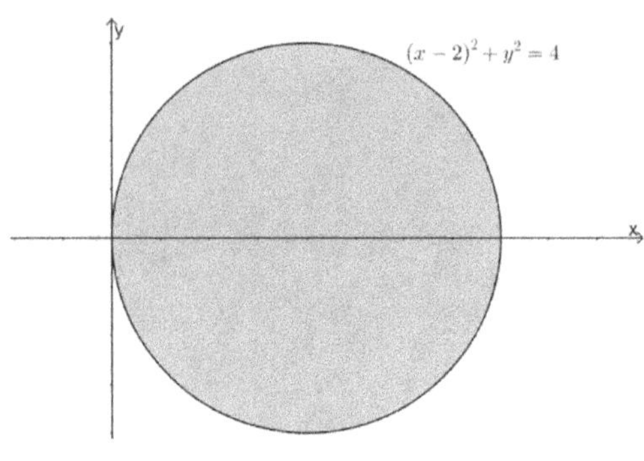


(2)

令 $R = \{(x,y): x^2 + (y-1)^2 \leq 1, y \geq 0\}$

则 $V = \iint_R \int_{-\sqrt{4-x^2-y^2}}^{\sqrt{4-x^2-y^2}} dzdA = \iint_{x^2+(y-1)^2\leq 1} 2\sqrt{4-x^2-y^2}dxdy$

令 $x = r\cos\theta, y = r\sin\theta$

则 $\{(x,y): x^2 + (y-1)^2 \leq 1, y \geq 0\} = \{(r,\theta): 0 \leq r \leq 2\sin\theta, 0 \leq \theta \leq \pi\}$

且 $dxdy = \left\| \begin{vmatrix} \frac{\partial x}{\partial r} & \frac{\partial x}{\partial \theta} \\ \frac{\partial y}{\partial r} & \frac{\partial y}{\partial \theta} \end{vmatrix} \right\| drd\theta = \left\| \begin{vmatrix} \cos\theta & -r\sin\theta \\ \sin\theta & r\cos\theta \end{vmatrix} \right\| drd\theta = rdrd\theta$

$$\therefore \iint_{x^2+(y-1)^2\leq 1} 2\sqrt{4-x^2-y^2}dxdy = 2\int_0^{\pi}\int_0^{2\sin\theta} \sqrt{4-r^2}rdrd\theta$$

$$= 2\int_0^{\pi} \left.\frac{(-1)(4-r^2)^{\frac{3}{2}}}{3}\right|_{r=0}^{r=2\sin\theta} d\theta = \frac{32}{3}\int_0^{\frac{\pi}{2}} 1-(1-\sin^2\theta)^{\frac{3}{2}}d\theta = \frac{32}{3}\int_0^{\frac{\pi}{2}} 1-\cos^3\theta \, d\theta$$

$$= \left(\frac{\pi}{2} - \frac{2}{3}\right)\frac{32}{3}$$

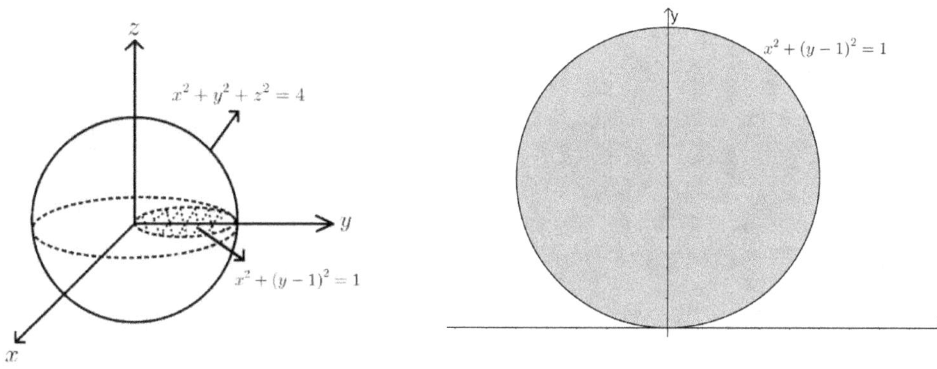

Example 21.

求由曲面 $z = x^2 + y^2$ 上方与 $z = y$ 下方所围区域的体积

【解】

令 $R = \left\{(x, y): x^2 + \left(y - \dfrac{1}{2}\right)^2 \leq \dfrac{1}{4}\right\}$

则 $V = \iint_R \int_{x^2+y^2}^{y} dz\,dA = \iint_R \dfrac{1}{4} - x^2 - \left(y - \dfrac{1}{2}\right)^2 dx\,dy$

令 $x = \dfrac{1}{2} r \cos\theta, y = \dfrac{1}{2} + \dfrac{1}{2} r \sin\theta$

则 $\left\{(x, y): x^2 + \left(y - \dfrac{1}{2}\right)^2 \leq \dfrac{1}{4}\right\} = \{(r, \theta): 0 \leq r \leq 1, 0 \leq \theta \leq 2\pi\}$

且 $dx\,dy = \left\|\begin{matrix} \dfrac{\partial x}{\partial r} & \dfrac{\partial x}{\partial \theta} \\ \dfrac{\partial y}{\partial r} & \dfrac{\partial y}{\partial \theta} \end{matrix}\right\| dr\,d\theta = \left\|\begin{matrix} \dfrac{1}{2}\cos\theta & \dfrac{1}{2} r \sin\theta \\ \dfrac{1}{2}\sin\theta & \dfrac{1}{2} r \cos\theta \end{matrix}\right\| dr\,d\theta = \dfrac{1}{4} r\,dr\,d\theta$

$\therefore \iint_R \dfrac{1}{4} - x^2 - \left(y - \dfrac{1}{2}\right)^2 dx\,dy = \int_0^{2\pi} \int_0^1 \left(\dfrac{1}{4} - \dfrac{1}{4} r^2\right) \dfrac{1}{4} r\,dr\,d\theta = \dfrac{1}{16} \int_0^{2\pi} d\theta \int_0^1 r - r^3 dr$

$= \dfrac{\pi}{32}$

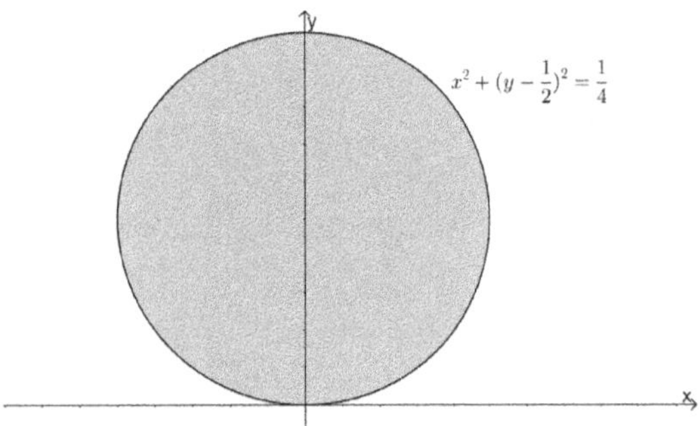



Example 22.

假设 V 为曲面 $z \geq x^2 + y^2,\ z \leq y,\ x > 0,\ y > 0,\ z > 0$ 所围区域,

求 $\iiint_V xdxdydz =?$

【解】

令 $R = \{(x, y): x^2 + (y - \frac{1}{2})^2 \leq \frac{1}{4}, x > 0, y > 0\}$

则 $\iiint_V xdxdydz = \iint_R \int_{x^2+y^2}^{y} xdzdA = \iint_R x\left(\frac{1}{4} - x^2 - \left(y - \frac{1}{2}\right)^2\right)dxdy$

令 $x = \frac{1}{2}r\cos\theta,\ y = \frac{1}{2} + \frac{1}{2}r\sin\theta$

则 $\{(x, y): x^2 + \left(y - \frac{1}{2}\right)^2 \leq \frac{1}{4}, x > 0, y > 0\} = \{(r, \theta): 0 \leq r \leq 1, 0 \leq \theta \leq \frac{\pi}{2}\}$

且 $dxdy = \left\|\begin{matrix} \dfrac{\partial x}{\partial r} & \dfrac{\partial x}{\partial \theta} \\ \dfrac{\partial y}{\partial r} & \dfrac{\partial y}{\partial \theta} \end{matrix}\right\| drd\theta = \left\|\begin{matrix} \dfrac{1}{2}\cos\theta & \dfrac{1}{2}r\sin\theta \\ \dfrac{1}{2}\sin\theta & \dfrac{1}{2}r\cos\theta \end{matrix}\right\| drd\theta = \dfrac{1}{4}rdrd\theta$

$\therefore \iint_R x\left(\frac{1}{4} - x^2 - \left(y - \frac{1}{2}\right)^2\right)dxdy$

$= \int_0^{\frac{\pi}{2}} \frac{1}{2}r\cos\theta \int_0^1 \left(\frac{1}{4} - \frac{1}{4}r^2\right)\frac{1}{4}rdrd\theta = \frac{1}{32}\int_0^{\frac{\pi}{2}}\cos\theta\, d\theta \int_0^1 r^2 - r^4 dr = \frac{1}{240}$

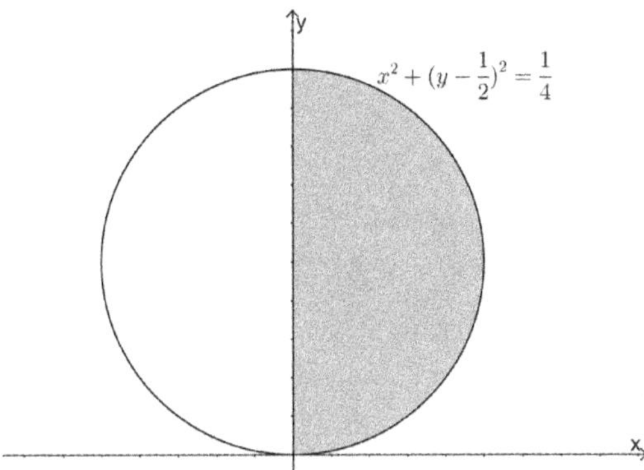


Example 23.

$(1)$计算$x^2 + y^2 = 1$ 与 $4x^2 + 4y^2 + z^2 = 64$ 交集区域的体积

$(2)$求由曲面 $z = \sqrt{4 - 3x^2 - 3y^2}$ 下方与 $z = \sqrt{x^2 + y^2}$ 上方所围区域的体积

【解】

(1)

令$R = \{(x, y): x^2 + y^2 \le 1\}$

则 $V = \iint_R \int_{-\sqrt{64-4x^2-4y^2}}^{\sqrt{64-4x^2-4y^2}} dzdA = \iint_{x^2+y^2 \le 1} 2\sqrt{64 - 4x^2 - 4y^2}\,dxdy$

令$x = r\cos\theta, y = r\sin\theta$ 则 $\{(x, y): x^2 + y^2 \le 1\} = \{(r, \theta): 0 \le r \le 1, 0 \le \theta \le 2\pi\}$

且 $dxdy = \left\| \begin{vmatrix} \frac{\partial x}{\partial r} & \frac{\partial x}{\partial \theta} \\ \frac{\partial y}{\partial r} & \frac{\partial y}{\partial \theta} \end{vmatrix} \right\| drd\theta = \left\| \begin{vmatrix} \cos\theta & -r\sin\theta \\ \sin\theta & r\cos\theta \end{vmatrix} \right\| drd\theta = rdrd\theta$

$\therefore \iint_{x^2+y^2 \le 1} 2\sqrt{64 - 4x^2 - 4y^2}\,dxdy = 2\int_0^{2\pi} \int_0^1 \sqrt{64 - 4r^2}\,rdrd\theta$

$= 4 \cdot (2\pi) \cdot \left. \frac{(-1)(16 - r^2)^{\frac{3}{2}}}{3} \right|_{r=0}^{r=1} = \frac{8\pi}{3}(64 - 15\sqrt{15})$

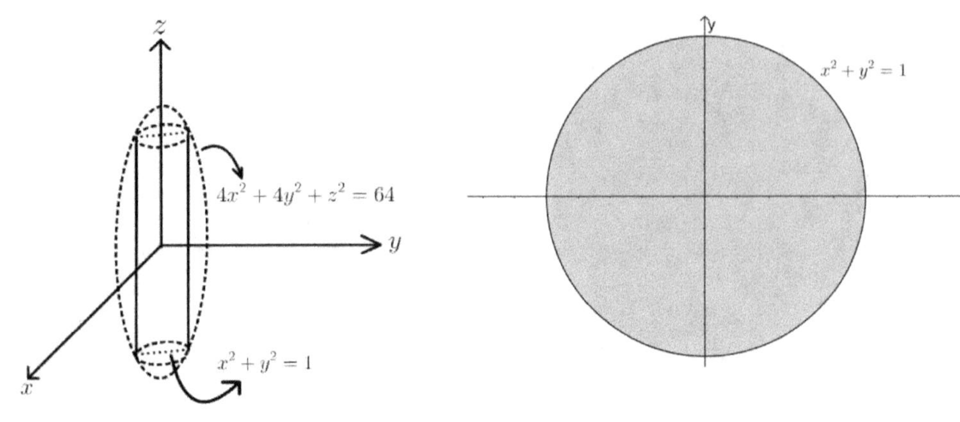

(2)

令$R = \{(x,y): x^2 + y^2 \leq 1\}$

则 $V = \displaystyle\iint_R \int_{\sqrt{x^2+y^2}}^{\sqrt{4-3x^2-3y^2}} dz\,dA = \iint_{x^2+y^2\leq 1} \sqrt{4 - 3x^2 - 3y^2} - \sqrt{x^2 + y^2}\,dxdy$

令$x = r\cos\theta, y = r\sin\theta$ 则 $\{(x,y): x^2 + y^2 \leq 1\} = \{(r,\theta): 0 \leq r \leq 1, 0 \leq \theta \leq 2\pi\}$

且 $dxdy = \left\|\begin{array}{cc} \dfrac{\partial x}{\partial r} & \dfrac{\partial x}{\partial \theta} \\[2mm] \dfrac{\partial y}{\partial r} & \dfrac{\partial y}{\partial \theta} \end{array}\right\| drd\theta = \left\|\begin{array}{cc} \cos\theta & -r\sin\theta \\ \sin\theta & r\cos\theta \end{array}\right\| drd\theta = r\,drd\theta$

$\therefore \displaystyle\iint_{x^2+y^2\leq 1} \sqrt{4 - 3x^2 - 3y^2} - \sqrt{x^2 + y^2}\,dxdy = \int_0^{2\pi}\int_0^1 (\sqrt{4 - 3r^2} - r)r\,drd\theta$

$= (2\pi)\cdot\left(\dfrac{(-1)(4 - 3r^2)^{\frac{3}{2}}}{9} - \dfrac{r^3}{3}\right)\Bigg|_{r=0}^{r=1} = \dfrac{8\pi}{9}$

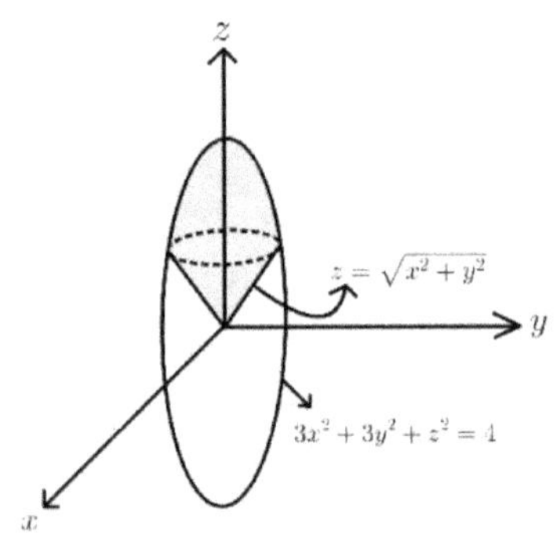

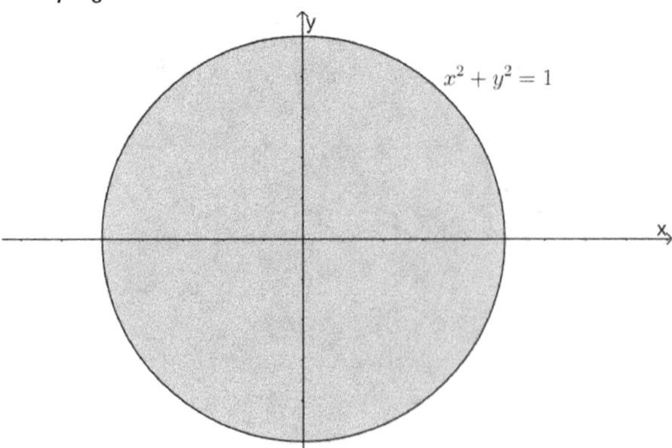

Example 24.

 (1)计算球体$x^2 + y^2 + z^2 \leq 100$ 被平面 $z = 6$ 切除后剩的较小块体积

 (2)计算球体$x^2 + y^2 + z^2 \leq 100$ 被平面 $z = 8$ 切除后剩的较小块体积

【解】

(1)

令$R = \{(x, y): x^2 + y^2 \leq 64\}$

则 $V = \iint_R \int_6^{\sqrt{100-x^2-y^2}} dz\,dA = \iint_{x^2+y^2\leq 64} \sqrt{100 - x^2 - y^2} - 6\,dxdy$

令$x = r\cos\theta, y = r\sin\theta$ 则 $\{(x,y): x^2 + y^2 \leq 64\} = \{(r,\theta): 0 \leq r \leq 8, 0 \leq \theta \leq 2\pi$

且 $dxdy = \left\|\begin{matrix} \dfrac{\partial x}{\partial r} & \dfrac{\partial x}{\partial \theta} \\ \dfrac{\partial y}{\partial r} & \dfrac{\partial y}{\partial \theta} \end{matrix}\right\| drd\theta = \left\|\begin{matrix} \cos\theta & -r\sin\theta \\ \sin\theta & r\cos\theta \end{matrix}\right\| drd\theta = r\,drd\theta$

$\therefore \iint_{x^2+y^2\leq 64} \sqrt{100 - x^2 - y^2} - 6\,dxdy = \int_0^{2\pi} \int_0^8 (\sqrt{100 - r^2} - 6) r\,drd\theta$

$= (2\pi) \cdot \left(\dfrac{(-1)(100 - r^2)^{\frac{3}{2}}}{3} - \dfrac{6r^2}{2}\right)\Bigg|_{r=0}^{r=8} = \dfrac{356\pi}{3}$

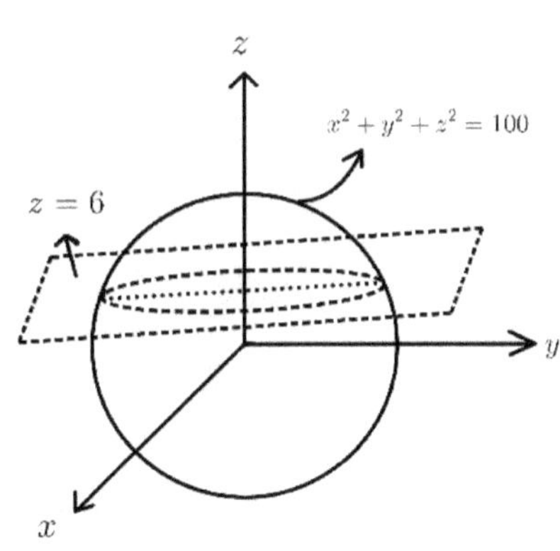


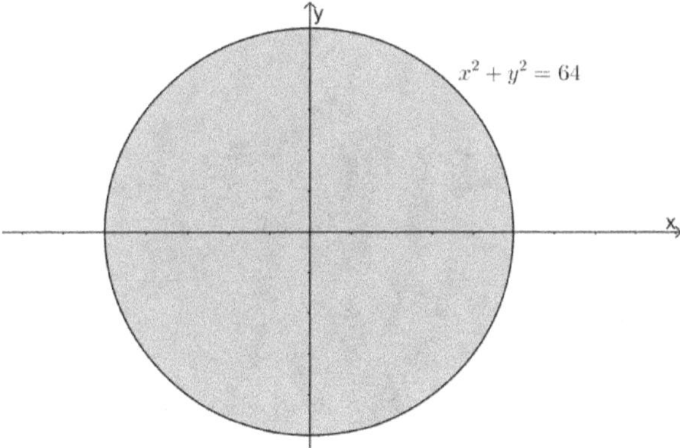


(2)

令$R = \{(x, y): x^2 + y^2 \leq 36\}$

则 $V = \iint_R \int_8^{\sqrt{100-x^2-y^2}} dz\,dA = \iint_{x^2+y^2\leq 36} \sqrt{100 - x^2 - y^2} - 8\,dxdy$

$\diamondsuit x = r\cos\theta, y = r\sin\theta$ 则 $\{(x,y): x^2 + y^2 \le 36\} = \{(r,\theta): 0 \le r \le 6, 0 \le \theta \le 2\pi\}$

且 $dxdy = \left\|\begin{vmatrix} \dfrac{\partial x}{\partial r} & \dfrac{\partial x}{\partial \theta} \\ \dfrac{\partial y}{\partial r} & \dfrac{\partial y}{\partial \theta} \end{vmatrix}\right\| drd\theta = \left\|\begin{vmatrix} \cos\theta & -r\sin\theta \\ \sin\theta & r\cos\theta \end{vmatrix}\right\| drd\theta = rdrd\theta$

$\therefore \iint_{x^2+y^2 \le 36} \sqrt{100 - x^2 - y^2} - 8dxdy = \int_0^{2\pi} \int_0^6 (\sqrt{100 - r^2} - 8) rdrd\theta$

$= (2\pi) \cdot \left( \dfrac{(-1)(100 - r^2)^{\frac{3}{2}}}{3} - \dfrac{8r^2}{2} \right)\Bigg|_{r=0}^{r=6} = \dfrac{112\pi}{3}$

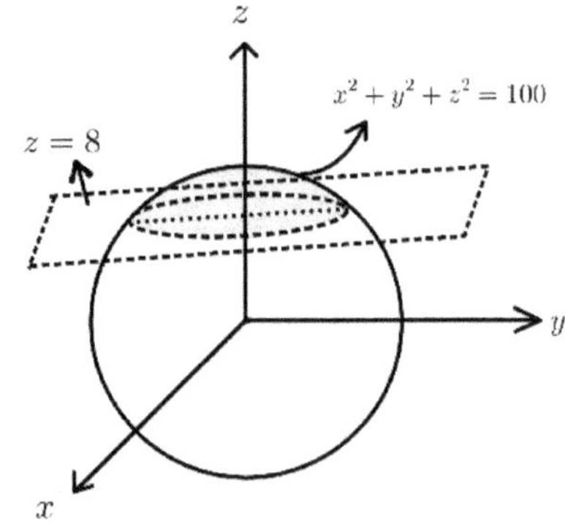


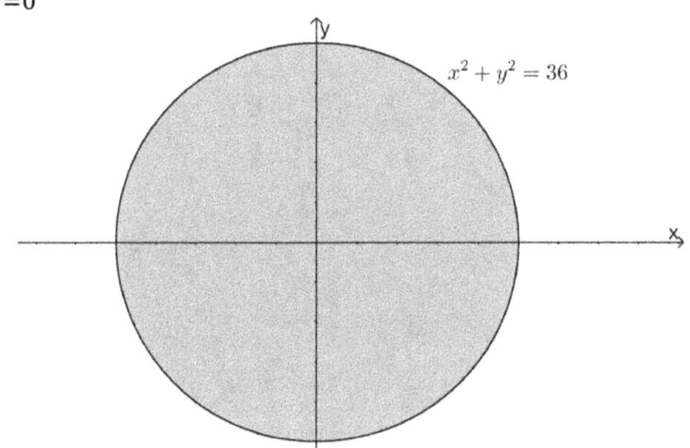


Example 25.

求 $\displaystyle\int_0^3 \int_0^{\sqrt{9-x^2}} \int_0^2 \sqrt{x^2 + y^2} \cos\sqrt{x^2 + y^2}\, dzdydx =?$

【解】

$\diamondsuit x = r\cos\theta, y = r\sin\theta$

则 $\{(x,y): 0 \le x \le 3, 0 \le y \le \sqrt{9 - x^2}\} = \{(r,\theta): 0 \le r \le 3, 0 \le \theta \le \dfrac{\pi}{2}\}$

且 $dxdy = \left\|\begin{vmatrix} \dfrac{\partial x}{\partial r} & \dfrac{\partial x}{\partial \theta} \\ \dfrac{\partial y}{\partial r} & \dfrac{\partial y}{\partial \theta} \end{vmatrix}\right\| drd\theta = \left\|\begin{vmatrix} \cos\theta & -r\sin\theta \\ \sin\theta & r\cos\theta \end{vmatrix}\right\| drd\theta = rdrd\theta$

$\therefore \displaystyle\int_0^3 \int_0^{\sqrt{9-x^2}} \int_0^2 \sqrt{x^2 + y^2} \cos\sqrt{x^2 + y^2}\, dzdydx = 2\int_0^3 \int_0^{\frac{\pi}{2}} r^2 \cos r\, d\theta dr$

$$= 2 \int_0^{\frac{\pi}{2}} d\theta \int_0^3 r^2 \cos r \, dr$$

令 $u = r^2, \; dv = \cos r \, dr$ 则 $du = 2r dr, \; v = \sin r$，藉由分部积分法

则 $\int r^2 \cos r \, dr = r^2 \sin r - 2 \int r \sin r \, dr + c$

令 $s = r, \; dt = \sin r \, dr$ 则 $ds = dr, \; t = -\cos r$，藉由分部积分法

则 $\int r \sin r \, dr = -r \cos r + \int \cos r \, dr = -r \cos r + \sin r$

$\therefore \int r^2 \cos r \, dr = r^2 \sin r - 2 \int r \sin r \, dr + c = r^2 \sin r + 2r \cos r - 2 \sin r + c$

令 $a, b \in R$

则 $\int_a^b r^2 \cos r \, dx = b^2 \sin b + 2b \cos b - 2 \sin b - (a^2 \sin a + 2a \cos a - 2 \sin a)$

$\therefore \int_0^3 \int_0^{\sqrt{9-x^2}} \int_0^2 \sqrt{x^2 + y^2} \cos \sqrt{x^2 + y^2} \, dz dy dx = 2 \int_0^{\frac{\pi}{2}} d\theta \int_0^3 r^2 \cos r \, dr$

$$= \pi(7 \sin 3 + 6 \cos 3)$$

Example 26.

    (1)计算球体$x^2 + y^2 + z^2 \leq 1$ 与 $z \geq \sqrt{3x^2 + 3y^2}$ 交集区域体积

    (2)计算球体$x^2 + y^2 + z^2 \leq 1$ 与 $\sqrt{3}z \geq \sqrt{x^2 + y^2}$ 交集区域体积

【解】

(1)

令 $R = \{(x, y): x^2 + y^2 \leq \frac{1}{4}\}$

则 $V = \iint_R \int_{\sqrt{3x^2+3y^2}}^{\sqrt{1-x^2-y^2}} dz dA = \iint_{x^2+y^2 \leq \frac{1}{4}} \sqrt{1 - x^2 - y^2} - \sqrt{3x^2 + 3y^2} \, dx dy$

令 $x = r \cos \theta, y = r \sin \theta$ 则 $\{(x, y): x^2 + y^2 \leq \frac{1}{4}\} = \{(r, \theta): 0 \leq r \leq \frac{1}{2}, 0 \leq \theta \leq 2\pi\}$

且 $dxdy = \left\| \begin{matrix} \dfrac{\partial x}{\partial r} & \dfrac{\partial x}{\partial \theta} \\ \dfrac{\partial y}{\partial r} & \dfrac{\partial y}{\partial \theta} \end{matrix} \right\| drd\theta = \left\| \begin{matrix} \cos\theta & -r\sin\theta \\ \sin\theta & r\cos\theta \end{matrix} \right\| drd\theta = rdrd\theta$

$\therefore \iint\limits_{x^2+y^2\leq\frac{1}{4}} \sqrt{1-x^2-y^2} - \sqrt{3x^2+3y^2}\, dxdy = \int_0^{2\pi} \int_0^{\frac{1}{2}} (\sqrt{1-r^2} - \sqrt{3r^2})\, rdrd\theta$

$= (2\pi) \cdot \left( \dfrac{(-1)(1-r^2)^{\frac{3}{2}}}{3} - \dfrac{\sqrt{3}r^3}{3} \right) \Bigg|_{r=0}^{r=\frac{1}{2}} = 2\pi\left( \dfrac{2-\sqrt{3}}{6} \right) = \pi\left( \dfrac{2-\sqrt{3}}{3} \right)$

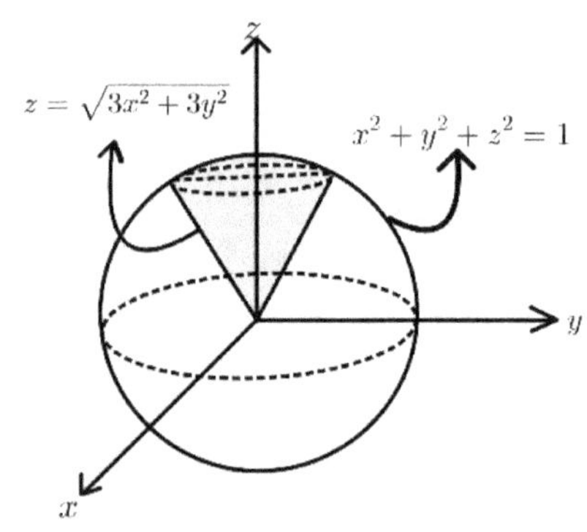


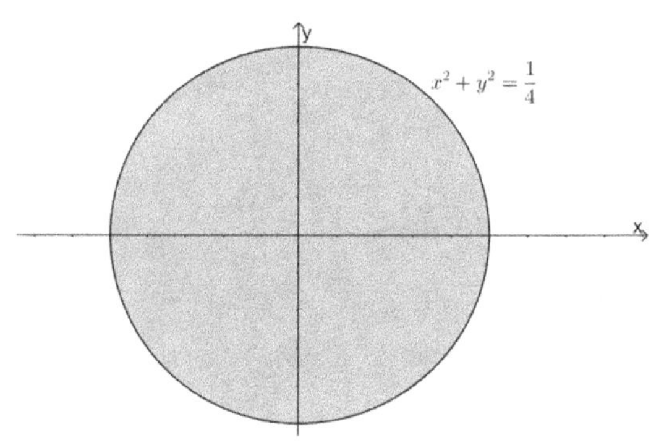


(2)

令 $R = \{(x,y): x^2+y^2 \leq \dfrac{3}{4}\}$

则 $V = \iint\limits_{R} \int_{\sqrt{\frac{x^2+y^2}{3}}}^{\sqrt{1-x^2-y^2}} dzdA = \iint\limits_{x^2+y^2\leq\frac{3}{4}} \sqrt{1-x^2-y^2} - \sqrt{\dfrac{x^2+y^2}{3}}\, dxdy$

令 $x = r\cos\theta, y = r\sin\theta$ 则 $\{(x,y): x^2+y^2 \leq \dfrac{3}{4}\} = \{(r,\theta): 0 \leq r \leq \sqrt{\dfrac{3}{4}}, 0 \leq \theta \leq 2\pi\}$

且 $dxdy = \left\| \begin{matrix} \dfrac{\partial x}{\partial r} & \dfrac{\partial x}{\partial \theta} \\ \dfrac{\partial y}{\partial r} & \dfrac{\partial y}{\partial \theta} \end{matrix} \right\| drd\theta = \left\| \begin{matrix} \cos\theta & -r\sin\theta \\ \sin\theta & r\cos\theta \end{matrix} \right\| drd\theta = rdrd\theta$

$$\therefore \iint_{x^2+y^2\leq\frac{3}{4}} \sqrt{1-x^2-y^2} - \sqrt{\frac{x^2+y^2}{3}}\, dxdy = \int_0^{2\pi}\int_0^{\sqrt{\frac{3}{4}}}\left(\sqrt{1-r^2}-\sqrt{\frac{r^2}{3}}\right)rdrd\theta$$

$$= (2\pi)\cdot\left(\frac{(-1)(1-r^2)^{\frac{3}{2}}}{3} - \frac{r^3}{3\sqrt{3}}\right)\Bigg|_{r=0}^{r=\sqrt{\frac{3}{4}}} = 2\pi\left(\frac{1}{6}\right) = \frac{\pi}{3}$$

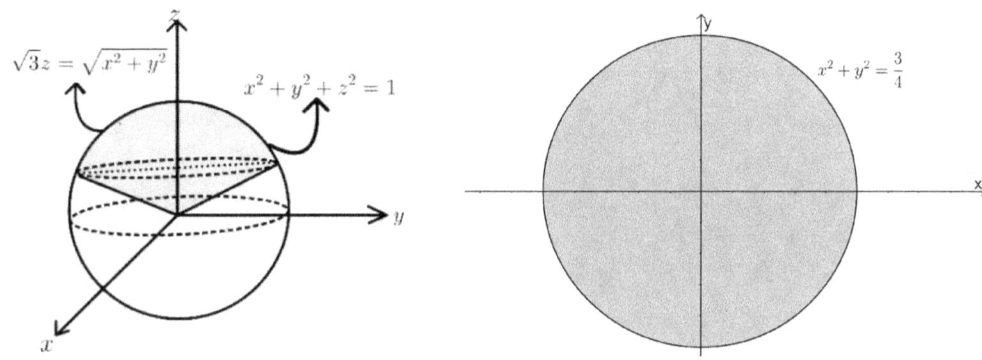


Example 27.

    (1)计算球体$x^2+y^2+(z-2)^2 \leq 4$ 与 $z\geq\sqrt{x^2+y^2}$交集区域体积

    (2)计算球体$x^2+y^2+z^2 \leq z$ 与 $z\geq\sqrt{x^2+y^2}$交集区域体积

【解】

(1)

令$R = \{(x,y)\colon x^2+y^2 \leq 4\}$

则 $V = \iint_R \int_{\sqrt{x^2+y^2}}^{2+\sqrt{4-x^2-y^2}} dzdA = \iint_{x^2+y^2\leq 4} 2+\sqrt{4-x^2-y^2}-\sqrt{x^2+y^2}\,dxdy$

令$x = r\cos\theta, y = r\sin\theta$ 则 $\{(x,y)\colon x^2+y^2 \leq 4\} = \{(r,\theta)\colon 0\leq r\leq 2, 0\leq\theta\leq 2\pi\}$

且 $dxdy = \left\|\begin{array}{cc}\frac{\partial x}{\partial r} & \frac{\partial x}{\partial\theta}\\ \frac{\partial y}{\partial r} & \frac{\partial y}{\partial\theta}\end{array}\right\| drd\theta = \left\|\begin{array}{cc}\cos\theta & -r\sin\theta\\ \sin\theta & r\cos\theta\end{array}\right\| drd\theta = rdrd\theta$

$$\therefore \iint_{x^2+y^2\leq4} 2 + \sqrt{4 - x^2 - y^2} - \sqrt{x^2 + y^2}\,dxdy = \int_0^{2\pi} \int_0^2 (2 + \sqrt{4 - r^2} - \sqrt{r^2})r\,drd\theta$$

$$= (2\pi) \cdot \left( r^2 + \frac{(-1)(4 - r^2)^{\frac{3}{2}}}{3} - \frac{r^3}{3} \right)\Bigg|_{r=0}^{r=2} = 8\pi$$

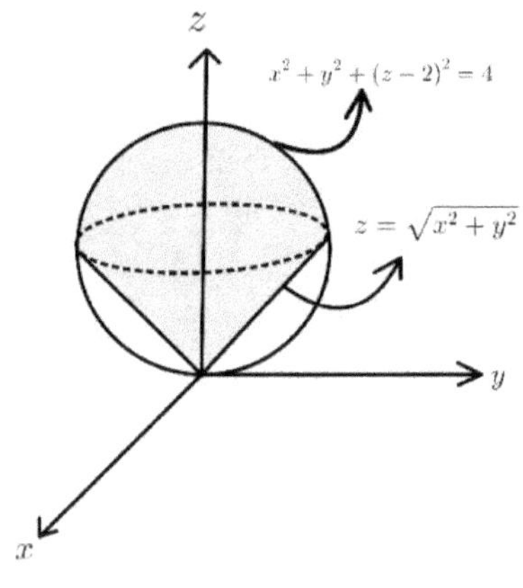


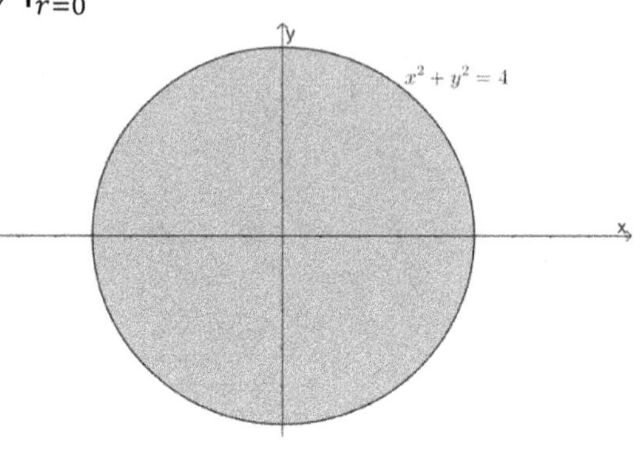


(2)

$$令 R = \left\{(x,y): x^2 + y^2 \leq \frac{1}{4}\right\}$$

$$则\ V = \iint_R \int_{\sqrt{x^2+y^2}}^{\frac{1}{2}+\sqrt{\frac{1}{4}-x^2-y^2}} dzdA = \iint_{x^2+y^2\leq\frac{1}{4}} \frac{1}{2} + \sqrt{\frac{1}{4} - x^2 - y^2} - \sqrt{x^2 + y^2}\,dxdy$$

$$令 x = r\cos\theta, y = r\sin\theta\ 则\ \left\{(x,y): x^2 + y^2 \leq \frac{1}{4}\right\} = \left\{(r,\theta): 0 \leq r \leq \frac{1}{2}, 0 \leq \theta \leq 2\pi\right\}$$

$$且\ dxdy = \left\|\begin{vmatrix} \dfrac{\partial x}{\partial r} & \dfrac{\partial x}{\partial \theta} \\ \dfrac{\partial y}{\partial r} & \dfrac{\partial y}{\partial \theta} \end{vmatrix}\right\| drd\theta = \left\|\begin{vmatrix} \cos\theta & -r\sin\theta \\ \sin\theta & r\cos\theta \end{vmatrix}\right\| drd\theta = rdrd\theta$$

$$\therefore \iint_{x^2+y^2\leq\frac{1}{4}} \frac{1}{2} + \sqrt{\frac{1}{4} - x^2 - y^2} - \sqrt{x^2 + y^2}\,dxdy = \int_0^{2\pi} \int_0^{\frac{1}{2}} \left( \frac{1}{2} + \sqrt{\frac{1}{4} - r^2} - r \right)r\,drd\theta$$

$$= 2\pi \left( \frac{r^2}{4} + \frac{(-1)\left(\frac{1}{4} - r^2\right)^{\frac{3}{2}}}{3} - \frac{r^3}{3} \right)\Bigg|_{r=0}^{r=\frac{1}{2}} = \frac{\pi}{8}$$

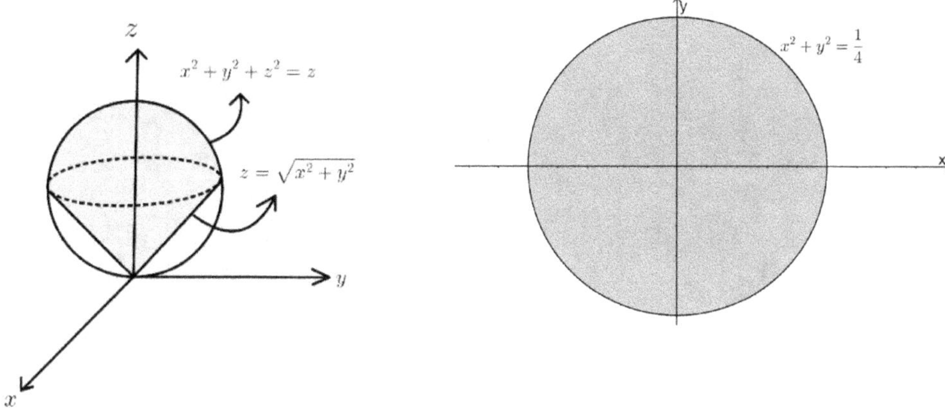

## Example 28.

求由曲面 $z = x^2 + 8y^2$ 上方与 $z = 9 - y^2$ 下方所围区域的体积

【解】

令 $R = \{(x,y): x^2 + 9y^2 \le 9\}$ 则 $V = \iint_R \int_{x^2+8y^2}^{9-y^2} dz\,dA = \iint_R 9 - x^2 - 9y^2\,dx\,dy$

令 $x = 3r\cos\theta, y = r\sin\theta$ 则 $\{(x,y): x^2 + 9y^2 \le 9\} = \{(r,\theta): 0 \le r \le 1, 0 \le \theta \le 2\pi\}$

且 $dx\,dy = \left\| \begin{vmatrix} \dfrac{\partial x}{\partial r} & \dfrac{\partial x}{\partial \theta} \\ \dfrac{\partial y}{\partial r} & \dfrac{\partial y}{\partial \theta} \end{vmatrix} \right\| dr\,d\theta = \left\| \begin{vmatrix} 2\cos\theta & -2r\sin\theta \\ \sin\theta & r\cos\theta \end{vmatrix} \right\| dr\,d\theta = 2r\,dr\,d\theta$

$\therefore \iint_R 9 - x^2 - 9y^2\,dx\,dy = \int_0^{2\pi} \int_0^1 (9 - 9r^2)2r\,dr\,d\theta = 18 \int_0^{2\pi} d\theta \int_0^1 r - r^3\,dr = 9\pi$

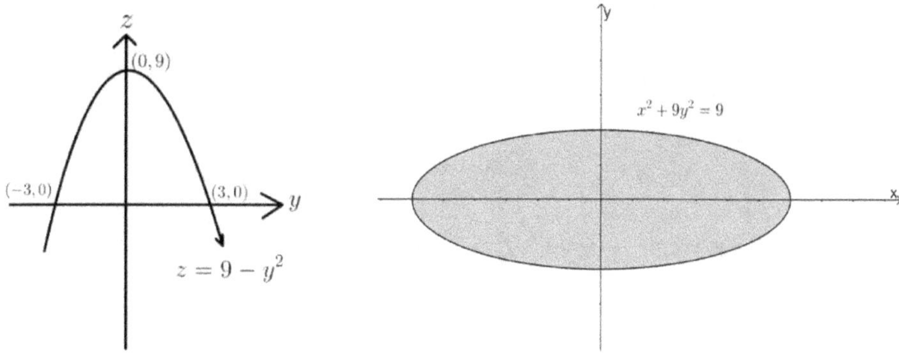

## Example 29.

假设 V 为曲面 $z \ge x^2 + 8y^2, z \le 9 - y^2, x > 0, y > 0, z > 0$ 所围区域

则 $\iiint_V xdxdydz =?$

【解】

令 $R = \{(x,y): x^2 + 9y^2 \le 9, x > 0, y > 0\}$

则 $\iiint_V xdxdydz = \iint_R \int_{x^2+8y^2}^{9-y^2} xdzdA = \iint_R x(9 - x^2 - 9y^2)dxdy$

令 $x = 3r\cos\theta$ , $y = r\sin\theta$

则 $\{(x,y): x^2 + 9y^2 \le 9, x > 0, y > 0\} = \{(r,\theta): 0 \le r \le 1, 0 \le \theta \le \frac{\pi}{2}\}$

且 $dxdy = \left\| \begin{vmatrix} \dfrac{\partial x}{\partial r} & \dfrac{\partial x}{\partial \theta} \\ \dfrac{\partial y}{\partial r} & \dfrac{\partial y}{\partial \theta} \end{vmatrix} \right\| drd\theta = \left\| \begin{vmatrix} 2\cos\theta & -2r\sin\theta \\ \sin\theta & r\cos\theta \end{vmatrix} \right\| drd\theta = 2rdrd\theta$

$\therefore \iint_R x(9 - x^2 - 9y^2)dxdy = \int_0^{\frac{\pi}{2}} 3r\cos\theta \int_0^1 (9 - 9r^2)2rdrd\theta$

$= 54 \int_0^{\frac{\pi}{2}} \cos\theta\, d\theta \int_0^1 r^2 - r^4 dr = \frac{36}{5}$

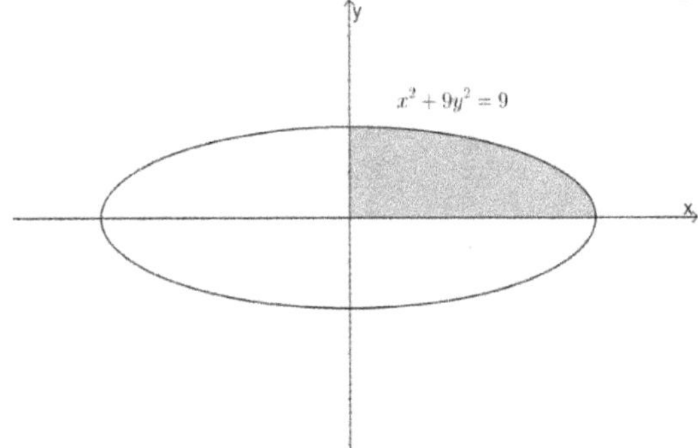


Example 30.

求由曲面 $z = 8 - 2y$ 下方与 $z = x^2 + y^2$ 上方所围区域的体积

【解】

令 $R = \{(x,y): x^2 + (y+1)^2 \le 9\}$

则 $V = \iint_R \int_{x^2+y^2}^{8-2y} dzdA = \iint_{x^2+(y+1)^2 \le 9} 9 - x^2 - (y+1)^2 dxdy$

令 $x = 3r\cos\theta, y = -1 + 3r\sin\theta$

则 $\{(x,y): x^2 + (y+1)^2 \le 9\} = \{(r,\theta): 0 \le r \le 1, 0 \le \theta \le 2\pi\}$

且 $dxdy = \left\| \begin{vmatrix} \dfrac{\partial x}{\partial r} & \dfrac{\partial x}{\partial \theta} \\ \dfrac{\partial y}{\partial r} & \dfrac{\partial y}{\partial \theta} \end{vmatrix} \right\| drd\theta = \left\| \begin{vmatrix} 3\cos\theta & 3r\sin\theta \\ 3\sin\theta & 3r\cos\theta \end{vmatrix} \right\| drd\theta = 9rdrd\theta$

$$\therefore \iint_{x^2+(y+1)^2 \le 9} 9 - x^2 - (y+1)^2 dxdy = \int_0^{2\pi}\int_0^1 (9-9r^2)9rdrd\theta = 81\int_0^{2\pi} d\theta \int_0^1 r - r^3 dr$$

$$= \frac{81\pi}{2}$$

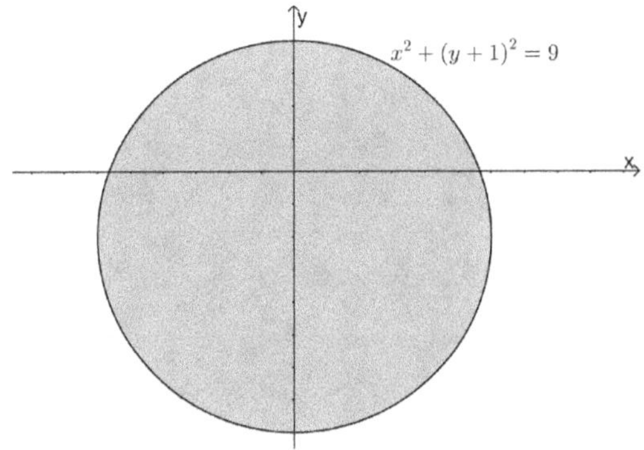

**Example 31.**

假设 V 为曲面 $z \le 8 - 2y,\ z \ge x^2 + y^2,\ x > 0,\ y > 0,\ z > 0$ 所围区域

求 $\iiint_V xdxdydz = ?$

【解】

令 $R = \{(x,y): x^2 + (y+1)^2 \le 9, x > 0, y > 0\}$

则 $\iiint_V xdxdydz = \iint_R \int_{x^2+y^2}^{8-2y} dzdA = \iint_R x(9 - x^2 - (y+1)^2)dxdy$

令 $x = 3r\cos\theta,\ y = -1 + 3r\sin\theta$

则 $\{(x,y): x^2 + (y+1)^2 \le 9\} = \{(r,\theta): 0 \le r \le 1, 0 \le \theta \le \dfrac{\pi}{2}\}$

且 $dxdy = \left\| \begin{vmatrix} \dfrac{\partial x}{\partial r} & \dfrac{\partial x}{\partial \theta} \\ \dfrac{\partial y}{\partial r} & \dfrac{\partial y}{\partial \theta} \end{vmatrix} \right\| drd\theta = \left\| \begin{vmatrix} 3\cos\theta & 3r\sin\theta \\ 3\sin\theta & 3r\cos\theta \end{vmatrix} \right\| drd\theta = 9rdrd\theta$

$$\therefore \iint_{x^2+(y+1)^2\leq 9, x>0, y>0} x(9-x^2-(y+1)^2)dxdy = \int_0^{\frac{\pi}{2}} 3r\cos\theta \int_0^1 (9-9r^2)9r\,dr\,d\theta$$

$$= 243 \int_0^{\frac{\pi}{2}} \cos\theta\, d\theta \int_0^1 r^2 - r^4 dr = \frac{162}{5}$$

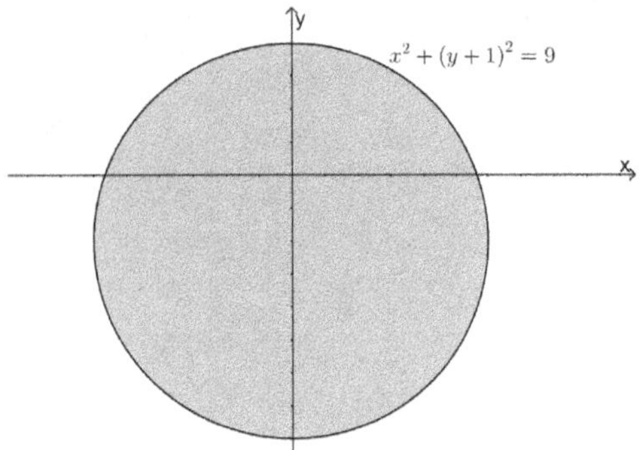


Example 32.

$$求 \int_0^1 \int_0^{\sqrt{1-x^2}} \int_0^1 \tan^{-1}\sqrt{x^2+y^2}\, dz\,dy\,dx = ?$$

【解】

令 $x = r\cos\theta, y = r\sin\theta$

则 $\{(x,y): 0\leq x\leq 1, 0\leq y\leq \sqrt{1-x^2}\} = \{(r,\theta): 0\leq r\leq 1, 0\leq \theta\leq \frac{\pi}{2}\}$

且 $dxdy = \left|\left|\begin{matrix}\dfrac{\partial x}{\partial r} & \dfrac{\partial x}{\partial \theta}\\ \dfrac{\partial y}{\partial r} & \dfrac{\partial y}{\partial \theta}\end{matrix}\right|\right| drd\theta = \left|\left|\begin{matrix}\cos\theta & -r\sin\theta\\ \sin\theta & r\cos\theta\end{matrix}\right|\right| drd\theta = rdrd\theta$

$$\therefore \int_0^1 \int_0^{\sqrt{1-x^2}} \int_0^1 \tan^{-1}\sqrt{x^2+y^2}\, dz\,dy\,dx = \int_0^1 \int_0^{\frac{\pi}{2}} r\tan^{-1}r\, d\theta\,dr = \int_0^{\frac{\pi}{2}} d\theta \int_0^1 r\tan^{-1}r\, dr$$

令 $u = \tan^{-1}r, \ dv = rdr$ 则 $du = \dfrac{dr}{1+r^2}, \ v = \dfrac{r^2}{2}$, 藉由分部积分法

则 $\displaystyle\int r\tan^{-1}r\, dr = \dfrac{r^2}{2}\tan^{-1}r - \dfrac{1}{2}\int \dfrac{r^2 dr}{1+r^2} = \dfrac{r^2}{2}\tan^{-1}r - \dfrac{1}{2}\int \dfrac{1+r^2-1 dr}{1+r^2}$

$$= \dfrac{r^2}{2}\tan^{-1}r - \dfrac{r}{2} + \dfrac{1}{2}\tan^{-1}r + c$$

$$\therefore \int_0^1 \int_0^{\sqrt{1-x^2}} \int_0^1 \tan^{-1}\sqrt{x^2+y^2}\,dzdydx = \int_0^{\frac{\pi}{2}} d\theta \int_0^1 r\tan^{-1}r\,dr = \frac{\pi}{2}\left(\frac{\pi}{4}-\frac{1}{2}\right)$$

## 8.4.3　使用球坐标转换求三重积分

使用球坐标转换的时机为三维空间的积分区域$V$出现 $x^2+y^2+z^2$

考试类型:

Type 1.

假设 $V = \{(x,y,z): x^2+y^2+z^2 \le a\},\ a>0,\ $ 求 $\iiint_V f(x,y,z)dV =?$

解题流程:

Step1.

令$x = \rho\sin\varphi\cos\theta\,, y = \rho\sin\varphi\sin\theta\,, \text{z} = \rho\cos\varphi$

$$\text{则}\,dxdydz = \begin{Vmatrix} \dfrac{\partial x}{\partial \rho} & \dfrac{\partial x}{\partial \varphi} & \dfrac{\partial x}{\partial \theta} \\[2mm] \dfrac{\partial y}{\partial \rho} & \dfrac{\partial y}{\partial \varphi} & \dfrac{\partial y}{\partial \theta} \\[2mm] \dfrac{\partial z}{\partial \rho} & \dfrac{\partial z}{\partial \varphi} & \dfrac{\partial z}{\partial \theta} \end{Vmatrix} d\rho d\varphi d\theta$$

$$= \begin{Vmatrix} \sin\varphi\cos\theta & \rho\cos\varphi\cos\theta & -\rho\sin\varphi\sin\theta \\ \sin\varphi\sin\theta & \rho\cos\varphi\sin\theta & -\rho\sin\varphi\cos\theta \\ \cos\varphi & -\rho\sin\varphi & 0 \end{Vmatrix} d\rho d\varphi d\theta = \rho^2\sin\varphi\,d\rho d\varphi d\theta$$

其中 $\{(\rho,\varphi,\theta): 0 \le \rho \le a, 0 \le \varphi \le \pi, 0 \le \theta \le 2\pi\}$

Step2.

$$\iiint_V f(x,y,z)dV = \int_0^\pi \int_0^{2\pi} \int_0^a f(\rho\sin\varphi\cos\theta\,, \rho\sin\varphi\sin\theta\,, \rho\cos\varphi)\rho^2\sin\varphi\,d\rho d\theta d\varphi$$

Step3.

求 $\displaystyle\int_0^\pi \int_0^{2\pi} \int_0^a f(\rho\sin\varphi\cos\theta\,, \rho\sin\varphi\sin\theta\,, \rho\cos\varphi)\rho^2\sin\varphi\,d\rho d\theta d\varphi =?$

范例说明:

求 $\iiint_V f(x,y,z)dV =?,\ \ V = \{(x,y,z): x^2 + y^2 + z^2 \le a\},\ \ a > 0$

令 $x = \rho \sin\varphi\cos\theta,\ y = \rho\sin\varphi\sin\theta,\ z = \rho\cos\varphi$ 则 $dxdydz = \rho^2\sin\varphi\, d\rho d\varphi d\theta$

其中 $0 \le \rho \le a,\ \ 0 \le \varphi \le \pi,\ \ 0 \le \theta \le 2\pi$

$$\iiint_V f(x,y,z)dV = \int_0^\pi \int_0^{2\pi}\int_0^a f(\rho\sin\varphi\cos\theta,\rho\sin\varphi\sin\theta,\rho\cos\varphi)\rho^2\sin\varphi\, d\rho d\theta d\varphi$$

(I) 当 $f(x,y,z) = \dfrac{1}{\sqrt{x^2+y^2+z^2}}$,

$$\iiint_V \frac{1}{\sqrt{x^2+y^2+z^2}}dV = \int_0^\pi\int_0^{2\pi}\int_0^a \rho\sin\varphi\, d\rho d\theta d\varphi$$

(II) 当 $f(x,y,z) = \dfrac{z^2}{\sqrt{x^2+y^2+z^2}}$,

$$\iiint_V \frac{z^2}{\sqrt{x^2+y^2+z^2}}dV = \int_0^\pi\int_0^{2\pi}\int_0^a \frac{\rho^2\cos^2\varphi\,\rho^2\sin\varphi}{\rho}\, d\rho d\theta d\varphi$$

(III) 当 $f(x,y,z) = \sqrt{1-(x^2+y^2+z^2)}$,

$$\iiint_V \sqrt{1-(x^2+y^2+z^2)}\,dV = \int_0^\pi\int_0^{2\pi}\int_0^a \sqrt{1-\rho^2}\,\rho^2\sin\varphi\, d\rho d\theta d\varphi$$

(IV) 当 $f(x,y,z) = \dfrac{1}{\sqrt{1-(x^2+y^2+z^2)}}$,

$$\iiint_V \frac{1}{\sqrt{1-(x^2+y^2+z^2)}}dV = \int_0^\pi\int_0^{2\pi}\int_0^a \frac{\rho^2}{\sqrt{1-\rho^2}}\sin\varphi\, d\rho d\theta d\varphi$$

(V) 当 $f(x,y,z) = xyz$,

$$\iiint_V xyz\,dV = \int_0^\pi \sin^3\varphi\cos\varphi\, d\varphi \int_0^{2\pi}\cos\theta\sin\theta\, d\theta \int_0^a \rho^5 d\rho$$

(VI) 当 $f(x,y,z) = (x^2+y^2+z^2)^{\frac{3}{2}}$,

$$\iiint_V (x^2+y^2+z^2)^{\frac{3}{2}}dV = \int_0^\pi\int_0^{2\pi}\int_0^a (\rho^2)^{\frac{3}{2}}\rho^2\sin\varphi\, d\rho d\theta d\varphi$$

(VII) 当 $f(x,y,z) = \dfrac{\cos\sqrt{x^2+y^2+z^2}}{x^2+y^2+z^2}$,

$$\iiint_V \frac{\cos\sqrt{x^2+y^2+z^2}}{x^2+y^2+z^2}\,dV = \int_0^\pi \int_0^{2\pi} \int_0^a \frac{\cos\rho}{\rho^2}\rho^2 \sin\varphi\,d\rho d\theta d\varphi$$

Example 1.

(1) 求由曲面 $x^{\frac{2}{3}} + y^{\frac{2}{3}} + z^{\frac{2}{3}} = r^2$ 围成区域的体积

(2) 求由曲面 $x^2 + 2y^2 + 3z^2 = 4$ 围成区域的体积

【解】

(1)

令 $x = u^3,\ y = v^3,\ z = s^3$

$$\text{则 } dxdydz = \left\| \begin{matrix} \dfrac{\partial x}{\partial u} & \dfrac{\partial x}{\partial v} & \dfrac{\partial x}{\partial s} \\[2mm] \dfrac{\partial y}{\partial u} & \dfrac{\partial y}{\partial v} & \dfrac{\partial y}{\partial s} \\[2mm] \dfrac{\partial z}{\partial u} & \dfrac{\partial z}{\partial v} & \dfrac{\partial z}{\partial s} \end{matrix} \right\| dudvds = \left\| \begin{matrix} 3u^2 & 0 & 0 \\ 0 & 3v^2 & 0 \\ 0 & 0 & 3s^2 \end{matrix} \right\| = 27u^2v^2s^2\,dudvds$$

$$\therefore\ V = \iiint_{x^{\frac{2}{3}}+y^{\frac{2}{3}}+z^{\frac{2}{3}}\le r^2} dxdydz = \iiint_{u^2+v^2+s^2\le r^2} 27u^2v^2s^2\,dudvds$$

令 $u = \rho\sin\varphi\cos\theta, v = \rho\sin\varphi\sin\theta, s = \rho\cos\varphi$ 则

$$dudvds = \left\| \begin{matrix} \dfrac{\partial u}{\partial \rho} & \dfrac{\partial u}{\partial \varphi} & \dfrac{\partial u}{\partial \theta} \\[2mm] \dfrac{\partial v}{\partial \rho} & \dfrac{\partial v}{\partial \varphi} & \dfrac{\partial v}{\partial \theta} \\[2mm] \dfrac{\partial s}{\partial \rho} & \dfrac{\partial s}{\partial \varphi} & \dfrac{\partial s}{\partial \theta} \end{matrix} \right\| d\rho d\varphi d\theta$$

$$= \left\| \begin{matrix} \sin\varphi\cos\theta & \rho\cos\varphi\cos\theta & -\rho\sin\varphi\sin\theta \\ \sin\varphi\sin\theta & \rho\cos\varphi\sin\theta & -\rho\sin\varphi\cos\theta \\ \cos\varphi & -\rho\sin\varphi & 0 \end{matrix} \right\| d\rho d\varphi d\theta = \rho^2 \sin\varphi\,d\rho d\varphi d\theta$$

且 $\{(u,v,s): u^2+v^2+s^2 \le r^2\} = \{(\rho,\varphi,\theta): 0\le\rho\le r, 0\le\varphi\le\pi, 0\le\theta\le 2\pi\}$

$$\therefore \iiint_{u^2+v^2+s^2 \le r^2} 27u^2v^2s^2\,dudvds$$

$$= \int_0^\pi \int_0^{2\pi} \int_0^r 27(\rho\sin\varphi\cos\theta)^2(\rho\sin\varphi\sin\theta)^2(\rho\cos\varphi)^2\rho^2\sin\varphi\,d\rho d\theta d\varphi$$

$$= 27\int_0^\pi \sin^5\varphi\cos^2\varphi\,d\varphi \int_0^{2\pi}\cos^2\theta\sin^2\theta d\theta \int_0^r \rho^8 d\rho$$

$$\because \int_0^\pi \sin^5\varphi\cos^2\varphi\,d\varphi = \int_0^\pi \sin\varphi(1-\cos^2\varphi)^2\cos^2\varphi\,d\varphi$$

$$= \left(\frac{(-1)\cos^3\varphi}{3} + \frac{2\cos^5\varphi}{5} + \frac{(-1)\cos^7\varphi}{7}\right)\Bigg|_0^\pi = \frac{16}{105}$$

$$\text{且}\int_0^{2\pi}\cos^2\theta\sin^2\theta d\theta = \int_0^{2\pi}\left(\frac{\sin 2\theta}{2}\right)^2 d\theta = \frac{1}{4}\int_0^{2\pi}\frac{1-\cos 4\theta}{2}d\theta = \frac{1}{4}\left(\frac{\theta}{2}-\frac{\sin 4\theta}{8}\right)\Bigg|_0^{2\pi} = \frac{\pi}{4}$$

$$\therefore V = 27\int_0^\pi \sin^5\varphi\cos^2\varphi\,d\varphi \int_0^{2\pi}\cos^2\theta\sin^2\theta d\theta \int_0^r \rho^8 d\rho = 27\cdot\frac{16}{105}\cdot\frac{\pi}{4}\cdot\frac{r^9}{9} = \frac{4\pi r^9}{35}$$

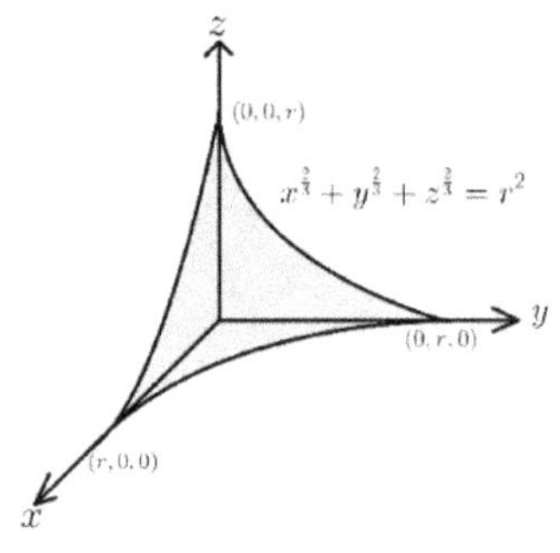


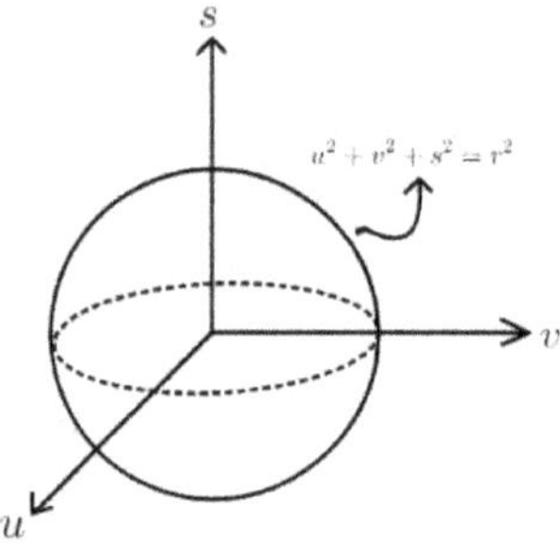


(2)

$$\text{令}x = u,\ y = \frac{v}{\sqrt{2}},\ z = \frac{s}{\sqrt{3}}$$

$$\text{则}\,dxdydz = \begin{Vmatrix}\dfrac{\partial x}{\partial u} & \dfrac{\partial x}{\partial v} & \dfrac{\partial x}{\partial s}\\[2mm] \dfrac{\partial y}{\partial u} & \dfrac{\partial y}{\partial v} & \dfrac{\partial y}{\partial s}\\[2mm] \dfrac{\partial z}{\partial u} & \dfrac{\partial z}{\partial v} & \dfrac{\partial z}{\partial s}\end{Vmatrix}dudvds = \begin{Vmatrix}1 & 0 & 0\\[1mm] 0 & \dfrac{1}{\sqrt{2}} & 0\\[1mm] 0 & 0 & \dfrac{1}{\sqrt{3}}\end{Vmatrix} = \frac{1}{\sqrt{6}}dudvds$$

$$\therefore \iiint_{x^2+2y^2+3z^2\le 4} dxdydz = \iiint_{u^2+v^2+s^2\le 4}\frac{1}{\sqrt{6}}dudvds$$

令 $u = \rho \sin\varphi \cos\theta\,,\, v = \rho \sin\varphi \sin\theta\,,\, s = \rho \cos\varphi$　则

$$dudvds = \begin{Vmatrix} \dfrac{\partial u}{\partial \rho} & \dfrac{\partial u}{\partial \varphi} & \dfrac{\partial u}{\partial \theta} \\[2mm] \dfrac{\partial v}{\partial \rho} & \dfrac{\partial v}{\partial \varphi} & \dfrac{\partial v}{\partial \theta} \\[2mm] \dfrac{\partial s}{\partial \rho} & \dfrac{\partial s}{\partial \varphi} & \dfrac{\partial s}{\partial \theta} \end{Vmatrix} d\rho d\varphi d\theta$$

$$= \begin{Vmatrix} \sin\varphi \cos\theta & \rho\cos\varphi \cos\theta & -\rho\sin\varphi \sin\theta \\ \sin\varphi \sin\theta & \rho\cos\varphi \sin\theta & -\rho\sin\varphi \cos\theta \\ \cos\varphi & -\rho\sin\varphi & 0 \end{Vmatrix} d\rho d\varphi d\theta = \rho^2 \sin\varphi\, d\rho d\varphi d\theta$$

且 $\{(u,v,s): u^2 + v^2 + s^2 \le 4\} = \{(\rho,\varphi,\theta): 0 \le \rho \le 2, 0 \le \varphi \le \pi, 0 \le \theta \le 2\pi\}$

$$\therefore \iiint_{u^2+v^2+s^2\le4} \frac{1}{\sqrt{6}} dudvds = \frac{1}{\sqrt{6}} \int_0^\pi \sin\varphi\, d\varphi \int_0^{2\pi} d\theta \int_0^2 \rho^2 d\rho = \frac{32\pi}{3\sqrt{6}}$$

Example 2.

$$求 \int_0^1 \int_0^{\sqrt{1-x^2}} \int_0^{\sqrt{1-(x^2+y^2)}} x^2 + y^2 + z^2\, dzdydx = ?$$

【解】

令 $x = \rho \sin\varphi \cos\theta\,,\, y = \rho \sin\varphi \sin\theta\,,\, z = \rho \cos\varphi$

$$则\ dxdydz = \begin{Vmatrix} \dfrac{\partial x}{\partial \rho} & \dfrac{\partial x}{\partial \varphi} & \dfrac{\partial x}{\partial \theta} \\[2mm] \dfrac{\partial y}{\partial \rho} & \dfrac{\partial y}{\partial \varphi} & \dfrac{\partial y}{\partial \theta} \\[2mm] \dfrac{\partial z}{\partial \rho} & \dfrac{\partial z}{\partial \varphi} & \dfrac{\partial z}{\partial \theta} \end{Vmatrix} d\rho d\varphi d\theta$$

$$= \begin{Vmatrix} \sin\varphi \cos\theta & \rho\cos\varphi \cos\theta & -\rho\sin\varphi \sin\theta \\ \sin\varphi \sin\theta & \rho\cos\varphi \sin\theta & -\rho\sin\varphi \cos\theta \\ \cos\varphi & -\rho\sin\varphi & 0 \end{Vmatrix} d\rho d\varphi d\theta = \rho^2 \sin\varphi\, d\rho d\varphi d\theta$$

令 $R = \{(\rho,\varphi,\theta): 0 \le \rho \le 1, 0 \le \varphi \le \dfrac{\pi}{2}, 0 \le \theta \le \dfrac{\pi}{2}\}$

$$\text{则} \int_0^1 \int_0^{\sqrt{1-x^2}} \int_0^{\sqrt{1-(x^2+y^2)}} x^2 + y^2 + z^2 \, dz\,dy\,dx = \iiint_R \rho^4 \sin\varphi \, dV$$

$$= \left(\frac{\rho^5}{5}\Big|_0^1\right) \int_0^{\frac{\pi}{2}} \sin\varphi \, d\varphi \int_0^{\frac{\pi}{2}} d\theta = \frac{\pi}{10}$$

Example 3.

$$\text{求} \int_{-2}^2 \int_{-\sqrt{4-x^2}}^{\sqrt{4-x^2}} \int_0^{\sqrt{4-(x^2+y^2)}} z^2\sqrt{x^2 + y^2 + z^2} \, dz\,dy\,dx = ?$$

【解】

令 $x = \rho \sin\varphi \cos\theta$ , $y = \rho \sin\varphi \sin\theta$ , $z = \rho \cos\varphi$

$$\text{则 } dxdydz = \left\| \begin{matrix} \dfrac{\partial x}{\partial \rho} & \dfrac{\partial x}{\partial \varphi} & \dfrac{\partial x}{\partial \theta} \\[2mm] \dfrac{\partial y}{\partial \rho} & \dfrac{\partial y}{\partial \varphi} & \dfrac{\partial y}{\partial \theta} \\[2mm] \dfrac{\partial z}{\partial \rho} & \dfrac{\partial z}{\partial \varphi} & \dfrac{\partial z}{\partial \theta} \end{matrix} \right\| d\rho d\varphi d\theta$$

$$= \left\| \begin{matrix} \sin\varphi\cos\theta & \rho\cos\varphi\cos\theta & -\rho\sin\varphi\sin\theta \\ \sin\varphi\sin\theta & \rho\cos\varphi\sin\theta & -\rho\sin\varphi\cos\theta \\ \cos\varphi & -\rho\sin\varphi & 0 \end{matrix} \right\| d\rho d\varphi d\theta = \rho^2 \sin\varphi \, d\rho d\varphi d\theta$$

令 $R = \{(\rho, \varphi, \theta): 0 \le \rho \le 2, 0 \le \varphi \le \dfrac{\pi}{2}, 0 \le \theta \le 2\pi\}$

$$\text{则} \int_{-2}^2 \int_{-\sqrt{4-x^2}}^{\sqrt{4-x^2}} \int_0^{\sqrt{4-(x^2+y^2)}} z^2\sqrt{x^2 + y^2 + z^2} \, dz\,dy\,dx = \iiint_R \rho^3 \cos^2\varphi \, \rho^2 \sin\varphi \, dV$$

$$= \left(\frac{\rho^6}{6}\Big|_0^2\right) \int_0^{\frac{\pi}{2}} \cos^2\varphi \sin\varphi \, d\varphi \int_0^{2\pi} d\theta = \frac{64}{6} \cdot \frac{1}{3} \cdot 2\pi = \frac{64\pi}{9}$$

Example 4.

$$\text{假设 } V = \left\{ (x, y, z): \frac{x^2}{a^2} + \frac{y^2}{b^2} + \frac{z^2}{c^2} \leq 1 \right\}, \quad \text{求 } \iiint_V x^2 y^2 \, dxdydz = ?$$

【解】

令 $x = a\rho \sin\varphi \cos\theta$, $y = b\rho \sin\varphi \sin\theta$, $z = c\rho \cos\varphi$

$$\text{则 } dxdydz = \begin{Vmatrix} \dfrac{\partial x}{\partial \rho} & \dfrac{\partial x}{\partial \varphi} & \dfrac{\partial x}{\partial \theta} \\ \dfrac{\partial y}{\partial \rho} & \dfrac{\partial y}{\partial \varphi} & \dfrac{\partial y}{\partial \theta} \\ \dfrac{\partial z}{\partial \rho} & \dfrac{\partial z}{\partial \varphi} & \dfrac{\partial z}{\partial \theta} \end{Vmatrix} d\rho d\varphi d\theta$$

$$= abc \begin{Vmatrix} \sin\varphi\cos\theta & \rho\cos\varphi\cos\theta & -\rho\sin\varphi\sin\theta \\ \sin\varphi\sin\theta & \rho\cos\varphi\sin\theta & -\rho\sin\varphi\cos\theta \\ \cos\varphi & -\rho\sin\varphi & 0 \end{Vmatrix} d\rho d\varphi d\theta = abc\rho^2 \sin\varphi \, d\rho d\varphi d\theta$$

令 $R = \{ (\rho, \varphi, \theta): 0 \leq \rho \leq 1, 0 \leq \varphi \leq \pi, 0 \leq \theta \leq 2\pi \}$

$$\text{则 } \iiint_V x^2 y^2 \, dxdydz = a^3 b^3 c \iiint_R \rho^6 \sin^5\varphi \sin^2\theta \cos^2\theta \, dV$$

$$= a^3 b^3 c \int_0^\pi \int_0^{2\pi} \int_0^1 \rho^6 \sin^5\varphi \sin^2\theta \cos^2\theta \, d\rho d\theta d\varphi$$

$$= a^3 b^3 c \left( \frac{\rho^7}{7} \Big|_0^1 \right) \int_0^\pi \sin^5\varphi \, d\varphi \int_0^{2\pi} \sin^2\theta \cos^2\theta \, d\theta$$

$$\because \int_0^\pi \sin^5\varphi \, d\varphi = \frac{4}{5} \cdot \frac{2}{3} \cdot \int_0^\pi \sin\varphi \, d\varphi = \frac{16}{15}$$

$$\int_0^{2\pi} \sin^2\theta \cos^2\theta \, d\theta = \int_0^{2\pi} \left( \frac{\sin 2\theta}{2} \right)^2 d\theta = \int_0^{2\pi} \frac{1 - \cos 4\theta}{8} \, d\theta = \frac{\pi}{4}$$

$$\therefore \iiint_V x^2 y^2 \, dxdydz = \frac{4\pi a^3 b^3 c}{105}$$

Example 5.

$$\text{假设 } R \text{ 为曲面 } x^2 + 2y^2 + 3z^2 = 4 \text{ 与 } z > 0 \text{ 围成的区域, } \quad \text{求 } \iiint_R z dV = ?$$

【解】

令 $x = u, y = \dfrac{v}{\sqrt{2}}, z = \dfrac{s}{\sqrt{3}}$

$$\text{则 } dxdydz = \left\| \begin{matrix} \dfrac{\partial x}{\partial u} & \dfrac{\partial x}{\partial v} & \dfrac{\partial x}{\partial s} \\[2mm] \dfrac{\partial y}{\partial u} & \dfrac{\partial y}{\partial v} & \dfrac{\partial y}{\partial s} \\[2mm] \dfrac{\partial z}{\partial u} & \dfrac{\partial z}{\partial v} & \dfrac{\partial z}{\partial s} \end{matrix} \right\| dudvds = \left\| \begin{matrix} 1 & 0 & 0 \\[2mm] 0 & \dfrac{1}{\sqrt{2}} & 0 \\[2mm] 0 & 0 & \dfrac{1}{\sqrt{3}} \end{matrix} \right\| = \dfrac{1}{\sqrt{6}} dudvds$$

$$\therefore \iiint_{x^2+2y^2+3z^2 \leq 4} z\,dxdydz = \frac{1}{3\sqrt{2}} \iiint_{u^2+v^2+s^2 \leq 4} s\,dudvds$$

令 $u = \rho \sin\varphi \cos\theta , v = \rho \sin\varphi \sin\theta , s = \rho \cos\varphi$ 则

$$dudvds = \left\| \begin{matrix} \dfrac{\partial u}{\partial \rho} & \dfrac{\partial u}{\partial \varphi} & \dfrac{\partial u}{\partial \theta} \\[2mm] \dfrac{\partial v}{\partial \rho} & \dfrac{\partial v}{\partial \varphi} & \dfrac{\partial v}{\partial \theta} \\[2mm] \dfrac{\partial s}{\partial \rho} & \dfrac{\partial s}{\partial \varphi} & \dfrac{\partial s}{\partial \theta} \end{matrix} \right\| d\rho d\varphi d\theta$$

$$= \left\| \begin{matrix} \sin\varphi\cos\theta & \rho\cos\varphi\cos\theta & -\rho\sin\varphi\sin\theta \\ \sin\varphi\sin\theta & \rho\cos\varphi\sin\theta & -\rho\sin\varphi\cos\theta \\ \cos\varphi & -\rho\sin\varphi & 0 \end{matrix} \right\| d\rho d\varphi d\theta = \rho^2 \sin\varphi\, d\rho d\varphi d\theta$$

且 $\{(u,v,s): u^2 + v^2 + s^2 \leq 4, s > 0\} = \{(\rho,\varphi,\theta): 0 \leq \rho \leq 2, 0 \leq \varphi \leq \dfrac{\pi}{2}, 0 \leq \theta \leq 2\pi\}$

$$\therefore \frac{1}{3\sqrt{2}} \iiint_{u^2+v^2+s^2 \leq 4} s\,dudvds = \frac{1}{\sqrt{6}} \int_0^{\frac{\pi}{2}} \sin\varphi\cos\varphi\, d\varphi \int_0^{2\pi} d\theta \int_0^2 \rho^3 d\rho$$

$$= \frac{\pi}{\sqrt{6}} \int_0^{\frac{\pi}{2}} \sin 2\varphi\, d\varphi \cdot \frac{\rho^4}{4}\bigg|_0^2 = \frac{\pi}{2\sqrt{6}} \cdot (-\cos 2\varphi)\big|_0^{\frac{\pi}{2}} \cdot 4 = \frac{4\pi}{\sqrt{6}}$$

Example 6.

$$\text{试求} \int_{-\infty}^{\infty} \int_{-\infty}^{\infty} \int_{-\infty}^{\infty} e^{-(x^2+y^2+z^2)} dxdydz = ?$$

【解】

令 $x = \rho \sin\varphi \cos\theta , y = \rho \sin\varphi \sin\theta , z = \rho \cos\varphi$

$$\text{则 } dxdydz = \begin{Vmatrix} \dfrac{\partial x}{\partial \rho} & \dfrac{\partial x}{\partial \varphi} & \dfrac{\partial x}{\partial \theta} \\[2mm] \dfrac{\partial y}{\partial \rho} & \dfrac{\partial y}{\partial \varphi} & \dfrac{\partial y}{\partial \theta} \\[2mm] \dfrac{\partial z}{\partial \rho} & \dfrac{\partial z}{\partial \varphi} & \dfrac{\partial z}{\partial \theta} \end{Vmatrix} d\rho d\varphi d\theta$$

$$= \begin{Vmatrix} \sin\varphi\cos\theta & \rho\cos\varphi\cos\theta & -\rho\sin\varphi\sin\theta \\ \sin\varphi\sin\theta & \rho\cos\varphi\sin\theta & -\rho\sin\varphi\cos\theta \\ \cos\varphi & -\rho\sin\varphi & 0 \end{Vmatrix} d\rho d\varphi d\theta = \rho^2 \sin\varphi \, d\rho d\varphi d\theta$$

令 $R = \{(\rho,\varphi,\theta): 0 \le \rho \le 1, 0 \le \varphi \le \pi, 0 \le \theta \le 2\pi\}$

$$\text{则 } \int_{-\infty}^{\infty}\int_{-\infty}^{\infty}\int_{-\infty}^{\infty} e^{-(x^2+y^2+z^2)}dxdydz = \iiint_R e^{-\rho^2}\rho^2 \sin\varphi \, d\rho d\theta d\varphi$$

$$= \int_0^{\pi}\int_0^{2\pi}\int_0^{\infty} e^{-\rho^2}\rho^2 \sin\varphi \, d\rho d\theta d\varphi = \int_0^{\pi} \sin\varphi \int_0^{2\pi} d\theta \int_0^{\infty} e^{-\rho^2}\rho^2 d\rho$$

藉由 integration by parts

$$\int_0^{\infty} e^{-\rho^2}\rho^2 d\rho = \frac{-\rho}{2}e^{-\rho^2}\Big|_0^{\infty} + \frac{1}{2}\int_0^{\infty} e^{-\rho^2} d\rho = \frac{\sqrt{\pi}}{4}$$

$$\therefore \int_{-\infty}^{\infty}\int_{-\infty}^{\infty}\int_{-\infty}^{\infty} e^{-(x^2+y^2+z^2)}dxdydz = \int_0^{\pi} \sin\varphi \int_0^{2\pi} d\theta \int_0^{\infty} e^{-\rho^2}\rho^2 d\rho = 2(2\pi)\frac{\sqrt{\pi}}{4} = \pi\sqrt{\pi}$$

Example 7.

$$\text{试求} \int_{-\infty}^{\infty}\int_{-\infty}^{\infty}\int_{-\infty}^{\infty} e^{-(x^2+2y^2+3z^2)}dxdydz = ?$$

【解】

令 $x = \rho\sin\varphi\cos\theta$, $y = \dfrac{\rho\sin\varphi\sin\theta}{\sqrt{2}}$, $z = \dfrac{\rho\cos\varphi}{\sqrt{3}}$

$$\text{则 } dxdydz = \begin{Vmatrix} \dfrac{\partial x}{\partial \rho} & \dfrac{\partial x}{\partial \varphi} & \dfrac{\partial x}{\partial \theta} \\[2mm] \dfrac{\partial y}{\partial \rho} & \dfrac{\partial y}{\partial \varphi} & \dfrac{\partial y}{\partial \theta} \\[2mm] \dfrac{\partial z}{\partial \rho} & \dfrac{\partial z}{\partial \varphi} & \dfrac{\partial z}{\partial \theta} \end{Vmatrix} d\rho d\varphi d\theta$$

$$= \frac{1}{\sqrt{6}} \begin{Vmatrix} \sin\varphi\cos\theta & \rho\cos\varphi\cos\theta & -\rho\sin\varphi\sin\theta \\ \sin\varphi\sin\theta & \rho\cos\varphi\sin\theta & -\rho\sin\varphi\cos\theta \\ \cos\varphi & -\rho\sin\varphi & 0 \end{Vmatrix} d\rho d\varphi d\theta = \frac{\rho^2\sin\varphi \, d\rho d\varphi d\theta}{\sqrt{6}}$$

令 $R = \{(\rho,\varphi,\theta): 0 \le \rho \le 1, 0 \le \varphi \le \pi, 0 \le \theta \le 2\pi\}$

则 $\displaystyle\int_{-\infty}^{\infty}\int_{-\infty}^{\infty}\int_{-\infty}^{\infty} e^{-(x^2+2y^2+3z^2)}dxdydz = \frac{1}{\sqrt{6}}\iiint_R e^{-\rho^2}\rho^2\sin\varphi \, d\rho d\theta d\varphi$

$$= \frac{1}{\sqrt{6}}\int_0^\pi \int_0^{2\pi}\int_0^\infty e^{-\rho^2}\rho^2\sin\varphi \, d\rho d\theta d\varphi = \frac{1}{\sqrt{6}}\int_0^\pi \sin\varphi \int_0^{2\pi} d\theta \int_0^\infty e^{-\rho^2}\rho^2 d\rho$$

藉由 integration by parts

$$\int_0^\infty e^{-\rho^2}\rho^2 d\rho = \frac{-\rho}{2}e^{-\rho^2}\Big|_0^\infty + \frac{1}{2}\int_0^\infty e^{-\rho^2}d\rho = \frac{\sqrt{\pi}}{4}$$

$$\therefore \frac{1}{\sqrt{6}}\int_0^\pi \sin\varphi \int_0^{2\pi} d\theta \int_0^\infty e^{-\rho^2}\rho^2 d\rho = \frac{1}{\sqrt{6}}\cdot 2(2\pi)\frac{\sqrt{\pi}}{4} = \pi\sqrt{\frac{\pi}{6}}$$

Example 8.

    (1)假设 $V = \{(x,y,z): x^2 + y^2 + z^2 \le 1\}$，求 $\displaystyle\iiint_V \frac{dV}{9-(x^2+y^2+z^2)} = ?$

    (2)假设 $V = \{(x,y,z): 1 \le x^2 + y^2 + z^2 \le 4, x \ge 0, y \ge 0\}$，求 $\displaystyle\iiint_V (3x+2z)dV = ?$

【解】

(1)

令 $x = \rho\sin\varphi\cos\theta$ , $y = \rho\sin\varphi\sin\theta$ , $z = \rho\cos\varphi$

则 $dxdydz = \begin{Vmatrix} \dfrac{\partial x}{\partial\rho} & \dfrac{\partial x}{\partial\varphi} & \dfrac{\partial x}{\partial\theta} \\[2mm] \dfrac{\partial y}{\partial\rho} & \dfrac{\partial y}{\partial\varphi} & \dfrac{\partial y}{\partial\theta} \\[2mm] \dfrac{\partial z}{\partial\rho} & \dfrac{\partial z}{\partial\varphi} & \dfrac{\partial z}{\partial\theta} \end{Vmatrix} d\rho d\varphi d\theta$

$$= \begin{Vmatrix} \sin\varphi\cos\theta & \rho\cos\varphi\cos\theta & -\rho\sin\varphi\sin\theta \\ \sin\varphi\sin\theta & \rho\cos\varphi\sin\theta & -\rho\sin\varphi\cos\theta \\ \cos\varphi & -\rho\sin\varphi & 0 \end{Vmatrix} d\rho d\varphi d\theta = \rho^2\sin\varphi \, d\rho d\varphi d\theta$$

令 $R = \{(\rho,\varphi,\theta): 0 \le \rho \le 1, 0 \le \varphi \le \pi, 0 \le \theta \le 2\pi\}$

则 $\displaystyle\iiint_V \frac{dV}{9-(x^2+y^2+z^2)} = \iiint_R \frac{\rho^2 \sin\varphi}{9-\rho^2} d\rho d\theta d\varphi = \int_0^\pi \int_0^{2\pi} \int_0^1 \frac{\rho^2 \sin\varphi}{9-\rho^2} d\rho d\theta d\varphi$

$\displaystyle = \int_0^\pi \sin\varphi\, d\varphi \int_0^{2\pi} d\theta \int_0^1 \frac{\rho^2}{9-\rho^2} d\rho = 2(2\pi)\left(\frac{3}{2}\ln\left|\frac{3+\rho}{3-\rho}\right| - \rho\right)\Big|_0^1 = 4\pi\left(\frac{3\ln 2}{2} - 1\right)$

(2)

令 $x = \rho \sin\varphi \cos\theta$, $y = \rho \sin\varphi \sin\theta$, $z = \rho\cos\varphi$

则 $dxdydz = \begin{Vmatrix} \dfrac{\partial x}{\partial \rho} & \dfrac{\partial x}{\partial \varphi} & \dfrac{\partial x}{\partial \theta} \\[2mm] \dfrac{\partial y}{\partial \rho} & \dfrac{\partial y}{\partial \varphi} & \dfrac{\partial y}{\partial \theta} \\[2mm] \dfrac{\partial z}{\partial \rho} & \dfrac{\partial z}{\partial \varphi} & \dfrac{\partial z}{\partial \theta} \end{Vmatrix} d\rho d\varphi d\theta$

$= \begin{Vmatrix} \sin\varphi \cos\theta & \rho\cos\varphi \cos\theta & -\rho\sin\varphi \sin\theta \\ \sin\varphi \sin\theta & \rho\cos\varphi \sin\theta & -\rho\sin\varphi \cos\theta \\ \cos\varphi & -\rho\sin\varphi & 0 \end{Vmatrix} d\rho d\varphi d\theta = \rho^2 \sin\varphi\, d\rho d\varphi d\theta$

令 $R = \{(\rho, \varphi, \theta): 1 \le \rho \le 2, 0 \le \varphi \le \pi, 0 \le \theta \le \frac{\pi}{2}\}$

则 $\displaystyle\iiint_V (3x+2z)dV = \iiint_R (3\rho\sin\varphi\cos\theta + 2\rho\cos\varphi)\rho^2 \sin\varphi\, d\rho d\theta d\varphi$

$\displaystyle = \int_0^\pi \int_0^{\frac{\pi}{2}} \int_1^2 (3\rho\sin\varphi\cos\theta + 2\rho\cos\varphi)\rho^2 \sin\varphi\, d\rho d\theta d\varphi$

$\displaystyle = \int_0^\pi \int_0^{\frac{\pi}{2}} \int_1^2 (3\rho^3 \sin^2\varphi\cos\theta + 2\rho^3 \cos\varphi\sin\varphi)d\rho d\theta d\varphi$

$\displaystyle = \int_0^\pi \int_0^{\frac{\pi}{2}} \sin^2\varphi\cos\theta \frac{3\rho^4}{4}\Big|_1^2 + \cos\varphi\sin\varphi \frac{\rho^4}{2}\Big|_1^2 d\rho d\theta d\varphi$

$\displaystyle = \int_0^\pi \int_0^{\frac{\pi}{2}} \sin^2\varphi\cos\theta \left(\frac{45}{4}\right) + \cos\varphi\sin\varphi \left(\frac{15}{2}\right) d\theta d\varphi$

$\displaystyle = \int_0^\pi \sin^2\varphi\sin\theta\Big|_{\theta=0}^{\theta=\frac{\pi}{2}} \left(\frac{45}{4}\right) + \left(\frac{\pi}{2}\right)\cos\varphi\sin\varphi \left(\frac{15}{2}\right) d\varphi$

$\displaystyle = \int_0^\pi \sin^2\varphi \left(\frac{45}{4}\right) + \left(\frac{\pi}{2}\right)\cos\varphi\sin\varphi \left(\frac{15}{2}\right)d\varphi = \frac{45}{4}\int_0^\pi \frac{1-\cos 2\varphi}{2} d\varphi + \frac{15\pi}{4}\int_0^\pi \cos\varphi\sin\varphi\, d\varphi$

$\displaystyle = \frac{45\pi}{8}$

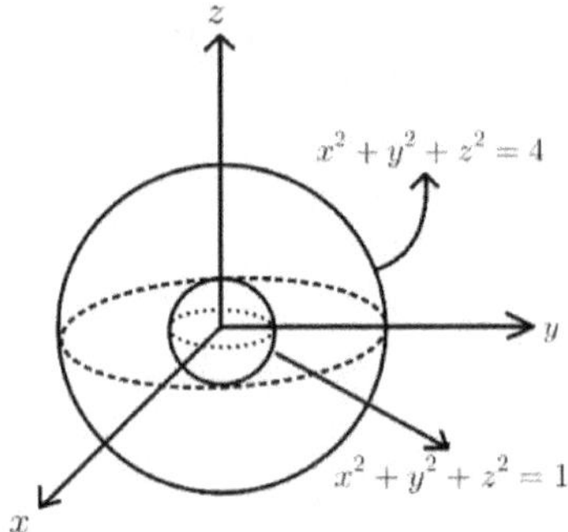


Example 9.

(1)假设 $V = \{(x, y, z): 1 \leq x^2 + y^2 + z^2 \leq 9, x \geq 0, y \geq 0, z \geq 0\}$，求 $\iiint_V z\, dV = ?$

(2)假设 $V = \{(x, y, z): 1 \leq x^2 + y^2 + z^2 \leq 3\}$，求 $\iiint_V \dfrac{\cos\sqrt{x^2 + y^2 + z^2}\, dxdydz}{x^2 + y^2 + z^2} = ?$

【解】

(1)

令 $x = \rho \sin\varphi \cos\theta$, $y = \rho \sin\varphi \sin\theta$, $z = \rho \cos\varphi$

则 $dxdydz = \begin{Vmatrix} \dfrac{\partial x}{\partial \rho} & \dfrac{\partial x}{\partial \varphi} & \dfrac{\partial x}{\partial \theta} \\[2mm] \dfrac{\partial y}{\partial \rho} & \dfrac{\partial y}{\partial \varphi} & \dfrac{\partial y}{\partial \theta} \\[2mm] \dfrac{\partial z}{\partial \rho} & \dfrac{\partial z}{\partial \varphi} & \dfrac{\partial z}{\partial \theta} \end{Vmatrix} d\rho d\varphi d\theta$

$= \begin{Vmatrix} \sin\varphi \cos\theta & \rho\cos\varphi \cos\theta & -\rho\sin\varphi \sin\theta \\ \sin\varphi \sin\theta & \rho\cos\varphi \sin\theta & -\rho\sin\varphi \cos\theta \\ \cos\varphi & -\rho\sin\varphi & 0 \end{Vmatrix} d\rho d\varphi d\theta = \rho^2 \sin\varphi\, d\rho d\varphi d\theta$

令 $R = \{(\rho, \varphi, \theta): 1 \leq \rho \leq 3, 0 \leq \varphi \leq \dfrac{\pi}{2}, 0 \leq \theta \leq \dfrac{\pi}{2}\}$

则 $\iiint_V z\, dV = \iiint_R \rho\cos\varphi\, \rho^2 \sin\varphi\, d\rho d\theta d\varphi = \int_0^{\frac{\pi}{2}} \int_0^{\frac{\pi}{2}} \int_1^3 \rho\cos\varphi\, \rho^2 \sin\varphi\, d\rho d\theta d\varphi$

$= \int_0^{\frac{\pi}{2}} \int_0^{\frac{\pi}{2}} \int_1^3 \rho^3 \cos\varphi \sin\varphi\, d\rho d\theta d\varphi = \int_0^{\frac{\pi}{2}} \cos\varphi \sin\varphi\, d\varphi \int_0^{\frac{\pi}{2}} d\theta \int_1^3 \rho^3 d\rho$

$$= \left( -\frac{\cos 2\varphi}{4} \Big|_0^{\frac{\pi}{2}} \right) \left( \frac{\pi}{2} \right) 20 = 5\pi$$

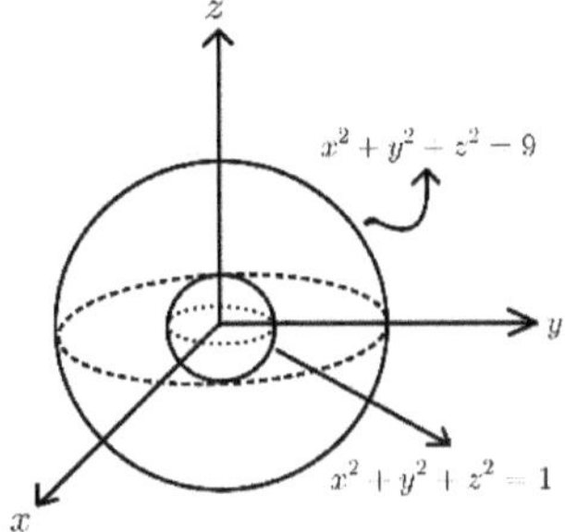


(2)

令 $x = \rho \sin\varphi \cos\theta$ , $y = \rho \sin\varphi \sin\theta$ , $\mathrm{z} = \rho \cos\varphi$

$$则\, dxdydz = \begin{Vmatrix} \dfrac{\partial x}{\partial \rho} & \dfrac{\partial x}{\partial \varphi} & \dfrac{\partial x}{\partial \theta} \\[2mm] \dfrac{\partial y}{\partial \rho} & \dfrac{\partial y}{\partial \varphi} & \dfrac{\partial y}{\partial \theta} \\[2mm] \dfrac{\partial z}{\partial \rho} & \dfrac{\partial z}{\partial \varphi} & \dfrac{\partial z}{\partial \theta} \end{Vmatrix} d\rho d\varphi d\theta$$

$$= \begin{Vmatrix} \sin\varphi \cos\theta & \rho\cos\varphi \cos\theta & -\rho\sin\varphi \sin\theta \\ \sin\varphi \sin\theta & \rho\cos\varphi \sin\theta & -\rho\sin\varphi \cos\theta \\ \cos\varphi & -\rho\sin\varphi & 0 \end{Vmatrix} d\rho d\varphi d\theta = \rho^2 \sin\varphi \, d\rho d\varphi d\theta$$

令 $R = \{(\rho, \varphi, \theta) : 1 \le \rho \le \sqrt{3}, 0 \le \varphi \le \pi, 0 \le \theta \le 2\pi\}$

$$则 \iiint_{1 \le x^2+y^2+z^2 \le 3} \frac{\cos\sqrt{x^2 + y^2 + z^2}\, dxdydz}{x^2 + y^2 + z^2}\, dV = \iiint_R \frac{\cos\rho}{\rho^2} \rho^2 \sin\varphi \, d\rho d\theta d\varphi$$

$$= \int_0^\pi \int_0^{2\pi} \int_1^{\sqrt{3}} \frac{\cos\rho}{\rho^2} \rho^2 \sin\varphi \, d\rho d\theta d\varphi = \int_0^\pi \sin\varphi \, d\varphi \int_0^{2\pi} d\theta \int_1^{\sqrt{3}} \cos\rho \, d\rho$$

$$= (-\cos\varphi|_0^\pi)(2\pi)\left( \sin\rho|_1^{\sqrt{3}} \right) = 4\pi(\sin\sqrt{3} - \sin 1)$$

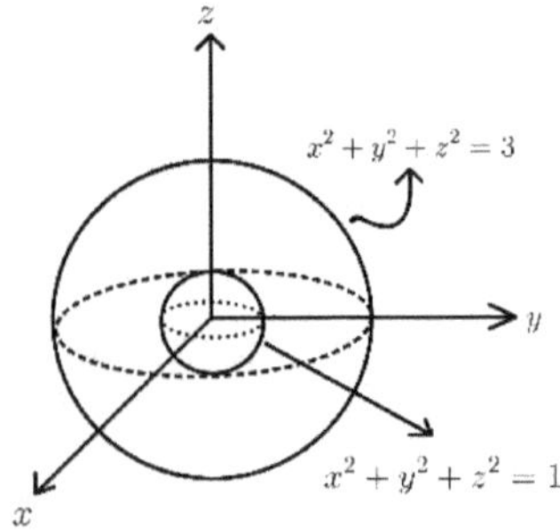


Example 10.

$$\text{假设 } V = \{(x, y, z): x^2 + y^2 + z^2 \le 1\}, \quad \text{求} \iiint_V \frac{z^2 dxdydz}{\sqrt{1 - x^2 - y^2 - z^2}} = ?$$

【解】

令 $x = \rho \sin \varphi \cos \theta$ , $y = \rho \sin \varphi \sin \theta$ , $z = \rho \cos \varphi$

$$\text{则 } dxdydz = \begin{Vmatrix} \dfrac{\partial x}{\partial \rho} & \dfrac{\partial x}{\partial \varphi} & \dfrac{\partial x}{\partial \theta} \\ \dfrac{\partial y}{\partial \rho} & \dfrac{\partial y}{\partial \varphi} & \dfrac{\partial y}{\partial \theta} \\ \dfrac{\partial z}{\partial \rho} & \dfrac{\partial z}{\partial \varphi} & \dfrac{\partial z}{\partial \theta} \end{Vmatrix} d\rho d\varphi d\theta$$

$$= \begin{Vmatrix} \sin \varphi \cos \theta & \rho\cos \varphi \cos \theta & -\rho\sin \varphi \sin \theta \\ \sin \varphi \sin \theta & \rho\cos \varphi \sin \theta & -\rho\sin \varphi \cos \theta \\ \cos \varphi & -\rho\sin \varphi & 0 \end{Vmatrix} d\rho d\varphi d\theta = \rho^2 \sin \varphi \, d\rho d\varphi d\theta$$

令 $R = \{(\rho, \varphi, \theta): 0 \le \rho \le 1, 0 \le \varphi \le \pi, 0 \le \theta \le 2\pi\}$

$$\text{则 } \iiint_V \frac{z^2 dxdydz}{\sqrt{1 - x^2 - y^2 - z^2}} = \iiint_R \frac{\rho^4 \sin \varphi \cos^2 \varphi}{\sqrt{1 - \rho^2}} d\rho d\theta d\varphi$$

$$= \int_0^\pi \int_0^{2\pi} \int_0^1 \frac{\rho^4 \sin \varphi \cos^2 \varphi}{\sqrt{1 - \rho^2}} d\rho d\theta d\varphi = \int_0^\pi \cos^2 \varphi \sin \varphi d\varphi \int_0^{2\pi} d\theta \int_0^1 \frac{\rho^4}{\sqrt{1 - \rho^2}} d\rho$$

$$= -\frac{\cos^3 \varphi}{3} \Big|_0^\pi (2\pi) \left( \int_0^1 \frac{\rho^4}{\sqrt{1 - \rho^2}} d\rho \right) = \frac{4\pi}{3} \left( \int_0^1 \frac{\rho^4}{\sqrt{1 - \rho^2}} d\rho \right)$$

令 $\rho = \sin \theta$ 则 $d\rho = \cos \theta \, d\theta$

$$\therefore \int_0^1 \frac{\rho^4}{\sqrt{1-\rho^2}}\,d\rho = \int_0^{\frac{\pi}{2}} \frac{\sin^4\theta\cos\theta}{\sqrt{1-\sin^2\theta}}\,d\theta = \int_0^{\frac{\pi}{2}} \sin^4\theta\,d\theta = \frac{1}{4}\left(\frac{3\theta}{2} - \sin 2\theta + \frac{\sin 4\theta}{8}\Big|_0^{\frac{\pi}{2}}\right) = \frac{3\pi}{16}$$

$$\therefore \iiint_V \frac{z^2\,dxdydz}{\sqrt{1-x^2-y^2-z^2}} = \frac{4\pi}{3}\left(\int_0^1 \frac{\rho^4}{\sqrt{1-\rho^2}}\,d\rho\right) = \frac{\pi^2}{4}$$

Example 11.

$$\text{假设 } V = \{(x,y,z): x^2+y^2+z^2 \le 1, z > 0\}, \quad \text{求} \iiint_V \frac{z\,dxdydz}{\sqrt{1-x^2-y^2-z^2}} = ?$$

【解】

令 $x = \rho\sin\varphi\cos\theta$, $y = \rho\sin\varphi\sin\theta$, $z = \rho\cos\varphi$

$$\text{则}\,dxdydz = \left\|\begin{vmatrix} \dfrac{\partial x}{\partial \rho} & \dfrac{\partial x}{\partial \varphi} & \dfrac{\partial x}{\partial \theta} \\[2mm] \dfrac{\partial y}{\partial \rho} & \dfrac{\partial y}{\partial \varphi} & \dfrac{\partial y}{\partial \theta} \\[2mm] \dfrac{\partial z}{\partial \rho} & \dfrac{\partial z}{\partial \varphi} & \dfrac{\partial z}{\partial \theta} \end{vmatrix}\right\| d\rho d\varphi d\theta$$

$$= \left\|\begin{vmatrix} \sin\varphi\cos\theta & \rho\cos\varphi\cos\theta & -\rho\sin\varphi\sin\theta \\ \sin\varphi\sin\theta & \rho\cos\varphi\sin\theta & -\rho\sin\varphi\cos\theta \\ \cos\varphi & -\rho\sin\varphi & 0 \end{vmatrix}\right\| d\rho d\varphi d\theta = \rho^2\sin\varphi\,d\rho d\varphi d\theta$$

令 $R = \{(\rho,\varphi,\theta): 0 \le \rho \le 1, 0 \le \varphi \le \dfrac{\pi}{2}, 0 \le \theta \le 2\pi\}$

$$\text{则} \iiint_V \frac{z\,dxdydz}{\sqrt{1-x^2-y^2-z^2}} = \iiint_R \frac{\rho^3\sin\varphi\cos\varphi}{\sqrt{1-\rho^2}}\,d\rho d\theta d\varphi$$

$$= \int_0^{\frac{\pi}{2}}\int_0^{2\pi}\int_0^1 \frac{\rho^3\sin\varphi\cos\varphi}{\sqrt{1-\rho^2}}\,d\rho d\theta d\varphi = \int_0^{\frac{\pi}{2}}\cos\varphi\sin\varphi d\varphi \int_0^{2\pi} d\theta \int_0^1 \frac{\rho^3}{\sqrt{1-\rho^2}}\,d\rho$$

$$= -\frac{\cos 2\varphi}{4}\Big|_0^{\frac{\pi}{2}} (2\pi)\left(\int_0^1 \frac{\rho^3}{\sqrt{1-\rho^2}}\,d\rho\right) = \pi\left(\int_0^1 \frac{\rho^3}{\sqrt{1-\rho^2}}\,d\rho\right)$$

令 $\rho = \sin\theta$ 则 $d\rho = \cos\theta\, d\theta$

$$\therefore \int_0^1 \frac{\rho^3}{\sqrt{1-\rho^2}}\, d\rho = \int_0^{\frac{\pi}{2}} \frac{\sin^3\theta\cos\theta}{\sqrt{1-\sin^2\theta}}\, d\theta = \int_0^{\frac{\pi}{2}} \sin^3\theta\, d\theta = -\cos\theta + \frac{\cos^3\theta}{3}\Big|_0^{\frac{\pi}{2}} = \frac{2}{3}$$

$$\therefore \iiint_V \frac{z\,dxdydz}{\sqrt{1-x^2-y^2-z^2}} = \frac{2\pi}{3}$$

Example 12.

假设 $V = \{(x,y,z): x^2 + y^2 + z^2 \le r^2\}$，求 $\iiint_V (x^2+y^2+z^2)^{\frac{3}{2}}\, dxdydz =?$

【解】

令 $x = \rho\sin\varphi\cos\theta\,, y = \rho\sin\varphi\sin\theta\,, z = \rho\cos\varphi$

$$\text{则}\ dxdydz = \begin{Vmatrix} \dfrac{\partial x}{\partial\rho} & \dfrac{\partial x}{\partial\varphi} & \dfrac{\partial x}{\partial\theta} \\[2mm] \dfrac{\partial y}{\partial\rho} & \dfrac{\partial y}{\partial\varphi} & \dfrac{\partial y}{\partial\theta} \\[2mm] \dfrac{\partial z}{\partial\rho} & \dfrac{\partial z}{\partial\varphi} & \dfrac{\partial z}{\partial\theta} \end{Vmatrix} d\rho d\varphi d\theta$$

$$= \begin{Vmatrix} \sin\varphi\cos\theta & \rho\cos\varphi\cos\theta & -\rho\sin\varphi\sin\theta \\ \sin\varphi\sin\theta & \rho\cos\varphi\sin\theta & -\rho\sin\varphi\cos\theta \\ \cos\varphi & -\rho\sin\varphi & 0 \end{Vmatrix} d\rho d\varphi d\theta = \rho^2\sin\varphi\, d\rho d\varphi d\theta$$

令 $R = \{(\rho,\varphi,\theta): 0 \le \rho \le r, 0 \le \varphi \le \pi, 0 \le \theta \le 2\pi\}$

$$\text{则}\ \iiint_V (x^2+y^2+z^2)^{\frac{3}{2}}\, dxdydz = \iiint_R (\rho^2)^{\frac{3}{2}}\rho^2\sin\varphi\, d\rho d\theta d\varphi$$

$$= \int_0^\pi \int_0^{2\pi} \int_0^r (\rho^2)^{\frac{3}{2}}\rho^2 \sin\varphi\, d\rho d\theta d\varphi = \int_0^\pi \sin\varphi d\varphi \int_0^{2\pi} d\theta \int_0^r \rho^5 d\rho = 2(2\pi)\left(\frac{r^6}{6}\right) = \frac{2\pi r^6}{3}$$

Example 13.

试求 $\displaystyle\int_{-3}^3 \int_0^{\sqrt{9-y^2}} \int_{-\sqrt{9-x^2-y^2}}^{\sqrt{9-x^2-y^2}} y^2\sqrt{x^2+y^2+z^2}\, dzdxdy =?$

【解】

令 $x = \rho \sin\varphi \cos\theta$, $y = \rho \sin\varphi \sin\theta$, $z = \rho \cos\varphi$

则 $dxdydz = \begin{Vmatrix} \dfrac{\partial x}{\partial \rho} & \dfrac{\partial x}{\partial \varphi} & \dfrac{\partial x}{\partial \theta} \\ \dfrac{\partial y}{\partial \rho} & \dfrac{\partial y}{\partial \varphi} & \dfrac{\partial y}{\partial \theta} \\ \dfrac{\partial z}{\partial \rho} & \dfrac{\partial z}{\partial \varphi} & \dfrac{\partial z}{\partial \theta} \end{Vmatrix} d\rho d\varphi d\theta$

$$= \begin{Vmatrix} \sin\varphi\cos\theta & \rho\cos\varphi\cos\theta & -\rho\sin\varphi\sin\theta \\ \sin\varphi\sin\theta & \rho\cos\varphi\sin\theta & -\rho\sin\varphi\cos\theta \\ \cos\varphi & -\rho\sin\varphi & 0 \end{Vmatrix} d\rho d\varphi d\theta = \rho^2 \sin\varphi \, d\rho d\varphi d\theta$$

令 $R = \{(\rho, \varphi, \theta): 0 \le \rho \le 3, 0 \le \varphi \le \pi, -\dfrac{\pi}{2} \le \theta \le \dfrac{\pi}{2}\}$

则 $\displaystyle\int_{-3}^{3} \int_{0}^{\sqrt{9-y^2}} \int_{-\sqrt{9-x^2-y^2}}^{\sqrt{9-x^2-y^2}} y^2\sqrt{x^2+y^2+z^2}\,dzdxdy = \iiint_R (\rho\sin\varphi\sin\theta)^2 \rho^3 \sin\varphi \, d\rho d\theta d\varphi$

$$= \int_{0}^{\pi} \int_{-\frac{\pi}{2}}^{\frac{\pi}{2}} \int_{0}^{3} (\rho\sin\varphi\sin\theta)^2 \rho^3 \sin\varphi \, d\rho d\theta d\varphi = \int_{0}^{\pi} \sin^3\varphi \, d\varphi \int_{-\frac{\pi}{2}}^{\frac{\pi}{2}} \sin^2\theta \, d\theta \int_{0}^{3} \rho^5 d\rho$$

$$= \frac{4}{3} \cdot \frac{\pi}{2} \cdot \frac{243}{2} = 81\pi$$

Example 14.

$$\text{试求} \int_{0}^{r} \int_{0}^{\sqrt{r^2-x^2}} \int_{0}^{\sqrt{r^2-x^2-y^2}} \frac{1}{x^2+y^2+z^2} \, dzdydx =?, \quad r > 0$$

【解】

令 $x = \rho \sin\varphi \cos\theta$, $y = \rho \sin\varphi \sin\theta$, $z = \rho \cos\varphi$

则 $dxdydz = \begin{Vmatrix} \dfrac{\partial x}{\partial \rho} & \dfrac{\partial x}{\partial \varphi} & \dfrac{\partial x}{\partial \theta} \\ \dfrac{\partial y}{\partial \rho} & \dfrac{\partial y}{\partial \varphi} & \dfrac{\partial y}{\partial \theta} \\ \dfrac{\partial z}{\partial \rho} & \dfrac{\partial z}{\partial \varphi} & \dfrac{\partial z}{\partial \theta} \end{Vmatrix} d\rho d\varphi d\theta$

$$= \begin{Vmatrix} \sin\varphi\cos\theta & \rho\cos\varphi\cos\theta & -\rho\sin\varphi\sin\theta \\ \sin\varphi\sin\theta & \rho\cos\varphi\sin\theta & -\rho\sin\varphi\cos\theta \\ \cos\varphi & -\rho\sin\varphi & 0 \end{Vmatrix} d\rho d\varphi d\theta = \rho^2\sin\varphi\, d\rho d\varphi d\theta$$

令 $R = \{(\rho,\varphi,\theta): 0 \le \rho \le r, 0 \le \varphi \le \dfrac{\pi}{2}, 0 \le \theta \le \dfrac{\pi}{2}\}$

则 $\displaystyle\int_0^r \int_0^{\sqrt{r^2-x^2}} \int_0^{\sqrt{r^2-x^2-y^2}} \frac{1}{x^2+y^2+z^2}\, dzdydx = \iiint_R \frac{1}{\rho^2}\rho^2\sin\varphi\, d\rho d\theta d\varphi$

$\displaystyle = \int_0^{\frac{\pi}{2}} \int_0^{\frac{\pi}{2}} \int_0^r \frac{1}{\rho^2}\rho^2\sin\varphi\, d\rho d\theta d\varphi = \int_0^{\frac{\pi}{2}} \sin\varphi\, d\varphi \int_0^{\frac{\pi}{2}} d\theta \int_0^r 1 d\rho = \frac{\pi r}{2}$

Example 15.

假设 $V = \{(x,y,z): x^2+y^2+z^2 \le 1, x \ge 0, y \ge 0, z \ge 0\}$, 求 $\displaystyle\iiint_V xyz\, dxdydz =?$

【解】

令 $x = \rho\sin\varphi\cos\theta$, $y = \rho\sin\varphi\sin\theta$, $z = \rho\cos\varphi$

则 $dxdydz = \begin{Vmatrix} \dfrac{\partial x}{\partial\rho} & \dfrac{\partial x}{\partial\varphi} & \dfrac{\partial x}{\partial\theta} \\ \dfrac{\partial y}{\partial\rho} & \dfrac{\partial y}{\partial\varphi} & \dfrac{\partial y}{\partial\theta} \\ \dfrac{\partial z}{\partial\rho} & \dfrac{\partial z}{\partial\varphi} & \dfrac{\partial z}{\partial\theta} \end{Vmatrix} d\rho d\varphi d\theta$

$$= \begin{Vmatrix} \sin\varphi\cos\theta & \rho\cos\varphi\cos\theta & -\rho\sin\varphi\sin\theta \\ \sin\varphi\sin\theta & \rho\cos\varphi\sin\theta & -\rho\sin\varphi\cos\theta \\ \cos\varphi & -\rho\sin\varphi & 0 \end{Vmatrix} d\rho d\varphi d\theta = \rho^2\sin\varphi\, d\rho d\varphi d\theta$$

令 $R = \{(\rho,\varphi,\theta): 0 \le \rho \le 1, 0 \le \varphi \le \dfrac{\pi}{2}, 0 \le \theta \le \dfrac{\pi}{2}\}$

则 $\displaystyle\iiint_V xyz\, dxdydz = \iiint_R (\rho\sin\varphi\cos\theta \cdot \rho\sin\varphi\sin\theta \cdot \rho\cos\varphi)\rho^2\sin\varphi\, d\rho d\theta d\varphi$

$\displaystyle = \int_0^{\frac{\pi}{2}} \int_0^{\frac{\pi}{2}} \int_0^1 (\rho\sin\varphi\cos\theta \cdot \rho\sin\varphi\sin\theta \cdot \rho\cos\varphi)\rho^2\sin\varphi\, d\rho d\theta d\varphi$

$\displaystyle = \left(\frac{\rho^6}{6}\bigg|_0^1\right) \int_0^{\frac{\pi}{2}} \sin^3\varphi\cos\varphi\, d\varphi \int_0^{\frac{\pi}{2}} \cos\theta\sin\theta\, d\theta$

$$\because \int_0^{\frac{\pi}{2}} \cos\theta \sin\theta \, d\theta = \frac{1}{2}\int_0^{\frac{\pi}{2}} \sin 2\theta \, d\theta = \frac{-1}{4}(\cos 2\theta)\Big|_0^{\frac{\pi}{2}} = \frac{1}{2}$$

$$\int_0^{\frac{\pi}{2}} \sin^3\varphi \cos\varphi \, d\varphi = \frac{\sin^4\varphi}{4}\Big|_0^{\frac{\pi}{2}} = \frac{1}{4}$$

$$\therefore \iiint_V xyz \, dxdydz = \frac{1}{6}\cdot\frac{1}{4}\cdot\frac{1}{2} = \frac{1}{48}$$

Example 16.

$$\text{假设 } V = \{(x,y,z): x^2 + y^2 + z^2 \leq 1\}, \quad \text{求} \iiint_V y^2 \, dV = ?$$

【解】

令 $x = \rho\sin\varphi\cos\theta$, $y = \rho\sin\varphi\sin\theta$, $z = \rho\cos\varphi$

$$\text{则 } dxdydz = \left\|\begin{vmatrix} \dfrac{\partial x}{\partial\rho} & \dfrac{\partial x}{\partial\varphi} & \dfrac{\partial x}{\partial\theta} \\[2mm] \dfrac{\partial y}{\partial\rho} & \dfrac{\partial y}{\partial\varphi} & \dfrac{\partial y}{\partial\theta} \\[2mm] \dfrac{\partial z}{\partial\rho} & \dfrac{\partial z}{\partial\varphi} & \dfrac{\partial z}{\partial\theta} \end{vmatrix}\right\| d\rho d\varphi d\theta$$

$$= \left\|\begin{vmatrix} \sin\varphi\cos\theta & \rho\cos\varphi\cos\theta & -\rho\sin\varphi\sin\theta \\ \sin\varphi\sin\theta & \rho\cos\varphi\sin\theta & -\rho\sin\varphi\cos\theta \\ \cos\varphi & -\rho\sin\varphi & 0 \end{vmatrix}\right\| d\rho d\varphi d\theta = \rho^2\sin\varphi \, d\rho d\varphi d\theta$$

令 $R = \{(\rho,\varphi,\theta): 0 \leq \rho \leq 1, 0 \leq \varphi \leq \pi, 0 \leq \theta \leq 2\pi\}$

$$\text{则} \iiint_V y^2 \, dV = \iiint_R \rho^4\sin^3\varphi\sin^2\theta \, d\rho d\theta d\varphi = \int_0^\pi \int_0^{2\pi} \int_0^1 \rho^4\sin^3\varphi\sin^2\theta \, d\rho d\theta d\varphi$$

$$= \left(\frac{\rho^5}{5}\Big|_0^1\right)\int_0^\pi \sin^3\varphi \, d\varphi \int_0^{2\pi} \sin^2\theta \, d\theta = \frac{1}{5}\cdot\frac{4}{3}\cdot\pi = \frac{4\pi}{15}$$

Example 17.

求封闭曲面$(x^2 + y^2 + z^2)^2 = 2z(x^2 + y^2)$所围区域体积

【解】

令$x = \rho \sin\varphi \cos\theta$ , $y = \rho \sin\varphi \sin\theta$ , $z = \rho \cos\varphi$

则 $dxdydz = \begin{Vmatrix} \dfrac{\partial x}{\partial \rho} & \dfrac{\partial x}{\partial \varphi} & \dfrac{\partial x}{\partial \theta} \\[2mm] \dfrac{\partial y}{\partial \rho} & \dfrac{\partial y}{\partial \varphi} & \dfrac{\partial y}{\partial \theta} \\[2mm] \dfrac{\partial z}{\partial \rho} & \dfrac{\partial z}{\partial \varphi} & \dfrac{\partial z}{\partial \theta} \end{Vmatrix} d\rho d\varphi d\theta$

$= \begin{Vmatrix} \sin\varphi \cos\theta & \rho\cos\varphi \cos\theta & -\rho\sin\varphi \sin\theta \\ \sin\varphi \sin\theta & \rho\cos\varphi \sin\theta & -\rho\sin\varphi \cos\theta \\ \cos\varphi & -\rho\sin\varphi & 0 \end{Vmatrix} d\rho d\varphi d\theta = \rho^2 \sin\varphi \, d\rho d\varphi d\theta$

令 $R = \{(\rho,\varphi,\theta): 0 \le \rho \le 2\cos\varphi \sin^2\varphi, 0 \le \varphi \le \dfrac{\pi}{2}, 0 \le \theta \le 2\pi\}$

则 $V = \iiint_R \; d\rho d\theta d\varphi = \int_0^{2\pi} \int_0^{\frac{\pi}{2}} \int_0^{2\cos\varphi \sin^2\varphi} \rho^2 \sin\varphi \, d\rho d\varphi d\theta$

$= 2\pi \int_0^{\frac{\pi}{2}} \dfrac{\rho^3}{3} \Big|_0^{2\cos\varphi \sin^2\varphi} \sin\varphi \, d\varphi = \dfrac{16\pi}{3} \int_0^{\frac{\pi}{2}} \cos^3\varphi \sin^7\varphi \, d\varphi$

$= \dfrac{16\pi}{3} \int_0^{\frac{\pi}{2}} \cos\varphi \, (1 - \sin^2\varphi) \sin^7\varphi \, d\varphi = \dfrac{16\pi}{3} \left( \dfrac{\sin^8\varphi}{8} - \dfrac{\sin^{10}\varphi}{10} \right) \Big|_0^{\frac{\pi}{2}} = \dfrac{2\pi}{15}$

Example 18.

    (1)计算球体$x^2 + y^2 + z^2 \le 100$ 被平面 $z = 4$ 切除后剩的较小块体积

    (2)计算球体$x^2 + y^2 + z^2 \le 100$ 被平面 $z = 2$ 切除后剩的较小块体积

【解】

(1)

令$x = \rho \sin\varphi \cos\theta$ , $y = \rho \sin\varphi \sin\theta$ , $z = \rho \cos\varphi$

$$\text{则 } dxdydz = \begin{Vmatrix} \dfrac{\partial x}{\partial \rho} & \dfrac{\partial x}{\partial \varphi} & \dfrac{\partial x}{\partial \theta} \\[2mm] \dfrac{\partial y}{\partial \rho} & \dfrac{\partial y}{\partial \varphi} & \dfrac{\partial y}{\partial \theta} \\[2mm] \dfrac{\partial z}{\partial \rho} & \dfrac{\partial z}{\partial \varphi} & \dfrac{\partial z}{\partial \theta} \end{Vmatrix} d\rho d\varphi d\theta$$

$$= \begin{Vmatrix} \sin\varphi\cos\theta & \rho\cos\varphi\cos\theta & -\rho\sin\varphi\sin\theta \\ \sin\varphi\sin\theta & \rho\cos\varphi\sin\theta & -\rho\sin\varphi\cos\theta \\ \cos\varphi & -\rho\sin\varphi & 0 \end{Vmatrix} d\rho d\varphi d\theta = \rho^2 \sin\varphi\, d\rho d\varphi d\theta$$

$$\text{令 } R = \left\{(\rho,\varphi,\theta): \frac{4}{\cos\varphi} \le \rho \le 10, 0 \le \varphi \le \cos^{-1}\frac{2}{5}, 0 \le \theta \le 2\pi\right\}$$

$$\text{则 } V = \iiint_R \rho^2 \sin\varphi\, d\rho d\theta d\varphi = \int_0^{\cos^{-1}\frac{2}{5}} \int_0^{2\pi} \int_{\frac{4}{\cos\varphi}}^{10} \rho^2 \sin\varphi\, d\rho d\theta d\varphi$$

$$= \int_0^{\cos^{-1}\frac{2}{5}} \int_0^{2\pi} \frac{\rho^3}{3}\Big|_{\frac{4}{\cos\varphi}}^{10} \sin\varphi\, d\theta d\varphi = \int_0^{\cos^{-1}\frac{2}{5}} \frac{1}{3}\left(1000 - \left(\frac{4}{\cos\varphi}\right)^3\right) \sin\varphi\, d\varphi \int_0^{2\pi} d\theta$$

$$= \frac{2\pi}{3}\left(-1000 \cdot \cos\varphi\Big|_0^{\cos^{-1}\frac{2}{5}} - \frac{4^3}{2} \cdot \cos^{-2}\varphi\Big|_0^{\cos^{-1}\frac{2}{5}}\right)$$

$$= \frac{2\pi}{3}\left(-1000 \cdot \left(\frac{2}{5} - 1\right) - \frac{4^3}{2} \cdot \left(\frac{25}{4} - 1\right)\right) = \frac{2\pi}{3}\left(1000 \cdot \left(\frac{3}{5}\right) - \frac{16}{2} \cdot 21\right) = 288\pi$$

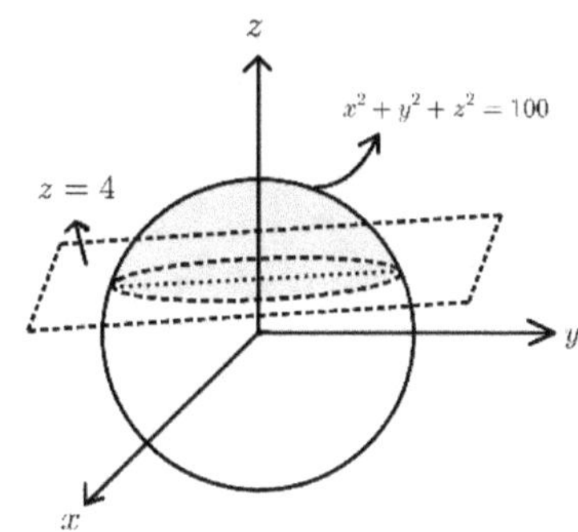


(2)

$$\text{令 } x = \rho\sin\varphi\cos\theta,\ y = \rho\sin\varphi\sin\theta,\ z = \rho\cos\varphi$$

$$\text{则 } dxdydz = \begin{Vmatrix} \dfrac{\partial x}{\partial \rho} & \dfrac{\partial x}{\partial \varphi} & \dfrac{\partial x}{\partial \theta} \\[2mm] \dfrac{\partial y}{\partial \rho} & \dfrac{\partial y}{\partial \varphi} & \dfrac{\partial y}{\partial \theta} \\[2mm] \dfrac{\partial z}{\partial \rho} & \dfrac{\partial z}{\partial \varphi} & \dfrac{\partial z}{\partial \theta} \end{Vmatrix} d\rho d\varphi d\theta$$

$$= \begin{Vmatrix} \sin\varphi\cos\theta & \rho\cos\varphi\cos\theta & -\rho\sin\varphi\sin\theta \\ \sin\varphi\sin\theta & \rho\cos\varphi\sin\theta & -\rho\sin\varphi\cos\theta \\ \cos\varphi & -\rho\sin\varphi & 0 \end{Vmatrix} d\rho d\varphi d\theta = \rho^2 \sin\varphi\, d\rho d\varphi d\theta$$

$$\text{令 } R = \left\{ (\rho, \varphi, \theta) : \frac{2}{\cos\varphi} \le \rho \le 10, 0 \le \varphi \le \cos^{-1}\frac{1}{5}, 0 \le \theta \le 2\pi \right\}$$

$$\text{则 } V = \iiint_R \rho^2 \sin\varphi\, d\rho d\theta d\varphi = \int_0^{\cos^{-1}\frac{1}{5}} \int_0^{2\pi} \int_{\frac{2}{\cos\varphi}}^{10} \rho^2 \sin\varphi\, d\rho d\theta d\varphi$$

$$= \int_0^{\cos^{-1}\frac{1}{5}} \int_0^{2\pi} \frac{\rho^3}{3}\Big|_{\frac{2}{\cos\varphi}}^{10} \sin\varphi\, d\theta d\varphi = \int_0^{\cos^{-1}\frac{1}{5}} \frac{1}{3}\left(1000 - \left(\frac{2}{\cos\varphi}\right)^3\right) \sin\varphi\, d\varphi \int_0^{2\pi} d\theta$$

$$= \frac{2\pi}{3}\left(-1000 \cdot \cos\varphi\Big|_0^{\cos^{-1}\frac{1}{5}} - \frac{2^3}{2} \cdot \cos^{-2}\varphi\Big|_0^{\cos^{-1}\frac{1}{5}}\right)$$

$$= \frac{2\pi}{3}\left(-1000 \cdot \left(\frac{1}{5} - 1\right) - \frac{2^3}{2} \cdot (25 - 1)\right) = \frac{2\pi}{3}\left(1000 \cdot \left(\frac{4}{5}\right) - 96\right) = \frac{1408\pi}{3}$$

Example 19.

(1)计算球体 $x^2 + y^2 + z^2 \le r^2$ 与 $z \ge \sqrt{3x^2 + 3y^2}$ 交集区域体积

(2)计算球体 $x^2 + y^2 + z^2 \le 1$ 与 $\sqrt{3}z \ge \sqrt{x^2 + y^2}$ 交集区域体积

【解】

(1)

令 $x = \rho\sin\varphi\cos\theta$, $y = \rho\sin\varphi\sin\theta$, $z = \rho\cos\varphi$

则 $dxdydz = \begin{Vmatrix} \dfrac{\partial x}{\partial \rho} & \dfrac{\partial x}{\partial \varphi} & \dfrac{\partial x}{\partial \theta} \\[2mm] \dfrac{\partial y}{\partial \rho} & \dfrac{\partial y}{\partial \varphi} & \dfrac{\partial y}{\partial \theta} \\[2mm] \dfrac{\partial z}{\partial \rho} & \dfrac{\partial z}{\partial \varphi} & \dfrac{\partial z}{\partial \theta} \end{Vmatrix} d\rho d\varphi d\theta$

$= \begin{Vmatrix} \sin\varphi\cos\theta & \rho\cos\varphi\cos\theta & -\rho\sin\varphi\sin\theta \\ \sin\varphi\sin\theta & \rho\cos\varphi\sin\theta & -\rho\sin\varphi\cos\theta \\ \cos\varphi & -\rho\sin\varphi & 0 \end{Vmatrix} d\rho d\varphi d\theta = \rho^2 \sin\varphi \, d\rho d\varphi d\theta$

令 $R = \{(\rho,\varphi,\theta) : 0 \le \rho \le r, 0 \le \varphi \le \dfrac{\pi}{6}, 0 \le \theta \le 2\pi\}$

则 $V = \iiint_R \rho^2 \sin\varphi \, d\rho d\theta d\varphi = \int_0^{\frac{\pi}{6}} \int_0^{2\pi} \int_0^r \rho^2 \sin\varphi \, d\rho d\theta d\varphi = \int_0^{\frac{\pi}{6}} \sin\varphi \, d\varphi \int_0^{2\pi} d\theta \int_0^r \rho^2 d\rho$

$= \left(\dfrac{1}{3} - \dfrac{\sqrt{3}}{6}\right) 2\pi r^3$

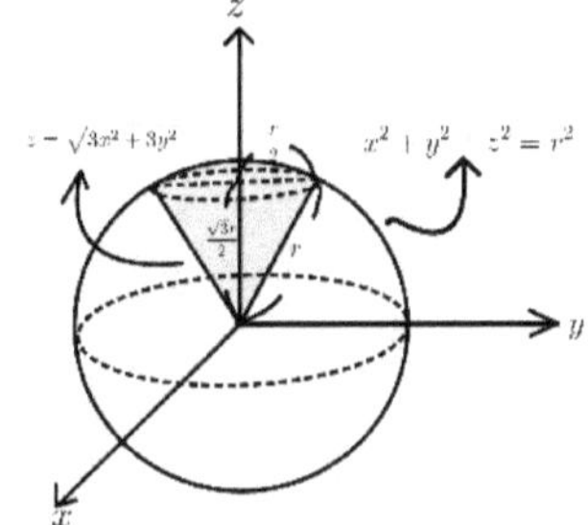


(2)

令 $x = \rho\sin\varphi\cos\theta$, $y = \rho\sin\varphi\sin\theta$, $z = \rho\cos\varphi$

则 $dxdydz = \begin{Vmatrix} \dfrac{\partial x}{\partial \rho} & \dfrac{\partial x}{\partial \varphi} & \dfrac{\partial x}{\partial \theta} \\[2mm] \dfrac{\partial y}{\partial \rho} & \dfrac{\partial y}{\partial \varphi} & \dfrac{\partial y}{\partial \theta} \\[2mm] \dfrac{\partial z}{\partial \rho} & \dfrac{\partial z}{\partial \varphi} & \dfrac{\partial z}{\partial \theta} \end{Vmatrix} d\rho d\varphi d\theta$

$= \begin{Vmatrix} \sin\varphi\cos\theta & \rho\cos\varphi\cos\theta & -\rho\sin\varphi\sin\theta \\ \sin\varphi\sin\theta & \rho\cos\varphi\sin\theta & -\rho\sin\varphi\cos\theta \\ \cos\varphi & -\rho\sin\varphi & 0 \end{Vmatrix} d\rho d\varphi d\theta = \rho^2 \sin\varphi \, d\rho d\varphi d\theta$

令 $R = \{(\rho, \varphi, \theta): 0 \le \rho \le 1, 0 \le \varphi \le \dfrac{\pi}{3}, 0 \le \theta \le 2\pi\}$

则 $V = \iiint_R \rho^2 \sin\varphi\, d\rho d\theta d\varphi = \displaystyle\int_0^{\frac{\pi}{3}} \int_0^{2\pi} \int_0^1 \rho^2 \sin\varphi\, d\rho d\theta d\varphi = \int_0^{\frac{\pi}{3}} \sin\varphi\, d\varphi \int_0^{2\pi} d\theta \int_0^1 \rho^2 d\rho$

$= \dfrac{\pi}{3}$

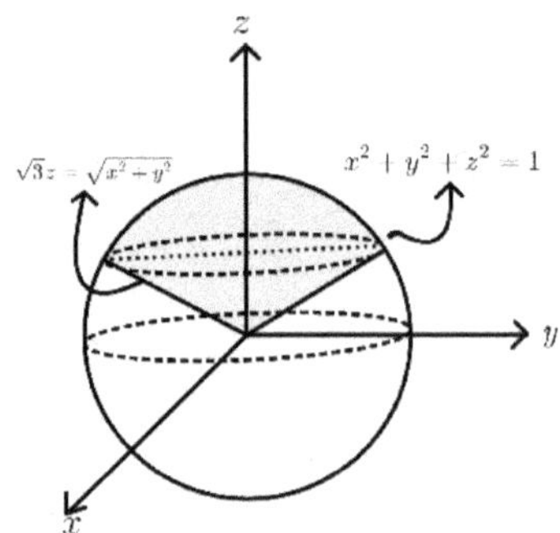


Example 20.

(1)计算球体$x^2 + y^2 + (z-1)^2 \le 1$ 与 $\sqrt{3}\,z \ge \sqrt{x^2+y^2}$ 交集区域体积

(2)计算球体$x^2 + y^2 + z^2 \le z$ 与 $\sqrt{3}\,z \ge \sqrt{x^2+y^2}$ 交集区域体积

【解】

(1)

令$x = \rho \sin\varphi \cos\theta\,, y = \rho \sin\varphi \sin\theta\,, \mathrm{z} = \rho \cos\varphi$

则$dxdydz = \begin{Vmatrix} \dfrac{\partial x}{\partial \rho} & \dfrac{\partial x}{\partial \varphi} & \dfrac{\partial x}{\partial \theta} \\[2mm] \dfrac{\partial y}{\partial \rho} & \dfrac{\partial y}{\partial \varphi} & \dfrac{\partial y}{\partial \theta} \\[2mm] \dfrac{\partial z}{\partial \rho} & \dfrac{\partial z}{\partial \varphi} & \dfrac{\partial z}{\partial \theta} \end{Vmatrix} d\rho d\varphi d\theta$

$= \begin{Vmatrix} \sin\varphi\cos\theta & \rho\cos\varphi\cos\theta & -\rho\sin\varphi\sin\theta \\ \sin\varphi\sin\theta & \rho\cos\varphi\sin\theta & -\rho\sin\varphi\cos\theta \\ \cos\varphi & -\rho\sin\varphi & 0 \end{Vmatrix} d\rho d\varphi d\theta = \rho^2 \sin\varphi\, d\rho d\varphi d\theta$

令 $R = \{(\rho, \varphi, \theta): 0 \le \rho \le 2\cos\varphi\,, 0 \le \varphi \le \dfrac{\pi}{3}, 0 \le \theta \le 2\pi\}$

则 $V = \iiint_R \rho^2 \sin\varphi \, d\rho d\theta d\varphi = \int_0^{\frac{\pi}{3}} \int_0^{2\pi} \int_0^{2\cos\varphi} \rho^2 \sin\varphi \, d\rho d\theta d\varphi$

$= 2\pi \int_0^{\frac{\pi}{3}} \int_0^{2\cos\varphi} \rho^2 \sin\varphi \, d\rho d\varphi = 2\pi \int_0^{\frac{\pi}{3}} \sin\varphi \int_0^{2\cos\varphi} \rho^2 \, d\rho d\varphi$

$\because \int_0^{2\cos\varphi} \rho^2 \, d\rho = \left.\frac{\rho^3}{3}\right|_0^{2\cos\varphi} = \frac{8\cos^3\varphi}{3}$

$\therefore 2\pi \int_0^{\frac{\pi}{3}} \sin\varphi \int_0^{2\cos\varphi} \rho^2 \, d\rho d\varphi = \frac{16\pi}{3} \int_0^{\frac{\pi}{3}} \sin\varphi \cos^3\varphi \, d\varphi = \frac{-16\pi}{3} \cdot \left.\frac{\cos^4\varphi}{4}\right|_0^{\frac{\pi}{3}}$

$= \frac{-4\pi}{3} \left(\frac{1}{16} - 1\right) = \frac{5\pi}{4}$

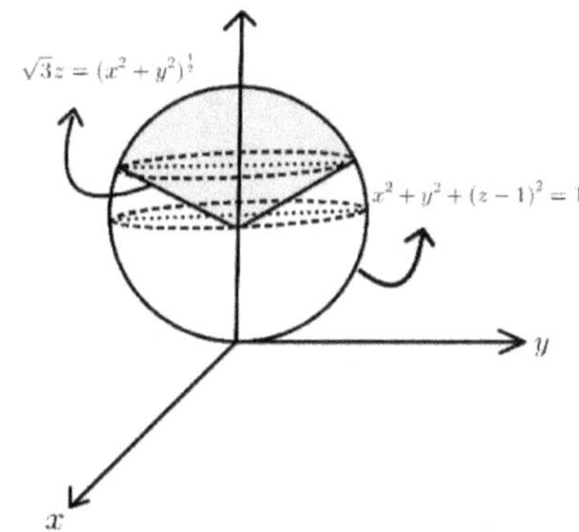


(2)

令 $x = \rho\sin\varphi\cos\theta \, , y = \rho\sin\varphi\sin\theta \, , z = \rho\cos\varphi$

则 $dxdydz = \begin{Vmatrix} \dfrac{\partial x}{\partial \rho} & \dfrac{\partial x}{\partial \varphi} & \dfrac{\partial x}{\partial \theta} \\[6pt] \dfrac{\partial y}{\partial \rho} & \dfrac{\partial y}{\partial \varphi} & \dfrac{\partial y}{\partial \theta} \\[6pt] \dfrac{\partial z}{\partial \rho} & \dfrac{\partial z}{\partial \varphi} & \dfrac{\partial z}{\partial \theta} \end{Vmatrix} d\rho d\varphi d\theta$

$= \begin{Vmatrix} \sin\varphi\cos\theta & \rho\cos\varphi\cos\theta & -\rho\sin\varphi\sin\theta \\ \sin\varphi\sin\theta & \rho\cos\varphi\sin\theta & -\rho\sin\varphi\cos\theta \\ \cos\varphi & -\rho\sin\varphi & 0 \end{Vmatrix} d\rho d\varphi d\theta = \rho^2 \sin\varphi \, d\rho d\varphi d\theta$

令 $R = \left\{ (\rho, \varphi, \theta): 0 \leq \rho \leq \cos\varphi, 0 \leq \varphi \leq \frac{\pi}{3}, 0 \leq \theta \leq 2\pi \right\}$

则 $V = \iiint_R \rho^2 \sin\varphi \, d\rho d\theta d\varphi = \int_0^{\frac{\pi}{3}} \int_0^{2\pi} \int_0^{\cos\varphi} \rho^2 \sin\varphi \, d\rho d\theta d\varphi$

$= 2\pi \int_0^{\frac{\pi}{3}} \int_0^{\cos\varphi} \rho^2 \sin\varphi \, d\rho d\varphi = 2\pi \int_0^{\frac{\pi}{3}} \sin\varphi \int_0^{\cos\varphi} \rho^2 \, d\rho d\varphi$

$\because \int_0^{\cos\varphi} \rho^2 \, d\rho = \left.\frac{\rho^3}{3}\right|_0^{\cos\varphi} = \frac{\cos^3\varphi}{3}$

$\therefore 2\pi \int_0^{\frac{\pi}{3}} \sin\varphi \int_0^{\cos\varphi} \rho^2 \, d\rho d\varphi = \frac{2\pi}{3} \int_0^{\frac{\pi}{3}} \sin\varphi \cos^3\varphi \, d\varphi = -\frac{2\pi}{3} \cdot \left.\frac{\cos^4\varphi}{4}\right|_0^{\frac{\pi}{3}} = -\frac{\pi}{6}\left(\frac{1}{16} - 1\right)$

$= \frac{5\pi}{32}$

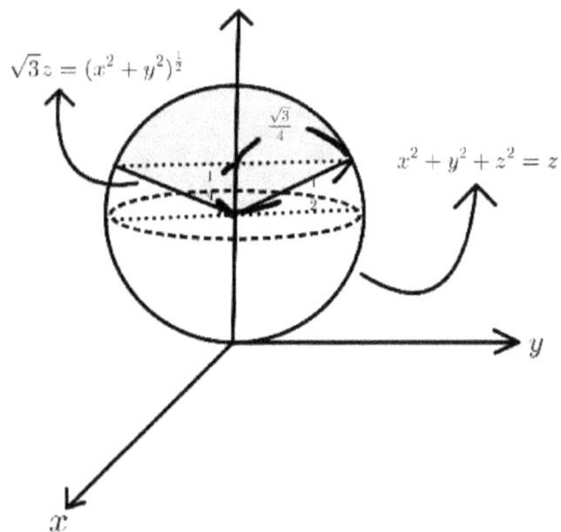


Example 21.

(1)试利用变数变换求 $\left(\dfrac{x^2}{a^2} + \dfrac{y^2}{b^2} + \dfrac{z^2}{c^2}\right)^{\frac{3}{2}} = \dfrac{x^2}{a^2} + \dfrac{y^2}{b^2}$ 所围成区域的体积

(2)假设 V 为 $\left(\dfrac{x^2}{a^2} + \dfrac{y^2}{b^2} + \dfrac{z^2}{c^2}\right)^{\frac{3}{2}} = \dfrac{x^2}{a^2} + \dfrac{y^2}{b^2}$ 与 $z > 0$ 所围区域, 试求 $\iiint_V zdxdydz =?$

【解】

(1)

令 $x = a\rho \sin\varphi \cos\theta \,, y = b\rho \sin\varphi \sin\theta \,, \mathrm{z} = c\rho \cos\varphi$

则 $dxdydz = \begin{Vmatrix} \dfrac{\partial x}{\partial \rho} & \dfrac{\partial x}{\partial \varphi} & \dfrac{\partial x}{\partial \theta} \\[2mm] \dfrac{\partial y}{\partial \rho} & \dfrac{\partial y}{\partial \varphi} & \dfrac{\partial y}{\partial \theta} \\[2mm] \dfrac{\partial z}{\partial \rho} & \dfrac{\partial z}{\partial \varphi} & \dfrac{\partial z}{\partial \theta} \end{Vmatrix} d\rho d\varphi d\theta$

$= abc \begin{Vmatrix} \sin\varphi\cos\theta & \rho\cos\varphi\cos\theta & -\rho\sin\varphi\sin\theta \\ \sin\varphi\sin\theta & \rho\cos\varphi\sin\theta & -\rho\sin\varphi\cos\theta \\ \cos\varphi & -\rho\sin\varphi & 0 \end{Vmatrix} d\rho d\varphi d\theta = abc\rho^2 \sin\varphi\, d\rho d\varphi d\theta$

令 $R = \{(\rho, \varphi, \theta): 0 \le \rho \le \sin^2\varphi, 0 \le \varphi \le \pi, 0 \le \theta \le 2\pi\}$

则 $V = abc \iiint_R \rho^2 \sin\varphi\, d\rho d\theta d\varphi = abc \int_0^\pi \int_0^{2\pi} \int_0^{\sin^2\varphi} \rho^2 \sin\varphi\, d\rho d\theta d\varphi$

$= abc \int_0^\pi \int_0^{2\pi} \dfrac{\rho^3}{3}\Big|_0^{\sin^2\varphi} \sin\varphi\, d\theta d\varphi = \dfrac{abc}{3} \cdot 2\pi \cdot \int_0^\pi \sin^7\varphi\, d\varphi = \dfrac{64abc\pi}{105}$

(2)

令 $x = a\rho\sin\varphi\cos\theta$ , $y = b\rho\sin\varphi\sin\theta$ , $z = c\rho\cos\varphi$

则 $dxdydz = \begin{Vmatrix} \dfrac{\partial x}{\partial \rho} & \dfrac{\partial x}{\partial \varphi} & \dfrac{\partial x}{\partial \theta} \\[2mm] \dfrac{\partial y}{\partial \rho} & \dfrac{\partial y}{\partial \varphi} & \dfrac{\partial y}{\partial \theta} \\[2mm] \dfrac{\partial z}{\partial \rho} & \dfrac{\partial z}{\partial \varphi} & \dfrac{\partial z}{\partial \theta} \end{Vmatrix} d\rho d\varphi d\theta$

$= abc \begin{Vmatrix} \sin\varphi\cos\theta & \rho\cos\varphi\cos\theta & -\rho\sin\varphi\sin\theta \\ \sin\varphi\sin\theta & \rho\cos\varphi\sin\theta & -\rho\sin\varphi\cos\theta \\ \cos\varphi & -\rho\sin\varphi & 0 \end{Vmatrix} d\rho d\varphi d\theta$

$= abc\rho^2 \sin\varphi\, d\rho d\varphi d\theta$

令 $R = \{(\rho, \varphi, \theta): 0 \le \rho \le \sin^2\varphi, 0 \le \varphi \le \dfrac{\pi}{2}, 0 \le \theta \le 2\pi\}$

则 $\iiint_V z\, dxdydz = abc^2 \iiint_R \rho^3 \sin\varphi\cos\varphi\, d\rho d\theta d\varphi$

$= abc^2 \int_0^{\frac{\pi}{2}} \int_0^{2\pi} \int_0^{\sin^2\varphi} \rho^3 \sin\varphi\cos\varphi\, d\rho d\theta d\varphi = abc^2 \int_0^{\frac{\pi}{2}} \int_0^{2\pi} \dfrac{\rho^4}{4}\Big|_0^{\sin^2\varphi} \sin\varphi\cos\varphi\, d\theta d\varphi$

$$= \frac{abc^2}{4} \cdot 2\pi \cdot \int_0^{\frac{\pi}{2}} \sin^9 \varphi \cos \varphi \, d\varphi = \frac{abc^2 \pi}{20}$$

Example 22.

假设 $V = \{(x,y,z): x^2 + y^2 + z^2 \le 1\}$, 求 $\iiint_V \dfrac{dxdydz}{\sqrt{1-x^2-y^2-z^2}} =?$

【解】

令 $x = \rho \sin \varphi \cos \theta \,, y = \rho \sin \varphi \sin \theta \,, \mathrm{z} = \rho \cos \varphi$

$$\text{则} \, dxdydz = \begin{Vmatrix} \dfrac{\partial x}{\partial \rho} & \dfrac{\partial x}{\partial \varphi} & \dfrac{\partial x}{\partial \theta} \\[2mm] \dfrac{\partial y}{\partial \rho} & \dfrac{\partial y}{\partial \varphi} & \dfrac{\partial y}{\partial \theta} \\[2mm] \dfrac{\partial z}{\partial \rho} & \dfrac{\partial z}{\partial \varphi} & \dfrac{\partial z}{\partial \theta} \end{Vmatrix} d\rho d\varphi d\theta$$

$$= \begin{Vmatrix} \sin \varphi \cos \theta & \rho \cos \varphi \cos \theta & -\rho \sin \varphi \sin \theta \\ \sin \varphi \sin \theta & \rho \cos \varphi \sin \theta & -\rho \sin \varphi \cos \theta \\ \cos \varphi & -\rho \sin \varphi & 0 \end{Vmatrix} d\rho d\varphi d\theta = \rho^2 \sin \varphi \, d\rho d\varphi d\theta$$

令 $\mathrm{R} = \{(\rho, \varphi, \theta): 0 \le \rho \le 1, 0 \le \varphi \le \pi, 0 \le \theta \le 2\pi\}$

$$\text{则} \iiint_V \frac{dxdydz}{\sqrt{1-x^2-y^2-z^2}} = \iiint_R \frac{\rho^2 \sin \varphi}{\sqrt{1-\rho^2}} d\rho d\theta d\varphi = \int_0^{\pi} \int_0^{2\pi} \int_0^1 \frac{\rho^2 \sin \varphi}{\sqrt{1-\rho^2}} d\rho d\theta d\varphi$$

$$= \int_0^{\pi} \sin \varphi d\varphi \int_0^{2\pi} d\theta \int_0^1 \frac{\rho^2}{\sqrt{1-\rho^2}} d\rho = 2(2\pi) \left( \int_0^1 \frac{\rho^2}{\sqrt{1-\rho^2}} d\rho \right)$$

令 $\rho = \sin \theta$ 则 $d\rho = \cos \theta \, d\theta$

$$\therefore \int_0^1 \frac{\rho^2}{\sqrt{1-\rho^2}} d\rho = \int_0^{\frac{\pi}{2}} \frac{\sin^2 \theta \cos \theta}{\sqrt{1-\sin^2 \theta}} d\theta = \int_0^{\frac{\pi}{2}} \sin^2 \theta \, d\theta = \int_0^{\frac{\pi}{2}} \frac{1-\cos 2\theta}{2} d\theta$$

$$= \frac{1}{2} \left( \frac{\pi}{2} - \frac{\sin 2\theta}{2} \Big|_0^{\frac{\pi}{2}} \right) = \frac{\pi}{4}$$

$$\therefore \iiint_V \frac{dxdydz}{\sqrt{1-x^2-y^2-z^2}} = 2(2\pi) \left( \int_0^1 \frac{\rho^2}{\sqrt{1-\rho^2}} d\rho \right) = \pi^2$$

Example 23.

假设 $V = \{(x, y, z): x^2 + y^2 + z^2 \leq r^2\}$, 求 $\iiint_V z^2(x^2 + y^2 + z^2)^{\frac{3}{2}} \, dxdydz =?$

【解】

令 $x = \rho \sin\varphi \cos\theta$, $y = \rho \sin\varphi \sin\theta$, $z = \rho \cos\varphi$

则 $dxdydz = \begin{Vmatrix} \dfrac{\partial x}{\partial \rho} & \dfrac{\partial x}{\partial \varphi} & \dfrac{\partial x}{\partial \theta} \\[2mm] \dfrac{\partial y}{\partial \rho} & \dfrac{\partial y}{\partial \varphi} & \dfrac{\partial y}{\partial \theta} \\[2mm] \dfrac{\partial z}{\partial \rho} & \dfrac{\partial z}{\partial \varphi} & \dfrac{\partial z}{\partial \theta} \end{Vmatrix} d\rho d\varphi d\theta$

$= \begin{Vmatrix} \sin\varphi \cos\theta & \rho\cos\varphi \cos\theta & -\rho\sin\varphi \sin\theta \\ \sin\varphi \sin\theta & \rho\cos\varphi \sin\theta & -\rho\sin\varphi \cos\theta \\ \cos\varphi & -\rho\sin\varphi & 0 \end{Vmatrix} d\rho d\varphi d\theta = \rho^2 \sin\varphi \, d\rho d\varphi d\theta$

令 $R = \{(\rho, \varphi, \theta): 0 \leq \rho \leq r, 0 \leq \varphi \leq \pi, 0 \leq \theta \leq 2\pi\}$

则 $\iiint_V z^2(x^2 + y^2 + z^2)^{\frac{3}{2}} \, dxdydz = \iiint_R \rho^2 \cos^2\varphi \, (\rho^2)^{\frac{3}{2}} \rho^2 \sin\varphi \, d\rho d\theta d\varphi$

$= \int_0^\pi \int_0^{2\pi} \int_0^r \rho^2 \cos^2\varphi \, (\rho^2)^{\frac{3}{2}} \rho^2 \sin\varphi \, d\rho d\theta d\varphi = \int_0^\pi \cos^2\varphi \sin\varphi d\varphi \int_0^{2\pi} d\theta \int_0^r \rho^7 d\rho$

$= -\dfrac{\cos^3\varphi}{3}\Big|_0^\pi (2\pi)\left(\dfrac{r^8}{8}\right) = \dfrac{\pi r^8}{6}$

Example 24.

假设 $V = \{(x, y, z): x^2 + y^2 + z^2 \leq r^2, z > 0\}$, 求 $\iiint_V z(x^2 + y^2 + z^2)^{\frac{3}{2}} \, dxdydz =?$

【解】

令 $x = \rho \sin\varphi \cos\theta$, $y = \rho \sin\varphi \sin\theta$, $z = \rho \cos\varphi$

则 $dxdydz = \begin{Vmatrix} \dfrac{\partial x}{\partial \rho} & \dfrac{\partial x}{\partial \varphi} & \dfrac{\partial x}{\partial \theta} \\[2mm] \dfrac{\partial y}{\partial \rho} & \dfrac{\partial y}{\partial \varphi} & \dfrac{\partial y}{\partial \theta} \\[2mm] \dfrac{\partial z}{\partial \rho} & \dfrac{\partial z}{\partial \varphi} & \dfrac{\partial z}{\partial \theta} \end{Vmatrix} d\rho d\varphi d\theta$

$$= \left\| \begin{matrix} \sin\varphi\cos\theta & \rho\cos\varphi\cos\theta & -\rho\sin\varphi\sin\theta \\ \sin\varphi\sin\theta & \rho\cos\varphi\sin\theta & -\rho\sin\varphi\cos\theta \\ \cos\varphi & -\rho\sin\varphi & 0 \end{matrix} \right\| d\rho d\varphi d\theta = \rho^2\sin\varphi\, d\rho d\varphi d\theta$$

令 $R = \{(\rho,\varphi,\theta): 0 \le \rho \le r, 0 \le \varphi \le \dfrac{\pi}{2}, 0 \le \theta \le 2\pi\}$

则 $\displaystyle\iiint_V z(x^2+y^2+z^2)^{\frac{3}{2}}\,dxdydz = \iiint_R \rho\cos\varphi\,(\rho^2)^{\frac{3}{2}}\rho^2\sin\varphi\,d\rho d\theta d\varphi$

$$= \int_0^{\frac{\pi}{2}}\int_0^{2\pi}\int_0^r \rho\cos\varphi\,(\rho^2)^{\frac{3}{2}}\rho^2\sin\varphi\,d\rho d\theta d\varphi = \int_0^{\frac{\pi}{2}}\cos\varphi\sin\varphi d\varphi \int_0^{2\pi} d\theta \int_0^r \rho^6 d\rho$$

$$= -\frac{\cos 2\varphi}{4}\Big|_0^{\frac{\pi}{2}}(2\pi)\left(\frac{r^7}{7}\right) = \frac{\pi r^7}{7}$$

Example 25.

假设 $V = \left\{(x,y,z): \dfrac{x^2}{a^2} + \dfrac{y^2}{b^2} + \dfrac{z^2}{c^2} \le r^2\right\}$，求 $\displaystyle\iiint_V xyz\,dV =?$

【解】

令 $x = a\rho\sin\varphi\cos\theta\,, y = b\rho\sin\varphi\sin\theta\,, z = c\rho\cos\varphi$

$$则\ dxdydz = \left\| \begin{matrix} \dfrac{\partial x}{\partial \rho} & \dfrac{\partial x}{\partial \varphi} & \dfrac{\partial x}{\partial \theta} \\[2mm] \dfrac{\partial y}{\partial \rho} & \dfrac{\partial y}{\partial \varphi} & \dfrac{\partial y}{\partial \theta} \\[2mm] \dfrac{\partial z}{\partial \rho} & \dfrac{\partial z}{\partial \varphi} & \dfrac{\partial z}{\partial \theta} \end{matrix} \right\| d\rho d\varphi d\theta$$

$$= abc \left\| \begin{matrix} \sin\varphi\cos\theta & \rho\cos\varphi\cos\theta & -\rho\sin\varphi\sin\theta \\ \sin\varphi\sin\theta & \rho\cos\varphi\sin\theta & -\rho\sin\varphi\cos\theta \\ \cos\varphi & -\rho\sin\varphi & 0 \end{matrix} \right\| d\rho d\varphi d\theta = abc\rho^2\sin\varphi\,d\rho d\varphi d\theta$$

令 $R = \{(\rho,\varphi,\theta): 0 \le \rho \le r, 0 \le \varphi \le \pi, 0 \le \theta \le 2\pi\}$

则 $\displaystyle\iiint_V xyz\,dV = a^2b^2c^2 \iiint_R (\rho\sin\varphi\cos\theta \cdot \rho\sin\varphi\sin\theta \cdot \rho\cos\varphi)\rho^2\sin\varphi\,d\rho d\theta d\varphi$

$$= a^2b^2c^2 \int_0^{\pi}\int_0^{2\pi}\int_0^r (\rho\sin\varphi\cos\theta \cdot \rho\sin\varphi\sin\theta \cdot \rho\cos\varphi)\rho^2\sin\varphi\,d\rho d\theta d\varphi$$

$$= a^2 b^2 c^2 \left( \left. \frac{\rho^6}{6} \right|_0^r \right) \int_0^\pi \sin^3 \varphi \cos \varphi \, d\varphi \int_0^{2\pi} \cos \theta \sin \theta \, d\theta$$

$$\because \int_0^{2\pi} \cos \theta \sin \theta \, d\theta = \frac{1}{2} \int_0^{2\pi} \sin 2\theta \, d\theta = \frac{-1}{2} (\cos 2\theta)|_0^{2\pi} = 0 \quad \therefore \iiint_V xyz \, dV = 0$$

Example 26.

$$假设 \, V = \left\{ (x, y, z) : \frac{x^2}{a^2} + \frac{y^2}{b^2} + \frac{z^2}{c^2} \leq r^2 \right\}, \quad 求 \iiint_V \sqrt{r^2 - \left( \frac{x^2}{a^2} + \frac{y^2}{b^2} + \frac{z^2}{c^2} \right)} \, dxdydz = ?$$

【解】

$$令 x = a\rho \sin \varphi \cos \theta \,, \, y = b\rho \sin \varphi \sin \theta \,, \, z = c\rho \cos \varphi$$

$$则 \, dxdydz = \begin{Vmatrix} \dfrac{\partial x}{\partial \rho} & \dfrac{\partial x}{\partial \varphi} & \dfrac{\partial x}{\partial \theta} \\ \dfrac{\partial y}{\partial \rho} & \dfrac{\partial y}{\partial \varphi} & \dfrac{\partial y}{\partial \theta} \\ \dfrac{\partial z}{\partial \rho} & \dfrac{\partial z}{\partial \varphi} & \dfrac{\partial z}{\partial \theta} \end{Vmatrix} d\rho d\varphi d\theta$$

$$= abc \begin{Vmatrix} \sin \varphi \cos \theta & \rho\cos \varphi \cos \theta & -\rho\sin \varphi \sin \theta \\ \sin \varphi \sin \theta & \rho\cos \varphi \sin \theta & -\rho\sin \varphi \cos \theta \\ \cos \varphi & -\rho\sin \varphi & 0 \end{Vmatrix} d\rho d\varphi d\theta = abc\rho^2 \sin \varphi \, d\rho d\varphi d\theta$$

$$令 \, R = \{ (\rho, \varphi, \theta) : 0 \leq \rho \leq r, 0 \leq \varphi \leq \pi, 0 \leq \theta \leq 2\pi \}$$

$$则 \iiint_V \sqrt{r^2 - \left( \frac{x^2}{a^2} + \frac{y^2}{b^2} + \frac{z^2}{c^2} \right)} \, dxdydz = abc \iiint_R \sqrt{r^2 - \rho^2} \rho^2 \sin \varphi \, d\rho d\theta d\varphi$$

$$= abc \int_0^\pi \int_0^{2\pi} \int_0^r \sqrt{r^2 - \rho^2} \rho^2 \sin \varphi \, d\rho d\theta d\varphi = abc \int_0^\pi \sin \varphi d\varphi \int_0^{2\pi} d\theta \int_0^r \sqrt{r^2 - \rho^2} \rho^2 d\rho$$

$$= abc(2)(2\pi) \int_0^r \sqrt{r^2 - \rho^2} \rho^2 d\rho$$

$$令 \rho = r \sin \theta \, 则 d\rho = r \cos \theta \, d\theta$$

$$\therefore \int_0^r \sqrt{r^2 - \rho^2} \rho^2 d\rho = r^4 \int_0^{\frac{\pi}{2}} \sin^2 \theta \cos^2 \theta \, d\theta = r^4 \int_0^{\frac{\pi}{2}} \left( \frac{\sin 2\theta}{2} \right)^2 d\theta = \frac{r^4}{4} \int_0^{\frac{\pi}{2}} \frac{1 - \cos 4\theta}{2} d\theta$$

$$= \frac{r^4}{8} \left( \frac{\pi}{2} - \left. \frac{\sin 4\theta}{4} \right|_0^{\frac{\pi}{2}} \right) = \frac{\pi r^4}{16}$$

$$\therefore \iiint_V \sqrt{r^2 - \left(\frac{x^2}{a^2} + \frac{y^2}{b^2} + \frac{z^2}{c^2}\right)}\, dxdydz = \frac{abc\pi^2 r^4}{4}$$

Example 27.

求封闭曲面 $(x^2 + y^2 + z^2)^3 = 27a^3 xyz (a > 0)$ 于第一象限内所围区域体积

【解】

令 $x = \rho \sin\varphi \cos\theta, y = \rho \sin\varphi \sin\theta, z = \rho \cos\varphi$

$$\text{则 } dxdydz = \begin{Vmatrix} \dfrac{\partial x}{\partial \rho} & \dfrac{\partial x}{\partial \varphi} & \dfrac{\partial x}{\partial \theta} \\[2mm] \dfrac{\partial y}{\partial \rho} & \dfrac{\partial y}{\partial \varphi} & \dfrac{\partial y}{\partial \theta} \\[2mm] \dfrac{\partial z}{\partial \rho} & \dfrac{\partial z}{\partial \varphi} & \dfrac{\partial z}{\partial \theta} \end{Vmatrix} d\rho d\varphi d\theta$$

$$= \begin{Vmatrix} \sin\varphi\cos\theta & \rho\cos\varphi\cos\theta & -\rho\sin\varphi\sin\theta \\ \sin\varphi\sin\theta & \rho\cos\varphi\sin\theta & -\rho\sin\varphi\cos\theta \\ \cos\varphi & -\rho\sin\varphi & 0 \end{Vmatrix} d\rho d\varphi d\theta = \rho^2 \sin\varphi\, d\rho d\varphi d\theta$$

令 $R = \{(\rho, \varphi, \theta): 0 \le \rho \le 3a\sqrt[3]{\sin^2\varphi\cos\varphi\cos\theta\sin\theta}, 0 \le \varphi \le \dfrac{\pi}{2}, 0 \le \theta \le \dfrac{\pi}{2}\}$

$$\text{则 } V = \iiint_R \rho^2 \sin\varphi\, d\rho d\theta d\varphi = \int_0^{\frac{\pi}{2}} \int_0^{\frac{\pi}{2}} \int_0^{3a\sqrt[3]{\sin^2\varphi\cos\varphi\cos\theta\sin\theta}} \rho^2 \sin\varphi\, d\rho d\theta d\varphi$$

$$= \int_0^{\frac{\pi}{2}} \int_0^{\frac{\pi}{2}} 9a^3 \sin^3\varphi\cos\varphi\cos\theta\sin\theta\, d\theta d\varphi = 9a^3 \left(\frac{\sin^4\varphi}{4}\right)\Bigg|_0^{\frac{\pi}{2}} \left(\frac{-\cos 2\theta}{4}\right)\Bigg|_0^{\frac{\pi}{2}}$$

$$= 9a^3 \cdot \frac{1}{4} \cdot \frac{1}{2} = \frac{9a^3}{8}$$

Example 28.

假设 $V = \{(x, y, z): 1 \le x^2 + y^2 + z^2 \le 9, z > 0\}$，求 $\iiint_V \dfrac{z^2 dV}{\sqrt{x^2 + y^2 + z^2}} = ?$

【解】

令 $x = \rho \sin\varphi \cos\theta, y = \rho \sin\varphi \sin\theta, z = \rho \cos\varphi$

$$\text{则 } dxdydz = \begin{Vmatrix} \dfrac{\partial x}{\partial \rho} & \dfrac{\partial x}{\partial \varphi} & \dfrac{\partial x}{\partial \theta} \\[2mm] \dfrac{\partial y}{\partial \rho} & \dfrac{\partial y}{\partial \varphi} & \dfrac{\partial y}{\partial \theta} \\[2mm] \dfrac{\partial z}{\partial \rho} & \dfrac{\partial z}{\partial \varphi} & \dfrac{\partial z}{\partial \theta} \end{Vmatrix} d\rho d\varphi d\theta$$

$$= \begin{Vmatrix} \sin\varphi\cos\theta & \rho\cos\varphi\cos\theta & -\rho\sin\varphi\sin\theta \\ \sin\varphi\sin\theta & \rho\cos\varphi\sin\theta & -\rho\sin\varphi\cos\theta \\ \cos\varphi & -\rho\sin\varphi & 0 \end{Vmatrix} d\rho d\varphi d\theta = \rho^2 \sin\varphi \, d\rho d\varphi d\theta$$

$$\text{令 } R = \left\{(\rho,\varphi,\theta): 1 \le \rho \le 3, 0 \le \varphi \le \frac{\pi}{2}, 0 \le \theta \le 2\pi\right\}$$

$$\text{则 } \iiint_V \frac{z^2 dV}{\sqrt{x^2+y^2+z^2}} = \iiint_R \frac{\rho^2\cos^2\varphi \, \rho^2\sin\varphi}{\rho} d\rho d\theta d\varphi$$

$$= \int_0^{\frac{\pi}{2}} \int_0^{2\pi} \int_1^3 \frac{\rho^2\cos^2\varphi \, \rho^2\sin\varphi}{\rho} d\rho d\theta d\varphi = \int_0^{\frac{\pi}{2}} \sin\varphi\cos^2\varphi \, d\varphi \int_0^{2\pi} d\theta \int_1^3 \rho^3 d\rho$$

$$= \frac{-\cos^3\varphi}{3}\Big|_0^{\frac{\pi}{2}} (2\pi)(20) = \frac{40\pi}{3}$$

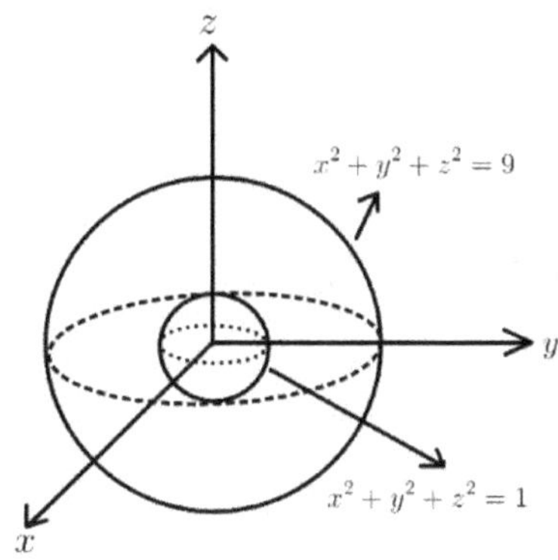


Example 29.

$$\text{假设 } V = \{(x,y,z): 1 \le x^2+y^2+z^2 \le 9, z>0\}, \quad \text{求} \iiint_V \frac{zdV}{\sqrt{x^2+y^2+z^2}} = ?$$

【解】

令 $x = \rho \sin\varphi\cos\theta$ , $y = \rho\sin\varphi\sin\theta$ , $\mathrm{z} = \rho\cos\varphi$

则 $dxdydz = \begin{Vmatrix} \dfrac{\partial x}{\partial \rho} & \dfrac{\partial x}{\partial \varphi} & \dfrac{\partial x}{\partial \theta} \\[6pt] \dfrac{\partial y}{\partial \rho} & \dfrac{\partial y}{\partial \varphi} & \dfrac{\partial y}{\partial \theta} \\[6pt] \dfrac{\partial z}{\partial \rho} & \dfrac{\partial z}{\partial \varphi} & \dfrac{\partial z}{\partial \theta} \end{Vmatrix} d\rho d\varphi d\theta$

$$= \begin{Vmatrix} \sin\varphi\cos\theta & \rho\cos\varphi\cos\theta & -\rho\sin\varphi\sin\theta \\ \sin\varphi\sin\theta & \rho\cos\varphi\sin\theta & -\rho\sin\varphi\cos\theta \\ \cos\varphi & -\rho\sin\varphi & 0 \end{Vmatrix} d\rho d\varphi d\theta = \rho^2\sin\varphi\, d\rho d\varphi d\theta$$

令 $\mathrm{R} = \{(\rho,\varphi,\theta): 1 \le \rho \le 3, 0 \le \varphi \le \dfrac{\pi}{2}, 0 \le \theta \le 2\pi\}$

则 $\iiint_V \dfrac{zdV}{\sqrt{x^2+y^2+z^2}} = \iiint_R \dfrac{\rho\cos\varphi\,\rho^2\sin\varphi}{\rho} d\rho d\theta d\varphi$

$$= \int_0^{\frac{\pi}{2}} \int_0^{2\pi} \int_1^3 \dfrac{\rho\cos\varphi\,\rho^2\sin\varphi}{\rho} d\rho d\theta d\varphi = \int_0^{\frac{\pi}{2}} \sin\varphi\cos\varphi\, d\varphi \int_0^{2\pi} d\theta \int_1^3 \rho^2 d\rho$$

$$= \dfrac{1}{2} \int_0^{\frac{\pi}{2}} \sin 2\varphi d\varphi \int_0^{2\pi} d\theta \int_1^3 \rho^2 d\rho = \dfrac{-\cos 2\varphi}{4}\Big|_0^{\frac{\pi}{2}} (2\pi)\dfrac{26}{3} = \dfrac{26\pi}{3}$$

Example 30.

假设 $\mathrm{V} = \{(x,y,z): x^2+y^2+z^2 \le 1\}$, 求 $\iiint_V \dfrac{z^2 dV}{9-(x^2+y^2+z^2)} = ?$

【解】

令 $x = \rho\sin\varphi\cos\theta$ , $y = \rho\sin\varphi\sin\theta$ , $\mathrm{z} = \rho\cos\varphi$

则 $dxdydz = \begin{Vmatrix} \dfrac{\partial x}{\partial \rho} & \dfrac{\partial x}{\partial \varphi} & \dfrac{\partial x}{\partial \theta} \\[6pt] \dfrac{\partial y}{\partial \rho} & \dfrac{\partial y}{\partial \varphi} & \dfrac{\partial y}{\partial \theta} \\[6pt] \dfrac{\partial z}{\partial \rho} & \dfrac{\partial z}{\partial \varphi} & \dfrac{\partial z}{\partial \theta} \end{Vmatrix} d\rho d\varphi d\theta$

令 $\mathrm{R} = \{(\rho,\varphi,\theta): 0 \le \rho \le 1, 0 \le \varphi \le \pi, 0 \le \theta \le 2\pi\}$

则 $\displaystyle\iiint_V \frac{z^2 dV}{9-(x^2+y^2+z^2)} = \iiint_R \frac{\rho^4 \sin\varphi \cos^2\varphi}{9-\rho^2} d\rho d\theta d\varphi$

$\displaystyle = \int_0^\pi \int_0^{2\pi} \int_0^1 \frac{\rho^4 \sin\varphi \cos^2\varphi}{9-\rho^2} d\rho d\theta d\varphi$

$\displaystyle = \int_0^\pi \cos^2\varphi \sin\varphi d\varphi \int_0^{2\pi} d\theta \int_0^1 \frac{\rho^2(\rho^2-9)+9(\rho^2-9)+81}{9-\rho^2} d\rho$

$\displaystyle = \frac{-\cos^3\varphi}{3}\Big|_0^\pi \cdot \int_0^{2\pi} d\theta \int_0^1 -\rho^2 - 9 + \frac{81}{9-\rho^2} d\rho = \frac{4\pi}{3} \cdot \left(-\frac{\rho^3}{3} - 9\rho + \frac{81}{6}\ln\frac{3+\rho}{3-\rho}\right)\Big|_0^1$

$\displaystyle = \frac{4\pi}{3} \cdot \left(\frac{28}{3} + \frac{81}{6}\ln 2\right)$

Example 31.

假设 $V = \{(x,y,z): x^2+y^2+z^2 \leq 1, z>0\}$, 求 $\displaystyle\iiint_V \frac{zdV}{9-(x^2+y^2+z^2)}$

【解】

令 $x = \rho\sin\varphi\cos\theta$, $y = \rho\sin\varphi\sin\theta$, $z = \rho\cos\varphi$

则 $\displaystyle dxdydz = \begin{Vmatrix} \dfrac{\partial x}{\partial\rho} & \dfrac{\partial x}{\partial\varphi} & \dfrac{\partial x}{\partial\theta} \\[2mm] \dfrac{\partial y}{\partial\rho} & \dfrac{\partial y}{\partial\varphi} & \dfrac{\partial y}{\partial\theta} \\[2mm] \dfrac{\partial z}{\partial\rho} & \dfrac{\partial z}{\partial\varphi} & \dfrac{\partial z}{\partial\theta} \end{Vmatrix} d\rho d\varphi d\theta$

令 $R = \{(\rho,\varphi,\theta): 0 \leq \rho \leq 1, 0 \leq \varphi \leq \dfrac{\pi}{2}, 0 \leq \theta \leq 2\pi\}$

则 $\displaystyle\iiint_V \frac{zdV}{9-(x^2+y^2+z^2)} = \iiint_R \frac{\rho^3 \sin\varphi \cos\varphi}{9-\rho^2} d\rho d\theta d\varphi$

$\displaystyle = \int_0^{\frac{\pi}{2}} \int_0^{2\pi} \int_0^1 \frac{\rho^3 \sin\varphi \cos\varphi}{9-\rho^2} d\rho d\theta d\varphi = \int_0^{\frac{\pi}{2}} \cos\varphi \sin\varphi d\varphi \int_0^{2\pi} d\theta \int_0^1 \frac{\rho(\rho^2-9)+9\rho}{9-\rho^2} d\rho$

$\displaystyle = \frac{1}{2}\int_0^{\frac{\pi}{2}} \sin 2\varphi d\varphi \int_0^{2\pi} d\theta \int_0^1 -\rho + \frac{9\rho}{9-\rho^2} d\rho = -\frac{\cos 2\varphi}{4}\Big|_0^{\frac{\pi}{2}} \cdot 2\pi \cdot \left(-\frac{\rho^2}{2} - \frac{9}{2}\ln(9-\rho^2)\right)\Big|_0^1$

$\displaystyle = \left(\frac{1}{2}\right)(2\pi)\left(-\frac{1}{2} - \frac{9}{2}\ln\frac{8}{9}\right) = \pi\left(-\frac{1}{2} + \frac{9}{2}(2\ln 3 - 3\ln 2)\right)$

Example 32.

$$假设\ V = \{(x, y, z): 1 \le x^2 + y^2 + z^2 \le 3\}$$

$$求 \iiint_V \frac{z^2 \cos\sqrt{x^2 + y^2 + z^2}\ dxdydz}{(x^2 + y^2 + z^2)^2} = ?$$

【解】

$$令 x = \rho\sin\varphi\cos\theta,\ y = \rho\sin\varphi\sin\theta,\ z = \rho\cos\varphi$$

$$则\ dxdydz = \begin{Vmatrix} \dfrac{\partial x}{\partial \rho} & \dfrac{\partial x}{\partial \varphi} & \dfrac{\partial x}{\partial \theta} \\[2mm] \dfrac{\partial y}{\partial \rho} & \dfrac{\partial y}{\partial \varphi} & \dfrac{\partial y}{\partial \theta} \\[2mm] \dfrac{\partial z}{\partial \rho} & \dfrac{\partial z}{\partial \varphi} & \dfrac{\partial z}{\partial \theta} \end{Vmatrix} d\rho d\varphi d\theta$$

$$= \begin{Vmatrix} \sin\varphi\cos\theta & \rho\cos\varphi\cos\theta & -\rho\sin\varphi\sin\theta \\ \sin\varphi\sin\theta & \rho\cos\varphi\sin\theta & -\rho\sin\varphi\cos\theta \\ \cos\varphi & -\rho\sin\varphi & 0 \end{Vmatrix} d\rho d\varphi d\theta = \rho^2\sin\varphi\ d\rho d\varphi d\theta$$

$$令\ R = \{(\rho, \varphi, \theta): 1 \le \rho \le \sqrt{3}, 0 \le \varphi \le \pi, 0 \le \theta \le 2\pi\}$$

$$则 \iiint_V \frac{z^2\cos\sqrt{x^2+y^2+z^2}\ dxdydz}{(x^2+y^2+z^2)^2}\ dV = \iiint_R \frac{\cos\rho}{\rho^4}\rho^4\sin\varphi\cos^2\varphi\ d\rho d\theta d\varphi$$

$$= \int_0^\pi \int_0^{2\pi} \int_1^{\sqrt{3}} \frac{\cos\rho}{\rho^4}\rho^4\sin\varphi\cos^2\varphi\ d\rho d\theta d\varphi = \frac{-\cos^3\varphi}{3}\bigg|_0^\pi \cdot 2\pi \cdot \int_1^{\sqrt{3}} \cos\rho\ d\rho$$

$$= \frac{4\pi}{3}\left(\sin\rho\big|_1^{\sqrt{3}}\right) = \frac{4\pi}{3}(\sin\sqrt{3} - \sin 1)$$

Example 33.

$$假设\ V = \{(x, y, z): 1 \le x^2 + y^2 + z^2 \le 3, z > 0\}$$

$$求 \iiint_V \frac{z\cos\sqrt{x^2 + y^2 + z^2}\ dxdydz}{(x^2 + y^2 + z^2)^{\frac{3}{2}}} = ?$$

【解】

令 $x = \rho \sin\varphi \cos\theta$ , $y = \rho \sin\varphi \sin\theta$ , $z = \rho \cos\varphi$

则 $dxdydz = \begin{Vmatrix} \dfrac{\partial x}{\partial \rho} & \dfrac{\partial x}{\partial \varphi} & \dfrac{\partial x}{\partial \theta} \\[2mm] \dfrac{\partial y}{\partial \rho} & \dfrac{\partial y}{\partial \varphi} & \dfrac{\partial y}{\partial \theta} \\[2mm] \dfrac{\partial z}{\partial \rho} & \dfrac{\partial z}{\partial \varphi} & \dfrac{\partial z}{\partial \theta} \end{Vmatrix} d\rho d\varphi d\theta$

$$= \begin{Vmatrix} \sin\varphi\cos\theta & \rho\cos\varphi\cos\theta & -\rho\sin\varphi\sin\theta \\ \sin\varphi\sin\theta & \rho\cos\varphi\sin\theta & -\rho\sin\varphi\cos\theta \\ \cos\varphi & -\rho\sin\varphi & 0 \end{Vmatrix} d\rho d\varphi d\theta = \rho^2 \sin\varphi \, d\rho d\varphi d\theta$$

令 $R = \{(\rho, \varphi, \theta) : 1 \le \rho \le \sqrt{3}, 0 \le \varphi \le \dfrac{\pi}{2}, 0 \le \theta \le 2\pi\}$

则 $\displaystyle\iiint_V \frac{z \cos\sqrt{x^2+y^2+z^2}\, dxdydz}{(x^2+y^2+z^2)^{\frac{3}{2}}} = \iiint_R \frac{\cos\rho}{\rho^3} \rho^3 \sin\varphi \cos\varphi \, d\rho d\theta d\varphi$

$$= \int_0^{\frac{\pi}{2}} \int_0^{2\pi} \int_1^{\sqrt{3}} \frac{\cos\rho}{\rho^3} \rho^3 \sin\varphi \cos\varphi \, d\rho d\theta d\varphi = \left. \frac{-\cos 2\varphi}{4} \right|_0^{\frac{\pi}{2}} \cdot 2\pi \cdot \int_1^{\sqrt{3}} \cos\rho \, d\rho = \pi \left( \sin\rho \Big|_1^{\sqrt{3}} \right)$$

$$= \pi(\sin\sqrt{3} - \sin 1)$$

Example 34.

试求 $\displaystyle\int_0^1 \int_0^{\sqrt{1-x^2}} \int_{\sqrt{x^2+y^2}}^{\sqrt{2-x^2-y^2}} dz dy dx = ?$

【解】

令 $x = \rho \sin\varphi \cos\theta$ , $y = \rho \sin\varphi \sin\theta$ , $z = \rho \cos\varphi$

则 $dxdydz = \begin{Vmatrix} \dfrac{\partial x}{\partial \rho} & \dfrac{\partial x}{\partial \varphi} & \dfrac{\partial x}{\partial \theta} \\[2mm] \dfrac{\partial y}{\partial \rho} & \dfrac{\partial y}{\partial \varphi} & \dfrac{\partial y}{\partial \theta} \\[2mm] \dfrac{\partial z}{\partial \rho} & \dfrac{\partial z}{\partial \varphi} & \dfrac{\partial z}{\partial \theta} \end{Vmatrix} d\rho d\varphi d\theta$

$$= \left\|\begin{matrix} \sin\varphi\cos\theta & \rho\cos\varphi\cos\theta & -\rho\sin\varphi\sin\theta \\ \sin\varphi\sin\theta & \rho\cos\varphi\sin\theta & -\rho\sin\varphi\cos\theta \\ \cos\varphi & -\rho\sin\varphi & 0 \end{matrix}\right\| d\rho d\varphi d\theta = \rho^2 \sin\varphi\, d\rho d\varphi d\theta$$

令 $R = \{(\rho,\varphi,\theta): 0 \le \rho \le \sqrt{2}, 0 \le \varphi \le \dfrac{\pi}{4}, 0 \le \theta \le \dfrac{\pi}{2}\}$

则 $\displaystyle\int_0^1 \int_0^{\sqrt{1-x^2}} \int_{\sqrt{x^2+y^2}}^{\sqrt{2-x^2-y^2}} dzdydx = \iiint_R \rho^2 \sin\varphi\, d\rho d\theta d\varphi = \int_0^{\frac{\pi}{4}} \int_0^{\frac{\pi}{2}} \int_0^{\sqrt{2}} \rho^2 \sin\varphi\, d\rho d\theta d\varphi$

$$= \frac{(\sqrt{2}-1)\pi}{3}$$

Example 35.

$$\text{假设 } V = \left\{(x,y,z): \frac{x^2}{a^2} + \frac{y^2}{b^2} + \frac{z^2}{c^2} \le 1\right\}, \quad \text{求} \iiint_V dxdydz = ?$$

【解】

令 $x = a\rho\sin\varphi\cos\theta\,, y = b\rho\sin\varphi\sin\theta\,, z = c\rho\cos\varphi$

$$\text{则 } dxdydz = \left\|\begin{matrix} \dfrac{\partial x}{\partial\rho} & \dfrac{\partial x}{\partial\varphi} & \dfrac{\partial x}{\partial\theta} \\ \dfrac{\partial y}{\partial\rho} & \dfrac{\partial y}{\partial\varphi} & \dfrac{\partial y}{\partial\theta} \\ \dfrac{\partial z}{\partial\rho} & \dfrac{\partial z}{\partial\varphi} & \dfrac{\partial z}{\partial\theta} \end{matrix}\right\| d\rho d\varphi d\theta$$

$$= abc \left\|\begin{matrix} \sin\varphi\cos\theta & \rho\cos\varphi\cos\theta & -\rho\sin\varphi\sin\theta \\ \sin\varphi\sin\theta & \rho\cos\varphi\sin\theta & -\rho\sin\varphi\cos\theta \\ \cos\varphi & -\rho\sin\varphi & 0 \end{matrix}\right\| d\rho d\varphi d\theta = abc\rho^2 \sin\varphi\, d\rho d\varphi d\theta$$

令 $R = \{(\rho,\varphi,\theta): 0 \le \rho \le 1, 0 \le \varphi \le \pi, 0 \le \theta \le 2\pi\}$

则 $\displaystyle\iiint_V dxdydz = abc\iiint_R \rho^2 \sin\varphi\, d\rho d\theta d\varphi = abc\int_0^{\pi}\int_0^{2\pi}\int_0^1 \rho^2 \sin\varphi\, d\rho d\theta d\varphi$

$$= \frac{4\pi abc}{3}$$

Example 36.

$$\text{求} \iiint_V \left(3x + \frac{1}{1+x^2+y^2+z^2}\right) dV = ?, \quad \text{其中 } V = \{(x,y,z): x^2+y^2+z^2 \le 4\}$$

【解】

令 $x = \rho \sin\varphi \cos\theta$ , $y = \rho \sin\varphi \sin\theta$ , $z = \rho \cos\varphi$

$$\text{则 } dxdydz = \begin{Vmatrix} \dfrac{\partial x}{\partial \rho} & \dfrac{\partial x}{\partial \varphi} & \dfrac{\partial x}{\partial \theta} \\[2mm] \dfrac{\partial y}{\partial \rho} & \dfrac{\partial y}{\partial \varphi} & \dfrac{\partial y}{\partial \theta} \\[2mm] \dfrac{\partial z}{\partial \rho} & \dfrac{\partial z}{\partial \varphi} & \dfrac{\partial z}{\partial \theta} \end{Vmatrix} d\rho d\varphi d\theta = \rho^2 \sin\varphi \, d\rho d\varphi d\theta$$

令 $R = \{(\rho, \varphi, \theta): 0 \le \rho \le 2, 0 \le \varphi \le \pi, 0 \le \theta \le 2\pi\}$

$$\text{则 } \iiint_V \left(3x + \frac{1}{1+x^2+y^2+z^2}\right) dV = \iiint_R \left(3\rho \sin\varphi \cos\theta + \frac{1}{1+\rho^2}\right) \rho^2 \sin\varphi \, d\rho d\theta d\varphi$$

$$= \int_0^\pi \int_0^{2\pi} \int_0^2 \left(3\rho \sin\varphi \cos\theta + \frac{1}{1+\rho^2}\right) \rho^2 \sin\varphi \, d\rho d\theta d\varphi$$

$$= \int_0^\pi \int_0^{2\pi} \int_0^2 3\rho^3 \sin^2\varphi \cos\theta \, d\rho d\theta d\varphi + \int_0^\pi \int_0^{2\pi} \int_0^2 \frac{\rho^2 \sin\varphi}{1+\rho^2} d\rho d\theta d\varphi$$

$$= \int_0^\pi \sin^2\varphi \, d\varphi \int_0^{2\pi} \cos\theta \, d\theta \int_0^2 3\rho^3 d\rho + 2\pi \int_0^\pi \sin\varphi \, d\varphi \int_0^2 \frac{\rho^2}{1+\rho^2} d\rho$$

$$= 2\pi \int_0^\pi \sin\varphi \, d\varphi \int_0^2 \frac{\rho^2}{1+\rho^2} d\rho = 4\pi \int_0^2 1 - \frac{1}{1+\rho^2} d\rho = 4\pi(2 - \tan^{-1} 2)$$

Example 37.

Find the mass of the solid bounded below by the half-cone $z = \sqrt{x^2+y^2}$ and above by the spherical surface $x^2+y^2+z^2 = 1$ given that the density function

$$f(x,y,z) = e^{(x^2+y^2+z^2)^{\frac{3}{2}}}$$

【解】

$$\because \text{mass} = \iiint_V f(x,y,z) \, dxdydz$$

令 $x = \rho \sin\varphi \cos\theta$ , $y = \rho \sin\varphi \sin\theta$ , $z = \rho \cos\varphi$

$$\text{则 } dxdydz = \begin{Vmatrix} \dfrac{\partial x}{\partial \rho} & \dfrac{\partial x}{\partial \varphi} & \dfrac{\partial x}{\partial \theta} \\[2mm] \dfrac{\partial y}{\partial \rho} & \dfrac{\partial y}{\partial \varphi} & \dfrac{\partial y}{\partial \theta} \\[2mm] \dfrac{\partial z}{\partial \rho} & \dfrac{\partial z}{\partial \varphi} & \dfrac{\partial z}{\partial \theta} \end{Vmatrix} d\rho d\varphi d\theta = \rho^2 \sin\varphi \, d\rho d\varphi d\theta$$

令 $R = \{(\rho, \varphi, \theta): 0 \leq \rho \leq 1, \dfrac{\pi}{4} \leq \varphi \leq \pi, 0 \leq \theta \leq 2\pi\}$

$$\text{则 } \iiint_V f(x,y,z)\, dxdydz = \iiint_V e^{(x^2+y^2+z^2)^{\frac{3}{2}}} \, dxdydz = \iiint_R e^{\rho^3} \rho^2 \sin\varphi \, d\rho d\theta d\varphi$$

$$= \int_{\frac{\pi}{4}}^{\pi} \int_0^{2\pi} \int_0^1 e^{\rho^3} \rho^2 \sin\varphi \, d\rho d\theta d\varphi = 2\pi \int_0^1 e^{\rho^3} \rho^2 d\rho \int_{\frac{\pi}{4}}^{\pi} \sin\varphi \, d\varphi = 2\pi \cdot \dfrac{e^{\rho^3}}{3}\Big|_0^1 \cdot (-\cos\varphi\,)\big|_{\frac{\pi}{4}}^{\pi}$$

$$= \dfrac{2\pi(e-1)(1+\dfrac{1}{\sqrt{2}})}{3}$$

# 第九章　　向量微积分

接下来依序介绍线积分、Green's Theorem、曲面积分、Stokes' Theorem、Gauss's Theorem (The Divergence Theorem)；线积分的被积分函数分为纯量函数与向量函数，曲线可能为二维平面曲线或三维空间曲线，无论是哪种情形皆是转成求单变量函数的定积分

当被积分函数的一阶偏导数存在且连续时，Green's Theorem 为计算二维平面封闭曲线积分与双重积分的重要工具；此外，Green's Theorem 是将微积分基本定理扩展到二维平面的一种形式，此定理将一个沿着简单封闭平面曲线的线积分与其所包围平面区域的双重积分链接起来，可以帮助将线积分与双重积分做双向的转换，如果线积分的计算是困难时，可藉由此定理转成双重积分，反之，当计算此双重积分是困难时，藉由此定理转成线积分做计算；Green's Theorem 要求双重积分中的区域 D 是简单连通，重要的是数个 disjoint 的简单连通集的联集为非简单连通集，因此，Green's Theorem 有时可以扩展应用在非简单连通集；读者应熟悉各种 Green's Theorem 的应用情境，当线积分的路径复杂或有复杂的被积分函数时，尝试使用 Green's Theorem 将问题转为求双重积分，转换后通常会搭配 Fubini's Theorem 求值；此外，也可能搭配极坐标转换或广义坐标转换求值，无论是哪种情形，精确找出积分边界是重要的，当被积分函数于所围区域并不满足一阶偏导数存在且连续的条件时，需另外绕出一个能逼近原先区域的封闭区域且在这新的封闭区域当中，原被积分函数满足一阶偏导数存在且连续的条件，在此情况下才能使用 Green's Theorem 求值，最后再逼近原先的封闭区域求解；反之，给封闭区域求面积时也可藉由 Green's Theorem 将问题转为求线积分，因此，求线积分的能力是重要的

曲面积分分为给空间曲面求封闭区域的表面积，给纯量函数求曲面$S$的积分以及给向量函数求通过曲面的通量，这三者最终皆是转换为双重积分的类型作计算，假设曲面$S$的参数式为$\vec{r}(s,t)$，上述这三者的计算皆与两偏导数的外积$\vec{r}_s \times \vec{r}_t$及其长度$|\vec{r}_s \times \vec{r}_t|$有关，因此，有效率且精确地求出这两者是重要的，曲面通常为圆柱、椭圆柱、圆锥体、抛物面..等，无论是哪种组合均透过投影的方法将求曲面积分的问题转成求双重积分的问题，并且也必须能精确且迅速求得投影后的积分边界；此外，投影后可能也须搭配极坐标转换或广义坐标转换

Stokes' Theorem 不仅是微积分基本定理在更高维度的推广，同时也是 Green's Theorem 的高维推广；Green's Theorem 是将平面区域的双重积分与平面上二维封闭曲线

之线积分做转换，Stokes' Theorem 则是将三维空间封闭曲线的线积分与非封闭曲面$S$的曲面积分做转换；Stokes' Theorem 说明可以将$\vec{F}$沿着非封闭曲面$S$其边界曲线 $C$ 的线积分转换为$\vec{F}$旋度在非封闭曲面$S$上的通量，相反地，也可藉由求$\vec{F}$沿着非封闭曲面$S$其边界曲线 $C$ 的线积分得到$\vec{F}$旋度在非封闭曲面$S$上的通量；Stokes' Theorem 有数种考试类型，其中重要成立条件是曲面是非封闭曲面且其边界是空间封闭曲线；再者，曲面$S$的定向与边界曲线 $C$ 是否为逆时针绕行有重要的关联；此外，当线积分的被积分函数是复杂的，可藉由 Stokes' Theorem 将问题转为求曲面积分，接着再透过投影方式转为求双重积分的问题，因此培养求双重积分的能力是重要的

　　整体而言，Gauss's Theorem (The Divergence Theorem)、Stokes' Theorem 与 Green's Theorem 都是微积分基本定理的扩展，此外，Gauss's Theorem 可以将向量场$\vec{F}$穿越封闭曲面$S$的通量(曲面积分)转换为$\vec{F}$的散度在封闭体积 V 的三重积分，例如，假设一个封闭曲面由$S_1, S_2$ 组成，原本求通过两曲面的方法是求各自的通量再相加，而藉由 Gauss's Theorem 则将问题转成纯量函数的体积分，相反地，也可藉由$\vec{F}$在封闭曲面$S$的通量(曲面积分)求得$\vec{F}$的散度在封闭体积 V 的三重积分；当向量函数于某封闭光滑曲面所围区域内的一阶偏导数存在且连续时，如果被积分的向量函数是复杂的，可藉由 Gauss's Theorem 将向量函数在曲面积分的问题转为求三重积分，因此，求三重积分的能力是重要的，此时有可能仍须搭配极坐标转换或广义坐标转换求值

## 9.1 参数化方程式(Parametric Equations)

【定义】

假设 $x$ 和 $y$ 是 $t$ 的连续函数, $t \in I$, 则方程式$x = x(t)$ 且 $y = y(t)$ 称为参数方程式;随着 $t$ 在区间$I$变化, 得到的点集合$(x, y)$称为参数曲线,用符号$C$表示

为了熟悉并理解以参数表示的曲线的图形，需练习如何将这两个方程式重新写成一个涉及变量$x$和$y$的单一方程式；另一方面，当给定变数 $x$ 和 $y$ 的单一方程式时，则对应的参数方程式并不唯一。

Example 1.

试将下列参数方程式用单一方程式表示

(1) $x(t) = t^2 - 5$, $y(t) = 2t + 1$, $-2 \leq t \leq 3$.

(2) $x(t) = \sqrt{2t + 3}$, $y(t) = 2t + 1, -2 \leq t \leq 6$.

(3) $x(t) = 2\cos t$, $y(t) = \sqrt{3}\sin t$, $-2 \le t \le 7$.

【解】

(1)

$$\because t = \frac{y-1}{2} \quad \therefore x = \left(\frac{y-1}{2}\right)^2 - 5 = \frac{y^2 - 2y + 1}{4} - 5 = \frac{y^2 - 2y - 19}{4}$$

$$\therefore x = \frac{y^2 - 2y - 19}{4}, \quad \forall -1 \le x \le 4$$

(2)

$$\because x = \sqrt{2t+3} \quad \therefore x^2 = 2t + 3 \Rightarrow t = \frac{x^2 - 3}{2}$$

$$\because y = 2t + 5 \quad \therefore y = 2\left(\frac{x^2 - 3}{2}\right) + 5 = x^2 + 2, \quad \forall 0 \le x \le \sqrt{15}$$

(3)

$$\because x(t) = 2\cos t \quad \therefore \cos t = \frac{x}{2}, \quad \because y(t) = \sqrt{3}\sin t \quad \therefore \sin t = \frac{y}{\sqrt{3}}$$

$$\because \cos^2 t + \sin^2 t = 1 \quad \therefore \left(\frac{x}{2}\right)^2 + \left(\frac{y}{\sqrt{3}}\right)^2 = 1 \Rightarrow \frac{x^2}{4} + \frac{y^2}{3} = 1$$

Example 2.

试将下列单一方程式找出所对应的两组不同参数方程式

(1)$y = 3x^2 - 1$ (2)$y = \ln x$ (3)$x = y^2 - 6y + 8$

【解】

(1)

Let $x(t) = t$ then $y(t) = 3t^2 - 1$

Let $x(t) = 2t - 1$ then $y(t) = 3x^2 - 1 = 3(2t - 1)^2 - 1 = 12t^2 - 12t + 2$

(2)

Let $x(t) = t$ then $y(t) = \ln t$

Let $x(t) = t^2$ then $y(t) = \ln t^2 = 2\ln t$

(3)

Let $y(t) = t$ then $x(t) = t^2 - 6t + 8$

Let $y(t) = t + 2$ then $x(t) = y^2 - 6y + 8 = (t + 2)^2 - 6(t + 2) + 8 = t^2 - 2t$

Example 3.

试将下列参数方程式用单一方程式表示

(1)$x(t) = t^5, \quad y(t) = 5\ln t$

(2)$x = e^{3t}, y = t + 3$

(3)$x(t) = 2t - 5, \quad y(t) = 4t - 7$

(4)$x = \sqrt{t}, \quad y = 2 - t$

(5)$x(t) = t^2 - 2, \quad y(t) = \dfrac{t}{3}, -2 \leq t \leq 6.$

(6)$x = t^2 - 2t, \quad y = t + 2, 0 \leq t \leq 5$

(7)$x(t) = 2\cos t, \quad y(t) = 3\sin t, 0 \leq t \leq 2\pi$

(8)$x(t) = 2 + \cos t, \quad y(t) = 4 - \sin t$

(9)$x = a\cos t, y = a\sin t, \quad 0 \leq t \leq 2\pi$

(10)$x(t) = \sin t, \quad y(t) = \cos 2t$

(11)$x(t) = \csc t, \quad y(t) = \cot t, \quad 0 < t < \dfrac{\pi}{2}$

(12)$x = \cos t, \quad y = \sec t, \quad 0 < t < \dfrac{\pi}{2}$

(13)$x = \sinh t, y = \cosh t$

(14)$x(t) = 3\cos 7t, \quad y(t) = 3\sin 7t$

(15)$x(t) = 3\cosh 4t, \quad y(t) = 4\sinh 4t$

(16)$x = \cos t, \quad y = \cos^2 t$

【解】

(1)

$\because x(t) = t^5 \quad \therefore 5\ln t = \ln x \Rightarrow y = \ln x, \quad \forall x \in [0, \infty)$

(2)

$\because x = e^{3t} \text{ and } y = t + 3 \quad \therefore t = \dfrac{\ln x}{3} = y - 3 \Rightarrow y = \dfrac{\ln x}{3} + 3$

$\therefore$ the curve given parametric equations is $\quad y = \dfrac{\ln x}{3} + 3$

(3)

$\because t = \dfrac{x + 5}{2} \quad \therefore y = 4t - 7 = 4\left(\dfrac{x + 5}{2}\right) - 7 = 2x + 3 \Rightarrow y = 2x + 3, \quad \forall x \in R$

(4)

$\because x^2 = t \text{ and } t = 2 - y$

$\therefore$ the curve given parametric equations is the parabola $y = 2 - x^2, \quad \forall x \geq 0$

(5)

$\because t = 3y \quad \therefore x = t^2 - 2 = (3y)^2 - 2 = 9y^2 - 2 \Rightarrow x = 9y^2 - 2$ where $-2 \leq x < \infty$

(6)

$\because t = y - 2, \quad \therefore x = t^2 - 2t = (y-2)^2 - 2(y-2) = y^2 - 6y + 8$

$\therefore$ the curve given parametric equations is the parabola $x = y^2 - 6y + 8, \quad -2 \leq y \leq 3$

(7)

$\because x(t) = 2\cos t \quad \therefore \cos t = \dfrac{x}{2}, \quad \because y(t) = 3\sin t \therefore \sin t = \dfrac{y}{3}$

$\because \cos^2 t + \sin^2 t = 1 \quad \therefore \left(\dfrac{x}{2}\right)^2 + \left(\dfrac{y}{3}\right)^2 = 1 \Rightarrow \dfrac{x^2}{4} + \dfrac{y^2}{9} = 1$ where $-2 \leq x \leq 2$

(8)

$\because x(t) = 2 + \cos t \therefore \cos t = x - 2, \quad \because y(t) = 4 - \sin t \quad \therefore \sin t = 4 - y$

$\because \cos^2 t + \sin^2 t = 1 \quad \therefore (x-2)^2 + (y-4)^2 = 1, \quad$ where $1 \leq x \leq 3$

(9)

$\because x^2 + y^2 = a^2\cos^2 t + a^2\sin^2 t = a^2$

$\therefore$ the curve given parametric equations is the circle with radius $a, \quad x^2 + y^2 = a^2$

(10)

$\because x(t) = \sin t, \ y(t) = \cos 2t$

$\because \sin^2 t = \dfrac{1 - \cos 2t}{2} \quad \therefore x^2 = \dfrac{1 - y}{2} \Rightarrow y = 1 - x^2, \quad$ where $-1 \leq x \leq 1$

(11)

$\because x(t) = \csc t, \ y(t) = \cot t$

$\because \csc^2 t = \cot^2 t + 1 \quad \therefore x^2 = y^2 + 1 \Rightarrow y = \sqrt{x^2 - 1}, \quad$ where $x > 1$

(12)

$\because x = \cos t, \ y = \sec t \quad \therefore y = \dfrac{1}{x}, \quad \forall y > 1$

$\therefore$ the curve given parametric equations is $y = \dfrac{1}{x}, \quad \forall y > 1$

(13)

$\because \cosh^2 t - \sinh^2 t = 1, \ x = \sinh t \ $ and $\ y = \cosh t$

$\therefore$ the curve given parametric equations is $y^2 - x^2 = 1$

(14)

$\because x(t) = 3\cos 7t, \ y(t) = 3\sin 7t,$

$$\because \cos^2 7t + \sin^2 7t = 1 \quad \therefore \left(\frac{x}{3}\right)^2 + \left(\frac{y}{3}\right)^2 = 1 \quad \therefore x^2 + y^2 = 9, \ \text{where} -3 \leq x \leq 3$$

(15)

$$\because x(t) = 3\cosh 4t, \ y(t) = 4\sinh 4t,$$

$$\because \cosh^2 4t - \sinh^2 4t = 1 \quad \therefore \left(\frac{x}{3}\right)^2 - \left(\frac{y}{4}\right)^2 = 1, \ \text{where} -3 \leq x \leq 3$$

$\therefore$ the curve is hyperbola

(16)

$$\because \ x = \cos t, \ y = \cos^2 t, \ \therefore y = \cos^2 t = x^2, \ \forall -1 \leq x \leq 1$$

$\therefore$ the curve given parametric equations is the parabola $y = x^2, \ \forall -1 \leq x \leq 1$


## 9.2 给定曲线的向量值表示式，求微分

### 【定义】

向量值函数为以下形式的函数

$$\vec{r}(t) = \big(f(t), g(t)\big) \ \text{或} \ \ \vec{r}(t) = (f(t), g(t), h(t))$$

其中, 分量函数$f, \ g$ 以及 $h$ 是参数$t$的实数值函数


上述定义当中, 前者是二维向量值函数, 后者是三维向量值函数


### 【定义】

給定向量值函数$\vec{r}(t)$,如果

$$\lim_{\Delta t \to 0} \frac{\vec{r}(t + \Delta t) - \vec{r}(t)}{\Delta t} \ 存在$$

则称 $\vec{r}(t)$ 于$t$可微分,用 $\vec{r}\,'(t)$ 表示 $\lim\limits_{\Delta t \to 0} \dfrac{\vec{r}(t + \Delta t) - \vec{r}(t)}{\Delta t}$, 即

$$\vec{r}\,'(t) = \lim_{\Delta t \to 0} \frac{\vec{r}(t + \Delta t) - \vec{r}(t)}{\Delta t}$$

再者,如果$\vec{r}\,'(t)$ 存在 $\forall t \in (a, b)$则称 $\vec{r}(t)$于区间$(a, b)$可微分


### 【定理】

假设 $\vec{r}(t) = (f(t), g(t), h(t))$ 其中 $f(t), g(t), h(t)$为可微分函数, 则

$$\vec{r}\,'(t) = (f'(t), g'(t), h'(t))$$

<u>Proof:</u>

$\because f(t), g(t), h(t)$ 为可微分函数

$\therefore \vec{r}'(t) = \lim\limits_{\Delta t \to 0} \dfrac{\vec{r}(t + \Delta t) - \vec{r}(t)}{\Delta t} = \lim\limits_{\Delta t \to 0} \dfrac{\big(f(t+\Delta t), g(t+\Delta t), h(t+\Delta t)\big) - \big(f(t), g(t), h(t)\big)}{\Delta t}$

$= \lim\limits_{\Delta t \to 0} \left( \dfrac{f(t+\Delta t) - f(t)}{\Delta t}, \dfrac{g(t+\Delta t) - g(t)}{\Delta t}, \dfrac{h(t+\Delta t) - h(t)}{\Delta t} \right)$

$= \left( \lim\limits_{\Delta t \to 0} \dfrac{f(t+\Delta t) - f(t)}{\Delta t}, \lim\limits_{\Delta t \to 0} \dfrac{g(t+\Delta t) - g(t)}{\Delta t}, \lim\limits_{\Delta t \to 0} \dfrac{h(t+\Delta t) - h(t)}{\Delta t} \right)$

$= (f'(t), g'(t), h'(t))$

## 【定义】

给定向量值函数 $\vec{r}(t)$，如果 $\vec{r}'(t)$ 连续且 $\vec{r}'(t) \neq 0, \forall t \in I$ 则称 $\vec{r}(t)$ 于 $I$ 为平滑向量值函数，如果曲线用平滑向量值函数所描述，则称该曲线为平滑曲线

底下先列出关于纯量函数线积分以及向量函数线积分的重要公式，细部推导请详见各节说明，第一个计算重点是给定曲线 $C$，求纯量函数 $f$ 的曲线积分，计算公式为

$$\int_C f(x, y, z)ds = \int_a^b f\big(\vec{r}(t)\big)|\vec{r}'(t)|dt$$

第二个计算重点是给定曲线 $C$，求向量函数 $\vec{F}$ 的曲线积分，计算公式为

$$\int_C \vec{F} \cdot d\vec{r} = \int_a^b \vec{F}(\vec{r}(t)) \cdot \vec{r}'(t)dt$$

其中 $C: \vec{r}(t) = \big(x(t), y(t), z(t)\big), \ \forall a \leq t \leq b.$

從上述观察得知，给定曲线 $C: \vec{r}(t) = \big(x(t), y(t), z(t)\big)$，如何有效率且快速地求得 $\vec{r}'(t)$ 及 $|\vec{r}'(t)|$ 是重要的

Example 1.

Assume $\vec{r}(t) = \big(\sqrt{t}, 2 - t\big)$. Find $\vec{r}'(t)$ and $\vec{r}'(1)$.

【解】

$\vec{r}'(t) = \left( \dfrac{1}{2\sqrt{t}}, -1 \right)$ and $\vec{r}'(1) = \left( \dfrac{1}{2\sqrt{2}}, -1 \right)$

Example 2.

试求下列各向量值函数的微分，$\vec{r}'(t) = ?$

(1) $\vec{r}(t) = (7t + 10, 5t^2 + 2t - 3)$　　(2) $\vec{r}(t) = (5\cos t, 7\sin t)$

(3) $\vec{r}(t) = (e^t \sin t, e^{-t} \cos t, -e^{3t})$  (4)$\vec{r}(t) = (t \ln t, 7e^t \sin t, \cos t + \sin t)$

(5) $\vec{r}(t) = (5 + t^4, te^{-t}, \cos 2t)$

【解】

(1)

$\vec{r}'(t) = (7, 10t + 2)$

(2)

$\vec{r}'(t) = (-5 \sin t, 7 \cos t)$

(3)

$\vec{r}'(t) = (e^t(\sin t + \cos t), -e^{-t}(\cos t + \sin t), -3e^{3t})$

(4)

$\vec{r}'(t) = (1 + \ln t, 7e^t(\sin t + \cos t), -\sin t + \cos t)$

(5)

$\vec{r}'(t) = (4t^3, (1 - t)e^{-t}, -2 \sin 2t)$


Example 3.

试求下列各向量值函数的 $\vec{r}'(t)$ 以及 $|\vec{r}'(t)| = ?$

(1)$\vec{r}(t) = (t, at),\ a > 0$　　(2)$\vec{r}(t) = (t, 4t, 3t)$　　　　(3)$\vec{r}(t) = (\cos t, \sin t, t)$

(4)$\vec{r}(t) = (e^t, e^{at}, e^{-t})$　　　(5)$\vec{r}(t) = (0, 3 \sin t, \sin^2 t)$　　(6)$\vec{r}(t) = (e^t \sin t, e^t \cos t)$

(7)$\vec{r}(t) = (r \cos t, r \sin t),\ r > 0$

【解】

(1)

$\vec{r}'(t) = (1, a),\ |\vec{r}'(t)| = \sqrt{1 + a^2}$

(2)

$\vec{r}'(t) = (1, 4, 3),\ |\vec{r}'(t)| = \sqrt{26}$

(3)

$\vec{r}'(t) = (-\sin t, \cos t, 1),\ |\vec{r}'(t)| = \sqrt{\cos^2 t + \sin^2 t + 1} = \sqrt{2}$

(4)

$\vec{r}'(t) = (e^t, ae^t, -e^{-t}),\ |\vec{r}'(t)| = \sqrt{e^{2t}(1 + a^2) + e^{-2t}}$

(5)

$\vec{r}'(t) = (0, 3 \cos t, 2 \sin t \cos t),\ |\vec{r}'(t)| = \sqrt{9 \cos^2 t + 4 \sin^2 t \cos^2 t}$

(6)

$\vec{r}'(t) = \left(e^t(\sin t + \cos t), e^t(\cos t - \sin t)\right),\ |\vec{r}'(t)| = \sqrt{2e^{2t}}$

(7)

$$\vec{r}'(t) = (-r\sin t, r\cos t), \quad |\vec{r}'(t)| = r\sqrt{\cos^2 t + \sin^2 t} = r$$

Example 4.

The curve $C$ is given by the vector- valued function $\vec{r}(t) = (t, t^2), \forall t \in R$. Is $C$ smooth ?

【解】

$\because \vec{r}'(t) = (1, 2t) \quad \therefore \vec{r}'(t) \neq \vec{0}, \forall t \in R$ 且 $\vec{r}'(t)$ 连续, $\forall t \quad \therefore C$ is a smooth curve

Example 5.

The curve $C$ is given by the vector- valued function $\vec{r}(t) = (t^4, t^5), \forall t \in R$. Is $C$ smooth ?

【解】

$\because \vec{r}'(t) = (4t^3, 5t^4), \quad \therefore \vec{r}'(0) = (0,0) \quad \therefore C$ is not a smooth curve

Example 6.

The curve $C$ is given by parametric equations $\vec{r}(t) = (t, |t|), \forall t \in R$. Is $C$ smooth ?

【解】

$\because \vec{r}'(0)$ doesn't exist $\therefore C$ is not a smooth curve

Example 7.

试判断下列曲线是否为平滑曲线

(1) $\vec{r}(t) = (1 + t^3, te^{-t}, \sin 2t), \forall t \in R$

(2) $\vec{r}(t) = (3\cos t, 4\sin t), \forall t \in R$

(3) $\vec{r}(t) = (e^t \sin t, e^t \cos t, -e^{2t}), \forall t \in R$

(4) $\vec{r}(t) = (t\ln t, 5e^t \cos t, \cos t - \sin t), \forall t \in R$

(5) $\vec{r}(t) = (0, 3\sin t, \sin^2 t), \forall t \in R$

(6) $\vec{r}(t) = (e^t \sin t, e^{-t} \cos t)$

(7) $\vec{r}(t) = (t^3 - 12t + 17, \ t^2 - 4t + 8), \forall t \in R$

(8) $\vec{r}(t) = \left(t, \ t^2 \sin\dfrac{1}{t}\right), \ \forall t \neq 0$ and $\vec{r}(0) = \vec{0}$.

【解】

(1)

$\because \vec{r}'(t) = (3t^2, (1-t)e^{-t}, 2\cos 2t) \neq \vec{0}$ 且 $\vec{r}'(t)$ 连续, $\forall t \in R \quad \therefore C$ is smooth

(2)

$\because \vec{r}'(t) = (-3\sin t, 4\cos t) \neq \vec{0}$ 且 $\vec{r}'(t)$ 连续, $\forall t \in R \quad \therefore C$ is smooth

(3)

$\because \vec{r}'(t) = (e^t(\sin t + \cos t), e^t(\cos t - \sin t), -2e^{2t}) \neq \vec{0}$ 且 $\vec{r}'(t)$ 连续, $\forall t \in R$

$\therefore C$ is smooth

(4)

$\because \vec{r}'(t) = (1 + \ln t, 5e^t(\cos t - \sin t), -\sin t - \cos t) \neq \vec{0}$ 且 $\vec{r}'(t)$ 连续, $\forall t \in R$

$\therefore C$ is smooth

(5)

$\because \vec{r}'(t) = (0, 3\cos t, 2\sin t \cos t) \quad \therefore \vec{r}'\left(\dfrac{\pi}{2}\right) = \vec{0} \quad \therefore C$ is not smooth

(6)

$\because \vec{r}'(t) = \left(e^t(\sin t + \cos t), -e^t(\sin t + \cos t)\right) = \left(e^t \sin\left(t + \dfrac{\pi}{4}\right), -e^t \sin\left(t + \dfrac{\pi}{4}\right)\right)$

$\therefore \vec{r}'\left(-\dfrac{\pi}{4}\right) = \vec{0} \quad \therefore C$ is not smooth

(7)

$\because x(t) = t^3 - 12t + 17, \ y(t) = t^2 - 4t + 8$

$\therefore x'(t) = 3t^2 - 12, \ y'(t) = 2t - 4 \Rightarrow x'(2) = y'(2) = 0$

$\therefore C$ is not a smooth curve

(8)

$\because y(t) = t^2 \sin\dfrac{1}{t} \quad \therefore y'(t) = -\cos\dfrac{1}{t} + 2t\sin\dfrac{1}{t}, \ \forall t \neq 0$

$\because y'(0) = 0 \ \text{ and } \ \lim\limits_{t \to 0} y'(t) = \lim\limits_{t \to 0} -\cos\dfrac{1}{t} + 2t\sin\dfrac{1}{t} \neq 0$

$\therefore y'(t)$ isn't continuous at $t = 0 \quad \therefore \vec{r}'(t)$ isn't continuous at $t = 0$

$\therefore C$ is not smooth


## 9.3 线积分

线积分的被积分函数可能为纯量函数或向量函数, 曲线 $C$ 可能为二维平面或三维空间的曲线, 类型通常包含: 参数式、圆、圆弧、椭圆..等; 无论是哪种情形, 关键步骤皆是将其转换为单变量的积分, 因此, 要清楚转换后被积分函数的样貌; 曲线 $C$ 从样貌上做区分又可分为封闭或非封闭, 当曲线为二维平面的封闭曲线时, 有时可藉由 Green's Theorem 将线积分转为双重积分, 当曲线是三维空间的封闭曲线时, 有时可藉由 Stokes' Theorem 将

线积分转为曲面积分

## 9.3.1　$f$ 为纯量函数, 給曲线 $C$ 求线积分

考虑一条空间曲线 $C$, 由参数方程式描述：

$$x = x(t), \quad y = y(t), \quad z = z(t), \quad \forall a \le t \le b$$

或者, 藉由向量值函数表示：

$$\vec{r}(t) = \big(x(t), y(t), z(t)\big), \quad \forall a \le t \le b$$

假设 $C$ 是一条光滑的空间曲线 $\big(\vec{r}'(t)$ 连续且 $\vec{r}'(t) \ne 0\big)$, 将定义域 $[a,b]$ 切成 $n$ 个等宽的子区间, 其中 $x_i = x(t_i)$, $y_i = y(t_i)$, $z_i = z(t_i)$, $\forall 0 \le i \le n$; 这些对应点 $(x_i, y_i, z_i)$ 将 $C$ 分成 $n$ 个子弧 $C_i$ 且长度分别为 $\Delta s_1, \Delta s_1 \ldots \Delta s_n \big(\Delta s_1, \Delta s_2, \ldots, \Delta s_n$ 不必然相等$\big)$; 在每个子弧中选任意点 $(x_i^*, y_i^*, z_i^*)$, 如果函数 $f$ 的定义域包含空间曲线 $C$, 则函数 $f$ 在点 $(x_i^*, y_i^*, z_i^*)$ 取值并乘以子弧的长度 $\Delta s_i$, 加总后则得到黎曼和的形式

$$\sum_{i=1}^{n} f(x_i^*, y_i^*, z_i^*) \Delta s_i,$$

藉由此黎曼和的极限可以定义纯量函数的线积分

【定义】

假设函数 $f$ 定义在平滑空间曲线 $C$ 上, 其中 $C: x = x(t)$, $y = y(t)$, $z = z(t), \forall a \le t \le b$, 如果极限

$$\lim_{n \to \infty} \sum_{i=1}^{n} f(x_i^*, y_i^*, z_i^*) \Delta s_i \ \text{存在},$$

则用 $\displaystyle\int_C f(x, y, z)\, ds$ 表示并称其为函数 $f$ 沿着空间曲线 $C$ 的线积分

平滑平面曲线的线积分也是类似, 考虑一条二维平面曲线 $C$, 由参数方程式描述：

$$x = x(t), \quad y = y(t), \quad \forall a \le t \le b$$

或者, 藉由向量值函数表示：

$$\vec{r}(t) = \big(x(t), y(t)\big), \quad \forall a \le t \le b$$

切割平面曲线 $C$ 的方式与三维空间类似(如下图), 函数 $f$ 在点 $(x_i^*, y_i^*)$ 取值并乘以子弧的长度 $\Delta s_i$, 加总后则得到黎曼和的形式

$$\sum_{i=1}^{n} f(x_i^*, y_i^*)\Delta s_i,$$

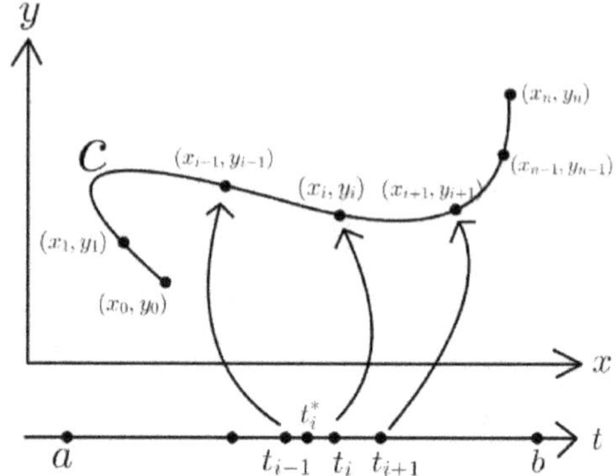

藉由此黎曼和的极限可定义纯量函数的线积分

## 【定义】

假设函数 $f$ 定义在平滑平面曲线 $C$ 上, 其中 $C: x = x(t),\ y = y(t),\ \forall a \le t \le b$, 如果极限

$$\lim_{n \to \infty} \sum_{i=1}^{n} f(x_i^*, y_i^*)\Delta s_i \ 存在,$$

则用 $\displaystyle\int_C f(x, y)\,ds$ 表示并称其为函数 $f$ 沿着平面曲线 $C$ 的线积分

值得注意的是, 因为任意定义域为闭区间的连续函数必为黎曼可积分函数, 因此, 如果上述定义的函数 $f$ 是一个连续函数, 则其上述黎曼和的极限值必定存在; 此外, 若曲线 $C(t)$ 为平滑曲线, $\forall t \in [a, b]$, 则 $\vec{r}'(t)$ 连续, $\forall a \le t \le b$, 藉由弧长公式,

$$\Delta s_i = \int_{t_{i-1}}^{t_i} |\vec{r}'(t)|\,dt.$$

令 $\Delta t_i = t_i - t_{i-1}$ 逼近零, 则

$$\Delta s_i = \int_{t_{i-1}}^{t_i} |\vec{r}'(t)|\,dt \ \approx \ |\vec{r}'(t_i^*)|\Delta t_i, \ \ 其中 t_i^* \in [t_i, t_{i-1}]$$

因此

$$\sum_{i=1}^{n} f(x_i^*, y_i^*)\Delta s_i = \sum_{i=1}^{n} f(\vec{r}(t_i^*))\Delta s_i \approx \sum_{i=1}^{n} f(\vec{r}(t_i^*))|\vec{r}'(t_i^*)|\Delta t_i$$

所以

$$\int_C f(x,y)ds = \lim_{n\to\infty}\sum_{i=1}^{n} f(x_i^*,y_i^*)\Delta s_i = \lim_{n\to\infty}\sum_{i=1}^{n} f(\vec{r}(t_i^*))|\vec{r}'(t_i^*)|\Delta t_i = \int_a^b f(\vec{r}(t))|\vec{r}'(t)|dt$$

上述的公式在说明计算函数 $f$ 沿着曲线 $C$ 的线积分相当于求单变量函数的黎曼积分, 特别的是单变量黎曼积分的被积分函数为

$$f(\vec{r}(t))|\vec{r}'(t)|$$

因此, 求曲线的线积分相当于先求得被积分函数之后, 再计算单变量黎曼积分; 底下针对曲线的线积分转换成单变数黎曼积分后, 有哪些不同类型作初步的说明; 掌握题型的重要关键需要熟悉曲线的样貌, 大方向可以区分为两类, 一者为二维平面的曲线, 另一者为三维空间的曲线; 再者又可细分为圆与非圆, 以及曲线是否用参数方程式表示, 在这些不同情境之下对于 $\int_a^b f(\vec{r}(t))|\vec{r}'(t)|dt$ 都会有一个明确对应的计算公式

假设 $\vec{r}(t) = (x(t),y(t))$ 则 $\vec{r}'(t) = (x'(t),y'(t)) \Rightarrow |\vec{r}'(t)| = \sqrt{(x'(t))^2 + (y'(t))^2}$

因此, 函数 $f$ 沿着曲线 $C$ 的线积分能改写为底下的公式并且用此公式求线积分

$$\int_C f(x,y)ds = \int_a^b f(\vec{r}(t))|\vec{r}'(t)|dt = \int_a^b f(x(t),y(t))\sqrt{(x'(t))^2 + (y'(t))^2}dt$$

再者, 当给定二维曲线 $C: y = g(x), \forall x \in [a,b]$, 即 $\vec{r}(x) = (x,g(x)) \Rightarrow \vec{r}'(x) = (1,g'(x))$

$$且 \quad |\vec{r}'(x)| = \sqrt{1 + (g'(x))^2}$$

因此, 纯量函数 $f$ 于曲线 $C$ 的线积分能改写为底下的公式, 并且用此公式求线积分

$$\int_C f(x,y)ds = \int_a^b f(\vec{r}(x))|\vec{r}'(x)|dx = \int_a^b f(x,g(x))\sqrt{1 + (g'(x))^2}dx$$

当给定圆 $C: x^2 + y^2 = r^2, r > 0$, 藉由极坐标参数化

$$x = r\cos\theta, y = r\sin\theta \quad 则 \quad \vec{r}(\theta) = (r\cos\theta, r\sin\theta), \quad 其中 \ 0 \le \theta \le 2\pi,$$

因为

$$\vec{r}'(\theta) = (-r\sin\theta, r\cos\theta) \quad 且 \quad |\vec{r}'(\theta)| = r\sqrt{\cos^2\theta + \sin^2\theta} = r$$

因此, 线积分则能改写为底下的公式, 并且用此公式求线积分

$$\int_C f(x,y)\,ds = \int_0^{2\pi} f(\vec{r}(\theta))|\vec{r}'(\theta)|\,d\theta = \int_0^{2\pi} f(r\cos\theta, r\sin\theta)\,r\,d\theta$$

上述为二维曲线在不同情况下, 计算线积分的方式, 接下来说明三维曲线的情形, 考虑一条参数化的平滑空间曲线 $C: \vec{r}(t) = (x(t), y(t), z(t))$, $\forall a \leq t \leq b$, 如果 $f$ 是一个定义于曲线 $C$ 的三变数连续函数, 使用与平面曲线相似的方法定义沿着空间曲线 $C$ 的 $f$ 的线积分

$$\int_C f(x,y,z)\,ds = \lim_{n\to\infty} \sum_{i=1}^{n} f(x_i^*, y_i^*, z_i^*)\Delta s_i$$

使用一个类似于对平面曲线应用的公式计算空间曲线 $C$ 的线积分,

$$\int_C f(x,y,z)\,ds = \int_a^b f(\vec{r}(t))|\vec{r}'(t)|\,dt$$

因为

$$\vec{r}'(t) = (x'(t), y'(t), z'(t)) \text{ 且 } |\vec{r}'(t)| = \sqrt{(x'(t))^2 + (y'(t))^2 + (z'(t))^2}$$

因此, 线积分能改写为底下的公式并且用此公式求线积分

$$\int_C f(x,y,z)\,ds = \int_a^b f(\vec{r}(t))|\vec{r}'(t)|\,dt$$
$$= \int_a^b f(x(t), y(t), z(t))\sqrt{(x'(t))^2 + (y'(t))^2 + (z'(t))^2}\,dt.$$

上述是针对单一光滑曲线推导明确的计算方式, 以考试类型而言, 曲线 $C$ 可能是一条分段光滑曲线(如下图), 接下来说明当线积分的曲线 $C$ 为分段光滑曲线时, 如何进行定义; 假设 $C$ 是一条分段光滑曲线, 即它由有限个光滑曲线 $C_1, ..., C_n$ 组成, 其中 $C_i$ 的终点与 $C_{i+1}$ 的起点重合, 则 $f$ 沿着 $C$ 的线积分 $= f$ 沿着每个光滑片段 $C_j$ 线积分之和, $\forall 1 \leq j \leq n$:

$$\int_C f(x,y,z)\,ds = \int_{C_1} f(x,y,z)\,ds + \cdots + \int_{C_n} f(x,y,z)\,ds$$

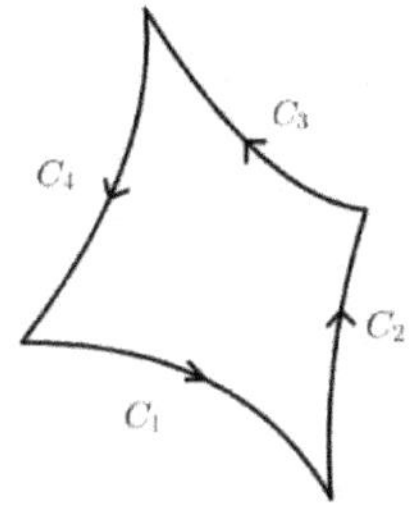 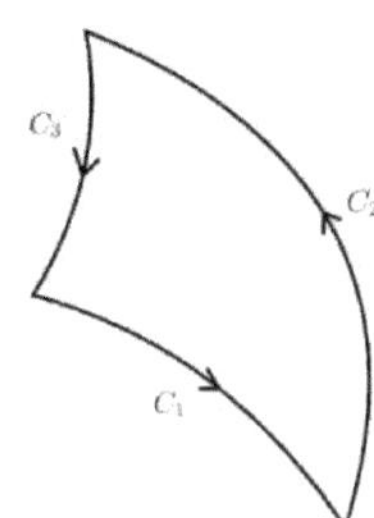 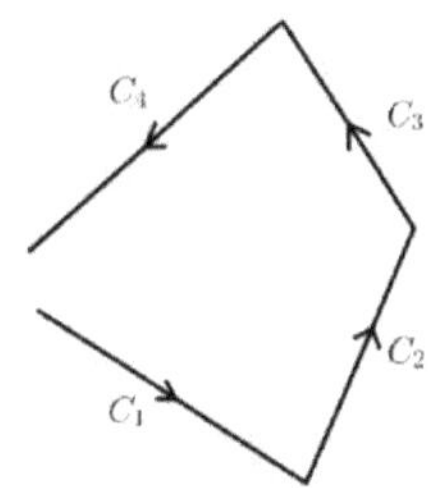

考试类型:

Type 1.

求 $\displaystyle\int_C f(x,y)ds =?$, 其中曲线 $C: y = g(x), \ \forall x \in [a,b]$

解题流程:

Step1.

令 $\vec{r}(x) = \big(x, g(x)\big)$ 则 $\vec{r}'(x) = (1, g'(x))$ 且 $|\vec{r}'(x)| = \sqrt{1 + \big(g'(x)\big)^2}$

Step2.

$$\int_C f(x,y)ds = \int_a^b f\big(\vec{r}(x)\big)|\vec{r}'(x)|dx = \int_a^b f\big(x, g(x)\big)\sqrt{1 + \big(g'(x)\big)^2}\,dx$$

Type 2.

求 $\displaystyle\int_C f(x,y)ds =?$, 其中曲线 $C: \big(x(t), y(t)\big), \ \forall a \le t \le b$ 为二维空间中的线段

求 $\displaystyle\int_C f(x,y,z)ds =?$, 其中曲线 $C: \big(x(t), y(t), z(t)\big), \ \forall a \le t \le b$ 为三维空间中的线段

解题流程:

Step1.

若曲线 $C$ 为二维空间中的线段,

令 $C: \vec{r}(t) = \big(x(t), y(t)\big), \ \forall a \le t \le b$ 则

$$\vec{r}'(t) = \big(x'(t), y'(t)\big) \quad 且 \quad |\vec{r}'(t)| = \sqrt{\big(x'(t)\big)^2 + \big(y'(t)\big)^2}$$

$$\therefore \int_C f(x,y)ds = \int_a^b f\big(\vec{r}(t)\big)|\vec{r}'(t)|dt = \int_a^b f(x(t), y(t))\sqrt{\big(x'(t)\big)^2 + \big(y'(t)\big)^2}\,dt$$

若曲线 $C$ 为三维空间中的线段,

令 $C: \vec{r}(t) = \big(x(t), y(t), z(t)\big), \ \forall a \le t \le b$ 则

$$\vec{r}'(t) = \big(x'(t), y'(t), z'(t)\big) \ \text{且} \ |\vec{r}'(t)| = \sqrt{\big(x'(t)\big)^2 + \big(y'(t)\big)^2 + \big(z'(t)\big)^2}$$

$$\therefore \int_C f(x,y,z)ds = \int_a^b f(\vec{r}(t))|\vec{r}'(t)|dt$$

$$= \int_a^b f(x(t), y(t), z(t))\sqrt{\big(x'(t)\big)^2 + \big(y'(t)\big)^2 + \big(z'(t)\big)^2}\, dt$$

Step2.

若曲线 $C$ 为二维空间中的线段, 求 $\displaystyle\int_a^b f(x(t), y(t))\sqrt{\big(x'(t)\big)^2 + \big(y'(t)\big)^2}\, dt$

若曲线 $C$ 为三维空间中的线段, 求 $\displaystyle\int_a^b f(x(t), y(t), z(t))\sqrt{\big(x'(t)\big)^2 + \big(y'(t)\big)^2 + \big(z'(t)\big)^2}\, dt$

<u>范例说明:</u>

(I) 求 $\displaystyle\int_C f(x,y)ds =?$, 其中 $f(x,y) = xy^3$, $C: \vec{r}(t) = (t, at)$, $-1 \le t \le 1$

$\because \vec{r}(t) = (t, at) \quad \therefore \vec{r}'(t) = (1, a)$ 且 $|\vec{r}'(t)| = \sqrt{1 + a^2}$

$$\int_C f(x,y)ds = \int_{-1}^1 f(\vec{r}(t))|\vec{r}'(t)|dt = \int_{-1}^1 t\,(at)^3 \cdot \sqrt{1 + a^2}\, dt = \frac{2a^3\sqrt{1 + a^2}}{5}$$

Type 3.

求 $\displaystyle\int_C f(x,y)ds =?$, 其中曲线 $C$ 为圆 $C: x^2 + y^2 = r^2$, $r > 0$

解题流程:

Step1.

令 $x = r\cos\theta$, $y = r\sin\theta$, $0 \le \theta \le 2\pi$ 则 $\vec{r}'(\theta) = (-r\sin\theta, r\cos\theta)$

$\therefore |\vec{r}'(\theta)| = r\sqrt{\cos^2\theta + \sin^2\theta} = r$

$$\therefore \int_C f(x,y)ds = \int_0^{2\pi} f(\vec{r}(\theta))|\vec{r}'(\theta)|d\theta = \int_0^{2\pi} f(r\cos\theta, r\sin\theta)r\,d\theta$$

Step2.

求 $\displaystyle\int_0^{2\pi} f(r\cos\theta, r\sin\theta)r\,d\theta =?$

<u>范例说明：</u>

求 $\displaystyle\int_C f(x,y)ds =?$，其中曲线 $C$ 为圆 $C: x^2 + y^2 = r^2,\ r > 0$

(I) 若 $f(x,y) = e^{x^2+y^2}$ 则 $\displaystyle\int_C e^{x^2+y^2}ds = \int_0^{2\pi} e^{r^2} r\,d\theta$

(II) 若 $f(x,y) = \ln(x^2 + y^2)$ 则 $\displaystyle\int_C \ln(x^2 + y^2)\,ds = \int_0^{2\pi} \ln(r^2)\,r\,d\theta$

(III) 若 $f(x,y) = \sin(x^2 + y^2)$ 则 $\displaystyle\int_C \sin(x^2 + y^2)\,ds = \int_0^{2\pi} \sin(r^2)\,r\,d\theta$

(IV) 若 $f(x,y) = \sin^{-1}(x^2 + y^2)$ 则 $\displaystyle\int_C \sin^{-1}(x^2 + y^2)\,ds = \int_0^{2\pi} \sin^{-1}(r^2)\,r\,d\theta$

(V) 若 $f(x,y) = \tan^{-1}(x^2 + y^2)$ 则 $\displaystyle\int_C \tan^{-1}(x^2 + y^2)\,ds = \int_0^{2\pi} \tan^{-1}(r^2)\,r\,d\theta$

(VI) 若 $f(x,y) = \dfrac{1}{\sqrt{x^2 + y^2}}$ 则 $\displaystyle\int_C \frac{1}{\sqrt{x^2 + y^2}}\,ds = \int_0^{2\pi} \int_0^a dr\,d\theta$

(VII) 若 $f(x,y) = \dfrac{1}{\sqrt{1 + x^2 + y^2}}$ 则 $\displaystyle\int_C \frac{1}{\sqrt{1 + x^2 + y^2}}\,ds = \int_0^{2\pi} \frac{r}{\sqrt{1 + r^2}}\,d\theta$

Example 1.

求 $\displaystyle\int_C xy^2 dt =?,\ \ C: x = 4t,\ y = e^t,\ 0 \le t \le 2$

【解】

$$\int_C xy^2 dt = \int_0^2 4t \cdot e^{2t}\,dt = 4\left(\frac{e^{2t}t}{2}\bigg|_0^2 - \frac{1}{2}\int_0^2 e^{2t}\,dt\right) = 4\left(e^4 - \frac{e^4 - 1}{4}\right) = 3e^4 + 1$$

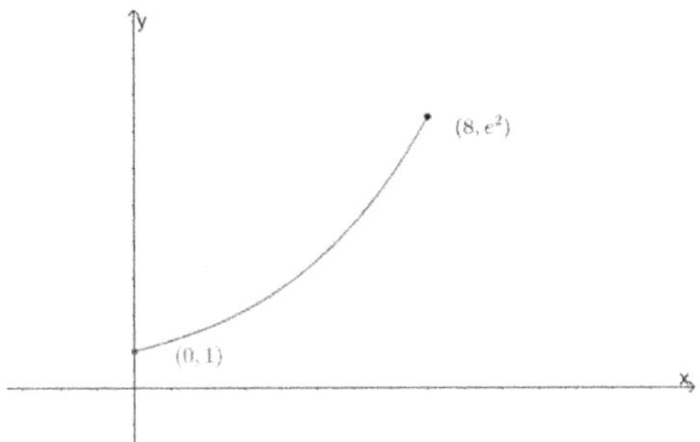

Example 2.

假设 $C: y = ax$ 且 $f(x,y) = xy^3$, 求 $f$ 沿着 $C$ 由点 $(-1,-a)$ 至点 $(1,a)$, $(a > 0)$ 的线积分

$$\int_C f(x,y)ds = ?$$

【解】

令 $\vec{r}(t) = (t, at), -1 \le t \le 1$ 则 $\vec{r}'(t) = (1, a)$ 且 $|\vec{r}'(t)| = \sqrt{1 + a^2}$

$$\therefore \int_C f(x,y)ds = \int_{-1}^{1} f(\vec{r}(t))|\vec{r}'(t)|dt = \int_{-1}^{1} t\,(at)^3 \cdot \sqrt{1 + a^2}\,dt = \frac{2a^3\sqrt{1 + a^2}}{5}$$

Example 3.

$$\text{求} \int_C 2 + x^2 y\, ds = ?, \quad C: x^2 + y^2 = r^2, \quad y \ge 0$$

【解】

令 $\vec{r}(\theta) = (r\cos\theta, r\sin\theta), \ 0 \le \theta \le \pi$ 则 $\vec{r}'(\theta) = (-r\sin\theta, r\cos\theta)$ 且 $|\vec{r}'(\theta)| = r$

$$\int_C 2 + x^2 y\, ds = \int_0^{\pi} (2 + r^3 \cos^2\theta \sin\theta) r\, d\theta = 2\pi r - r^4 \cdot \left.\frac{\cos^3\theta}{3}\right|_0^{\pi} = 2\pi r + \frac{2r^4}{3}$$

Example 4.

$$\text{求} \int_C x - 3y^2 + z\, ds = ?, \quad C \text{为} (0,0,0) \text{至} (1,4,3) \text{的直线线段}$$

【解】

令 $\vec{r}(t) = (t, 4t, 3t), \ 0 \le t \le 1$ 则 $\vec{r}'(t) = (1,4,3)$ 且 $|\vec{r}'(t)| = \sqrt{26}$

$$\therefore \int_C x - 3y^2 + z\, ds = \int_0^1 (t - 3(4t)^2 + 3t) \cdot \sqrt{26}\,dt = -14\sqrt{26}$$

Example 5.

$$\text{求} \int_C 2x + y\, ds = ?, \quad C \text{为圆}: x^2 + y^2 = 25 \text{ 从} (4,3) \text{至} (3,4) \text{的圆弧}$$

【解】

令 $\vec{r}(\theta) = (5\cos\theta, 5\sin\theta)$ 则 $\cos^{-1}\frac{4}{5} \le \theta \le \cos^{-1}\frac{3}{5}$ 且 $\sin^{-1}\frac{3}{5} \le \theta \le \sin^{-1}\frac{4}{5}$

則 $\vec{r}'(\theta) = (-5\sin\theta, 5\cos\theta)$ 且 $|\vec{r}'(\theta)| = 5\sqrt{\cos^2\theta + \sin^2\theta}\, d\theta = 5$

$$\int_C 2x + y\,ds = \int_{\sin^{-1}\frac{3}{5}}^{\sin^{-1}\frac{4}{5}} 2\cdot 5\cos\theta \cdot |\vec{r}'(\theta)|\,d\theta + \int_{\cos^{-1}\frac{4}{5}}^{\cos^{-1}\frac{3}{5}} 5\sin\theta \cdot |\vec{r}'(\theta)|\,d\theta$$

$$= 5\int_{\sin^{-1}\frac{3}{5}}^{\sin^{-1}\frac{4}{5}} 10\cos\theta\,d\theta + 5\int_{\cos^{-1}\frac{4}{5}}^{\cos^{-1}\frac{3}{5}} 5\sin\theta\,d\theta = -25\cos\theta\Big|_{\cos^{-1}\frac{4}{5}}^{\cos^{-1}\frac{3}{5}} + 50\sin\theta\Big|_{\sin^{-1}\frac{3}{5}}^{\sin^{-1}\frac{4}{5}} = 15$$

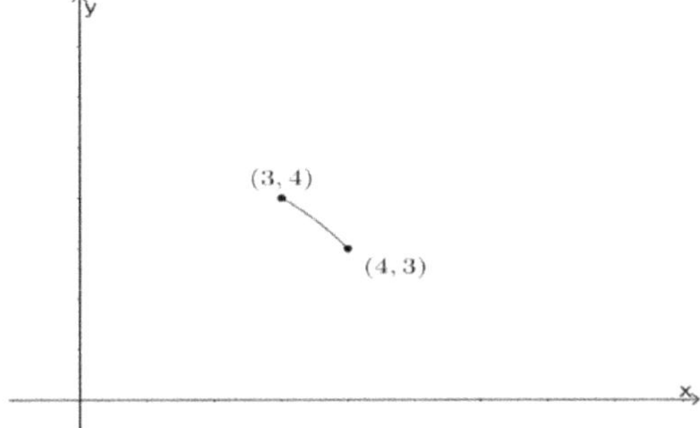

Example 6.

$\quad$ 求 $\displaystyle\int_C 3 + x^2 y\,ds =?$, 其中曲线 $C$ 是单位圆 $x^2 + y^2 = 1$ 于第一象限的圆弧

【解】

令 $\vec{r}(t) = (\cos t, \sin t)$ 則 $\vec{r}'(t) = (-\sin t, \cos t)$ 且 $|\vec{r}'(t)| = \sqrt{\sin^2 t + \cos^2 t} = 1\,, 0 \leq t \leq \dfrac{\pi}{2}$

$$\therefore \int_C 3 + x^2 y\,ds = \int_0^{\frac{\pi}{2}} (3 + \cos^2 t \sin t)|\vec{r}'(t)|\,dt = \int_0^{\frac{\pi}{2}} (3 + \cos^2 t \sin t)\,dt = \left(3t - \frac{\cos^3 t}{3}\right)\Bigg|_0^{\frac{\pi}{2}}$$

$$= \frac{3\pi}{2} + \frac{1}{3}$$

Example 7.

$\quad$ 求 $\displaystyle\int_C 2x\,ds =?$, 其中 $C$ 包含抛物线 $C_1$ 沿着 $y = x^2$ 從 $(0,0)$ 至 $(2,4)$ 以及垂直线 $C_2$ 從 $(2,4)$ 至 $(2,5)$

【解】

$\because C_1: y = x^2,\ 0 \leq x \leq 2$

$$\therefore \int_{C_1} 2xds = \int_0^2 2x\sqrt{\left(\frac{dx}{dx}\right)^2 + \left(\frac{dy}{dx}\right)^2}\,dx = \int_0^2 2x\sqrt{1+4x^2}\,dx = \frac{17\sqrt{17}-1}{6}$$

$$\because C_2: x = 2, \ 4 \le y \le 5 \quad \therefore \int_{C_2} 2xds = \int_4^5 2\sqrt{\left(\frac{dx}{dy}\right)^2 + \left(\frac{dy}{dy}\right)^2}\,dy = 2$$

$$\therefore \int_C 2xds = \int_{C_1} 2xds + \int_{C_2} 2xds = \frac{17\sqrt{17}-1}{6} + 2$$

Example 8.

$$求 \int_C y^2dx + xdy, \quad (a)\, C = C_1: 從 (-5,-3) 至 (0,2) 的直线线段 \quad (b)\, C = C_2: 沿着拋物$$
线 $x = 4 - y^2$ 從 $(-5,-3)$ 至 $(0,2)$

【解】

$(a)$

$\because C_1: x(t) = 5t - 5, \ y(t) = 5t - 3, \ \forall 0 \le t \le 1$

$$\therefore \int_{C_1} y^2dx + xdy = \int_0^1 (5t-3)^2 5dt + (5t-5)5dt = -\frac{5}{6}$$

$(b)$

$\because C_2: x = 4 - y^2, \ \forall -3 \le y \le 2$

$$\therefore \int_{C_2} y^2dx + xdy = \int_{-3}^2 y^2(-2y)dy + (4-y^2)dy = \frac{245}{6}$$

Example 9.

$$求 \int_C xydx + ydy, \quad 其中 \ C: x = 2t, y = 10t, \ 0 \le t \le 2$$

【解】

令 $C: \vec{r}(t) = (2t, 10t), \ \forall 0 \le t \le 2$

$$則 \int_C xydx + ydy = \int_0^2 20t^2 \cdot 2dt + 10t \cdot 10dt = \left(\frac{40t^3}{3} + 50t^2\right)\Big|_0^2 = \frac{320}{3} + 200 = \frac{920}{3}$$

Example 10.

$$求 \int_C \frac{y}{2x^2 - y^2} ds, \ 其中 \ C(t) = (t,t), \ 1 \le t \le 3$$

【解】

令 $C: \vec{r}(t) = (t,t), \ \forall \, 0 \le t \le 3$ 則 $\vec{r}'(t) = (1,1)$ 且 $|\vec{r}'(t)| = \sqrt{2}$

$$\therefore \int_C \frac{y}{2x^2 - y^2} ds = \int_1^3 \frac{y(t)|\vec{r}'(t)|}{2x^2(t) - y^2(t)} dt = \sqrt{2} \int_1^3 \frac{t}{t^2} dt = \sqrt{2} \int_1^3 \frac{1}{t} dt = \sqrt{2} \ln t \big|_1^3 = \sqrt{2} \ln 3$$

Example 11.

$$求 \int_C xyz \, ds = ?, \ 其中 \ C: x = 2\sin t, y = t, z = -2\cos t, \ \forall 0 \le t \le \pi$$

【解】

令 $C: \vec{r}(t) = (2\sin t, t, -2\cos t), \ \forall 0 \le t \le \pi$

則 $\vec{r}'(t) = (2\cos t, 1, 2\sin t)$ 且 $|\vec{r}'(t)| = \sqrt{(2\cos t)^2 + (1)^2 + (2\sin t)^2} = \sqrt{5}$

$$\therefore \int_C xyz \, ds = -4 \int_0^\pi t \sin t \cos t \cdot |\vec{r}'(t)| dt = -4\sqrt{5} \int_0^\pi t \sin t \cos t \, dt = -2\sqrt{5} \int_0^\pi t \sin 2t \, dt$$

$$= 2\sqrt{5} \cdot \frac{t \cos 2t}{2} \bigg|_0^\pi = \sqrt{5}\pi$$

Example 12.

Evaluate $\int_C x \cos z \ ds = ?$, where

(1) $C$ is the circular helix given by the equation $x = \cos t, y = \sin t, z = t, 0 \le t \le 2\pi$

(2) $C$ is the circular helix given by the equation $x = \cos t, y = \sin t, z = 0, 0 \le t \le 2\pi$

【解】

(1)

$$\int_C x \cos z \ ds = \int_0^{2\pi} \cos t \cos t \sqrt{\left(x'(t)\right)^2 + \left(y'(t)\right)^2 + \left(z'(t)\right)^2} dt$$

$$= \int_0^{2\pi} \cos^2 t \sqrt{\sin^2 t + \cos^2 t + 1} dt = \sqrt{2} \int_0^{2\pi} \frac{1 + \cos 2t}{2} dt = \sqrt{2}\pi$$

(2)

$$\int_C x\cos z \; ds = \int_0^{2\pi} \cos t \sqrt{\left(x'(t)\right)^2 + \left(y'(t)\right)^2 + \left(z'(t)\right)^2}\, dt = \int_0^{2\pi} \cos t \sqrt{\sin^2 t + \cos^2 t}\, dt$$
$$= 0$$

Example 13.

求 $\int_C ydx + zdy + xdz$,其中 $C$ 包含 $C_1$ 從 $(3,0,0)$ 至 $(4,4,5)$ 的直線線段, 以及從 $(4,4,5)$ 至 $(4,4,0)$ 的垂直線段.

【解】

$\because C_1 : \vec{r}(t) = (1-t)(3,0,0) + t(4,4,5) = (3+t, 4t, 5t), 0 \leq t \leq 1$

$\therefore \int_{C_1} ydx + zdy + xdz = \int_0^1 4tdt + 5t(4)dt + (3+t)5dt = \dfrac{59}{2}$

$\because C_2 : \vec{r}(t) = (1-t)(4,4,5) + t(4,4,0) = (4,4,5-5t), 0 \leq t \leq 1$

$\therefore \int_{C_2} ydx + zdy + xdz = \int_0^1 -20dt = -20$

$\therefore \int_C ydx + zdy + xdz = \dfrac{59}{2} - 20 = \dfrac{19}{2}$

Example 14.

求 $\int_C x - yds =?$, 其中 $\vec{r}(t) = (4t, 3t), \ \forall \, 0 \leq t \leq 4$

【解】

$\because C : \vec{r}(t) = (4t, 3t), \ \forall \, 0 \leq t \leq 4 \quad \therefore \vec{r}'(t) = (4,3)$ 且 $|\vec{r}'(t)| = 5, \ \forall \, 0 \leq t \leq 4$

$\therefore \int_C x - yds = \int_0^4 (x(t) - y(t))|\vec{r}'(t)|dt = \int_0^4 t \cdot 5dt = \dfrac{5t^2}{2}\Big|_0^4 = 40$

Example 15.

求 $\int_C xy^4 ds =?$,其中 $C$ 为沿着 $x^2 + y^2 = 16$ 逆时针绕行的右半圆

【解】

令 $\vec{r}(t) = (4\cos t, 4\sin t)$ 則 $\vec{r}'(t) = (-4\sin t, 4\cos t)$ 且 $|\vec{r}'(t)| = 4, -\dfrac{\pi}{2} \leq t \leq \dfrac{\pi}{2}$

$$\therefore \int_C xy^4 ds = \int_{-\frac{\pi}{2}}^{\frac{\pi}{2}} x(t)y^4(t)|\vec{r}'(t)|dt = 4^5 \int_{-\frac{\pi}{2}}^{\frac{\pi}{2}} (\cos t \sin^4 t) \cdot 4dt = 4^6 \int_{-\frac{\pi}{2}}^{\frac{\pi}{2}} \cos t \sin^4 t \, dt$$

$$= 4^6 \left(\frac{\sin^5 t}{5}\right)\Big|_{-\frac{\pi}{2}}^{\frac{\pi}{2}} = \frac{8192}{5}$$

Example 16.

$$\text{求} \int_C yzdx + xzdy + xydz = ?, \quad C \text{为從 } (1,1,1) \text{ 至 } (3,2,0) \text{的直线线段}$$

【解】

令 $C: \vec{r}(t) = (1 - t)(1,1,1) + t(3,2,0) = (1 + 2t, t + 1, 1 - t), \quad 0 \le t \le 1$

$$\therefore \int_C yzdx + xzdy + xydz = \int_0^1 (t + 1)(1 - t)2 + (1 + 2t)(1 - t) - (1 + 2t)(t + 1)dt$$

$$= \int_0^1 2 - 2t - 6t^2 dt = 2t - t^2 - 2t^3|_0^1 = -1$$

Example 17.

$$\text{Evaluate} \int_C x + y + z \, ds = ?, \quad \text{where } C \text{ is parameterized by } (\cos t, \sin t, t), t \in [0, 2\pi]$$

【解】

令 $\vec{r}(t) = \big(x(t), y(t), z(t)\big) = (\cos t, \sin t, t), \quad \forall 0 \le t \le 2\pi$

則 $\vec{r}'(t) = (-\sin t, \cos t, 1)$ 且 $|\vec{r}'(t)| = \sqrt{2}, \quad \forall 0 \le t \le 2\pi$

$$\therefore \int_C x + y + z \, ds = \int_0^{2\pi} \big(x(t) + y(t) + z(t)\big)|\vec{r}'(t)| \, dt = \int_0^{2\pi} (\cos t + \sin t + t)\sqrt{2} \, dt$$

$$= 2\sqrt{2}\pi^2$$

Example 18.

$$\text{求} \int_C (x^2 + y^2 + z^2)^2 ds = ?, \text{其中曲线 } C \text{ 为沿着空间螺旋线} = (\cos t, \sin t, 3t) \text{從}$$

$(1,0,0)$ 至 $(1,0,6\pi)$ 的线段

【解】

令 $\vec{r}(t) = \big(x(t), y(t), z(t)\big) = (\cos t, \sin t, 3t), \quad \forall 0 \le t \le 2\pi$

则 $\vec{r}'(t) = (-\sin t, \cos t, 3)$ 且 $|\vec{r}'(t)| = \sqrt{10}, \quad \forall 0 \le t \le 2\pi$

$$\therefore \int_C (x^2 + y^2 + z^2)^2 ds = \int_0^{2\pi} \left(x^2(t) + y^2(t) + z^2(t)\right)^2 |\vec{r}'(t)|\, dt$$

$$= \int_0^{2\pi} ((\cos t)^2 + (\sin t)^2 + (3t)^2)^2 \sqrt{10}\, dt = \sqrt{10} \int_0^{2\pi} (1 + 9t^2)^2\, dt$$

$$= \sqrt{10} \left( 2\pi + 48\pi^3 + \frac{2592\pi^5}{5} \right)$$

Example 19.

$$求 \int_C x e^{yz} ds = ?, 其中 C 是從 (0,0,0) 至 (1,2,3) 的直线线段$$

【解】

令 $C: \vec{r}(t) = (1-t)(0,0,0) + t(1,2,3) = (t, 2t, 3t), \quad 0 \le t \le 1$

则 $\vec{r}'(t) = (1,2,3)$ 且 $|\vec{r}'(t)| = \sqrt{14}$

$$\therefore \int_C x e^{yz} ds = \int_0^1 x(t) e^{y(t)z(t)} |\vec{r}'(t)| dt = \int_0^1 t e^{6t^2} \sqrt{14} dt = \frac{\sqrt{14} e^{6t^2}}{12} \Big|_0^1 = \frac{\sqrt{14}(e^6 - 1)}{12}$$

Example 20.

$$求 \int_C z^2 dx + x^2 dy + y^2 dz = ?, 其中 C 是從 (1,0,0) 至 (4,1,2) 的直线线段$$

【解】

令 $C: \vec{r}(t) = (1-t)(1,0,0) + t(4,1,2) = (1 + 3t, t, 2t), \quad 0 \le t \le 1$

则 $\int_C z^2 dx + x^2 dy + y^2 dz = \int_0^1 4t^2 \cdot 3 + (1 + 3t)^2 + t^2 \cdot 2 dt$

$$= 4t^3 + (t + 3t^2 + 3t^3) + \frac{2t^3}{3} \Big|_0^1 = \frac{35}{3}$$

## 9.3.2    $\vec{F}$为向量函数, 给曲线 $C$ 求线积分

假设 $\vec{F}(x,y,z) = \left(f_1(x,y,z), f_2(x,y,z), f_3(x,y,z)\right)$ 是一个三维空间的连续向量场, 其定义域包含一条平滑的空间曲线 $C: \vec{r}(t) = \left(x(t), y(t), z(t)\right), \forall t \in [a,b]$, 将参数区间 $[a,b]$ 切成 $n$

个等宽的子区间，其中 $x_i = x(t_i)$, $y_i = y(t_i)$, $z_i = z(t_i), \forall 0 \le i \le n$; 这些对应点 $(x_i, y_i, z_i)$ 将 $C$ 分成 $n$ 个子弧 $C_i$ 且长度分别为 $\Delta s_1, \Delta s_1 \dots \Delta s_n \left(\Delta s_1, \Delta s_2, \dots, \Delta s_n$ 不必然相等$\right)$; 在第 $i$ 个子弧段上取一点 $(x_i^*, y_i^*, z_i^*)$, 对应参数值为 $t_i^*$, 以物理的角度来看, 当粒子沿着曲线 $C_i$ 经过时, 如果 $\Delta s_i$ 很小，将非常接近单位切向量 $\vec{u}(t_i^*)$ 的方向 $($如下图$)$, 因此, 力 $\vec{\mathbf{F}}$ 在沿着曲线 $C_i$ 将粒子从起点移动到终点的过程中所做的功，近似为：

$$\vec{\mathbf{F}}(x_i^*, y_i^*, z_i^*) \cdot \Delta s_i \vec{u}(t_i^*) = \vec{\mathbf{F}}\big(\vec{r}(t_i^*)\big) \cdot \Delta s_i \vec{u}(t_i^*)$$

因为曲线的切线向量为 $\vec{r}'(t) = \dfrac{d\vec{r}(t)}{dt}$, 所以单位切向量 $\vec{u}(t) = \dfrac{\vec{r}'(t)}{|\vec{r}'(t)|}$, 因此

$$\vec{\mathbf{F}}\big(\vec{r}(t_i^*)\big) \cdot \Delta s_i \vec{u}(t_i^*) = \vec{\mathbf{F}}\big(\vec{r}(t_i^*)\big) \cdot \Delta s_i \left(\frac{\vec{r}'(t_i^*)}{|\vec{r}'(t_i^*)|}\right)$$

所以沿着曲线移动粒子所做的总功近似等于

$$\sum_{i=1}^{n} \vec{\mathbf{F}}\big(\vec{r}(t_i^*)\big) \cdot \Delta s_i \left(\frac{\vec{r}'(t_i^*)}{|\vec{r}'(t_i^*)|}\right).$$

因为 $\Delta s_i = \displaystyle\int_{t_{i-1}}^{t_i} |\vec{r}'(t)| dt$, 当 $\Delta t_i = t_i - t_{i-1}$ 逼近零，则 $\Delta s_i = \displaystyle\int_{t_{i-1}}^{t_i} |\vec{r}'(t)| dt \approx |\vec{r}'(t_i^*)| \Delta t_i$

沿着曲线移动粒子所做的总功近似等于

$$\sum_{i=1}^{n} \vec{\mathbf{F}}\big(\vec{r}(t_i^*)\big) \cdot |\vec{r}'(t_i^*)| \Delta t_i \left(\frac{\vec{r}'(t_i^*)}{|\vec{r}'(t_i^*)|}\right) = \sum_{i=1}^{n} \left(\vec{\mathbf{F}}\big(\vec{r}(t_i^*)\big) \cdot \vec{r}'(t_i^*)\right) \Delta t_i.$$

所以，藉由力场 $\vec{\mathbf{F}}$ 所作的功 $W$ 定义为上述黎曼和的极限：

$$W = \lim_{n\to\infty} \sum_{i=1}^{n} \left(\vec{\mathbf{F}}\big(\vec{r}(t_i^*)\big) \cdot \vec{r}'(t_i^*)\right) \Delta t_i = \int_{a}^{b} \vec{\mathbf{F}}\big(\vec{r}(t)\big) \cdot \vec{r}'(t) dt$$

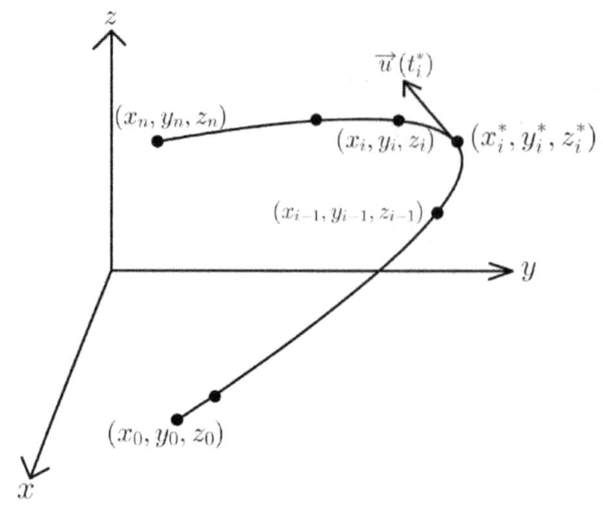


藉由上述的观察, 底下说明向量函数线积分的定义

## 【定义】

假设 $\vec{\mathbf{F}}$ 是一个连续向量场且定义于光滑曲线 $C: \vec{r}(t), \forall t \in [a,b]$则$\vec{\mathbf{F}}$沿着曲线$C$的线积分表示为

$$\int_C \vec{\mathbf{F}} \cdot d\vec{r} = \int_a^b \vec{\mathbf{F}}(\vec{r}(t)) \cdot \vec{r}'(t)dt$$

上述的公式说明计算向量函数的线积分相当于求单变量函数的黎曼积分, 特别的是单变量黎曼积分的被积分函数为

$$\vec{\mathbf{F}}(\vec{r}(t)) \cdot \vec{r}'(t)$$

因此, 求向量函数的线积分相当于要先求得被积分函数之后, 再计算单变量黎曼积分; 底下针对曲线的向量函数线积分转换成单变量黎曼积分后, 有哪些不同类型作初步说明; 值得注意的是向量函数的线积分以及纯量函数线积分的关联性, 假设 $\vec{F} = (f_1, f_2, f_3)$且曲线 $C: \vec{r}(t) = (x, y, z)$则 $d\vec{r} = (dx, dy, dz)$, 因此,

$$\int_C \vec{\mathbf{F}} \cdot d\vec{r} = \int_C (f_1, f_2, f_3) \cdot (dx, dy, dz) = \int_C f_1 dx + f_2 dy + f_3 dz.$$

掌握题型的重要关键与上一节相同, 需要熟悉曲线的样貌, 大方向可以区分为两类, 一者为二维平面的曲线, 另一者为三维空间的曲线; 再者又可细分为圆与非圆, 以及是否用参数方程式表示, 在这些不同情境下对于 $\int_a^b \vec{\mathbf{F}}(\vec{r}(t)) \cdot \vec{r}'(t)dt$ 需要一个明确对应的计算公式

给定向量函数 $\vec{\mathbf{F}} = \big(f_1(x, y), f_2(x, y)\big)$, 且曲线 $C$ 为二维曲线: $y = g(x)$, $\forall a \leq x \leq b$,

目标求线积分 $\int_C \vec{\mathbf{F}} \cdot d\vec{r}$, 首先对曲线做参数化,

$$\text{令 } \vec{r}(x) = (x, g(x)), \quad \text{则 } \vec{r}'(x) = (1, g'(x)),$$

因此线积分可表示为

$$\int_C \vec{\mathbf{F}} \cdot d\vec{r} = \int_a^b \vec{\mathbf{F}}(\vec{r}(x)) \cdot \vec{r}'(x)dx = \int_a^b \vec{\mathbf{F}}(\vec{r}(x)) \cdot (1, g'(x))dx$$

$$= \int_a^b f_1(x, g(x))dx + \int_a^b f_2(x, g(x))g'(x)dx$$

特别的是如果曲线 $C$ 为一个圆: $x^2 + y^2 = r^2$, 藉由极坐标转换

$$\text{令 } x = r\cos\theta, \quad y = r\sin\theta \quad \text{则 } \vec{r}(\theta) = (r\cos\theta, r\sin\theta), \quad \forall\, 0 \leq \theta \leq 2\pi$$

因为

$$\vec{r}'(\theta) = (-r\sin\theta, r\cos\theta)$$

所以, 线积分 $\int_C \vec{\mathbf{F}} \cdot d\vec{r}$ 可以表示为两个积分的和,

$$\int_C \vec{\mathbf{F}} \cdot d\vec{r} = \int_0^{2\pi} \vec{\mathbf{F}}(\vec{r}(\theta)) \cdot \vec{r}'(\theta)d\theta = \int_0^{2\pi} \big(f_1(\vec{r}(\theta)), f_2(\vec{r}(\theta))\big) \cdot (-r\sin\theta, r\cos\theta)d\theta$$

$$= \int_0^{2\pi} f_1(r\cos\theta, r\sin\theta)(-r\sin\theta)d\theta + \int_0^{2\pi} f_2(r\cos\theta, r\sin\theta)(r\cos\theta)d\theta$$

给定三维空间的向量函数 $\vec{\mathbf{F}} = \big(f_1(x, y, z), f_2(x, y, z), f_3(x, y, z)\big)$, 目标求 $\int_C \vec{\mathbf{F}} \cdot d\vec{r}$

其中曲线 $C$ 的向量函数为 $\vec{r}(t) = (x(t), y(t), z(t))$, $\forall a \leq t \leq b$, 因为

$$\vec{r}'(t) = (x'(t), y'(t), z'(t)),$$

所以

$$\int_C \vec{\mathbf{F}} \cdot d\vec{r} = \int_a^b \vec{\mathbf{F}}(\vec{r}(t)) \cdot \vec{r}'(t)dt$$

$$= \int_a^b f_1(\vec{r}(t))x'(t)dt + \int_a^b f_2(\vec{r}(t))y'(t)dt + \int_a^b f_3(\vec{r}(t))z'(t)dt$$

给定向量值函数 $\vec{\mathbf{F}} = \big(f_1(x, y, z), f_2(x, y, z), f_3(x, y, z)\big)$ 以及三维空间中的封闭椭圆曲

线 $C: \dfrac{x^2}{a^2} + \dfrac{y^2}{b^2} = 1$ 且 $z = c$, 因为曲线 $C = \left\{(x, y, z): \dfrac{x^2}{a^2} + \dfrac{y^2}{b^2} = 1, z = c\right\}$, 藉由极坐标转换

$x = a\cos\theta$, $y = b\sin\theta$ 则 $\vec{r}(\theta) = (a\cos\theta, b\sin\theta, c)$ 且 $\vec{r}'(\theta) = (-a\sin\theta, b\cos\theta, 0)$, 因此线积分可表示为

$$\oint_C \vec{\mathbf{F}} \cdot d\vec{r} = \int_0^{2\pi} \vec{\mathbf{F}}(\vec{r}(\theta)) \cdot \vec{r}'(\theta) d\theta$$

$$= \int_0^{2\pi} f_1(a\cos\theta, b\sin\theta)(-a\sin\theta)d\theta + \int_0^{2\pi} f_2(a\cos\theta, b\sin\theta) b\cos\theta\, d\theta$$

上述给出了当 $\vec{\mathbf{F}}$ 为向量值函数时, 曲线 $C$ 在不同的情境之下的计算公式, 然而, 针对不同考试类型而言, 必须给出详细的解题流程

考试类型:

Type 1.

给定向量值函数 $\vec{\mathbf{F}} = (f_1(x,y), f_2(x,y))$, 假设曲线 $C$ 为沿着 $y = g(x)$ 的二维平面线段, 其中 $a \le x \le b$, 求 $\displaystyle\int_C \vec{\mathbf{F}} \cdot d\vec{r} =?$

解题流程:

Step1.

令 $\vec{r}(x) = (x, g(x))$ 则 $\vec{r}'(x) = (1, g'(x))$

$$\therefore \int_C \vec{\mathbf{F}} \cdot d\vec{r} = \int_a^b \vec{\mathbf{F}}(\vec{r}(x)) \cdot \vec{r}'(x)dx = \int_a^b f_1(x, g(x))dx + \int_a^b f_2(x, g(x))g'(x)dx$$

Step2.

求 $\displaystyle\int_a^b f_1(x, g(x))dx + \int_a^b f_2(x, g(x))g'(x)dx =?$

<u>范例说明:</u>

(I) 求 $\displaystyle\int_C \vec{\mathbf{F}} \cdot d\vec{r} =?$, 其中 $\vec{\mathbf{F}} = (y\cos x, x\sin y)$ 且曲线 $C = \{(x,y): 0 \le x \le a, y = x\}$

$$\int_C \vec{\mathbf{F}} \cdot d\vec{r} = \int_C y\cos x\, dx + x\sin y\, dy = \int_a^0 x\cos x\, dx + x\sin x\, dx$$

$$= -(a+1)\sin a + (a-1)\cos a + 1$$

Type 2.

给定向量函数 $\vec{\mathbf{F}} = (f_1(x,y,z), f_2(x,y,z), f_3(x,y,z))$, 求 $\displaystyle\int_C \vec{\mathbf{F}} \cdot d\vec{r} =?$, 其中曲线 $C$ 为:

$$\vec{r}(t) = \big(x(t), y(t), z(t)\big), \ \forall \, a \leq t \leq b$$

解题流程:

Step1.

$$\int_C \vec{F} \cdot d\vec{r} = \int_a^b f_1\big(\vec{r}(t)\big)x'(t)dt + \int_a^b f_2\big(\vec{r}(t)\big)y'(t)dt + \int_a^b f_3\big(\vec{r}(t)\big)z'(t)dt$$

Step2.

$$求 \int_a^b f_1\big(\vec{r}(t)\big)x'(t)dt + \int_a^b f_2\big(\vec{r}(t)\big)y'(t)dt + \int_a^b f_3\big(\vec{r}(t)\big)z'(t)dt = ?$$

<u>范例说明:</u>

$$求 \int_C \vec{F} \cdot d\vec{r} = ?$$

(I)若 $\vec{F} = (x^2y, x - z, xyz)$ 且 $\vec{r}(t) = (t, t^2, 2), \ 0 \leq t \leq 2$

$$则 \int_C \vec{F} \cdot d\vec{r} = \int_0^2 \vec{F}(\vec{r}(t)) \cdot \vec{r}'(t)dt = \int_0^2 (t^2 \cdot t^2 + (t - 2) \cdot 2t + 2t^3 \cdot 0) \ dt = \frac{56}{15}$$

(II)若 $\vec{F} = (x^3, 3zy^2, x^2y)$ 且 $\vec{r}(t) = (3t, 2t, t), \ -1 \leq t \leq 0$

$$则 \int_C \vec{F} \cdot d\vec{r} = \int_{-1}^0 \vec{F}(\vec{r}(t)) \cdot \vec{r}'(t)dt = \int_{-1}^0 (27t^3 \cdot 3 + 3t(2t)^2 \cdot 2 + 9t^2 \cdot 2t) \ dt = \frac{-123}{4}$$

(III)若 $\vec{F} = (xy, yz, xz)$ 且 $\vec{r}(t) = (t, t^2, t^3), \ 0 \leq t \leq 1$

$$则 \int_C \vec{F} \cdot d\vec{r} = \int_0^1 \vec{F}(\vec{r}(t)) \cdot \vec{r}'(t)dt = \int_{-1}^0 t^3 + 5t^6 \ dt = \frac{27}{28}$$

(IV)若 $\vec{F} = (x^2, -xy)$ 且 $\vec{r}(t) = (\cos t, \sin t), 0 \leq t \leq \frac{\pi}{2}$

$$\int_C \vec{F} \cdot d\vec{r} = \int_0^{\frac{\pi}{2}} \vec{F}(\vec{r}(t)) \cdot \vec{r}'(t)dt = \int_0^{\frac{\pi}{2}} (\cos^2 t, -\cos t \sin t) \cdot (-\sin t, \cos t)dt$$

$$= -2\int_0^{\frac{\pi}{2}} \sin t \cos^2 t \ dt = -\frac{2}{3}$$

Type 3.

$$求 \oint_C \vec{F} \cdot d\vec{r} = ?, \ \ 其中 \vec{F} = \big(f_1(x,y), f_2(x,y)\big), \ \ 曲线C为半径r的圆: x^2 + y^2 = r^2$$

解题流程:

Step1.

令 $x = r\cos\theta$, $y = r\sin\theta$, $0 \le \theta \le 2\pi$ 则 $\oint_C \vec{F}\cdot d\vec{r} = \int_C f_1(x,y)dx + f_2(x,y)dy$

Step2.

$$\int_C f_1(x,y)dx = \int_0^{2\pi} f_1(r\cos\theta, r\sin\theta)(-r\sin\theta)d\theta$$

$$\int_C f_2(x,y)dy = \int_0^{2\pi} f_2(r\cos\theta, r\sin\theta)(r\cos\theta)d\theta$$

Step3.

求 $\int_0^{2\pi} f_1(r\cos\theta, r\sin\theta)(-r\sin\theta)d\theta + \int_0^{2\pi} f_2(r\cos\theta, r\sin\theta)(r\cos\theta)d\theta =?$

<u>范例说明:</u>

求 $\oint_C \vec{F}\cdot d\vec{r} =?$, 其中 $\vec{F} = \big(f_1(x,y), f_2(x,y)\big)$ 且曲线 $C: x^2 + y^2 = r^2$

(I) 若 $f_1(x,y) = \dfrac{-y}{x^2+y^2}$, $f_2(x,y) = \dfrac{x}{x^2+y^2}$

则 $\oint_C \vec{F}\cdot d\vec{r} = \int_0^{2\pi} \cos^2\theta + \sin^2\theta \, d\theta = 2\pi$

(II) 若 $f_1(x,y) = \dfrac{x^2}{x^2+y^2}$, $f_2(x,y) = \dfrac{y^2}{x^2+y^2}$

则 $\oint_C \vec{F}\cdot d\vec{r} = r\int_0^{2\pi} -\cos^2\theta\sin\theta + \sin^2\theta\cos\theta \, d\theta = 0$

(III) 若 $f_1(x,y) = y^3$, $f_2(x,y) = -x^3$

则 $\oint_C \vec{F}\cdot d\vec{r} = \int_0^{2\pi} r^3\sin^3\theta\cdot(-r\sin\theta) - r^3\cos^3\theta\cdot(r\cos\theta)d\theta = -\dfrac{3r^4\pi}{2}$

Type 4.

求 $\oint_C \vec{F}\cdot d\vec{r} =?$, 其中 $\vec{F} = \big(f_1(x,y,z), f_2(x,y,z), f_3(x,y,z)\big)$ 且曲线 $C$ 为三维空间中的

封闭圆: $x^2 + y^2 = r^2$, $z = c$

解题流程:

Step1.

令 $C = \{(x,y,z): x^2+y^2 = r^2, z = c\}$

令 $x = r\cos\theta$, $y = r\sin\theta$ 则 $\vec{r}(\theta) = (r\cos\theta, r\sin\theta, c)$ $\therefore \vec{r}'(\theta) = (-a\sin\theta, b\cos\theta, 0)$

Step2.

$$\oint_C \vec{\mathbf{F}} \cdot d\vec{r} = \int_0^{2\pi} \vec{\mathbf{F}}(\vec{r}(\theta)) \cdot \vec{r}'(\theta)d\theta = \int_C (f_1, f_2, f_3) \cdot (-a\sin\theta, b\cos\theta, 0)d\theta$$

$$= \int_0^{2\pi} f_1(r\cos\theta, r\sin\theta)(-r\sin\theta)d\theta + \int_0^{2\pi} f_2(r\cos\theta, r\sin\theta)\, r\cos\theta\, d\theta$$

Step3.

$$求 \int_0^{2\pi} f_1(r\cos\theta, r\sin\theta)(-r\sin\theta)d\theta + \int_0^{2\pi} f_2(r\cos\theta, r\sin\theta)\, r\cos\theta\, d\theta = ?$$

<u>范例说明:</u>

求 $\oint_C \vec{\mathbf{F}} \cdot d\vec{r} = ?$, 其中 $\vec{\mathbf{F}} = (f_1(x,y,z), f_2(x,y,z), f_3(x,y,z))$ 且曲线 $C$ 为三维空间中的封闭

圆, $x^2 + y^2 = r^2$, $z = c$

(I)若 $\vec{\mathbf{F}} = (-y^3\cos z, x^3 e^z, -e^z)$

则 $\oint_C \vec{\mathbf{F}} \cdot d\vec{r} = \int_0^{2\pi} \vec{\mathbf{F}}(\vec{r}(\theta)) \cdot \vec{r}'(\theta)d\theta$

$$= \int_0^{2\pi} (-(a\sin\theta)^3, (a\cos\theta)^3, 1) \cdot (-a\sin\theta, a\cos\theta, 0)d\theta = \frac{3a^4\pi}{2}$$

Type 5.

求 $\oint_C \vec{\mathbf{F}} \cdot d\vec{r} = ?$, 其中 $\vec{\mathbf{F}} = (f_1(x,y,z), f_2(x,y,z), f_3(x,y,z))$ 且曲线 $C$ 为三维空间中的封闭

椭圆, $\dfrac{x^2}{a^2} + \dfrac{y^2}{b^2} = 1$, $z = c$

解题流程:

Step1.

令 $C = \{(x,y,z): \dfrac{x^2}{a^2} + \dfrac{y^2}{b^2} = 1, z = c\}$

令 $x = a\cos\theta$, $y = b\sin\theta$ 则 $\vec{r}(\theta) = (a\cos\theta, b\sin\theta, c)$ $\therefore \vec{r}'(\theta) = (-a\sin\theta, b\cos\theta, 0)$

Step2.

$$\oint_C \vec{\mathbf{F}} \cdot d\vec{r} = \int_0^{2\pi} \vec{\mathbf{F}}(\vec{r}(\theta)) \cdot \vec{r}'(\theta)d\theta = \int_C (f_1, f_2, f_3) \cdot (-a\sin\theta, b\cos\theta, 0)\, d\theta$$

$$= \int_0^{2\pi} f_1(a\cos\theta, b\sin\theta)(-a\sin\theta)d\theta + \int_0^{2\pi} f_2(a\cos\theta, b\sin\theta)\,b\cos\theta\,d\theta$$

Step3.

$$\text{求} \int_0^{2\pi} f_1(a\cos\theta, b\sin\theta)(-a\sin\theta)d\theta + \int_0^{2\pi} f_2(a\cos\theta, b\sin\theta)\,b\cos\theta\,d\theta = ?$$

<u>范例说明:</u>

求 $\displaystyle\oint_C \vec{\mathbf{F}}\cdot d\vec{r} = ?$, 其中 $\vec{\mathbf{F}} = \big(f_1(x,y,z), f_2(x,y,z), f_3(x,y,z)\big)$ 且曲线 $C$ 为三维空间中的封闭

椭圆, $\dfrac{x^2}{a^2} + \dfrac{y^2}{b^2} = 1$, $z = c$

(I)若 $\vec{\mathbf{F}} = (2x - y, 2y + z, xyz)$

则 $\displaystyle\oint_C \vec{\mathbf{F}}\cdot d\vec{r} = \int_0^{2\pi} \vec{\mathbf{F}}(\vec{r}(\theta))\cdot \vec{r}'(\theta)d\theta$

$$= \int_0^{2\pi} (2a\cos\theta - b\sin\theta, 2b\sin\theta, abc\cos\theta\sin\theta)\cdot(-a\sin\theta, b\cos\theta, 0)d\theta = ab\pi$$

Type 6.

空间中二次曲面与平面 $z = c$ 的交集常为椭圆或圆, 底下整理数种交集为圆或椭圆的情形

(i)$C$ 为椭球面 $\dfrac{x^2}{a^2} + \dfrac{y^2}{b^2} + \dfrac{z^2}{c^2} = 1$ 与 $z = 0$ 交线 $\Rightarrow C:\dfrac{x^2}{a^2} + \dfrac{y^2}{b^2} = 1, z = 0$

(ii)$C$ 为椭球面 $\dfrac{x^2}{a^2} + \dfrac{y^2}{b^2} + \dfrac{z^2}{c^2} = c_1$ 与 $\dfrac{x^2}{a^2} + \dfrac{y^2}{b^2} = c_2$ 交线 $(c_1 > c_2 > 0)$

$\Rightarrow C:\dfrac{x^2}{a^2} + \dfrac{y^2}{b^2} = c_2$, $z = c\sqrt{c_1 - c_2}$

(iii)$C$ 为椭圆抛物面 $\dfrac{x^2}{a^2} + \dfrac{y^2}{b^2} = \dfrac{z}{c}$ 与 $z = c$ 交线 $\Rightarrow C:\dfrac{x^2}{a^2} + \dfrac{y^2}{b^2} = 1$, $z = c$

(iv)$C$ 为椭圆椎 $\dfrac{x^2}{a^2} + \dfrac{y^2}{b^2} = \dfrac{z^2}{c^2}$ 与 $z = c$ 交线 $\Rightarrow C:\dfrac{x^2}{a^2} + \dfrac{y^2}{b^2} = 1$, $z = c$

(v)$C$ 为双曲面 $\dfrac{x^2}{a^2} + \dfrac{y^2}{b^2} - \dfrac{z^2}{c^2} = 1$ 与 $z = 0$ 交线 $\Rightarrow C:\dfrac{x^2}{a^2} + \dfrac{y^2}{b^2} = 1$, $z = 0$

求 $\displaystyle\oint_C \vec{\mathbf{F}}\cdot d\vec{r} = ?$, 其中 $\vec{\mathbf{F}} = \big(f_1(x,y,z), f_2(x,y,z), f_3(x,y,z)\big)$ 且曲线 $C$ 为上述三维空间

当中的封闭椭圆

Example 1.

$$求 \oint (6y + x)dx + (y + 2x)dy =?, \quad C: (x - 2)^2 + (y - 3)^2 = r^2$$

【解】

令$x = 2 + r\cos\theta, y = 3 + r\sin\theta, \quad 0 \leq \theta \leq 2\pi$

则 $\oint (6y + x)dx + (y + 2x)dy$

$$= \int_0^{2\pi} (6(3 + r\sin\theta) + 2 + r\cos\theta)(-r\sin\theta) + (3 + r\sin\theta + 2(2 + r\cos\theta))r\cos\theta \, d\theta$$

$$= \int_0^{2\pi} -6r^2\sin^2\theta + 2r^2\cos^2\theta \, d\theta = \int_0^{2\pi} -6r^2\sin^2\theta + 2r^2(1 - \sin^2\theta)d\theta$$

$$= \int_0^{2\pi} -8r^2\sin^2\theta + 2r^2 d\theta = 4\pi r^2 - 8r^2 \int_0^{2\pi} \frac{1 - \cos 2\theta}{2} d\theta = -4\pi r^2$$

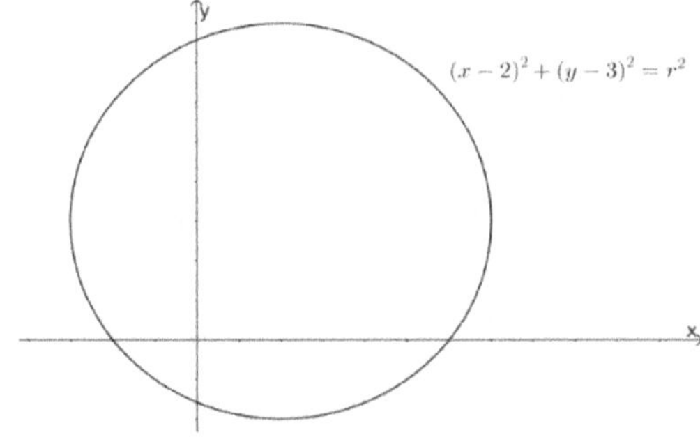


Example 2.

$$求 \int_C x^2ydx + (x - z)dy + xyzdz =?, \quad C为沿 y = x^2, \ z = 2, 從点(0,0,2)至点(2,4,2)$$
的线段

【解】

令$\vec{r}(t) = (t, t^2, 2), 0 \leq t \leq 2$ 则 $\vec{r}'(t) = (1, 2t, 0)$

$$\therefore \int_C x^2ydx + (x - z)dy + xyzdz = \int_0^2 (t^2 \cdot t^2, t - 2, 2t^3) \cdot (1, 2t, 0)dt = \int_0^2 t^4 + 2t^2 - 4t dt$$

$$= \frac{t^5}{5} + \frac{2t^3}{3} - 2t^2 \bigg|_0^2 = \frac{56}{15}$$

Example 3.

$$求 \int_C x^3 dx + 3zy^2 dy + x^2 y dz =?，\quad C 为 (-3,-2,-1) 至 (0,0,0) 的线段$$

【解】

令 $\vec{r}(t) = (3t, 2t, t),\quad -1 \le t \le 0$ 则 $\vec{r}'(t) = (3,2,1)$

$$\therefore \int_C x^3 dx + 3zy^2 dy + x^2 y dz = \int_{-1}^{0} 27t^3 \cdot 3dt + 3t(2t)^2 \cdot 2dt + 9t^2 \cdot 2t dt = \frac{-123}{4}$$

Example 4.

$$求 \oint y^3 dx - x^3 \, dy =?，\quad C: x^2 + y^2 = r^2$$

【解】

令 $x = r\cos\theta, y = r\sin\theta, 0 \le \theta \le 2\pi$

$$则 \oint y^3 dx - x^3 \, dy = \int_0^{2\pi} r^3 \sin^3\theta \cdot (-r\sin\theta) - r^3 \cos^3\theta \cdot (r\cos\theta) d\theta$$

$$= -r^4 \int_0^{2\pi} \sin^4\theta + \cos^4\theta \, d\theta = -r^4 \int_0^{2\pi} (\sin^2\theta + \cos^2\theta)^2 - 2\sin^2\theta\cos^2\theta \, d\theta$$

$$= -r^4 \int_0^{2\pi} 1 - 2\left(\frac{\sin 2\theta}{2}\right)^2 d\theta = -r^4 \int_0^{2\pi} 1 - \frac{\sin^2 2\theta}{2} d\theta = -r^4 \int_0^{2\pi} 1 - \frac{1-\cos 4\theta}{4} d\theta$$

$$= -r^4 \cdot \frac{3}{4} \cdot 2\pi = -\frac{3r^4\pi}{2}$$

Example 5.

$$假设曲线 C: \vec{r}(t) = (\cos t, \sin t),\quad 0 \le t \le \frac{\pi}{2},\quad 求力场 \vec{F}(x,y) = (x, -y^2) 沿 C 所作的功$$

【解】

$$\because \vec{r}(t) = (\cos t, \sin t) \quad \therefore \vec{r}'(t) = (-\sin t, \cos t),\quad 0 \le t \le \frac{\pi}{2}$$

$$\therefore \int_C \vec{F} \cdot d\vec{r} = \int_0^{\frac{\pi}{2}} (\cos t, -\sin^2 t) \cdot \vec{r}'(t) \, dt = \int_0^{\frac{\pi}{2}} \cos t(-\sin t) - \sin^2 t \cdot \cos t \, dt = -\frac{1}{2} - \frac{1}{3}$$

$$= -\frac{5}{6}$$

Example 6.

$$\text{求} \int_C \frac{-y}{x^2+y^2}\,dx + \frac{x}{x^2+y^2}\,dy = ?, \quad C: \text{圆} \, x^2+y^2 = r^2$$

【解】

令 $x = r\cos\theta$, $y = r\sin\theta$, $0 \le \theta \le 2\pi$

则 $\displaystyle\int_C \frac{-y}{x^2+y^2}\,dx + \frac{x}{x^2+y^2}\,dy = \int_0^{2\pi} (-\sin\theta)(-\sin\theta) + \cos\theta \cdot \cos\theta\,d\theta$

$$= \int_0^{2\pi} \sin^2\theta + \cos^2\theta\,d\theta = \int_0^{2\pi} d\theta = 2\pi$$

Example 7.

$$\text{求} \oint 2y\,dx + (x^2+y^2)\,dy = ?, \quad C: x^2 + (y-3)^2 = r^2$$

【解】

令 $x = r\cos\theta$, $y = 3 + r\sin\theta$, $0 \le \theta \le 2\pi$

则 $\displaystyle\oint 2y\,dx + (x^2+y^2)\,dy = \int_0^{2\pi} 2(3 + r\sin\theta)(-r\sin\theta) + \big((9+r^2) + 6r\sin\theta\big)r\cos\theta\,d\theta$

$$= \int_0^{2\pi} -6r\sin\theta - 2r^2\sin^2\theta + (9+r^2)r\cos\theta + 6r^2\sin\theta\cos\theta\,d\theta$$

$$= \int_0^{2\pi} -2r^2\sin^2\theta + 6r^2\sin\theta\cos\theta\,d\theta = \int_0^{2\pi} -2r^2\left(\frac{1-\cos 2\theta}{2}\right) + 3r^2\sin 2\theta\,d\theta = -2r^2\pi$$

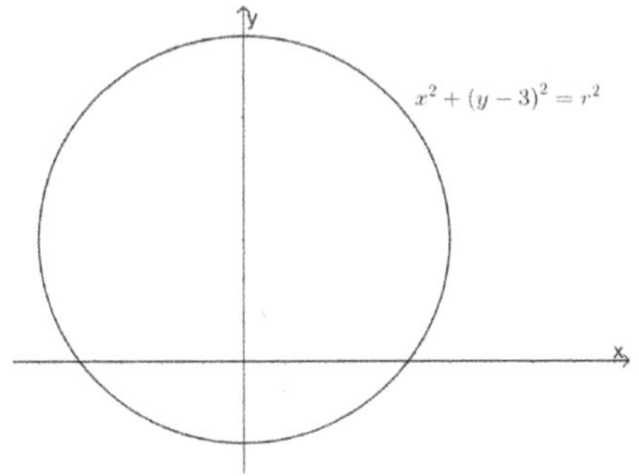


Example 8.

$$\vec{\mathbf{F}} = \left(-z, x, \frac{y^2 z}{2}\right), \quad \text{曲线} \, C: z = a, \, x^2 + y^2 = a^2, \quad \text{求} \oint_C \vec{\mathbf{F}} \cdot d\vec{r} = ?$$

【解】

令 $\vec{r}(t) = (a\cos t, a\sin t, a)$, $0 \le t \le 2\pi$ 则 $\vec{r}'(t) = (-a\sin t, a\cos t, 0)$

$$\oint_C \vec{\mathbf{F}} \cdot d\vec{r} = \int_0^{2\pi} \vec{\mathbf{F}}(\vec{r}(t)) \cdot \vec{r}'(t)\, dt = \int_0^{2\pi} \left(-a, a\cos t, \frac{a^3\sin^2 t}{2}\right) \cdot (-a\sin t, a\cos t, 0)\, dt$$

$$= \int_0^{2\pi} a^2\sin t + a^2\cos^2 t\, dt = 2\pi \cdot \frac{a^2}{2} = \pi a^2$$

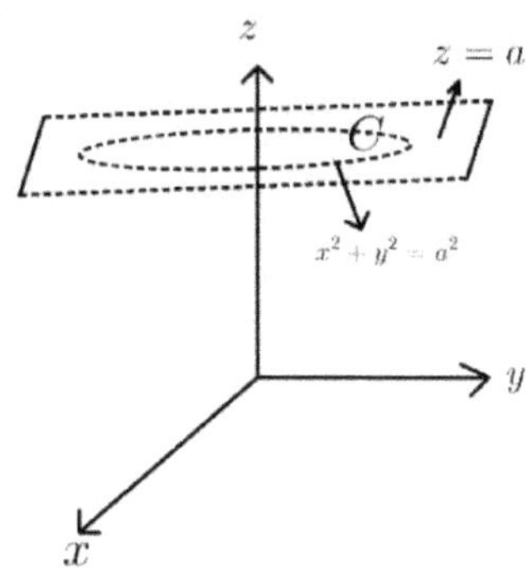


Example 9.

$$\vec{\mathbf{F}} = (0, -xz, -xy), \quad 曲线 C : x^2 + z^2 \le 1, \ y = 1, \quad 求 \oint_C \vec{\mathbf{F}} \cdot d\vec{r} = ?$$

【解】

令 $\vec{r}(t) = (\cos t, 1, \sin t)$, $0 \le t \le 2\pi$ 则 $\vec{r}'(t) = (-\sin t, 0, \cos t)$

$$\therefore \oint_C \vec{\mathbf{F}} \cdot d\vec{r} = \int_0^{2\pi} \vec{\mathbf{F}}(\vec{r}(t)) \cdot \vec{r}'(t)\, dt = \int_0^{2\pi} (0, -\cos t\sin t, -\cos t) \cdot (-\sin t, 0, \cos t)\, dt$$

$$= -\int_0^{2\pi} \cos^2 t\, dt = -\pi$$

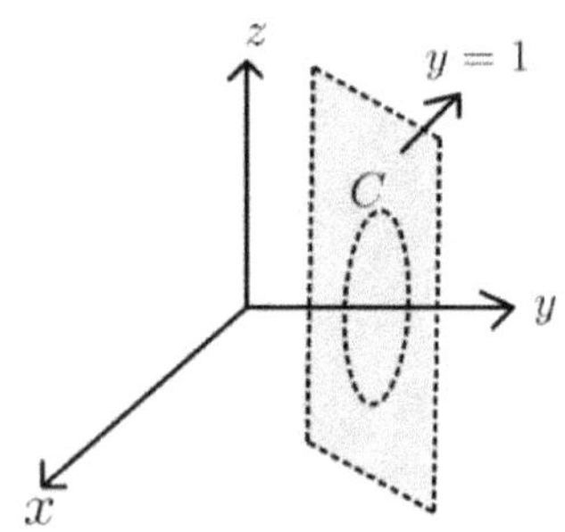


Example 10.

求 $\int_C \dfrac{-y}{x^2+y^2}dx + \dfrac{x}{x^2+y^2}dy =?$, $C$ 为沿着圆 $x^2+y^2=4$ 从 $(0,2)$ 至 $\left(-\sqrt{2},-\sqrt{2}\right)$ 的圆弧

【解】

令 $x=2\cos\theta$, $y=2\sin\theta$, $\dfrac{\pi}{2} \leq \theta \leq \dfrac{5\pi}{4}$

则 $\int_C \dfrac{-y}{x^2+y^2}dx + \dfrac{x}{x^2+y^2}dy = \int_{\frac{\pi}{2}}^{\frac{5\pi}{4}} \dfrac{-2\sin\theta}{4} \cdot (-2\sin\theta) + \dfrac{2\cos\theta}{4} \cdot 2\cos\theta\, d\theta$

$= \int_{\frac{\pi}{2}}^{\frac{5\pi}{4}} \cos^2\theta + \sin^2\theta\, d\theta = \dfrac{3\pi}{4}$

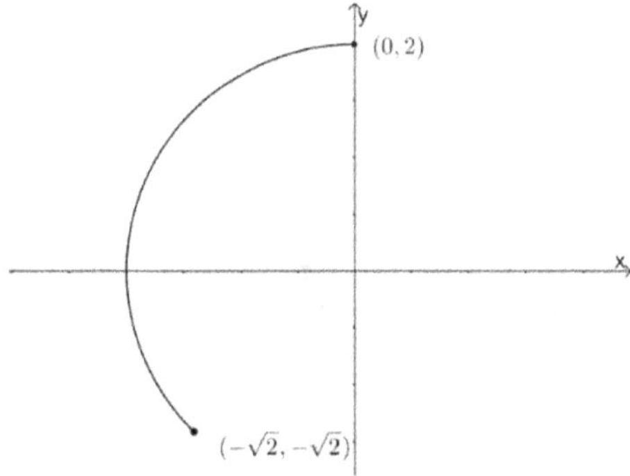


Example 11.

求 $\int_C \dfrac{x^2}{x^2+y^2}dx + \dfrac{y^2}{x^2+y^2}dy =?$, $C$ 为沿着圆 $x^2+y^2=r^2$ 從 $(r,0)$ 至 $\left(-\dfrac{r}{\sqrt{2}}, \dfrac{r}{\sqrt{2}}\right)$ 的圆弧, $r>0$

【解】

令 $x=r\cos\theta$, $y=r\sin\theta$, $0 \leq \theta \leq \dfrac{3\pi}{4}$

则 $\int_C \dfrac{x^2}{x^2+y^2}dx + \dfrac{y^2}{x^2+y^2}dy = \int_0^{\frac{3\pi}{4}} \dfrac{r^2\cos^2\theta}{r^2} \cdot (-r\sin\theta) + \dfrac{r^2\sin^2\theta}{r^2} \cdot r\cos\theta\, d\theta$

$= r\int_0^{\frac{3\pi}{4}} -\cos^2\theta\sin\theta + \sin^2\theta\cos\theta\, d\theta = \dfrac{r}{3}(\cos^3\theta + \sin^3\theta)\Big|_0^{\frac{3\pi}{4}} = -\dfrac{r}{3}$

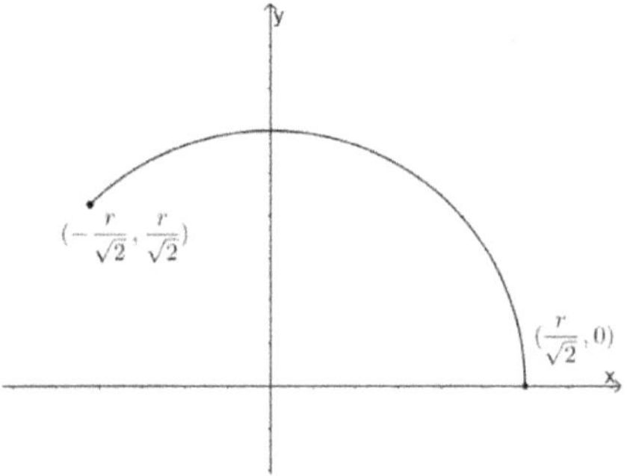


Example 12.

$$求 \int_C yzdx - xzdy + xydz =?, \quad C: \vec{r}(t) = (e^t, e^{at}, e^{-t}), \ 0 \le t \le 1$$

【解】

$\because \vec{r}(t) = (e^t, e^{at}, e^{-t}) \quad \therefore \vec{r}'(t) = (e^t, ae^{at}, -e^{-t})$

$$\therefore \int_C yzdx - xzdy + xydz = \int_0^1 e^{(a-1)t} \cdot e^t - ae^{at} - e^{at}dt = 1 - e^a$$

Example 13.

$$求 \int_C \vec{F} \cdot d\vec{r} =?, \quad \vec{F} = (z, x, y), \ C: \vec{r}(t) = (0, 3\sin t, \sin^2 t), \ 0 \le t \le \frac{\pi}{2}$$

【解】

$\because \vec{r}(t) = (0, 3\sin t, \sin^2 t) \quad \therefore \vec{r}'(t) = (0, 3\cos t, 2\sin t \cos t)$

$$\therefore \int_C \vec{F} \cdot d\vec{r} = \int_0^{\frac{\pi}{2}} (\sin^2 t, 0, 3\sin t) \cdot (0, 3\cos t, 2\sin t \cos t)dt = \int_0^{\frac{\pi}{2}} 6\sin^2 t \cos t \, dt$$

$$= 2\sin^3 t \Big|_0^{\frac{\pi}{2}} = 2$$

Example 14.

$$\vec{F} = (3x - 4y + 2z, 4x + 2y - 3z^2, 2xz - 4y^2 + z^3), 椭圆曲线 C: \frac{x^2}{16} + \frac{y^2}{9} = 1, \ z = 0,$$

$$求沿着椭圆 C 上半部绕一圈所作的功 \int_C \vec{F} \cdot d\vec{r} =?$$

【解】

$令 \vec{r}(t) = (4\cos t, 3\sin t, 0), \ 0 \le t \le \pi \ 则 \ \vec{r}'(t) = (-4\sin t, 3\cos t, 0)$

$$\therefore \int_C \vec{F} \cdot d\vec{r} = \int_0^\pi \vec{F}(t) \cdot \vec{r}'(t) dt$$

$$= \int_0^\pi (12\cos t - 12\sin t, 16\cos t + 6\sin t)(-4\sin t, 3\cos t) dt$$

$$= \int_0^\pi (12\cos t - 12\sin t)(-4\sin t) + (16\cos t + 6\sin t)(3\cos t) dt$$

$$= 48 \int_0^\pi \sin^2 t + \cos^2 t \, dt = 48\pi$$

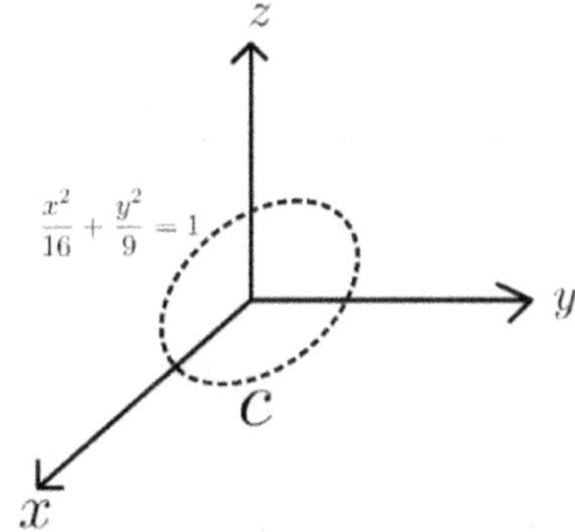

Example 15.

$$\vec{\mathbf{F}} = \left(0, \frac{2x^3}{3}, 2y^3\right), \quad S \text{是} x + y + z = 1 \text{ 在第一卦限所围成的平面, } C \text{为} S \text{的逆时针边}$$

界, 求 $\oint_C \vec{\mathbf{F}} \cdot d\vec{r} =?$

【解】

令 $C_1: \vec{r}_1(t) = (1-t, t, 0), \quad C_2: \vec{r}_2(t) = (0, 1-t, t), \quad C_3: \vec{r}_3(t) = (t, 0, 1-t), \quad 0 \le t \le 1$

则 $\vec{r}_1'(t) = (-1, 1, 0), \quad \vec{r}_2'(t) = (0, -1, 1), \quad \vec{r}_3'(t) = (1, 0, -1)$

$$\therefore \oint_C \vec{\mathbf{F}} \cdot d\vec{r} = \int_{C_1} \vec{\mathbf{F}}(\vec{r}_1(t)) \cdot \vec{r}_1'(t) dt + \int_{C_2} \vec{\mathbf{F}}(\vec{r}_2(t)) \cdot \vec{r}_2'(t) dt + \int_{C_3} \vec{\mathbf{F}}(\vec{r}_3(t)) \cdot \vec{r}_3'(t) dt$$

$$= \frac{2}{3} \int_0^1 (1-t)^3 \, dt + 2 \int_0^1 (1-t)^3 \, dt = \frac{2}{3}$$

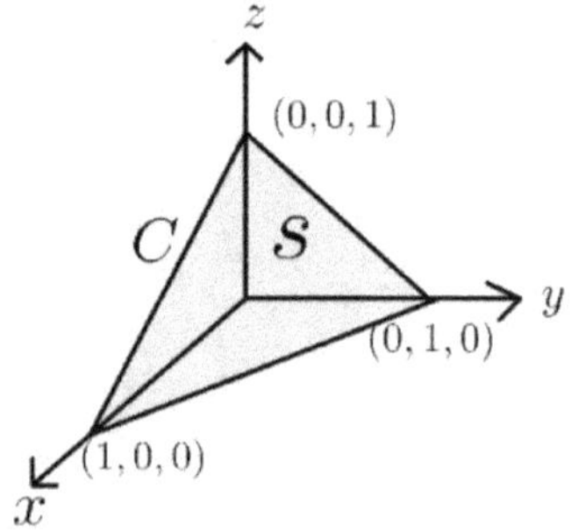

**Example 16.**

$$\vec{\mathbf{F}} = \left(\frac{z^2}{2}, \frac{x^2}{2}, \frac{y^2}{2}\right), \quad S为(a,0,0),(0,a,0),(0,0,a)于第一象限所围的平面, C为S的逆时$$

针边界, 求 $\displaystyle\oint_C \vec{\mathbf{F}} \cdot d\vec{r} = ?$

【解】

令 $C_1: \vec{r}_1(t) = (a - at, at, 0), C_2: \vec{r}_2(t) = (0, a - at, at), C_3: \vec{r}_3(t) = (at, 0, a - at), 0 \leq t \leq 1$

则 $\vec{r}_1'(t) = (-a, a, 0), \quad \vec{r}_2'(t) = (0, -a, a), \quad \vec{r}_3'(t) = (a, 0, -a)$

$$\therefore \oint_C \vec{\mathbf{F}} \cdot d\vec{r} = \int_{C_1} \vec{\mathbf{F}}(\vec{r}_1(t)) \cdot \vec{r}_1'(t) dt + \int_{C_2} \vec{\mathbf{F}}(\vec{r}_2(t)) \cdot \vec{r}_2'(t) dt + \int_{C_3} \vec{\mathbf{F}}(\vec{r}_3(t)) \cdot \vec{r}_3'(t) dt$$

$$= \frac{a}{2} \int_0^1 (a - at)^3 \, dt \cdot 3 = \frac{a^3}{2}$$

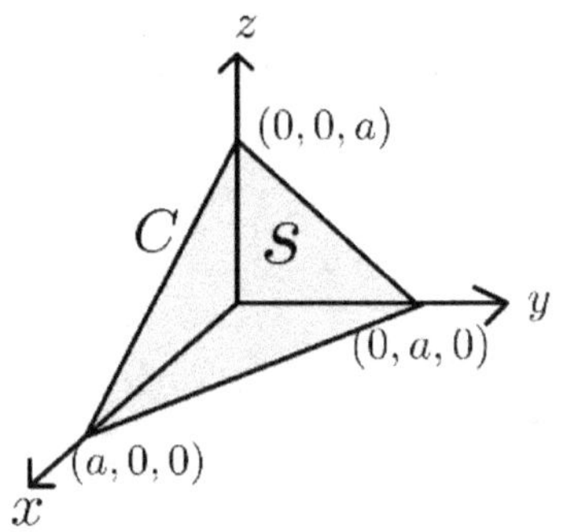

**Example 17.**

假设 $\vec{\mathbf{F}} = (0, e^x + 16xy, e^y + 6y^2), \quad S$ 是 $x + y + z = 6$ 在第一卦限的部分, $C$ 为 $S$ 的

逆时针边界, 求 $\oint_C \vec{\mathbf{F}} \cdot d\vec{r} = ?$

【解】

令 $C_1 : \vec{r}_1(t) = (6 - 6t, 6t, 0)$,  $C_2 : \vec{r}_2(t) = (0, 6 - 6t, 6t)$,  $C_3 : \vec{r}_3(t) = (6t, 0, 6 - 6t)$

则 $\vec{r}_1'(t) = (-6, 6, 0)$,  $\vec{r}_2'(t) = (0, -6, 6)$,  $\vec{r}_3'(t) = (6, 0, -6), 0 \le t \le 1$

$\because \oint_C \vec{\mathbf{F}} \cdot d\vec{r} = \int_{C_1} \vec{\mathbf{F}}\big(\vec{r}_1(t)\big) \cdot \vec{r}_1'(t) dt + \int_{C_2} \vec{\mathbf{F}}\big(\vec{r}_2(t)\big) \cdot \vec{r}_2'(t) dt + \int_{C_3} \vec{\mathbf{F}}\big(\vec{r}_3(t)\big) \cdot \vec{r}_3'(t) dt$

$\because \int_{C_1} \vec{\mathbf{F}}\big(\vec{r}_1(t)\big) \cdot \vec{r}_1'(t) dt = 6 \int_0^1 e^{6-6t} + 96t(6 - 6t)\, dt = e^6 - 1 + 576$

$\int_{C_2} \vec{\mathbf{F}}\big(\vec{r}_2(t)\big) \cdot \vec{r}_2'(t) dt = -6 \int_0^1 dt + 6 \int_0^1 e^{6-6t} + 6(6 - 6t)^2\, dt = -6 + e^6 - 1 + 72$

$= e^6 + 65$

且 $\int_{C_3} \vec{\mathbf{F}}\big(\vec{r}_3(t)\big) \cdot \vec{r}_3'(t) dt = -6 \int_0^1 dt = -6$

$\therefore \oint_C \vec{\mathbf{F}} \cdot d\vec{r} = 634 + 2e^6$

Example 18.

求 $\int_C y \cos x\, dx + x \sin y\, dy = ?$,  $C$ 为 $(0,0), (a, 0), (a, a)$ 所围成逆时针方向的三角形路径 $(a > 0)$

【解】

令 $C_1 = \{(x, y) : 0 \le x \le a, y = 0\}$,  $C_2 = \{(x, y) : 0 \le y \le a, x = a\}$
且 $C_3 = \{(x, y) : 0 \le x \le a, y = x\}$

则 $\int_C y \cos x\, dx + x \sin y\, dy$

$= \int_{C_1} y \cos x\, dx + x \sin y\, dy + \int_{C_2} y \cos x\, dx + x \sin y\, dy + \int_{C_3} y \cos x\, dx + x \sin y\, dy$

$$\because \int_{C_1} y\cos x\, dx + x\sin y\, dy = 0,$$

$$\int_{C_2} y\cos x\, dx + x\sin y\, dy = a\int_0^a \sin y\, dy = a(-\cos a + 1),$$

$$\int_{C_3} y\cos x\, dx + x\sin y\, dy = \int_a^0 x\cos x\, dx + x\sin x\, dx = -(a+1)\sin a + (a-1)\cos a + 1$$

$$\therefore \oint y\cos x\, dx + x\sin y\, dy = -(a+1)\sin a - \cos a + a + 1$$

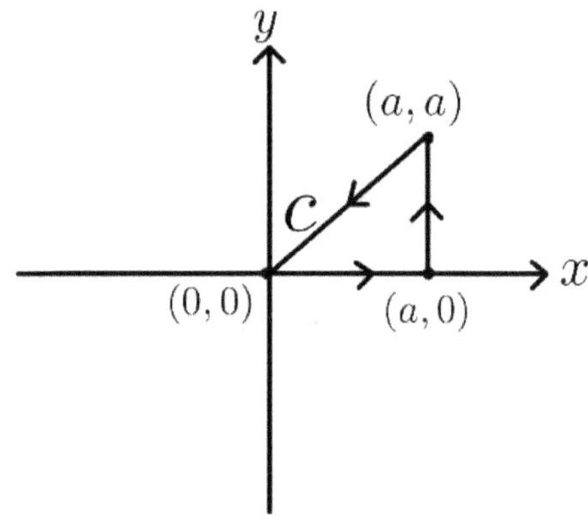


Example 19.

$$\text{求} \int_C -y^4 dx + xy^2 dy = ?, \quad C \text{为} (0,0), (1,0), (1,1) \text{所围成逆时针方向的三角形路径}$$

【解】

令 $C_1 = \{(x,y): 0 \le x \le 1, y = 0\}, \quad C_2 = \{(x,y): 0 \le y \le 1, x = 1\}$

且 $C_3 = \{(x,y): 0 \le x \le 1, y = x\}$

则 $\displaystyle\int_C -y^4 dx + xy^2 dy = \int_{C_1} -y^4 dx + xy^2 dy + \int_{C_2} -y^4 dx + xy^2 dy + \int_{C_3} -y^4 dx + xy^2 dy$

$$\because \int_{C_1} -y^4 dx + xy^2 dy = 0, \quad \int_{C_2} -y^4 dx + xy^2 dy = \int_0^1 y^2 dy = \frac{1}{3}$$

$$\text{且} \int_{C_3} -y^4 dx + xy^2 dy = \int_1^0 -x^4 dx + x^3 dy = \frac{-1}{20}$$

$$\therefore \int_C -y^4 dx + xy^2 dy = \frac{1}{3} - \frac{1}{20} = \frac{17}{60}$$

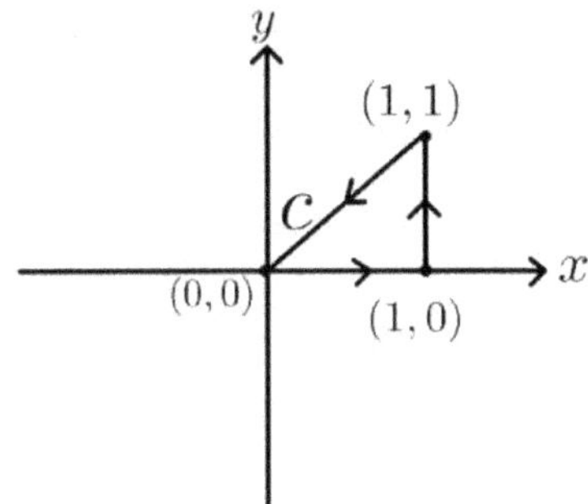

Example 20.

$$\text{假设 } \vec{F} = \left( x(x^2 + y^2)^{-\frac{3}{2}}, y(x^2 + y^2)^{-\frac{3}{2}} \right), \ C: \vec{r}(t) = (e^t \sin t, e^t \cos t), \ 0 \le t \le 2,$$

$$\text{求} \int_C \vec{F} \cdot d\vec{r} = ?$$

【解】

$\because \vec{r}(t) = (e^t \sin t, e^t \cos t) \quad \therefore \vec{r}'(t) = (e^t(\sin t + \cos t), e^t(\cos t - \sin t))$

$\because \vec{F} = \left( x(x^2 + y^2)^{-\frac{3}{2}}, y(x^2 + y^2)^{-\frac{3}{2}} \right) \ \text{且} \int_C \vec{F} \cdot d\vec{r} = \int_0^2 \vec{F}(\vec{r}(t)) \cdot \vec{r}'(t) dt$

$\because \vec{F}(\vec{r}(t)) \cdot \vec{r}'(t) = e^t \sin t \cdot e^{-3t} \cdot e^t(\sin t + \cos t) + e^t \cos t \cdot e^{-3t} \cdot e^t(\cos t - \sin t) = e^{-t}$

$\therefore \int_0^2 \vec{F}(\vec{r}(t)) \cdot \vec{r}'(t) dt = \int_0^2 e^{-t} dt = 1 - e^{-2}$

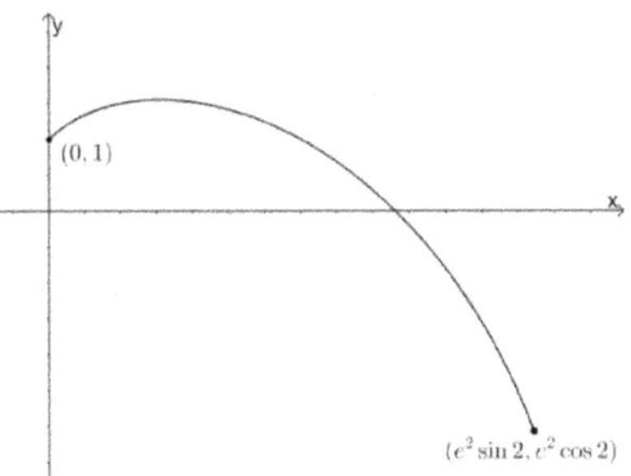


Example 21.

$$\text{假设曲线 } C: \vec{r}(t) = (t, t^2), \ 0 \le t \le 1, \ \text{求力场} \vec{F}(x, y) = (x^2, -y) \text{沿着} C \text{所作的功}$$

【解】

$\because \vec{r}(t) = (t, t^2) \quad \therefore \vec{r}'(t) = (1, 2t)$

$$\therefore \int_C \vec{F} \cdot d\vec{r} = \int_0^2 \vec{F}(\vec{r}(t)) \cdot \vec{r}'(t)dt = \int_0^2 (t^2, -t^2) \cdot (1, 2t)dt = \int_0^1 t^2 - 2t^3 \, dt = \left(\frac{t^3}{3} - \frac{t^4}{2}\right)\Big|_0^1$$

$$= -\frac{1}{6}$$

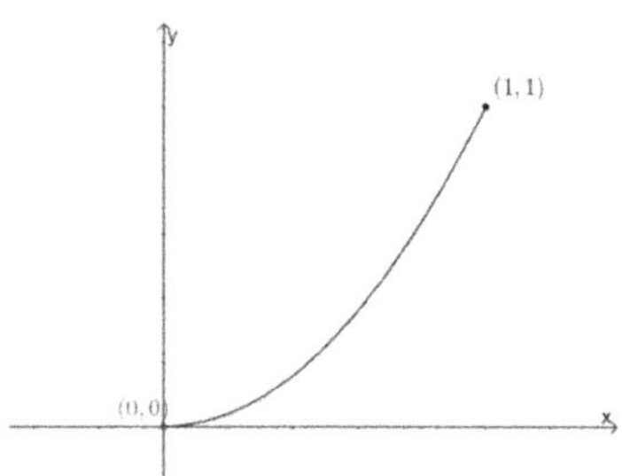


Example 22.

假设 $\vec{F}(x, y, z) = (3x^2 + 6y, -14yz, 20xz^2)$, 求從 $(0,0,0)$ 至 $(1,1,1)$ 沿下列路径的线积分 $(1)C: x = t, y = t^2, z = t^3$ $(2)C:$ 由 $(0,0,0)$ 至 $(1,1,1)$ 的直线

【解】

(1)

令 $\vec{r}(t) = (t, t^2, t^3)$ 则 $\vec{r}'(t) = (1, 2t, 3t^2)$

$$\therefore \int_C \vec{F} \cdot d\vec{r} = \int_0^1 \vec{F}(\vec{r}(t)) \cdot \vec{r}'(t)dt = \int_C \vec{F}(t, t^2, t^3) \cdot (1, 2t, 3t^2)dt$$

$$= \int_0^1 (3t^2 + 6t^2) - 14t^5 \cdot 2t + 20t^7 \cdot 3t^2 dt = \int_0^1 9t^2 - 28t^6 + 60t^9 dt = 5$$

(2)

令 $\vec{r}(t) = (t, t, t)$ 则 $\vec{r}'(t) = (1, 1, 1)$

$$\int_C \vec{F} \cdot d\vec{r} = \int_0^1 \vec{F}(\vec{r}(t)) \cdot \vec{r}'(t)dt = \int_C \vec{F}(t, t, t) \cdot (1, 1, 1)dt = \int_0^1 (3t^2 + 6t) - 14t^2 + 20t^3 dt$$

$$= \int_0^1 20t^3 - 11t^2 + 6t dt = \frac{13}{3}$$

Example 23.

求 $\int_C (x^2 + y^2)dx - xdy = ?$, 其中 $C$ 为圆 $x^2 + y^2 = r^2$ 從 $(r, 0)$ 至 $(0, r)$ 的圆弧

【解】

令 $x = r\cos\theta$, $y = r\sin\theta$, $0 \le \theta \le \dfrac{\pi}{2}$ 则 $x'(\theta) = -r\sin\theta$, $y'(\theta) = r\cos\theta$

$\therefore \displaystyle\int_C (x^2 + y^2)dx - xdy = \int_0^{\frac{\pi}{2}} r^2(\cos^2\theta + \sin^2\theta)(-r\sin\theta) - r^2\cos^2\theta \, d\theta$

$= \displaystyle\int_0^{\frac{\pi}{2}} -r^3\sin\theta - r^2\cos^2\theta \, d\theta = -r^3 - \dfrac{\pi r^2}{4}$

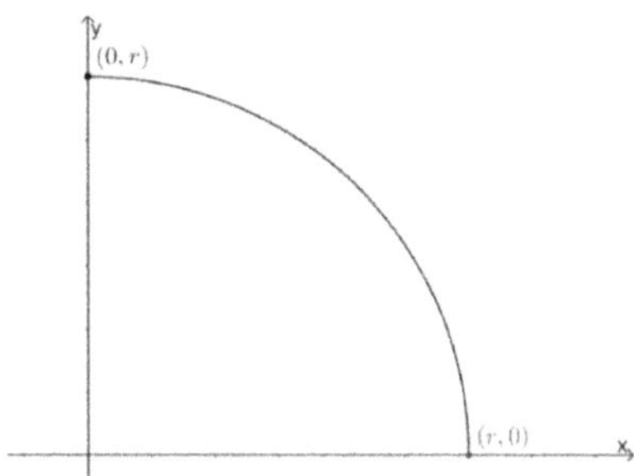

## Example 24.

$\qquad$ 求 $\displaystyle\int_C xdx - zdy + ydz = ?$, 其中 $C: \vec{r}(t) = (t^2, -t, t^2)$, $\forall 1 \le t \le 3$

【解】

令 $\vec{r}(t) = (t^2, -t, t^2)$ 则 $\vec{r}'(t) = (2t, -1, 2t)$

$\therefore \displaystyle\int_C xdx - zdy + ydz = \int_1^3 t^2 \cdot 2tdt - t^2 \cdot (-1)dt - t \cdot 2t \, dt = \left(\dfrac{t^4}{2} - \dfrac{t^3}{3}\right)\bigg|_1^3 = \dfrac{81}{2} - 9 - \dfrac{1}{6}$

$= \dfrac{94}{3}$

## Example 25.

$\qquad$ 求 $\displaystyle\int_C yz^2dx + (xz^2 + ze^{yz})dy + \left(2xyz + ye^{yz} + \dfrac{1}{1+z}\right)dz = ?$, 其中曲线 $C$:

$\qquad \vec{r}(t) = (t, t^2, t^3), 0 \le t \le 2$

【解】

令 $\vec{r}(t) = (t, t^2, t^3)$ 则 $\vec{r}'(t) = (1, 2t, 3t^2)$

$$\therefore \int_C yz^2\,dx + (xz^2 + ze^{yz})\,dy + \left(2xyz + ye^{yz} + \frac{1}{1+z}\right)dz$$

$$= \int_0^2 t^8 + (t^7 + t^3 e^{t^5})\,2t + \left(2t^6 + t^2 e^{t^5} + \frac{1}{1+t^3}\right)3t^2\,dt$$

$$= \int_0^2 9t^8\,dt + 5\int_0^2 t^4 e^{t^5}\,dt + \int_0^2 \frac{3t^2}{1+t^3}\,dt = \left(t^9 + e^{t^5} + \ln(1+t^3)\right)\Big|_0^2 = 511 + e^{32} + 2\ln 3$$

Example 26.

求 $\int_C x^2 y\,dx + xy\,dy =?$, 其中曲线 $C$ 为沿着 $x = \sqrt{1-y^2}$ 從 $(1,0)$ 至 $(0,1)$ 的线段

【解】

$$\because x = \sqrt{1-y^2} \quad \therefore dx = \frac{1}{2}(1-y^2)^{-\frac{1}{2}}(-2y)\,dy$$

$$\therefore \int_C x^2 y\,dx + xy\,dy = \int_0^1 (1-y^2)y \cdot \frac{1}{2}(1-y^2)^{-\frac{1}{2}}(-2y) + y\sqrt{1-y^2}\,dy$$

$$= \int_0^1 -y^2 \cdot (1-y^2)^{\frac{1}{2}} + y\sqrt{1-y^2}\,dy$$

$$\because \int_0^1 y\sqrt{1-y^2}\,dy = \frac{-(1-y^2)^{\frac{3}{2}}}{3}\bigg|_0^1 = \frac{1}{3}$$

令 $y = \sin\theta$ 则 $dy = \cos\theta\,d\theta$

$$\therefore \int_0^1 -y^2 \cdot (1-y^2)^{\frac{1}{2}}\,dy = -\int_0^{\frac{\pi}{2}} \sin^2\theta \cdot \cos^2\theta\,d\theta = -\int_0^{\frac{\pi}{2}} \left(\frac{\sin 2\theta}{2}\right)^2 d\theta = -\frac{1}{4}\int_0^{\frac{\pi}{2}} \sin^2 2\theta\,d\theta$$

$$= -\frac{1}{4}\int_0^{\frac{\pi}{2}} \frac{1 - \cos 4\theta}{2}\,d\theta = -\frac{\pi}{16}$$

$$\therefore \int_C x^2 y\,dx + xy\,dy = -\frac{\pi}{16} + \frac{1}{3}$$

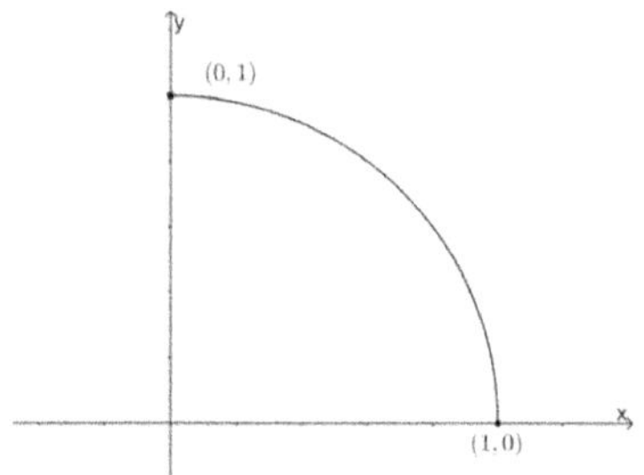

Example 27.

$\vec{\mathbf{F}} = (-y^3 \cos z, x^3 e^z, -e^z)$, $C$为曲面 $S: x^2 + y^2 + (z-a)^2 = 2a^2$ 与 $z = 0$ 交集的

逆时针封闭曲线，求$\oint_C \vec{\mathbf{F}} \cdot d\vec{r} =?$

【解】

令$C$为曲面 $S$ 与 $z = 0$ 交集的逆时针封闭曲线则 $C = \{(x, y, z): x^2 + y^2 = a^2, z = 0\}$

令 $\vec{r}(\theta) = (a\cos\theta, a\sin\theta, 0)$ 则 $\vec{r}'(\theta) = (-a\sin\theta, a\cos\theta, 0)$

$$\therefore \oint_C \vec{\mathbf{F}} \cdot d\vec{r} = \int_0^{2\pi} \vec{\mathbf{F}}(\vec{r}(\theta)) \cdot \vec{r}'(\theta)\, d\theta$$

$$= \int_0^{2\pi} (-(a\sin\theta)^3, (a\cos\theta)^3, -1) \cdot (-a\sin\theta, a\cos\theta, 0)\, d\theta = a^4 \int_0^{2\pi} \sin^4\theta + \cos^4\theta\, d\theta$$

$$\because \sin^4\theta + \cos^4\theta = (\sin^2\theta + \cos^2\theta)^2 - 2\sin^2\theta\cos^2\theta = 1 - 2(\sin\theta\cos\theta)^2$$

$$= 1 - 2\left(\frac{\sin 2\theta}{2}\right)^2 = 1 - \frac{\sin^2 2\theta}{2} = 1 - \frac{1 - \cos 4\theta}{4} = \frac{3 + \cos 4\theta}{4}$$

$$\therefore \int_0^{2\pi} \sin^4\theta + \cos^4\theta\, d\theta = \int_0^{2\pi} \frac{3 + \cos 4\theta}{4}\, d\theta = \frac{3\pi}{2}$$

$$\therefore \oint_C \vec{\mathbf{F}} \cdot d\vec{r} = \frac{3a^4\pi}{2}$$

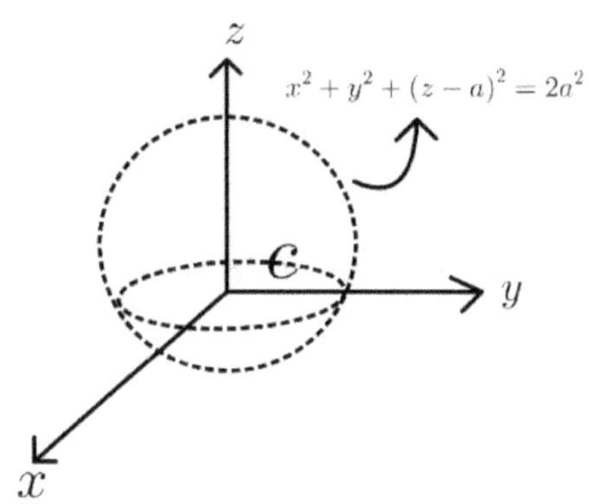


Example 28.

$\vec{\mathbf{F}} = (2x - y, 2y + z, xyz)$, $C$为曲面 $S: \dfrac{x^2}{a^2} + \dfrac{y^2}{b^2} + \dfrac{z^2}{c^2} = 1$ 与 $z = 0$ 交集的逆时针封闭

$$\text{曲线, 求} \oint_C \vec{\mathbf{F}} \cdot d\vec{r} = ?$$

【解】

令 $C$ 为曲面 $S$ 与 $z = 0$ 交集的逆时针封闭曲线则 $C = \{(x, y, z): \dfrac{x^2}{a^2} + \dfrac{y^2}{b^2} = 1, z = 0\}$

令 $\vec{r}(\theta) = (a\cos\theta, b\sin\theta, 0)$ 则 $\vec{r}'(\theta) = (-a\sin\theta, b\cos\theta, 0)$

$$\oint_C \vec{\mathbf{F}} \cdot d\vec{r} = \int_0^{2\pi} \vec{\mathbf{F}}(\vec{r}(\theta)) \cdot \vec{r}'(\theta)\, d\theta$$

$$= \int_0^{2\pi} (2a\cos\theta - b\sin\theta, 2b\sin\theta, 0) \cdot (-a\sin\theta, b\cos\theta, 0)\, d\theta$$

$$= \int_0^{2\pi} -2a^2 \sin\theta \cos\theta + ab\sin^2\theta + 2b^2 \sin\theta \cos\theta\, d\theta$$

$$= \int_0^{2\pi} 2(b^2 - a^2)\sin\theta\cos\theta + ab\sin^2\theta\, d\theta = \int_0^{2\pi} (b^2 - a^2)\sin 2\theta + ab\left(\dfrac{1 - \cos 2\theta}{2}\right) d\theta$$

$$= ab\pi$$

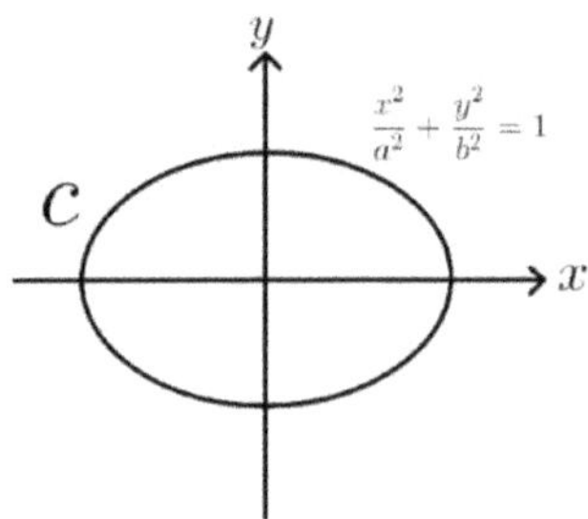


Example 29.

$\vec{\mathbf{F}} = (xz, yz, xy)$, $C$ 为曲面 $S: x^2 + y^2 + z^2 = 4$ $(z > 0)$ 与 $x^2 + y^2 = 1$ 交集的逆时针封闭曲线, 求 $\oint_C \vec{\mathbf{F}} \cdot d\vec{r} = ?$

【解】

令 $C$ 为 $S$ 与 $x^2 + y^2 = 1$ 交集的封闭曲线则 $C = \{(x, y, z): x^2 + y^2 = 1, z = \sqrt{3}\}$

令 $\vec{r}(\theta) = (\cos\theta, \sin\theta, \sqrt{3})$ 则 $\vec{r}'(\theta) = (-\sin\theta, \cos\theta, 0)$

$$\therefore \oint_C \vec{\mathbf{F}} \cdot d\vec{r} = \int_0^{2\pi} \vec{\mathbf{F}}(\vec{r}(\theta)) \cdot \vec{r}'(\theta)\, d\theta$$

$$= \int_0^{2\pi} (\sqrt{3}\cos\theta, \sqrt{3}\sin\theta, \sin\theta\cos\theta) \cdot (-\sin\theta, \cos\theta, 0)\, d\theta = \sqrt{3}\int_0^{2\pi} 0\, d\theta = 0$$

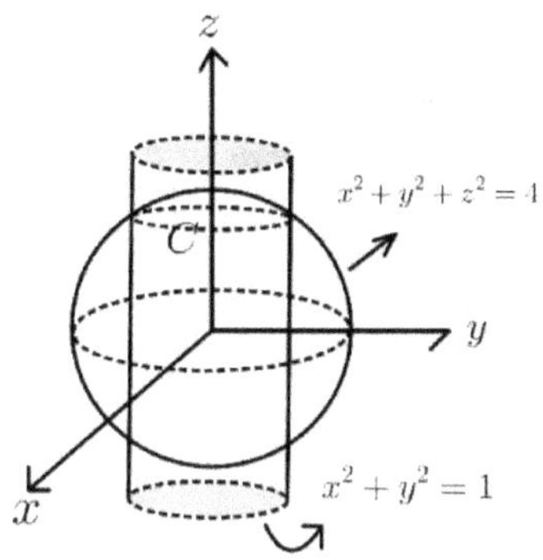


Example 30.

$\vec{\mathbf{F}} = (2x - y, -yz^2, -zy^2)$, $C$为球面$x^2 + y^2 + z^2 = 1$ 与 $z = 0$ 交集的逆时针封闭

曲线, 求 $\oint_C \vec{\mathbf{F}} \cdot d\vec{r} =?$

【解】

令$C$为球面与 $z = 0$ 交集的封闭曲线则$C = \{(x, y, z): x^2 + y^2 = 1, z = 0\}$

令 $\vec{r}(\theta) = (\cos\theta, \sin\theta, 0)$则 $\vec{r}'(\theta) = (-\sin\theta, \cos\theta, 0)$

$$\therefore \oint_C \vec{\mathbf{F}} \cdot d\vec{r} = \int_0^{2\pi} \vec{\mathbf{F}}(\vec{r}(\theta)) \cdot \vec{r}'(\theta)\, d\theta = \int_0^{2\pi} (2\cos\theta - \sin\theta, 0, 0) \cdot (-\sin\theta, \cos\theta, 0)\, d\theta$$

$$= \int_0^{2\pi} -2\sin\theta\cos\theta + \sin^2\theta\, d\theta = \int_0^{2\pi} -2\sin\theta\cos\theta + \frac{1 - \cos 2\theta}{2}\, d\theta = \pi$$

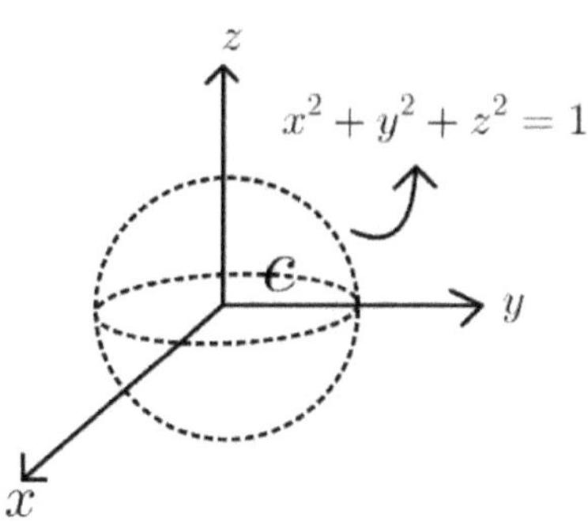


Example 31.

$\vec{\mathbf{F}} = (y^2, x^2, 0)$, $C$为曲面$S: x^2 + y^2 \leq 1$ 与 $z = 0$ 交集的逆时针封闭曲线, 求

$\oint_C \vec{\mathbf{F}} \cdot d\vec{r} =?$

【解】

令 $C$ 为曲面 $S$ 与 $z=0$ 交集的逆时封闭曲线则 $C=\{(x,y,z): x^2+y^2=1, z=0\}$

令 $\vec{r}(\theta)=(\cos\theta,\sin\theta,0)$则 $\vec{r}'(\theta)=(-\sin\theta,\cos\theta,0)$

$$\oint_C \vec{F}\cdot d\vec{r}=\int_0^{2\pi}\vec{F}(\vec{r}(\theta))\cdot\vec{r}'(\theta)\,d\theta=\int_0^{2\pi}(\sin^2,\cos^2,0)\cdot(-\sin\theta,\cos\theta,0)$$

$$=\int_0^{2\pi}-\sin^3\theta+\cos^3\theta\,d\theta$$

$\because -\sin^3\theta+\cos^3\theta=(\cos\theta-\sin\theta)(\cos^2\theta+\sin\theta\cos\theta+\sin^2\theta)$

$=(\cos\theta-\sin\theta)(1+\sin\theta\cos\theta)=(\cos\theta-\sin\theta)-\sin^2\theta\cos\theta+\sin\theta\cos^2\theta$

$$\therefore \int_0^{2\pi}-\sin^3\theta+\cos^3\theta\,d\theta=\int_0^{2\pi}(\cos\theta-\sin\theta)-\sin^2\theta\cos\theta+\sin\theta\cos^2\theta\,d\theta$$

$$=-\frac{\sin^3\theta}{3}+\frac{\cos^3\theta}{3}\bigg|_0^{2\pi}=0$$

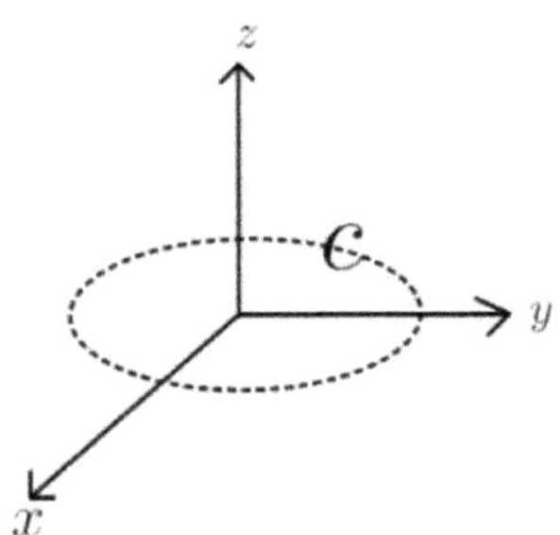


## 9.4  Green's Theorem $\Delta := \partial g/\partial x(x,y) - \partial f/\partial y(x,y)$

Green's Theorem 的条件成立与简单连通集有关，底下先介绍简单连通集的定义

【定义】

(i)若曲线$\vec{r}(t)$, $\forall a\le t\le b$满足 $\vec{r}(t_1)\ne\vec{r}(t_2)$, $\forall a<t_1<t_2<b$ 则称作 simple curve

(ii)再者,若 simple curve 满足 $\vec{r}(a)=\vec{r}(b)$, 则称作 simple closed curve

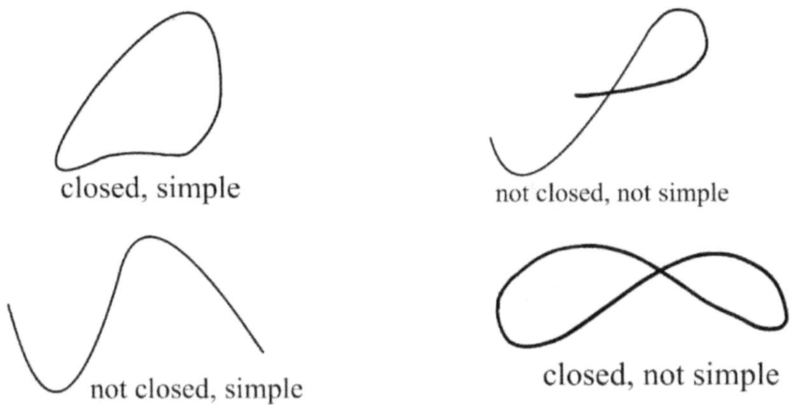



## 【定义】

如果一个连通集 D 内的任意简单封闭曲线所围成的区域都包含在 D 之内则 D 是简单连通

　　藉由上述定义可以推论两个 disjoint 的开集合的联集为非简单连通，或者当一个集合 D 里存在一个开集合不属于集合 D 则为非简单连通，如下图

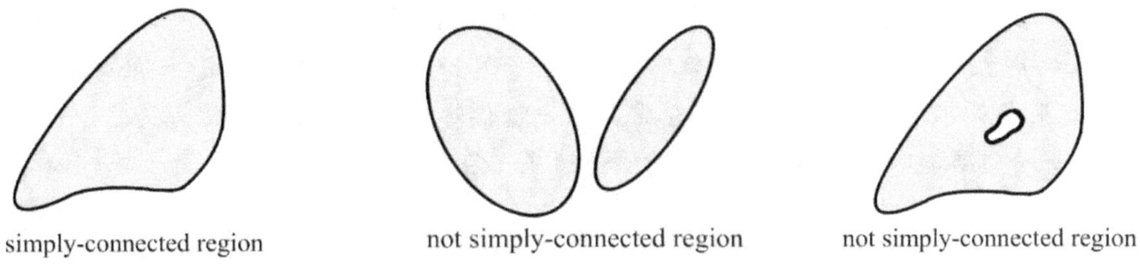



　　接下来介绍 Green's Theorem，当被积分函数的一阶偏导数存在且连续时，Green's Theorem 是计算二维封闭曲线积分与双重积分的重要工具，此外，Green's Theorem 是将微积分基本定理扩展到二维平面的一种形式，此定理将一个沿着简单封闭平面曲线 C 的线积分与其所包围平面区域 R（如下图）的双重积分连结起来，可以帮助将线积分与双重积分做双向的转换，如果线积分的计算是困难时，可藉由此定理转成双重积分，反之，当计算双重积分是困难时，藉由此定理可转成线积分做计算；Green's Theorem 要求双重积分中的区域 D 是简单连通，重要的是数个 disjoint 的简单连通集的联集为非简单连通集，因此，Green's Theorem 可以扩展应用在非简单连通集，实作上可以先拆成数个在简单连通集的积分，个别使用 Green's Theorem 之后再相加即可。Green's Theorem 有两种形式：旋度形式和通量形式，两者改写后的表示式对于 Stokes' Theorem 以及 The Divergence Theorem 提供了直观的说明，此外，Green's Theorem 当中封闭曲线的正方向是指逆时针方

向

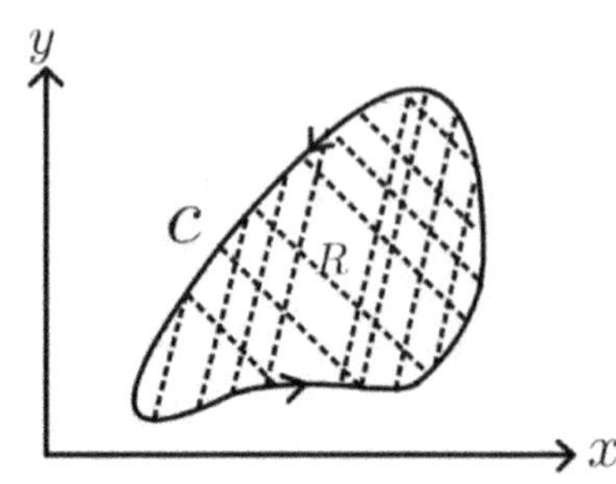

  Green's Theorem 的考试类型包含：说明线积分与积分路径无关、给定封闭区域 R 求面积、将封闭线积转成面积分、将封闭线积分转成面积分再搭配坐标转换、将面积分转成封闭线积分、造出不含奇异点的封闭区域再用 Green's Theorem, 值得注意的是除了第一种情况, 其余的类型皆为

$$\frac{\partial g}{\partial x}(x,y) - \frac{\partial f}{\partial y}(x,y) \neq 0,$$

整体而言, 若题目出现二维封闭曲线时, 可能就是 Green's Theorem 的应用时机, Green's Theorem 的成立条件之一是当中的两函数 $f, g$ 在所围封闭区域 $R$ 上需要是一阶偏导数存在且连续, 如果条件不满足时, 例如: $f, g$ 在原点的一阶偏导数不存在, 则通常先造出 $R'$ 使得 $f, g$ 在 $R'$ 上是一阶偏导数存在且连续, 接着让 $R'$ 逼近 $R$, 如下图所示

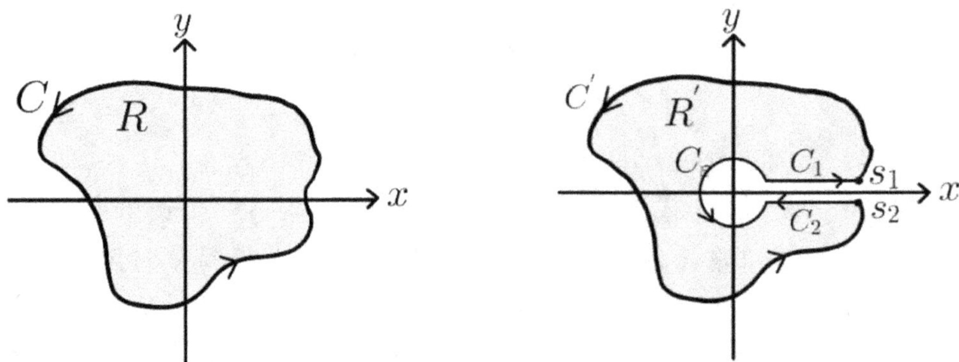

【**定理**】Green's Theorem

假设

1. $C$ 为二维空间中逆时针的封闭曲线且所围区域为 $R, R$ 简单连通
2. $f(x,y), g(x,y)$ 定义于 open set $B, R \subseteq B$ 且 $f, g$ 于 $B$ 的一阶偏导数存在且连续

则 $\displaystyle\iint_R \frac{\partial g}{\partial x}(x,y) - \frac{\partial f}{\partial y}(x,y)\,dxdy = \oint_C f(x,y)\,dx + g(x,y)\,dy$

<u>Proof:</u>

Let $C$ be made up of $C_1$ and $C_2$, $C_1: y = f_1(x), a \leq x \leq b$ and $C_2: y = f_2(x), \; a \leq x \leq b$

then $\displaystyle\oint_C f(x,y)\,dx = \oint_{C_1} f(x,y)\,dx - \oint_{C_2} f(x,y)\,dx$

Claim: $\displaystyle\iint_R \frac{\partial f}{\partial y}(x,y)\,dxdy = -\oint_C f(x,y)\,dx$

$\because \displaystyle\iint_R \frac{\partial f}{\partial y}(x,y)\,dxdy = \int_a^b \int_{f_1(x)}^{f_2(x)} \frac{\partial f}{\partial y}(x,y)\,dydx$

藉由微积分基本定理，$\displaystyle\int_{f_1(x)}^{f_2(x)} \frac{\partial f}{\partial y}(x,y)\,dy = f\big(x, f_2(x)\big) - f(x, f_1(x))$

$\therefore \displaystyle\iint_R \frac{\partial f}{\partial y}(x,y)\,dxdy = \int_a^b \int_{f_1(x)}^{f_2(x)} \frac{\partial f}{\partial y}(x,y)\,dydx = \int_a^b f\big(x, f_2(x)\big) - f(x, f_1(x))\,dx$

$\displaystyle = \int_a^b f\big(x, f_2(x)\big)\,dx - \int_a^b f(x, f_1(x))\,dx = -\left(\oint_{C_1} f(x,y)\,dx - \oint_{C_2} f(x,y)\,dx\right)$

$\displaystyle = -\oint_C f(x,y)\,dx$

Let $C$ be made up of $C_3$ and $C_4$, $C_3: y = g_1(y), \; c \leq y \leq d$ and $C_4: y = g_2(y), \; c \leq y \leq d$

then $\displaystyle\oint_C g(x,y)\,dx = \oint_{C_3} g(x,y)\,dx - \oint_{C_4} g(x,y)\,dx$

Claim: $\displaystyle\iint_R \frac{\partial g}{\partial x}(x,y)\,dxdy = \oint_C g(x,y)\,dy$

$\because \displaystyle\iint_R \frac{\partial g}{\partial x}(x,y)\,dxdy = \int_c^d \int_{g_2(y)}^{g_1(y)} \frac{\partial g}{\partial x}(x,y)\,dxdy$

藉由微积分基本定理，$\displaystyle\int_{g_2(y)}^{g_1(y)} \frac{\partial g}{\partial x}(x,y)\,dx = g(g_1(y), y) - g(g_2(y), y)$

$\therefore \displaystyle\iint_R \frac{\partial g}{\partial x}(x,y)\,dxdy = \int_c^d \int_{g_2(y)}^{g_1(y)} \frac{\partial g}{\partial x}(x,y)\,dxdy = \int_c^d g(g_1(y), y) - g(g_2(y), y)\,dy$

$\displaystyle = \int_c^d g(g_1(y), y)\,dy - \int_c^d g(g_2(y), y)\,dy = \oint_{C_3} g(x,y)\,dy - \oint_{C_4} g(x,y)\,dy = \oint_C g(x,y)\,dy$

因此

$$\iint_R \frac{\partial g}{\partial x}(x,y) - \frac{\partial f}{\partial y}(x,y)dxdy = \oint_C f(x,y)\,dx + g(x,y)dy$$

Green's Theorem 可以藉由旋度$(\nabla \times \vec{F})$以及散度$(\operatorname{div}\vec{F})$重新改写，改写后的表示式对于 Stokes' Theorem 以及 The Divergence Theorem 提供了直观的说明， 假设平面封闭曲线为 $C$，所围区域为 R，函数$f$及$g$两者的偏导数存在且连续，考虑向量场$\vec{F} = (f,g)$

$$\because \operatorname{curl}\vec{F} = \nabla \times \vec{F} = \left\|\begin{matrix} \boldsymbol{i} & \boldsymbol{j} & \boldsymbol{k} \\ \frac{\partial}{\partial x} & \frac{\partial}{\partial y} & \frac{\partial}{\partial z} \\ f & g & 0 \end{matrix}\right\| = \left(\frac{\partial g}{\partial x} - \frac{\partial f}{\partial y}\right)\boldsymbol{k} \qquad \therefore \frac{\partial g}{\partial x} - \frac{\partial f}{\partial y} = (\operatorname{curl}\vec{F}) \cdot \boldsymbol{k}$$

另一方面， 线积分表示为

$$\oint_C f\,dx + g\,dy = \oint_C \vec{F} \cdot d\vec{r}, \ \ \text{其中}\ \vec{F} = (f(x,y), g(x,y), 0).$$

By Green's Theorem

$$\oint_C \vec{F} \cdot d\vec{r} = \oint_C f\,dx + g\,dy = \iint_R \frac{\partial g}{\partial x}(x,y) - \frac{\partial f}{\partial y}(x,y)dxdy = \iint_R (\nabla \times \vec{F}) \cdot \boldsymbol{k}\,dA$$

因此， Green's Theorem 的向量形式的表示为

$$\oint_C \vec{F} \cdot d\vec{r} = \iint_R (\nabla \times \vec{F}) \cdot \boldsymbol{k}\,dA$$

如果是三维空间时， 直观的推论为

$$\oint_C \vec{F} \cdot d\vec{r} = \iint_S (\nabla \times \vec{F}) \cdot \vec{n}\,dA$$

其中$S$为空间中非封闭的曲面且$C$为$S$的边界(空间中的封闭曲线)

接下来说明 Green's Theorem 的散度形式， 假设曲线 $C$ 由向量方程式表示
$$\vec{r}(t) = (x(t), y(t)), \ \ a \le t \le b$$

$\because$ 单位切向量 $= \left(\dfrac{x'(t)}{|\vec{r}'(t)|}, \dfrac{y'(t)}{|\vec{r}'(t)|}\right)$ $\quad \therefore$ 向外的单位法向量 $\vec{n} = \left(\dfrac{y'(t)}{|\vec{r}'(t)|}, -\dfrac{x'(t)}{|\vec{r}'(t)|}\right)$

因此

$$\oint_C \vec{F} \cdot \vec{n}\,ds = \int_a^b (\vec{F} \cdot \vec{n})(t)|\vec{r}'(t)|dt = \int_a^b \left(\frac{f(x(t),y(t))y'(t)}{|\vec{r}'(t)|} - \frac{g(x(t),y(t))x'(t)}{|\vec{r}'(t)|}\right)|\vec{r}'(t)|dt$$

$$= \int_a^b f(x(t),y(t))y'(t) - g(x(t),y(t))x'(t)dt = \int_C f\,dy - g\,dx$$

根据 Green's Theorem

$$\int_C f\,dy - g\,dx = \iint_R \frac{\partial f}{\partial x} + \frac{\partial g}{\partial y}\,dA = \iint_R \operatorname{div}\vec{F}\,dA$$

因此

$$\oint_C \vec{F}\cdot\vec{n}\,ds = \iint_R \operatorname{div}\vec{F}\,dA$$

如果是三维空间时，直观的推论为

$$\oiint_S \vec{F}\cdot\vec{n}\,dA = \iiint_V \nabla\cdot\vec{F}\,dV$$

其中 $S$ 为空间中封闭的曲面且 V 为 $S$ 所围体积

后续 Stokes' Theorem 以及 The Divergence Theorem 的说明，皆是以上述 Green's Theorem 改写后的形式做说明

## 9.4.1　$\Delta = 0$, 线积分与积分路径无关

Green's Theorem 当中的两函数 $f, g$ 满足一阶偏导数存在且连续的条件,藉由

$$\frac{\partial g}{\partial x}(x,y) - \frac{\partial f}{\partial y}(x,y) = 0,$$

试证线积分与积分路径无关

考试类型：

Type 1.

给定函数 $f(x,y), g(x,y)$ 与两点 $(x_0, y_0), (x_1, y_1)$，试证 $\displaystyle\int_{(x_0,y_0)}^{(x_1,y_1)} f(x,y)\,dx + g(x,y)\,dy$ 与路径

无关并求此积分

解题流程：

Step1.

令 $C_1$、$C_2$ 为从 $(x_0, y_0)$ 至 $(x_1, y_1)$ 的任意两条曲线

令 $C$ 为 $C_1$、$C_2$ 所围成的逆时针封闭曲线且 R 为所围封闭区域

Step2.

Claim: $\dfrac{\partial g}{\partial x}(x,y) - \dfrac{\partial f}{\partial y}(x,y) = 0$

Step3.

藉由 Green's Theorem,　$\displaystyle\oint_C fdx+gdy=\iint_R \frac{\partial g}{\partial x}(x,y)-\frac{\partial f}{\partial y}(x,y)dxdy=0$

$\therefore \displaystyle\int_{C_1} fdx+gdy-\int_{C_2} fdx+gdy=0 \Rightarrow \int_{C_1} fdx+gdy=\int_{C_2} fdx+gdy$

$\therefore \displaystyle\int_{(x_0,y_0)}^{(x_1,y_1)} f(x,y)dx+g(x,y)dy$ 与路径无关

Step4.

$\therefore \displaystyle\int_{(x_0,y_0)}^{(x_1,y_1)} f(x,y)dx+g(x,y)dy$ 与路径无关

$\therefore \exists G(x,y)$ 使得 $\nabla G(x,y,z)=(f(x,y),g(x,y))$

且 $\displaystyle\int_{(x_0,y_0)}^{(x_1,y_1)} fdx+gdy=\int_{(x_0,y_0)}^{(x_1,y_1)} \nabla G\cdot(dx,dy)=\int_{(x_0,y_0)}^{(x_1,y_1)} dG=G(x_1,y_1)-G(x_0,y_0)$

Step5.

找 $G(x,y)=$? 且求 $G(x_1,y_1)-G(x_0,y_0)=$?


<u>范例说明:</u>

給定平面上两点 $(x_0,y_0),(x_1,y_1)$,　试证 $\displaystyle\int_{(x_0,y_0)}^{(x_1,y_1)} f(x,y)dx+g(x,y)dy$ 与路径无关

令 $C_1$、$C_2$ 为從 $(x_0,y_0)$ 至 $(x_1,y_1)$ 的任意两条曲线

令 $C$ 为 $C_1$、$C_2$ 所围成的逆时针封闭曲线且 R 为所围封闭区域

(I)若 $f(x,y)=6xy^2-y^3$,　$g(x,y)=6x^2y-3xy^2$

则 $\displaystyle\frac{\partial g}{\partial x}(x,y)-\frac{\partial f}{\partial y}(x,y)=12xy-3y^2-(12xy-3y^2)=0$

(II)若 $f(x,y)=2xy-y^4+3$,　$g(x,y)=x^2-4xy^3$

则 $\displaystyle\frac{\partial g}{\partial x}(x,y)-\frac{\partial f}{\partial y}(x,y)=2x-4y^3-(2x-4y^3)=0$

(III)若 $f(x,y)=4x^3y$,　$g(x,y)=x^4$ 则 $\displaystyle\frac{\partial g}{\partial x}(x,y)-\frac{\partial f}{\partial y}(x,y)=4x^3-(4x^3)=0$

藉由 Green's Theorem,　$\displaystyle\oint_C fdx+gdy=\iint_R \frac{\partial g}{\partial x}(x,y)-\frac{\partial f}{\partial y}(x,y)dxdy=0$

$\therefore \displaystyle\int_{C_1} fdx+gdy-\int_{C_2} fdx+gdy=0 \Rightarrow \int_{C_1} fdx+gdy=\int_{C_2} fdx+gdy$

$$\therefore \int_{(x_0,y_0)}^{(x_1,y_1)} f(x,y)dx + g(x,y)dy \text{ 与路径无关}$$

Example 1.

$$试证 \int_{(1,1)}^{(2,2)} (6xy^2 - y^3)dx + (6x^2y - 3xy^2)dy \text{ 与路径无关并求此积分}$$

【解】

令 $C_1$、$C_2$ 为從 $(1,1)$ 至 $(2,2)$ 的任意两条曲线

令 $C$ 为 $C_1$、$C_2$ 所围成的逆时针封闭曲线且 R 为所围封闭区域

令 $f(x,y) = 6xy^2 - y^3$, $g(x,y) = 6x^2y - 3xy^2$

则 $\dfrac{\partial g}{\partial x}(x,y) - \dfrac{\partial f}{\partial y}(x,y) = 12xy - 3y^2 - (12xy - 3y^2) = 0$

$\because f(x,y), g(x,y)$ 的一阶偏导数存在且连续

藉由 Green's Theorem, $\displaystyle\oint_C fdx + gdy = \iint_R \dfrac{\partial g}{\partial x}(x,y) - \dfrac{\partial f}{\partial y}(x,y)dxdy = 0$

$$\therefore \int_{C_1} fdx + gdy - \int_{C_2} fdx + gdy = 0 \Rightarrow \int_{C_1} fdx + gdy = \int_{C_2} fdx + gdy$$

$$\therefore \int_{(1,1)}^{(2,2)} (6xy^2 - y^3)dx + (6x^2y - 3xy^2)dy \text{ 与积分路径无关}$$

$\therefore \exists G(x,y)$ 使得 $\nabla G(x,y) = (f(x,y), g(x,y))$

且 $\displaystyle\int_{(1,1)}^{(2,2)} fdx + gdy = \int_{(1,1)}^{(2,2)} \nabla G \cdot (dx,dy) = \int_{(1,1)}^{(2,2)} dG = G(3,4) - G(1,2)$

$$\therefore \begin{cases} \dfrac{\partial G}{\partial x} = 6xy^2 - y^3 \\[2mm] \dfrac{\partial G}{\partial y} = 6x^2y - 3xy^2 \end{cases} \quad \therefore \begin{cases} G(x,y,z) = 3x^2y^2 - xy^3 + c \\ G(x,y,z) = 3x^2y^2 - xy^3 + c \end{cases}$$

$\therefore G(x,y,z) = 3x^2y^2 - xy^3 + c$

$$\therefore \int_{(1,1)}^{(2,2)} (6xy^2 - y^3)dx + (6x^2y - 3xy^2)dy = 3x^2y^2 - xy^3 \Big|_{(1,1)}^{(2,2)} = 30$$

Example 2.

求 $\oint_C (x^2 y \cos x + 2xy \sin x - y^2 e^x)dx + (x^2 \sin x - 2ye^x)dy = ?$,

$C: x^{\frac{2}{3}} + y^{\frac{2}{3}} = a^{\frac{2}{3}}$ 为逆时针封闭曲线

【解】

令 $f(x,y) = x^2 y \cos x + 2xy \sin x - y^2 e^x$, $g(x,y) = x^2 \sin x - 2ye^x$

则 $\dfrac{\partial g}{\partial x}(x,y) - \dfrac{\partial f}{\partial y}(x,y) = 2x \sin x + x^2 \cos x - 2ye^x - (x^2 \cos x + 2x \sin x - 2ye^x) = 0$

$\because f(x,y), g(x,y)$ 的一阶偏导数存在且连续

藉由 Green's Theorem, $\oint_C f dx + g dy = \iint_R \dfrac{\partial g}{\partial x}(x,y) - \dfrac{\partial f}{\partial y}(x,y) dxdy = 0$

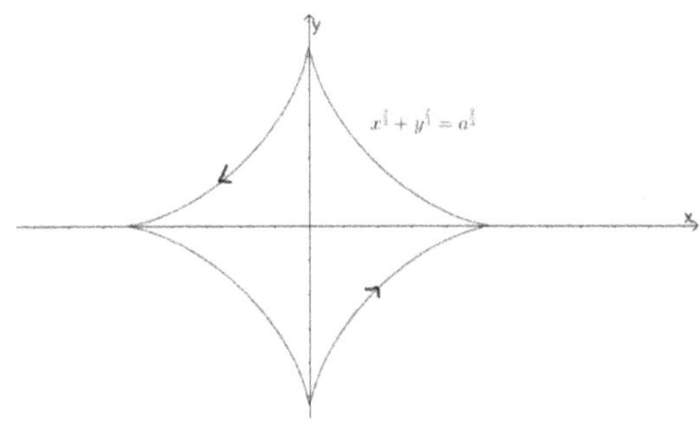


Example 3.

试证 $\displaystyle\int_{(0,0)}^{(3,1)} (2xy - y^4 + 3)dx + (x^2 - 4xy^3)dy$ 与 $(0,0),(3,1)$ 的路径无关并求此积分

【解】

令 $C_1$、$C_2$ 为從 $(0,0)$ 至 $(3,1)$ 的任意两条曲线

令 $C$ 为 $C_1$、$C_2$ 所围成的逆时针封闭曲线且 R 为所围封闭区域

令 $f(x,y) = 2xy - y^4 + 3$, $g(x,y) = x^2 - 4xy^3$

则 $\dfrac{\partial g}{\partial x}(x,y) - \dfrac{\partial f}{\partial y}(x,y) = 2x - 4y^3 - (2x - 4y^3) = 0$

$\because f(x,y), g(x,y)$ 的一阶偏导数存在且连续

藉由 Green's Theorem, $\oint_C f dx + g dy = \iint_R \dfrac{\partial g}{\partial x}(x,y) - \dfrac{\partial f}{\partial y}(x,y) dxdy = 0$

$$\therefore \int_{C_1} fdx + gdy - \int_{C_2} fdx + gdy = 0 \Rightarrow \int_{C_1} fdx + gdy = \int_{C_2} fdx + gdy$$

$$\therefore \int_{(0,0)}^{(3,1)} (2xy - y^4 + 3)dx + (x^2 - 4xy^3)dy \text{与积分路径无关}$$

$$\therefore \exists G(x,y) \text{使得} \nabla G(x,y) = (f(x,y), g(x,y))$$

$$\text{且} \int_{(0,0)}^{(3,1)} fdx + gdy = \int_{(0,0)}^{(3,1)} \nabla G \cdot (dx, dy) = \int_{(0,0)}^{(3,1)} dG = G(3,1) - G(0,0)$$

$$\therefore \begin{cases} \dfrac{\partial G}{\partial x} = 2xy - y^4 + 3 \\ \dfrac{\partial G}{\partial y} = x^2 - 4xy^3 \end{cases} \qquad \therefore \begin{cases} G(x,y,z) = x^2 y - xy^4 + 3x + c \\ G(x,y,z) = x^2 y - xy^4 + c \end{cases}$$

$$\therefore G(x,y,z) = x^2 y - xy^4 + 3x + c$$

$$\therefore \int_{(0,0)}^{(3,1)} (2xy - y^4 + 3)dx + (x^2 - 4xy^3)dy = x^2 y - xy^4 + 3x \Big|_{(0,0)}^{(3,1)} = 15$$

Example 4.

$$\text{试证} \int_{(0,0)}^{(a,b)} 4x^3 ydx + x^4 dy \text{与}(0,0),(a,b) \text{的路径无关并求此积分}$$

【解】

令$C_1$、$C_2$为从$(0,0)$至$(a,b)$的任意两条曲线

令$C$为$C_1$、$C_2$所围成的逆时针封闭曲线且 R 为所围封闭区域

令$f(x,y) = 4x^3 y$,  $g(x,y) = x^4$  则 $\dfrac{\partial g}{\partial x}(x,y) - \dfrac{\partial f}{\partial y}(x,y) = 4x^3 - (4x^3) = 0$

$\because f(x,y), g(x,y)$的一阶偏导数存在且连续

藉由Green's Theorem,  $\displaystyle\oint_C fdx + gdy = \iint_R \dfrac{\partial g}{\partial x}(x,y) - \dfrac{\partial f}{\partial y}(x,y)dxdy = 0$

$$\therefore \int_{C_1} fdx + gdy - \int_{C_2} fdx + gdy = 0 \Rightarrow \int_{C_1} fdx + gdy = \int_{C_2} fdx + gdy$$

$$\therefore \int_{(0,0)}^{(a,b)} 4x^3 ydx + x^4 dy \text{与积分路径无关}$$

$$\therefore \exists G(x,y) \text{使得} \nabla G(x,y) = (f(x,y), g(x,y))$$

$$\text{且} \int_{(0,0)}^{(a,b)} fdx + gdy = \int_{(0,0)}^{(a,b)} \nabla G \cdot (dx, dy) = \int_{(0,0)}^{(a,b)} dG = G(a,b) - G(0,0)$$

$$\therefore \begin{cases} \dfrac{\partial G}{\partial x} = 4x^3y \\ \dfrac{\partial G}{\partial y} = x^4 \end{cases} \qquad \therefore \begin{cases} G(x,y,z) = x^4y + c \\ G(x,y,z) = x^4y + c \end{cases}$$

$$\therefore G(x,y,z) = x^4y + c \qquad \therefore \int_{(0,0)}^{(a,b)} 4x^3y\,dx + x^4\,dy = x^4y\big|_{(0,0)}^{(a,b)} = a^4b$$

Example 5.

$$求 \int_{(0,0)}^{(-2,-1)} (10x^4 - 2xy^3)dx - 3x^2y^2\,dy = ?, \quad 其中积分路径为 x^4 - 6xy^3 = 4y^2$$

【解】

令 $C_1$、$C_2$ 为從 $(0,0)$ 至 $(-2,-1)$ 的任意两条曲线

令 $C$ 为 $C_1$、$C_2$ 所围成的逆时针封闭曲线且 R 为所围封闭区域

令 $f(x,y) = 10x^4 - 2xy^3, g(x,y) = -3x^2y^2$

则 $\dfrac{\partial g}{\partial x}(x,y) - \dfrac{\partial f}{\partial y}(x,y) = -6xy^2 - (-6xy^2) = 0$

$\because f(x,y), \ g(x,y)$ 的一阶偏导数存在且连续

藉由 Green's Theorem, $\displaystyle\oint_C f\,dx + g\,dy = \iint_R \dfrac{\partial g}{\partial x}(x,y) - \dfrac{\partial f}{\partial y}(x,y)\,dxdy = 0$

$$\therefore \int_{C_1} f\,dx + g\,dy - \int_{C_2} f\,dx + g\,dy = 0 \Rightarrow \int_{C_1} f\,dx + g\,dy = \int_{C_2} f\,dx + g\,dy$$

$$\therefore \int_{(0,0)}^{(-2,-1)} (10x^4 - 2xy^3)dx - 3x^2y^2\,dy \text{ 与积分路径无关}$$

$\therefore \exists G(x,y)$ 使得 $\nabla G(x,y) = (f(x,y), g(x,y))$ 且

$$\int_{(0,0)}^{(-2,-1)} f\,dx + g\,dy = \int_{(0,0)}^{(-2,-1)} \nabla G \cdot (dx, dy) = \int_{(0,0)}^{(-2,-1)} dG = G(-2,-1) - G(0,0)$$

$$\therefore \begin{cases} \dfrac{\partial G}{\partial x} = 10x^4 - 2xy^3 \\ \dfrac{\partial G}{\partial y} = -3x^2y^2 \end{cases} \qquad \therefore \begin{cases} G(x,y,z) = 2x^5 - x^2y^3 + c \\ G(x,y,z) = -x^2y^3 + c \end{cases} \quad \therefore G(x,y,z) = 2x^5 - x^2y^3 + c$$

$$\therefore \int_{(0,0)}^{(-2,-1)} (10x^4 - 2xy^3)dx - (3x^2y^2)dy = 2x^5 - x^2y^3\big|_{(0,0)}^{(-2,-1)} = -60$$

Example 6.

$$求 \int_{(0,0)}^{\left(\frac{\pi}{2},1\right)} (x^2 + 6xy - 2y^2)dx + (3x^2 - 4xy + 2y)dy =?, \quad 其中路径为 y = \sin x$$

【解】

令 $C_1$、$C_2$ 为從 $(0,0)$ 至 $\left(\frac{\pi}{2}, 1\right)$ 的任意两条曲线

令 $C$ 为 $C_1$、$C_2$ 所围成的逆时针封闭曲线且 R 为所围封闭区域

令 $f(x,y) = x^2 + 6xy - 2y^2, \quad g(x,y) = 3x^2 - 4xy + 2y$

则 $\dfrac{\partial g}{\partial x}(x,y) - \dfrac{\partial f}{\partial y}(x,y) = 6x - 4y - (6x - 4y) = 0$

$\because f(x,y), g(x,y)$ 的一阶偏导数存在且连续

藉由 Green's Theorem, $\displaystyle\oint_C f dx + g dy = \iint_R \frac{\partial g}{\partial x}(x,y) - \frac{\partial f}{\partial y}(x,y)dxdy = 0$

$\therefore \displaystyle\int_{C_1} f dx + g dy - \int_{C_2} f dx + g dy = 0 \Rightarrow \int_{C_1} f dx + g dy = \int_{C_2} f dx + g dy$

$\therefore \displaystyle\int_{(0,0)}^{\left(\frac{\pi}{2},1\right)} (x^2 + 6xy - 2y^2)dx + (3x^2 - 4xy + 2y)dy$ 与积分路径无关

$\therefore \exists G(x,y)$ 使得 $\nabla G(x,y) = (f(x,y), g(x,y))$

且 $\displaystyle\int_{(0,0)}^{\left(\frac{\pi}{2},1\right)} f dx + g dy = \int_{(0,0)}^{\left(\frac{\pi}{2},1\right)} \nabla G \cdot (dx, dy) = \int_{(0,0)}^{\left(\frac{\pi}{2},1\right)} dG = G\left(\frac{\pi}{2}, 1\right) - G(0,0)$

$$\because \begin{cases} \dfrac{\partial G}{\partial x} = x^2 + 6xy - 2y^2 \\[2mm] \dfrac{\partial G}{\partial y} = 3x^2 - 4xy + 2y \end{cases} \qquad \therefore \begin{cases} G(x,y,z) = \dfrac{x^3}{3} + 3x^2 y - 2xy^2 + c \\[2mm] G(x,y,z) = 3x^2 y - 2xy^2 + y^2 + c \end{cases}$$

$\therefore G(x,y,z) = 3x^2 y - 2xy^2 + y^2 + \dfrac{x^3}{3} + c$

$$\therefore \int_{(0,0)}^{\left(\frac{\pi}{2},1\right)} (x^2 + 6xy - 2y^2)dx + (3x^2 - 4xy + 2y)dy = 3x^2 y - 2xy^2 + y^2 + \frac{x^3}{3} \Bigg|_{(0,0)}^{\left(\frac{\pi}{2},1\right)}$$

$$= 3\left(\frac{\pi}{2}\right)^2 - 2\left(\frac{\pi}{2}\right) + 1 + \frac{\left(\frac{\pi}{2}\right)^3}{3}$$

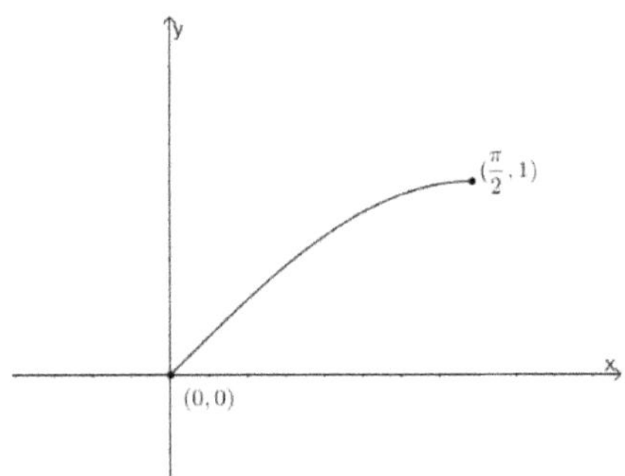


## 9.4.2　　$\Delta \neq 0$, 給封閉区域R求面积

考试类型:

Type 1.

假设$C$为封闭区域$R$的边界, 求$R$所围面积

解题流程:

令 $f(x,y) = -y$,　$g(x,y) = x$, 藉由 Green's Theorem 则

$$R\text{ 所围面积} = \iint_R dxdy = \frac{1}{2}\iint_R \frac{\partial x}{\partial x} - \frac{\partial(-y)}{\partial y} dxdy = \frac{1}{2}\oint_C -y\,dx + xdy$$

求 $\dfrac{1}{2}\displaystyle\oint_C -y\,dx + xdy =?$

<u>補充说明:</u>

将求封闭区域$R$面积的问题转成求 $\dfrac{1}{2}\displaystyle\oint_C -y\,dx + xdy =?$, 其中$C$为$R$的边界

<u>范例说明:</u>

假设$R$为封闭区域且$C$为$R$的边界, 求$R$所围面积 $\displaystyle\iint_R dxdy =?$

令 $f(x,y) = -y$,　$g(x,y) = x$, 藉由 Green's Theorem,

$$\iint_R dxdy = \frac{1}{2}\iint_R \frac{\partial x}{\partial x} - \frac{\partial(-y)}{\partial y} dxdy = \frac{1}{2}\oint_C -y\,dx + xdy$$

(I)若 $C$: $\dfrac{x^2}{a^2} + \dfrac{y^2}{b^2} = 1$

则 $\dfrac{1}{2}\oint_C -y\,dx + x\,dy = \dfrac{1}{2}\int_0^{2\pi} (-b\sin\theta)(-a\sin\theta) + (a\cos\theta)(b\cos\theta)d\theta = ab\pi$

(II)若 $C$: $r(\theta) = a(1 \pm \cos\theta)$

则 $\dfrac{1}{2}\oint_C -y\,dx + x\,dy = \dfrac{1}{2}\oint_C r^2(\theta)\,d\theta = \dfrac{1}{2}\left(\int_0^{2\pi} a^2(1+\cos\theta)^2 d\theta\right) = \dfrac{3a^2\pi}{2}$

(III)若 $C$: $r(\theta) = a(1 \pm \sin\theta)$

则 $\dfrac{1}{2}\oint_C -y\,dx + x\,dy = \dfrac{1}{2}\oint_C r^2(\theta)\,d\theta = \dfrac{1}{2}\left(\int_0^{2\pi} a^2(1\pm\sin\theta)^2 d\theta\right) = \dfrac{3a^2\pi}{2}$


**Example 1.**

假设 $R$ 为封闭区域且 $C$ 为 $R$ 逆时针方向的边界, 使用 Green's Theorem 求 $R$ 所围面积

【解】

令 $f(x,y) = -y,\ g(x,y) = x,\ \because f(x,y), g(x,y)$ 的一阶偏导数存在且连续

藉由 Green's Theorem

则 R 所围面积 $= \iint_R dx\,dy = \dfrac{1}{2}\iint_R \dfrac{\partial x}{\partial x} - \dfrac{\partial(-y)}{\partial y} dx\,dy = \dfrac{1}{2}\oint_C -y\,dx + x\,dy$


**Example 2.**

使用 Green's Theorem, 求 $\dfrac{x^2}{a^2} + \dfrac{y^2}{b^2} = 1$ 所围面积

【解】

令 R: $\dfrac{x^2}{a^2} + \dfrac{y^2}{b^2} \leq 1$ 且 $C$ 为 R 的逆时针边界则 R 所围面积 $= \dfrac{1}{2}\oint_C -y\,dx + x\,dy$

令 $x = a\cos\theta,\ y = b\sin\theta$ 且 $0 \leq \theta \leq 2\pi$

则 $\dfrac{1}{2}\oint_C -y\,dx + x\,dy = \dfrac{1}{2}\int_0^{2\pi} (-b\sin\theta)(-a\sin\theta) + a\cos\theta\, b\cos\theta\, d\theta = ab\pi$

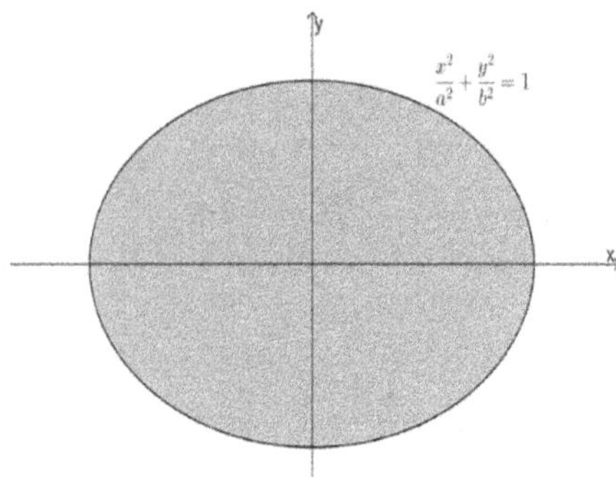

## Example 3.

使用 Green's Theorem，　求 $y = x^r$、$x = y^r (r > 1)$ 所围面积

【解】

令 $C_1: y = x^r$、$C_2: x = y^r$ 且 $0 \leq x, y \leq 1$

令 $C$ 为 $C_1$、$C_2$ 所围逆时针封闭曲线且 R 为 $C$ 所围面积,藉由 Green's Theorem

则 R 所围面积 $= \dfrac{1}{2} \oint_C -y\,dx + x\,dy = \dfrac{1}{2} \int_{C_1} -y\,dx + x\,dy + \dfrac{1}{2} \int_{C_2} -y\,dx + x\,dy$

$= \dfrac{1}{2} \int_0^1 -x^r dx + x(rx^{r-1})dx + \dfrac{1}{2} \int_1^0 -y(ry^{r-1})dy + y^r dy$

$= \dfrac{1}{2} \int_0^1 (r-1)x^r dx + \dfrac{1}{2} \int_1^0 (1-r)y^r dy = \dfrac{1}{2}\left(\dfrac{r-1}{r+1} \cdot x^{r+1}\big|_0^1 + \dfrac{1-r}{r+1} \cdot y^{r+1}\big|_1^0\right)$

$= \dfrac{1}{2} \cdot \dfrac{r-1-(1-r)}{r+1} = \dfrac{r-1}{r+1}$

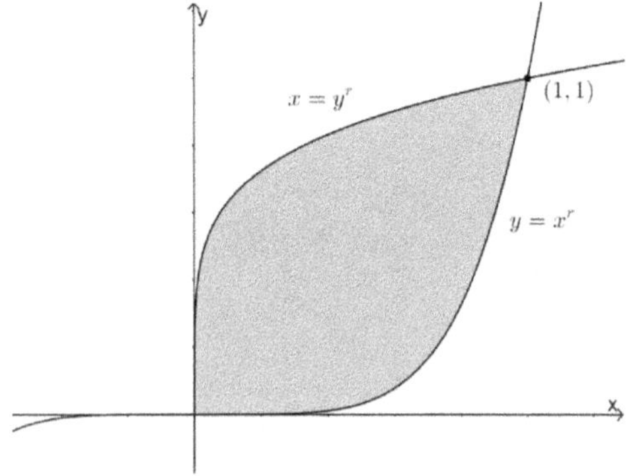


## Example 4.

使用 Green's Theorem，　求 $r(\theta) = a(1 + \cos\theta)$ 的封闭面积

【解】

$$
\text{所围面积} = \frac{1}{2}\oint_C -y\,dx + x\,dy = \frac{1}{2}\oint_C r^2(\theta)\,d\theta = \frac{1}{2}\int_0^{2\pi} a^2(1+\cos\theta)^2\,d\theta
$$

$$
= \frac{a^2}{2}\int_0^{2\pi} 1 + \frac{1+\cos 2\theta}{2} + 2\cos\theta\,d\theta = \frac{a^2}{2}\left(\theta + 2\sin\theta + \frac{\theta}{2} + \frac{\sin 2\theta}{4}\right)\Big|_0^{2\pi} = \frac{3a^2\pi}{2}
$$

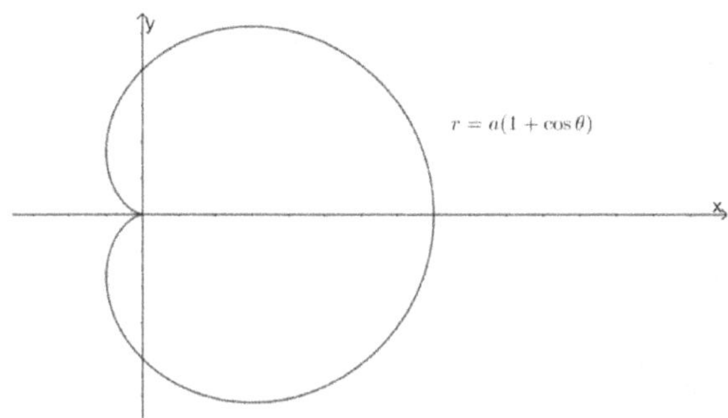


Example 5.

使用 Green's Theorem, 求 $\vec{r}(t) = (\cos^3 t, \sin^3 t)$ 所围面积, $0 \le t \le 2\pi$

【解】

藉由 Green's Theorem

令 $C$ 为 $x = \cos^3 t, y = \sin^3 t$ 逆时针封闭边界且 $0 \le \theta \le 2\pi$,所围区域为 R

则 $dx = 3\cos^2 t\,(-\sin t)dt$ 且 $dy = 3\sin^2 t\cos t\,dt$ 且 $\iint_R dxdy = \frac{1}{2}\oint_C -y\,dx + x\,dy$

$$
\frac{1}{2}\oint_C -y\,dx + x\,dy = \frac{1}{2}\int_0^{2\pi}(-\sin^3 t)\,3\cos^2 t\,(-\sin t) + (\cos^3 t)(3\sin^2 t\cos t)dt
$$

$$
= \frac{3}{2}\int_0^{2\pi}(\sin^2 t + \cos^2 t)\cos^2 t\sin^2 t\,dt = \frac{3}{2}\int_0^{2\pi}\cos^2 t\sin^2 t\,dt = \frac{3}{2}\int_0^{2\pi}\left(\frac{\sin 2t}{2}\right)^2 dt
$$

$$
= \frac{3}{2}\int_0^{2\pi}\frac{1-\cos 4t}{8}\,dt = \frac{3\pi}{8}
$$

### 9.4.3　Δ ≠ 0, 将封闭线积转成面积分

应用时机为线积分的被积分函数较为复杂，使用 Green's Theorem 将封闭线积分转成面积分后，被积分函数变得较为干净；藉由 Green's Theorem,

$$\oint_C f(x,y)\,dx + g(x,y)dy = \iint_R \frac{\partial g}{\partial x}(x,y) - \frac{\partial f}{\partial y}(x,y)dxdy$$

考试类型:

Type 1.

求 $\oint_C f(x,y)\,dx + g(x,y)dy =?$, 其中 $\dfrac{\partial g}{\partial x}(x,y) - \dfrac{\partial f}{\partial y}(x,y) = c\,(c$为常数$)$, 曲线$C$为逆时针的封闭圆: $x^2 + y^2 = r^2$ 且 $f,g$的一阶偏导数存在且连续

解题流程:

令 $R = \{(x,y): x^2 + y^2 \leq r^2\}$ 藉由 Green's Theorem 则

$$\oint_C f(x,y)\,dx + g(x,y)dy = \iint_R \frac{\partial g}{\partial x}(x,y) - \frac{\partial f}{\partial y}(x,y)dxdy = \iint_{x^2+y^2\leq r^2} c\,dxdy = c\pi r^2$$

<u>范例说明:</u>

求 $\oint_C f(x,y)\,dx + g(x,y)dy =?$, $C$为圆: $x^2 + y^2 = r^2, r > 0$

(I)若 $f(x,y) = 3x^2y + \cos x + e^y$, $g(x,y) = x^3 + xe^y + x$ 则 $\dfrac{\partial g}{\partial x}(x,y) - \dfrac{\partial f}{\partial y}(x,y) = 1$

$$\therefore \oint_C f(x,y)\,dx + g(x,y)dy = \iint_R \frac{\partial g}{\partial x}(x,y) - \frac{\partial f}{\partial y}(x,y)dxdy = \iint_{x^2+y^2\leq r^2} dxdy = \pi r^2$$

(II)若 $f(x,y) = x + 6y$, $g(x,y) = 2x + y$ 则 $\dfrac{\partial g}{\partial x}(x,y) - \dfrac{\partial f}{\partial y}(x,y) = 2 - 6 = -4$

$$\therefore \oint_C f(x,y)\,dx + g(x,y)dy = \iint_R \frac{\partial g}{\partial x}(x,y) - \frac{\partial f}{\partial y}(x,y)dxdy = \iint_{x^2+y^2\leq r^2} -4dxdy$$
$$= -4\pi r^2$$

<u>补充说明</u>

有时候积分区域不一定刚好是个圆可能是椭圆, 方法也是类似

Type 2.

藉由 Green's Theorem 和 Fubini's Theorem, 求 $\oint_C f(x,y)\,dx + g(x,y)dy =?$,

其中$C$ 为 $R = [a,b] \times [c,d]$的逆时针封闭边界 且 $f,g$的一阶偏导数存在且连续

解题流程:

Step1.

藉由 Green's Theorem

$$\oint_C f(x,y)\,dx + g(x,y)dy = \iint_R \frac{\partial g}{\partial x}(x,y) - \frac{\partial f}{\partial y}(x,y)dxdy = \iint_R h(x,y)dxdy$$

其中 $\dfrac{\partial g}{\partial x}(x,y) - \dfrac{\partial f}{\partial y}(x,y) = h(x,y)$

Step2.

藉由 Fubini'sTheorem, $\quad \iint_R h(x,y)dxdy = \int_a^b \int_c^d h(x,y)dydx$

Step3.

求 $\displaystyle\int_a^b \int_c^d h(x,y)dydx =?$

<u>范例说明：</u>

求 $\displaystyle\oint_C f(x,y)\,dx + g(x,y)dy =?$

(I)若 $f(x,y) = x^2$, $g(x,y) = 2xy$, $C$ 为 $(0,0)$、$(0,5)$、$(12,5)$、$(12,0)$ 所围的顺时针长方形边界

藉由 Green's Theorem

$$-\oint_C x^2\,dx + 2xydy = -\int_0^5 \int_0^{12} 2y\,dxdy = -\int_0^5 24ydy = -12y^2|_0^5 = -300$$

<u>补充说明</u>
有时候积分区域不一定是个长方形，但方法也是类似

Example 1.

藉由 Green's Theorem 求 $\displaystyle\oint_C (3x^2y + \cos x + e^y)\,dx + (x^3 + xe^y + x)dy =?$，其中

$C$ 为 $x^2 + y^2 = 1$ 的逆时针封闭圆

【解】

藉由 Green's Theorem

则 $\displaystyle\oint_C (3x^2y + \cos x + e^y)\,dx + (x^3 + xe^y + x)dy = \iint_{x^2+y^2 \leq 1} dxdy = \pi$

Example 2.

藉由 Green's Theorem 求 $\oint_C (6y + x)dx + (y + 2x)dy = ?$，$C$ 为逆时针封闭圆：

$(x - 2)^2 + (y - 3)^2 = r^2$

【解】

令 $R = \{(x, y): (x - 2)^2 + (y - 3)^2 \leq r^2\}$

藉由 Green's Theorem 则 $\oint_C (6y + x)dx + (y + 2x)dy = \iint_R 2 - 6\,dxdy = -4\pi r^2$

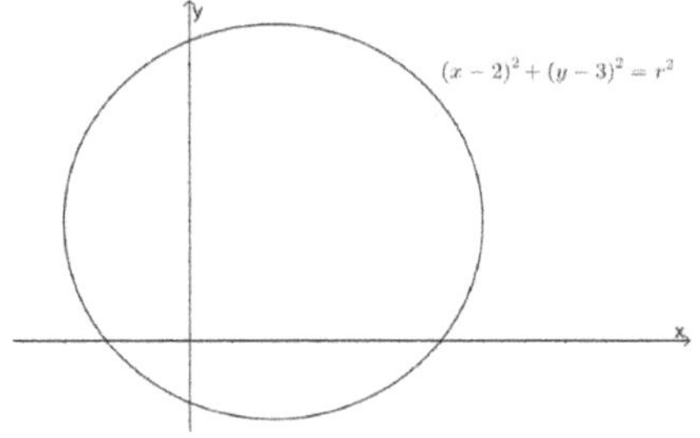


Example 3.

藉由 Green's Theorem 求 $\oint_C 3x - 4ydx + (4x + 2y)dy = ?$，$C$ 为上半椭圆的逆时针封

闭边界：$\dfrac{x^2}{16} + \dfrac{y^2}{9} = 1, y \geq 0$

【解】

令 $R = \{(x, y): \dfrac{x^2}{16} + \dfrac{x^2}{9} \leq 1, y \geq 0\}$

藉由 Green's Theorem 则 $\oint_C 3x - 4ydx + (4x + 2y)dy = \iint_R 8\,dxdy = \dfrac{8 \cdot 4 \cdot 3 \cdot \pi}{2} = 48\pi$

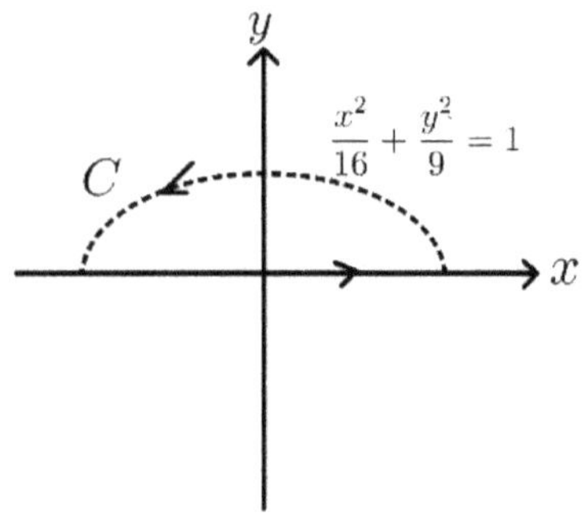

Example 4.

藉由 Green's Theorem 求 $\oint_C x^4\,dx + xy\,dy = ?$，$C$为 $(0,0)$、$(2,0)$、$(0,2)$ 所围的逆时针三角形边界

【解】

令 $R = \{(x,y): 0 \le x \le 2, 0 \le y \le x\}$

藉由 Green's Theorem 则 $\oint_C x^4\,dx + xy\,dy = \iint_R y\,dx\,dy$

藉由 Fubini's Theorem 则 $\iint_R y\,dx\,dy = \int_0^2 \int_0^x y\,dy\,dx = \int_0^2 \left.\frac{y^2}{2}\right|_0^x dx = \int_0^2 \frac{x^2}{2}\,dx = \left.\frac{x^3}{6}\right|_0^2 = \frac{4}{3}$

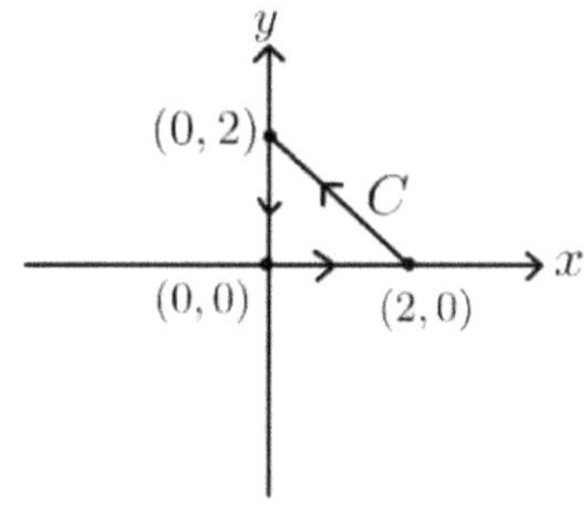


Example 5.

藉由 Green's Theorem 求 $\oint_C (x + 6y)\,dx + (2x + y)\,dy = ?$，其中 $C$为逆时针封闭圆：$(x - 2)^2 + (y - 3)^2 = 9$

【解】

藉由 Green's Theorem

则 $\oint_C (x + 6y)\,dx + (2x + y)\,dy = \iint_{(x-2)^2+(y-3)^2 \le 9} 2 - 6\,dx\,dy = -36\pi$

Example 6.

藉由 Green's Theorem 求 $\oint_C x^2\,dx + 2xy\,dy = ?$，$C$为 $(0,0)$、$(0,5)$、$(12,5)$、$(12,0)$ 所围的顺时针长方形边界

【解】

藉由 Green's Theorem 则

$$-\oint_C x^2\,dx + 2xydy = -\int_0^5 \int_0^{12} 2y\,dxdy = -\int_0^5 24ydy = -12y^2\big|_0^5 = -300$$

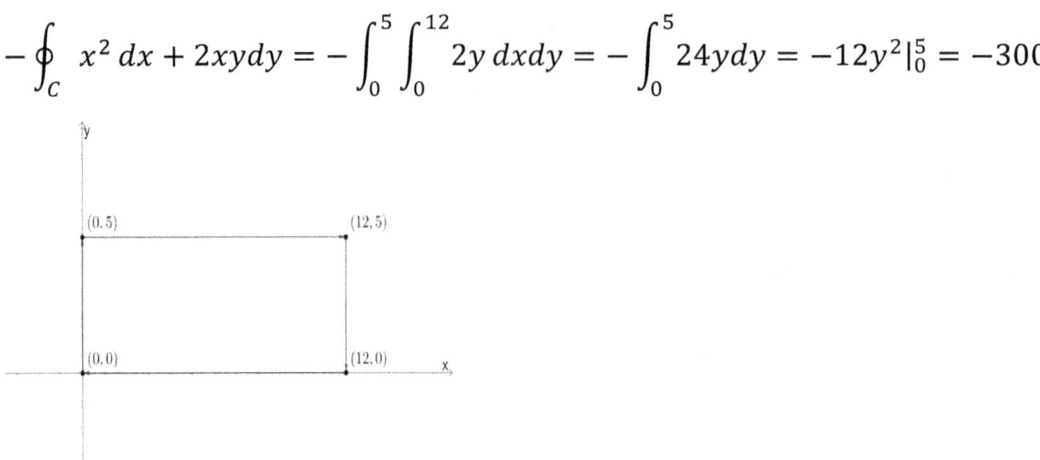

## Example 7.

藉由 Green's Theorem 求 $\oint_C (x^2 - xy^3)\,dx + (y^2 - 2xy)dy =?$, $C$ 为绕 $(0,0), (r,0),$

$(r,r), (0,r)$ 的逆时针封闭曲线, $r > 0$

【解】

令 $R = \{(x,y): 0 \le x \le r, 0 \le y \le r\}$

藉由 Green's Theorem 则 $\oint_C (x^2 - xy^3)\,dx + (y^2 - 2xy)dy = \iint_R -2y + 3xy^2 dxdy$

藉由 Fubini's Theorem

则 $\iint_R -2y + 3xy^2 dxdy = \int_0^r \int_0^r -2y + 3xy^2\,dydx = \int_0^r -y^2 + xy^3\big|_0^r dx$

$$= \int_0^r -r^2 + xr^3 dx = -r^3 + \frac{r^5}{2}$$

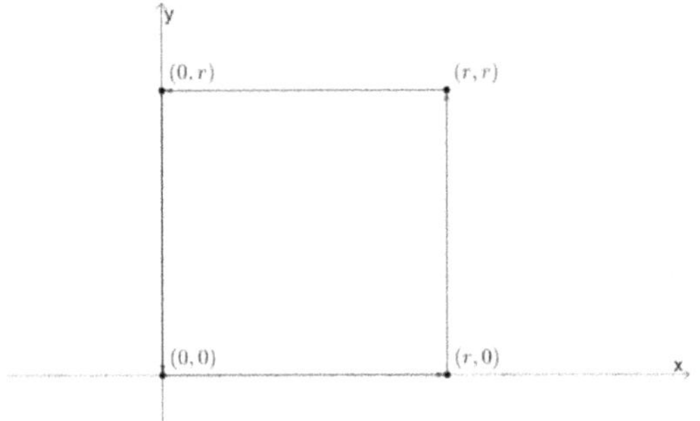

## Example 8.

藉由 Green's Theorem 求 $\displaystyle\oint_C \left(\frac{y^2}{2} - 4y\right) dx + \left(-\frac{x^2}{2}\right) dy =?$，其中 $C$ 为

$R = \{(x, y) : 0 \le x \le 2, 0 \le y \le 2\}$ 的逆时针封闭边界

【解】

藉由 Green's Theorem，$\displaystyle\oint_C \left(\frac{y^2}{2} - 4y\right) dx + \left(-\frac{x^2}{2}\right) dy = \iint_R 4 - x - y\, dA$

$\because f(x, y) = 4 - x - y$ 在 $R$ 区域为连续函数，藉由 Fubinis Theorem

$$\iint_R 4 - x - y\, dA = \int_0^2 \int_0^2 4 - x - y\, dy\, dx = \int_0^2 4y - xy - \frac{y^2}{2}\Big|_{y=0}^{y=2}\, dx = \int_0^2 8 - 2x - 2\, dx$$

$$= (6x - x^2)\big|_{x=0}^{x=2} = 12 - 4 = 8$$

Example 9.

藉由 Green's Theorem 求 $\displaystyle\oint_C xy^2\, dx + (5x^2y + x) dy =?$，其中 $C$ 为

$R = \{(x, y) : 1 \le x \le 2, 0 \le y \le 2\}$ 的逆时针封闭边界

【解】

藉由 Green's Theorem，$\displaystyle\oint_C xy^2\, dx + (5x^2y + x) dy = \iint_R 1 + 8xy\, dA$

$\because f(x, y) = 1 + 8xy$ 在 $R$ 区域为连续函数，藉由 Fubinis Theorem

$$\iint_R 1 + 8xy\, dA = \int_0^2 \int_1^2 1 + 8xy\, dx\, dy = \int_0^2 (x + 4x^2y)\big|_{x=1}^{x=2}\, dy = \int_0^2 1 + 12y\, dy$$

$$= (y + 6y^2)\big|_{y=0}^{y=2} = 2 + 6 \cdot 4 = 26$$

Example 10.

求 $\displaystyle\oint_C (y - \sin x)\, dx + (\cos x + \sqrt{2}) dy =?$，$C$ 为 $(0,0)$、$\left(\frac{\pi}{2}, 0\right)$、$\left(\frac{\pi}{2}, 1\right)$ 所围的逆

时针三角形边界 (1)直接计算 (2)使用 Green's Theorem

【解】

(1)

令 $C_1 = \left\{(x, y) : 0 \le x \le \frac{\pi}{2}, y = 0\right\}$，$C_2 = \left\{(x, y) : 0 \le y \le 1, x = \frac{\pi}{2}\right\}$

$$C_3 = \left\{(x,y): y = \frac{2x}{\pi}, 0 \le x \le \frac{\pi}{2}\right\} \text{ 则 } C = C_1 \cup C_2 \cup C_3$$

$$\because \int_{C_1} (y - \sin x)\, dx + (\cos x + \sqrt{2})dy = \int_0^{\frac{\pi}{2}} -\sin x\, dx = -1$$

$$\int_{C_2} (y - \sin x)\, dx + (\cos x + \sqrt{2})dy = \int_0^1 \cos\frac{\pi}{2} + \sqrt{2}\,dy = \sqrt{2}$$

$$\int_{C_3} (y - \sin x)\, dx + (\cos x + \sqrt{2})dy = \int_{\frac{\pi}{2}}^0 \left(\frac{2x}{\pi} - \sin x\right) + \frac{2}{\pi}(\cos x + \sqrt{2})dx$$

$$= \left(\frac{x^2}{\pi} + \cos x + \frac{2}{\pi}\sin x + \frac{2\sqrt{2}}{\pi}x\right)\Bigg|_{\frac{\pi}{2}}^0 = 1 - \frac{\pi}{4} - \frac{2}{\pi} - \sqrt{2}$$

$$\therefore \oint_C (y - \sin x)\, dx + \cos x\, dy = -\frac{\pi}{4} - \frac{2}{\pi}$$

(2)

$$\text{令 } R = \left\{(x,y): 0 \le x \le \frac{\pi}{2}, 0 \le y \le \frac{2x}{\pi}\right\}$$

藉由 Green's Theorem 则 $\displaystyle\oint_C (y - \sin x)\, dx + (\cos x + \sqrt{2})dy = \iint_R -\sin x - 1\, dxdy$

藉由 Fubini's Theorem

则 $\displaystyle\iint_R -\sin x - 1\, dxdy = \int_0^{\frac{\pi}{2}}\int_0^{\frac{2x}{\pi}} -\sin x - 1\, dydx = \int_0^{\frac{\pi}{2}}(-y\sin x - y)\Big|_{y=0}^{y=\frac{2x}{\pi}} dx$

$$= -\frac{2}{\pi}\left(\int_0^{\frac{\pi}{2}} x\sin x + x\,dx\right)$$

藉由 integration by parts 则 $\displaystyle\int_0^{\frac{\pi}{2}} x\sin x + x\,dx = \left(-x\cos x + \sin x + \frac{x^2}{2}\right)\Bigg|_0^{\frac{\pi}{2}} = 1 + \frac{\pi^2}{8}$

$$\therefore \oint_C (y - \sin x)\, dx + (\cos x + \sqrt{2})dy = -\frac{2}{\pi}\left(1 + \frac{\pi^2}{8}\right) = -\frac{2}{\pi} - \frac{\pi}{4}$$

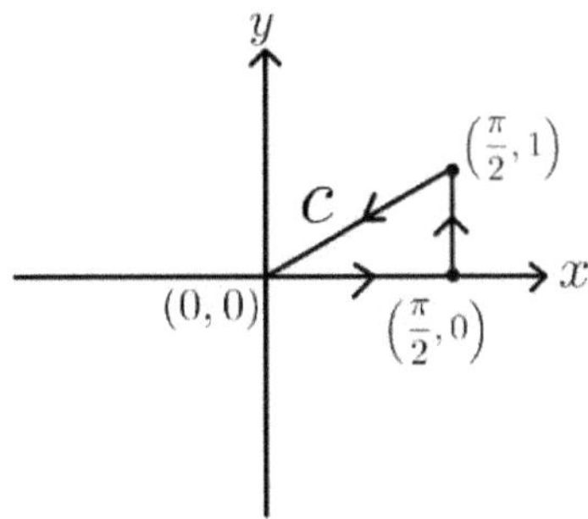

## Example 11.

藉由 Green's Theorem 求 $\oint_C y\cos x\,dx + x\sin y\,dy = ?$，$C$ 为 $(0,0),(a,0),(a,a)$

为顶点的三角形逆时针封闭路径 $(a > 0)$

【解】

令 $R = \{(x,y): 0 \le x \le a, 0 \le y \le x\}$

藉由 Green's Theorem 则 $\oint_C y\cos x\,dx + x\sin y\,dy = \iint_R \sin y - \cos x\,dxdy$

藉由 Fubini's Theorem 则

$$\iint_R \sin y - \cos x\,dxdy = \int_0^a \int_0^x \sin y - \cos x\,dydx = \int_0^a (-\cos y)\big|_0^x - x\cos x\,dx$$

$$= \int_0^a 1 - \cos x - x\cos x\,dx = a - \sin a - a\sin a - \cos a + 1$$

$$= -(a+1)\sin a - \cos a + a + 1$$

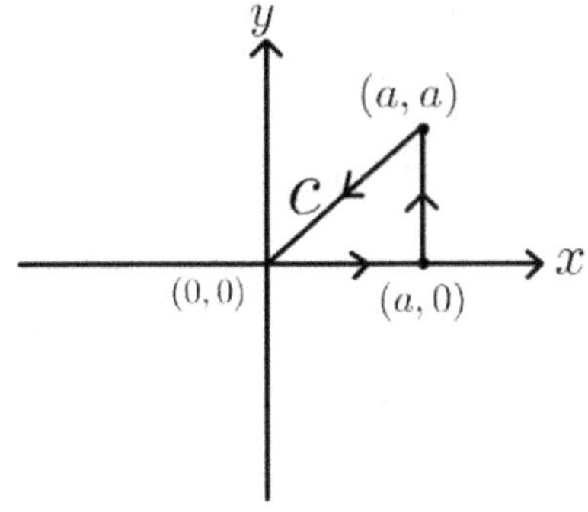

## Example 12.

藉由 Green's Theorem 求 $\oint_C -y^4dx + xy^2dy =?$，$C$ 为 $(0,0),(1,0),(1,1)$ 为顶点的三角形逆时针封闭路径

【解】

令 $R = \{(x,y): 0 \le x \le 1, 0 \le y \le x\}$

藉由 Green's Theorem 则 $\oint_C -y^4dx + xy^2dy = \iint_R y^2 + 4y^3 dxdy$

藉由 Fubini's Theorem 则

$$\iint_R y^2 + 4y^3 dxdy = \int_0^1 \int_0^x y^2 + 4y^3\, dydx = \int_0^1 \frac{x^3}{3} + x^4 dx = \frac{x^4}{12} + \frac{x^5}{5} = \frac{17}{60}$$

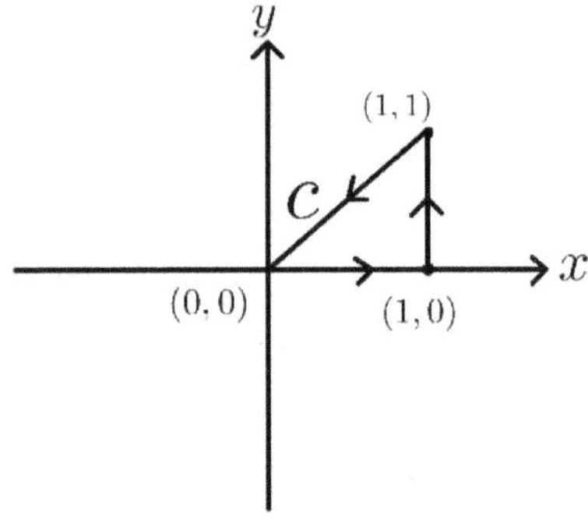

Example 13.

藉由 Green's Theorem 求 $\oint_C (2xy - x^2)\, dx + (x + y^2)dy =?$，$C$ 为從 $(0,0)$ 沿着 $y = x^2$ 至 $(1,1)$ 再沿着 $x = y^2$ 绕回 $(0,0)$ 的封闭曲线

【解】

令 $R = \{(x,y): 0 \le x \le 1, x^2 \le y \le \sqrt{x}\}$

藉由 Green's Theorem 则 $\oint_C (2xy - x^2)\, dx + (x + y^2)dy = \iint_R 1 - 2x dxdy$

藉由 Fubini's Theorem

则 $\displaystyle \iint_R 1 - 2x dxdy = \int_0^1 \int_{x^2}^{\sqrt{x}} 1 - 2x\, dydx = \int_0^1 x^{\frac{1}{2}} - 2x^{\frac{3}{2}} - x^2 + 2x^3 dx = \frac{1}{30}$

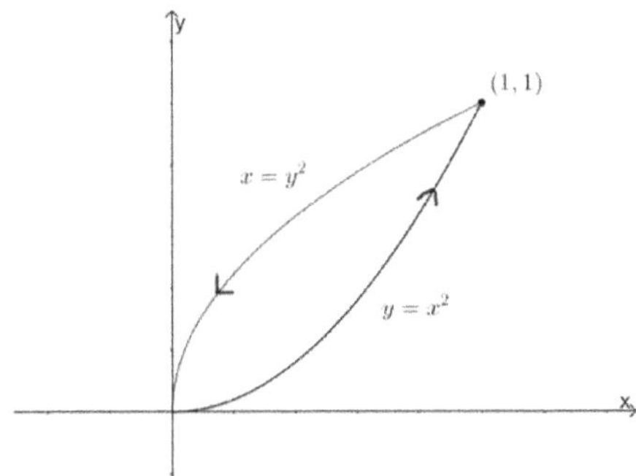


Example 14.

　　$C$ 为從 $(0,0)$ 沿着 $y^2 = 2x$ 至 $(2,2)$，　再沿着 $y^3 = 4x$ 绕回 $(0,0)$ 的逆时针封闭路径, 求

$$\oint_C y\,dx + x^2 y\,dy = ? \quad (1)直接计算 \quad (2)\ 使用\ Green's\ Theorem$$

【解】

(1)

令 $C_1 = \{(x,y): 0 \le x \le 2, y^2 = 2x\}, C_2 = \{(x,y): 0 \le x \le 2, y^3 = 4x\}$ 则 $C = C_1 \cup C_2$

$$\because \int_{C_1} y\,dx + x^2 y\,dy = \int_0^2 \sqrt{2x}\,dx + x^2\,dx = \frac{16}{3}$$

$$\int_{C_2} y\,dx + x^2 y\,dy = \int_2^0 (4x)^{\frac{1}{3}}\,dx + \frac{4}{3}(4x)^{\frac{-1}{3}}\,dx = -3 - 2 = -5$$

$$\therefore \oint_C y\,dx + x^2 y\,dy = \frac{16}{3} - 5 = \frac{1}{3}$$

(2)

令 $R = \{(x,y): \dfrac{y^3}{4} \le x \le \dfrac{y^2}{2}, 0 \le y \le 2\}$

藉由 Green's Theorem 则 $\displaystyle \oint_C y\,dx + x^2 y\,dy = \iint_R 2xy - 1\,dxdy$

藉由 Fubini's Theorem 则

$$\iint_R 2xy - 1\,dxdy = \int_0^2 \int_{\frac{y^3}{4}}^{\frac{y^2}{2}} 2xy - 1\,dxdy = \int_0^2 (x^2 y - x)\Big|_{\frac{y^3}{4}}^{\frac{y^2}{2}}\,dy$$

$$= \int_0^2 \frac{y^5}{4} - \frac{y^2}{2} - \left(\frac{y^7}{16} - \frac{y^3}{4}\right)\,dy = \frac{1}{3}$$

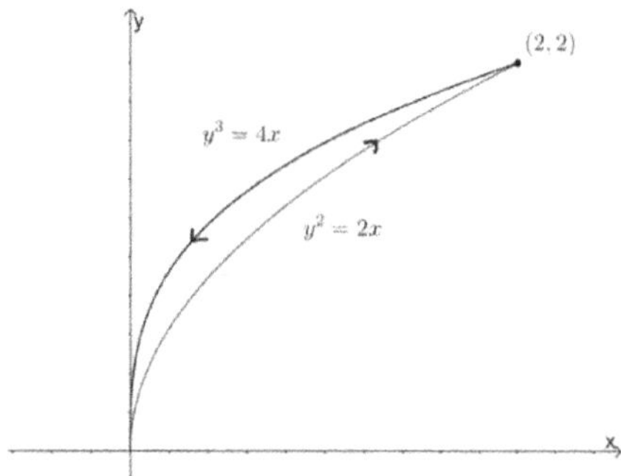


Example 15.

藉由 Green's Theorem 求 $\displaystyle\oint_C \frac{xy^2}{2}dx + (x^2y)dy =?$，其中 $C$ 为

$R = \{(x,y): 0 \leq x \leq 2, 2x-4 \leq y \leq 0\}$ 的逆时针封闭边界

【解】

藉由 Green's Theorem， $\displaystyle\oint_C \frac{xy^2}{2}dx + (x^2y)dy = \iint_R xy\,dA$

$\because f(x,y) = xy$ 在 $R$ 区域为连续函数，藉由 Fubinis Theorem

$$\iint_R xy\,dA = \int_0^2 \int_{2x-4}^0 xy\,dy\,dx = \int_0^2 \frac{xy^2}{2}\bigg|_{y=2x-4}^{y=0} dx = -\int_0^2 \frac{x(2x-4)^2}{2}dx$$

$$= -2\int_0^2 x(x-2)^2 dx = -2\int_0^2 x^3 - 4x^2 + 4x\,dx = -2\left(\frac{x^4}{4} - \frac{4x^3}{3} + 2x^2\right)\bigg|_{x=0}^{x=2}$$

$$= -2\left(4 - \frac{32}{3} + 8\right) = \frac{-8}{3}$$

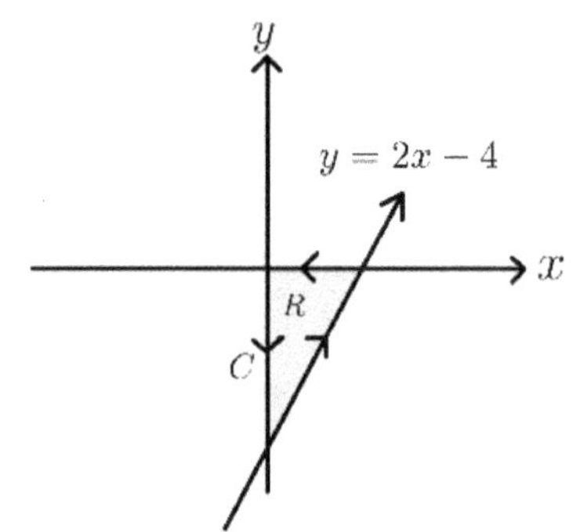

Example 16.

藉由 Green's Theorem 求 $\oint_C \dfrac{y^3}{3}dx + x^2dy =?$，其中 $C$ 为

$R = \{(x,y): x - y + 1 \leq 0, y \leq 2, x + y - 1 \geq 0\}$ 的逆时针封闭边界

【解】

藉由 Green's Theorem, $\oint_C \dfrac{y^3}{3}dx + x^2dy = \iint_R 2x - y^2 dA$

$\because f(x,y) = 2x - y^2$ 在 $R$ 区域为连续函数, 藉由 Fubinis Theorem

则 $\iint_R 2x - y^2 dA = \int_1^2 \int_{1-y}^{y-1} 2x - y^2 dxdy = \int_1^2 x^2 - y^2 x\big|_{x=1-y}^{x=y-1} dy$

$= \int_1^2 (y-1)^2 - y^2(y-1) - ((1-y)^2 - y^2(1-y))\, dy = \int_1^2 -2y^3 + 2y^2\, dy$

$= \dfrac{-y^4}{2}\bigg|_{y=1}^{y=2} + \dfrac{2y^3}{3}\bigg|_{y=1}^{y=2} = \dfrac{-17}{6}$

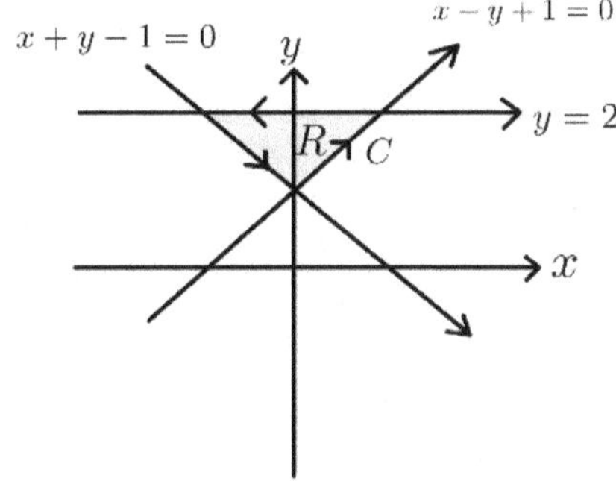


Example 17.

藉由 Green's Theorem 求 $\oint_C y\, dx + 2xdy =?$，其中 $C$ 为

$R = \{(x,y): -3x - y + 6 \leq 0, 4x - x^2 - y \geq 0, y \geq 0\}$ 的逆时针封闭边界

【解】

藉由 Green's Theorem,

$\oint_C y\, dx + 2xdy = \iint_R 1dA = \int_1^2 4x - x^2 - (6 - 3x)\, dx + \int_2^4 4x - x^2\, dx$

$$= \left(\frac{7x^2}{2} - \frac{x^3}{3} - 6x\right)\bigg|_1^2 + \left(2x^2 - \frac{x^3}{3}\right)\bigg|_2^4 = \frac{21}{2} - \frac{7}{3} - 6 + 2\cdot 12 - \frac{56}{3} = \frac{15}{2}$$

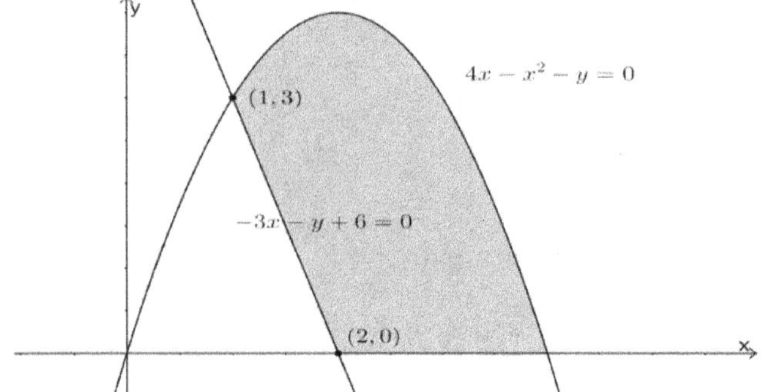


Example 18.

藉由 Green's Theorem 求 $\oint_C \dfrac{y^3}{3}dx + \dfrac{2x^{\frac{3}{2}}}{3}dy = ?$, $C$ 为 $R = \left\{(x,y): y \geq x^2, y \leq x^{\frac{1}{4}}\right\}$

的逆时针封闭边界

【解】

藉由 Green's Theorem, $\quad \oint_C \dfrac{y^3}{3}dx + \dfrac{2x^{\frac{3}{2}}}{3}dy = \iint_R \sqrt{x} - y^2\, dxdy$

令 $x^2 = x^{\frac{1}{4}}$ 则 $x = 0$ 或 $1 \Rightarrow 0 \leq x \leq 1 \Rightarrow R = \{(x,y): 0 \leq x \leq 1, y \geq x^2, y \leq x^{\frac{1}{4}}\}$

$\because f(x,y) = \sqrt{x} - y^2$ 在 $R$ 区域为连续函数, 藉由 Fubinis Theorem

则 $\displaystyle\iint_R \sqrt{x} - y^2\, dxdy = \int_0^1 \int_{x^2}^{x^{\frac{1}{4}}} \sqrt{x} - y^2\, dydx = \int_0^1 \sqrt{x}y - \frac{y^3}{3}\bigg|_{y=x^2}^{y=x^{\frac{1}{4}}}dx$

$\displaystyle = \int_0^1 \sqrt{x}(x^{\frac{1}{4}}) - \frac{x^{\frac{3}{4}}}{3} - \left(\sqrt{x}(x^2) - \frac{x^6}{3}\right)dx = \int_0^1 x^{\frac{3}{4}} - \frac{x^{\frac{3}{4}}}{3} - x^{\frac{5}{2}} + \frac{x^6}{3}dx$

$\displaystyle = \left(\frac{4x^{\frac{7}{4}}}{7} - \frac{4x^{\frac{7}{4}}}{21} - \frac{2x^{\frac{7}{2}}}{7} + \frac{x^7}{21}\right)\bigg|_{x=0}^{x=1} = \frac{2}{7} - \frac{3}{21} = \frac{1}{7}$

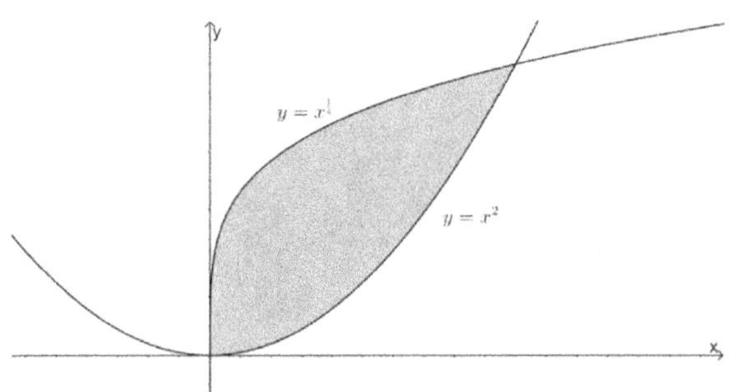


## 9.4.4 $\quad \Delta \neq 0$,将封闭线积分转成面积分再搭配坐标转换

如果封闭线积分的被积分函数较为复杂且使用 Green's Theorem 将封闭线积分转成面积分
之后无法直接求值，得再搭配极坐标转换或广义坐标转换

考试类型：

Type 1.

求 $\oint_C f(x,y)\,dx + g(x,y)dy =?$，其中 $C$ 为逆时针封闭圆：$x^2 + y^2 = a^2$ 且 $f,g$ 的一阶偏导
数存在且连续

解题流程：

Step1.

令 $R = \{(x,y): x^2 + y^2 \leq a^2\}$，藉由 Green's Theoremm

$$\oint_C f(x,y)\,dx + g(x,y)dy = \iint_R \frac{\partial g}{\partial x}(x,y) - \frac{\partial f}{\partial y}(x,y)dxdy = \iint_R h(x,y)dxdy$$

其中 $\dfrac{\partial g}{\partial x}(x,y) - \dfrac{\partial f}{\partial y}(x,y) = h(x,y)$

Step2.

令 $x = r\cos\theta$，$y = r\sin\theta$ 则 $\{(x,y): x^2 + y^2 \leq a^2\} = \{(r,\theta): 0 \leq r \leq a, 0 \leq \theta \leq 2\pi\}$

且 $dxdy = \left\| \begin{vmatrix} \dfrac{\partial x}{\partial r} & \dfrac{\partial x}{\partial \theta} \\ \dfrac{\partial y}{\partial r} & \dfrac{\partial y}{\partial \theta} \end{vmatrix} \right\| drd\theta = \left| \begin{matrix} \cos\theta & -r\sin\theta \\ \sin\theta & r\cos\theta \end{matrix} \right| drd\theta = rdrd\theta$

Step3.

$$\iint_R h(x,y)dxdy = \int_0^{2\pi} \int_0^a h(r\cos\theta, r\sin\theta)\, rdrd\theta$$

Step4.

$$求 \int_0^{2\pi} \int_0^a h(r\cos\theta, r\sin\theta)\, rdrd\theta = ?$$

<u>范例说明:</u>

(I)假设 $f(x,y) = -\dfrac{y^3}{3}$ 且 $g(x,y) = \dfrac{x^3}{3}$, $C$为逆时针封闭圆: $x^2 + y^2 = a^2$

$$求 \oint_C f(x,y)\, dx + g(x,y)dy = ?$$

令$R = \{(x,y): x^2 + y^2 \leq a^2\}$, 藉由 Green's Theorem

$$\oint_C f(x,y)\, dx + g(x,y)dy = \iint_R \frac{\partial g}{\partial x}(x,y) - \frac{\partial f}{\partial y}(x,y)dxdy = \iint_R x^2 + y^2 dxdy$$

令$x = r\cos\theta, y = r\sin\theta$ 则 $\{(x,y): x^2 + y^2 \leq a^2\} = \{(r,\theta): 0 \leq r \leq a, 0 \leq \theta \leq 2\pi\}$

$$且\ dxdy = \left\| \begin{matrix} \dfrac{\partial x}{\partial r} & \dfrac{\partial x}{\partial \theta} \\ \dfrac{\partial y}{\partial r} & \dfrac{\partial y}{\partial \theta} \end{matrix} \right\| drd\theta = \left\| \begin{matrix} \cos\theta & -r\sin\theta \\ \sin\theta & r\cos\theta \end{matrix} \right\| drd\theta = rdrd\theta$$

$$\iint_R x^2 + y^2 dxdy = \int_0^{2\pi} \int_0^a r^3\, drd\theta = \frac{a^4\pi}{2}$$

<u>补充说明:</u>
有时候积分区域是个椭圆, 方法也是类似

Example 1.

藉由Green's Theorem 求 $\oint_C y^3\, dx - x^3 dy = ?$, $C$为逆时针封闭圆: $x^2 + y^2 = 25$

【解】

令 $R = \{(x,y): x^2 + y^2 \leq 25\}$

藉由Green's Theorem 则 $\oint_C y^3\, dx - x^3 dy = -\iint_R 3x^2 + 3y^2 dxdy$

令$x = r\cos\theta, y = r\sin\theta$ 则 $\{(x,y): x^2 + y^2 \leq 25\} = \{(r,\theta): 0 \leq r \leq 5, 0 \leq \theta \leq 2\pi\}$

$$\text{且 } dxdy = \left\| \begin{vmatrix} \dfrac{\partial x}{\partial r} & \dfrac{\partial x}{\partial \theta} \\[2mm] \dfrac{\partial y}{\partial r} & \dfrac{\partial y}{\partial \theta} \end{vmatrix} \right\| drd\theta = \left\| \begin{vmatrix} \cos\theta & -r\sin\theta \\ \sin\theta & r\cos\theta \end{vmatrix} \right\| drd\theta = rdrd\theta$$

$$\therefore -3\iint_R x^2 + y^2 dxdy = -3\int_0^{2\pi}\int_0^5 r^3\, drd\theta = -6\pi \cdot \left.\frac{r^4}{4}\right|_0^5 = -\frac{3\pi}{2}\cdot 625 = -\frac{1875\pi}{2}$$

Example 2.

藉由 Green's Theorem 求 $\displaystyle\oint_C (6y + x)\, dx + (y + 2x)dy$,　$C$ 为逆时针封闭圆：

$(x-2)^2 + (y-3)^2 = 9$

【解】

令 $R = \{(x,y): (x-2)^2 + (y-3)^2 \le 9\}$

藉由 Green's Theorem 则 $\displaystyle\oint_C (6y + x)\, dx + (y + 2x)dy = \iint_R 2 - 6dxdy$

令 $x = 2 + r\cos\theta, y = 3 + r\sin\theta$

则 $\{(x,y): (x-2)^2 + (y-3)^2 \le 9\} = \{(r,\theta): 0 \le r \le 3, 0 \le \theta \le 2\pi\}$

$$\text{且 } dxdy = \left\| \begin{vmatrix} \dfrac{\partial x}{\partial r} & \dfrac{\partial x}{\partial \theta} \\[2mm] \dfrac{\partial y}{\partial r} & \dfrac{\partial y}{\partial \theta} \end{vmatrix} \right\| drd\theta = \left\| \begin{vmatrix} \cos\theta & -r\sin\theta \\ \sin\theta & r\cos\theta \end{vmatrix} \right\| drd\theta = rdrd\theta$$

$$\therefore -4\iint_R dxdy = -4\int_0^{2\pi}\int_0^3 r\, drd\theta = -8\pi \cdot \left.\frac{r^2}{2}\right|_0^3 = -36\pi$$

Example 3.

藉由 Green's Theorem 求 $\displaystyle\oint_C 2y\, dx + (x^2 + y^2)dy = ?$,　$C$ 为逆时针封闭圆：

$x^2 + (y-3)^2 = 16$

【解】

令 $R = \{(x,y): x^2 + (y-3)^2 \le 16\}$

藉由 Green's Theorem 则 $\oint_C 2y\,dx + (x^2 + y^2)dy = \iint_R 2x - 2\,dxdy$

令 $x = r\cos\theta, y = 3 + r\sin\theta$

则 $\{(x, y): x^2 + (y-3)^2 \leq 16\} = \{(r, \theta): 0 \leq r \leq 4, 0 \leq \theta \leq 2\pi\}$

且 $dxdy = \left\|\begin{matrix} \dfrac{\partial x}{\partial r} & \dfrac{\partial x}{\partial \theta} \\ \dfrac{\partial y}{\partial r} & \dfrac{\partial y}{\partial \theta} \end{matrix}\right\| drd\theta = \left\|\begin{matrix} \cos\theta & -r\sin\theta \\ \sin\theta & r\cos\theta \end{matrix}\right\| drd\theta = r\,drd\theta$

$\therefore \iint_R 2x - 2\,dxdy = 2\int_0^{2\pi}\int_0^4 r(r\cos\theta - 1)\,drd\theta = 2\int_0^{2\pi}\left(\dfrac{r^3}{3}\cos\theta - \dfrac{r^2}{2}\right)\Big|_0^4 d\theta = -32\pi$

**Example 4.**

藉由 Green's Theorem 求 $\oint_C 2ydx + (x^2 + y^2)dy = ?$，$C$ 为逆时针封闭圆：

$x^2 + (y-3)^2 = a^2, (a > 3)$

【解】

令 $R = \{(x, y): x^2 + (y-3)^2 \leq a^2\}$

藉由 Green's Theorem 则 $\oint_C 2ydx + (x^2 + y^2)dy = \iint_R 2x - 2\,dxdy$

令 $x = r\cos\theta, y = r\sin\theta, 0 \leq \theta \leq 2\pi$

则 $\iint_R 2x - 2\,dxdy = \int_0^{2\pi}\int_0^a 2\,r^2\cos\theta - 2r\,drd\theta = \int_0^{2\pi}\int_0^a -2r\,drd\theta = -2a^2\pi$

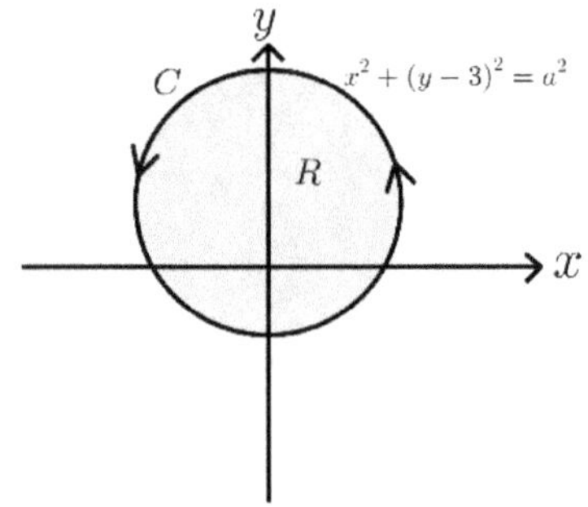


**Example 5.**

$C$ 为 $R = \{(x,y): y \geq 0,\ x^2 + y^2 \leq 4\}$ 的逆时针封闭边界, 藉由 Green's Theorem

求 $\displaystyle\oint_C x^2 y^2\, dx + x^3 y dy =?$

【解】

藉由 Green's Theorem, $\displaystyle\oint_C x^2 y^2\, dx + x^3 y dy = \iint_R x^2 y\, dxdy$

令 $x = r\cos\theta,\ y = r\sin\theta$ 则 $\{(x,y): y \geq 0, x^2 + y^2 \leq 4\} = \{(r,\theta): 0 \leq r \leq 2, 0 \leq \theta \leq \pi\}$

且 $dxdy = \left\|\begin{array}{cc} \dfrac{\partial x}{\partial r} & \dfrac{\partial x}{\partial \theta} \\[2mm] \dfrac{\partial y}{\partial r} & \dfrac{\partial y}{\partial \theta} \end{array}\right\| drd\theta = \left\|\begin{array}{cc} \cos\theta & -r\sin\theta \\ \sin\theta & r\cos\theta \end{array}\right\| drd\theta = rdrd\theta$

$\therefore \displaystyle\iint_R x^2 y\, dxdy = \int_0^\pi \int_0^2 (r\cos\theta)^2 (r\sin\theta) r drd\theta = \int_0^\pi \sin\theta \cos^2\theta\, d\theta \int_0^2 r^4 dr$

$\displaystyle = \frac{(-1)\cos^3\theta}{3}\bigg|_0^\pi \cdot \frac{r^5}{5}\bigg|_0^2 = \frac{2}{3} \cdot \frac{32}{5} = \frac{64}{15}$

Example 6.

藉由 Green's Theorem 求 $\displaystyle\oint_C -\frac{y^3}{3} dx + \frac{x^3}{3} dy =?$, $C$ 为 $R = \left\{(x,y): \dfrac{x^2}{a^2} + \dfrac{y^2}{b^2} \leq 4\right\}$

的逆时针封闭边界

【解】

藉由 Green's Theorem, $\displaystyle\oint_C -\frac{y^3}{3} dx + \frac{x^3}{3} dy = \iint_R x^2 + y^2 dA$

令 $x = ar\cos\theta,\ y = br\sin\theta$ 则 $\left\{(x,y): \dfrac{x^2}{a^2} + \dfrac{y^2}{b^2} \leq 4\right\} = \{(r,\theta): 0 \leq r \leq 2, 0 \leq \theta \leq 2\pi\}$

且 $dxdy = \left\|\begin{array}{cc} \dfrac{\partial x}{\partial r} & \dfrac{\partial x}{\partial \theta} \\[2mm] \dfrac{\partial y}{\partial r} & \dfrac{\partial y}{\partial \theta} \end{array}\right\| drd\theta = \left\|\begin{array}{cc} a\cos\theta & -ra\sin\theta \\ b\sin\theta & rb\cos\theta \end{array}\right\| drd\theta = abrdrd\theta$

$\therefore \displaystyle\iint_R x^2 + y^2 dA = \int_0^{2\pi} \int_0^2 ((ar)^2 \cos^2\theta + (br)^2 \sin^2\theta) abr\, drd\theta$

$\displaystyle = a^3 b \int_0^{2\pi} \cos^2\theta\, d\theta \int_0^2 r^3 dr + ab^3 \int_0^{2\pi} \sin^2\theta\, d\theta \int_0^2 r^3 dr$

$$\because \int_0^{2\pi} \cos^2\theta\, d\theta = \int_0^{2\pi} \frac{1+\cos 2\theta}{2}\, d\theta = \pi$$

$$\because \int_0^{2\pi} \sin^2\theta\, d\theta = \int_0^{2\pi} \frac{1-\cos 2\theta}{2}\, d\theta = \pi \quad \text{且} \quad \int_0^{2} r^3\, dr = \frac{16}{4} = 4$$

$$\therefore \iint_R x^2 + y^2\, dA = a^3 b(4\pi) + ab^3(4\pi) = 4ab\pi(a^2+b^2)$$

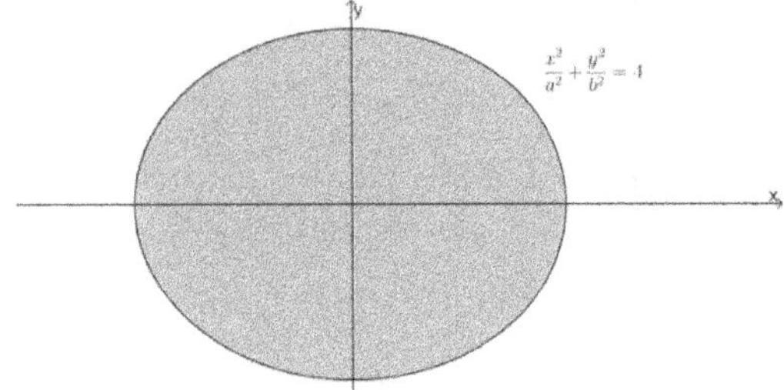


Example 7.

藉由Green's Theorem 求 $\oint_C (y^3 - 2y)\, dx - (2x^3 + x)\, dy =?$，$C$为逆时针封闭圆：

$$6x^2 + 3y^2 = 1$$

【解】

令 $R = \{(x,y): 6x^2 + 3y^2 \le 1\}$

藉由Green's Theorem 则 $\oint_C (y^3 - 2y)\, dx - (2x^3 + x)\, dy = \iint_R 1 - 6x^2 - 3y^2\, dxdy$

令 $x = \dfrac{r\cos\theta}{\sqrt{6}},\ y = \dfrac{r\sin\theta}{\sqrt{3}}$ 则$\{(x,y): 6x^2 + 3x^2 \le 1\} = \{(r,\theta): 0 \le r \le 1, 0 \le \theta \le 2\pi\}$

且 $dxdy = \left\| \begin{Vmatrix} \dfrac{\partial x}{\partial r} & \dfrac{\partial x}{\partial \theta} \\ \dfrac{\partial y}{\partial r} & \dfrac{\partial y}{\partial \theta} \end{Vmatrix} \right\| drd\theta = \left\| \begin{Vmatrix} \dfrac{\cos\theta}{\sqrt{6}} & \dfrac{-r\sin\theta}{\sqrt{6}} \\ \dfrac{\sin\theta}{\sqrt{3}} & \dfrac{r\cos\theta}{\sqrt{3}} \end{Vmatrix} \right\| drd\theta = \dfrac{1}{3\sqrt{2}} r\, drd\theta$

$$\therefore \iint_R 1 - 6x^2 - 3y^2\, dxdy = \int_0^{2\pi}\int_0^1 (1-r^2)\frac{1}{3\sqrt{2}} r\, drd\theta = \frac{\pi}{6\sqrt{2}}$$

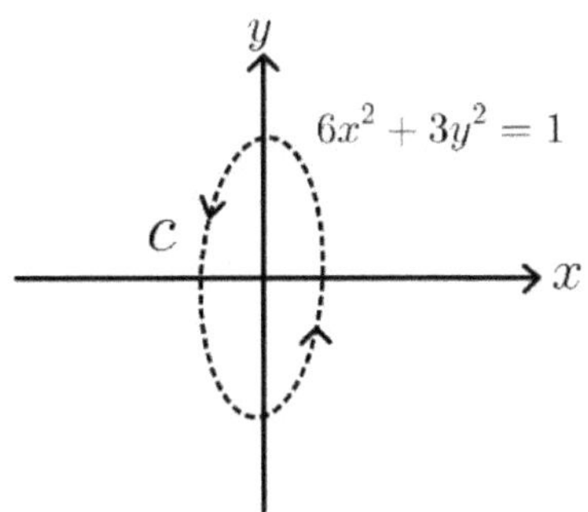

Example 8.

藉由 Green's Theorem 求 $\displaystyle\oint_C (y^3 + e^{-x^2} + e^x)\,dx + (e^{-y^2} - x^3 + y^2 + 6xy)\,dy =?$,

C 为逆时针封闭圆：$x^2 + (y-1)^2 = 1$

【解】

令 $R = \{(x,y): x^2 + (y-1)^2 \leq 1\}$，藉由 Green's Theorem

则 $\displaystyle\oint_C (y^3 + e^{-x^2} + e^x)\,dx + (e^{-y^2} - x^3 + y^2 + 6xy)\,dy = \iint_R -3x^2 + 6y - 3y^2\,dxdy$

$$= 3\iint_R 1 - x^2 - (y-1)^2\,dxdy$$

令 $x = r\cos\theta,\ y = r\sin\theta + 1$

则 $\{(x,y): x^2 + (y-1)^2 \leq 1\} = \{(r,\theta): 0 \leq r \leq 1, 0 \leq \theta \leq 2\pi\}$

且 $dxdy = \left\| \begin{vmatrix} \dfrac{\partial x}{\partial r} & \dfrac{\partial x}{\partial \theta} \\ \dfrac{\partial y}{\partial r} & \dfrac{\partial y}{\partial \theta} \end{vmatrix} \right\| drd\theta = \left\| \begin{vmatrix} \cos\theta & -r\sin\theta \\ \sin\theta & r\cos\theta \end{vmatrix} \right\| drd\theta = rdrd\theta$

$\therefore 3\iint_R 1 - x^2 - (y-1)^2\,dxdy = 3\int_0^{2\pi}\int_0^1 (1-r^2)r\,drd\theta = \dfrac{3\pi}{2}$

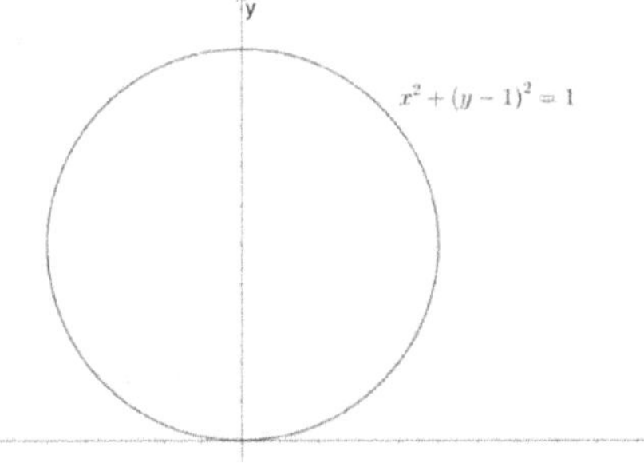

Example 9.

藉由 Green's Theorem 求 $\oint_C x^2y^2\,dx + x^3y\,dy = ?$，其中 $C$ 为

$\qquad R = \{(x,y): y \geq 0,\ (x-1)^2 + y^2 \leq 1\}$ 的逆时针封闭边界

【解】

藉由 Green's Theorem, $\oint_C x^2y^2\,dx + x^3y\,dy = \iint_R x^2y\,dA$

令 $x = r\cos\theta + 1,\ y = r\sin\theta$

则 $\{(x,y): (x-1)^2 + y^2 \leq 1, y \geq 0\} = \{(r,\theta): 0 \leq r \leq 1, 0 \leq \theta \leq \pi\}$

且 $dxdy = \left\|\begin{vmatrix} \dfrac{\partial x}{\partial r} & \dfrac{\partial x}{\partial \theta} \\[2mm] \dfrac{\partial y}{\partial r} & \dfrac{\partial y}{\partial \theta} \end{vmatrix}\right\| drd\theta = \left\|\begin{vmatrix} \cos\theta & -r\sin\theta \\ \sin\theta & r\cos\theta \end{vmatrix}\right\| drd\theta = rdrd\theta$

$\therefore \iint_R x^2y\,dA = \int_0^\pi \int_0^1 (r\cos\theta + 1)^2 r\sin\theta\,rdrd\theta$

$= \int_0^\pi \int_0^1 (r^2\cos^2\theta + 2r\cos\theta + 1)r\sin\theta\,rdrd\theta$

$= \int_0^\pi \cos^2\theta\sin\theta\,d\theta \int_0^1 r^4dr + \int_0^\pi 2\cos\theta\sin\theta\,d\theta \int_0^1 r^3dr + \int_0^\pi \sin\theta\,d\theta \int_0^1 r^2dr$

$= (-1)\dfrac{\cos^3\theta}{3}\Big|_0^\pi \cdot \dfrac{r^5}{5}\Big|_0^1 - \dfrac{\cos 2\theta}{2}\Big|_0^\pi \cdot \dfrac{r^4}{4}\Big|_0^1 - \cos\theta|_0^\pi \cdot \dfrac{r^3}{3}\Big|_0^1 = \dfrac{4}{5}$

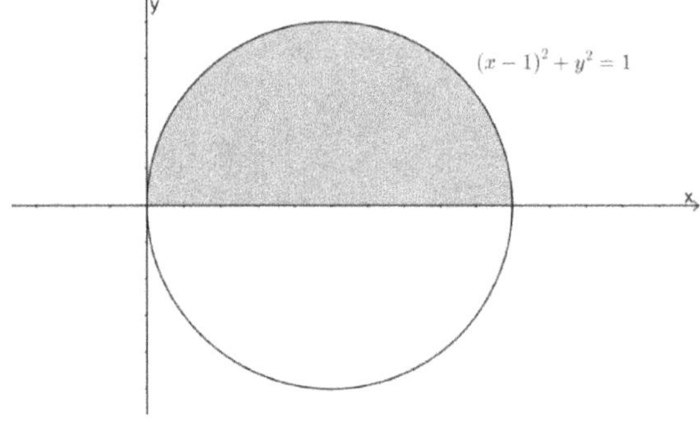


Example 10.

假设 $C = \{(x,y): x = 2\cos t, y = 2\sin t\},\ 0 \leq t \leq 2\pi$, 藉由 Green's Theorem

求 $\oint_C e^x y - y^3 dx + (e^x + x^3)dy = ?$

【解】

令 $R = \{(x, y): x^2 + y^2 \leq 4\}$

藉由 Green's Theorem 则 $\oint_C e^x y - y^3 dx + (e^x + x^3)dy = \iint_R 3x^2 + 3y^2 dxdy$

令 $x = r\cos\theta$, $y = r\sin\theta$ 则 $\{(x, y): x^2 + y^2 \leq 4\} = \{(r, \theta): 0 \leq r \leq 2, 0 \leq \theta \leq 2\pi\}$

且 $dxdy = \left\|\begin{matrix} \dfrac{\partial x}{\partial r} & \dfrac{\partial x}{\partial \theta} \\ \dfrac{\partial y}{\partial r} & \dfrac{\partial y}{\partial \theta} \end{matrix}\right\| drd\theta = \left\|\begin{matrix} \cos\theta & -r\sin\theta \\ \sin\theta & r\cos\theta \end{matrix}\right\| drd\theta = rdrd\theta$

$\therefore 3\iint_R x^2 + y^2 dxdy = 3\int_0^{2\pi}\int_0^2 r^3 \, drd\theta = 24\pi$

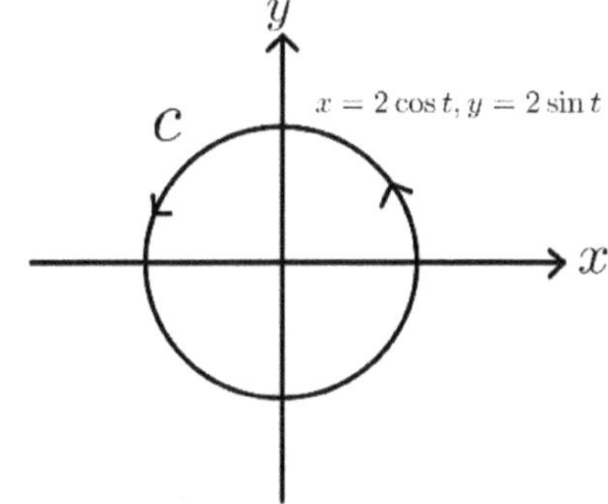


Example 11.

藉由 Green's Theorem 求 $\oint_C -\dfrac{2y^{\frac{3}{2}}}{3} dx + \dfrac{2x^{\frac{3}{2}}}{3} dy = ?$, 其中 $R$ 为 $\sqrt{x} + \sqrt{y} = 1$,

$x = 0$, $y = 0$ 所围封闭区域, $C$ 为 $R$ 的逆时针边界

【解】

藉由 Green's Theorem, $\oint_C -\dfrac{2y^{\frac{3}{2}}}{3} dx + \dfrac{2x^{\frac{3}{2}}}{3} dy = \iint_R \sqrt{x} + \sqrt{y} \, dxdy$

令 $x = r\cos^4\theta$, $y = r\sin^4\theta$

則 $dxdy = \begin{Vmatrix} \dfrac{\partial x}{\partial r} & \dfrac{\partial x}{\partial \theta} \\ \dfrac{\partial y}{\partial r} & \dfrac{\partial y}{\partial \theta} \end{Vmatrix} drd\theta = \begin{Vmatrix} \cos^4 \theta & -4r\cos^3 \theta \sin\theta \\ \sin^4 \theta & 4r\sin^3 \theta \cos\theta \end{Vmatrix} drd\theta = 4r\sin^3 \theta \cos^3 \theta \, drd\theta$

$\because R = \{(x,y): x \geq 0,\ y \geq 0, \sqrt{x} + \sqrt{y} \leq 1\} = \{(r,\theta): 0 \leq r \leq 1, 0 \leq \theta \leq \dfrac{\pi}{2}\}$

$\therefore \iint_R \sqrt{x} + \sqrt{y}\, dxdy = \int_0^{\frac{\pi}{2}} \int_0^1 \sqrt{r}(4r\sin^3 \theta \cos^3 \theta)\, drd\theta = \int_0^1 4r^{\frac{3}{2}}\, dr \int_0^{\frac{\pi}{2}} \sin^3 \theta \cos^3 \theta\, d\theta$

$= \int_0^1 4r^{\frac{3}{2}}\, dr \int_0^{\frac{\pi}{2}} \left(\dfrac{\sin 2\theta}{2}\right)^3 d\theta = \dfrac{2}{15}$

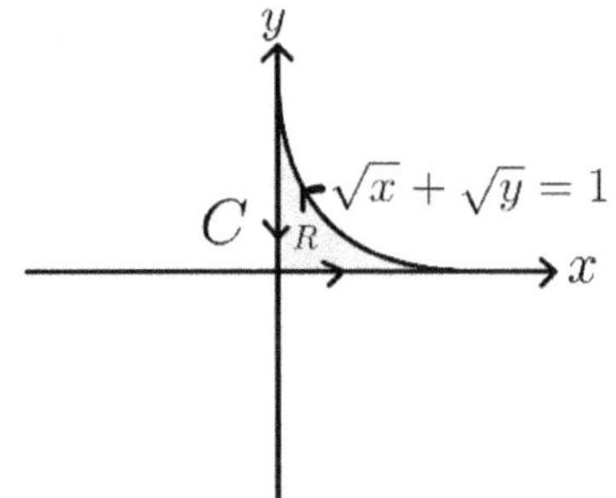


Example 12.

藉由 Green's Theorem 求 $\oint_C (6y + x)\, dx + (y + 2x)dy,$  $C$ 为逆时针封闭圆:
$(x - x_0)^2 + (y - y_0)^2 = a^2$

【解】

令 $R = \{(x,y): (x - x_0)^2 + (y - y_0)^2 \leq a^2\}$

藉由 Green's Theorem 則 $\oint_C (6y + x)\, dx + (y + 2x)dy = \iint_R 2 - 6\, dxdy$

令 $x = x_0 + r\cos\theta,\ y = y_0 + r\sin\theta$

則 $\{(x,y): (x - x_0)^2 + (y - y_0)^2 \leq a^2\} = \{(r,\theta): 0 \leq r \leq a, 0 \leq \theta \leq 2\pi\}$

且 $dxdy = \begin{Vmatrix} \dfrac{\partial x}{\partial r} & \dfrac{\partial x}{\partial \theta} \\ \dfrac{\partial y}{\partial r} & \dfrac{\partial y}{\partial \theta} \end{Vmatrix} drd\theta = \begin{Vmatrix} \cos\theta & -r\sin\theta \\ \sin\theta & r\cos\theta \end{Vmatrix} drd\theta = r\, drd\theta$

$$\therefore -4 \iint_R dxdy = -4 \int_0^{2\pi} \int_0^a r\, drd\theta = -4a^2\pi$$

Example 13.

$C$ 为逆时针封闭椭圆：$\dfrac{x^2}{a^2} + \dfrac{y^2}{b^2} = 1$ 且 $f, g$ 满足 $\dfrac{\partial g}{\partial x}(x,y) - \dfrac{\partial f}{\partial y}(x,y) = 1 - \left(\dfrac{x^2}{a^2} + \dfrac{y^2}{b^2}\right)$,

藉由 Green's Theorem 求 $\displaystyle\oint_C f(x,y)\, dx + g(x,y)dy = ?$

【解】

令 $R = \left\{(x,y): \dfrac{x^2}{a^2} + \dfrac{y^2}{b^2} \leq 1\right\}$，藉由 Green's Theorem 则

$$\oint_C f(x,y)\, dx + g(x,y)dy = \iint_R \frac{\partial g}{\partial x}(x,y) - \frac{\partial f}{\partial y}(x,y)dxdy = \iint_R 1 - \left(\frac{x^2}{a^2} + \frac{y^2}{b^2}\right)dxdy$$

令 $x = ra\cos\theta$, $y = rb\sin\theta$ 则 $\{(x,y): \dfrac{x^2}{a^2} + \dfrac{y^2}{b^2} \leq 1\} = \{(r,\theta): 0 \leq r \leq 1, 0 \leq \theta \leq 2\pi\}$

且 $dxdy = \left\|\begin{vmatrix} \dfrac{\partial x}{\partial r} & \dfrac{\partial x}{\partial \theta} \\ \dfrac{\partial y}{\partial r} & \dfrac{\partial y}{\partial \theta} \end{vmatrix}\right\| drd\theta = \left\|\begin{vmatrix} a\cos\theta & -ra\sin\theta \\ b\sin\theta & rb\cos\theta \end{vmatrix}\right\| drd\theta = abr\, drd\theta$

$$\therefore \iint_R 1 - \left(\frac{x^2}{a^2} + \frac{y^2}{b^2}\right)dxdy = \int_0^{2\pi} \int_0^1 (1 - r^2)abr\, drd\theta = \frac{ab\pi}{2}$$

Example 14.

藉由 Green's Theorem 求 $\displaystyle\oint_C \frac{e^{x+y}}{3} dx + \frac{4e^{x+y}}{3} dy = ?$, $C$ 为 $R = \{(x,y): |x| + |y| \leq 2\}$

的逆时针封闭边界

【解】

藉由 Green's Theorem, $\displaystyle\oint_C \frac{e^{x+y}}{3} dx + \frac{4e^{x+y}}{3} dy = \iint_R e^{x+y} dA$

令 $x + y = u$, $x - y = v$ 则 $\{(x,y): |x| + |y| \leq 2\} = \{(u,v): -2 \leq u \leq 2, -2 \leq v \leq 2\}$

且 $x = \dfrac{u+v}{2}$, $y = \dfrac{u-v}{2}$, $dxdy = \left\|\begin{vmatrix} \dfrac{\partial x}{\partial u} & \dfrac{\partial x}{\partial v} \\ \dfrac{\partial y}{\partial u} & \dfrac{\partial y}{\partial v} \end{vmatrix}\right\| dudv = \left\|\begin{vmatrix} \dfrac{1}{2} & \dfrac{1}{2} \\ \dfrac{1}{2} & \dfrac{-1}{2} \end{vmatrix}\right\| dudv = |\dfrac{-1}{2}| dudv$

$$\therefore \iint_R e^{x+y}\, dA = \int_{-2}^{2}\int_{-2}^{2} \frac{e^u}{2}\, du\, dv = \frac{1}{2}\int_{-2}^{2} dv \int_{-2}^{2} e^u\, du = 2(e^2 - e^{-2})$$

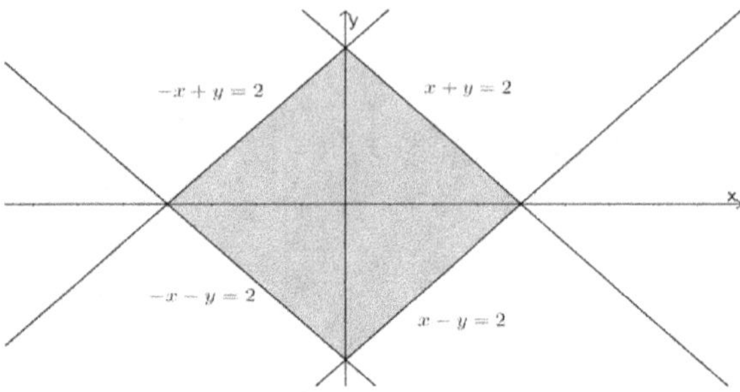


**Example 15.**

藉由 Green's Theorem 求 $\displaystyle\oint_C \frac{y}{2}\, dx + \frac{3x}{2}\, dy =?$, $C$ 为 $R = \{(x,y): ax^2 + bxy + cy^2 \leq \alpha^2\}$

的逆时针封闭边界

【解】

藉由 Green's Theorem $\displaystyle\oint_C \frac{y}{2}\, dx + \frac{3x}{2}\, dy = \iint_R\, dA$

$$\because ax^2 + bxy + cy^2 = a\left(x + \frac{by}{2a}\right)^2 + \frac{4ac - b^2}{4a}y^2$$

$$\diamondsuit \sqrt{a}\left(x + \frac{by}{2a}\right) = u,\quad \sqrt{\frac{4ac - b^2}{4a}}\, y = v$$

$$\text{則}\, dxdy = \left\|\begin{matrix} \dfrac{\partial x}{\partial u} & \dfrac{\partial x}{\partial v} \\ \dfrac{\partial y}{\partial u} & \dfrac{\partial y}{\partial v} \end{matrix}\right\| dudv = \left\|\begin{matrix} \dfrac{1}{\sqrt{a}} & 0 \\ \dfrac{2\sqrt{a}}{b} & \sqrt{\dfrac{4a}{4ac - b^2}} \end{matrix}\right\| dudv = \frac{2}{\sqrt{4ac - b^2}}\, dudv$$

$$\therefore \iint_R\, dA = \iint_{u^2+v^2 \leq \alpha^2} \frac{2}{\sqrt{4ac - b^2}}\, dudv$$

$$\diamondsuit u = r\cos\theta,\ v = r\sin\theta\ \text{則}\{(u,v): u^2 + v^2 \leq \alpha^2\} = \{(r,\theta): 0 \leq r \leq \alpha, 0 \leq \theta \leq 2\pi\}$$

$$\text{且}\, dudv = \left\|\begin{matrix} \dfrac{\partial u}{\partial r} & \dfrac{\partial u}{\partial \theta} \\ \dfrac{\partial v}{\partial r} & \dfrac{\partial v}{\partial \theta} \end{matrix}\right\| drd\theta = \left\|\begin{matrix} \cos\theta & -r\sin\theta \\ \sin\theta & r\cos\theta \end{matrix}\right\| drd\theta = r\, drd\theta$$

$$\therefore \iint_{u^2+v^2 \leq \alpha^2} \frac{2}{\sqrt{4ac-b^2}}\, dudv = \int_0^{2\pi} \int_0^\alpha \frac{2}{\sqrt{4ac-b^2}}\, rdrd\theta = \frac{2}{\sqrt{4ac-b^2}} \int_0^{2\pi} d\theta \int_0^\alpha rdr$$

$$= \frac{2\pi\alpha^2}{\sqrt{4ac-b^2}}$$

Example 16.

藉由 Green's Theorem 求 $\oint_C \left(\frac{xy^2}{4} - \frac{y^3}{6}\right) dx + \left(\frac{x^3}{6}\right) dy = ?$,其中

$C$ 为 $R = \{(x,y): x^2 - xy + y^2 \leq 2\}$ 的逆时针封闭边界

【解】

藉由 Green's Theorem 則 $\oint_C \left(\frac{xy^2}{4} - \frac{y^3}{6}\right) dx + \left(\frac{x^3}{6}\right) dy = \iint_R \frac{x^2 - xy + y^2}{2}\, dA$

$\because \dfrac{x^2 - xy + y^2}{2} = \dfrac{1}{2}\left(x - \dfrac{y}{2}\right)^2 + \dfrac{3}{8}y^2$, 令 $\sqrt{\dfrac{1}{2}}\left(x - \dfrac{y}{2}\right) = u$, $\sqrt{\dfrac{3}{8}}\, y = v$

$$則\ dxdy = \left\| \begin{matrix} \frac{\partial x}{\partial u} & \frac{\partial x}{\partial v} \\ \frac{\partial y}{\partial u} & \frac{\partial y}{\partial v} \end{matrix} \right\| dudv = \left\| \begin{matrix} \sqrt{2} & 0 \\ -2\sqrt{2} & \sqrt{\frac{8}{3}} \end{matrix} \right\| dudv = \frac{4}{\sqrt{3}} dudv$$

$$\therefore \iint_R \frac{x^2 - xy + y^2}{2}\, dA = \iint_{u^2+v^2 \leq 1} (u^2 + v^2)\frac{4}{\sqrt{3}} dudv$$

令 $u = r\cos\theta$, $v = r\sin\theta$ 則 $\{(u,v): u^2 + v^2 \leq 1\} = \{(r,\theta): 0 \leq r \leq 1, 0 \leq \theta \leq 2\pi\}$

$$且\ dudv = \left\| \begin{matrix} \frac{\partial u}{\partial r} & \frac{\partial u}{\partial \theta} \\ \frac{\partial v}{\partial r} & \frac{\partial v}{\partial \theta} \end{matrix} \right\| drd\theta = \left\| \begin{matrix} \cos\theta & -r\sin\theta \\ \sin\theta & r\cos\theta \end{matrix} \right\| drd\theta = rdrd\theta$$

$$\therefore \iint_{u^2+v^2 \leq 1} (u^2 + v^2)\frac{4}{\sqrt{3}} dudv = \frac{4}{\sqrt{3}} \int_0^{2\pi} \int_0^1 r^2 \cdot rdrd\theta = \frac{4}{\sqrt{3}} \int_0^{2\pi} d\theta \int_0^1 r^3 dr$$

$$= \frac{4}{\sqrt{3}} \cdot 2\pi \cdot \frac{r^4}{4}\Big|_{r=0}^{r=1} = \frac{2\pi}{\sqrt{3}}$$

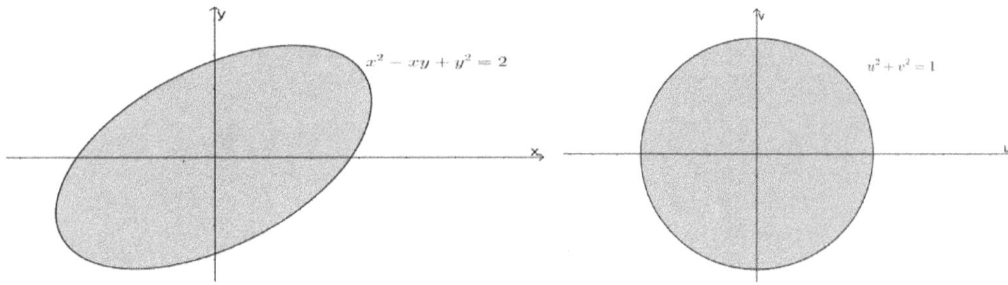


Example 17.

藉由 Green's Theorem 求 $\oint_C \dfrac{y}{2}dx + \dfrac{3x}{2}dy = ?$, $R$ 为 $y = x^2, y = 3x^2, x = y^2, x = 4y^2$

所围封闭区域, $C$ 为 R 的逆时针边界

【解】

藉由 Green's Theorem, $\oint_C \dfrac{y}{2}dx + \dfrac{3x}{2}dy = \iint_R dA$

令 $u = \dfrac{y}{x^2}$, $v = \dfrac{x}{y^2}$ 則 $x = u^{-\frac{2}{3}}v^{-\frac{1}{3}}$, $y = u^{-\frac{1}{3}}v^{-\frac{2}{3}}$

且 $dxdy = \left\| \begin{matrix} \dfrac{\partial x}{\partial u} & \dfrac{\partial x}{\partial v} \\ \dfrac{\partial y}{\partial u} & \dfrac{\partial y}{\partial v} \end{matrix} \right\| dudv = \left\| \begin{matrix} \dfrac{-2u^{-\frac{5}{3}}v^{-\frac{1}{3}}}{3} & \dfrac{-u^{-\frac{2}{3}}v^{-\frac{4}{3}}}{3} \\ \dfrac{-u^{-\frac{4}{3}}v^{-\frac{2}{3}}}{3} & \dfrac{-2u^{-\frac{1}{3}}v^{-\frac{5}{3}}}{3} \end{matrix} \right\| dudv = \dfrac{u^{-2}v^{-2}}{3} dudv$

令 $R = \left\{ (x, y) : 1 \leq \dfrac{y}{x^2} \leq 3, 1 \leq \dfrac{x}{y^2} \leq 4 \right\}$ 則 $R = \{(u, v) : 1 \leq u \leq 3, 1 \leq v \leq 4\}$

$\therefore$ 面积 $= \iint_R 1dA = \int_1^3 \int_1^4 \dfrac{u^{-2}v^{-2}}{3} dvdu = \dfrac{1}{6}$

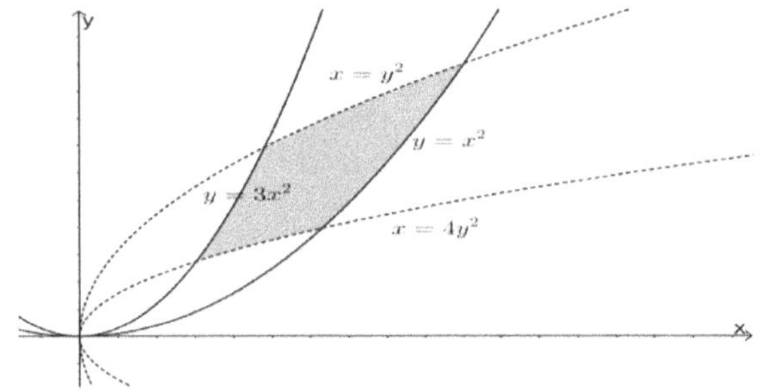


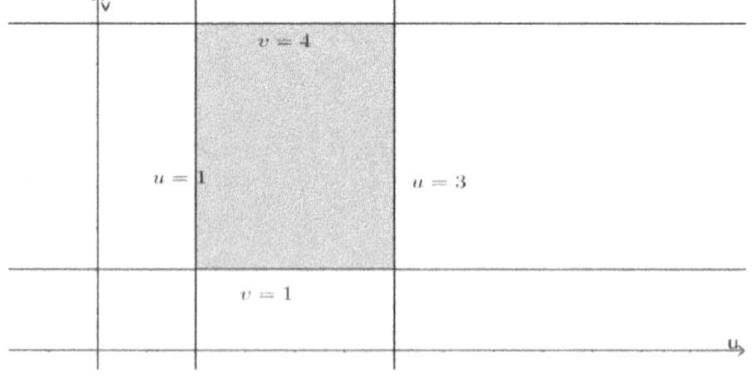

Example 18.

藉由 Green's Theorem 求 $\oint_C -\dfrac{y^3}{3}dx + \dfrac{x^3}{3}dy =?$，其中 $R$ 为 $x^2 - y^2 = 1$,

$x^2 - y^2 = 9,\ xy = 2,\ xy = 4$ 所围封闭区域,$C$ 为 R 的逆时针边界

【解】

藉由 Green's Theorem, $\oint_C -\dfrac{y^3}{3}dx + \dfrac{x^3}{3}dy = \iint_R x^2 + y^2\, dxdy$

令 $x^2 - y^2 = u,\ 2xy = v$

则 $dudv = \left\|\begin{matrix} \dfrac{\partial u}{\partial x} & \dfrac{\partial u}{\partial y} \\ \dfrac{\partial v}{\partial x} & \dfrac{\partial v}{\partial y} \end{matrix}\right\| dxdy = \left\|\begin{matrix} 2x & -2y \\ 2y & 2x \end{matrix}\right\| dxdy = 4(x^2 + y^2)dxdy$

$\because (x^2 + y^2)^2 = (x^2 - y^2)^2 + (2xy)^2 = u^2 + v^2 \quad \therefore x^2 + y^2 = (u^2 + v^2)^{\frac{1}{2}}$

$\therefore dxdy = \dfrac{dudv}{4(u^2 + v^2)^{\frac{1}{2}}}$

$\because R = \{(x,y): 1 \le x^2 - y^2 \le 9, 4 \le 2xy \le 8\} \quad \therefore R = \{(u,v): 1 \le u \le 9, 4 \le v \le 8\}$

$\therefore \iint_R x^2 + y^2\, dxdy = \int_4^8 \int_1^9 \dfrac{(u^2 + v^2)^{\frac{1}{2}}}{4(u^2 + v^2)^{\frac{1}{2}}}\, dudv = 8$

Example 19.

(1) 藉由 Green's Theorem 求 $\oint_C -\dfrac{y^3}{3}dx + \dfrac{x^3}{3}dy =?$，其中 $R$ 为 $-x = y, y = -x + 3,$

$x - 3 = y,\ y = x$ 包围区域,$C$ 为 R 的逆时针边界

(2) 藉由 Green's Theorem 求 $\oint_C \dfrac{xy^2}{2}dx + x^2ydy =?$，其中 $R$ 为 $x^2 + y^2 = a,$

$x^2 + y^2 = b,\ x^2 - y^2 = c,\ x^2 - y^2 = d, (b > a > 0, d > c > 0)$ 包围于第一象限区域,

$C$ 为 $R$ 的逆时针边界

【解】

(1)

藉由 Green's Theorem, $\oint_C -\dfrac{y^3}{3}dx + \dfrac{x^3}{3}dy = \iint_R x^2 + y^2\, dxdy$

令 $x + y = u,\ x - y = v$ 則 $x = \dfrac{u+v}{2},\ \ y = \dfrac{u-v}{2}$

且 $dxdy = \left\| \begin{matrix} \dfrac{\partial x}{\partial u} & \dfrac{\partial x}{\partial v} \\ \dfrac{\partial y}{\partial u} & \dfrac{\partial y}{\partial v} \end{matrix} \right\| dudv = \left\| \begin{matrix} \dfrac{1}{2} & \dfrac{1}{2} \\ \dfrac{1}{2} & \dfrac{-1}{2} \end{matrix} \right\| dudv = \dfrac{1}{2} dudv$

$\because R = \{(x,y) : 0 \le x+y \le 3, 0 \le x-y \le 3\}$　$\therefore R = \{(u,v) : 0 \le u \le 3, 0 \le v \le 3\}$

$\therefore \displaystyle\iint_R x^2 + y^2 \, dA = \int_0^3 \int_0^3 \left( \left(\frac{u+v}{2}\right)^2 + \left(\frac{u-v}{2}\right)^2 \right) \frac{1}{2} dudv = \frac{1}{4} \int_0^3 \int_0^3 u^2 + v^2 \, dudv = \frac{27}{2}$

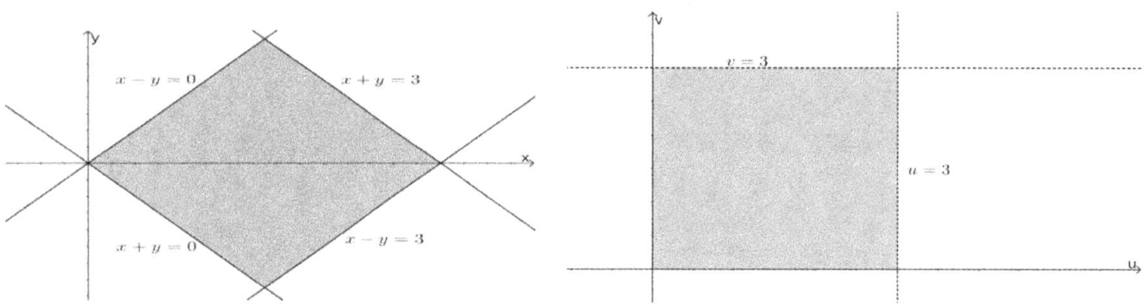

(2)

藉由 Green's Theorem, $\displaystyle\oint_C \frac{xy^2}{2} dx + x^2 y \, dy = \iint_R xy \, dxdy$

令 $x^2 + y^2 = u,\ x^2 - y^2 = v$ 則 $x = \dfrac{(u+v)^{\frac{1}{2}}}{\sqrt{2}},\ \ y = \dfrac{(u-v)^{\frac{1}{2}}}{\sqrt{2}}$

且 $dxdy = \left\| \begin{matrix} \dfrac{\partial x}{\partial u} & \dfrac{\partial x}{\partial v} \\ \dfrac{\partial y}{\partial u} & \dfrac{\partial y}{\partial v} \end{matrix} \right\| dudv = \left\| \begin{matrix} \dfrac{(u+v)^{-\frac{1}{2}}}{2\sqrt{2}} & \dfrac{(u+v)^{-\frac{1}{2}}}{2\sqrt{2}} \\ \dfrac{(u-v)^{-\frac{1}{2}}}{2\sqrt{2}} & \dfrac{(u-v)^{-\frac{1}{2}}}{2\sqrt{2}} \end{matrix} \right\| dudv = \dfrac{(u^2 - v^2)^{-\frac{1}{2}}}{4} dudv$

$\because R = \{(x,y) : a \le x^2 + y^2 \le b, c \le x^2 - y^2 \le d\}$　$\therefore R = \{(u,v) : a \le u \le b, c \le v \le d\}$

$\therefore \displaystyle\iint_R xy \, dA = \int_c^d \int_a^b \frac{(u^2 - v^2)^{\frac{1}{2}}}{2} \cdot \frac{(u^2 - v^2)^{-\frac{1}{2}}}{4} dudv = \frac{(b-a)(c-d)}{8}$

Example 20.

藉由Green's Theorem 求 $\oint_C -\dfrac{y^2}{2}dx + x^2dy =?$，$R$为$x^2 - 2xy + y^2 + x + y = 0$

与 $x + y + 4 = 0$ 包围区域，$C$为 R 的逆时针边界

【解】

藉由Green's Theorem，$\oint_C -\dfrac{y^2}{2}dx + x^2dy = \iint_R 2x + y\,dxdy$

$\because x^2 - 2xy + y^2 + x + y = 0 \Leftrightarrow (x - y)^2 = -(x + y)$

令 $x + y = u$，$x - y = v$ 则 $x = \dfrac{u+v}{2}$，$y = \dfrac{u-v}{2}$

且 $dxdy = \left\| \begin{vmatrix} \dfrac{\partial x}{\partial u} & \dfrac{\partial x}{\partial v} \\ \dfrac{\partial y}{\partial u} & \dfrac{\partial y}{\partial v} \end{vmatrix} \right\| dudv = \left\| \begin{vmatrix} \dfrac{1}{2} & \dfrac{1}{2} \\ \dfrac{1}{2} & \dfrac{1}{2} \end{vmatrix} \right\| dudv = \dfrac{1}{2}dudv$

$\because R = \{(u,v): -v^2 \leq u \leq 0, -2 \leq v \leq 2\}$

$\therefore \iint_R 2x + y\,dxdy = \int_{-2}^{2} \int_{-v^2}^{0} 3u + v\,dudv = -\dfrac{96}{5}$

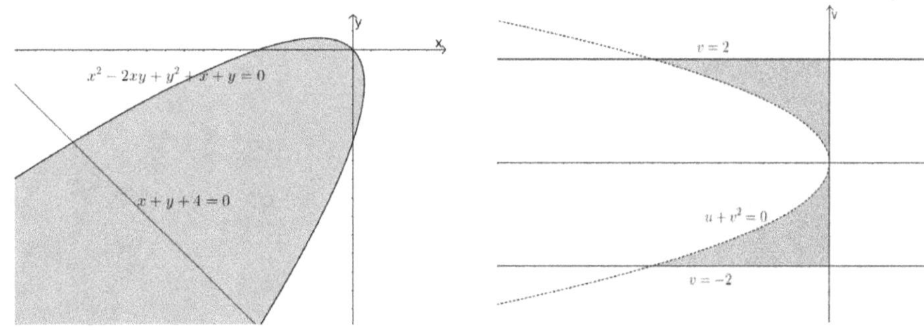


## 9.4.5　$\Delta \neq 0$, 将面积分转成封闭线积分

如果面积分的被积分函数较为复杂，则使用 Green's Theorem 将面积分转成封闭线积分
Type 1.

藉由 Green's Theorem 求 $\iint_R \dfrac{\partial g}{\partial x}(x,y) - \dfrac{\partial f}{\partial y}(x,y)\,dxdy =?$，$R = \left\{(x,y): \dfrac{x^2}{a^2} + \dfrac{y^2}{b^2} \leq r^2\right\}$，

$C$为$R$的逆时针封闭边界且$f,g$的一阶偏导数存在且连续

解题流程：

Step1.

藉由 Green's Theorem, $\displaystyle\iint_R \frac{\partial g}{\partial x}(x,y) - \frac{\partial f}{\partial y}(x,y)dxdy = \oint_C f(x,y)\,dx + g(x,y)dy$

Step2.

令$x = ra\cos\theta$, $y = rb\sin\theta$ 则

$$\oint_C f(x,y)\,dx + g(x,y)dy$$

$$= \int_0^{2\pi} f(ra\cos\theta, rb\sin\theta)(-ra\sin\theta)\,d\theta + \int_0^{2\pi} g(ra\cos\theta, rb\sin\theta)(rb\cos\theta)\,d\theta$$

Type 2.

藉由 Green's Theorem 求 $\displaystyle\iint_R \frac{\partial g}{\partial x}(x,y) - \frac{\partial f}{\partial y}(x,y)\,dxdy =?$, $C$为$R$的逆时针封闭边界且

$$C = \bigcup_{i=1}^{k} C_i, k \in N, \ C_i为平滑线段且f,g的一阶偏导数存在且连续$$

解题流程：

Step1.

藉由 Green's Theorem, $\displaystyle\iint_R \frac{\partial g}{\partial x}(x,y) - \frac{\partial f}{\partial y}(x,y)dxdy = \oint_C f(x,y)\,dx + g(x,y)dy$

Step2.

$$\oint_C f(x,y)\,dx + g(x,y)dy = \sum_{i=1}^{k} \int_{C_i} f(x,y)\,dx + g(x,y)dy$$

Example 1.

藉由 Green's Theorem 求 $\displaystyle\iint_R \frac{\partial g}{\partial x}(x,y) - \frac{\partial f}{\partial y}(x,y)\,dxdy =?$, 其中

$$f(x,y) = \frac{x^2 y^2}{2}, \ g(x,y) = \frac{2x^3 y}{3} \ 且\ R = \{(x,y): y \geq 0, \ x^2 + y^2 \leq 4\}$$

【解】

令 $C_1: \{(x,y): x^2 + y^2 = 4, y > 0\}$, $C_2: \{(x,y): -2 \leq x \leq 2, y = 0\}$ 且 $C = C_1 \cup C_2$

藉由 Green's Theorem, $\displaystyle\iint_R \frac{\partial g}{\partial x}(x,y) - \frac{\partial f}{\partial y}(x,y)\, dxdy = \oint_C \frac{x^2 y^2}{2}\, dx + \frac{2x^3 y}{3}\, dy$

令 $x = 2\cos\theta$, $y = 2\sin\theta$, $0 \leq \theta \leq \pi$

则 $\displaystyle\int_{C_1} f(x,y)dx + g(x,y)dy = \int_{C_1} \frac{x^2 y^2}{2}\, dx + \frac{2x^3 y}{3}\, dy$

$$= \int_0^\pi \frac{16 \cdot \cos^2\theta \cdot \sin^2\theta}{2}(-2\sin\theta)d\theta + \frac{2}{3}\int_0^\pi 16 \cdot \cos^3\theta \cdot \sin\theta\,(2\cos\theta)\,d\theta$$

$$= -16\int_0^\pi (1 - \sin^2\theta) \cdot \sin^3\theta\, d\theta + \frac{64}{3}\int_0^\pi \cos^4\theta \cdot \sin\theta\, d\theta$$

$$= -16\left(-\cos\theta + \frac{\cos^3\theta}{3} + \frac{\cos^5\theta}{5} - \frac{2\cos^3\theta}{3} + \cos\theta\right)\Big|_0^\pi - \frac{64}{3} \cdot \frac{\cos^5\theta}{5}\Big|_0^\pi$$

$$= -16\left(2 - \frac{2}{3} - \frac{2}{5} + \frac{4}{3} - 2\right) - \frac{64}{3} \cdot \left(\frac{-2}{5}\right) = \frac{64}{15}$$

$$\because \int_{C_2} \frac{x^2 y^2}{2}\, dx + \frac{2x^3 y}{3}\, dy = 0 \quad \therefore \oint_C \frac{x^2 y^2}{2}\, dx + \frac{2x^3 y}{3}\, dy = \frac{64}{15}$$

Example 2.

藉由 Green's Theorem 求 $\displaystyle\iint_R \frac{\partial g}{\partial x}(x,y) - \frac{\partial f}{\partial y}(x,y)\, dxdy =?$, 其中

$$f(x,y) = -\frac{y^3}{3}, \quad g(x,y) = \frac{x^3}{3}, \quad R = \left\{(x,y): \frac{x^2}{a^2} + \frac{y^2}{b^2} \leq 4\right\}$$

【解】

令 $C$ 为 $R$ 的逆时针封闭边界

藉由 Green's Theorem, $\displaystyle\iint_R \frac{\partial g}{\partial x}(x,y) - \frac{\partial f}{\partial y}(x,y)\, dxdy = \oint_C -\frac{y^3}{3}\, dx + \frac{x^3}{3}\, dy$

令 $x = 2a\cos\theta$, $y = 2b\sin\theta$ 则

$$\oint_C -\frac{y^3}{3}\, dx + \frac{x^3}{3}\, dy = \frac{-1}{3}\int_0^{2\pi} 8b^3 \cdot \sin^3\theta \cdot (-2a\sin\theta)\, d\theta + \frac{1}{3}\int_0^{2\pi} 8a^3 \cdot \cos^3\theta \cdot 2b\cos\theta\, d\theta$$

$$= \frac{16b^3 a}{3}\int_0^{2\pi} \sin^4\theta\, d\theta + \frac{16a^3 b}{3}\int_0^{2\pi} \cos^4\theta\, d\theta = \frac{16b^3 a}{3} \cdot \frac{3\pi}{4} + \frac{16a^3 b}{3} \cdot \frac{3\pi}{4} = 4ab\pi(a^2 + b^2)$$

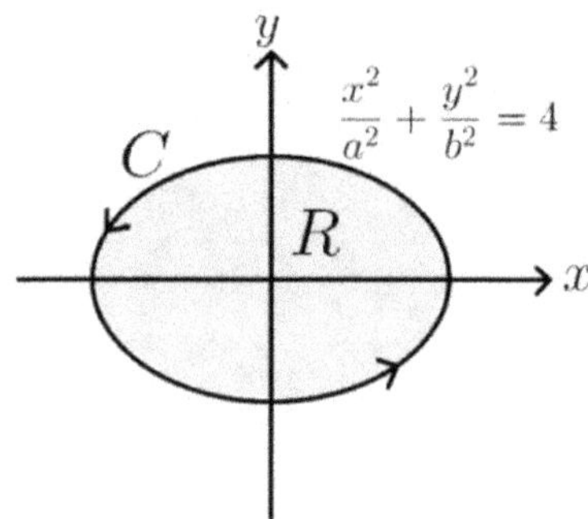


Example 3.

藉由 Green's Theorem 求 $\iint_R \frac{\partial g}{\partial x}(x,y) - \frac{\partial f}{\partial y}(x,y)\,dxdy =?$ 其中 $f(x,y) = e^{x+y}$, $g(x,y) = 2e^{x+y}$,  $R = \{(x,y): |x| + |y| \le 2\}$

【解】

令 $C_1: \{(x,y): x - y = 2\}$, $C_2: \{(x,y): x + y = 2\}$, $C_3: \{(x,y): -x + y = 2\}$, $C_4: \{(x,y): -x - y = 2\}$ 且 $C = C_1 \cup C_2 \cup C_3 \cup C_4$

藉由 Green's Theorem, $\iint_R \frac{\partial g}{\partial x}(x,y) - \frac{\partial f}{\partial y}(x,y)\,dxdy = \oint_C e^{x+y}\,dx + 2e^{x+y}dy$

$\because \int_{C_1} f(x,y)dx + g(x,y)dy = \int_{C_1} e^{x+y}dx + 2e^{x+y}dy = \int_0^2 e^{x+y}\,dx + 2e^{x+y}dx$

$= 3\int_0^2 e^{2x-2}\,dx = \frac{3}{2}(e^2 - e^{-2})$

$\int_{C_2} f(x,y)dx + g(x,y)dy = \int_2^0 e^2 - 2e^2\,dx = 2e^2$

$\int_{C_4} f(x,y)dx + g(x,y)dy = \int_{-2}^0 e^{x+y} - 2e^{x+y}\,dx = -\int_{-2}^0 e^{-2}\,dx = -2e^2$

$\int_{C_3} f(x,y)dx + g(x,y)dy = \int_{C_3} e^{x+y}dx + 2e^{x+y}dy = \int_0^{-2} e^{x+y}\,dx + 2e^{x+y}dx$

$= 3\int_0^{-2} e^{2x+2}\,dx = \frac{3}{2}(e^{-2} - e^2)$

$\therefore \oint_C f\,dx + g\,dy = \frac{3}{2}(e^2 - e^{-2}) + 2e^2 + \frac{3}{2}(e^{-2} - e^2) - 2e^2 = 2(e^2 - e^{-2})$

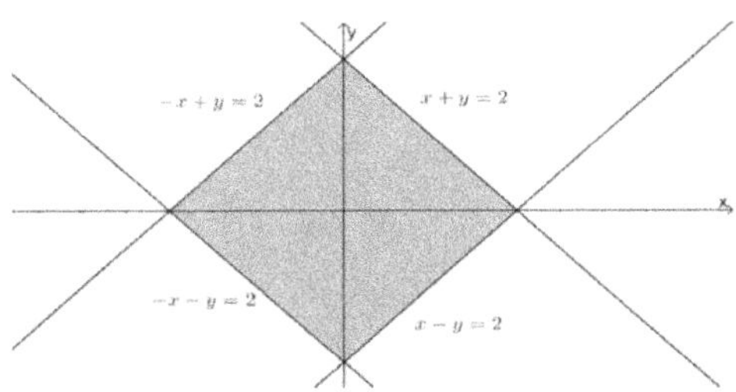

## Example 4.

藉由 Green's Theorem 求 $\displaystyle\iint_R \frac{\partial g}{\partial x}(x,y) - \frac{\partial f}{\partial y}(x,y)\,dxdy =?$ 其中

$$f(x,y) = -\frac{2y^{\frac{3}{2}}}{3}, \quad g(x,y) = \frac{2x^{\frac{3}{2}}}{3}, \quad R \text{ 为 } \sqrt{x} + \sqrt{y} = 1, \ x = 0, \ y = 0 \text{ 所围区域}$$

【解】

令 $C_1 : \{(x,y): \sqrt{x} + \sqrt{y} = 1, 0 \le x \le 1\}, C_2 : \{(x,y): x = 0, 0 \le y \le 2\}$,

$C_3 : \{(x,y): y = 0, 0 \le x \le 1\}$ 且 $C = C_1 \cup C_2 \cup C_3$

藉由 Green's Theorem

$$\iint_R \frac{\partial g}{\partial x}(x,y) - \frac{\partial f}{\partial y}(x,y)\,dxdy = \iint_R \left(\sqrt{x} + \sqrt{y}\right) dxdy = \oint_C -\frac{2y^{\frac{3}{2}}}{3}dx + \frac{2x^{\frac{3}{2}}}{3}dy$$

$$\because \int_{C_1} -\frac{2y^{\frac{3}{2}}}{3}dx + \frac{2x^{\frac{3}{2}}}{3}dy = \int_1^0 -\frac{2}{3}\left(1 - t^{\frac{1}{2}}\right)^3 dt + \frac{2}{3}\int_1^0 t^{\frac{3}{2}}\left(1 - t^{-\frac{1}{2}}\right) dt$$

令 $u = 1 - t^{\frac{1}{2}}$ 则

$$\int_1^0 -\frac{2}{3}\left(1 - t^{\frac{1}{2}}\right)^3 dt = \frac{4}{3}\int_0^1 u^3(1-u)\,du = \frac{1}{15} \quad \text{且} \quad \frac{2}{3}\int_1^0 t^{\frac{3}{2}}\left(1 - t^{-\frac{1}{2}}\right) dt = \frac{1}{15}$$

$$\because \int_{C_2} -\frac{2y^{\frac{3}{2}}}{3}dx + \frac{2x^{\frac{3}{2}}}{3}dy = 0 \quad \text{且} \quad \int_{C_3} -\frac{2y^{\frac{3}{2}}}{3}dx + \frac{2x^{\frac{3}{2}}}{3}dy = 0$$

$$\therefore \oint_C -\frac{2y^{\frac{3}{2}}}{3}dx + \frac{2x^{\frac{3}{2}}}{3}dy = \frac{2}{15}$$

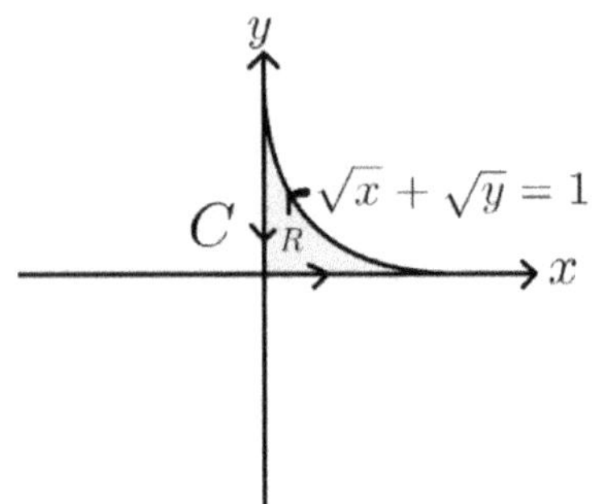


Example 5.

$$f(x,y) = \left(\frac{y^2}{2} - 4y\right), \quad g(x,y) = -\frac{x^2}{2}, \quad R = \{(x,y): 0 \leq x \leq 2, 0 \leq y \leq 2\},$$

藉由 Green's Theorem 求 $\displaystyle\iint_R \frac{\partial g}{\partial x}(x,y) - \frac{\partial f}{\partial y}(x,y)\, dxdy =?$

【解】

令 $C_1: \{(x,y): 0 \leq x \leq 2, y = 0\}, C_2: \{(x,y): x = 2, 0 \leq y \leq 2\},$

$C_3: \{(x,y): 0 \leq x \leq 2, y = 2\}, C_4: \{(x,y): x = 0, 0 \leq y \leq 2\}$ 且 $C = C_1 \cup C_2 \cup C_3 \cup C_4$

藉由 Green's Theorem

$$\iint_R \frac{\partial g}{\partial x}(x,y) - \frac{\partial f}{\partial y}(x,y)\, dxdy = \iint_R 4 - x - y\, dA = \oint_C \left(\frac{y^2}{2} - 4y\right) dx + \left(-\frac{x^2}{2}\right) dy$$

$$\because \int_{C_1} \left(\frac{y^2}{2} - 4y\right) dx + \left(-\frac{x^2}{2}\right) dy = 0, \quad \int_{C_2} \left(\frac{y^2}{2} - 4y\right) dx + \left(-\frac{x^2}{2}\right) dy = \int_0^2 -2\, dx = -4$$

$$\int_{C_3} \left(\frac{y^2}{2} - 4y\right) dx + \left(-\frac{x^2}{2}\right) dy = \int_2^0 2 - 8\, dx = 12 \text{ 且} \int_{C_4} \left(\frac{y^2}{2} - 4y\right) dx + \left(-\frac{x^2}{2}\right) dy = 0$$

$$\therefore \oint_C \left(\frac{y^2}{2} - 4y\right) dx + \left(-\frac{x^2}{2}\right) dy = 8$$

Example 6.

藉由 Green's Theorem 求 $\displaystyle\iint_R \frac{\partial g}{\partial x}(x,y) - \frac{\partial f}{\partial y}(x,y)\, dxdy =?$ 其中

$$f(x,y) = \frac{xy^2}{2}, \quad g(x,y) = x^2 y, \quad R = \{(x,y): 0 \leq x \leq 2, 2x - 4 \leq y \leq 0\},$$

【解】

令 $C_1:\{(x,y): 0 \leq x \leq 2, y = 2x - 4\}, C_2:\{(x,y): 0 \leq x \leq 2, y = 0\},$

$C_3:\{(x,y): x = 0, -4 \leq y \leq 0\}$ 且 $C = C_1 \cup C_2 \cup C_3$

藉由 Green's Theorem

$$\iint_R \frac{\partial g}{\partial x}(x,y) - \frac{\partial f}{\partial y}(x,y)\,dxdy = \iint_R xy\,dA = \oint_C \frac{xy^2}{2}dx + (x^2 y)dy$$

$$\because \int_{C_1} \frac{xy^2}{2}dx + (x^2 y)dy = \int_0^2 \frac{x(2x-4)^2}{2}dx + 2x^2(2x-4)dx = \int_0^2 (2x^2 - 4x)(3x - 2)dx$$

$$= \int_0^2 6x^3 - 16x^2 + 8x\,dx = \frac{-8}{3},$$

$$\int_{C_2} \frac{xy^2}{2}dx + (x^2 y)dy = 0 \quad \text{且} \quad \int_{C_3} \frac{xy^2}{2}dx + (x^2 y)dy = 0$$

$$\therefore \oint_C \frac{xy^2}{2}dx + (x^2 y)dy = \frac{-8}{3}$$

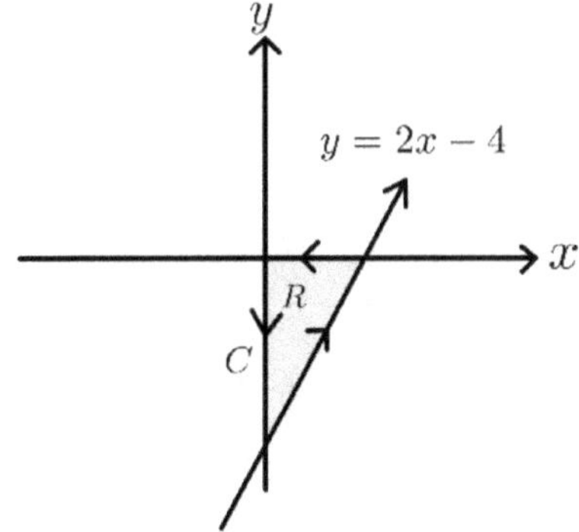


Example 7.

$$f(x,y) = \frac{y^3}{3}, \quad g(x,y) = x^2, \quad R = \{(x,y): x - y + 1 \leq 0, y \leq 2, x + y - 1 \geq 0\},$$

藉由 Green's Theorem 求 $\iint_R \frac{\partial g}{\partial x}(x,y) - \frac{\partial f}{\partial y}(x,y)\,dxdy =?$

【解】

令 $C_1:\{(x,y): 0 \leq x \leq 1, x - y + 1 = 0\}, C_2:\{(x,y): -1 \leq x \leq 1, y = 2\},$

$C_3:\{(x,y): x + y - 1 = 0, -1 \leq x \leq 0\}$ 且 $C = C_1 \cup C_2 \cup C_3$

藉由 Green's Theorem

$$\iint_R \frac{\partial g}{\partial x}(x,y) - \frac{\partial f}{\partial y}(x,y)\,dxdy = \iint_R 2x - y^2\,dA = \oint_C \frac{y^3}{3}dx + x^2\,dy$$

$$\because \int_{C_1} \frac{y^3}{3}\,dx + x^2\,dy = \int_0^1 \frac{(x+1)^3}{3}\,dx + x^2\,dx = \int_0^1 \frac{x^3 + 6x^2 + 3x + 1}{3}\,dx = \frac{19}{12},$$

$$\int_{C_2} \frac{y^3}{3}\,dx + x^2\,dy = \int_1^{-1} \frac{8}{3}\,dx = -\frac{16}{3},$$

$$\int_{C_3} \frac{y^3}{3}\,dx + x^2\,dy = \int_{-1}^0 \frac{(1-x)^3}{3}\,dx - x^2\,dx = \int_{-1}^0 \frac{1 - 3x - x^3}{3}\,dx = \frac{11}{12}$$

$$\therefore \oint_C \frac{y^3}{3}\,dx + x^2\,dy = \frac{19}{12} - \frac{16}{3} + \frac{11}{12} = \frac{-17}{6}$$

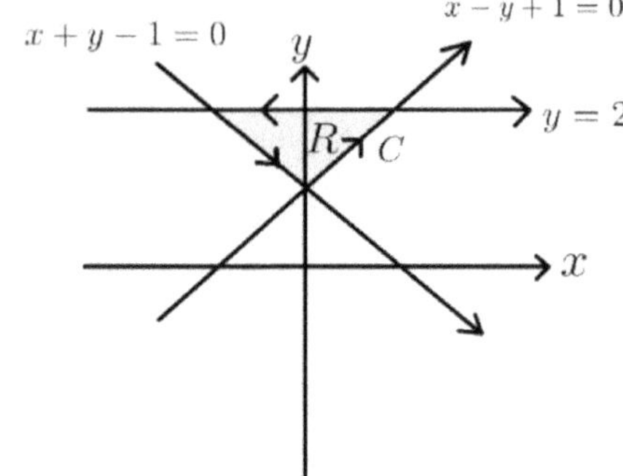


Example 8.

$$f(x,y) = y,\; g(x,y) = 2x,\quad R = \{(x,y): -3x - y + 6 \leq 0, 4x - x^2 - y \geq 0, y \geq 0\},$$

藉由 Green's Theorem 求 $\displaystyle\iint_R \frac{\partial g}{\partial x}(x,y) - \frac{\partial f}{\partial y}(x,y)\,dxdy =\,?$

【解】

令 $C_1: \{(x,y): 1 \leq x \leq 4, 4x - x^2 - y = 0\}$, $C_2: \{(x,y): 1 \leq x \leq 2, -3x - y + 6 = 0\}$,

$C_3: \{(x,y): 2 \leq x \leq 4, y = 0\}$ 且 $C = C_1 \cup C_2 \cup C_3$

藉由 Green's Theorem, $\displaystyle\iint_R \frac{\partial g}{\partial x}(x,y) - \frac{\partial f}{\partial y}(x,y)\,dxdy = \iint_R dA = \oint_C y\,dx + 2x\,dy$

$$\because \int_{C_1} y\,dx + 2x\,dy = \int_4^1 4x - x^2\,dx + 2x(4 - 2x)\,dx = \int_4^1 -5x^2 + 12x\,dx = \frac{-5x^3}{3} + 6x^2 \Big|_4^1$$

$$= 15,$$

$$\int_{C_2} y\,dx + 2x\,dy = \int_1^2 6 - 3x\,dx + 2x(-3)\,dx = \int_1^2 -9x + 6\,dx = \frac{-9x^2}{2} + 6x \Big|_1^2 = -\frac{15}{2}$$

且 $\displaystyle\int_{C_3} ydx + 2xdy = 0$

$\therefore \displaystyle\oint_C y\,dx + 2xdy = \dfrac{15}{2}$

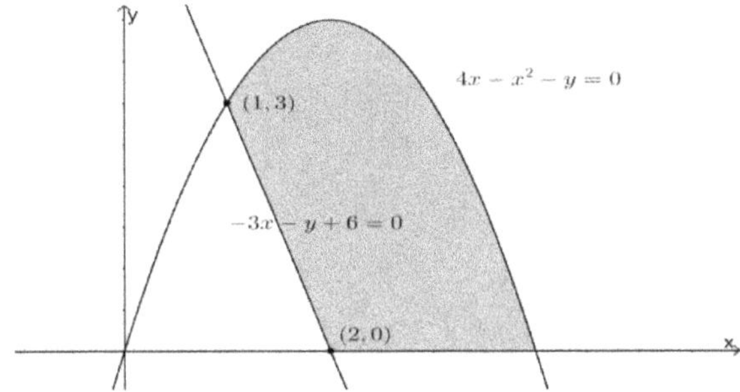


Example 9.

藉由 Green's Theorem 求 $\displaystyle\iint_R \dfrac{\partial g}{\partial x}(x,y) - \dfrac{\partial f}{\partial y}(x,y)\,dxdy =?$ 其中 $f(x,y) = \dfrac{y^3}{3}$,

$g(x,y) = \dfrac{2x^{\frac{3}{2}}}{3}$, $R = \left\{(x,y): y \geq x^2, y \leq x^{\frac{1}{4}}\right\}$,

【解】

令 $C_1: \{(x,y): 0 \leq x \leq 1, y = x^2\}$, $C_2: \left\{(x,y): 0 \leq x \leq 1, y = x^{\frac{1}{4}}\right\}$ 且 $C = C_1 \cup C_2$

藉由 Green's Theorem

$$\iint_R \dfrac{\partial g}{\partial x}(x,y) - \dfrac{\partial f}{\partial y}(x,y)\,dxdy = \iint_R \sqrt{x} - y^2\,dxdy = \oint_C \dfrac{y^3}{3}dx + \dfrac{2x^{\frac{3}{2}}}{3}dy$$

$$\because \int_{C_1} \dfrac{y^3}{3}dx + \dfrac{2x^{\frac{3}{2}}}{3}dy = \int_0^1 \dfrac{x^6}{3}dx + \dfrac{2x^{\frac{3}{2}}}{3}\cdot 2xdx = \dfrac{3}{7},$$

$$\int_{C_2} \dfrac{y^3}{3}dx + \dfrac{2x^{\frac{3}{2}}}{3}dy = \int_1^0 \dfrac{x^{\frac{3}{4}}}{3}dx + \dfrac{x^{\frac{3}{4}}}{6}dx = -\dfrac{2}{7}$$

$$\therefore \oint_C \dfrac{y^3}{3}dx + \dfrac{2x^{\frac{3}{2}}}{3}dy = \dfrac{1}{7}$$

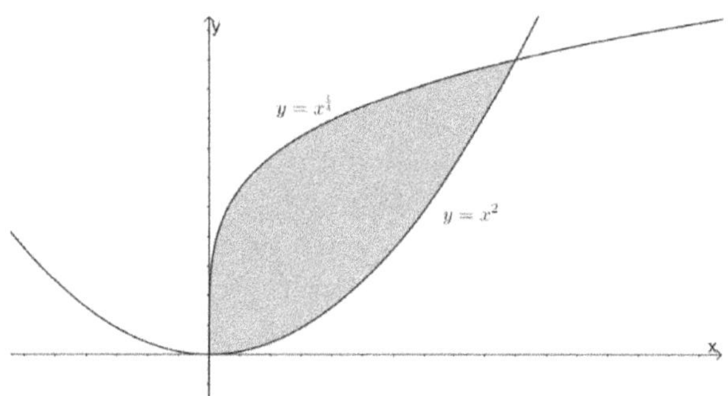


Example 10.

$$f(x,y) = 6y + x, \quad g(x,y) = y + 2x, \quad R = \{(x,y): (x-2)^2 + (y-3)^2 \le 9\},$$

藉由 Green's Theorem 求 $\displaystyle\iint_R \frac{\partial g}{\partial x}(x,y) - \frac{\partial f}{\partial y}(x,y)\,dxdy = ?$

【解】

令 $C$ 为逆时针封闭圆：$\{(x,y): (x-2)^2 + (y-3)^2 = 9\}$，藉由 Green's Theorem

$$\iint_R \frac{\partial g}{\partial x}(x,y) - \frac{\partial f}{\partial y}(x,y)\,dxdy = \iint_R 2 - 6\,dxdy = \oint_C (6y + x)\,dx + (y + 2x)dy$$

令 $x = 2 + 3\cos\theta, y = 3 + 3\sin\theta$ 则

$$\oint_C (6y + x)\,dx + (y + 2x)dy$$

$$= \int_0^{2\pi} \big(6(3 + 3\sin\theta) + (2 + 3\cos\theta)\big)(-3\sin\theta)d\theta$$

$$+ \int_0^{2\pi} \big(3 + 3\sin\theta + (4 + 6\cos\theta)\big)3\cos\theta\,d\theta$$

$$= \int_0^{2\pi} -\frac{54}{2} + 18 \cdot \frac{1}{2}\,d\theta = 2\pi \cdot (-18) = -36\pi$$

Example 11.

$$f(x,y) = y^3 - 2y, \quad g(x,y) = -(2x^3 + x), \quad R = \{(x,y): 6x^2 + 3y^2 \le 1\},$$

藉由 Green's Theorem 求 $\displaystyle\iint_R \frac{\partial g}{\partial x}(x,y) - \frac{\partial f}{\partial y}(x,y)\,dxdy = ?$

【解】

令 $C$ 为逆时针封闭圆：$\{(x,y): 6x^2 + 3y^2 = 1\}$，藉由 Green's Theorem

$$\iint_R \frac{\partial g}{\partial x}(x,y) - \frac{\partial f}{\partial y}(x,y)\,dxdy = \iint_R 1 - 6x^2 - 3y^2\,dxdy = \oint_C y^3 - 2y\,dx - (2x^3 + x)dy$$

令 $x = \dfrac{\cos\theta}{\sqrt{6}}, \quad y = \dfrac{\sin\theta}{\sqrt{3}}$ 则

$$\oint_C (y^3 - 2y)\,dx - (2x^3 + x)dy$$

$$= \int_0^{2\pi}\left(\frac{\sin^3\theta}{3\sqrt{3}} - \frac{2\sin\theta}{\sqrt{3}}\right)\left(-\frac{\sin\theta}{\sqrt{6}}\right)d\theta - \int_0^{2\pi}\left(\frac{\cos^3\theta}{3\sqrt{6}} + \frac{\cos\theta}{\sqrt{6}}\right)\left(\frac{\cos\theta}{\sqrt{3}}\right)d\theta$$

$$= -\frac{1}{9\sqrt{2}}\int_0^{2\pi}\sin^4\theta\,d\theta + \frac{\sqrt{2}}{3}\int_0^{2\pi}\sin^2\theta\,d\theta - \frac{1}{9\sqrt{2}}\int_0^{2\pi}\cos^4\theta\,d\theta - \frac{1}{3\sqrt{2}}\int_0^{2\pi}\cos^2\theta\,d\theta$$

$$= -\frac{1}{9\sqrt{2}}\cdot\frac{3}{8}\cdot 2\pi + \frac{\sqrt{2}}{3}\cdot\frac{1}{\sqrt{2}}\cdot 2\pi - \frac{1}{9\sqrt{2}}\cdot\frac{3}{8}\cdot 2\pi - \frac{1}{3\sqrt{2}}\cdot\frac{1}{2}\cdot 2\pi = \frac{\pi}{6\sqrt{2}}$$

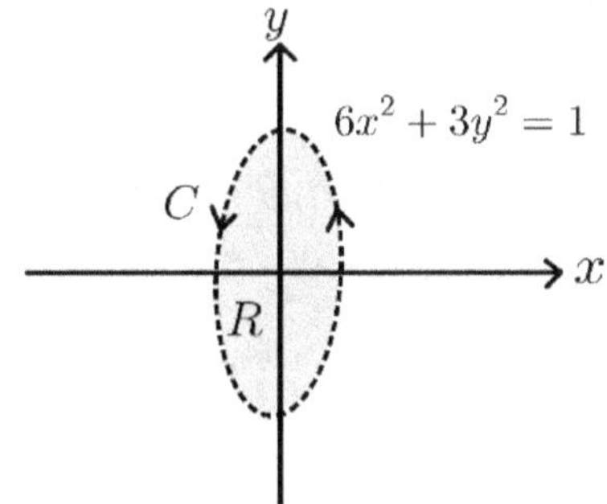


Example 12.

$$f(x,y) = y^3, \quad g(x,y) = -x^3, \quad R:\{(x,y): x^2 + y^2 \leq r^2\}, \text{ 藉由 Green's Theorem}$$

求 $\displaystyle\iint_R \frac{\partial g}{\partial x}(x,y) - \frac{\partial f}{\partial y}(x,y)\,dxdy = ?$

【解】

令 $C$ 为逆时针封闭圆: $\{(x,y): x^2 + y^2 = r^2\}$, 藉由 Green's Theorem

$$\iint_R \frac{\partial g}{\partial x}(x,y) - \frac{\partial f}{\partial y}(x,y)\,dxdy = \iint_R -3x^2 - 3y^2\,dxdy = \oint_C y^3\,dx - x^3\,dy$$

令 $x = r\cos\theta, \quad y = r\sin\theta, \quad 0 \leq \theta \leq 2\pi$

则 $\displaystyle\oint_C y^3\,dx - x^3\,dy = \int_0^{2\pi} r^3\sin^3\theta\cdot(-r\sin\theta) - r^3\cos^3\theta\cdot(r\cos\theta)d\theta$

$$= -r^4\int_0^{2\pi}\sin^4\theta + \cos^4\theta\,d\theta = -r^4\int_0^{2\pi}(\sin^2\theta + \cos^2\theta)^2 - 2\sin^2\theta\cos^2\theta\,d\theta$$

$$= -r^4 \int_0^{2\pi} 1 - 2\left(\frac{\sin 2\theta}{2}\right)^2 d\theta = -r^4 \int_0^{2\pi} 1 - \frac{\sin^2 2\theta}{2} d\theta = -r^4 \int_0^{2\pi} 1 - \frac{1 - \cos 4\theta}{4} d\theta$$

$$= -r^4 \cdot \frac{3}{4} \cdot 2\pi = -\frac{3r^4\pi}{2}$$

Example 13.

$$f(x,y) = 2y, \quad g(x,y) = x^2 + y^2, \quad R:\{(x,y): x^2 + (y-3)^2 \le r^2\}, \quad r > 3, \quad \text{藉由}$$

Green's Theorem 求 $\iint_R \frac{\partial g}{\partial x}(x,y) - \frac{\partial f}{\partial y}(x,y)\, dxdy = ?$

【解】

令 $C$ 为逆时针封闭圆：$\{(x,y): x^2 + (y-3)^2 = r^2\}$，藉由 Green's Theorem

$$\iint_R \frac{\partial g}{\partial x}(x,y) - \frac{\partial f}{\partial y}(x,y)\, dxdy = \iint_R 2x - 2\, dxdy = \int_C 2y\, dx + (x^2 + y^2)\, dy$$

令 $x = r\cos\theta$, $y = 3 + r\sin\theta$, $0 \le \theta \le 2\pi$ 则

$$\int_C 2y\, dx + (x^2 + y^2)\, dy = \int_0^{2\pi} 2(3 + r\sin\theta)(-r\sin\theta) + ((9 + r^2) + 6r\sin\theta) r\cos\theta\, d\theta$$

$$= \int_0^{2\pi} -6r\sin\theta - 2r^2\sin^2\theta + (9 + r^2)r\cos\theta + 6r^2\sin\theta\cos\theta\, d\theta$$

$$= \int_0^{2\pi} -2r^2\sin^2\theta + 6r^2\sin\theta\cos\theta\, d\theta = \int_0^{2\pi} -2r^2\left(\frac{1 - \cos 2\theta}{2}\right) + 3r^2\sin 2\theta\, d\theta = -2r^2\pi$$

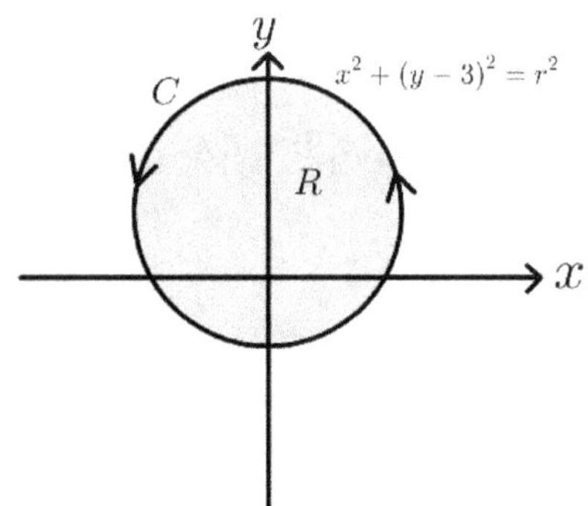


Example 14.

$$f(x,y) = y\cos x, \quad g(x,y) = x\sin y, \quad R \text{为} (0,0), (a,0), (a,a) \text{所围逆时针封闭三角形}$$

区域 $(a > 0)$, 藉由 Green's Theorem 求 $\iint_R \frac{\partial g}{\partial x}(x,y) - \frac{\partial f}{\partial y}(x,y)\, dxdy = ?$

【解】

令 $C_1 = \{(x,y): 0 \le x \le a, y = 0\}$,　$C_2 = \{(x,y): 0 \le y \le a, x = a\}$,

$C_3 = \{(x,y): 0 \le x \le a, y = x\}$ 且 $C = C_1 \cup C_2 \cup C_3$

藉由 Green's Theorem

$$\iint_R \frac{\partial g}{\partial x}(x,y) - \frac{\partial f}{\partial y}(x,y)\, dxdy = \iint_R \sin y - \cos x\, dxdy = \oint_C y \cos x\, dx + x \sin y\, dy$$

則 $\displaystyle\oint_C y \cos x\, dx + x \sin y\, dy$

$$= \int_{C_1} y \cos x\, dx + x \sin y\, dy + \int_{C_2} y \cos x\, dx + x \sin y\, dy + \int_{C_3} y \cos x\, dx + x \sin y\, dy$$

$\because \displaystyle\int_{C_1} y \cos x\, dx + x \sin y\, dy = 0$

$$\int_{C_2} y \cos x\, dx + x \sin y\, dy = a \int_0^a \sin y\, dy = a(-\cos a + 1)$$

$$\int_{C_3} y \cos x\, dx + x \sin y\, dy = \int_a^0 x \cos x\, dx + x \sin x\, dx = -(a+1)\sin a + (a-1)\cos a + 1$$

$\therefore \displaystyle\oint_C y \cos x\, dx + x \sin y\, dy = -(a+1)\sin a - \cos a + a + 1$

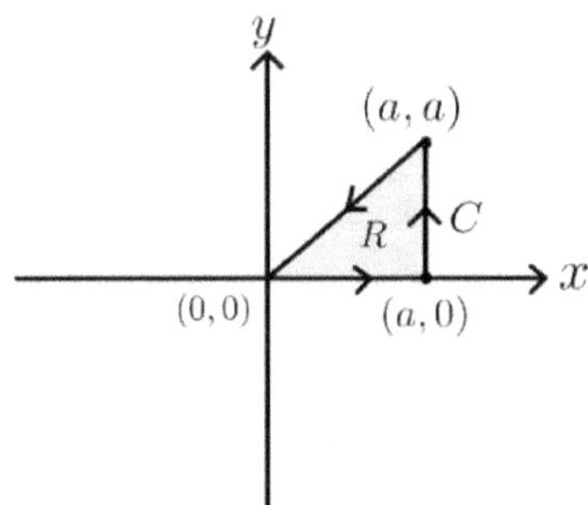


Example 15.

$f(x,y) = -y^4$,　$g(x,y) = xy^2$,　$R$ 为 $(0,0), (1,0), (1,1)$ 所围逆时针封闭三角形区域,

藉由 Green's Theorem 求 $\displaystyle\iint_R \frac{\partial g}{\partial x}(x,y) - \frac{\partial f}{\partial y}(x,y)\, dxdy =$?

【解】

令 $C_1 = \{(x,y): 0 \le x \le 1, y = 0\}$,　$C_2 = \{(x,y): 0 \le y \le 1, x = 1\}$

$C_3 = \{(x, y): 0 \le x \le 1, y = x\}$ 且 $C = C_1 \cup C_2 \cup C_3$

藉由 Green's Theorem

$$\iint_R \frac{\partial g}{\partial x}(x, y) - \frac{\partial f}{\partial y}(x, y)\, dxdy = \iint_R 2xy + 4y^3\, dxdy = \oint_C -y^4 dx + xy^2 dy$$

则 $\displaystyle\oint_C -y^4 dx + xy^2 dy = \sum_{j=1}^{3} \int_{C_j} -y^4 dx + xy^2 dy$

$\because \displaystyle\int_{C_1} -y^4 dx + xy^2 dy = 0, \quad \int_{C_2} -y^4 dx + xy^2 dy = \int_0^1 y^2 dy = \frac{1}{3}$

$\displaystyle\int_{C_3} -y^4 dx + xy^2 dy = \int_1^0 -x^4 dx + x^3 dy = \frac{-1}{20}$

$\therefore \displaystyle\oint_C -y^4 dx + xy^2 dy = \frac{1}{3} - \frac{1}{20} = \frac{17}{60}$

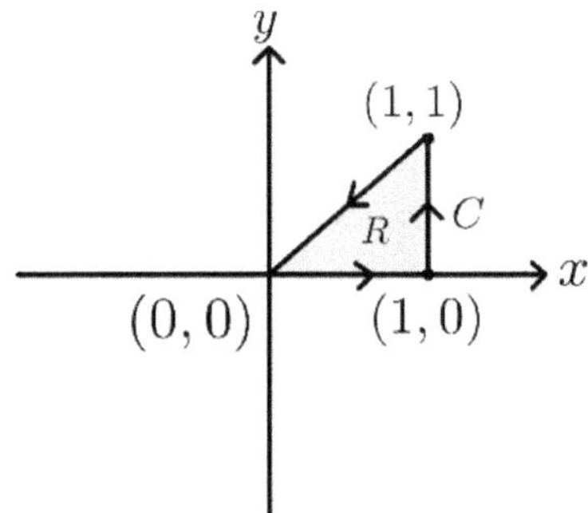


## 9.4.6  $\Delta \neq 0$, 造出不含奇异点的封闭区域再用 Green's Theorem

如果 Green's Theorem 当中两函数 $f, g$ 无法满足在封闭区域内一阶偏导数存在且连续的条件, 则无法直接使用 Green's Theorem, 得先造出不包含奇异点(奇异点为使得函数 $f, g$ 不连续的点或不可微分的点)的封闭区域再使用 Green's Theorem

考试类型:
Type 1.

求 $\displaystyle\oint_C f(x, y)\, dx + g(x, y) dy = ?$, 其中 $C$ 为任意包含原点的封闭曲线(如下图), $f, g$ 在原点

一阶偏导数不存在或者不连续

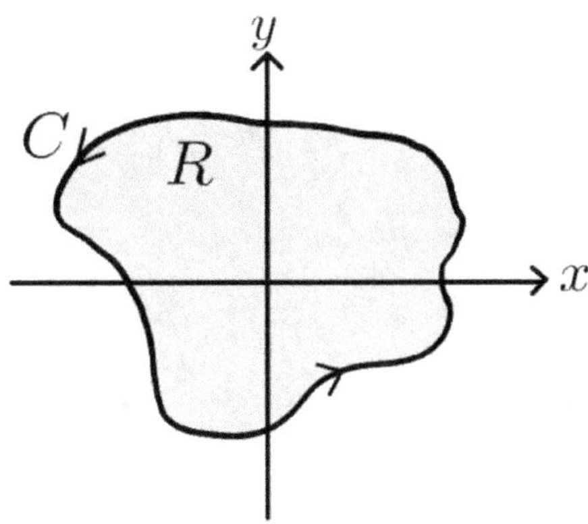


补充说明：

常考封闭曲线为长方形、正方形、椭圆、圆、不规则形，无论是哪种曲线证明叙述皆类似

解题流程：

Step1.

令$O_\varepsilon$为包含原点半径为 $\varepsilon$ 的圆

且取圆上两点$(\varepsilon\cos\theta,\varepsilon\sin\theta)(\varepsilon\cos(-\theta),\varepsilon\sin(-\theta)),\theta>0$

假设过$(\varepsilon\cos\theta,\varepsilon\sin\theta)$的水平直线与$C$交于$s_1$

过$(\varepsilon\cos(-\theta),\varepsilon\sin(-\theta))$的水平直线与$C$交于$s_2$

令$C_\varepsilon$为$O_\varepsilon$圆上从$(\varepsilon\cos\theta,\varepsilon\sin\theta)$至$(\varepsilon\cos(-\theta),\varepsilon\sin(-\theta))$的逆时针弧线

令$C_1$为$(\varepsilon\cos\theta,\varepsilon\sin\theta)$至$s_1$的直线

令$C_2$为$s_2$至$(\varepsilon\cos(-\theta),\varepsilon\sin(-\theta))$的直线

令$C'$为曲线$C$上从$s_1$至$s_2$的逆时针弧线

令$R'$为$C_\varepsilon$、$C_1$、$C_2$、$C'$所围不包含原点的封闭区域

Step2.

藉由 Green's Theorem

$$\int_{C'}fdx+gdy+\int_{C_1}fdx+gdy-\int_{C_\varepsilon}fdx+gdy+\int_{C_2}fdx+gdy$$

$$=\iint_{R'}\frac{\partial g}{\partial x}(x,y)-\frac{\partial f}{\partial y}(x,y)dxdy$$

Step3.

Claim: $\lim\limits_{\theta\to0}\int_{C'}fdx+gdy=\lim\limits_{\theta\to0}\int_{C_\varepsilon}fdx+gdy$

$\because\dfrac{\partial g}{\partial x}(x,y)-\dfrac{\partial f}{\partial y}(x,y)=0\quad\therefore\iint_{R'}\dfrac{\partial g}{\partial x}(x,y)-\dfrac{\partial f}{\partial y}(x,y)dxdy=0$

$$\Rightarrow \int_{C'} f\,dx + g\,dy + \int_{C_1} f\,dx + g\,dy - \int_{C_\varepsilon} f\,dx + g\,dy + \int_{C_2} f\,dx + g\,dy = 0$$

$$\because \lim_{\theta \to 0} \int_{C_1} f\,dx + g\,dy + \int_{C_2} f\,dx + g\,dy = 0 \quad \therefore \lim_{\theta \to 0} \int_{C'} f\,dx + g\,dy - \int_{C_\varepsilon} f\,dx + g\,dy = 0$$

$$\therefore \lim_{\theta \to 0} \int_{C'} f\,dx + g\,dy = \lim_{\theta \to 0} \int_{C_\varepsilon} f\,dx + g\,dy$$

Step4.

令 $x = \varepsilon \cos t, \; y = \varepsilon \sin t$ 则 $dx = -\varepsilon \sin t, \; dy = \varepsilon \cos t$

$$\because \lim_{\theta \to 0} \int_{C_\varepsilon} f\,dx + g\,dy = \int_0^{2\pi} f(\varepsilon \cos t, \varepsilon \sin t)(-\varepsilon \sin t) + g(\varepsilon \cos t, \varepsilon \sin t)\varepsilon \cos t\,dt$$

$$\because \lim_{\theta \to 0} \int_{C'} f\,dx + g\,dy = \oint_C f\,dx + g\,dy$$

$$\therefore \oint_C f\,dx + g\,dy = \int_0^{2\pi} f(\varepsilon \cos t, \varepsilon \sin t)(-\varepsilon \sin t) + g(\varepsilon \cos t, \varepsilon \sin t)\varepsilon \cos t\,dt$$

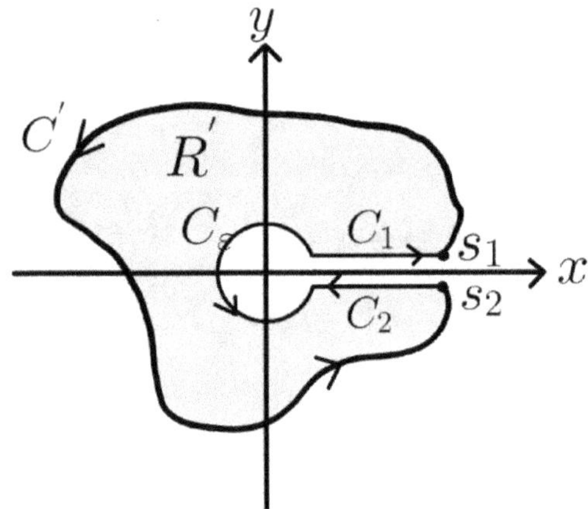


范例说明：

求 $\oint_C f(x,y)\,dx + g(x,y)\,dy = ?$,

(I) 若 $f(x,y) = \dfrac{-y}{x^2 + y^2}$, $g(x,y) = \dfrac{x}{x^2 + y^2}$ 且 $C$ 为任意包含原点的封闭曲线

则 $f, g$ 在原点并非一阶偏导数存在且连续

令 $(x,y) \neq (0,0)$

则 $\dfrac{\partial g}{\partial x}(x,y) = \dfrac{x^2 + y^2 - 2x^2}{(x^2 + y^2)^2} = \dfrac{-x^2 + y^2}{(x^2 + y^2)^2}$, $\dfrac{\partial f}{\partial y}(x,y) = \dfrac{-(x^2 + y^2) + 2y^2}{(x^2 + y^2)^2} = \dfrac{-x^2 + y^2}{(x^2 + y^2)^2}$

$$\therefore \frac{\partial g}{\partial x}(x,y) - \frac{\partial f}{\partial y}(x,y) = 0, \quad \forall (x,y) \neq (0,0)$$

(II)若 $f(x,y) = \dfrac{x-y}{x^2+y^2}$，$g(x,y) = \dfrac{x+y}{x^2+y^2}$ 且 $C$ 为任意包含原点的封闭曲线

则 $f,g$ 在原点并非一阶偏导数存在且连续

令 $(x,y) \neq (0,0)$

则 $\dfrac{\partial g}{\partial x}(x,y) = \dfrac{x^2+y^2-2x^2-2xy}{(x^2+y^2)^2} = \dfrac{-x^2+y^2-2xy}{(x^2+y^2)^2}$

且 $\dfrac{\partial f}{\partial y}(x,y) = \dfrac{-(x^2+y^2)+2y^2-2xy}{(x^2+y^2)^2} = \dfrac{-x^2+y^2-2xy}{(x^2+y^2)^2}$

$$\therefore \frac{\partial g}{\partial x}(x,y) - \frac{\partial f}{\partial y}(x,y) = 0, \quad \forall (x,y) \neq (0,0)$$

(III)若 $f(x,y) = \dfrac{-y^3}{(x^2+y^2)^2}$，$g(x,y) = \dfrac{xy^2}{(x^2+y^2)^2}$，$C$：任意包含原点的封闭曲线

则 $f,g$ 在原点并非一阶偏导数存在且连续

令 $(x,y) \neq (0,0)$

则 $\dfrac{\partial g}{\partial x}(x,y) = \dfrac{y^2(x^2+y^2)^2 - xy^2 \cdot 2(x^2+y^2)2x}{(x^2+y^2)^4} = \dfrac{y^2(x^2+y^2)-4x^2y^2}{(x^2+y^2)^3}$

且 $\dfrac{\partial f}{\partial y}(x,y) = \dfrac{-3y^2(x^2+y^2)^2 + y^3 \cdot 2(x^2+y^2)2y}{(x^2+y^2)^4} = \dfrac{-3y^2(x^2+y^2)+4y^4}{(x^2+y^2)^3}$

$$\therefore \frac{\partial g}{\partial x}(x,y) - \frac{\partial f}{\partial y}(x,y) = 0, \quad \forall (x,y) \neq (0,0)$$

令 $O_\varepsilon$ 为包含原点半径为 $\varepsilon$ 的圆

且取圆上两点 $(\varepsilon\cos\theta, \varepsilon\sin\theta)(\varepsilon\cos(-\theta), \varepsilon\sin(-\theta)), \theta > 0$

假设过 $(\varepsilon\cos\theta, \varepsilon\sin\theta)$ 的水平直线与 $C$ 交于 $s_1$

过 $(\varepsilon\cos(-\theta), \varepsilon\sin(-\theta))$ 的水平直线与 $C$ 交于 $s_2$

令 $C_\varepsilon$ 为 $O_\varepsilon$ 圆上从 $(\varepsilon\cos\theta, \varepsilon\sin\theta)$ 至 $(\varepsilon\cos(-\theta), \varepsilon\sin(-\theta))$ 的逆时针弧线

令 $C_1$ 为 $(\varepsilon\cos\theta, \varepsilon\sin\theta)$ 至 $s_1$ 的直线

令 $C_2$ 为 $s_2$ 至 $(\varepsilon\cos(-\theta), \varepsilon\sin(-\theta))$ 的直线

令 $C'$ 为曲线 $C$ 上从 $s_1$ 至 $s_2$ 的逆时针弧线

藉由 Green's Theorem

$$\int_{C'} f dx + g dy + \int_{C_1} f dx + g dy - \int_{C_\varepsilon} f dx + g dy + \int_{C_2} f dx + g dy$$

$$= \iint_{R'} \frac{\partial g}{\partial x}(x,y) - \frac{\partial f}{\partial y}(x,y)\,dxdy$$

Claim: $\displaystyle\lim_{\theta\to 0}\int_{C'} f\,dx + g\,dy = \lim_{\theta\to 0}\int_{C_\varepsilon} f\,dx + g\,dy$

$$\because \frac{\partial g}{\partial x}(x,y) - \frac{\partial f}{\partial y}(x,y) = 0 \qquad \therefore \iint_{R'} \frac{\partial g}{\partial x}(x,y) - \frac{\partial f}{\partial y}(x,y)\,dxdy = 0$$

$$\Rightarrow \int_{C'} f\,dx + g\,dy + \int_{C_1} f\,dx + g\,dy - \int_{C_\varepsilon} f\,dx + g\,dy + \int_{C_2} f\,dx + g\,dy = 0$$

$$\because \lim_{\theta\to 0}\int_{C_1} f\,dx + g\,dy + \int_{C_2} f\,dx + g\,dy = 0 \qquad \therefore \lim_{\theta\to 0}\int_{C'} f\,dx + g\,dy - \int_{C_\varepsilon} f\,dx + g\,dy = 0$$

$$\therefore \lim_{\theta\to 0}\int_{C'} f\,dx + g\,dy = \lim_{\theta\to 0}\int_{C_\varepsilon} f\,dx + g\,dy$$

(I)

Claim: $\displaystyle\oint_C f\,dx + g\,dy = 2\pi$

令 $x = \varepsilon\cos t,\ y = \varepsilon\sin t$ 则 $dx = -\varepsilon\sin t,\ dy = \varepsilon\cos t$

$$\because \lim_{\theta\to 0}\int_{C_\varepsilon} f\,dx + g\,dy = \int_0^{2\pi} \frac{-\varepsilon\sin t(-\varepsilon\sin t) + \varepsilon^2\cos^2 t}{\varepsilon^2}\,dt = 2\pi$$

$$\therefore \lim_{\theta\to 0}\int_{C'} f\,dx + g\,dy = \lim_{\theta\to 0}\int_{C_\varepsilon} f\,dx + g\,dy = 2\pi$$

$$\because \lim_{\theta\to 0}\int_{C'} f\,dx + g\,dy = \oint_C f\,dx + g\,dy \qquad \therefore \oint_C f\,dx + g\,dy = 2\pi$$

(II)

Claim: $\displaystyle\oint_C f\,dx + g\,dy = 2\pi$

令 $x = \varepsilon\cos t,\ y = \varepsilon\sin t$ 则 $dx = -\varepsilon\sin t,\ dy = \varepsilon\cos t$

$$\because \lim_{\theta\to 0}\int_{C_\varepsilon} f\,dx + g\,dy = \int_0^{2\pi} \frac{(\varepsilon\cos t - \varepsilon\sin t)(-\varepsilon\sin t) + (\varepsilon\cos t + \varepsilon\sin t)\varepsilon\cos t}{\varepsilon^2}\,dt = 2\pi$$

$$\therefore \lim_{\theta\to 0}\int_{C'} f\,dx + g\,dy = \lim_{\theta\to 0}\int_{C_\varepsilon} f\,dx + g\,dy = 2\pi$$

$$\because \lim_{\theta\to 0}\int_{C'} f\,dx + g\,dy = \oint_C f\,dx + g\,dy \qquad \therefore \oint_C f\,dx + g\,dy = 2\pi$$

(III)

Claim: $\displaystyle\oint_C fdx + gdy = \pi$

令$x = \varepsilon\cos t, \ y = \varepsilon\sin t$ 则$dx = -\varepsilon\sin t, \ dy = \varepsilon\cos t$

$$\because \lim_{\theta\to 0}\int_{C_\varepsilon} fdx + gdy = \int_0^{2\pi} \frac{-(\varepsilon\sin t)^3(-\varepsilon\sin t) + (\varepsilon\cos t\cdot\varepsilon^2\sin^2 t)\varepsilon\cos t}{\varepsilon^4} dt$$

$$= \int_0^{2\pi} \sin^2 t\, dt = \pi$$

$$\therefore \lim_{\theta\to 0}\int_{C'} fdx + gdy = \lim_{\theta\to 0}\int_{C_\varepsilon} fdx + gdy = \pi$$

$$\because \lim_{\theta\to 0}\int_{C'} fdx + gdy = \oint_C fdx + gdy \quad \therefore \oint_C fdx + gdy = \pi$$


Example 1.

假设$f(x,y) = \dfrac{-y}{x^2 + y^2}$、$g(x,y) = \dfrac{x}{x^2 + y^2}$，$C$为任意包含原点的封闭曲线，

求 $\displaystyle\oint_C f(x,y)\,dx + g(x,y)dy =?$

【解】

令$O_\varepsilon$为包含原点半径为 $\varepsilon$ 的圆且取圆上两点$(\varepsilon\cos\theta, \varepsilon\sin\theta)(\varepsilon\cos(-\theta), \varepsilon\sin(-\theta)), \theta > 0$

假设过$(\varepsilon\cos\theta, \varepsilon\sin\theta)$的水平直线与$C$交于$s_1$

过$(\varepsilon\cos(-\theta), \varepsilon\sin(-\theta))$的水平直线与$C$交于$s_2$

令$C_\varepsilon$为$O_\varepsilon$圆上从$(\varepsilon\cos\theta, \varepsilon\sin\theta)$至$(\varepsilon\cos(-\theta), \varepsilon\sin(-\theta))$的逆时针弧线

令$C_1$为$(\varepsilon\cos\theta, \varepsilon\sin\theta)$至$s_1$的直线

令$C_2$为$s_2$至$(\varepsilon\cos(-\theta), \varepsilon\sin(-\theta))$的直线

令$C'$为曲线$C$上从$s_1$至$s_2$的逆时针弧线

令$R'$为$C_\varepsilon$、$C_1$、$C_2$、$C'$所围的不包含原点的封闭区域

藉由 Green's Theorem

$$\int_{C'} fdx + gdy + \int_{C_1} fdx + gdy - \int_{C_\varepsilon} fdx + gdy + \int_{C_2} fdx + gdy$$

$$= \iint_{R'} \frac{\partial g}{\partial x}(x,y) - \frac{\partial f}{\partial y}(x,y)dxdy$$

令 $(x,y) \neq (0,0)$

则 $\dfrac{\partial g}{\partial x}(x,y) = \dfrac{x^2+y^2-2x^2}{(x^2+y^2)^2} = \dfrac{-x^2+y^2}{(x^2+y^2)^2}, \quad \dfrac{\partial f}{\partial y}(x,y) = \dfrac{-(x^2+y^2)+2y^2}{(x^2+y^2)^2} = \dfrac{-x^2+y^2}{(x^2+y^2)^2}$

$\therefore \dfrac{\partial g}{\partial x}(x,y) - \dfrac{\partial f}{\partial y}(x,y) = 0 \Rightarrow \iint_{R'} \dfrac{\partial g}{\partial x}(x,y) - \dfrac{\partial f}{\partial y}(x,y)\,dxdy = 0$

$\Rightarrow \displaystyle\int_{C'} f\,dx + g\,dy + \int_{C_1} f\,dx + g\,dy - \int_{C_\varepsilon} f\,dx + g\,dy + \int_{C_2} f\,dx + g\,dy = 0$

$\because \displaystyle\lim_{\theta\to 0}\int_{C_1} f\,dx + g\,dy + \int_{C_2} f\,dx + g\,dy = 0 \quad \therefore \lim_{\theta\to 0}\int_{C'} f\,dx + g\,dy - \int_{C_\varepsilon} f\,dx + g\,dy = 0$

$\therefore \displaystyle\lim_{\theta\to 0}\int_{C'} f\,dx + g\,dy = \lim_{\theta\to 0}\int_{C_\varepsilon} f\,dx + g\,dy$

令 $x = \varepsilon\cos t, y = \varepsilon\sin t$ 则 $dx = -\varepsilon\sin t, dy = \varepsilon\cos t$

$\therefore \displaystyle\lim_{\theta\to 0}\int_{C_\varepsilon} f\,dx + g\,dy = \int_0^{2\pi} \dfrac{-\varepsilon\sin t(-\varepsilon\sin t) + \varepsilon^2\cos^2 t}{\varepsilon^2}\,dt = 2\pi$

$\therefore \displaystyle\lim_{\theta\to 0}\int_{C'} f\,dx + g\,dy = \lim_{\theta\to 0}\int_{C_\varepsilon} f\,dx + g\,dy = 2\pi$

$\because \displaystyle\lim_{\theta\to 0}\int_{C'} f\,dx + g\,dy = \oint_C f\,dx + g\,dy \quad \therefore \oint_C f\,dx + g\,dy = 2\pi$

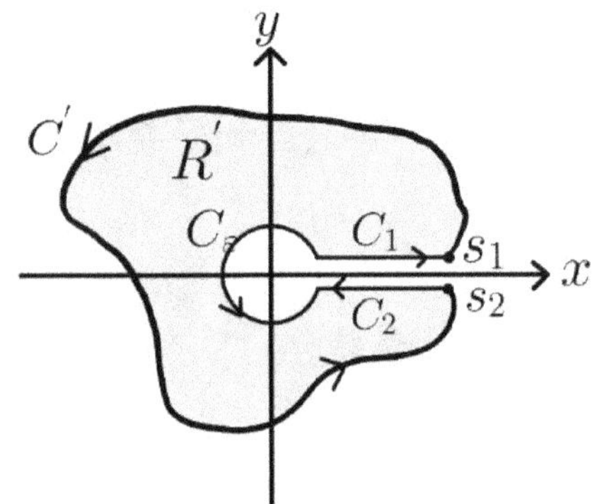


Example 2.

假设 $f(x,y) = \dfrac{x-y}{x^2+y^2}$ 、$g(x,y) = \dfrac{x+y}{x^2+y^2}$, $C$ 为 $(1,0)$、$(1,1)$、$(-1,1)$、$(-1,0)$

所围的逆时针长方形边界, 求 $\displaystyle\oint_C f(x,y)\,dx + g(x,y)\,dy = ?$

【解】

令 $s_1 = (-\varepsilon, 0), \; s_2 = (\varepsilon, 0) \in C$ 且 $\varepsilon > 0$

令 $O_\varepsilon$ 为包含原点半径为 $\varepsilon$ 的圆

令 $C_\varepsilon$ 为 $O_\varepsilon$ 圆上從 $(\varepsilon, 0)$ 至 $(-\varepsilon, 0)$ 的逆时针上半圆弧线

令 $C'$ 为曲线 $C$ 上 $s_2$ 至 $s_1$ 的逆时针路径

令 $R'$ 为 $C_\varepsilon$、$C'$ 所围的封闭区域

藉由 Green's Theorem $\displaystyle\int_{C'} f\,dx + g\,dy - \int_{C_\varepsilon} f\,dx + g\,dy = \iint_{R'} \frac{\partial g}{\partial x}(x, y) - \frac{\partial f}{\partial y}(x, y)\,dxdy$

令 $(x, y) \neq (0, 0)$

则 $\dfrac{\partial g}{\partial x}(x, y) = \dfrac{x^2 + y^2 - 2x^2 - 2xy}{(x^2 + y^2)^2} = \dfrac{-x^2 + y^2 - 2xy}{(x^2 + y^2)^2}$

$\dfrac{\partial f}{\partial y}(x, y) = \dfrac{-(x^2 + y^2) + 2y^2 - 2xy}{(x^2 + y^2)^2} = \dfrac{-x^2 + y^2 - 2xy}{(x^2 + y^2)^2}$

$\therefore \dfrac{\partial g}{\partial x}(x, y) - \dfrac{\partial f}{\partial y}(x, y) = 0 \Rightarrow \iint_{R'} \dfrac{\partial g}{\partial x}(x, y) - \dfrac{\partial f}{\partial y}(x, y)\,dxdy = 0$

$\Rightarrow \displaystyle\int_{C'} f\,dx + g\,dy - \int_{C_\varepsilon} f\,dx + g\,dy = 0 \Rightarrow \int_{C'} f\,dx + g\,dy = \int_{C_\varepsilon} f\,dx + g\,dy$

令 $x = \varepsilon\cos\theta, \; y = \varepsilon\sin\theta$ 则 $dx = -\varepsilon\sin\theta, \; dy = \varepsilon\cos\theta$

$\displaystyle\lim_{\varepsilon\to 0}\int_{C_\varepsilon} f\,dx + g\,dy = \lim_{\varepsilon\to 0}\int_0^\pi \frac{\varepsilon(\cos\theta - \sin\theta)(-\varepsilon\sin\theta) + \varepsilon(\cos\theta + \sin\theta)(\varepsilon\cos\theta)}{\varepsilon^2}\,d\theta = \pi$

$\therefore \displaystyle\lim_{\varepsilon\to 0}\int_{C'} f\,dx + g\,dy = \lim_{\varepsilon\to 0}\int_{C_\varepsilon} f\,dx + g\,dy = \pi$

$\because \displaystyle\lim_{\varepsilon\to 0}\int_{C'} f\,dx + g\,dy = \oint_C f\,dx + g\,dy \qquad \therefore \oint_C f\,dx + g\,dy = \pi$

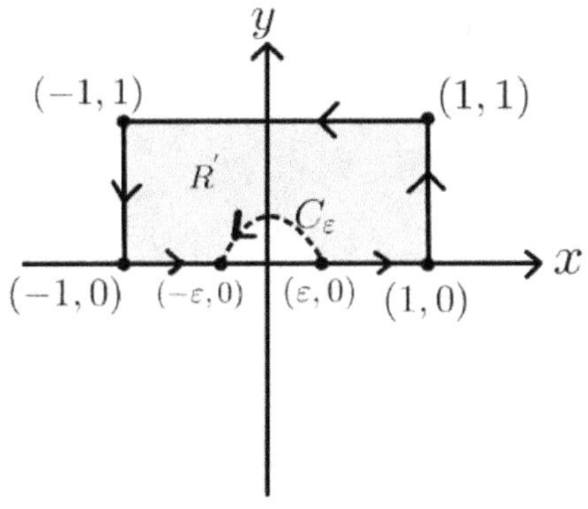

Example 3.

$$假设 f(x,y) = \frac{x-y}{x^2+y^2} \text{、} g(x,y) = \frac{x+y}{x^2+y^2}, \quad C 为 (1,1) \text{、} (-1,1) \text{、} (-1,-1) \text{、} (1,-1)$$

$$所围逆时针正方形, \quad 求 \oint_C f(x,y)\,dx + g(x,y)\,dy = ?$$

【解】

令 $O_\varepsilon$ 为包含原点半径为 $\varepsilon$ 的圆且取圆上两点 $(\varepsilon\cos\theta, \varepsilon\sin\theta)(\varepsilon\cos(-\theta), \varepsilon\sin(-\theta)), \theta > 0$

假设过 $(\varepsilon\cos\theta, \varepsilon\sin\theta)$ 的水平直线与 $C$ 交于 $s_1$

过 $(\varepsilon\cos(-\theta), \varepsilon\sin(-\theta))$ 的水平直线与 $C$ 交于 $s_2$

令 $C_\varepsilon$ 为 $O_\varepsilon$ 圆上从 $(\varepsilon\cos\theta, \varepsilon\sin\theta)$ 至 $(\varepsilon\cos(-\theta), \varepsilon\sin(-\theta))$ 的逆时针弧线

令 $C_1$ 为 $(\varepsilon\cos\theta, \varepsilon\sin\theta)$ 至 $s_1$ 的直线

令 $C_2$ 为 $s_2$ 至 $(\varepsilon\cos(-\theta), \varepsilon\sin(-\theta))$ 的直线

令 $C'$ 为曲线 $C$ 上从 $s_1$ 至 $s_2$ 的逆时针线

令 $R'$ 为 $C_\varepsilon$、$C_1$、$C_2$、$C'$ 所围的不包含原点的封闭区域

藉由 Green's Theorem

$$\int_{C'} f\,dx + g\,dy + \int_{C_1} f\,dx + g\,dy - \int_{C_\varepsilon} f\,dx + g\,dy + \int_{C_2} f\,dx + g\,dy$$

$$= \iint_{R'} \frac{\partial g}{\partial x}(x,y) - \frac{\partial f}{\partial y}(x,y)\,dxdy$$

令 $(x,y) \neq (0,0)$

则 $\dfrac{\partial g}{\partial x}(x,y) = \dfrac{x^2+y^2-2x^2-2xy}{(x^2+y^2)^2} = \dfrac{-x^2+y^2-2xy}{(x^2+y^2)^2}$

$\dfrac{\partial f}{\partial y}(x,y) = \dfrac{-(x^2+y^2)+2y^2-2xy}{(x^2+y^2)^2} = \dfrac{-x^2+y^2-2xy}{(x^2+y^2)^2}$

$\therefore \dfrac{\partial g}{\partial x}(x,y) - \dfrac{\partial f}{\partial y}(x,y) = 0 \Rightarrow \iint_{R'} \dfrac{\partial g}{\partial x}(x,y) - \dfrac{\partial f}{\partial y}(x,y)\,dxdy = 0$

$$\Rightarrow \int_{C'} f\,dx + g\,dy + \int_{C_1} f\,dx + g\,dy - \int_{C_\varepsilon} f\,dx + g\,dy + \int_{C_2} f\,dx + g\,dy = 0$$

$$\because \lim_{\theta\to 0}\int_{C_1} f\,dx + g\,dy + \int_{C_2} f\,dx + g\,dy = 0 \quad \therefore \lim_{\theta\to 0}\int_{C'} f\,dx + g\,dy - \int_{C_\varepsilon} f\,dx + g\,dy = 0$$

$$\therefore \lim_{\theta\to 0}\int_{C'} f\,dx + g\,dy = \lim_{\theta\to 0}\int_{C_\varepsilon} f\,dx + g\,dy$$

令 $x = \varepsilon\cos t, y = \varepsilon\sin t$ 则 $dx = -\varepsilon\sin t, dy = \varepsilon\cos t$

$$\therefore \lim_{\theta \to 0} \int_{C_\varepsilon} f\,dx + g\,dy = \int_0^{2\pi} \frac{\varepsilon(\cos t - \sin t)(-\varepsilon \sin t) + \varepsilon(\cos t + \sin t)(\varepsilon \cos t)}{\varepsilon^2}\,dt = 2\pi$$

$$\therefore \lim_{\theta \to 0} \int_{C'} f\,dx + g\,dy = \lim_{\theta \to 0} \int_{C_\varepsilon} f\,dx + g\,dy = 2\pi$$

$$\because \lim_{\theta \to 0} \int_{C'} f\,dx + g\,dy = \oint_C f\,dx + g\,dy \quad \therefore \oint_C f\,dx + g\,dy = 2\pi$$

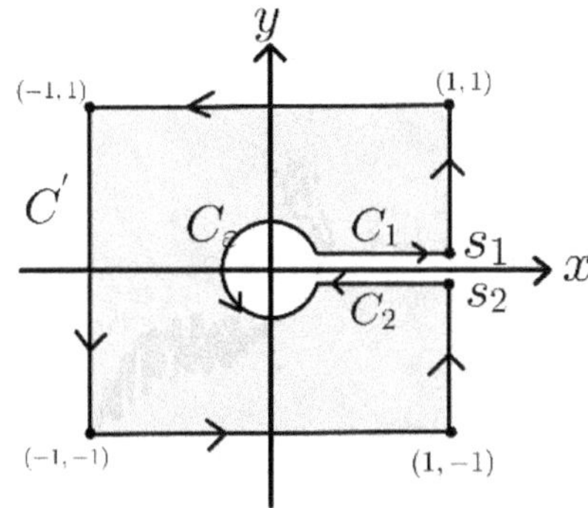

Example 4.

$$\text{假设} f(x,y) = \frac{x-y}{x^2+y^2} \,\text{、}\, g(x,y) = \frac{x+y}{x^2+y^2}, \quad C\text{为椭圆} \frac{x^2}{a^2} + \frac{y^2}{b^2} = \text{的逆时针边界，}$$

$$\text{求} \oint_C f(x,y)\,dx + g(x,y)\,dy = ?$$

【解】

令 $O_\varepsilon$ 为包含原点半径为 $\varepsilon$ 的圆且取圆上两点$(\varepsilon \cos\theta, \varepsilon \sin\theta)(\varepsilon \cos(-\theta), \varepsilon \sin(-\theta))$，$\theta > 0$

假设过$(\varepsilon \cos\theta, \varepsilon \sin\theta)$的水平直线与$C$交于$s_1$

过$(\varepsilon \cos(-\theta), \varepsilon \sin(-\theta))$的水平直线与$C$交于$s_2$

令 $C_\varepsilon$ 为 $O_\varepsilon$ 圆上从$(\varepsilon \cos\theta, \varepsilon \sin\theta)$至$(\varepsilon \cos(-\theta), \varepsilon \sin(-\theta))$的逆时针弧线

令 $C_1$ 为$(\varepsilon \cos\theta, \varepsilon \sin\theta)$至$s_1$的直线

令 $C_2$ 为$s_2$至$(\varepsilon \cos(-\theta), \varepsilon \sin(-\theta))$的直线

令 $C'$ 为曲线$C$上从$s_1$至$s_2$的逆时针弧线

令 $R'$ 为 $C_\varepsilon$、$C_1$、$C_2$、$C'$ 所围的不包含原点的封闭区域

藉由 Green's Theorem

$$\int_{C'} f\,dx + g\,dy + \int_{C_1} f\,dx + g\,dy - \int_{C_\varepsilon} f\,dx + g\,dy + \int_{C_2} f\,dx + g\,dy$$

$$= \iint_{R'} \frac{\partial g}{\partial x}(x,y) - \frac{\partial f}{\partial y}(x,y)\,dxdy$$

令 $(x,y) \neq (0,0)$

则 $\dfrac{\partial g}{\partial x}(x,y) = \dfrac{x^2 + y^2 - 2x^2 - 2xy}{(x^2 + y^2)^2} = \dfrac{-x^2 + y^2 - 2xy}{(x^2 + y^2)^2}$

$\dfrac{\partial f}{\partial y}(x,y) = \dfrac{-(x^2 + y^2) + 2y^2 - 2xy}{(x^2 + y^2)^2} = \dfrac{-x^2 + y^2 - 2xy}{(x^2 + y^2)^2}$

$\therefore \dfrac{\partial g}{\partial x}(x,y) - \dfrac{\partial f}{\partial y}(x,y) = 0 \Rightarrow \iint_{R'} \dfrac{\partial g}{\partial x}(x,y) - \dfrac{\partial f}{\partial y}(x,y)\,dxdy = 0$

$\Rightarrow \int_{C'} f\,dx + g\,dy + \int_{C_1} f\,dx + g\,dy - \int_{C_\varepsilon} f\,dx + g\,dy + \int_{C_2} f\,dx + g\,dy = 0$

$\because \lim_{\theta \to 0} \int_{C_1} f\,dx + g\,dy + \int_{C_2} f\,dx + g\,dy = 0 \quad \therefore \lim_{\theta \to 0} \int_{C'} f\,dx + g\,dy - \int_{C_\varepsilon} f\,dx + g\,dy = 0$

$\therefore \lim_{\theta \to 0} \int_{C'} f\,dx + g\,dy = \lim_{\theta \to 0} \int_{C_\varepsilon} f\,dx + g\,dy$

令 $x = \varepsilon \cos t, y = \varepsilon \sin t$ 则 $dx = -\varepsilon \sin t, dy = \varepsilon \cos t$

$\therefore \lim_{\theta \to 0} \int_{C_\varepsilon} f\,dx + g\,dy = \int_0^{2\pi} \dfrac{\varepsilon(\cos t - \sin t)(-\varepsilon \sin t) + \varepsilon(\cos t + \sin t)(\varepsilon \cos t)}{\varepsilon^2}\,dt = 2\pi$

$\therefore \lim_{\theta \to 0} \int_{C'} f\,dx + g\,dy = \lim_{\theta \to 0} \int_{C_\varepsilon} f\,dx + g\,dy = 2\pi$

$\because \lim_{\theta \to 0} \int_{C'} f\,dx + g\,dy = \oint_C f\,dx + g\,dy \quad \therefore \oint_C f\,dx + g\,dy = 2\pi$

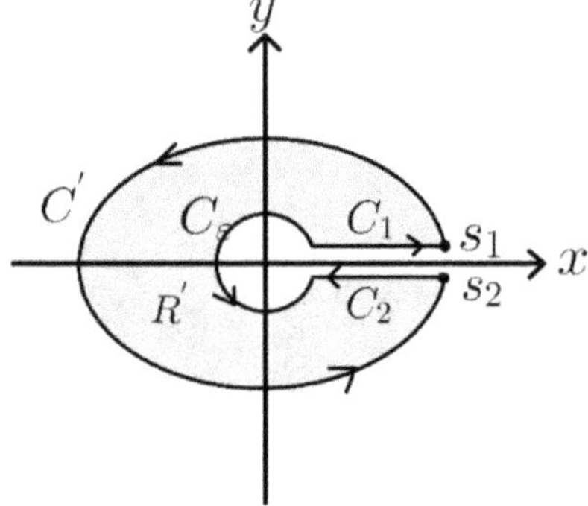


Example 5.

假设 $f(x,y) = \dfrac{-y}{x^2 + y^2}$ ，$g(x,y) = \dfrac{x}{x^2 + y^2}$, $C$ 为圆 $x^2 + y^2 = 4$ 的逆时针边界,

求 $\displaystyle\oint_C f(x,y)\,dx + g(x,y)\,dy = ?$

【解】

令 $O_\varepsilon$ 为包含原点半径为 $\varepsilon$ 的圆且取圆上两点 $(\varepsilon\cos\theta, \varepsilon\sin\theta)(\varepsilon\cos(-\theta), \varepsilon\sin(-\theta)), \theta > 0$

假设过 $(\varepsilon\cos\theta, \varepsilon\sin\theta)$ 的水平直线与 $C$ 交于 $s_1$

过 $(\varepsilon\cos(-\theta), \varepsilon\sin(-\theta))$ 的水平直线与 $C$ 交于 $s_2$

令 $C_\varepsilon$ 为 $O_\varepsilon$ 圆上从 $(\varepsilon\cos\theta, \varepsilon\sin\theta)$ 至 $(\varepsilon\cos(-\theta), \varepsilon\sin(-\theta))$ 的逆时针弧线

令 $C_1$ 为 $(\varepsilon\cos\theta, \varepsilon\sin\theta)$ 至 $s_1$ 的直线

令 $C_2$ 为 $s_2$ 至 $(\varepsilon\cos(-\theta), \varepsilon\sin(-\theta))$ 的直线

令 $C'$ 为曲线 $C$ 上从 $s_1$ 至 $s_2$ 的逆时针弧线

令 $R'$ 为 $C_\varepsilon$、$C_1$、$C_2$、$C'$ 所围的不包含原点的封闭区域

藉由 Green's Theorem

$$\int_{C'} f\,dx + g\,dy + \int_{C_1} f\,dx + g\,dy - \int_{C_\varepsilon} f\,dx + g\,dy + \int_{C_2} f\,dx + g\,dy$$

$$= \iint_{R'} \frac{\partial g}{\partial x}(x,y) - \frac{\partial f}{\partial y}(x,y)\,dxdy$$

令 $(x,y) \neq (0,0)$

则 $\dfrac{\partial g}{\partial x}(x,y) = \dfrac{x^2 + y^2 - 2x^2}{(x^2+y^2)^2} = \dfrac{-x^2+y^2}{(x^2+y^2)^2}$, $\quad \dfrac{\partial f}{\partial y}(x,y) = \dfrac{-(x^2+y^2)+2y^2}{(x^2+y^2)^2} = \dfrac{-x^2+y^2}{(x^2+y^2)^2}$

$\therefore \dfrac{\partial g}{\partial x}(x,y) - \dfrac{\partial f}{\partial y}(x,y) = 0 \Rightarrow \iint_{R'} \dfrac{\partial g}{\partial x}(x,y) - \dfrac{\partial f}{\partial y}(x,y)\,dxdy = 0$

$\Rightarrow \displaystyle\int_{C'} f\,dx + g\,dy + \int_{C_1} f\,dx + g\,dy - \int_{C_\varepsilon} f\,dx + g\,dy + \int_{C_2} f\,dx + g\,dy = 0$

$\because \displaystyle\lim_{\theta\to 0}\int_{C_1} f\,dx + g\,dy + \int_{C_2} f\,dx + g\,dy = 0 \quad \therefore \lim_{\theta\to 0}\int_{C'} f\,dx + g\,dy - \int_{C_\varepsilon} f\,dx + g\,dy = 0$

$\therefore \displaystyle\lim_{\theta\to 0}\int_{C'} f\,dx + g\,dy = \lim_{\theta\to 0}\int_{C_\varepsilon} f\,dx + g\,dy$

令 $x = \varepsilon\cos t, y = \varepsilon\sin t$ 则 $dx = -\varepsilon\sin t, dy = \varepsilon\cos t$

$\therefore \displaystyle\lim_{\theta\to 0}\int_{C_\varepsilon} f\,dx + g\,dy = \int_0^{2\pi} \frac{-\varepsilon\sin t(-\varepsilon\sin t) + \varepsilon^2\cos^2 t}{\varepsilon^2}\,dt = 2\pi$

$\therefore \displaystyle\lim_{\theta\to 0}\int_{C'} f\,dx + g\,dy = \lim_{\theta\to 0}\int_{C_\varepsilon} f\,dx + g\,dy = 2\pi$

$$\because \lim_{\theta \to 0} \int_{C'} f\,dx + g\,dy = \oint_C f\,dx + g\,dy \qquad \therefore \oint_C f\,dx + g\,dy = 2\pi$$

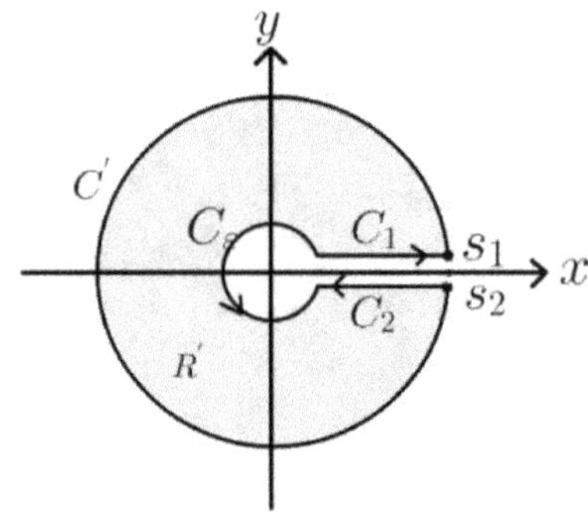


Example 6.

$$求 \oint_C \frac{-y}{x^2 + y^2}\,dx + \frac{x}{x^2 + y^2}\,dy =? 沿下列路径的值$$

(1)$C$为沿 $x^2 + y^2 = 1$ 的逆时针封闭路径

(2)$C$为沿 $x^2 + y^2 = r\,(r > 1)$的逆时针封闭路径

(3)$C$为沿 $\left(\dfrac{x}{4}\right)^2 + \left(\dfrac{y}{3}\right)^2 = 1$ 的逆时针封闭路径

(4)$C$为沿 $(1,1), (-1,1), (-1,-1), (1,-1), (1,1)$的逆时针封闭路径

(5)$C$为沿 $(x-5)^2 + (y-5)^2 = 16$ 的逆时针封闭路径

(6)$C$为沿 $(x-1)^2 + y^2 = 1$ 的逆时针封闭路径

(7)$C$为沿 $(x-1)^2 + y^2 = a^2, a \in \mathrm{R}$ 的逆时针封闭路径

【解】

(1)(2)(3)(4)

令$O_\varepsilon$为包含原点半径为 $\varepsilon$ 的圆且取圆上两点$(\varepsilon\cos\theta, \varepsilon\sin\theta)(\varepsilon\cos(-\theta), \varepsilon\sin(-\theta)), \theta > 0$

假设过$(\varepsilon\cos\theta, \varepsilon\sin\theta)$的水平直线与$C$交于$s_1$

过$(\varepsilon\cos(-\theta), \varepsilon\sin(-\theta))$的水平直线与$C$交于$s_2$

令$C_\varepsilon$为$O_\varepsilon$圆上从$(\varepsilon\cos\theta, \varepsilon\sin\theta)$至$(\varepsilon\cos(-\theta), \varepsilon\sin(-\theta))$的逆时针弧线

令$C_1$为$(\varepsilon\cos\theta, \varepsilon\sin\theta)$至$s_1$的直线

令$C_2$为$s_2$至$(\varepsilon\cos(-\theta), \varepsilon\sin(-\theta))$的直线

令$C'$为曲线$C$上从$s_1$至$s_2$的逆时针弧线

令$R'$为$C_\varepsilon$、$C_1$、$C_2$、$C'$所围的不包含原点的封闭区域

藉由Green's Theorem

$$\int_{C'} f dx + g dy + \int_{C_1} f dx + g dy - \int_{C_\varepsilon} f dx + g dy + \int_{C_2} f dx + g dy$$

$$= \iint_{R'} \frac{\partial g}{\partial x}(x,y) - \frac{\partial f}{\partial y}(x,y) dx dy$$

令 $(x,y) \neq (0,0)$

则 $\dfrac{\partial g}{\partial x}(x,y) = \dfrac{x^2 + y^2 - 2x^2}{(x^2+y^2)^2} = \dfrac{-x^2+y^2}{(x^2+y^2)^2}, \quad \dfrac{\partial f}{\partial y}(x,y) = \dfrac{-(x^2+y^2)+2y^2}{(x^2+y^2)^2} = \dfrac{-x^2+y^2}{(x^2+y^2)^2}$

$\therefore \dfrac{\partial g}{\partial x}(x,y) - \dfrac{\partial f}{\partial y}(x,y) = 0 \Rightarrow \iint_{R'} \dfrac{\partial g}{\partial x}(x,y) - \dfrac{\partial f}{\partial y}(x,y) dx dy = 0$

$\Rightarrow \displaystyle\int_{C'} f dx + g dy + \int_{C_1} f dx + g dy - \int_{C_\varepsilon} f dx + g dy + \int_{C_2} f dx + g dy = 0$

$\because \displaystyle\lim_{\theta \to 0} \int_{C_1} f dx + g dy + \int_{C_2} f dx + g dy = 0 \quad \therefore \lim_{\theta \to 0} \int_{C'} f dx + g dy - \int_{C_\varepsilon} f dx + g dy = 0$

$\therefore \displaystyle\lim_{\theta \to 0} \int_{C'} f dx + g dy = \lim_{\theta \to 0} \int_{C_\varepsilon} f dx + g dy$

令 $x = \varepsilon \cos t, y = \varepsilon \sin t$ 则 $dx = -\varepsilon \sin t, dy = \varepsilon \cos t$

$\therefore \displaystyle\lim_{\theta \to 0} \int_{C_\varepsilon} f dx + g dy = \int_0^{2\pi} \frac{-\varepsilon \sin t(-\varepsilon \sin t) + \varepsilon^2 \cos^2 t}{\varepsilon^2} dt = 2\pi$

$\therefore \displaystyle\lim_{\theta \to 0} \int_{C'} f dx + g dy = \lim_{\theta \to 0} \int_{C_\varepsilon} f dx + g dy = 2\pi$

$\therefore \displaystyle\lim_{\theta \to 0} \int_{C'} f dx + g dy = \oint_{C} f dx + g dy \quad \therefore \oint_{C} f dx + g dy = 2\pi$

(5)

令 $R'$ 为 $C$ 所围封闭区域则 $f, g$ 于 $R'$ 的一阶偏导数存在且连续

藉由 Green's Theorem, $\displaystyle\oint_C f dx + g dy = \iint_{R'} \frac{\partial g}{\partial x}(x,y) - \frac{\partial f}{\partial y}(x,y) dx dy$

$\because \dfrac{\partial g}{\partial x}(x,y) = \dfrac{x^2 + y^2 - 2x^2}{(x^2+y^2)^2} = \dfrac{-x^2+y^2}{(x^2+y^2)^2}, \quad \dfrac{\partial f}{\partial y}(x,y) = \dfrac{-(x^2+y^2)+2y^2}{(x^2+y^2)^2} = \dfrac{-x^2+y^2}{(x^2+y^2)^2}$

$\therefore \dfrac{\partial g}{\partial x}(x,y) - \dfrac{\partial f}{\partial y}(x,y) = 0 \Rightarrow \iint_{R'} \dfrac{\partial g}{\partial x}(x,y) - \dfrac{\partial f}{\partial y}(x,y) dx dy = 0 \quad \therefore \oint_C f dx + g dy = 0$

(6)

令 $O_\varepsilon$ 为包含原点半径为 $\varepsilon$ 的圆

且与 $C$ 交于 $s_1 = (\varepsilon \cos \theta_1, \varepsilon \sin \theta_1) \text{、} s_2 = (\varepsilon \cos \theta_2, \varepsilon \sin \theta_2), \theta_2 > \theta_1$

令 $C_\varepsilon$ 为 $O_\varepsilon$ 圆上从 $(\varepsilon\cos\theta_1,\varepsilon\sin\theta_1)$ 至 $(\varepsilon\cos\theta_2,\varepsilon\sin\theta_2)$ 的顺时针弧线

令 $C'$ 为曲线 $C$ 上从 $s_2$ 至 $s_1$ 的逆时针弧线

令 $R'$ 为 $C_\varepsilon$、$C'$ 所围的封闭区域

藉由 Green's Theorem，$\displaystyle\int_{C'} f dx + g dy + \int_{C_\varepsilon} f dx + g dy = \iint_{R'} \frac{\partial g}{\partial x}(x,y) - \frac{\partial f}{\partial y}(x,y) dx dy$

令 $(x,y) \neq (0,0)$

则 $\dfrac{\partial g}{\partial x}(x,y) = \dfrac{x^2+y^2-2x^2}{(x^2+y^2)^2} = \dfrac{-x^2+y^2}{(x^2+y^2)^2}$，$\dfrac{\partial f}{\partial y}(x,y) = \dfrac{-(x^2+y^2)+2y^2}{(x^2+y^2)^2} = \dfrac{-x^2+y^2}{(x^2+y^2)^2}$

$\therefore \dfrac{\partial g}{\partial x}(x,y) - \dfrac{\partial f}{\partial y}(x,y) = 0 \Rightarrow \iint_{R'} \dfrac{\partial g}{\partial x}(x,y) - \dfrac{\partial f}{\partial y}(x,y) dx dy = 0$

$\Rightarrow \displaystyle\int_{C'} f dx + g dy + \int_{C_\varepsilon} f dx + g dy = 0 \Rightarrow \int_{C'} f dx + g dy = -\int_{C_\varepsilon} f dx + g dy$

令 $x = \varepsilon\cos\theta, y = \varepsilon\sin\theta$ 则 $dx = -\varepsilon\sin\theta, dy = \varepsilon\cos\theta$

$$\lim_{\varepsilon\to 0}\int_{C_\varepsilon} f dx + g dy = \lim_{\varepsilon\to 0}\int_{\frac{\pi}{2}}^{-\frac{\pi}{2}} \frac{\varepsilon\sin\theta(\varepsilon\sin\theta) + \varepsilon^2\cos^2\theta}{\varepsilon^2} d\theta = -\pi$$

$$\lim_{\varepsilon\to 0}\int_{C'} f dx + g dy = -\lim_{\varepsilon\to 0}\int_{C_\varepsilon} f dx + g dy = \pi$$

$\because \displaystyle\lim_{\varepsilon\to 0}\int_{C'} f dx + g dy = \oint_C f dx + g dy \quad \therefore \oint_C f dx + g dy = \pi$

(7)

$$a < 1 \Rightarrow \oint_C f dx + g dy = 0$$

$$a = 1 \Rightarrow \oint_C f dx + g dy = \pi$$

$$a > 1 \Rightarrow \oint_C f dx + g dy = 2\pi$$

## 9.5　曲面积分

### 9.5.1　空间曲面 $S$ 的表示式与光滑曲面

曲面积分的考试类型包含给曲面$S$求封闭区域的表面积、给定纯量函数$f$与曲面$S$求质量、给定向量函数$\vec{F}$与曲面$S$求通量。三维空间的曲面可用两个参数来表示（如下图）：

$$\vec{r}(s,t) = \big(x(s,t),y(s,t),z(s,t)\big)$$

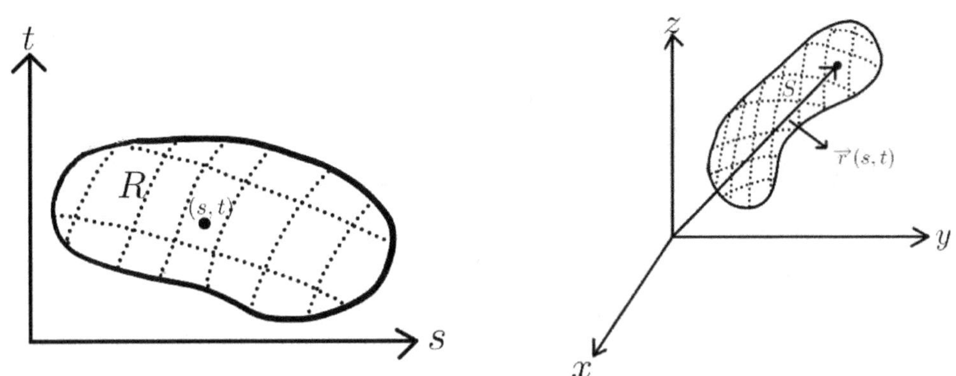

假设函数$f$为一个双变数函数，曲面$S$的方程式为$z = f(x,y)$，则曲面参数化的表示式为$\vec{r}(x,y) = \big(x,y,f(x,y)\big)$，其中$\forall\,(x,y) \in D_f$，$D_f$为$f$的定义域；同理，如果曲面$S$的方程式为

$$x = g(y,z),\ \forall\,(y,z) \in D_g \quad 或 \quad y = h(x,z),\ \forall(x,z) \in D_h$$

则参数化表示式分别为

$$\vec{r}(y,z) = (g(y,z),y,z),\ \forall\,(y,z) \in D_g \quad 或 \quad \vec{r}(x,z) = (x,h(x,z),z),\ \forall\,(x,z) \in D_h$$

【定义】

假设曲面的参数式为$\vec{r}(s,t) = \big(x(s,t),y(s,t),z(s,t)\big)$，则

$$\vec{r}_s = \frac{\partial \vec{r}}{\partial s} := \left(\frac{\partial x}{\partial s},\frac{\partial y}{\partial s},\frac{\partial z}{\partial s}\right) \ 且 \ \vec{r}_t = \frac{\partial \vec{r}}{\partial t} := \left(\frac{\partial x}{\partial t},\frac{\partial y}{\partial t},\frac{\partial z}{\partial t}\right)$$

【定义】

假设曲面的参数式为$\vec{r}(s,t) = \big(x(s,t),y(s,t),z(s,t)\big)$，如果存在定义域的某点$(s,t)$使得$\vec{r}_s \times \vec{r}_t \neq 0$则称$\vec{r}(s,t)$为正则参数化(regular parameterization)

接下来皆只考虑正则参数化的曲面，参数化曲面是否平滑与曲线是否平滑的定义类似，如果曲线没有尖锐的转角，则参数化曲线是平滑的。如果曲面没有尖锐的转角，则称参数化曲面是平滑的。

【定义】

假设曲面的参数化为$\vec{r}(s,t) = \big(x(s,t),y(s,t),z(s,t)\big)$，如果$\vec{r}_s \times \vec{r}_t \neq 0,\ \forall\,(s,t) \in D$

则称曲面为平滑曲面

底下列出计算曲面面积、纯量函数曲面积分以及向量函数曲面积分的重要公式, 细部推导请见各节说明, 假设曲面$S$的参数式为$\vec{r}(s,t)$, 第一个计算重点是给定曲面$S$求封闭区域的表面积, 公式为

$$\iint_S dA = \iint_D |\vec{r}_s \times \vec{r}_t| dA$$

第二个计算重点是给定曲面$S$, 求纯量函数$f$的曲面积分, 公式为

$$\iint_S f(x,y,z)dS = \iint_D f(\vec{r}(s,t))|\vec{r}_s \times \vec{r}_t| dA$$

第三个计算重点是给定曲面$S$, 求向量函数$\vec{F}$的曲面积分, 公式为

$$\iint_S \vec{F} \cdot d\vec{S} \iint_D \vec{F}(\vec{r}(s,t)) \cdot (\vec{r}_s \times \vec{r}_t) ds dt$$

其中 $\vec{r}_s = \left(\dfrac{\partial x}{\partial s}, \dfrac{\partial y}{\partial s}, \dfrac{\partial z}{\partial s}\right)$ 且 $\vec{r}_t = \left(\dfrac{\partial x}{\partial t}, \dfrac{\partial y}{\partial t}, \dfrac{\partial z}{\partial t}\right)$

從上述观察得知, 给定参数化曲面$\vec{r}(s,t) = (x(s,t), y(s,t), z(s,t))$, 如何有效率且精确地求得两偏导数的外积$\vec{r}_s \times \vec{r}_t$及其长度$|\vec{r}_s \times \vec{r}_t|$ 是重要的

Example 1.

假设曲面的参数表示式为$\vec{r}(s,t) = (3\cos s, t, 3\sin s)$, 求曲面的单一方程式表示式

【解】

$\because$ 曲面的参数式为 $x = 3\cos s, y = t, z = 3\sin s$

$\therefore x^2 + z^2 = 9\cos^2 s + 9\sin^2 s = 9.$

Example 2.

找出一个向量函数, 通过具有位置向量$\vec{r}_0$的点$P_0$, 并包含两个非平行向量$\vec{u}$和$\vec{v}$的平面

【解】

令 P 点为平面上的任意点则存在纯量$s, t$使得$\overrightarrow{P_0 P} = s\vec{u} + t\vec{v}$

令 $\vec{r}$ 为 $P$ 点的位置向量 则 $\vec{r} = \overrightarrow{OP_0} + \overrightarrow{P_0 P} = \vec{r}_0 + s\vec{u} + t\vec{v}$

$\therefore \vec{r}(s,t) = r_0 + s\vec{u} + t\vec{v}$

Example 3.

求下列各个曲面方程式的参数表示式

(1)球体方程式表示 $x^2 + y^2 + z^2 = r^2,\ r > 0$

(2)圆柱体方程式表示$x^2 + y^2 = a^2$，其中 $0 \leq z \leq 1, a > 0$

(3)锥体方程式表示 $z = 5\sqrt{x^2 + y^2}$

(4)椭圆抛物面方程式 $z = x^2 + 2y^2$

(5)锥体$x^2 + y^2 = z^2$在平面 $z = -3$ 的上方曲面

(6)椭球面方程式 $\dfrac{x^2}{a^2} + \dfrac{y^2}{b^2} + \dfrac{z^2}{c^2} = 1$

【解】

(1)

$\vec{r}(\theta, \varphi) = r\cos\theta\sin\varphi,\, y = r\sin\theta\sin\varphi,\, z = r\cos\varphi$

(2)

$\vec{r}(\theta, z) = (a\cos\theta,\, a\sin\theta,\, z),\ 0 \leq \theta \leq 2\pi\ \text{且}\ 0 \leq z \leq 1$

(3)

参数表示方式为 $\vec{r}(x, y) = (x, y, 5\sqrt{x^2 + y^2})$

另一个参数表示方式可用极坐标表示

令 $x = r\cos\theta,\, y = r\sin\theta$ 则 $z = 5\sqrt{x^2 + y^2} = 5r$

$\therefore$ 极坐标参数表示方式为 $\vec{r}(r, \theta) = (r\cos\theta,\, r\sin\theta,\, 5r)$

(4)

参数表示方式为 $\vec{r}(x, y) = (x, y, x^2 + 2y^2)$

(5)

参数表示方式为 $\vec{r}(s, t) = (s\cos t,\, s\sin t,\, s)$ 其中 $-3 \leq s < \infty, 0 \leq t < 2\pi$

(6)

$\vec{r}(\theta, \varphi) = (a\cos\theta\sin\varphi,\, b\sin\theta\sin\varphi,\, c\cos\varphi)$ 其中 $0 \leq \theta \leq 2\pi,\ 0 \leq \varphi \leq \pi$

Example 4.

假设曲面的参数表示为 $\vec{r}(s, t) = (s\cos t,\, s\sin t,\, s^2), 0 \leq s < \infty, 0 \leq t < 2\pi$，求其单一方程式表示式

【解】

$\because x^2 + y^2 = (s\cos t)^2 + (s\sin t)^2 = s^2 = z$

$\therefore$ 单一方程式表示式为椭圆抛物面 $x^2 + y^2 = z$

Example 5.

假设曲面$S$的参数式为 $\vec{r}(s,t) = \big((3 + \cos t)\cos s, (3 + \cos t)\sin s, \sin t\big)$，其中 $0 \leq s < 2\pi$, $0 \leq t < 2\pi$, 判断曲面是否为平滑曲面

【解】

$\vec{r}_s = (-(3 + \cos t)\sin s, (3 + \cos t)\cos s, 0)$, and $\vec{r}_t = (-\sin t \cos s, -\sin t \sin s, \cos t)$

$\because \vec{r}_s \times \vec{r}_t = \big((3 + \cos t)\cos s \cos t, (3 + \cos t)\sin s \cos t, (3 + \cos t)\sin t\big)$

Let $\vec{r}_s \times \vec{r}_t = \vec{0}$ then $(3 + \cos t)\sin t = 0 \Rightarrow t = 0$ or $\pi$

$\because t = 0$ 或 $\pi$ 且 $(3 + \cos t)\sin s \cos t = 0$ $\quad \therefore s = 0$ or $\pi$

$\because (3 + \cos t)\cos s \cos t \neq 0$, for $s = 0$ or $\pi$ and $t = 0$ or $\pi$

$\therefore$ 曲面为平滑曲面

Example 6.

求 $\dfrac{\partial \vec{r}}{\partial s} \times \dfrac{\partial \vec{r}}{\partial t} =?$ 以及 $\left|\dfrac{\partial \vec{r}}{\partial s} \times \dfrac{\partial \vec{r}}{\partial t}\right| =?$, 并判断是否为平滑曲面

(1)假设曲面的参数化方程式为$\vec{r}(s,t) = (kt\cos s, kt\sin s, t)$,

(2)假设曲面的参数化方程式为$\vec{r}(s,t) = (a\cos s\cos t, a\sin s\cos t, a\sin t)$

【解】

(1)

$\because \dfrac{\partial \vec{r}}{\partial s} = (-kt\sin s, kt\cos s, 0)$, $\dfrac{\partial \vec{r}}{\partial t} = (k\cos s, k\sin s, 1)$

$\therefore \dfrac{\partial \vec{r}}{\partial s} \times \dfrac{\partial \vec{r}}{\partial t} = (kt\cos s, kt\sin s, -k^2 t)$ 且 $\left|\dfrac{\partial \vec{r}}{\partial s} \times \dfrac{\partial \vec{r}}{\partial t}\right| = kt\sqrt{1 + k^2}$

$\because \dfrac{\partial \vec{r}}{\partial s} \times \dfrac{\partial \vec{r}}{\partial t} = (0,0,0)$, as $t = 0$ $\quad \therefore$ 曲面不是平滑曲面

(2)

$\because \dfrac{\partial \vec{r}}{\partial s} \times \dfrac{\partial \vec{r}}{\partial t} = a\cos t\,(a\cos s\cos t, a\sin s\cos t, a\sin t) = a\cos t\,\vec{r}(s,t)$ and

$\left|\dfrac{\partial \vec{r}}{\partial s} \times \dfrac{\partial \vec{r}}{\partial t}\right| = a\cos^2 t$

$\because \dfrac{\partial \vec{r}}{\partial s} \times \dfrac{\partial \vec{r}}{\partial t} = (0,0,0)$, as $t = \dfrac{\pi}{2}$ $\quad \therefore$ 曲面不是平滑曲面

Example 7.

曲面的参数化方程式为$\vec{r}(x,\theta) = (x, x^2\cos\theta, x^2\sin\theta)$, 求(1)$\dfrac{\partial\vec{r}}{\partial x}\times\dfrac{\partial\vec{r}}{\partial\theta}$=? 以及 $\left|\dfrac{\partial\vec{r}}{\partial x}\times\dfrac{\partial\vec{r}}{\partial\theta}\right|$=?

(2)判断是否为平滑曲面

【解】

(1)

$\because \dfrac{\partial\vec{r}}{\partial x} = (1, 2x\cos\theta, 2x\sin\theta), \quad \dfrac{\partial\vec{r}}{\partial\theta} = (0, -x^2\sin\theta, x^2\cos\theta)$

$\therefore \dfrac{\partial\vec{r}}{\partial x}\times\dfrac{\partial\vec{r}}{\partial\theta} = (2x^3, -x^2\sin\theta, -x^2\sin\theta)$ 且 $\left|\dfrac{\partial\vec{r}}{\partial x}\times\dfrac{\partial\vec{r}}{\partial\theta}\right| = x^2\sqrt{4x^2+1}$

(2)

$\because \dfrac{\partial\vec{r}}{\partial x}\times\dfrac{\partial\vec{r}}{\partial\theta} = (0,0,0), \quad \text{as } x = 0 \quad \therefore$ 曲面不是平滑曲面

Example 8.

求 $\dfrac{\partial\vec{r}}{\partial x}\times\dfrac{\partial\vec{r}}{\partial y}$=? 以及 $\left|\dfrac{\partial\vec{r}}{\partial x}\times\dfrac{\partial\vec{r}}{\partial y}\right|$=? 并判断是否为平滑曲面

(1)假设曲面为空间平面$Ax + By + Cz = D$

(2)假设空间曲面为 $z = \dfrac{2}{3}y^{\frac{3}{2}}$,

(3)空间曲面为 $z = x + ay + b$

(4)假设空间曲面为 $z = xy$

(5)假设空间曲面为 $z = y^2 + bx + c$

(6)假设空间曲面为 $z = x^2 + y^2$

(7)假设曲面为 $z = 3 - y^2$

(8)假设曲面为 $z = x^3$

【解】

(1)

令 $\vec{r}(x,y) = \left(x, y, \dfrac{D - (Ax + By)}{C}\right)$

$$\because \frac{\partial \vec{r}}{\partial x} = \left(1, 0, -\frac{A}{C}\right), \quad \frac{\partial \vec{r}}{\partial y} = \left(0, 1, -\frac{B}{C}\right)$$

$$\therefore \frac{\partial \vec{r}}{\partial x} \times \frac{\partial \vec{r}}{\partial y} = \left(\frac{A}{C}, \frac{B}{C}, 1\right) \Rightarrow \left|\frac{\partial \vec{r}}{\partial x} \times \frac{\partial \vec{r}}{\partial y}\right| = \sqrt{1 + \frac{A^2 + B^2}{C^2}} = \sqrt{\frac{A^2 + B^2 + C^2}{C^2}}$$

$$\because \frac{\partial \vec{r}}{\partial x} \times \frac{\partial \vec{r}}{\partial y} \neq (0, 0, 0), \ \forall x, y \in R \quad \therefore 曲面是平滑曲面$$

(2)

$$令 \vec{r}(x, y) = \left(x, y, \frac{2}{3} y^{\frac{3}{2}}\right)$$

$$\because \frac{\partial \vec{r}}{\partial x} = (1, 0, 0), \quad \frac{\partial \vec{r}}{\partial y} = \left(0, 1, y^{\frac{1}{2}}\right) \quad \therefore \frac{\partial \vec{r}}{\partial x} \times \frac{\partial \vec{r}}{\partial y} = \left(0, -y^{\frac{1}{2}}, 1\right) \Rightarrow \left|\frac{\partial \vec{r}}{\partial x} \times \frac{\partial \vec{r}}{\partial y}\right| = (y + 1)^{\frac{1}{2}}$$

$$\because \frac{\partial \vec{r}}{\partial x} \times \frac{\partial \vec{r}}{\partial y} \neq (0, 0, 0), \ \forall x, y \in R \quad \therefore 曲面是平滑曲面$$

(3)

$$令 \ \vec{r}(x, y) = (x, y, x + ay + b)$$

$$\because \frac{\partial \vec{r}}{\partial x} = (1, 0, 1), \quad \frac{\partial \vec{r}}{\partial y} = (0, 1, a) \quad \therefore \frac{\partial \vec{r}}{\partial x} \times \frac{\partial \vec{r}}{\partial y} = (-1, -a, 1) \Rightarrow \left|\frac{\partial \vec{r}}{\partial x} \times \frac{\partial \vec{r}}{\partial y}\right| = \sqrt{a^2 + 2}$$

$$\because \frac{\partial \vec{r}}{\partial x} \times \frac{\partial \vec{r}}{\partial y} \neq (0, 0, 0), \ \forall x, y \in R \quad \therefore 曲面是平滑曲面$$

(4)

$$令 \ \vec{r}(x, y) = (x, y, xy)$$

$$\because \frac{\partial \vec{r}}{\partial x} = (1, 0, y), \quad \frac{\partial \vec{r}}{\partial y} = (0, 1, x) \quad \therefore \frac{\partial \vec{r}}{\partial x} \times \frac{\partial \vec{r}}{\partial y} = (-y, -x, 1) \Rightarrow \left|\frac{\partial \vec{r}}{\partial x} \times \frac{\partial \vec{r}}{\partial y}\right| = (x^2 + y^2 + 1)^{\frac{1}{2}}$$

$$\because \frac{\partial \vec{r}}{\partial x} \times \frac{\partial \vec{r}}{\partial y} \neq (0, 0, 0), \ \forall x, y \in R \quad \therefore 曲面是平滑曲面$$

(5)

令 $\vec{r}(x,y) = (x, y, y^2 + bx + c)$

$\because \dfrac{\partial \vec{r}}{\partial x} = (1,0,b), \quad \dfrac{\partial \vec{r}}{\partial y} = (0,1,2y) \quad \therefore \dfrac{\partial \vec{r}}{\partial x} \times \dfrac{\partial \vec{r}}{\partial y} = (-b, -2y, 1) \Rightarrow \left| \dfrac{\partial \vec{r}}{\partial x} \times \dfrac{\partial \vec{r}}{\partial y} \right| = (b^2 + 1 + 4y^2)^{\frac{1}{2}}$

$\because \dfrac{\partial \vec{r}}{\partial x} \times \dfrac{\partial \vec{r}}{\partial y} \neq (0,0,0), \ \forall x, y \in R \quad \therefore$ 曲面是平滑曲面

(6)

令 $\vec{r}(x,y) = (x, y, x^2 + y^2) \quad \because \dfrac{\partial \vec{r}}{\partial x} = (1,0,2x), \quad \dfrac{\partial \vec{r}}{\partial y} = (0,1,2y)$

$\therefore \dfrac{\partial \vec{r}}{\partial x} \times \dfrac{\partial \vec{r}}{\partial y} = (-2x, -2y, 1) \Rightarrow \left| \dfrac{\partial \vec{r}}{\partial x} \times \dfrac{\partial \vec{r}}{\partial y} \right| = (4x^2 + 4y^2 + 1)^{\frac{1}{2}}$

$\because \dfrac{\partial \vec{r}}{\partial x} \times \dfrac{\partial \vec{r}}{\partial y} \neq (0,0,0), \ \forall x, y \in R \quad \therefore$ 曲面是平滑曲面

(7)

令 $\vec{r}(x,y) = (x, y, 3 - y^2)$

$\because \dfrac{\partial \vec{r}}{\partial x} = (1,0,0), \quad \dfrac{\partial \vec{r}}{\partial y} = (0,1,-2y) \quad \therefore \dfrac{\partial \vec{r}}{\partial x} \times \dfrac{\partial \vec{r}}{\partial y} = (0, 2y, 1) \Rightarrow \left| \dfrac{\partial \vec{r}}{\partial x} \times \dfrac{\partial \vec{r}}{\partial y} \right| = \sqrt{1 + 4y^2}$

$\because \dfrac{\partial \vec{r}}{\partial x} \times \dfrac{\partial \vec{r}}{\partial y} \neq (0,0,0), \ \forall x, y \in R \quad \therefore$ 曲面是平滑曲面

(8)

令 $\vec{r}(x,y) = (x, y, x^3)$

$\because \dfrac{\partial \vec{r}}{\partial x} = (1,0,3x^2), \quad \dfrac{\partial \vec{r}}{\partial y} = (0,1,0) \quad \therefore \dfrac{\partial \vec{r}}{\partial x} \times \dfrac{\partial \vec{r}}{\partial y} = (-3x^2, 0, -1)$

$\left| \dfrac{\partial \vec{r}}{\partial x} \times \dfrac{\partial \vec{r}}{\partial y} \right| = \sqrt{1 + 9x^4}$

$\because \dfrac{\partial \vec{r}}{\partial x} \times \dfrac{\partial \vec{r}}{\partial y} \neq (0,0,0), \ \forall x, z \in R \quad \therefore$ 曲面是平滑曲面

Example 9.

假设球面$S = \{(x, y, z): x^2 + y^2 + z^2 = r^2, z \geq 0\}$，$\vec{r}(\theta, \varphi)$为曲面参数化方程式，

求$(1)\dfrac{\partial \vec{r}}{\partial \varphi} \times \dfrac{\partial \vec{r}}{\partial \theta} = ?$ 以及 $\left|\dfrac{\partial \vec{r}}{\partial \varphi} \times \dfrac{\partial \vec{r}}{\partial \theta}\right| = ?$

【解】

令$\vec{r}(\theta, \varphi) = (r \cos\theta \sin\varphi, r \sin\theta \sin\varphi, r\cos\varphi)$

$\because \dfrac{\partial \vec{r}}{\partial \theta} = (-r \sin\theta \sin\varphi, r \cos\theta \sin\varphi, 0)$ 且 $\dfrac{\partial \vec{r}}{\partial \varphi} = (r \cos\theta \cos\varphi, r \sin\theta \cos\varphi, -r\sin\varphi)$

$\therefore \dfrac{\partial \vec{r}}{\partial \varphi} \times \dfrac{\partial \vec{r}}{\partial \theta} = (r^2 \cos\theta \sin^2\varphi, r^2 \sin\theta \sin^2\varphi, r^2 \sin\varphi \cos\varphi) \Rightarrow \left|\dfrac{\partial \vec{r}}{\partial \varphi} \times \dfrac{\partial \vec{r}}{\partial \theta}\right| = r^2 \sin\varphi$

Example 10.

求$\dfrac{\partial \vec{r}}{\partial y} \times \dfrac{\partial \vec{r}}{\partial z} = ?$ 以及 $\left|\dfrac{\partial \vec{r}}{\partial y} \times \dfrac{\partial \vec{r}}{\partial z}\right| = ?$ 并判断是否为平滑曲面

(1)曲面为 $x = \sqrt{z^2 - y^2}$　　(2)曲面为 $x - y^2 - z = 2$

【解】

(1)

令$\vec{r}(x, y) = \left(\sqrt{z^2 - y^2}, y, z\right)$

$\because \dfrac{\partial \vec{r}}{\partial y} = \left(-y(z^2 - y^2)^{-\frac{1}{2}}, 1, 0\right), \quad \dfrac{\partial \vec{r}}{\partial z} = \left(z(z^2 - y^2)^{-\frac{1}{2}}, 0, 1\right),$

$\therefore \dfrac{\partial \vec{r}}{\partial y} \times \dfrac{\partial \vec{r}}{\partial z} = \left(1, y(z^2 - y^2)^{-\frac{1}{2}}, -z(z^2 - y^2)^{-\frac{1}{2}}\right)$ 且

$\left|\dfrac{\partial \vec{r}}{\partial y} \times \dfrac{\partial \vec{r}}{\partial z}\right| = \sqrt{1 + (z^2 + y^2)(z^2 - y^2)^{-1}}$

$\because \dfrac{\partial \vec{r}}{\partial y} \times \dfrac{\partial \vec{r}}{\partial z} \neq (0, 0, 0), \ \forall x, y \in R \quad \therefore$ 曲面是平滑曲面

(2)

令 $\vec{r}(y,z) = (y^2 + z + 2, y, z)$

$\because \dfrac{\partial \vec{r}}{\partial y} = (2y, 1, 0),\quad \dfrac{\partial \vec{r}}{\partial z} = (1,0,1)\quad \therefore \dfrac{\partial \vec{r}}{\partial y} \times \dfrac{\partial \vec{r}}{\partial z} = (1, -2y, -1) \Rightarrow \left|\dfrac{\partial \vec{r}}{\partial y} \times \dfrac{\partial \vec{r}}{\partial z}\right| = \sqrt{2 + 4y^2}$

$\because \dfrac{\partial \vec{r}}{\partial y} \times \dfrac{\partial \vec{r}}{\partial z} \neq (0,0,0),\ \forall x, y \in R\quad \therefore$ 曲面是平滑曲面

Example 11.

假设曲面为 $y = x^2$, 求 (1) $\dfrac{\partial \vec{r}}{\partial x} \times \dfrac{\partial \vec{r}}{\partial z} =?$ 以及 $\left|\dfrac{\partial \vec{r}}{\partial x} \times \dfrac{\partial \vec{r}}{\partial z}\right| =?$ (2) 判断是否为平滑曲面

【解】

令 $\vec{r}(x,z) = (x, x^2, z)$

$\because \dfrac{\partial \vec{r}}{\partial x} = (1, 2x, 0),\quad \dfrac{\partial \vec{r}}{\partial z} = (0,0,1)\quad \therefore \dfrac{\partial \vec{r}}{\partial x} \times \dfrac{\partial \vec{r}}{\partial z} = (2x, -1, 0)$ 且 $\left|\dfrac{\partial \vec{r}}{\partial x} \times \dfrac{\partial \vec{r}}{\partial z}\right| = \sqrt{4x^2 + 1}$

(2)

$\because \dfrac{\partial \vec{r}}{\partial x} \times \dfrac{\partial \vec{r}}{\partial z} \neq (0,0,0),\ \forall x, z \in R\quad \therefore$ 曲面是平滑曲面

## 9.5.2　給曲面S求封闭区域的表面积

給定一个参数曲面 $S$: $\vec{r}(s,t) = \big(x(s,t), y(s,t), z(s,t)\big)$, 定义域为一个矩形参数区域$D$, 将定义域$D$分割为数个小矩形 $D_{ij}$, $D_{ij}$的中心点表示为$\big(s_i^*, t_j^*\big)$, 假设$D_{ij}$对应曲面 $S$ 上的区域为$S_{ij}$且$\big(s_i^*, t_j^*\big)$对应曲面$S_{ij}$ 上的点为 $E_{ij}$(如下图), 其位置向量为$\vec{r}\big(s_i^*, t_j^*\big)$; 由于向量

$$\vec{r}_s\big(s_i^*, t_j^*\big) \times \vec{r}_t\big(s_i^*, t_j^*\big)$$

是垂直于曲面在点$\vec{r}\big(s_i^*, t_j^*\big)$的向量，可以藉由线性函数对此$S_{ij}$的切平面进行参数化

$$令\, f(s,t) = s\,\vec{r}_s\big(s_i^*, t_j^*\big) + t\vec{r}_t\big(s_i^*, t_j^*\big) + \big(\vec{r}\big(s_i^*, t_j^*\big) - s_i^*\vec{r}_s\big(s_i^*, t_j^*\big) - t_j^*\vec{r}_t\big(s_i^*, t_j^*\big)\big)$$

因为 $S_{ij}$可以藉由两向量 $f(s_1 + \Delta s, t_1) - f(s_1, t_1)$ 且 $f(s_1, t_1 + \Delta t) - f(s_1, t_1)$所围的平行

四边形作逼近并且

$$f(s_1 + \Delta s, t_1) - f(s_1, t_1) = \Delta s \vec{r}_s(s_i^*, t_j^*) \text{ 且 } f(s_1, t_1 + \Delta t) - f(s_1, t_1) = \Delta t \vec{r}_t(s_i^*, t_j^*)$$

因此，这个平行四边形的面积为

$$\left| \Delta s\, \vec{r}_s(s_i^*, t_j^*) \times \Delta t \vec{r}_t(s_i^*, t_j^*) \right| = \left| \vec{r}_s(s_i^*, t_j^*) \times \vec{r}_t(s_i^*, t_j^*) \right| \Delta s \Delta t$$

所以

$$\Delta S_{ij} \approx \left| \vec{r}_s(s_i^*, t_j^*) \times \vec{r}_t(s_i^*, t_j^*) \right| \Delta s \Delta t$$

令 $\vec{r}_s^* = \vec{r}_s(s_i^*, t_j^*)$ 且 $\vec{r}_t^* = \vec{r}_t(s_i^*, t_j^*)$ 则

$$\text{曲面}S \text{ 的近似面积} \approx \sum_{i=1}^{m} \sum_{j=1}^{n} |\vec{r}_s^* \times \vec{r}_t^*| \Delta s \Delta t$$

将此双重和当作黎曼积分 $\displaystyle\iint_D |\vec{r}_s \times \vec{r}_t| ds dt$ 的黎曼和，视为定义曲面面积的背后動机

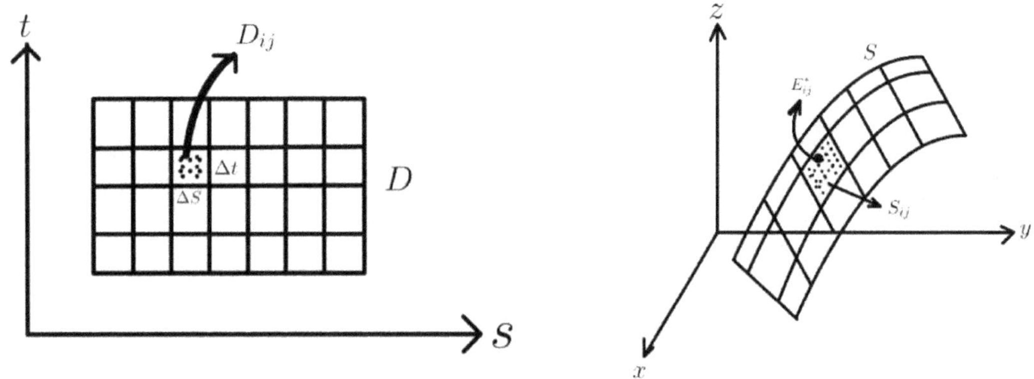

## 【定义】

考虑一个光滑参数曲面$S$: $\vec{r}(s,t) = (x(s,t), y(s,t), z(s,t)), (s,t) \in D$ 并且 $S$ 在整个参数域 $D$ 当中被正好覆盖一次，则

$$S\text{的曲面积} = \iint_S dA = \iint_D |\vec{r}_s \times \vec{r}_t| ds dt \text{ 其中 } \vec{r}_s = \left( \frac{\partial x}{\partial s}, \frac{\partial y}{\partial s}, \frac{\partial z}{\partial s} \right) \text{ 且 } \vec{r}_t = \left( \frac{\partial x}{\partial t}, \frac{\partial y}{\partial t}, \frac{\partial z}{\partial t} \right)$$

由上述观察可知，计算曲面的面积相当于求双重积分，特别的是双重积分的被积分函数为

$$|\vec{r}_s(s,t) \times \vec{r}_t(s,t)|$$

因此，求曲面的面积相当于先求得被积分函数之后，再求双重积分；底下针对曲面面积转换成双重积分后，有哪些不同类型作初步的说明；

假定曲面 $S: z = f(x,y)$，其中 $a \le x \le b, \ c \le y \le d$，目标是计算表面积 $\iint_S dA$，将曲面 $S$ 参数化，$S: \vec{r}(x,y) = (x,y,f(x,y))$，令 $R = \{(x,y): a \le x \le b, c \le y \le d\}$ 则曲面表面积为

$$\iint_S dA = \iint_R \left| \frac{\partial \vec{r}}{\partial x} \times \frac{\partial \vec{r}}{\partial y} \right| dxdy$$

因为 $\dfrac{\partial \vec{r}}{\partial x} = (1,0,f_x(x,y))$ 且 $\dfrac{\partial \vec{r}}{\partial y} = \left(0,1,f_y(x,y)\right)$，两者向量外积为

$$\frac{\partial \vec{r}}{\partial x} \times \frac{\partial \vec{r}}{\partial y} = (-f_x(x,y), -f_y(x,y), 1)$$

且

$$\left| \frac{\partial \vec{r}}{\partial x} \times \frac{\partial \vec{r}}{\partial y} \right| = \sqrt{1 + (f_x(x,y))^2 + \left(f_y(x,y)\right)^2}$$

表面积公式可改写为

$$\iint_S dA = \iint_R \left| \frac{\partial \vec{r}}{\partial x} \times \frac{\partial \vec{r}}{\partial y} \right| dxdy = \int_a^b \int_c^d \sqrt{1 + (f_x(x,y))^2 + \left(f_y(x,y)\right)^2}\, dydx$$

如果曲面为 $z = f(x,y)$ 且投影至 $xy$ 平面 $= \{(x,y): x^2 + y^2 \le a\}$，藉由极坐标转换 $x = r\cos\theta, \ y = r\sin\theta$ 则

$$\iint_S dA = \iint_R \left| \frac{\partial \vec{r}}{\partial x} \times \frac{\partial \vec{r}}{\partial y} \right| dxdy$$
$$= \int_0^{2\pi} \int_0^a \sqrt{1 + (f_x(r\cos\theta, r\sin\theta))^2 + \left(f_y(r\cos\theta, r\sin\theta)\right)^2}\, rdrd\theta$$

假设曲面 $S$ 为一球体，$S = \{(x,y,z): x^2 + y^2 + z^2 = r^2, r > 0\}$，目标是计算 $S$ 的表面积 $\iint_S dA$；曲面 $S$ 的参数方程式为

$$\vec{r}(\theta,\varphi) = (r\cos\theta\sin\varphi, r\sin\theta\sin\varphi, r\cos\varphi)$$

令 $R = \{(\theta,\varphi): 0 \le \theta \le 2\pi, 0 \le \varphi \le \pi\}$ 则 $\iint_S dA = \iint_R \left| \dfrac{\partial \vec{r}}{\partial \varphi} \times \dfrac{\partial \vec{r}}{\partial \theta} \right| d\theta d\varphi$.

因为 $\dfrac{\partial \vec{r}}{\partial \theta} = (-r\sin\theta\sin\varphi, r\cos\theta\sin\varphi, 0)$ 且 $\dfrac{\partial \vec{r}}{\partial \varphi} = (r\cos\theta\cos\varphi, r\sin\theta\cos\varphi, -r\sin\varphi)$，两向量外积为

$$\frac{\partial \vec{r}}{\partial \varphi} \times \frac{\partial \vec{r}}{\partial \theta} = (r^2 \cos\theta \sin^2\varphi, r^2 \sin\theta \sin^2\varphi, r^2 \sin\varphi \cos\varphi)$$

且

$$\left| \frac{\partial \vec{r}}{\partial \varphi} \times \frac{\partial \vec{r}}{\partial \theta} \right| = r^2 \sin\varphi$$

因此,

$$\iint_S dA = \iint_R \left| \frac{\partial \vec{r}}{\partial \varphi} \times \frac{\partial \vec{r}}{\partial \theta} \right| d\theta d\varphi = r^2 \int_0^{2\pi} \int_0^{\pi} \sin\varphi \, d\varphi d\theta = -r^2 \int_0^{2\pi} \cos\varphi |_0^{\pi} \, d\theta = 4r^2\pi$$

考试类型:

Type 1.

給定一参数化曲面 $S$: $\vec{r}(s,t)$, $\forall a \leq s \leq b$, $c \leq t \leq d$, 求表面积 $\iint_S dA =?$

解题流程:

Step1.

$$\because \iint_S dA = \iint_D \left| \frac{\partial \vec{r}}{\partial s} \times \frac{\partial \vec{r}}{\partial t} \right| dsdt$$

Step2.

求 $\dfrac{\partial \vec{r}}{\partial s}$, $\dfrac{\partial \vec{r}}{\partial t}$, $\dfrac{\partial \vec{r}}{\partial s} \times \dfrac{\partial \vec{r}}{\partial t}$ 以及 $\left| \dfrac{\partial \vec{r}}{\partial s} \times \dfrac{\partial \vec{r}}{\partial t} \right|$

Step3.

$$\because \iint_S dA = \iint_R \left| \frac{\partial \vec{r}}{\partial s} \times \frac{\partial \vec{r}}{\partial t} \right| dsdt = \int_a^b \int_c^d \left| \frac{\partial \vec{r}}{\partial s} \times \frac{\partial \vec{r}}{\partial t} \right| dtds, \quad 求 \int_a^b \int_c^d \left| \frac{\partial \vec{r}}{\partial s} \times \frac{\partial \vec{r}}{\partial t} \right| dtds =?$$

Type 2.

給定一参数化曲面 $S$: $z = f(x,y)$, $\forall a \leq x \leq b$, $c \leq y \leq d$, 求表面积 $\iint_S dA =?$

解题流程:

Step1.

令 $\vec{r}(x,y) = (x, y, f(x,y))$ 且 $R = \{(x,y): a \leq x \leq b, c \leq y \leq d\}$

则 $\iint_S dA = \iint_R \left| \dfrac{\partial \vec{r}}{\partial x} \times \dfrac{\partial \vec{r}}{\partial y} \right| dxdy$

Step2.

$$\because \frac{\partial \vec{r}}{\partial x} = (1,0,f_x(x,y)), \quad \frac{\partial \vec{r}}{\partial y} = \left(0,1,f_y(x,y)\right) \quad \therefore \frac{\partial \vec{r}}{\partial x} \times \frac{\partial \vec{r}}{\partial y} = \left(-f_x(x,y), -f_y(x,y), 1\right)$$

$$\Rightarrow \left| \frac{\partial \vec{r}}{\partial x} \times \frac{\partial \vec{r}}{\partial y} \right| = \sqrt{1 + \left(f_x(x,y)\right)^2 + \left(f_y(x,y)\right)^2}$$

Step3.

$$\iint_S dA = \iint_R \left| \frac{\partial \vec{r}}{\partial x} \times \frac{\partial \vec{r}}{\partial y} \right| dxdy = \int_a^b \int_c^d \sqrt{1 + \left(f_x(x,y)\right)^2 + \left(f_y(x,y)\right)^2} \, dydx$$

Type 3.

給定一參數化曲面 $S: z = f(x,y)$，假設曲面 $S$ 于 $xy$ 平面投影 $= \{(x,y): x^2 + y^2 \leq a^2\}$，

求 $\displaystyle\iint_S dA = ?$

解題流程:

Step1.

令 $\vec{r}(x,y) = \left(x, y, f(x,y)\right)$ 且 $R$ 为曲面 $S$ 投影至 $xy$ 平面的封閉区域

則 $R = \{(x,y): x^2 + y^2 \leq a^2\}$ 且 $\displaystyle\iint_S dA = \iint_R \left| \frac{\partial \vec{r}}{\partial x} \times \frac{\partial \vec{r}}{\partial y} \right| dxdy$

Step2.

$$\because \frac{\partial \vec{r}}{\partial x} = (1,0,f_x(x,y)), \quad \frac{\partial \vec{r}}{\partial y} = \left(0,1,f_y(x,y)\right) \quad \therefore \frac{\partial \vec{r}}{\partial x} \times \frac{\partial \vec{r}}{\partial y} = \left(-f_x(x,y), -f_y(x,y), 1\right)$$

$$\Rightarrow \left| \frac{\partial \vec{r}}{\partial x} \times \frac{\partial \vec{r}}{\partial y} \right| = \sqrt{1 + \left(f_x(x,y)\right)^2 + \left(f_y(x,y)\right)^2}$$

Step3.

$$\therefore \iint_S dA = \iint_R \left| \frac{\partial \vec{r}}{\partial x} \times \frac{\partial \vec{r}}{\partial y} \right| dxdy = \iint_R \sqrt{1 + \left(f_x(x,y)\right)^2 + \left(f_y(x,y)\right)^2} \, dxdy$$

Step4.

令 $x = r\cos\theta, y = r\sin\theta$ 則 $R = \{(x,y): x^2 + y^2 \leq a^2\} = \{(r,\theta): 0 \leq r \leq a, 0 \leq \theta \leq 2\pi\}$

$$\iint_R \sqrt{1 + \left(f_x(x,y)\right)^2 + \left(f_y(x,y)\right)^2} \, dxdy$$

$$= \int_0^{2\pi} \int_0^a \sqrt{1 + \left(f_x(r\cos\theta, r\sin\theta)\right)^2 + \left(f_y(r\cos\theta, r\sin\theta)\right)^2} \, r \, dr \, d\theta$$

Type 4.

假设曲面 $S: \{(x,y,z): x^2 + y^2 + z^2 = r^2, r > 0\}$，　求$S$的表面积 $\iint_S dA =?$

解题流程：

Step1.

令 $\vec{r}(\theta,\varphi) = (r\cos\theta\sin\varphi, r\sin\theta\sin\varphi, r\cos\varphi)$且 $R = \{(\theta,\varphi): 0 \le \theta \le 2\pi, 0 \le \varphi \le \pi\}$

则 $\iint_S dA = \iint_R \left|\dfrac{\partial\vec{r}}{\partial\varphi} \times \dfrac{\partial\vec{r}}{\partial\theta}\right| d\theta d\varphi$

Step2.

$\because \dfrac{\partial\vec{r}}{\partial\theta} = (-r\sin\theta\sin\varphi, r\cos\theta\sin\varphi, 0)$ 且 $\dfrac{\partial\vec{r}}{\partial\varphi} = (r\cos\theta\cos\varphi, r\sin\theta\cos\varphi, -r\sin\varphi)$

$\therefore \dfrac{\partial\vec{r}}{\partial\varphi} \times \dfrac{\partial\vec{r}}{\partial\theta} = (r^2\cos\theta\sin^2\varphi, r^2\sin\theta\sin^2\varphi, r^2\sin\varphi\cos\varphi) \Rightarrow \left|\dfrac{\partial\vec{r}}{\partial\varphi} \times \dfrac{\partial\vec{r}}{\partial\theta}\right| = r^2\sin\varphi$

则 $\iint_S dA = \iint_R \left|\dfrac{\partial\vec{r}}{\partial\varphi} \times \dfrac{\partial\vec{r}}{\partial\theta}\right| d\theta d\varphi = r^2 \int_0^{2\pi} \int_0^{\pi} \sin\varphi\, d\varphi d\theta = -r^2 \int_0^{2\pi} \cos\varphi\big|_0^{\pi}\, d\theta = 4r^2\pi$

Type 5.

给定空间中的平面$S: Ax + By + Cz = D$

(1)求在 $0 \le ax + by \le s,\ 0 \le cx + dy \le t,\ \forall t, s > 0$ 所围区域的表面积

(2)求在$y = x^{\alpha},\ y = \beta x^{\alpha},\ x = y^{\alpha},\ x = \gamma y^{\alpha}$，所围区域的表面积，其中 $\alpha \ne 1, \beta, \gamma > 1$

(3)求在 $ax^2 + bxy + cy^2 = \alpha^2$，所围区域的表面积（$\alpha > 0,\ b^2 - 4ac < 0$）

(4)求在 $y = ax, y = bx, xy = c, xy = d$，所围区域的表面积，其中 $a < b, c < d$

解题流程：

令 $\vec{r}(x,y) = \left(x, y, \dfrac{D - (Ax + By)}{C}\right)$　且$R$为投影至$xy$平面区域

则 $\iint_S dA = \iint_R \left|\dfrac{\partial\vec{r}}{\partial x} \times \dfrac{\partial\vec{r}}{\partial y}\right| dxdy$

$\because \dfrac{\partial\vec{r}}{\partial x} = \left(1, 0, -\dfrac{A}{C}\right)$ 且 $\dfrac{\partial\vec{r}}{\partial y} = \left(0, 1, -\dfrac{B}{C}\right)$　$\therefore \dfrac{\partial\vec{r}}{\partial x} \times \dfrac{\partial\vec{r}}{\partial y} = \left(\dfrac{A}{C}, \dfrac{B}{C}, 1\right)$

$$\Rightarrow \left| \frac{\partial \vec{r}}{\partial x} \times \frac{\partial \vec{r}}{\partial y} \right| = \sqrt{1 + \frac{A^2 + B^2}{C^2}} = \sqrt{\frac{A^2 + B^2 + C^2}{C^2}}$$

$$\therefore \iint_S dA = \iint_R \left| \frac{\partial \vec{r}}{\partial x} \times \frac{\partial \vec{r}}{\partial y} \right| dxdy = \sqrt{\frac{A^2 + B^2 + C^2}{C^2}} \iint_R dxdy$$

(1)

Step1.

令 $u = ax + by$, $v = cx + dy$ 則 $x = \dfrac{du - bv}{ad - bc}$ 且 $y = \dfrac{av - cu}{ad - bc}$

$$\because dxdy = \left\| \begin{matrix} \frac{\partial x}{\partial u} & \frac{\partial x}{\partial v} \\ \frac{\partial y}{\partial u} & \frac{\partial y}{\partial v} \end{matrix} \right\| dudv = \frac{\left| \begin{matrix} d & -b \\ -c & a \end{matrix} \right|}{|ad - bc|^2} dudv = \frac{1}{|ad - bc|} dudv$$

$$\therefore \iint_R dxdy = \int_0^t \int_0^s \frac{1}{|ad - bc|} dudv$$

Step2.

計算此積分 $\displaystyle\int_0^t \int_0^s \frac{1}{|ad - bc|} dudv = ?$

$$\therefore \iint_R dxdy = \int_0^t \int_0^s \frac{1}{|ad - bc|} dudv = \frac{ts}{|ad - bc|}$$

$$\therefore \iint_S dA = \sqrt{\frac{A^2 + B^2 + C^2}{C^2}} \iint_R dxdy = \frac{ts}{|ad - bc|} \sqrt{\frac{A^2 + B^2 + C^2}{C^2}}$$

(2)

Step1.

令 $u = \dfrac{y}{x^\alpha}$, $v = \dfrac{x}{y^\alpha}$ 則 $x = u^{\frac{\alpha}{1-\alpha^2}} v^{\frac{1}{1-\alpha^2}}$, $y = u^{\frac{1}{1-\alpha^2}} v^{\frac{\alpha}{1-\alpha^2}}$

$$\because dxdy = \left\| \begin{matrix} \frac{\partial x}{\partial u} & \frac{\partial x}{\partial v} \\ \frac{\partial y}{\partial u} & \frac{\partial y}{\partial v} \end{matrix} \right\| dudv = \left\| \begin{matrix} \dfrac{\alpha u^{\left(\frac{\alpha}{1-\alpha^2}-1\right)} v^{\frac{1}{1-\alpha^2}}}{1 - \alpha^2} & \dfrac{u^{\frac{\alpha}{1-\alpha^2}} v^{\left(\frac{1}{1-\alpha^2}-1\right)}}{1 - \alpha^2} \\ \dfrac{u^{\left(\frac{1}{1-\alpha^2}-1\right)} v^{\frac{\alpha}{1-\alpha^2}}}{1 - \alpha^2} & \dfrac{\alpha u^{\frac{1}{1-\alpha^2}} v^{\left(\frac{\alpha}{1-\alpha^2}-1\right)}}{1 - \alpha^2} \end{matrix} \right\| dudv$$

$$= \frac{\alpha^2 u^{\frac{\alpha-1+\alpha^2+1}{1-\alpha^2}} v^{\frac{1+\alpha-1+\alpha^2}{1-\alpha^2}} - u^{\frac{\alpha-1+\alpha^2+1}{1-\alpha^2}} v^{\frac{1+\alpha-1+\alpha^2}{1-\alpha^2}}}{(1-\alpha^2)^2} \, dudv$$

$$= \frac{\alpha^2 u^{\frac{\alpha}{1-\alpha}} v^{\frac{\alpha}{1-\alpha}} - u^{\frac{\alpha}{1-\alpha}} v^{\frac{\alpha}{1-\alpha}}}{(1-\alpha^2)^2} \, dudv = \frac{u^{\frac{\alpha}{1-\alpha}} v^{\frac{\alpha}{1-\alpha}}}{\alpha^2-1} \, dudv$$

Step2.

令 $R = \left\{(x,y): 1 \le \dfrac{y}{x^\alpha} \le \beta, 1 \le \dfrac{x}{y^\alpha} \le \gamma\right\}$ 則 $R = \{(u,v): 1 \le u \le \beta, 1 \le v \le \gamma\}$

Step3.

$$面积 = \iint_R dxdy = \int_1^\beta \int_1^\gamma \frac{u^{\frac{\alpha}{1-\alpha}} v^{\frac{\alpha}{1-\alpha}}}{\alpha^2-1} \, dvdu$$

$$\therefore \iint_S dA = \sqrt{\frac{A^2+B^2+C^2}{C^2}} \iint_R dxdy = \sqrt{\frac{A^2+B^2+C^2}{C^2}} \int_1^\beta \int_1^\gamma \frac{u^{\frac{\alpha}{1-\alpha}} v^{\frac{\alpha}{1-\alpha}}}{\alpha^2-1} \, dvdu$$

求 $\displaystyle\int_1^\beta \int_1^\gamma \frac{u^{\frac{\alpha}{1-\alpha}} v^{\frac{\alpha}{1-\alpha}}}{\alpha^2-1} \, dvdu$

(3)

Step1.

$$\because ax^2 + bxy + cy^2 = a\left(x + \frac{by}{2a}\right)^2 + \frac{4ac-b^2}{4a} y^2$$

令 $\sqrt{a}\left(x + \dfrac{by}{2a}\right) = u, \quad \sqrt{\dfrac{4ac-b^2}{4a}}\, y = v$

$$則 \ dxdy = \left\| \begin{matrix} \dfrac{\partial x}{\partial u} & \dfrac{\partial x}{\partial v} \\ \dfrac{\partial y}{\partial u} & \dfrac{\partial y}{\partial v} \end{matrix} \right\| dudv = \left\| \begin{matrix} \dfrac{1}{\sqrt{a}} & 0 \\ \dfrac{2\sqrt{a}}{b} & \sqrt{\dfrac{4a}{4ac-b^2}} \end{matrix} \right\| dudv = \frac{2}{\sqrt{4ac-b^2}} \, dudv$$

$$\therefore \iint_R dxdy = \iint_{u^2+v^2 \le a^2} \frac{2}{\sqrt{4ac-b^2}} \, dudv$$

Step2.

令 $u = r\cos\theta, v = r\sin\theta$ 則 $\{(u,v): u^2 + v^2 \leq \alpha^2\} = \{(r,\theta): 0 \leq r \leq \alpha, 0 \leq \theta \leq 2\pi\}$

且 $dudv = \left\|\begin{vmatrix} \dfrac{\partial u}{\partial r} & \dfrac{\partial u}{\partial \theta} \\ \dfrac{\partial v}{\partial r} & \dfrac{\partial v}{\partial \theta} \end{vmatrix}\right\| drd\theta = \left\|\begin{vmatrix} \cos\theta & -r\sin\theta \\ \sin\theta & r\cos\theta \end{vmatrix}\right\| drd\theta = rdrd\theta$

Step3.

$$\iint_{u^2+v^2\leq\alpha^2} \frac{2}{\sqrt{4ac-b^2}}\,dudv = \int_0^{2\pi}\int_0^{\alpha} \frac{2}{\sqrt{4ac-b^2}}\,rdrd\theta = \frac{2}{\sqrt{4ac-b^2}}\int_0^{2\pi} d\theta \int_0^{\alpha} rdr$$

$$= \frac{2\pi\alpha^2}{\sqrt{4ac-b^2}}$$

$$\therefore \iint_S dA = \frac{2\pi\alpha^2}{\sqrt{4ac-b^2}}\sqrt{\frac{A^2+B^2+C^2}{C^2}}$$

(4)

Step1.

令 $\dfrac{y}{x} = u, \quad xy = v, \quad$ 則 $x = \sqrt{\dfrac{v}{u}}, \quad y = \sqrt{uv}$

$$\because dxdy = \left\|\begin{vmatrix} \dfrac{\partial x}{\partial u} & \dfrac{\partial x}{\partial v} \\ \dfrac{\partial y}{\partial u} & \dfrac{\partial y}{\partial v} \end{vmatrix}\right\| dudv = \left\|\begin{vmatrix} \dfrac{1}{2}\left(\dfrac{v}{u}\right)^{-\frac{1}{2}}\left(-\dfrac{v}{u^2}\right) & \dfrac{1}{2}\left(\dfrac{v}{u}\right)^{-\frac{1}{2}}\left(\dfrac{1}{u}\right) \\ \dfrac{1}{2}(uv)^{-\frac{1}{2}}v & \dfrac{1}{2}(uv)^{-\frac{1}{2}}u \end{vmatrix}\right\| dudv = \frac{1}{2u}\,dudv$$

Step2.

令 $R = \left\{(x,y): a \leq \dfrac{y}{x} \leq b, c \leq xy \leq d\right\}$ 則 $R = \{(u,v): a \leq u \leq b, c \leq v \leq d\}$

Step3.

$$\because \iint_R dxdy = \int_a^b \int_c^d \frac{1}{2u}\,dvdu = \frac{(d-c)\ln\dfrac{b}{a}}{2}$$

$$\therefore \iint_S dA = \sqrt{\frac{A^2+B^2+C^2}{C^2}}\iint_R dxdy = \sqrt{\frac{A^2+B^2+C^2}{C^2}}\int_a^b \int_c^d \frac{1}{2u}\,dvdu$$

$$= \frac{(d-c)\ln\dfrac{b}{a}\sqrt{\dfrac{A^2+B^2+C^2}{C^2}}}{2}$$

Example 1.

　　假设曲面$S$的参数式为$\vec{r}(s,t) = (kt\cos s, kt\sin s, t), 0 \le s \le 2\pi,\ 0 \le t \le h$, 求曲面$S$
的表面积

【解】

$$\because \frac{\partial \vec{r}}{\partial s} = (-kt\sin s, kt\cos s, 0)\ \text{且}\ \frac{\partial \vec{r}}{\partial t} = (k\cos s, k\sin s, 1)$$

$$\therefore \frac{\partial \vec{r}}{\partial s} \times \frac{\partial \vec{r}}{\partial t} = (kt\cos s, kt\sin s, -k^2 t)\ \text{且}\ \left|\frac{\partial \vec{r}}{\partial s} \times \frac{\partial \vec{r}}{\partial t}\right| = kt\sqrt{1+k^2}$$

$$\therefore 表面积 = \iint_S dA = \int_0^h \int_0^{2\pi} kt\sqrt{1+k^2}\,dsdt = \pi k h^2 \sqrt{1+k^2}$$

Example 2.

　　假设曲面$S$的参数式为$\vec{r}(s,t) = (s^2, s^3, 2t), \forall 0 \le s \le 1, 0 \le t \le 2$, 求曲面表面积

【解】

$$\because \frac{\partial \vec{r}}{\partial s} = (2s, 3s^2, 0)\ \text{且}\ \frac{\partial \vec{r}}{\partial t} = (0,0,2)$$

$$\therefore \frac{\partial \vec{r}}{\partial s} \times \frac{\partial \vec{r}}{\partial t} = (6s^2, -4s, 0)\ \text{且}\ \left|\frac{\partial \vec{r}}{\partial s} \times \frac{\partial \vec{r}}{\partial t}\right| = \sqrt{16s^2 + 36s^4} = 4s\sqrt{1+\frac{9s^2}{4}}$$

$$\therefore 表面积 = \iint_S dA = \int_0^2 \int_0^1 4s\sqrt{1+\frac{9s^2}{4}}\,dsdt = 8\int_0^1 s\sqrt{1+\frac{9s^2}{4}}\,ds = \frac{32}{27}\left(1+\frac{9s^2}{4}\right)^{\frac{3}{2}}\Bigg|_0^1$$

$$= \frac{32}{27}\left(\frac{13\sqrt{13}}{8} - 1\right)$$

Example 3.

　　求平面$acx + bcy + abz = abc$ 于第一卦限所围面积

【解】

$$令\ \vec{r}(x,y) = \left(x, y, \frac{abc - (acx + bcy)}{ab}\right) 且 R 为平面投影至 xy 平面的封闭区域$$

則 $R = \{(x,y): acx + bcy \le abc, x \ge 0, y \ge 0\}$ 且表面积 $= \iint_S dA = \iint_R \left|\dfrac{\partial \vec{r}}{\partial x} \times \dfrac{\partial \vec{r}}{\partial y}\right| dxdy$

$\because \dfrac{\partial \vec{r}}{\partial x} = \left(1, 0, -\dfrac{c}{b}\right)$ 且 $\dfrac{\partial \vec{r}}{\partial y} = \left(0, 1, -\dfrac{b}{a}\right)$ $\therefore \dfrac{\partial \vec{r}}{\partial x} \times \dfrac{\partial \vec{r}}{\partial y} = \left(\dfrac{c}{b}, \dfrac{c}{a}, 1\right)$

$\Rightarrow \left|\dfrac{\partial \vec{r}}{\partial x} \times \dfrac{\partial \vec{r}}{\partial y}\right| = \sqrt{1 + \dfrac{a^2c^2 + b^2c^2}{a^2b^2}} = \sqrt{\dfrac{a^2c^2 + b^2c^2 + a^2b^2}{a^2b^2}}$

$\therefore$ 表面积 $= \iint_S dA = \iint_R \left|\dfrac{\partial \vec{r}}{\partial x} \times \dfrac{\partial \vec{r}}{\partial y}\right| dxdy = \sqrt{\dfrac{a^2c^2 + b^2c^2 + a^2b^2}{a^2b^2}} \iint_R dxdy$

$= \sqrt{\dfrac{a^2c^2 + b^2c^2 + a^2b^2}{a^2b^2}} \times \dfrac{ab}{2} = \dfrac{\sqrt{a^2c^2 + b^2c^2 + a^2b^2}}{2}$

Example 4.

假设曲面的参数式为 $\vec{r}(s,t) = (a\cos s \cos t, a \sin s \cos t, a \sin t), 0 \le s \le 2\pi,$

$-\dfrac{\pi}{2} \le t \le \dfrac{\pi}{2}$，求曲面表面积

【解】

$\because \dfrac{\partial \vec{r}}{\partial s} = (-a \sin s \cos t, a \cos s \cos t, 0)$ 且 $\dfrac{\partial \vec{r}}{\partial t} = (-a \cos s \sin t, -a \sin s \sin t, a \cos t)$

$\because \dfrac{\partial \vec{r}}{\partial s} \times \dfrac{\partial \vec{r}}{\partial t} = a\cos t\,(a\cos s \cos t, a \sin s \cos t, a \sin t) = a\cos t\,\vec{r}(s,t)$ 且

$\left|\dfrac{\partial \vec{r}}{\partial s} \times \dfrac{\partial \vec{r}}{\partial t}\right| = a\cos^2 t$

$\therefore \iint_S dA = \int_0^{2\pi} \int_{-\frac{\pi}{2}}^{\frac{\pi}{2}} \left|\dfrac{\partial \vec{r}}{\partial s} \times \dfrac{\partial \vec{r}}{\partial t}\right| dtds = \int_0^{2\pi} \int_{-\frac{\pi}{2}}^{\frac{\pi}{2}} a\cos^2 t\, dtds = 2\pi a^2 \int_{-\frac{\pi}{2}}^{\frac{\pi}{2}} a\cos^2 t\, dt = 4\pi a^2$

Example 5.

Find the area of the surface, $z = a^2 - (x^2 + y^2)$ with $\dfrac{a^2}{4} \le x^2 + y^2 \le a^2$

【解】

令 $S$: $z = a^2 - (x^2 + y^2)$, $\dfrac{a^2}{4} \leq x^2 + y^2 \leq a^2$

令 $\vec{r}(x, y) = \left(x, y, a^2 - (x^2 + y^2)\right)$ 且 $R$ 为曲面$S$投影至$xy$平面的封闭区域

则 $R = \left\{(x, y): \dfrac{a^2}{4} \leq x^2 + y^2 \leq a^2\right\}$ 且 表面积 $= \iint_S dA = \iint_R \left|\dfrac{\partial \vec{r}}{\partial x} \times \dfrac{\partial \vec{r}}{\partial y}\right| dxdy$

$\because \dfrac{\partial \vec{r}}{\partial x} = (1, 0, -2x)$ 且 $\dfrac{\partial \vec{r}}{\partial y} = (0, 1, -2y)$ $\quad \therefore \dfrac{\partial \vec{r}}{\partial x} \times \dfrac{\partial \vec{r}}{\partial y} = (2x, 2y, 1)$

$\Rightarrow \left|\dfrac{\partial \vec{r}}{\partial x} \times \dfrac{\partial \vec{r}}{\partial y}\right| = (4x^2 + 4y^2 + 1)^{\frac{1}{2}}$

$\therefore \iint_S dA = \iint_R \left|\dfrac{\partial \vec{r}}{\partial x} \times \dfrac{\partial \vec{r}}{\partial y}\right| dxdy = \iint_{\frac{a^2}{4} \leq x^2 + y^2 \leq a^2} (4x^2 + 4y^2 + 1)^{\frac{1}{2}} dxdy$

令 $x = r\cos\theta$, $y = r\sin\theta$ 则 $\left\{(x, y): \dfrac{a^2}{4} \leq x^2 + y^2 \leq a^2\right\} = \left\{(r, \theta): \dfrac{a}{2} \leq r \leq a, 0 \leq \theta \leq 2\pi\right\}$

$\therefore \iint_S dA = \iint_{\frac{a^2}{4} \leq x^2 + y^2 \leq a^2} (4x^2 + 4y^2 + 1)^{\frac{1}{2}} dxdy = \int_0^{2\pi} \int_{\frac{a}{2}}^{a} (4r^2 + 1)^{\frac{1}{2}} r\, dr\, d\theta$

$= \dfrac{\pi}{6}(4r^2 + 1)^{\frac{3}{2}} \Big|_{\frac{a}{2}}^{a} = \dfrac{\pi}{6}\left((4a^2 + 1)^{\frac{3}{2}} - (a^2 + 1)^{\frac{3}{2}}\right)$

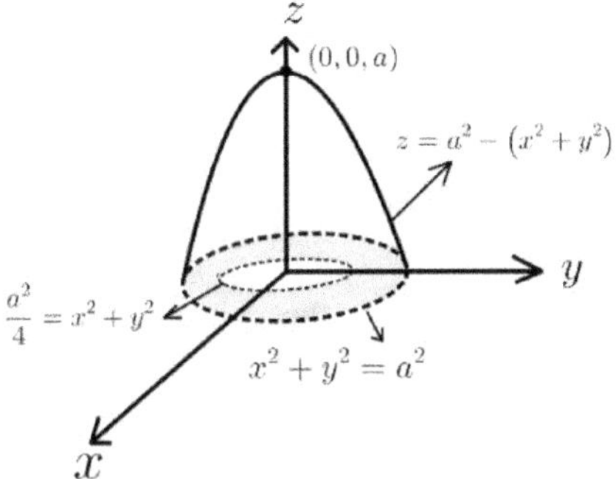


## Example 6.

Find the area of the surface $z = \dfrac{2}{3}\left(x^{\frac{3}{2}} + y^{\frac{3}{2}}\right)$ with $0 \le x \le 1, 0 \le y \le 1$

【解】

令 $S$: $z = \dfrac{2}{3}\left(x^{\frac{3}{2}} + y^{\frac{3}{2}}\right)$, $\ 0 \le x \le 1, 0 \le y \le 1$

令 $\vec{r}(x,y) = \left(x, y, \dfrac{2}{3}\left(x^{\frac{3}{2}} + y^{\frac{3}{2}}\right)\right)$ 且 $R$ 为曲面 $S$ 投影至 $xy$ 平面的封闭区域

则 $R = \{(x,y): 0 \le x \le 1, 0 \le y \le 1\}$ 且 $\displaystyle\iint_S dA = \iint_R \left|\dfrac{\partial \vec{r}}{\partial x} \times \dfrac{\partial \vec{r}}{\partial y}\right| dx\,dy$

$\because \dfrac{\partial \vec{r}}{\partial x} = \left(1, 0, x^{\frac{1}{2}}\right)$ 且 $\dfrac{\partial \vec{r}}{\partial y} = \left(0, 1, y^{\frac{1}{2}}\right)$ $\quad \therefore \dfrac{\partial \vec{r}}{\partial x} \times \dfrac{\partial \vec{r}}{\partial y} = \left(x^{\frac{1}{2}}, y^{\frac{1}{2}}, 1\right) \Rightarrow \left|\dfrac{\partial \vec{r}}{\partial x} \times \dfrac{\partial \vec{r}}{\partial y}\right| = (x + y + 1)^{\frac{1}{2}}$

$\displaystyle\iint_S dA = \iint_R \left|\dfrac{\partial \vec{r}}{\partial x} \times \dfrac{\partial \vec{r}}{\partial y}\right| dx\,dy = \int_0^1 \int_0^1 (x + y + 1)^{\frac{1}{2}} dy\,dx = \dfrac{2}{3}\int_0^1 (2 + y)^{\frac{3}{2}} - (1 + y)^{\frac{3}{2}} dx$

$= \dfrac{2}{3}\left(\dfrac{2}{5}\left((2 + y)^{\frac{5}{2}} - (1 + y)^{\frac{5}{2}}\right)\right)\Big|_{y=0}^{y=1} = \dfrac{4}{15}\left(3^{\frac{5}{2}} - 2^{\frac{7}{2}} + 1\right)$

## Example 7.

Find the area of the surface $x^2 + y^2 + z^2 - 4z = 0$ with $0 \le 3(x^2 + y^2) \le z^2, z \ge 2$

**【解】**

令 $S: x^2 + y^2 + z^2 = 4z,\ 0 \le 3(x^2 + y^2) \le z^2,\ z \ge 2$ 则表面积 $= \iint_S dA$

令 $\vec{r}(x, y) = \left(x, y, 2 + \sqrt{4 - (x^2 + y^2)}\right)$ 且 $R$ 为球面$S$投影至$xy$平面的封闭区域

则 $R = \{(x, y): 0 \le x^2 + y^2 \le 3\}$ 且 $\iint_S dA = \iint_R \left|\dfrac{\partial \vec{r}}{\partial x} \times \dfrac{\partial \vec{r}}{\partial y}\right| dx\, dy$

$\because \dfrac{\partial \vec{r}}{\partial x} = \left(1, 0, -x(4 - (x^2 + y^2))^{-\frac{1}{2}}\right)$ 且 $\dfrac{\partial \vec{r}}{\partial y} = \left(0, 1, -y(4 - (x^2 + y^2))^{-\frac{1}{2}}\right),$

$\therefore \dfrac{\partial \vec{r}}{\partial x} \times \dfrac{\partial \vec{r}}{\partial y} = \left(x(4 - (x^2 + y^2))^{-\frac{1}{2}}, y(4 - (x^2 + y^2))^{-\frac{1}{2}}, 1\right)$

$\Rightarrow \left|\dfrac{\partial \vec{r}}{\partial x} \times \dfrac{\partial \vec{r}}{\partial y}\right| = (x^2(4 - (x^2 + y^2))^{-1} + y^2(4 - (x^2 + y^2))^{-1} + 1)^{\frac{1}{2}} = 2(4 - (x^2 + y^2))^{-\frac{1}{2}}$

$\therefore \iint_S dA = \iint_R \left|\dfrac{\partial \vec{r}}{\partial x} \times \dfrac{\partial \vec{r}}{\partial y}\right| dx\, dy = \iint_{x^2 + y^2 \le 3} 2(4 - (x^2 + y^2))^{-\frac{1}{2}} dx\, dy$

令 $x = r \cos\theta,\ y = r \sin\theta$ 则 $\{(x, y): x^2 + y^2 \le 3\} = \{(r, \theta): 0 \le r \le \sqrt{3}, 0 \le \theta \le 2\pi\}$

$\displaystyle \iint_{x^2 + y^2 \le 3} 2(4 - (x^2 + y^2))^{-\frac{1}{2}} dx\, dy = \int_0^{2\pi} \int_0^{\sqrt{3}} 2(4 - r^2)^{-\frac{1}{2}} r\, dr\, d\theta = 4\pi(4 - r^2)^{\frac{1}{2}}\Big|_{\sqrt{3}}^{0} = 4\pi$

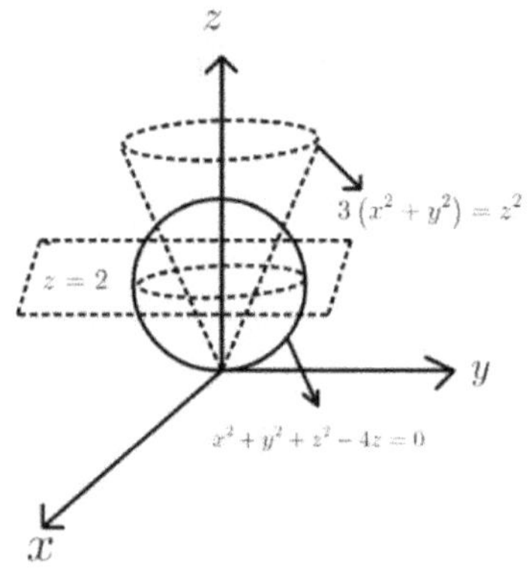

Example 8.

Find the  area of the part of the plane $x + 2y + 3z = 1$ that lies within  the cylinder $x^2 + y^2 = 3$.

【解】

令 $\vec{r}(x,y) = \left(x, y, \dfrac{1-(x+2y)}{3}\right)$ 且 $R$ 为平面投影至 $xy$ 平面的封闭区域

则 $R = \{(x,y): x^2 + y^2 \leq 3\}$ 且 表面积 $= \displaystyle\iint_S dA = \iint_R \left|\dfrac{\partial\vec{r}}{\partial x} \times \dfrac{\partial\vec{r}}{\partial y}\right| dxdy$

$\because \dfrac{\partial\vec{r}}{\partial x} = \left(1, 0, -\dfrac{1}{3}\right)$ 且 $\dfrac{\partial\vec{r}}{\partial y} = \left(0, 1, -\dfrac{2}{3}\right)$ $\therefore \dfrac{\partial\vec{r}}{\partial x} \times \dfrac{\partial\vec{r}}{\partial y} = \left(\dfrac{1}{3}, \dfrac{2}{3}, 1\right) \Rightarrow \left|\dfrac{\partial\vec{r}}{\partial x} \times \dfrac{\partial\vec{r}}{\partial y}\right| = \sqrt{1 + \dfrac{5}{9}} = \sqrt{\dfrac{14}{9}}$

$\therefore \displaystyle\iint_S dA = \iint_R \left|\dfrac{\partial\vec{r}}{\partial x} \times \dfrac{\partial\vec{r}}{\partial y}\right| dxdy = \sqrt{\dfrac{14}{9}} \iint_{x^2+y^2\leq 3} dxdy = \sqrt{\dfrac{14}{9}} \times 3\pi = \sqrt{14}\pi$

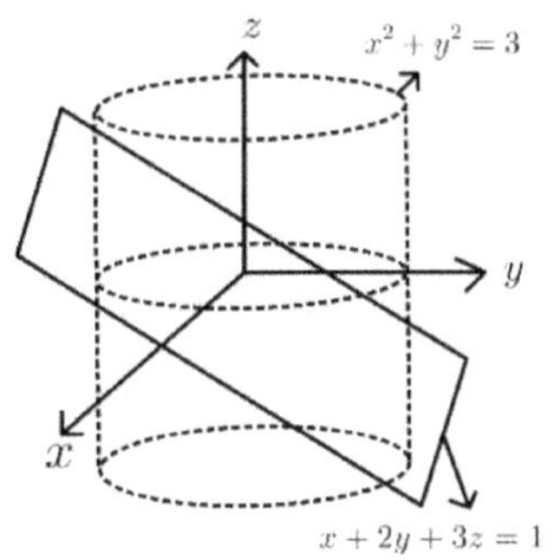


Example 9.

Find the  surface area of that part of the parabolic cylinder $z = y^2$ that lies  over the triangle with vertices $(0,0), (0,1), (1,1)$ in the $xy$– plane.

【解】

令 $S: z = y^2$, $0 \leq x \leq 1, 0 \leq y \leq 1$ 则表面积 $= \displaystyle\iint_S dA$

令 $\vec{r}(x,y) = (x, y, y^2)$ 且 $R$ 为曲面 $S$ 投影至 $xy$ 平面的封闭区域

則 $R = \{(x, y): 0 \le x \le y, 0 \le y \le 1\}$ 且表面积 $= \iint_S dA = \iint_R \left| \dfrac{\partial \vec{r}}{\partial x} \times \dfrac{\partial \vec{r}}{\partial y} \right| dxdy$

$\because \dfrac{\partial \vec{r}}{\partial x} = (1,0,0)$ 且 $\dfrac{\partial \vec{r}}{\partial y} = (0,1,2y)$ $\therefore \dfrac{\partial \vec{r}}{\partial x} \times \dfrac{\partial \vec{r}}{\partial y} = (0,2y,1) \Rightarrow \left| \dfrac{\partial \vec{r}}{\partial x} \times \dfrac{\partial \vec{r}}{\partial y} \right| = (1 + 4y^2)^{\frac{1}{2}}$

$$\iint_R \left| \dfrac{\partial \vec{r}}{\partial x} \times \dfrac{\partial \vec{r}}{\partial y} \right| dxdy = \int_0^1 \int_0^y (1 + 4y^2)^{\frac{1}{2}} dxdy = \int_0^1 y(1 + 4y^2)^{\frac{1}{2}} dy = \dfrac{(1 + 4y^2)^{\frac{3}{2}}}{12} \Bigg|_0^1$$

$$= \dfrac{5\sqrt{5} - 1}{12}$$

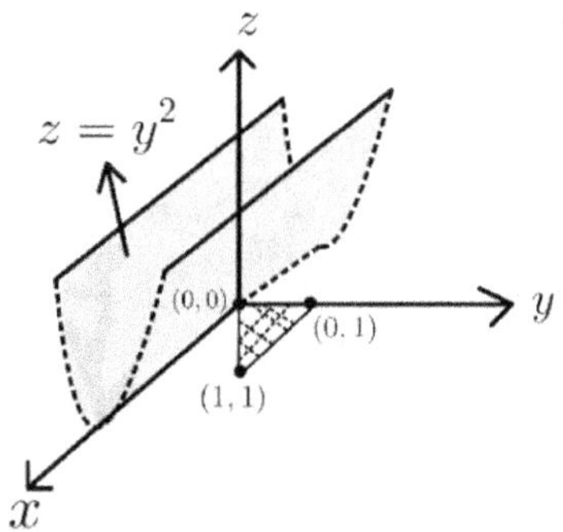


Example 10.

Find the area of the surface with parametric equation $x = s^2$, $y = st$, $z = \dfrac{t^2}{2}$, $0 \le s \le 1, 0 \le t \le 2$.

【解】

令 $\vec{r}(s,t) = \left( s^2, st, \dfrac{t^2}{2} \right)$, $\forall 0 \le s \le 1, 0 \le t \le 2$ 則 $\dfrac{\partial \vec{r}}{\partial s} = (2s, t, 0)$ 且 $\dfrac{\partial \vec{r}}{\partial t} = (0, s, t)$

$\therefore \dfrac{\partial \vec{r}}{\partial s} \times \dfrac{\partial \vec{r}}{\partial t} = (t^2, 2st, 2s^2)$ 且 $\left| \dfrac{\partial \vec{r}}{\partial s} \times \dfrac{\partial \vec{r}}{\partial t} \right| = \sqrt{t^4 + 4s^2t^2 + 4s^4} = t^2 + 2s^2$

$\therefore \iint_S dA = \iint_{[0,1]\times[0,2]} \left| \dfrac{\partial \vec{r}}{\partial s} \times \dfrac{\partial \vec{r}}{\partial t} \right| dsdt = \int_0^2 \int_0^1 t^2 + 2s^2 dsdt = \int_0^2 t^2 + \dfrac{1}{3} dt = 4$

Example 11.

Find the area of the surface $y = 4x + z^2$ that lies between the planes $x = 0, x = 1,$

$z = 0, \text{and } z = 1.$

【解】

令 $S: y = 4x + z^2, \ 0 \le x \le 1, \ 0 \le z \le 1$ 则表面积 $= \iint_S dA$

令 $\vec{r}(x,y) = (x, 4x + z^2, z)$ 则 $R$ 为曲面 $S$ 投影至 $xz$ 平面的封闭区域

则 $R = \{(x,z) : 0 \le x \le 1, 0 \le z \le 1\}$ 且 $\iint_S dA = \iint_R \left| \dfrac{\partial \vec{r}}{\partial x} \times \dfrac{\partial \vec{r}}{\partial z} \right| dxdz$

$\because \dfrac{\partial \vec{r}}{\partial x} = (1,4,0)$ 且 $\dfrac{\partial \vec{r}}{\partial z} = (0,2z,1)$ $\therefore \dfrac{\partial \vec{r}}{\partial x} \times \dfrac{\partial \vec{r}}{\partial z} = (4,-1,2z) \Rightarrow \left| \dfrac{\partial \vec{r}}{\partial x} \times \dfrac{\partial \vec{r}}{\partial z} \right| = (17 + 4z^2)^{\frac{1}{2}}$

$\therefore \iint_S dA = \iint_R \left| \dfrac{\partial \vec{r}}{\partial x} \times \dfrac{\partial \vec{r}}{\partial z} \right| dxdz = \int_0^1 \int_0^1 (17 + 4z^2)^{\frac{1}{2}} dxdz = \int_0^1 (17 + 4z^2)^{\frac{1}{2}} dz$

令 $t = 2z$ 则 $\displaystyle\int_0^1 (17 + 4z^2)^{\frac{1}{2}} dz = \frac{1}{2}\int_0^2 (17 + t^2)^{\frac{1}{2}} dt$

令 $t = \sqrt{17}\tan\theta$ 则 $dt = \sqrt{17}\sec^2\theta\, d\theta$

$\therefore \dfrac{1}{2}\int_0^2 (17 + t^2)^{\frac{1}{2}} dt = \dfrac{17}{2}\int_0^{\tan^{-1} 2} \sec^3\theta\, d\theta = \dfrac{1}{2}\left( \dfrac{t\sqrt{17 + t^2}}{2} + \dfrac{17}{2}\ln\left( \dfrac{t}{\sqrt{17}} + \dfrac{\sqrt{17 + t^2}}{\sqrt{17}} \right) \right)\Bigg|_{t=0}^{t=2}$

$= \dfrac{\sqrt{21}}{2} + \dfrac{17}{4}\left( \ln(2 + \sqrt{21}) - \ln\sqrt{17} \right)$

Example 12.

求平面 $Ax + By + Cz = D$ 在椭圆柱 $\dfrac{x^2}{a^2} + \dfrac{y^2}{b^2} = 1$ 內的表面积

【解】

令 $\vec{r}(x,y) = \left( x, y, \dfrac{D - (Ax + By)}{C} \right)$ 且 $R$ 为所围平面投影至 $xy$ 平面的封闭区域

則 $R = \left\{(x,y): \dfrac{x^2}{a^2} + \dfrac{y^2}{b^2} \leq 1\right\}$ 且 表面积 $= \iint_S dA = \iint_R \left|\dfrac{\partial \vec{r}}{\partial x} \times \dfrac{\partial \vec{r}}{\partial y}\right| dxdy$

$\because \dfrac{\partial \vec{r}}{\partial x} = \left(1,0,-\dfrac{A}{C}\right)$ 且 $\dfrac{\partial \vec{r}}{\partial y} = \left(0,1,-\dfrac{B}{C}\right)$ $\quad \therefore \dfrac{\partial \vec{r}}{\partial x} \times \dfrac{\partial \vec{r}}{\partial y} = \left(\dfrac{A}{C}, \dfrac{B}{C}, 1\right)$

$\Rightarrow \left|\dfrac{\partial \vec{r}}{\partial x} \times \dfrac{\partial \vec{r}}{\partial y}\right| = \sqrt{1 + \dfrac{A^2+B^2}{C^2}} = \sqrt{\dfrac{A^2+B^2+C^2}{C^2}}$

$\therefore$ 表面积 $= \iint_S dA = \iint_R \left|\dfrac{\partial \vec{r}}{\partial x} \times \dfrac{\partial \vec{r}}{\partial y}\right| dxdy = \sqrt{\dfrac{A^2+B^2+C^2}{C^2}} \iint_{\frac{x^2}{a^2}+\frac{y^2}{b^2}\leq 1} dxdy$

$令 x = ra\cos\theta, \ y = rb\sin\theta$ 則 $\left\{(x,y): \dfrac{x^2}{a^2} + \dfrac{y^2}{b^2} \leq 1\right\} = \{(r,\theta): 0 \leq r \leq 1, 0 \leq \theta \leq 2\pi\}$

$\therefore \iint_{\frac{x^2}{a^2}+\frac{y^2}{b^2}\leq 1} dxdy = \int_0^{2\pi} \int_0^1 abr\,dr\,d\theta = ab\pi \quad \therefore$ 表面积 $= ab\pi \sqrt{\dfrac{A^2+B^2+C^2}{C^2}}$

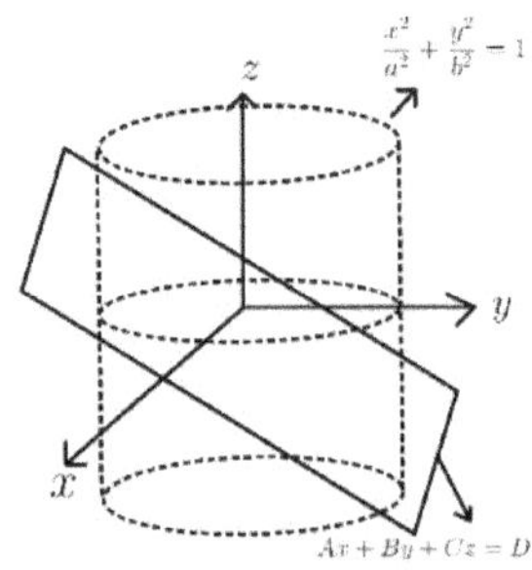


Example 13.

求平面 $Ax + By + Cz = D$ 在 $y = x^2, \ y = 3x^2, \ x = y^2, \ x = 4y^2$ 所围区域的表面积

【解】

$令 \ \vec{r}(x,y) = \left(x, y, \dfrac{D-(Ax+By)}{C}\right)$ 且 $R = \{(x,y): x^2 \leq y \leq 3x^2, 4y^2 \leq x \leq y^2\}$

$$\text{表面积} = \iint_S dA = \iint_R \left|\frac{\partial \vec{r}}{\partial x} \times \frac{\partial \vec{r}}{\partial y}\right| dxdy$$

$$\because \frac{\partial \vec{r}}{\partial x} = \left(1,0,-\frac{A}{C}\right) \text{ 且 } \frac{\partial \vec{r}}{\partial y} = \left(0,1,-\frac{B}{C}\right) \quad \therefore \frac{\partial \vec{r}}{\partial x} \times \frac{\partial \vec{r}}{\partial y} = \left(\frac{A}{C},\frac{B}{C},1\right)$$

$$\Rightarrow \left|\frac{\partial \vec{r}}{\partial x} \times \frac{\partial \vec{r}}{\partial y}\right| = \sqrt{1+\frac{A^2+B^2}{C^2}} = \sqrt{\frac{A^2+B^2+C^2}{C^2}}$$

$$\therefore \text{表面积} = \iint_S dA = \iint_R \left|\frac{\partial \vec{r}}{\partial x} \times \frac{\partial \vec{r}}{\partial y}\right| dxdy = \sqrt{\frac{A^2+B^2+C^2}{C^2}} \iint_R dxdy$$

$$\text{令 } u = \frac{y}{x^2}, \quad v = \frac{x}{y^2} \text{ 則 } x = u^{-\frac{2}{3}}v^{-\frac{1}{3}}, \quad y = u^{-\frac{1}{3}}v^{-\frac{2}{3}}$$

$$\text{且 } dxdy = \left\|\begin{matrix} \dfrac{\partial x}{\partial u} & \dfrac{\partial x}{\partial v} \\ \dfrac{\partial y}{\partial u} & \dfrac{\partial y}{\partial v} \end{matrix}\right\| dudv = \left\|\begin{matrix} \dfrac{-2u^{-\frac{5}{3}}v^{-\frac{1}{3}}}{3} & \dfrac{-u^{-\frac{2}{3}}v^{-\frac{4}{3}}}{3} \\ \dfrac{-u^{-\frac{4}{3}}v^{-\frac{2}{3}}}{3} & \dfrac{-2u^{-\frac{1}{3}}v^{-\frac{5}{3}}}{3} \end{matrix}\right\| dudv = \frac{u^{-2}v^{-2}}{3} dudv$$

$$\because R = \left\{(x,y): 1 \le \frac{y}{x^2} \le 3, 1 \le \frac{x}{y^2} \le 4\right\} \quad \therefore R = \{(u,v): 1 \le u \le 3, 1 \le v \le 4\}$$

$$\therefore \iint_R 1dA = \int_1^3 \int_1^4 \frac{u^{-2}v^{-2}}{3} dvdu = \frac{1}{6} \quad \therefore \iint_S dA = \frac{1}{6}\sqrt{\frac{A^2+B^2+C^2}{C^2}}$$

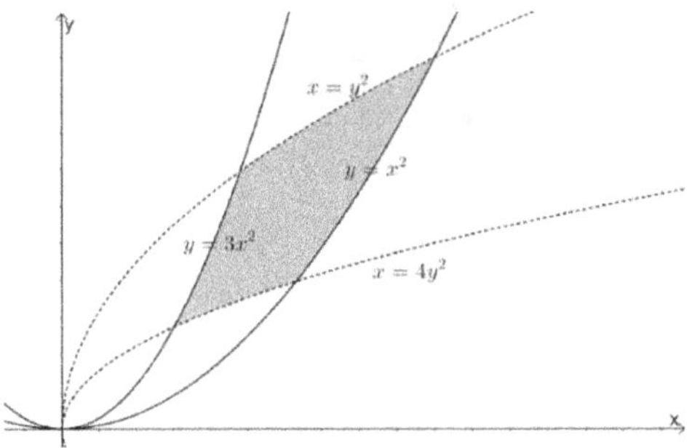


Example 14.

求平面$Ax + By + Cz = D$ 在 $x - 2y = -4,\ x - 2y = 1,\ 2x - y = 0,\ 2x - y = 3$
所圍區域內的表面積

【解】

令 $\vec{r}(x,y) = \left(x, y, \dfrac{D - (Ax + By)}{C}\right)$ 且 $R = \{(x,y): -4 \le x - 2y \le 1, 0 \le 2x - y \le 3\}$

則表面積 $= \displaystyle\iint_S dA = \iint_R \left|\dfrac{\partial \vec{r}}{\partial x} \times \dfrac{\partial \vec{r}}{\partial y}\right| dxdy$

$\because \dfrac{\partial \vec{r}}{\partial x} = \left(1, 0, -\dfrac{A}{C}\right)$ 且 $\dfrac{\partial \vec{r}}{\partial y} = \left(0, 1, -\dfrac{B}{C}\right)$ $\quad \therefore \dfrac{\partial \vec{r}}{\partial x} \times \dfrac{\partial \vec{r}}{\partial y} = \left(\dfrac{A}{C}, \dfrac{B}{C}, 1\right)$

$\Rightarrow \left|\dfrac{\partial \vec{r}}{\partial x} \times \dfrac{\partial \vec{r}}{\partial y}\right| = \sqrt{1 + \dfrac{A^2 + B^2}{C^2}} = \sqrt{\dfrac{A^2 + B^2 + C^2}{C^2}}$

$\therefore \displaystyle\iint_R \left|\dfrac{\partial \vec{r}}{\partial x} \times \dfrac{\partial \vec{r}}{\partial y}\right| dxdy = \sqrt{\dfrac{A^2 + B^2 + C^2}{C^2}} \iint_R dxdy$

令 $x - 2y = u,\ 2x - y = v$ 則 $x = -\dfrac{u}{3} + \dfrac{2v}{3},\ y = -\dfrac{2u}{3} + \dfrac{v}{3}$

且 $dxdy = \left\|\begin{array}{cc} \dfrac{\partial x}{\partial u} & \dfrac{\partial x}{\partial v} \\ \dfrac{\partial y}{\partial u} & \dfrac{\partial y}{\partial v} \end{array}\right\| dudv = \left\|\begin{array}{cc} \dfrac{-1}{3} & \dfrac{2}{3} \\ \dfrac{-2}{3} & \dfrac{1}{3} \end{array}\right\| dudv = \dfrac{1}{3} dudv$

$\because R = \{(x,y): -4 \le x - 2y \le 1, 0 \le 2x - y \le 3\}$ $\quad \therefore R = \{(u,v): -4 \le u \le 1, 0 \le v \le 3\}$

$\therefore \displaystyle\iint_R 1\, dA = \int_{-4}^{1} \int_{0}^{3} \dfrac{1}{3} dudv = 5$ $\quad \therefore$ 表面積 $= \displaystyle\iint_S dA = 5\sqrt{\dfrac{A^2 + B^2 + C^2}{C^2}}$

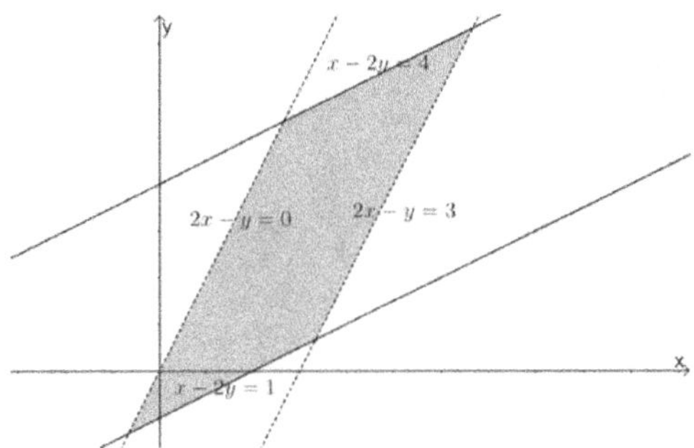


Example 15.

求平面 $Ax + By + Cz = D$ 在 $y = x,\ y = 2x,\ xy = 1,\ xy = 4$ 所围区域内的表面积

【解】

令 $\vec{r}(x,y) = \left(x, y, \dfrac{D - (Ax + By)}{C}\right)$ 且 $R$ 为 $y = x,\ y = 2x,\ xy = 1,\ xy = 4$ 所围区域

则表面积 $= \displaystyle\iint_S dA = \iint_R \left|\dfrac{\partial \vec{r}}{\partial x} \times \dfrac{\partial \vec{r}}{\partial y}\right| dx\,dy$

$\because \dfrac{\partial \vec{r}}{\partial x} = \left(1, 0, -\dfrac{A}{C}\right)$ 且 $\dfrac{\partial \vec{r}}{\partial y} = \left(0, 1, -\dfrac{B}{C}\right)$ $\quad \therefore \dfrac{\partial \vec{r}}{\partial x} \times \dfrac{\partial \vec{r}}{\partial y} = \left(\dfrac{A}{C}, \dfrac{B}{C}, 1\right)$

$\Rightarrow \left|\dfrac{\partial \vec{r}}{\partial x} \times \dfrac{\partial \vec{r}}{\partial y}\right| = \sqrt{1 + \dfrac{A^2 + B^2}{C^2}} = \sqrt{\dfrac{A^2 + B^2 + C^2}{C^2}}$

$\therefore \displaystyle\iint_R \left|\dfrac{\partial \vec{r}}{\partial x} \times \dfrac{\partial \vec{r}}{\partial y}\right| dx\,dy = \sqrt{\dfrac{A^2 + B^2 + C^2}{C^2}} \iint_R dx\,dy$

令 $\dfrac{y}{x} = u,\ xy = v$ 则 $x = \sqrt{\dfrac{v}{u}},\ y = \sqrt{uv}$

且 $dx\,dy = \left\|\begin{matrix} \dfrac{\partial x}{\partial u} & \dfrac{\partial x}{\partial v} \\[2mm] \dfrac{\partial y}{\partial u} & \dfrac{\partial y}{\partial v} \end{matrix}\right\| du\,dv = \left\|\begin{matrix} \dfrac{1}{2}\left(\dfrac{v}{u}\right)^{-\frac{1}{2}}\left(-\dfrac{v}{u^2}\right) & \dfrac{1}{2}\left(\dfrac{v}{u}\right)^{-\frac{1}{2}}\left(\dfrac{1}{u}\right) \\[3mm] \dfrac{1}{2}(uv)^{-\frac{1}{2}}v & \dfrac{1}{2}(uv)^{-\frac{1}{2}}u \end{matrix}\right\| du\,dv = \dfrac{1}{2u} du\,dv$

$\because R = \left\{(x,y): 1 \le \dfrac{y}{x} \le 2, 1 \le xy \le 4\right\}$ $\quad \therefore R = \left\{(u,v): 1 \le u \le 2, 1 \le v \le 4\right\}$

$$\therefore \iint_R dxdy = 2\int_1^4 \int_1^2 \frac{1}{2u}\, du\, dv = 3\ln 2 \quad \therefore \text{表面积} = \iint_S dA = 3\ln 2 \sqrt{\frac{A^2 + B^2 + C^2}{C^2}}$$

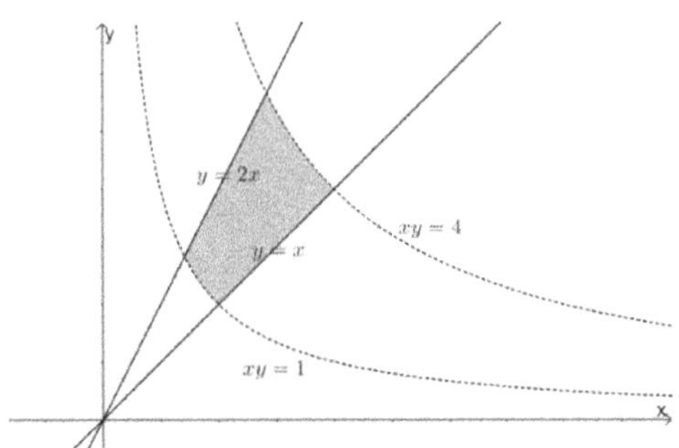


Example 16.

　　求平面$Ax + By + Cz = D$ 在 $ax^2 + bxy + cy^2 = \alpha^2$所围区域內的表面积

【解】

令 $\vec{r}(x,y) = \left(x, y, \dfrac{D - (Ax + By)}{C}\right)$ 且 $R = \{(x,y): ax^2 + bxy + cy^2 \leq \alpha^2\}$

则 $\displaystyle\iint_S dA = \iint_R \left|\frac{\partial \vec{r}}{\partial x} \times \frac{\partial \vec{r}}{\partial y}\right| dxdy$

$\because \dfrac{\partial \vec{r}}{\partial x} = \left(1, 0, -\dfrac{A}{C}\right)$ 且 $\dfrac{\partial \vec{r}}{\partial y} = \left(0, 1, -\dfrac{B}{C}\right)$ $\quad \therefore \dfrac{\partial \vec{r}}{\partial x} \times \dfrac{\partial \vec{r}}{\partial y} = \left(\dfrac{A}{C}, \dfrac{B}{C}, 1\right)$

$\Rightarrow \left|\dfrac{\partial \vec{r}}{\partial x} \times \dfrac{\partial \vec{r}}{\partial y}\right| = \sqrt{1 + \dfrac{A^2 + B^2}{C^2}} = \sqrt{\dfrac{A^2 + B^2 + C^2}{C^2}}$

$\therefore \displaystyle\iint_R \left|\dfrac{\partial \vec{r}}{\partial x} \times \dfrac{\partial \vec{r}}{\partial y}\right| dxdy = \sqrt{\dfrac{A^2 + B^2 + C^2}{C^2}} \iint_R dxdy$

$\because ax^2 + bxy + cy^2 = a\left(x + \dfrac{by}{2a}\right)^2 + \left(\dfrac{4ac - b^2}{4a}\right)y^2$

$$\diamondsuit \sqrt{a}\left(x + \frac{by}{2a}\right) = u, \quad \sqrt{\frac{4ac - b^2}{4a}}\, y = v$$

$$\text{則}\, dxdy = \left\|\begin{vmatrix} \dfrac{\partial x}{\partial u} & \dfrac{\partial x}{\partial v} \\[2mm] \dfrac{\partial y}{\partial u} & \dfrac{\partial y}{\partial v} \end{vmatrix}\right\| dudv = \left\|\begin{vmatrix} \dfrac{1}{\sqrt{a}} & 0 \\[2mm] 0 & \sqrt{\dfrac{4a}{4ac - b^2}} \end{vmatrix}\right\| dudv = \frac{2}{\sqrt{4ac - b^2}}\, dudv$$

$$\therefore \iint_R dA = \iint_{u^2 + v^2 \le \alpha^2} \frac{2}{\sqrt{4ac - b^2}}\, dudv$$

$$\diamondsuit u = r\cos\theta, v = r\sin\theta \ \ \text{則}\, \{(u,v): u^2 + v^2 \le \alpha^2\} = \{(r,\theta): 0 \le r \le \alpha, 0 \le \theta \le 2\pi\}$$

$$\text{且}\, dudv = \left\|\begin{vmatrix} \dfrac{\partial u}{\partial r} & \dfrac{\partial u}{\partial \theta} \\[2mm] \dfrac{\partial v}{\partial r} & \dfrac{\partial v}{\partial \theta} \end{vmatrix}\right\| drd\theta = \left\|\begin{matrix} \cos\theta & -r\sin\theta \\ \sin\theta & r\cos\theta \end{matrix}\right\| drd\theta = rdrd\theta$$

$$\therefore \iint_{u^2+v^2\le\alpha^2} \frac{2}{\sqrt{4ac - b^2}}\, dudv = \int_0^{2\pi}\int_0^\alpha \frac{2}{\sqrt{4ac - b^2}}\, rdrd\theta = \frac{2}{\sqrt{4ac - b^2}}\int_0^{2\pi}d\theta\int_0^\alpha rdr$$

$$= \frac{2\pi\alpha^2}{\sqrt{4ac - b^2}}$$

$$\therefore \text{表面积} = \iint_S dA = \iint_R \left|\frac{\partial \vec{r}}{\partial x} \times \frac{\partial \vec{r}}{\partial y}\right| dxdy = \frac{2\pi\alpha^2}{\sqrt{4ac - b^2}}\sqrt{\frac{A^2 + B^2 + C^2}{C^2}}$$

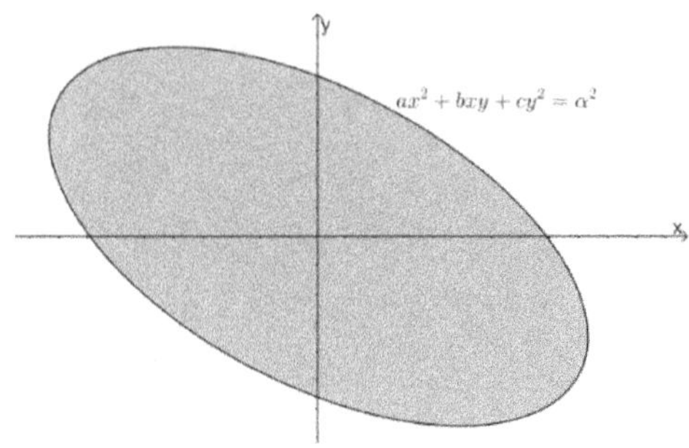


Example 17.

$$\text{求曲面}\, z = \frac{2}{3}y^{\frac{3}{2}}\, \text{在}\, 0 < x < 2, 0 < y < 2\, \text{內的表面积}$$

【解】

令 $S\colon z = \dfrac{2}{3}y^{\frac{3}{2}},\ 0 < x < 2,\ 0 < y < 2$ 则表面积 $= \displaystyle\iint_S dA$

令 $\vec{r}(x,y) = \left(x, y, \dfrac{2}{3}y^{\frac{3}{2}}\right)$ 且 $R$ 为曲面 $S$ 投影至 $xy$ 平面的封闭区域

则 $R = \{(x,y)\colon 0 < x < 2, 0 < y < 2\}$ 且 $\displaystyle\iint_S dA = \iint_R \left|\dfrac{\partial \vec{r}}{\partial x} \times \dfrac{\partial \vec{r}}{\partial y}\right| dxdy$

$\because \dfrac{\partial \vec{r}}{\partial x} = (1,0,0)$ 且 $\dfrac{\partial \vec{r}}{\partial y} = \left(0,1,y^{\frac{1}{2}}\right)$ $\therefore \dfrac{\partial \vec{r}}{\partial x} \times \dfrac{\partial \vec{r}}{\partial y} = \left(0, -y^{\frac{1}{2}}, 1\right) \Rightarrow \left|\dfrac{\partial \vec{r}}{\partial x} \times \dfrac{\partial \vec{r}}{\partial y}\right| = (y+1)^{\frac{1}{2}}$

$\therefore \displaystyle\iint_S dA = \iint_R \left|\dfrac{\partial \vec{r}}{\partial x} \times \dfrac{\partial \vec{r}}{\partial y}\right| dxdy = \int_0^2 \int_0^2 (y+1)^{\frac{1}{2}} dxdy = 2 \cdot \left. \dfrac{2(y+1)^{\frac{3}{2}}}{3}\right|_0^2 = \dfrac{4}{3}\left(3\sqrt{3} - 1\right)$

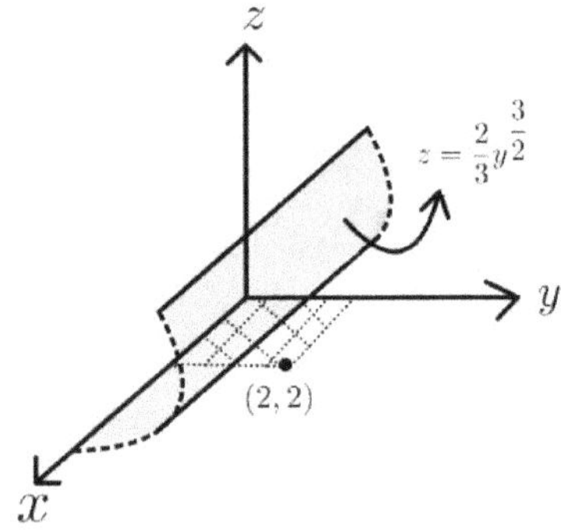


Example 18.

　　求曲面 $z = x + ay + b$ 在 $y > x^2,\ y < 1, x > 0$ 內的表面积

【解】

令 $S\colon z = x + ay + b,\ y > x^2,\ y < 1,\ x > 0$ 则表面积 $= \displaystyle\iint_S dA$

令 $\vec{r}(x,y) = (x, y, x + ay + b)$ 且 $R$ 为曲面 $S$ 投影至 $xy$ 平面的封闭区域

則 $R = \{(x, y): y > x^2, y < 1, x > 0\}$ 且 $\displaystyle\iint_S dA = \iint_R \left|\frac{\partial \vec{r}}{\partial x} \times \frac{\partial \vec{r}}{\partial y}\right| dxdy$

$\because \dfrac{\partial \vec{r}}{\partial x} = (1,0,1)$ 且 $\dfrac{\partial \vec{r}}{\partial y} = (0,1,a)$ $\therefore \dfrac{\partial \vec{r}}{\partial x} \times \dfrac{\partial \vec{r}}{\partial y} = (-1,-a,1) \Rightarrow \left|\dfrac{\partial \vec{r}}{\partial x} \times \dfrac{\partial \vec{r}}{\partial y}\right| = \sqrt{a^2 + 2}$

$\therefore \displaystyle\iint_S dA = \iint_R \left|\frac{\partial \vec{r}}{\partial x} \times \frac{\partial \vec{r}}{\partial y}\right| dxdy = \int_0^1 \int_{x^2}^1 \sqrt{a^2 + 2}\, dydx = \frac{2\sqrt{a^2 + 2}}{3}$

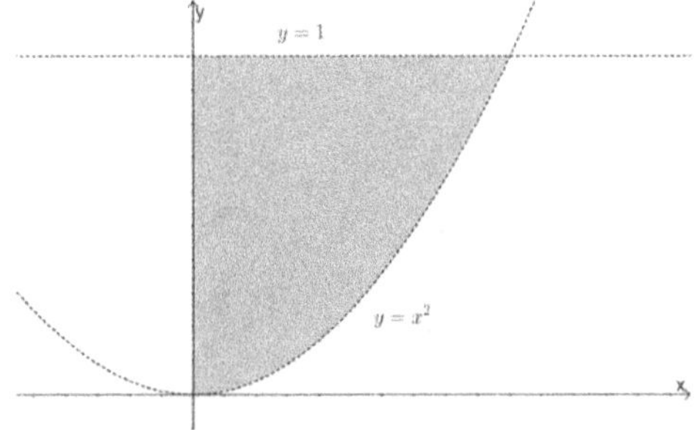


Example 19.

　　求曲面 $z = xy$ 在柱面 $x^2 + y^2 = a^2$, $x \geq 0$, $y \geq 0$ 之內的曲面面積

【解】

令 $S$: $z = xy$, $x^2 + y^2 \leq a^2$ 則 表面積 $= \displaystyle\iint_S dA$

令 $\vec{r}(x,y) = (x, y, xy)$ 且 $R$ 为曲面 $S$ 投影至 $xy$ 平面的封闭区域

則 $R = \{(x, y): x^2 + y^2 \leq a^2\}$ 且 $\displaystyle\iint_S dA = \iint_R \left|\frac{\partial \vec{r}}{\partial x} \times \frac{\partial \vec{r}}{\partial y}\right| dxdy$

$\because \dfrac{\partial \vec{r}}{\partial x} = (1,0,y)$ 且 $\dfrac{\partial \vec{r}}{\partial y} = (0,1,x)$ $\therefore \dfrac{\partial \vec{r}}{\partial x} \times \dfrac{\partial \vec{r}}{\partial y} = (-y,-x,1) \Rightarrow \left|\dfrac{\partial \vec{r}}{\partial x} \times \dfrac{\partial \vec{r}}{\partial y}\right| = (x^2 + y^2 + 1)^{\frac{1}{2}}$

$\therefore \displaystyle\iint_S dA = \iint_R \left|\frac{\partial \vec{r}}{\partial x} \times \frac{\partial \vec{r}}{\partial y}\right| dxdy = \iint_{x^2+y^2 \leq a^2} (x^2 + y^2 + 1)^{\frac{1}{2}} dxdy$

令 $x = r\cos\theta$, $y = r\sin\theta$

則 $\{(x,y): x^2 + y^2 \leq a^2, x \geq 0, y \geq 0\} = \{(r,\theta): 0 \leq r \leq a, 0 \leq \theta \leq \frac{\pi}{2}\}$

$$\therefore \iint_{x^2+y^2 \leq a^2} (x^2 + y^2 + 1)^{\frac{1}{2}} dxdy = \int_0^{\frac{\pi}{2}} \int_0^a (r^2 + 1)^{\frac{1}{2}} r \, dr d\theta = \frac{\pi}{6}(r^2 + 1)^{\frac{3}{2}}\Big|_0^a$$

$$= \frac{\pi}{6}((a^2 + 1)^{\frac{3}{2}} - 1)$$

Example 20.

求曲面 $z = y^2 + bx + c$ 在 $xy$ 平面 $(0,0)$、$(0,a)$、$(a,a)$ 所圍三角形內的区域面积

【解】

令 $S: z = y^2 + bx + c, \ 0 \leq x \leq y, \ 0 \leq y \leq a$ 則表面積 $= \iint_S dA$

令 $\vec{r}(x,y) = (x, y, y^2 + bx + c)$ 且 $R$ 为曲面 $S$ 投影至 $xy$ 平面的封闭区域

則 $R = \{(x,y): 0 \leq x \leq y, 0 \leq y \leq a\}$ 且 $\iint_S dA = \iint_R \left|\frac{\partial \vec{r}}{\partial x} \times \frac{\partial \vec{r}}{\partial y}\right| dxdy$

$$\because \frac{\partial \vec{r}}{\partial x} = (1,0,b), \ \frac{\partial \vec{r}}{\partial y} = (0,1,2y) \ \therefore \frac{\partial \vec{r}}{\partial x} \times \frac{\partial \vec{r}}{\partial y} = (-b, -2y, 1) \Rightarrow \left|\frac{\partial \vec{r}}{\partial x} \times \frac{\partial \vec{r}}{\partial y}\right| = (b^2 + 1 + 4y^2)^{\frac{1}{2}}$$

$$\therefore \iint_S dA = \iint_R \left|\frac{\partial \vec{r}}{\partial x} \times \frac{\partial \vec{r}}{\partial y}\right| dxdy = \int_0^a \int_0^y (b^2 + 1 + 4y^2)^{\frac{1}{2}} dxdy = \int_0^a (b^2 + 1 + 4y^2)^{\frac{1}{2}} y \, dy$$

$$= \frac{(b^2 + 1 + 4y^2)^{\frac{3}{2}}}{12}\Big|_0^a = \frac{1}{12}\left((1 + 4a^2 + b^2)\sqrt{1 + 4a^2 + b^2} - (b^2 + 1)\sqrt{b^2 + 1}\right)$$

Example 21.

求抛物体 $z = x^2 + y^2$ 在 $0 \leq z \leq 3$ 內的表面積

【解】

令 $S: z = x^2 + y^2, \ 0 \le z \le 3$ 則表面积 $= \iint_S dA$

令 $\vec{r}(x,y) = (x, y, x^2 + y^2)$ 且 $R$ 为曲面投影至 $xy$ 平面的封闭区域

则 $R = \{(x,y): x^2 + y^2 \le 3\}$ 且 $\displaystyle\iint_S dA = \iint_R \left| \frac{\partial \vec{r}}{\partial x} \times \frac{\partial \vec{r}}{\partial y} \right| dxdy$

$\because \dfrac{\partial \vec{r}}{\partial x} = (1, 0, 2x)$ 且 $\dfrac{\partial \vec{r}}{\partial y} = (0, 1, 2y)$ $\quad \therefore \dfrac{\partial \vec{r}}{\partial x} \times \dfrac{\partial \vec{r}}{\partial y} = (-2x, -2y, 1)$

$\Rightarrow \left| \dfrac{\partial \vec{r}}{\partial x} \times \dfrac{\partial \vec{r}}{\partial y} \right| = (4x^2 + 4y^2 + 1)^{\frac{1}{2}}$

$\therefore \displaystyle\iint_R \left| \frac{\partial \vec{r}}{\partial x} \times \frac{\partial \vec{r}}{\partial y} \right| dxdy = \iint_{x^2 + y^2 \le 3} (4x^2 + 4y^2 + 1)^{\frac{1}{2}} dxdy$

令 $x = r\cos\theta, \ y = r\sin\theta$ 则 $\{(x,y): x^2 + y^2 \le 3\} = \{(r, \theta): 0 \le r \le \sqrt{3}, 0 \le \theta \le 2\pi\}$

$\therefore \displaystyle\iint_{x^2 + y^2 \le 3} (4x^2 + 4y^2 + 1)^{\frac{1}{2}} dxdy = \int_0^{2\pi} \int_0^{\sqrt{3}} (4r^2 + 1)^{\frac{1}{2}} r\, dr\, d\theta = 2\pi \cdot \frac{1}{12} \cdot (4r^2 + 1)^{\frac{3}{2}} \Big|_0^{\sqrt{3}}$

$= \dfrac{\pi}{6}(13\sqrt{13} - 1)$

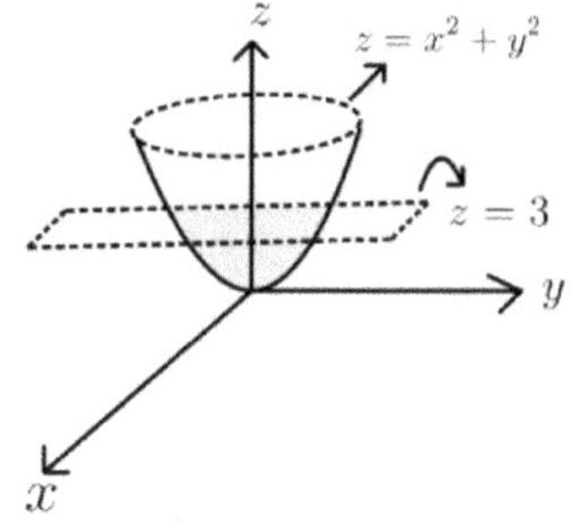


Example 22.

　　求平面 $ax + by + cz = 1$ 在第一卦限被 $x = 0$、$y = 0$、$x^2 + y^2 = r^2$ 所截的表面积

【解】

令 $S$: $ax + by + cz = 1$, $x \geq 0$, $y \geq 0$, $x^2 + y^2 \leq r^2$ 則表面积 $= \dfrac{1}{4}\iint_S dA$

令 $\vec{r}(x, y) = \left(x, y, \dfrac{1}{c} - \dfrac{ax}{c} - \dfrac{by}{c}\right)$ 且 $R$ 为平面 $S$ 投影至 $xy$ 平面的封闭区域

則 $R = \{(x, y): x \geq 0 、 y \geq 0 、 x^2 + y^2 \leq r^2\}$ 且 $\displaystyle\iint_S dA = \iint_R \left|\dfrac{\partial \vec{r}}{\partial x} \times \dfrac{\partial \vec{r}}{\partial y}\right| dxdy$

$\because \dfrac{\partial \vec{r}}{\partial x} = \left(1, 0, -\dfrac{a}{c}\right)$ 且 $\dfrac{\partial \vec{r}}{\partial y} = \left(0, 1, -\dfrac{b}{c}\right)$ $\therefore \dfrac{\partial \vec{r}}{\partial x} \times \dfrac{\partial \vec{r}}{\partial y} = \left(\dfrac{a}{c}, \dfrac{b}{c}, 1\right) \Rightarrow \left|\dfrac{\partial \vec{r}}{\partial x} \times \dfrac{\partial \vec{r}}{\partial y}\right| = \dfrac{\sqrt{a^2 + b^2 + c^2}}{c}$

$\therefore \dfrac{1}{4}\iint_S dA = \dfrac{1}{4}\iint_R \left|\dfrac{\partial \vec{r}}{\partial x} \times \dfrac{\partial \vec{r}}{\partial y}\right| dxdy = \dfrac{1}{4}\iint_{x^2+y^2 \leq r^2} \dfrac{\sqrt{a^2 + b^2 + c^2}}{c} dxdy$

$= \dfrac{r^2 \pi \sqrt{a^2 + b^2 + c^2}}{4c}$

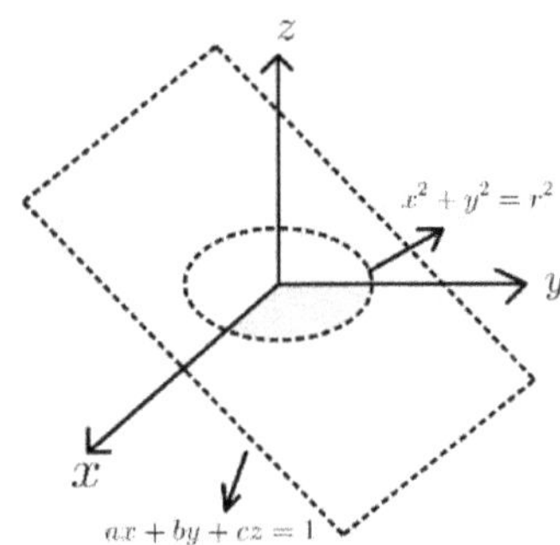


Example 23.

    (1) 求球面 $x^2 + y^2 + z^2 = 8$ 位于圆锥面 $z = \sqrt{x^2 + y^2}$ 內的表面积

    (2) 求球面 $x^2 + y^2 + z^2 = 2a (a > 0)$ 位于圆锥面 $z = \sqrt{x^2 + y^2}$ 內的表面积

【解】

(1)

令 $S: x^2 + y^2 + z^2 = 8,\ z \le \sqrt{x^2 + y^2}$ 則表面积 $= \iint_S dA$

令 $\vec{r}(x, y) = \left(x, y, \sqrt{8 - (x^2 + y^2)}\right)$ 且 $R$ 为球面 $S$ 投影至 $xy$ 平面的封闭区域

則 $R = \{(x, y): x^2 + y^2 \le 4\}$ 且 $\iint_S dA = \iint_R \left|\dfrac{\partial \vec{r}}{\partial x} \times \dfrac{\partial \vec{r}}{\partial y}\right| dxdy$

$\because \dfrac{\partial \vec{r}}{\partial x} = \left(1, 0, -x(8 - (x^2 + y^2))^{-\frac{1}{2}}\right)$ 且 $\dfrac{\partial \vec{r}}{\partial y} = \left(0, 1, -y(8 - (x^2 + y^2))^{-\frac{1}{2}}\right)$

$\therefore \dfrac{\partial \vec{r}}{\partial x} \times \dfrac{\partial \vec{r}}{\partial y} = \left(x(8 - (x^2 + y^2))^{-\frac{1}{2}}, y(8 - (x^2 + y^2))^{-\frac{1}{2}}, 1\right)$

$\Rightarrow \left|\dfrac{\partial \vec{r}}{\partial x} \times \dfrac{\partial \vec{r}}{\partial y}\right| = (x^2(8 - (x^2 + y^2))^{-1} + y^2(8 - (x^2 + y^2))^{-1} + 1)^{\frac{1}{2}} = \sqrt{8}(8 - (x^2 + y^2))^{-\frac{1}{2}}$

$\therefore \iint_S dA = \iint_R \left|\dfrac{\partial \vec{r}}{\partial x} \times \dfrac{\partial \vec{r}}{\partial y}\right| dxdy = \iint_{x^2 + y^2 \le 4} \sqrt{8}(8 - (x^2 + y^2))^{-\frac{1}{2}} dxdy$

令 $x = r\cos\theta,\ y = r\sin\theta$ 則 $\{(x, y): x^2 + y^2 \le 4\} = \{(r, \theta): 0 \le r \le 2, 0 \le \theta \le 2\pi\}$

$\therefore \iint_{x^2 + y^2 \le 4} \sqrt{8}(8 - (x^2 + y^2))^{-\frac{1}{2}} dxdy = \int_0^{2\pi} \int_0^2 \sqrt{8}(8 - r^2)^{-\frac{1}{2}} r\, dr\, d\theta = 4\sqrt{2}\pi(8 - r^2)^{\frac{1}{2}}\Big|_2^0$

$= \pi(16 - 8\sqrt{2})$

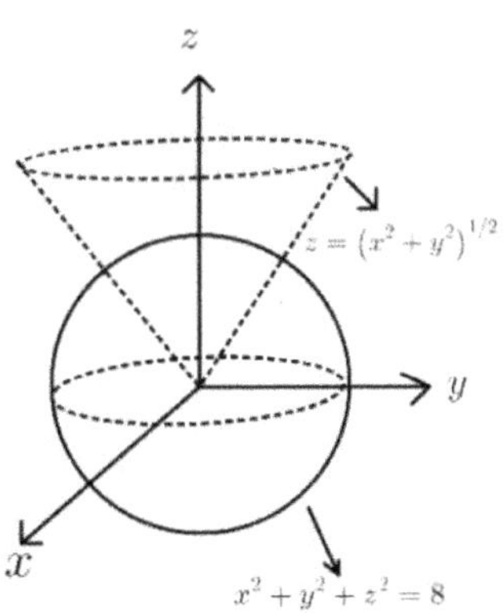

(2)

令 $S: x^2 + y^2 + z^2 = 2a, \ z \leq \sqrt{x^2 + y^2}$ 则表面积 $= \iint_S dA$

令 $\vec{r}(x,y) = \left(x, y, \sqrt{2a - (x^2 + y^2)}\right)$ 且 $R$ 为球面 $S$ 投影至 $xy$ 平面的封闭区域

则 $R = \{(x,y): x^2 + y^2 \leq a\}$ 且 $\iint_S dA = \iint_R \left|\dfrac{\partial \vec{r}}{\partial x} \times \dfrac{\partial \vec{r}}{\partial y}\right| dxdy$

$\because \dfrac{\partial \vec{r}}{\partial x} = \left(1, 0, -x(2a - (x^2 + y^2))^{-\frac{1}{2}}\right)$ 且 $\dfrac{\partial \vec{r}}{\partial y} = \left(0, 1, -y(2a - (x^2 + y^2))^{-\frac{1}{2}}\right)$

$\therefore \dfrac{\partial \vec{r}}{\partial x} \times \dfrac{\partial \vec{r}}{\partial y} = \left(x(2a - (x^2 + y^2))^{-\frac{1}{2}}, y(2a - (x^2 + y^2))^{-\frac{1}{2}}, 1\right)$

$\Rightarrow \left|\dfrac{\partial \vec{r}}{\partial x} \times \dfrac{\partial \vec{r}}{\partial y}\right| = (x^2(2a - (x^2 + y^2))^{-1} + y^2(2a - (x^2 + y^2))^{-1} + 1)^{\frac{1}{2}}$

$= \sqrt{2a}(2a - (x^2 + y^2))^{-\frac{1}{2}}$

$\therefore \iint_R \left|\dfrac{\partial \vec{r}}{\partial x} \times \dfrac{\partial \vec{r}}{\partial y}\right| dxdy = \iint_{x^2 + y^2 \leq a} \sqrt{2a}(2a - (x^2 + y^2))^{-\frac{1}{2}} dxdy$

令 $x = r\cos\theta, y = r\sin\theta$ 则 $\{(x,y): x^2 + y^2 \leq a\} = \{(r,\theta): 0 \leq r \leq \sqrt{a}, 0 \leq \theta \leq 2\pi\}$

$\therefore \iint_{x^2 + y^2 \leq a} \sqrt{2a}(2a - (x^2 + y^2))^{-\frac{1}{2}} dxdy = \int_0^{2\pi} \int_0^{\sqrt{a}} \sqrt{2a}(2a - r^2)^{-\frac{1}{2}} r \, dr \, d\theta$

$= 2\sqrt{2a}\pi(2a - r^2)^{\frac{1}{2}}\Big|_{\sqrt{a}}^{0} = 4a\pi - 2a\sqrt{2}\pi$

Example 24.

(1)求球面$x^2 + y^2 + z^2 = 6z$ 在抛物面 $z = x^2 + y^2$ 内的表面积

(2)求球面$x^2 + y^2 + z^2 = 2az, (a > \dfrac{1}{2})$在抛物面$z = x^2 + y^2$内的表面积

【解】

(1)

令 $S: x^2 + y^2 + z^2 = 6z, \ z \le x^2 + y^2$ 则表面积 $= \iint_S dA$

令 $\vec{r}(x,y) = \left(x, y, 3 + \sqrt{9 - (x^2 + y^2)}\right)$ 且 $R$ 为球面$S$投影至$xy$平面的封闭区域

则$R = \{(x,y): x^2 + y^2 \le 5\}$ 且 $\iint_S dA = \iint_R \left|\dfrac{\partial \vec{r}}{\partial x} \times \dfrac{\partial \vec{r}}{\partial y}\right| dxdy$

$\because \dfrac{\partial \vec{r}}{\partial x} = \left(1, 0, -x(9 - (x^2 + y^2))^{-\frac{1}{2}}\right)$ 且 $\dfrac{\partial \vec{r}}{\partial y} = \left(0, 1, -y(9 - (x^2 + y^2))^{-\frac{1}{2}}\right)$

$\therefore \dfrac{\partial \vec{r}}{\partial x} \times \dfrac{\partial \vec{r}}{\partial y} = \left(x(9 - (x^2 + y^2))^{-\frac{1}{2}}, y(9 - (x^2 + y^2))^{-\frac{1}{2}}, 1\right)$

$\Rightarrow \left|\dfrac{\partial \vec{r}}{\partial x} \times \dfrac{\partial \vec{r}}{\partial y}\right| = (x^2(9 - (x^2 + y^2))^{-1} + y^2(9 - (x^2 + y^2))^{-1} + 1)^{\frac{1}{2}} = 3(9 - (x^2 + y^2))^{-\frac{1}{2}}$

$\therefore \iint_R \left|\dfrac{\partial \vec{r}}{\partial x} \times \dfrac{\partial \vec{r}}{\partial y}\right| dxdy = \iint_{x^2 + y^2 \le 5} 3(9 - (x^2 + y^2))^{-\frac{1}{2}} dxdy$

令$x = r\cos\theta, \ y = r\sin\theta$ 则 $\{(x,y): x^2 + y^2 \le 5\} = \{(r, \theta): 0 \le r \le \sqrt{5}, 0 \le \theta \le 2\pi\}$

$\therefore \iint_{x^2 + y^2 \le 5} 3(9 - (x^2 + y^2))^{-\frac{1}{2}} dxdy = \int_0^{2\pi} \int_0^{\sqrt{5}} 3(9 - r^2)^{-\frac{1}{2}} r\, dr\, d\theta = 6\pi(9 - r^2)^{\frac{1}{2}} \Big|_{\sqrt{5}}^{0} = 6\pi$

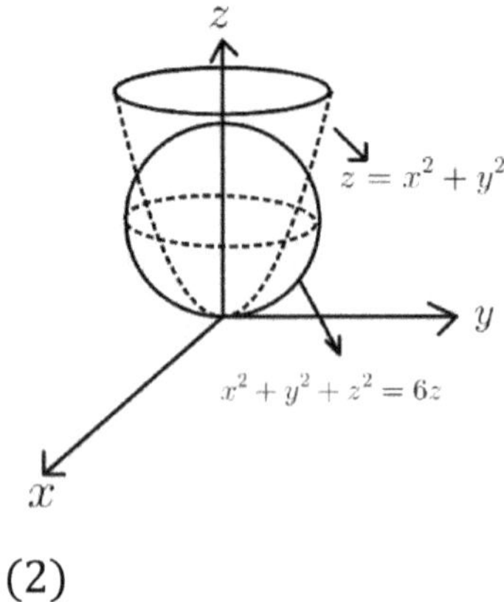


(2)

令 $S: x^2 + y^2 + z^2 = 2az,\ z \le x^2 + y^2$ 則 表面积 $= \displaystyle\iint_S dA$

令 $\vec{r}(x,y) = \left(x, y, a + \sqrt{a^2 - (x^2 + y^2)}\right)$ 且 $R$为球面$S$投影至$xy$平面的封闭区域

則$R = \{(x,y): x^2 + y^2 \le 2a - 1\}$ 且 $\displaystyle\iint_S dA = \iint_R \left|\dfrac{\partial \vec{r}}{\partial x} \times \dfrac{\partial \vec{r}}{\partial y}\right| dxdy$

$\because \dfrac{\partial \vec{r}}{\partial x} = \left(1, 0, -x(a^2 - (x^2 + y^2))^{-\frac{1}{2}}\right)$ 且 $\dfrac{\partial \vec{r}}{\partial y} = \left(0, 1, -y(a^2 - (x^2 + y^2))^{-\frac{1}{2}}\right)$

$\therefore \dfrac{\partial \vec{r}}{\partial x} \times \dfrac{\partial \vec{r}}{\partial y} = \left(x(a^2 - (x^2 + y^2))^{-\frac{1}{2}}, y(a^2 - (x^2 + y^2))^{-\frac{1}{2}}, 1\right)$

$\Rightarrow \left|\dfrac{\partial \vec{r}}{\partial x} \times \dfrac{\partial \vec{r}}{\partial y}\right| = (x^2(a^2 - (x^2 + y^2))^{-1} + y^2(a^2 - (x^2 + y^2))^{-1} + 1)^{\frac{1}{2}}$

$= a(a^2 - (x^2 + y^2))^{-\frac{1}{2}}$

$\therefore \displaystyle\iint_R \left|\dfrac{\partial \vec{r}}{\partial x} \times \dfrac{\partial \vec{r}}{\partial y}\right| dxdy = \iint_{x^2 + y^2 \le 2a - 1} a(a^2 - (x^2 + y^2))^{-\frac{1}{2}} dxdy$

令$x = r\cos\theta,\ y = r\sin\theta$

則 $\{(x,y): x^2 + y^2 \le 2a - 1\} = \{(r, \theta): 0 \le r \le \sqrt{2a - 1}, 0 \le \theta \le 2\pi\}$

$$\therefore \iint_{x^2+y^2 \leq 2a-1} a(a^2-(x^2+y^2))^{-\frac{1}{2}} dxdy = \int_0^{2\pi}\int_0^{\sqrt{2a-1}} a(a^2-r^2)^{-\frac{1}{2}} rdrd\theta$$

$$= 2a\pi(a^2-r^2)^{\frac{1}{2}}\Big|_{\sqrt{2a-1}}^{0} = 2a\pi(a-(a-1)) = 2a\pi$$

Example 25.

(1) 半径为 $r$ 的球面被两个相距为 $h$ 的平行平面所截, 试证:所截球的表面积仅与 $h$ 有关, 与所截位置无关

(2) 假设 $-r < a < b < r$, 求单位球面 $x^2 + y^2 + z^2 = r^2$ 在 $z = a$ 与 $z = b$ 之间的表面积 $(r > 0)$

【解】

(1)

设球心离较近的平面距离 $= b$

令 $S: \vec{r}(\theta, \varphi) = (r\cos\theta\sin\varphi, r\sin\theta\sin\varphi, r\cos\varphi), \cos^{-1}\dfrac{b+h}{r} \leq \varphi \leq \cos^{-1}\dfrac{b}{r}, 0 \leq \theta \leq 2\pi$

则 表面积 $= \displaystyle\iint_S dA$

$\because \dfrac{\partial\vec{r}}{\partial\theta} = (-r\sin\theta\sin\varphi, r\cos\theta\sin\varphi, 0)$ 且 $\dfrac{\partial\vec{r}}{\partial\varphi} = (r\cos\theta\cos\varphi, r\sin\theta\cos\varphi, -r\sin\varphi)$

$\therefore \dfrac{\partial\vec{r}}{\partial\varphi} \times \dfrac{\partial\vec{r}}{\partial\theta} = (r^2\cos\theta\sin^2\varphi, r^2\sin\theta\sin^2\varphi, r^2\sin\varphi\cos\varphi) \Rightarrow \left|\dfrac{\partial\vec{r}}{\partial\varphi} \times \dfrac{\partial\vec{r}}{\partial\theta}\right| = r^2\sin\varphi$

$\therefore \displaystyle\iint_S dA = \iint_R \left|\dfrac{\partial\vec{r}}{\partial\varphi} \times \dfrac{\partial\vec{r}}{\partial\theta}\right| d\theta d\varphi = \int_0^{2\pi}\int_{\cos^{-1}\frac{b+h}{r}}^{\cos^{-1}\frac{b}{r}} r^2\sin\varphi \, d\varphi d\theta$

$$= -r^2\int_0^{2\pi} \cos\varphi\Big|_{\cos^{-1}\frac{b+h}{r}}^{\cos^{-1}\frac{b}{r}} d\theta = 2\pi rh$$

$\therefore$ 所截球的表面积仅与 $h$ 有关

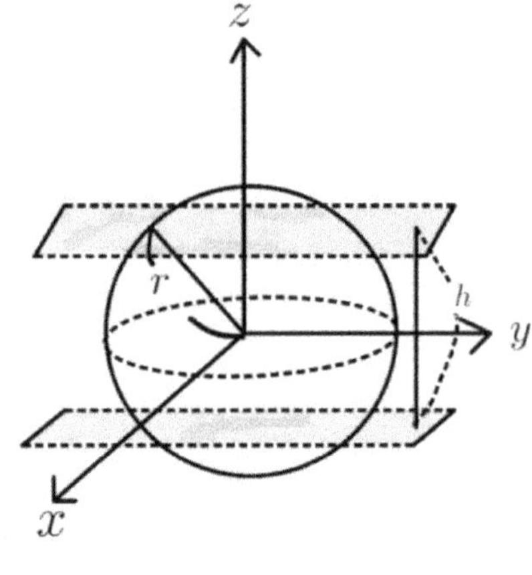


(2)

令 $S: x^2 + y^2 + z^2 = r^2, \ a \leq z \leq b$ 则表面积 $= \displaystyle\iint_S dA$

令 $\vec{r}(\theta, \varphi) = (r\cos\theta\sin\varphi, r\sin\theta\sin\varphi, r\cos\varphi)$

令 $R = \left\{(\theta, \varphi): 0 \leq \theta \leq 2\pi, \cos^{-1}\dfrac{b}{r} \leq \varphi \leq \cos^{-1}\dfrac{a}{r}\right\}$ 则 $\displaystyle\iint_S dA = \iint_R \left|\dfrac{\partial\vec{r}}{\partial\varphi} \times \dfrac{\partial\vec{r}}{\partial\theta}\right| d\theta d\varphi$

$\because \dfrac{\partial\vec{r}}{\partial\theta} = (-r\sin\theta\sin\varphi, r\cos\theta\sin\varphi, 0)$ 且 $\dfrac{\partial\vec{r}}{\partial\varphi} = (r\cos\theta\cos\varphi, r\sin\theta\cos\varphi, -r\sin\varphi)$

$\therefore \dfrac{\partial\vec{r}}{\partial\varphi} \times \dfrac{\partial\vec{r}}{\partial\theta} = (r^2\cos\theta\sin^2\varphi, r^2\sin\theta\sin^2\varphi, r^2\sin\varphi\cos\varphi) \Rightarrow \left|\dfrac{\partial\vec{r}}{\partial\varphi} \times \dfrac{\partial\vec{r}}{\partial\theta}\right| = r^2\sin\varphi$

$\therefore \displaystyle\iint_S dA = \iint_R \left|\dfrac{\partial\vec{r}}{\partial\varphi} \times \dfrac{\partial\vec{r}}{\partial\theta}\right| d\theta d\varphi = r^2 \int_0^{2\pi} \int_{\cos^{-1}\frac{b}{r}}^{\cos^{-1}\frac{a}{r}} \sin\varphi\, d\varphi d\theta = -r^2 \int_0^{2\pi} \cos\varphi\Big|_{\cos^{-1}\frac{b}{r}}^{\cos^{-1}\frac{a}{r}} d\theta$

$= 2\pi r(b - a)$

Example 26.

假设 $S_1 = \left\{(x, y, z): x^2 + y^2 + z^2 = 1, z \geq \dfrac{\sqrt{2}}{2}\right\}, S_2 = \left\{(x, y, z): x^2 + y^2 \leq 1, z = \dfrac{\sqrt{2}}{2}\right\}$

求 $S_1 \cup S_2$ 的表面积

【解】

令 $S_1 = \left\{(x, y, z): x^2 + y^2 + z^2 = 1, z \geq \dfrac{\sqrt{2}}{2}\right\}$ 且 $S_2 = \left\{(x, y, z): x^2 + y^2 \leq 1, z = \dfrac{\sqrt{2}}{2}\right\}$

则表面积 $= \iint_{S_1} dA + \iint_{S_2} dA$

令 $\vec{r}(\theta, \varphi) = (\cos\theta \sin\varphi, \sin\theta \sin\varphi, \cos\varphi)$ 且 $R = \left\{(\theta, \varphi): 0 \leq \theta \leq 2\pi, 0 \leq \varphi \leq \dfrac{\pi}{4}\right\}$

则 $\displaystyle\iint_{S_1} dA = \iint_{R} \left|\dfrac{\partial\vec{r}}{\partial\varphi} \times \dfrac{\partial\vec{r}}{\partial\theta}\right| d\theta d\varphi$

$\because \dfrac{\partial\vec{r}}{\partial\theta} = (-\sin\theta \sin\varphi, \cos\theta \sin\varphi, 0)$ 且 $\dfrac{\partial\vec{r}}{\partial\varphi} = (\cos\theta \cos\varphi, \sin\theta \cos\varphi, -\sin\varphi)$

$\therefore \dfrac{\partial\vec{r}}{\partial\varphi} \times \dfrac{\partial\vec{r}}{\partial\theta} = (\cos\theta \sin^2\varphi, \sin\theta \sin^2\varphi, \sin\varphi \cos\varphi) \Rightarrow \left|\dfrac{\partial\vec{r}}{\partial\varphi} \times \dfrac{\partial\vec{r}}{\partial\theta}\right| = \sin\varphi$

$\therefore \displaystyle\iint_{S_1} dA = \iint_{R} \left|\dfrac{\partial\vec{r}}{\partial\varphi} \times \dfrac{\partial\vec{r}}{\partial\theta}\right| d\theta d\varphi = \int_0^{2\pi} \int_0^{\frac{\pi}{4}} \sin\varphi \, d\varphi d\theta = 2\pi\left(1 - \dfrac{1}{\sqrt{2}}\right)$

$\because S_2 = \left\{(x, y, z): x^2 + y^2 \leq \dfrac{1}{2}, z = \dfrac{\sqrt{2}}{2}\right\} \quad \therefore \displaystyle\iint_{S_2} dA = \dfrac{\pi}{2} \Rightarrow$ 表面积 $= \dfrac{\pi}{2} + 2\pi\left(1 - \dfrac{1}{\sqrt{2}}\right)$

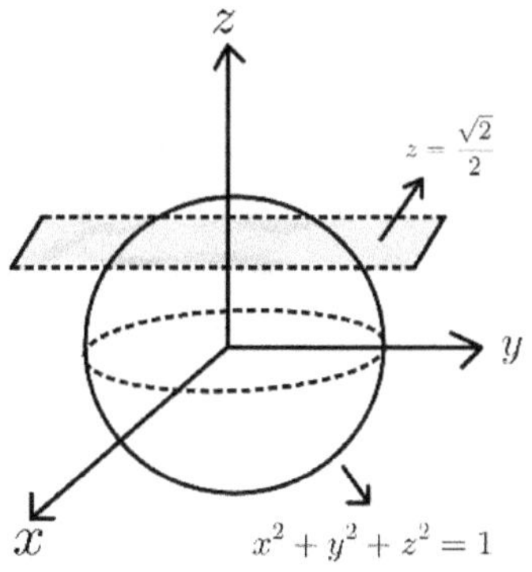

Example 27.

$$S = \{(x, y, z): x^2 + y^2 + z^2 = r^2, z \geq 0\} \text{ 求}S\text{的表面积}$$

【解】

令$\vec{r}(\theta, \varphi) = (r \cos \theta \sin \varphi, r \sin \theta \sin \varphi, r\cos \varphi)$ 且 $R = \left\{(\theta, \varphi): 0 \leq \theta \leq 2\pi, 0 \leq \varphi \leq \frac{\pi}{2}\right\}$

则 $\iint_S dA = \iint_R \left|\frac{\partial \vec{r}}{\partial \varphi} \times \frac{\partial \vec{r}}{\partial \theta}\right| d\theta d\varphi$

$\because \dfrac{\partial \vec{r}}{\partial \theta} = (-r \sin \theta \sin \varphi, r \cos \theta \sin \varphi, 0)$ 且 $\dfrac{\partial \vec{r}}{\partial \varphi} = (r \cos \theta \cos \varphi, r \sin \theta \cos \varphi, -r\sin \varphi)$

$\therefore \dfrac{\partial \vec{r}}{\partial \varphi} \times \dfrac{\partial \vec{r}}{\partial \theta} = (r^2 \cos \theta \sin^2 \varphi, r^2 \sin \theta \sin^2 \varphi, r^2 \sin \varphi \cos \varphi) \Rightarrow \left|\dfrac{\partial \vec{r}}{\partial \varphi} \times \dfrac{\partial \vec{r}}{\partial \theta}\right| = r^2 \sin \varphi$

$\therefore$ 表面积 $= \displaystyle\iint_S dA = \iint_R \left|\frac{\partial \vec{r}}{\partial \varphi} \times \frac{\partial \vec{r}}{\partial \theta}\right| d\theta d\varphi = \int_0^{2\pi} \int_0^{\frac{\pi}{2}} r^2 \sin \varphi \, d\varphi d\theta = 2\pi r^2$

Example 28.

$$S = \left\{(x, y, z): x^2 + y^2 + z^2 = r^2, \frac{r^2}{4} \leq x^2 + y^2 \leq \frac{3r^2}{4}, z > 0\right\}, \text{ 求}S\text{的表面积}$$

【解】

令 $\vec{r}(\theta, \varphi) = (r \cos \theta \sin \varphi, r \sin \theta \sin \varphi, r\cos \varphi)$

令 $R = \left\{(\theta, \varphi): 0 \leq \theta \leq 2\pi, \cos^{-1} \frac{\sqrt{3}}{2} \leq \varphi \leq \cos^{-1} \frac{1}{2}\right\}$ 则 $\displaystyle\iint_S dA = \iint_R \left|\frac{\partial \vec{r}}{\partial \varphi} \times \frac{\partial \vec{r}}{\partial \theta}\right| d\theta d\varphi$

$\because \dfrac{\partial \vec{r}}{\partial \theta} = (-r \sin \theta \sin \varphi, r \cos \theta \sin \varphi, 0)$ 且 $\dfrac{\partial \vec{r}}{\partial \varphi} = (r \cos \theta \cos \varphi, r \sin \theta \cos \varphi, -r\sin \varphi)$

$\therefore \dfrac{\partial \vec{r}}{\partial \varphi} \times \dfrac{\partial \vec{r}}{\partial \theta} = (r^2 \cos \theta \sin^2 \varphi, r^2 \sin \theta \sin^2 \varphi, r^2 \sin \varphi \cos \varphi) \Rightarrow \left|\dfrac{\partial \vec{r}}{\partial \varphi} \times \dfrac{\partial \vec{r}}{\partial \theta}\right| = r^2 \sin \varphi$

則 $\displaystyle\iint_S dA = \iint_R \left|\frac{\partial\vec{r}}{\partial\varphi}\times\frac{\partial\vec{r}}{\partial\theta}\right| d\theta d\varphi = r^2\int_0^{2\pi}\int_{\cos^{-1}\frac{\sqrt{3}}{2}}^{\cos^{-1}\frac{1}{2}}\sin\varphi\, d\varphi d\theta = -r^2\int_0^{2\pi}\cos\varphi\Big|_{\cos^{-1}\frac{\sqrt{3}}{2}}^{\cos^{-1}\frac{1}{2}}\, d\theta$

$= r^2\pi(\sqrt{3}-1)$

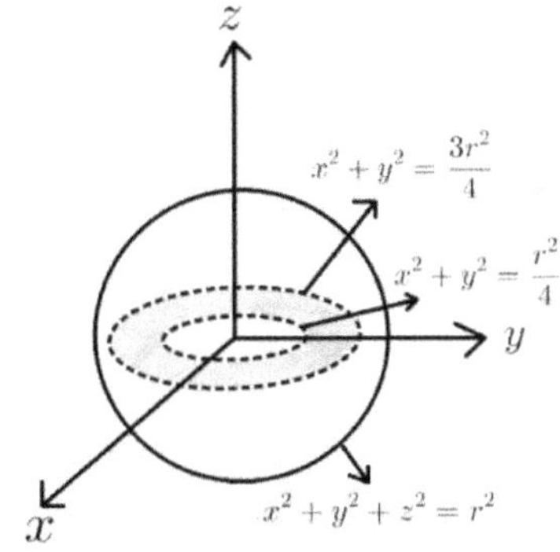


Example 29.

    (1) 求球面 $x^2+y^2+z^2=a^2$ 在圓柱 $x^2+y^2=b^2$ 內且位于 $xy$ 平面上方的表面積, $a>b>0$

    (2) 求球面 $x^2+y^2+z^2=16$ 在圓柱 $(x-2)^2+y^2=4$ 內的表面積

    (3) 求球面 $x^2+y^2+z^2=4a^2$ 在圓柱 $x^2+y^2=2ay$ 內的表面積

【解】

(1)

令 $S: x^2+y^2+z^2=a^2,\ x^2+y^2\le b^2,\ z\ge 0$ 則表面積 $=\displaystyle\iint_S dA$

令 $\vec{r}(x,y)=\left(x,y,\sqrt{a^2-(x^2+y^2)}\right)$ 且 $R$ 为球面投影至 $xy$ 平面的封閉區域

則 $R=\{(x,y): x^2+y^2\le b^2\}$ 且 $\displaystyle\iint_S dA = \iint_R \left|\frac{\partial\vec{r}}{\partial x}\times\frac{\partial\vec{r}}{\partial y}\right| dxdy$

$\because \dfrac{\partial\vec{r}}{\partial x}=\left(1,0,-x(a^2-(x^2+y^2))^{-\frac{1}{2}}\right)$ 且 $\dfrac{\partial\vec{r}}{\partial y}=\left(0,1,-y(a^2-(x^2+y^2))^{-\frac{1}{2}}\right)$

$\therefore \dfrac{\partial\vec{r}}{\partial x}\times\dfrac{\partial\vec{r}}{\partial y}=\left(x(a^2-(x^2+y^2))^{-\frac{1}{2}}, y(a^2-(x^2+y^2))^{-\frac{1}{2}}, 1\right)$

$$\Rightarrow \left|\frac{\partial \vec{r}}{\partial x} \times \frac{\partial \vec{r}}{\partial y}\right| = (x^2(a^2 - (x^2 + y^2))^{-1} + y^2(a^2 - (x^2 + y^2))^{-1} + 1)^{\frac{1}{2}}$$

$$= a(a^2 - (x^2 + y^2))^{-\frac{1}{2}}$$

$$\therefore \iint_S dA = \iint_R \left|\frac{\partial \vec{r}}{\partial x} \times \frac{\partial \vec{r}}{\partial y}\right| dxdy = \iint_{x^2+y^2\leq b^2} a(a^2 - (x^2 + y^2))^{-\frac{1}{2}}dxdy$$

令$x = r\cos\theta, y = r\sin\theta$ 則 $\{(x,y): x^2 + y^2 \leq b^2\} = \{(r,\theta): 0 \leq r \leq b, 0 \leq \theta \leq 2\pi\}$

$$\therefore \iint_{x^2+y^2\leq b^2} a(a^2 - (x^2 + y^2))^{-\frac{1}{2}}dxdy = \int_0^{2\pi} \int_0^b a(a^2 - r^2)^{-\frac{1}{2}} rdrd\theta = 2\pi a(a - \sqrt{a^2 - b^2})$$

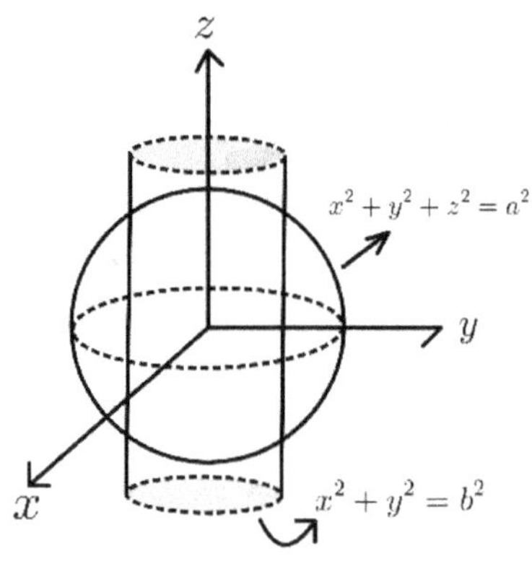


(2)

令$S: x^2 + y^2 + z^2 = 16, \ (x - 2)^2 + y^2 \leq 4, \ z \geq 0$ 則表面積 $= 2\iint_S dA$

令$\vec{r}(x,y) = \left(x, y, \sqrt{16 - (x^2 + y^2)}\right)$ 且 $R$为球面投影至$xy$平面的封闭区域

則 $R = \{(x,y): (x - 2)^2 + y^2 \leq 4\}$ 且 $\iint_S dA = \iint_R \left|\frac{\partial \vec{r}}{\partial x} \times \frac{\partial \vec{r}}{\partial y}\right| dxdy$

$$\therefore \frac{\partial \vec{r}}{\partial x} = \left(1, 0, -x(16 - (x^2 + y^2))^{-\frac{1}{2}}\right) \text{ 且 } \frac{\partial \vec{r}}{\partial y} = \left(0, 1, -y(16 - (x^2 + y^2))^{-\frac{1}{2}}\right)$$

$$\therefore \frac{\partial \vec{r}}{\partial x} \times \frac{\partial \vec{r}}{\partial y} = \left( x(16 - (x^2 + y^2))^{-\frac{1}{2}}, y(16 - (x^2 + y^2))^{-\frac{1}{2}}, 1 \right)$$

$$\Rightarrow \left| \frac{\partial \vec{r}}{\partial x} \times \frac{\partial \vec{r}}{\partial y} \right| = (x^2(16 - (x^2 + y^2))^{-1} + y^2(16 - (x^2 + y^2))^{-1} + 1)^{\frac{1}{2}}$$

$$= \left( \frac{x^2 + y^2}{16 - (x^2 + y^2)} + 1 \right)^{\frac{1}{2}} = \left( \frac{x^2 + y^2 + 16 - (x^2 + y^2)}{16 - (x^2 + y^2)} \right)^{\frac{1}{2}} = 4(16 - (x^2 + y^2))^{-\frac{1}{2}}$$

$$\therefore 2 \iint_S dA = 2 \iint_R \left| \frac{\partial \vec{r}}{\partial x} \times \frac{\partial \vec{r}}{\partial y} \right| dxdy = 2 \iint_R 4(16 - (x^2 + y^2))^{-\frac{1}{2}} dxdy$$

令 $x = r\cos\theta , y = r\sin\theta$

則 $\{(x,y): (x-2)^2 + y^2 \le 4\} = \{(r,\theta): 0 \le r \le 4\cos\theta , -\frac{\pi}{2} \le \theta \le \frac{\pi}{2}\}$

$$\therefore 2 \iint_R 4(16 - (x^2 + y^2))^{-\frac{1}{2}} dxdy = 8 \int_{\frac{-\pi}{2}}^{\frac{\pi}{2}} \int_0^{4\cos\theta} (16 - r^2)^{-\frac{1}{2}} r\, dr\, d\theta$$

$$= 16 \int_0^{\frac{\pi}{2}} (-1)(16 - r^2)^{\frac{1}{2}} \Big|_0^{4\cos\theta} d\theta = 16 \int_0^{\frac{\pi}{2}} 4 - 4\sin\theta \, d\theta = 32(\pi - 2)$$

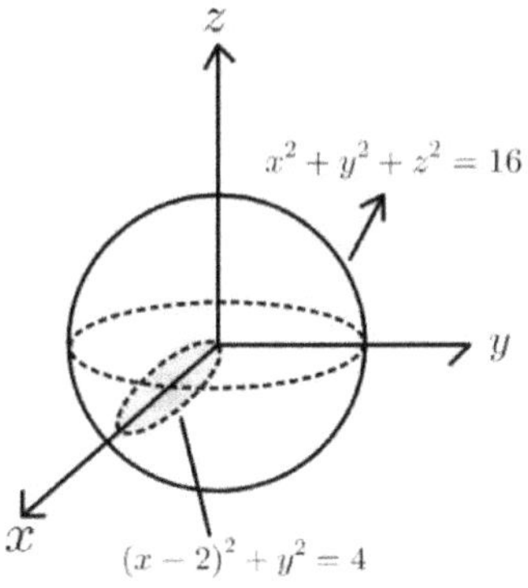



(3)

令 $S: x^2 + y^2 + z^2 = 4a^2, \ x^2 + y^2 \leq 2ay, \ z \geq 0$ 則表面積 $= 2\iint_S dA$

令 $\vec{r}(x,y) = \left(x, y, \sqrt{4a^2 - (x^2 + y^2)}\right)$ 且 $R$ 為球面投影至 $xy$ 平面的封閉區域

則 $R = \{(x,y): x^2 + y^2 \leq 2ay\}$ 且 $\iint_S dA = \iint_R \left|\dfrac{\partial \vec{r}}{\partial x} \times \dfrac{\partial \vec{r}}{\partial y}\right| dxdy$

$\because \dfrac{\partial \vec{r}}{\partial x} = \left(1, 0, -x(4a^2 - (x^2 + y^2))^{-\frac{1}{2}}\right)$ 且 $\dfrac{\partial \vec{r}}{\partial y} = \left(0, 1, -y(4a^2 - (x^2 + y^2))^{-\frac{1}{2}}\right)$

$\therefore \dfrac{\partial \vec{r}}{\partial x} \times \dfrac{\partial \vec{r}}{\partial y} = \left(x(4a^2 - (x^2 + y^2))^{-\frac{1}{2}}, y(4a^2 - (x^2 + y^2))^{-\frac{1}{2}}, 1\right)$

$\Rightarrow \left|\dfrac{\partial \vec{r}}{\partial x} \times \dfrac{\partial \vec{r}}{\partial y}\right| = (x^2(4a^2 - (x^2 + y^2))^{-1} + y^2(4a^2 - (x^2 + y^2))^{-1} + 1)^{\frac{1}{2}}$

$= 2a(4a^2 - (x^2 + y^2))^{-\frac{1}{2}}$

$\therefore 2\iint_S dA = 2\iint_R \left|\dfrac{\partial \vec{r}}{\partial x} \times \dfrac{\partial \vec{r}}{\partial y}\right| dxdy = 2\iint_R 2a(4a^2 - (x^2 + y^2))^{-\frac{1}{2}}dxdy$

令 $x = r\cos\theta, y = r\sin\theta$ 則 $\{(x,y): x^2 + y^2 \leq 2ay\} = \{(r,\theta): 0 \leq r \leq 2a\sin\theta, 0 \leq \theta \leq \pi\}$

$\therefore 2\iint_R 2a(4a^2 - (x^2 + y^2))^{-\frac{1}{2}}dxdy = 2\int_0^{\pi} \int_0^{2a\sin\theta} 2a(4a^2 - r^2)^{-\frac{1}{2}} r\, dr\, d\theta$

$= 8a\int_0^{\frac{\pi}{2}} \int_0^{2a\sin\theta} (4a^2 - r^2)^{-\frac{1}{2}} r\, dr\, d\theta = -8a\int_0^{\frac{\pi}{2}} (4a^2 - r^2)^{\frac{1}{2}}\Big|_0^{2a\sin\theta} d\theta$

$= 16a^2 \int_0^{\frac{\pi}{2}} 1 - \cos\theta\, d\theta = 8a^2(\pi - 2)$

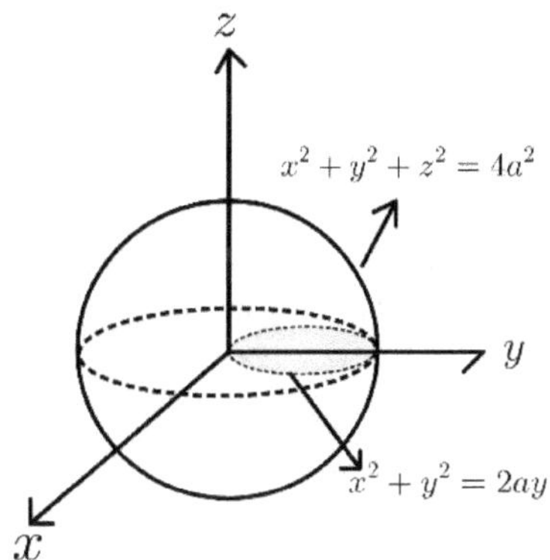


Example 30.

求曲面 $z^2 = x^2 + y^2,\ z \geq 0$ 在 $y^2 + z^2 \leq 2$ 內的表面积

【解】

令 $S: z^2 = x^2 + y^2,\ z \geq 0,\ y^2 + z^2 \leq 2$ 則表面积 $= \iint_S dA$

令 $\vec{r}(x,y) = \left(\sqrt{z^2 - y^2}, y, z\right)$ 且 $R$ 为曲面投影至 $yz$ 平面的封闭区域

則 $R = \left\{(y,z): 0 \leq y \leq 1, y \leq z \leq \sqrt{2 - y^2}\right\}$ 且 $\iint_S dA = \iint_R \left|\dfrac{\partial \vec{r}}{\partial y} \times \dfrac{\partial \vec{r}}{\partial z}\right| dydz$

$\because \dfrac{\partial \vec{r}}{\partial y} = \left(-y(z^2 - y^2)^{-\frac{1}{2}}, 1, 0\right)$ 且 $\dfrac{\partial \vec{r}}{\partial z} = \left(z(z^2 - y^2)^{-\frac{1}{2}}, 0, 1\right)$

$\therefore \dfrac{\partial \vec{r}}{\partial y} \times \dfrac{\partial \vec{r}}{\partial z} = \left(1, y(z^2 - y^2)^{-\frac{1}{2}}, -z(z^2 - y^2)^{-\frac{1}{2}}\right)$

$\Rightarrow \left|\dfrac{\partial \vec{r}}{\partial y} \times \dfrac{\partial \vec{r}}{\partial z}\right| = (1 + y^2(z^2 - y^2)^{-1} + z^2(z^2 - y^2)^{-1})^{\frac{1}{2}} = \sqrt{2}z(z^2 - y^2)^{-\frac{1}{2}}$

$\therefore \iint_S dA = \iint_R \left|\dfrac{\partial \vec{r}}{\partial y} \times \dfrac{\partial \vec{r}}{\partial z}\right| dydz = 4\int_0^1 \int_y^{\sqrt{2-y^2}} \sqrt{2}z(z^2 - y^2)^{-\frac{1}{2}} dzdy$

$$= 4\sqrt{2} \int_0^1 (z^2 - y^2)^{\frac{1}{2}} \Big|_y^{\sqrt{2-y^2}} \, dy = 4\sqrt{2} \int_0^1 (2 - 2y^2)^{\frac{1}{2}} \, dy = 8 \int_0^1 (1 - y^2)^{\frac{1}{2}} \, dy$$

令 $y = \sin\theta$ 則 $8 \int_0^1 (1 - y^2)^{\frac{1}{2}} \, dy = 8 \int_0^{\frac{\pi}{2}} \cos^2\theta \, d\theta = 8 \int_0^{\frac{\pi}{2}} \frac{1 + \cos 2\theta}{2} \, d\theta = 2\pi$





Example 31.

求曲面 $z^2 = x^2 + y^2$ 在圓柱 $x^2 + y^2 = 2ay$ 內的表面积

【解】

令 $S: z^2 = x^2 + y^2, \ x^2 + y^2 \leq 2ay, \ z \geq 0$ 則表面积 $= 2 \iint_S dA$

令 $\vec{r}(x, y) = \left(x, y, \sqrt{x^2 + y^2}\right)$ 且 $R$ 为曲面 $S$ 投影至 $xy$ 平面的封閉区域

則 $R = \{(x, y): x^2 + (y - a)^2 \leq a^2\}$ 且 $\iint_S dA = \iint_R \left|\frac{\partial \vec{r}}{\partial x} \times \frac{\partial \vec{r}}{\partial y}\right| dxdy$

$\because \ \frac{\partial \vec{r}}{\partial x} = \left(1, 0, x(x^2 + y^2)^{-\frac{1}{2}}\right)$ 且 $\frac{\partial \vec{r}}{\partial y} = \left(0, 1, y(x^2 + y^2)^{-\frac{1}{2}}\right)$

$\therefore \ \frac{\partial \vec{r}}{\partial x} \times \frac{\partial \vec{r}}{\partial y} = \left(x(x^2 + y^2)^{-\frac{1}{2}}, y(x^2 + y^2)^{-\frac{1}{2}}, 1\right)$

$$\Rightarrow \left| \frac{\partial \vec{r}}{\partial x} \times \frac{\partial \vec{r}}{\partial y} \right| = \left( (x^2 + y^2)(x^2 + y^2)^{-1} + 1 \right)^{\frac{1}{2}} = \sqrt{2}$$

$$\therefore 2 \iint_S dA = 2 \iint_R \left| \frac{\partial \vec{r}}{\partial x} \times \frac{\partial \vec{r}}{\partial y} \right| dxdy = 2\sqrt{2} \iint_{x^2+(y-a)^2 \le a^2} dxdy = 2\sqrt{2}\pi a^2$$

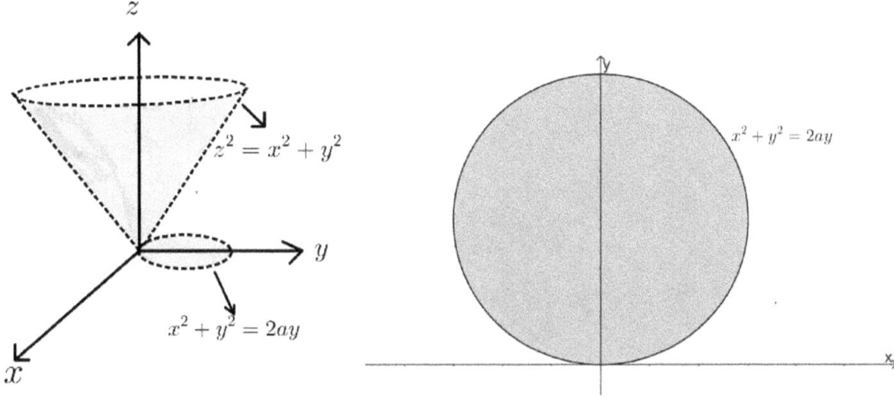


Example 32.

　　求曲面 $z^2 = x^2 + y^2$ 在橢圓柱 $\dfrac{x^2}{a^2} + \dfrac{y^2}{b^2} = 1$ 內的表面积

【解】

令 $S: z^2 = x^2 + y^2$,  $\dfrac{x^2}{a^2} + \dfrac{y^2}{b^2} \le 1$,  $z \ge 0$ 則 表面积 $= 2 \iint_S dA$

令 $\vec{r}(x,y) = \left( x, y, \sqrt{x^2 + y^2} \right)$ 且 $R$ 为曲面 $S$ 投影至 $xy$ 平面的封閉区域

則 $R = \left\{ (x,y) : \dfrac{x^2}{a^2} + \dfrac{y^2}{b^2} \le 1 \right\}$ 且 $\iint_S dA = \iint_R \left| \dfrac{\partial \vec{r}}{\partial x} \times \dfrac{\partial \vec{r}}{\partial y} \right| dxdy$

$\because \dfrac{\partial \vec{r}}{\partial x} = \left( 1, 0, x(x^2 + y^2)^{-\frac{1}{2}} \right)$ 且 $\dfrac{\partial \vec{r}}{\partial y} = \left( 0, 1, y(x^2 + y^2)^{-\frac{1}{2}} \right)$

$\therefore \dfrac{\partial \vec{r}}{\partial x} \times \dfrac{\partial \vec{r}}{\partial y} = \left( x(x^2 + y^2)^{-\frac{1}{2}}, y(x^2 + y^2)^{-\frac{1}{2}}, 1 \right)$

$$\Rightarrow \left|\frac{\partial \vec{r}}{\partial x} \times \frac{\partial \vec{r}}{\partial y}\right| = \left((x^2 + y^2)(x^2 + y^2)^{-1} + 1\right)^{\frac{1}{2}} = \sqrt{2}$$

$$\therefore 2\iint_S dA = 2\iint_R \left|\frac{\partial \vec{r}}{\partial x} \times \frac{\partial \vec{r}}{\partial y}\right| dxdy = 2\sqrt{2}\iint_{\frac{x^2}{a^2}+\frac{y^2}{b^2}\leq 1} dxdy = 2\sqrt{2}ab\pi$$

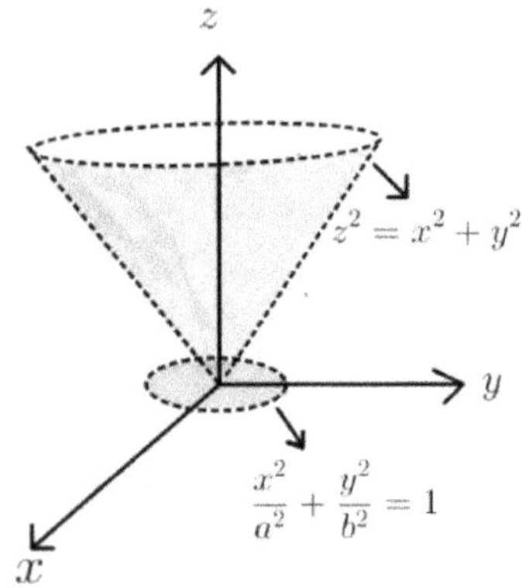


Example 33.

　　求曲面 $z^2 = x^2 + y^2$ 在 $y = x^2$，$y = 3x^2$，$x = y^2$，$x = 4y^2$ 所围区域内的表面积

【解】

令 $S: z^2 = x^2 + y^2, y \geq x^2$，$y \leq 3x^2$，$x \geq y^2$，$x \leq 4y^2$ 则表面积 $= 2\iint_S dA$

令 $\vec{r}(x, y) = \left(x, y, \sqrt{x^2 + y^2}\right)$ 且 $R = \left\{(x, y): 1 \leq \frac{y}{x^2} \leq 3, 1 \leq \frac{x}{y^2} \leq 4\right\}$

则 $\iint_S dA = \iint_R \left|\frac{\partial \vec{r}}{\partial x} \times \frac{\partial \vec{r}}{\partial y}\right| dxdy$

$\because \frac{\partial \vec{r}}{\partial x} = \left(1, 0, x(x^2 + y^2)^{-\frac{1}{2}}\right)$ 且 $\frac{\partial \vec{r}}{\partial y} = \left(0, 1, y(x^2 + y^2)^{-\frac{1}{2}}\right)$

$\therefore \frac{\partial \vec{r}}{\partial x} \times \frac{\partial \vec{r}}{\partial y} = \left(x(x^2 + y^2)^{-\frac{1}{2}}, y(x^2 + y^2)^{-\frac{1}{2}}, 1\right)$

$$\Rightarrow \left|\frac{\partial \vec{r}}{\partial x} \times \frac{\partial \vec{r}}{\partial y}\right| = \left((x^2+y^2)(x^2+y^2)^{-1}+1\right)^{\frac{1}{2}} = \sqrt{2}$$

$$\therefore 2\iint_S dA = 2\iint_R \left|\frac{\partial \vec{r}}{\partial x}\times\frac{\partial \vec{r}}{\partial y}\right| dxdy = 2\sqrt{2}\iint_R dxdy$$

令 $u = \dfrac{y}{x^2}$, $v = \dfrac{x}{y^2}$ 則 $x = u^{-\frac{2}{3}}v^{-\frac{1}{3}}$, $y = u^{-\frac{1}{3}}v^{-\frac{2}{3}}$

$$\text{且 } dxdy = \left\|\begin{matrix}\dfrac{\partial x}{\partial u} & \dfrac{\partial x}{\partial v}\\[2mm] \dfrac{\partial y}{\partial u} & \dfrac{\partial y}{\partial v}\end{matrix}\right\| dudv = \left\|\begin{matrix}\dfrac{-2u^{-\frac{5}{3}}v^{-\frac{1}{3}}}{3} & \dfrac{-u^{-\frac{2}{3}}v^{-\frac{4}{3}}}{3}\\[3mm] \dfrac{-u^{-\frac{4}{3}}v^{-\frac{2}{3}}}{3} & \dfrac{-2u^{-\frac{1}{3}}v^{-\frac{5}{3}}}{3}\end{matrix}\right\| dudv = \dfrac{u^{-2}v^{-2}}{3}dudv$$

$$\because R = \left\{(x,y): 1\le\frac{y}{x^2}\le 3, 1\le\frac{x}{y^2}\le 4\right\} \quad \therefore R = \{(u,v): 1\le u\le 3, 1\le v\le 4\}$$

$$\therefore \iint_R 1\,dA = \int_1^3\int_1^4 \frac{u^{-2}v^{-2}}{3}dvdu = \frac{1}{6} \quad \therefore \text{表面积} = 2\iint_S dA = 2\sqrt{2}\iint_R dxdy = \frac{\sqrt{2}}{3}$$

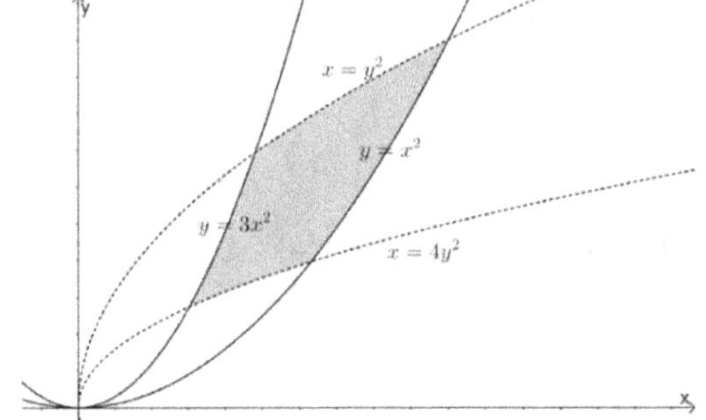


Example 34.

　　求曲面 $z^2 = x^2+y^2$ 在 $y=x$, $y=2x$, $xy=1$, $xy=4$ 所围区域內的表面积

【解】

令 $S: z^2 = x^2+y^2$, $y\ge x, y\le 2x$, $xy\ge 1$, $xy\le 4$ 則表面積 $= 2\displaystyle\iint_S dA$

令 $\vec{r}(x,y) = \left(x, y, \sqrt{x^2+y^2}\right)$ 且 $R = \left\{(x,y): 1\le\dfrac{y}{x}\le 2, 1\le xy\le 4\right\}$

則 $\displaystyle\iint_S dA = \iint_R \left|\frac{\partial \vec{r}}{\partial x} \times \frac{\partial \vec{r}}{\partial y}\right| dxdy$

$\because \dfrac{\partial \vec{r}}{\partial x} = \left(1, 0, x(x^2+y^2)^{-\frac{1}{2}}\right)$ 且 $\dfrac{\partial \vec{r}}{\partial y} = \left(0, 1, y(x^2+y^2)^{-\frac{1}{2}}\right)$

$\therefore \dfrac{\partial \vec{r}}{\partial x} \times \dfrac{\partial \vec{r}}{\partial y} = \left(x(x^2+y^2)^{-\frac{1}{2}}, y(x^2+y^2)^{-\frac{1}{2}}, 1\right)$

$\Rightarrow \left|\dfrac{\partial \vec{r}}{\partial x} \times \dfrac{\partial \vec{r}}{\partial y}\right| = \left((x^2+y^2)(x^2+y^2)^{-1} + 1\right)^{\frac{1}{2}} = \sqrt{2}$

$\therefore 2\displaystyle\iint_S dA = 2\iint_R \left|\frac{\partial \vec{r}}{\partial x} \times \frac{\partial \vec{r}}{\partial y}\right| dxdy = 2\sqrt{2}\iint_R dxdy$

令 $\dfrac{y}{x} = u, \ xy = v$ 則 $x = \sqrt{\dfrac{v}{u}}, \ y = \sqrt{uv}$

且 $dxdy = \left\|\begin{matrix} \dfrac{\partial x}{\partial u} & \dfrac{\partial x}{\partial v} \\ \dfrac{\partial y}{\partial u} & \dfrac{\partial y}{\partial v} \end{matrix}\right\| dudv = \left\|\begin{matrix} \dfrac{1}{2}\left(\dfrac{v}{u}\right)^{-\frac{1}{2}}\left(-\dfrac{v}{u^2}\right) & \dfrac{1}{2}\left(\dfrac{v}{u}\right)^{-\frac{1}{2}}\left(\dfrac{1}{u}\right) \\ \dfrac{1}{2}(uv)^{-\frac{1}{2}}v & \dfrac{1}{2}(uv)^{-\frac{1}{2}}u \end{matrix}\right\| dudv = \dfrac{1}{2u}dudv$

$\because R = \left\{(x,y): 1 \leq \dfrac{y}{x} \leq 2, 1 \leq xy \leq 4\right\}$  $\therefore R = \{(u,v): 1 \leq u \leq 2, 1 \leq v \leq 4\}$

$\therefore \displaystyle\iint_R dxdy = 2\int_1^4 \int_1^2 \frac{1}{2u}dudv = 3\ln 2$  $\therefore$ 表面積 $= 2\iint_S dA = 2\sqrt{2}\iint_R dxdy = 6\sqrt{2}\ln 2$

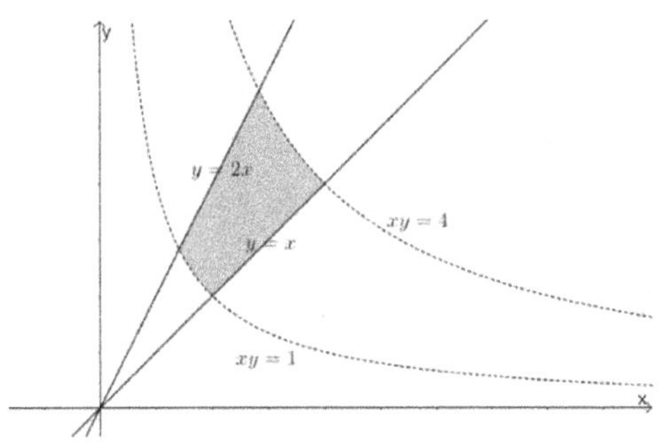

### 9.5.3    $f$为纯量函数, 給曲面$S$求质量

表面积分和表面积的关联如同线积分和弧长的关联, 考虑一个参数化曲面$S$:
$$\vec{r}(s,t) = \big(x(s,t), y(s,t), z(s,t)\big), \ \forall (s,t) \in D$$
假设参数定义域$D$为矩形, 将其分割为数个小矩形 $D_{ij}$, 长为 $\Delta s$, 宽为 $\Delta t$, $D_{ij}$的中心点为 $\big(s_i^*, t_j^*\big)$, $D_{ij}$对应曲面 $S$ 上的区域为 $S_{ij}$并且$\big(s_i^*, t_j^*\big)$对应曲面$S_{ij}$上的点为$E_{ij}^*$, 其位置向量为 $\vec{r}\big(s_i^*, t_j^*\big)$; 因此, 曲面$S$被分割成数个片段$S_{ij}$, 对于每个片段, $f$在点$E_{ij}^*$取值, 乘以片段的面积$\Delta S_{ij}$产生了黎曼和

$$\sum_{i=1}^{m} \sum_{j=1}^{n} f\big(E_{ij}^*\big) \Delta S_{ij}$$

随着分割数量的增加则可以藉由上述黎曼和的极限定义纯量函数$f$在曲面$S$上的表面积

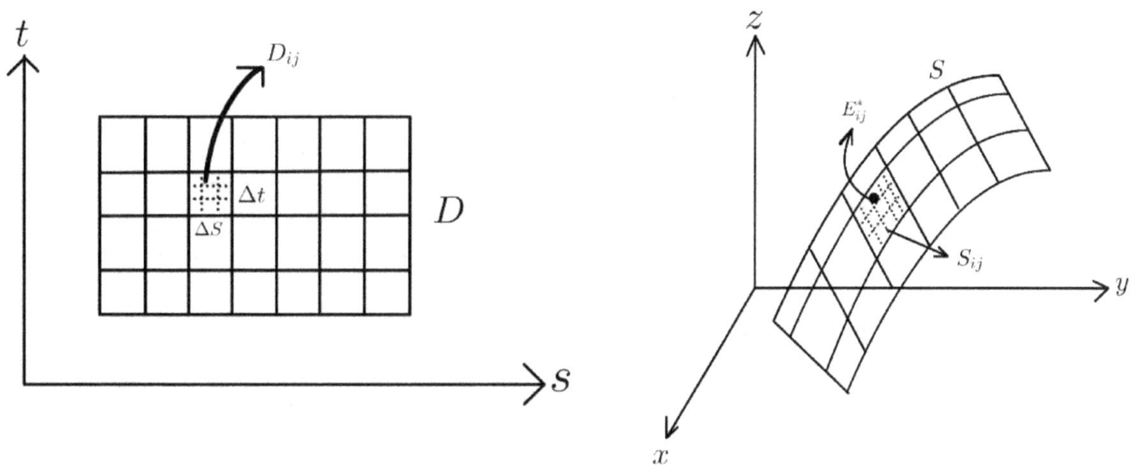


### 【定义】

纯量函数$f$于片段光滑曲面$S$的曲面积分定义如下

$$\iint_S f(x,y,z)\,dS = \lim_{m \to \infty, n \to \infty} \sum_{i=1}^{m} \sum_{j=1}^{n} f\big(E_{ij}^*\big) \Delta S_{ij},$$

其中$E_{ij}^*$, $\Delta S_{ij}$如上述

    值得注意的是, 因为任意定义域为闭区间的连续函数皆为黎曼可积分函数, 因此, 如果上述定义的函数$f$是一个连续函数, 则上述黎曼和的极限值必定存在; 为了计算上述定义

中的表面积分, 考虑位于切平面内的一个近似平行四边形的面积来逼近区域 $\Delta S_{ij}$, 之前在表面积的讨论中, $\Delta S_{ij}$ 有一个近似估计

$$\Delta S_{ij} \approx \left|\vec{r}_s(s_i^*, t_j^*) \times \vec{r}_t(s_i^*, t_j^*)\right|\Delta s \Delta t$$

所以

$$\iint_S f(x, y, z)dS = \lim_{m\to\infty, n\to\infty} \sum_{i=1}^{m} \sum_{j=1}^{n} f(E_{ij}^*)\Delta S_{ij}$$

$$= \lim_{m\to\infty, n\to\infty} \sum_{i=1}^{m} \sum_{j=1}^{n} f(E_{ij}^*)\left|\vec{r}_s(s_i^*, t_j^*) \times \vec{r}_t(s_i^*, t_j^*)\right|\Delta s \Delta t = \iint_D f(\vec{r}(s, t))\left|\vec{r}_s \times \vec{r}_t\right|dsdt$$

藉由上述观察可知, 计算曲面积分相当于求双重积分, 特别的是双重积分的被积分函数为

$$f(\vec{r}(s, t))\left|\vec{r}_s(s, t) \times \vec{r}_t(s, t)\right|$$

因此求曲面的积分相当于先求得被积分函数之后, 再计算双重积分; 底下针对曲面的积分转换成双重积分后, 有哪些不同类型作初步的说明

给定曲面 $S: z = g(x, y)$, $\forall a \leq x \leq b, c \leq y \leq d$, 目标是求表面积分 $\iint_S f(x, y, z)dA$, 令 $\vec{r}(x, y) = (x, y, g(x, y))$ 且 $R = \{(x, y): a \leq x \leq b, c \leq y \leq d\}$ 则表面积分为

$$\iint_S f(x, y, z)dS = \iint_R f(x, y, g(x, y))\left|\frac{\partial \vec{r}}{\partial x} \times \frac{\partial \vec{r}}{\partial y}\right|dxdy$$

因为 $\dfrac{\partial \vec{r}}{\partial x} = (1, 0, g_x(x, y))$ 且 $\dfrac{\partial \vec{r}}{\partial y} = (0, 1, g_y(x, y))$ 则两者的外积

$$\frac{\partial \vec{r}}{\partial x} \times \frac{\partial \vec{r}}{\partial y} = (-g_x(x, y), -g_y(x, y), 1)$$

且

$$\left|\frac{\partial \vec{r}}{\partial x} \times \frac{\partial \vec{r}}{\partial y}\right| = \sqrt{1 + (g_x(x, y))^2 + (g_y(x, y))^2}$$

因此, 表面积分可改写为

$$\iint_S f(x, y, z)dA = \int_a^b \int_c^d f(x, y, g(x, y))\sqrt{1 + (g_x(x, y))^2 + (g_y(x, y))^2}\, dydx$$

特别的是如果曲面为 $z = g(x, y)$, 投影至$xy$平面 $= \{(x, y): x^2 + y^2 \leq a\}$, 藉由极坐标转换, 令$x = r\cos\theta, y = r\sin\theta$ 则

$$\iint_S f(x, y, z)dA$$

$$= \int_0^{2\pi} \int_0^a f(x, y, g(x, y))\sqrt{1 + (g_x(x, y))^2 + (g_y(x, y))^2}\Bigg|_{x=r\cos\theta, y=r\sin\theta} r\, dr d\theta$$

假设曲面$S$为球面$x^2 + y^2 + z^2 = r^2$, 其中 $r > 0$, 将曲面$S$作参数化：
$$\vec{r}(\theta, \varphi) = (r\cos\theta\sin\varphi, r\sin\theta\sin\varphi, r\cos\varphi),$$
其中定义域为$R = \{(\theta, \varphi): 0 \leq \theta \leq 2\pi, 0 \leq \varphi \leq \pi\}$, 则表面积分转换为

$$\iint_S f(x, y, z)dA = \iint_D f(\vec{r}(\theta, \varphi))|\vec{r}_\theta \times \vec{r}_\varphi|d\theta d\varphi = \iint_R f(\vec{r}(\theta, \varphi))\left|\frac{\partial\vec{r}}{\partial\varphi} \times \frac{\partial\vec{r}}{\partial\theta}\right|d\theta d\varphi$$

因为

$$\frac{\partial\vec{r}}{\partial\theta} = (-r\sin\theta\sin\varphi, r\cos\theta\sin\varphi, 0) \text{ 且 } \frac{\partial\vec{r}}{\partial\varphi} = (r\cos\theta\cos\varphi, r\sin\theta\cos\varphi, -r\sin\varphi)$$

所以

$$\frac{\partial\vec{r}}{\partial\varphi} \times \frac{\partial\vec{r}}{\partial\theta} = (r^2\cos\theta\sin^2\varphi, r^2\sin\theta\sin^2\varphi, r^2\sin\varphi\cos\varphi) \text{ 且 } \left|\frac{\partial\vec{r}}{\partial\varphi} \times \frac{\partial\vec{r}}{\partial\theta}\right| = r^2\sin\varphi$$

因此, 球体曲面积分的公式可改写为

$$\iint_S f(x, y, z)dS = \iint_R f(r\cos\theta\sin\varphi, r\sin\theta\sin\varphi, r\cos\varphi)r^2\sin\varphi\, d\theta d\varphi$$

考试类型:

Type 1.

给定曲面$S$: $\vec{r}(s, t)$, $\forall a \leq s \leq b, c \leq t \leq d$, 求 $\iint_S f(x, y, z)dA = ?$

解题流程:

Step1.

令 $R = \{(x, y): a \leq s \leq b, c \leq t \leq d\}$则

$$\iint_S f(x, y, z)dA = \iint_R f(\vec{r}(s, t))|\vec{r}_s \times \vec{r}_t|dsdt = \int_a^b \int_c^d f(\vec{r}(s, t))|\vec{r}_s \times \vec{r}_t|\, dtds$$

Step2.

求 $\vec{r}_s$, $\vec{r}_s$, $\vec{r}_s \times \vec{r}_t$, 以及 $|\vec{r}_s \times \vec{r}_t|$

求 $\displaystyle\int_a^b \int_c^d f(\vec{r}(s,t)))|\vec{r}_s \times \vec{r}_t|\, dt ds =$?

Type 2.

给定曲面 $S$: $z = g(x,y)$, $\forall a \le x \le b, c \le y \le d$, 求 $\displaystyle\iint_S f(x,y,z)dA =$?

解题流程:

Step1.

令 $\vec{r}(x,y) = (x,y,g(x,y))$ 且 $R = \{(x,y): a \le x \le b, c \le y \le d\}$

则 $\displaystyle\iint_S f(x,y,z)dA = \iint_R f(x,y,g(x,y))\left|\frac{\partial \vec{r}}{\partial x} \times \frac{\partial \vec{r}}{\partial y}\right| dxdy$

Step2.

$$\because \frac{\partial \vec{r}}{\partial x} = (1,0,g_x(x,y)), \quad \frac{\partial \vec{r}}{\partial y} = \left(0,1,g_y(x,y)\right) \quad \therefore \frac{\partial \vec{r}}{\partial x} \times \frac{\partial \vec{r}}{\partial y} = (-g_x(x,y), -g_y(x,y), 1)$$

$$\Rightarrow \left|\frac{\partial \vec{r}}{\partial x} \times \frac{\partial \vec{r}}{\partial y}\right| = \sqrt{1 + \left(g_x(x,y)\right)^2 + \left(g_y(x,y)\right)^2}$$

Step3.

$$\iint_S f(x,y,z)dA = \int_a^b \int_c^d f(x,y,g(x,y))\sqrt{1 + \left(g_x(x,y)\right)^2 + \left(g_y(x,y)\right)^2}\, dydx$$

同理, 如果给定曲面 $S$: $y = g(x,z)$, $\forall a \le x \le b, c \le z \le d$ 则

$$\iint_S f(x,y,z)dA = \int_a^b \int_c^d f(x,g(x,z),z)\sqrt{1 + \left(g_x(x,z)\right)^2 + \left(g_z(x,z)\right)^2}\, dzdx$$

如果给定曲面 $S$: $x = g(y,z)$, $\forall a \le y \le b, c \le z \le d$

$$\iint_S f(x,y,z)dA = \int_a^b \int_c^d f(g(y,z),y,z)\sqrt{1 + \left(g_y(y,z)\right)^2 + \left(g_z(y,z)\right)^2}\, dzdy$$

<u>范例说明:</u>

求 $\displaystyle\iint_S f(x,y,z)dA =$?

(I) 若 $f(x,y,z) = x + y + z - 1$, $S:\{(x,y,z): z = x + y + 1, \ 0 \le y \le x, \ 0 \le x \le 2\}$

令 $\vec{r}(x,y) = (x,y,x+y+z-1)$

$$\because \frac{\partial \vec{r}}{\partial x} = (1,0,1) \text{ 且 } \frac{\partial \vec{r}}{\partial y} = (0,1,1) \quad \therefore \frac{\partial \vec{r}}{\partial x} \times \frac{\partial \vec{r}}{\partial y} = (-1,-1,1) \Rightarrow \left| \frac{\partial \vec{r}}{\partial x} \times \frac{\partial \vec{r}}{\partial y} \right| = \sqrt{3}$$

$$\therefore \iint_S f(x,y,z)\, dA = \int_0^2 \int_0^x (2x + 2y) \sqrt{3}\, dy dx = 8\sqrt{3}$$

(II)若 $f(x,y,z) = (1 - x^2)y,\ \ S:\{(x,y,z): z = 3 - y^2,\ \ 0 \leq x \leq 1,\ \ 0 \leq y \leq 1\}$

令$\vec{r}(x,y) = (x, y, (1 - x^2)y)$

$$\because \frac{\partial \vec{r}}{\partial x} = (1,0,0) \text{ 且 } \frac{\partial \vec{r}}{\partial y} = (0,1,-2y) \quad \therefore \frac{\partial \vec{r}}{\partial x} \times \frac{\partial \vec{r}}{\partial y} = (0,2y,1) \Rightarrow \left| \frac{\partial \vec{r}}{\partial x} \times \frac{\partial \vec{r}}{\partial y} \right| = \sqrt{1 + 4y^2}$$

$$\text{则} \iint_S f(x,y,z)\, dA = \int_0^1 \int_0^1 y(1 - x^2) \sqrt{1 + 4y^2}\, dy dx = \frac{5\sqrt{5} - 1}{18}$$

Type 3.

給定曲面$S$: $z = g(x,y)$,若$S$投影至$xy$平面 $= \{(x,y): x^2 + y^2 \leq a^2\}$, 求 $\iint_S f(x,y,z)dA = ?$

<u>補充说明:</u>

如: 圆柱、椭圆柱、圆锥体、抛物面

解题流程:

Step1.

令$\vec{r}(x,y) = (x, y, g(x,y))$且 $R = \{(x,y): x^2 + y^2 \leq a^2\}$ 则 $\iint_S dA = \iint_R \left| \frac{\partial \vec{r}}{\partial x} \times \frac{\partial \vec{r}}{\partial y} \right| dxdy$

Step2.

$$\because \frac{\partial \vec{r}}{\partial x} = (1,0,g_x(x,y)) \text{ 且 } \frac{\partial \vec{r}}{\partial y} = \left(0,1,g_y(x,y)\right) \quad \therefore \frac{\partial \vec{r}}{\partial x} \times \frac{\partial \vec{r}}{\partial y} = \left(-g_x(x,y), -g_y(x,y), 1\right)$$

$$\Rightarrow \left| \frac{\partial \vec{r}}{\partial x} \times \frac{\partial \vec{r}}{\partial y} \right| = \sqrt{1 + \left(g_x(x,y)\right)^2 + \left(g_y(x,y)\right)^2}$$

Step3.

$$\therefore \iint_S f(x,y,z)\, dA = \iint_R f(x,y,g(x,y)) \left| \frac{\partial \vec{r}}{\partial x} \times \frac{\partial \vec{r}}{\partial y} \right| dxdy$$

$$= \iint_R f(x,y,g(x,y)) \sqrt{1 + \left(g_x(x,y)\right)^2 + \left(g_y(x,y)\right)^2}\, dxdy$$

Step4.

令$x = r \cos\theta,\, y = r \sin\theta$ 则 $\{(x,y): x^2 + y^2 \leq a^2\} = \{(r,\theta): 0 \leq r \leq a, 0 \leq \theta \leq 2\pi\}$

$$\iint_R f(x,y,g(x,y))\sqrt{1+\left(g_x(x,y)\right)^2+\left(g_y(x,y)\right)^2}\,dxdy$$

$$=\int_0^{2\pi}\int_0^a f(x,y,g(x,y))\sqrt{1+\left(g_x(x,y)\right)^2+\left(g_y(x,y)\right)^2}\,\Bigg|_{x=r\cos\theta,y=r\sin\theta}\,rdrd\theta$$

Type 4.

給定球面 $S: x^2+y^2+z^2=r^2,\ r>0,$ 求 $\iint_S f(x,y,z)dA=?$

解题流程:

Step1.

令 $\vec{r}(\theta,\varphi)=(r\cos\theta\sin\varphi,r\sin\theta\sin\varphi,r\cos\varphi)$ 且 $R=\{(\theta,\varphi):0\le\theta\le 2\pi,0\le\varphi\le\pi\}$

则 $\iint_S f(x,y,z)dA=\iint_R f(r\cos\theta\sin\varphi,r\sin\theta\sin\varphi,r\cos\varphi)\left|\dfrac{\partial\vec{r}}{\partial\varphi}\times\dfrac{\partial\vec{r}}{\partial\theta}\right|d\theta d\varphi$

Step2.

$\because\dfrac{\partial\vec{r}}{\partial\theta}=(-r\sin\theta\sin\varphi,r\cos\theta\sin\varphi,0)$ 且 $\dfrac{\partial\vec{r}}{\partial\varphi}=(r\cos\theta\cos\varphi,r\sin\theta\cos\varphi,-r\sin\varphi)$

$\therefore\dfrac{\partial\vec{r}}{\partial\varphi}\times\dfrac{\partial\vec{r}}{\partial\theta}=(r^2\cos\theta\sin^2\varphi,r^2\sin\theta\sin^2\varphi,r^2\sin\varphi\cos\varphi)\Rightarrow\left|\dfrac{\partial\vec{r}}{\partial\varphi}\times\dfrac{\partial\vec{r}}{\partial\theta}\right|=r^2\sin\varphi$

Step3.

$$\iint_S f(x,y,z)dA=\iint_R f(r\cos\theta\sin\varphi,r\sin\theta\sin\varphi,r\cos\varphi)\,r^2\sin\varphi\,d\theta d\varphi$$

求 $\iint_R f(r\cos\theta\sin\varphi,r\sin\theta\sin\varphi,r\cos\varphi)\,r^2\sin\varphi\,d\theta d\varphi=?$

<u>范例说明:</u>

求 $\iint_S f(x,y,z)dA=?$, 其中 $S: x^2+y^2+z^2=r^2,\ r>0$

(I)若 $f(x,y,z)=x^2$ 则 $\iint_S f(x,y,z)dA=\iint_R r^2\cos^2\theta\sin^2\varphi\,r^2\sin\varphi\,d\theta d\varphi=\dfrac{4\pi r^4}{3}$

(II)若 $f(x,y,z)=z^2$ 则 $\iint_S f(x,y,z)dA=\iint_R r^2\cos^2\varphi\,r^2\sin\varphi\,d\theta d\varphi=\dfrac{4\pi r^4}{3}$

Example 1.

$$求 \iint_S 3xy \, dA =?, \quad S: z = xy + \sqrt{7}, \quad 0 \le x \le 1, \quad 0 \le y \le 1$$

【解】

令 $\vec{r}(x,y) = \left(x, y, xy + \sqrt{7}\right)$ 且 $R$ 为曲面 $S$ 投影至 $xy$ 平面的封闭区域

则 $R = \{(x,y): 0 \le x \le 1, 0 \le y \le 1\}$ 且 $\iint_S 3xy \, dA = \iint_R 3xy \left| \dfrac{\partial \vec{r}}{\partial x} \times \dfrac{\partial \vec{r}}{\partial y} \right| dxdy$

$\because \dfrac{\partial \vec{r}}{\partial x} = (1,0,y)$ 且 $\dfrac{\partial \vec{r}}{\partial y} = (0,1,x)$ $\quad \therefore \dfrac{\partial \vec{r}}{\partial x} \times \dfrac{\partial \vec{r}}{\partial y} = (-y, -x, 1) \Rightarrow \left| \dfrac{\partial \vec{r}}{\partial x} \times \dfrac{\partial \vec{r}}{\partial y} \right| = (x^2 + y^2 + 1)^{\frac{1}{2}}$

$\therefore 3 \iint_S xy \, dA = \iint_R 3xy \left| \dfrac{\partial \vec{r}}{\partial x} \times \dfrac{\partial \vec{r}}{\partial y} \right| dxdy = 3 \iint_R xy(x^2 + y^2 + 1)^{\frac{1}{2}} dxdy$

令 $u = x^2$, $v = y^2$ 则 $R = \{(x,y): 0 \le x \le 1, 0 \le y \le 1\} = \{(u,v): 0 \le u \le 1, 0 \le v \le 1\}$

$\therefore 3 \iint_R xy(x^2 + y^2 + 1)^{\frac{1}{2}} dxdy = \dfrac{3}{4} \int_0^1 \int_0^1 (u + v + 1)^{\frac{1}{2}} \, dudv = \dfrac{9\sqrt{3} - 8\sqrt{2} + 1}{5}$

Example 2.

$$假设 f(x,y,z) = x + y + z - 1, \quad S: z = x + y + 1, \quad 0 \le y \le x, \quad 0 \le x \le 2,$$

$$求 \iint_S f(x,y,z) \, dA =?$$

【解】

令 $\vec{r}(x,y) = (x, y, x + y + 1)$ 且 $R$ 为曲面 $S$ 投影至 $xy$ 平面的封闭区域

则 $R = \{(x,y): 0 \le y \le x, 0 \le x \le 2\}$

且 $\iint_S f(x,y,z) \, dA = \iint_S f(x,y,x+y+1) \, dA = \iint_R f(x,y,x+y+1) \left| \dfrac{\partial \vec{r}}{\partial x} \times \dfrac{\partial \vec{r}}{\partial y} \right| dxdy$

$\because \dfrac{\partial \vec{r}}{\partial x} = (1,0,1)$ 且 $\dfrac{\partial \vec{r}}{\partial y} = (0,1,1)$ $\quad \therefore \dfrac{\partial \vec{r}}{\partial x} \times \dfrac{\partial \vec{r}}{\partial y} = (-1, -1, 1) \Rightarrow \left| \dfrac{\partial \vec{r}}{\partial x} \times \dfrac{\partial \vec{r}}{\partial y} \right| = \sqrt{3}$

$$\therefore \iint_S f(x,y,z)\, dA = \iint_R f(x,y,x+y+1)\left|\frac{\partial \vec{r}}{\partial x} \times \frac{\partial \vec{r}}{\partial y}\right| dxdy = \int_0^2 \int_0^x (2x+2y)\sqrt{3}\, dydx$$

$$= \sqrt{3}\int_0^2 2xy + y^2 \big|_0^x\, dx = \sqrt{3}\int_0^2 3x^2\, dx = 8\sqrt{3}$$

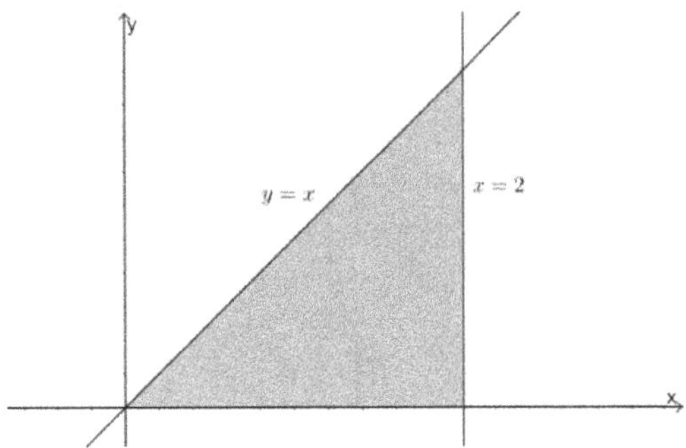


Example 3.

Evaluate $\displaystyle\iint_S \sqrt{x^2+y^2}\, dA = ?$, over the surface $S: z = xy,\ \ 0 \le x^2 + y^2 \le 1$.

【解】

令 $\vec{r}(x,y) = (x,y,xy)$ 且 $R$ 为曲面 $S$ 投影至 $xy$ 平面的封闭区域

则 $R = \{(x,y): 0 \le x^2 + y^2 \le 1\}$ 且 $\displaystyle\iint_S \sqrt{x^2+y^2}\, dA = \iint_R \sqrt{x^2+y^2}\left|\frac{\partial \vec{r}}{\partial x} \times \frac{\partial \vec{r}}{\partial y}\right| dxdy$

$\because \dfrac{\partial \vec{r}}{\partial x} = (1,0,y)$ 且 $\dfrac{\partial \vec{r}}{\partial y} = (0,1,x)$ $\quad \therefore \dfrac{\partial \vec{r}}{\partial x} \times \dfrac{\partial \vec{r}}{\partial y} = (-y,-x,1) \Rightarrow \left|\dfrac{\partial \vec{r}}{\partial x} \times \dfrac{\partial \vec{r}}{\partial y}\right| = \sqrt{1+x^2+y^2}$

$$\therefore \iint_R \sqrt{x^2+y^2}\left|\frac{\partial \vec{r}}{\partial x} \times \frac{\partial \vec{r}}{\partial y}\right| dxdy = \iint_{0 \le \sqrt{x^2+y^2} \le 1} \sqrt{x^2+y^2}\sqrt{1+x^2+y^2}\, dxdy$$

令 $x = r\cos\theta,\ y = r\sin\theta$ 且 $R = \{(r,\theta): 0 \le r \le 1, 0 \le \theta \le 2\pi\}$

$$\iint_{0 \le \sqrt{x^2+y^2} \le 1} \sqrt{x^2+y^2}\sqrt{1+x^2+y^2}\, dxdy = \int_0^{2\pi}\int_0^1 r^2\sqrt{1+r^2}\, drd\theta = 2\pi\int_0^1 r^2\sqrt{1+r^2}\, dr$$

令 $r = \tan\alpha$ 则 $\displaystyle 2\pi\int_0^1 r^2\sqrt{1+r^2}\, dr = 2\pi\int_0^{\frac{\pi}{4}} \tan^2\alpha\, \sec^3\alpha\, d\alpha = 2\pi\int_0^{\frac{\pi}{4}} \sec^5\alpha - \sec^3\alpha\, d\alpha$

$$= \frac{\pi}{4}\left(3\sqrt{2} - \ln(1+\sqrt{2})\right)$$

Example 4.

Evaluate $\iint_S xy\, dA =?$, over the surface $\vec{r}(s,t) = s\vec{a} + t\vec{b}$,  where $0 \le s \le 1$,

$0 \le t \le 1$,  $\vec{a} = (a_1, a_2, a_3)$,  $\vec{b} = (b_1, b_2, b_3)$

【解】

令 $\vec{r}(s,t) = s\vec{a} + t\vec{b} = (sa_1 + tb_1, sa_2 + tb_2, sa_3 + tb_3)$ 且 $R = \{(s,t): 0 \le s \le 1, 0 \le t \le 1\}$

$\because \dfrac{\partial \vec{r}}{\partial s} = (a_1, a_2, a_3)$ 且 $\dfrac{\partial \vec{r}}{\partial t} = (b_1, b_2, b_3)$   $\therefore \dfrac{\partial \vec{r}}{\partial s} \times \dfrac{\partial \vec{r}}{\partial t} = \vec{a} \times \vec{b}$ 且 $\left|\dfrac{\partial \vec{r}}{\partial s} \times \dfrac{\partial \vec{r}}{\partial t}\right| = |\vec{a} \times \vec{b}|$

$$\iint_S xy\, dA = \iint_R (sa_1 + tb_1)(sa_2 + tb_2)\left|\frac{\partial \vec{r}}{\partial s} \times \frac{\partial \vec{r}}{\partial t}\right| ds\, dt$$

$$= |\vec{a} \times \vec{b}| \int_0^1 \int_0^1 (sa_1 + tb_1)(sa_2 + tb_2)\, ds\, dt$$

$$= |\vec{a} \times \vec{b}| \int_0^1 \int_0^1 s^2 a_1 a_2 + st(a_1 b_2 + a_2 b_1) + t^2 b_1 b_2\, ds\, dt$$

$$= |\vec{a} \times \vec{b}|\left(\frac{a_1 a_2}{3} + \frac{a_1 b_2 + a_2 b_1}{4} + \frac{b_1 b_2}{3}\right)$$

Example 5.

求 $\iint_S x^2\, dA =?$,  $S: x^2 + y^2 + z^2 = r^2$

【解】

令 $\vec{r}(\theta, \varphi) = (r\cos\theta \sin\varphi, r\sin\theta \sin\varphi, r\cos\varphi)$ 且 $R = \{(\theta, \varphi): 0 \le \theta \le 2\pi, 0 \le \varphi \le \pi\}$

则 $\iint_S x^2\, dA = \iint_R r^2\cos^2\theta \sin^2\varphi \left|\dfrac{\partial \vec{r}}{\partial \varphi} \times \dfrac{\partial \vec{r}}{\partial \theta}\right| d\theta\, d\varphi$

$$\because \frac{\partial \vec{r}}{\partial \theta} = (-r\sin\theta\sin\varphi, r\cos\theta\sin\varphi, 0) \ \text{且} \ \frac{\partial \vec{r}}{\partial \varphi} = (r\cos\theta\cos\varphi, r\sin\theta\cos\varphi, -r\sin\varphi)$$

$$\therefore \frac{\partial \vec{r}}{\partial \varphi} \times \frac{\partial \vec{r}}{\partial \theta} = (r^2\cos\theta\sin^2\varphi, r^2\sin\theta\sin^2\varphi, r^2\sin\varphi\cos\varphi) \Rightarrow \left|\frac{\partial \vec{r}}{\partial \varphi} \times \frac{\partial \vec{r}}{\partial \theta}\right| = r^2\sin\varphi$$

$$\therefore \iint_S x^2\, dA = \iint_R r^2\cos^2\theta\sin^2\varphi \left|\frac{\partial \vec{r}}{\partial \varphi} \times \frac{\partial \vec{r}}{\partial \theta}\right| d\theta d\varphi = r^4 \int_0^{2\pi} \int_0^{\pi} \cos^2\theta\sin^2\varphi\sin\varphi\, d\varphi d\theta$$

$$= r^4 \int_0^{2\pi} \cos^2\theta\, d\theta \int_0^{\pi} \sin^3\varphi\, d\varphi = r^4 \cdot \pi \cdot \frac{4}{3} = \frac{4\pi r^4}{3}$$

Example 6.

$$求 \iint_S y\, dA =?, \quad S: x - y^2 - z = 2, \ 0 \le y \le 1, \ 0 \le z \le 1$$

【解】

令 $\vec{r}(x, y) = (y^2 + z + 2, y, z)$ 且 $R$ 为曲面$S$投影至$yz$平面的封闭区域

则 $R = \{(y, z): 0 \le y \le 1, 0 \le z \le 1\}$ 且 $\iint_S y\, dA = \iint_R y \left|\frac{\partial \vec{r}}{\partial y} \times \frac{\partial \vec{r}}{\partial z}\right| dydz$

$$\because \frac{\partial \vec{r}}{\partial y} = (2y, 1, 0) \ \text{且} \ \frac{\partial \vec{r}}{\partial z} = (1, 0, 1) \quad \therefore \frac{\partial \vec{r}}{\partial y} \times \frac{\partial \vec{r}}{\partial z} = (1, -2y, -1) \Rightarrow \left|\frac{\partial \vec{r}}{\partial y} \times \frac{\partial \vec{r}}{\partial z}\right| = \sqrt{2 + 4y^2}$$

$$\therefore \iint_S y\, dA = \iint_R y \left|\frac{\partial \vec{r}}{\partial y} \times \frac{\partial \vec{r}}{\partial z}\right| dxdy = \int_0^1 \int_0^1 y\sqrt{2 + 4y^2}\, dydz = \frac{3\sqrt{6} - \sqrt{2}}{6}$$

Example 7.

$$\text{Evaluate} \iint_S \sqrt{x^2 + y^2}\, dA =?, \ \text{over the surface} \ \vec{r}(s, t) = (s\cos wt, s\sin wt, bt),$$

$$\text{where } 0 \le s \le k, 0 \le t \le \frac{2\pi}{w}$$

【解】

$$\because \frac{\partial \vec{r}}{\partial s} = (\cos wt, \sin wt, 0) \;\; \text{且} \;\; \frac{\partial \vec{r}}{\partial t} = (-ws\sin wt, ws\cos wt, b)$$

$$\therefore \frac{\partial \vec{r}}{\partial s} \times \frac{\partial \vec{r}}{\partial t} = (b\sin wt, -b\cos wt, ws) \;\; \text{且} \;\; \left|\frac{\partial \vec{r}}{\partial s} \times \frac{\partial \vec{r}}{\partial t}\right| = \sqrt{b^2 + w^2 s^2}$$

$$\text{令} \; R = \left\{(s,t): 0 \le s \le k, 0 \le t \le \frac{2\pi}{w}\right\} \text{则}$$

$$\therefore \iint_S \sqrt{x^2 + y^2}\, dA = \iint_R \sqrt{(s\cos wt)^2 + (s\sin wt)^2} \left|\frac{\partial \vec{r}}{\partial s} \times \frac{\partial \vec{r}}{\partial t}\right| ds\,dt$$

$$= \iint_R \sqrt{(s\cos wt)^2 + (s\sin wt)^2} \sqrt{b^2 + w^2 s^2}\, dA = \iint_R s\sqrt{b^2 + w^2 s^2}\, dA$$

$$= \int_0^{\frac{2\pi}{w}} \int_0^k s\sqrt{b^2 + w^2 s^2}\, ds\,dt = \frac{2\pi}{w} \int_0^k s\sqrt{b^2 + w^2 s^2}\, ds = \frac{2\pi}{3w^3}\left((b^2 + w^2 k^2)^{\frac{3}{2}} - b^3\right)$$

Example 8.

Evaluate $\iint_S 3y\, dA = ?$, over the surface $S$: $z = \dfrac{y^2}{2}$, $0 \le x \le 1$, $0 \le y \le 1$

【解】

$$\text{令} \; \vec{r}(x,y) = \left(x, y, \frac{y^2}{2}\right) \text{且} \; R \text{为} S \text{曲面投影至} xy \text{平面的封闭区域}$$

$$\text{则} \; R = \{(x,y): 0 \le x \le 1, 0 \le y \le 1\} \;\; \text{且} \;\; \iint_S 3y\, dA = \iint_R 3y\left|\frac{\partial \vec{r}}{\partial x} \times \frac{\partial \vec{r}}{\partial y}\right| dx\,dy$$

$$\because \frac{\partial \vec{r}}{\partial x} = (1,0,0) \;\; \text{且} \;\; \frac{\partial \vec{r}}{\partial y} = (0,1,y) \;\; \therefore \frac{\partial \vec{r}}{\partial x} \times \frac{\partial \vec{r}}{\partial y} = (0,-y,1) \Rightarrow \left|\frac{\partial \vec{r}}{\partial x} \times \frac{\partial \vec{r}}{\partial y}\right| = (y^2 + 1)^{\frac{1}{2}}$$

$$\iint_R 3y\left|\frac{\partial \vec{r}}{\partial x} \times \frac{\partial \vec{r}}{\partial y}\right| dx\,dy = 3\iint_R y(y^2 + 1)^{\frac{1}{2}} dx\,dy = (y^2 + 1)^{\frac{3}{2}}\Big|_0^1 = 2\sqrt{2} - 1$$

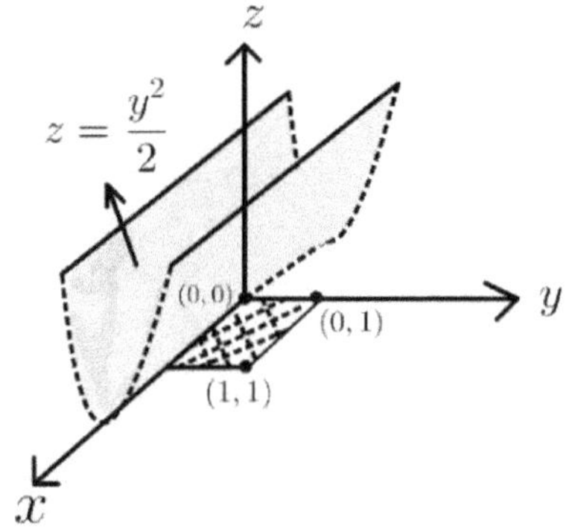

Example 9.

假设 $f(x,y,z) = (1-x^2)y, \quad S: z = 3 - y^2, \quad 0 \le x \le 1, \quad 0 \le y \le 1$

求 $\displaystyle\iint_S f(x,y,z)\, dA =?$

【解】

令 $\vec{r}(x,y) = (x,y,3-y^2)$ 且 $R$ 为曲面 $S$ 投影至 $xy$ 平面的封闭区域

则 $R = \{(x,y): 0 \le x \le 1, 0 \le y \le 1\}$ 且 $\displaystyle\iint_S f(x,y,z)\, dA = \iint_R f(x,y,z)\left|\frac{\partial \vec{r}}{\partial x} \times \frac{\partial \vec{r}}{\partial y}\right| dxdy$

$\because \dfrac{\partial \vec{r}}{\partial x} = (1,0,0)$ 且 $\dfrac{\partial \vec{r}}{\partial y} = (0,1,-2y)$ $\quad \therefore \dfrac{\partial \vec{r}}{\partial x} \times \dfrac{\partial \vec{r}}{\partial y} = (0,2y,1) \Rightarrow \left|\dfrac{\partial \vec{r}}{\partial x} \times \dfrac{\partial \vec{r}}{\partial y}\right| = \sqrt{1+4y^2}$

$\therefore \displaystyle\iint_S f(x,y,z)\, dA = \iint_R f(x,y,z)\left|\frac{\partial \vec{r}}{\partial x} \times \frac{\partial \vec{r}}{\partial y}\right| dxdy = \int_0^1 \int_0^1 y(1-x^2)\sqrt{1+4y^2}\, dy\, dx$

$= \displaystyle\int_0^1 y\sqrt{1+4y^2}\, dy \int_0^1 (1-x^2)\, dx$

$\because \displaystyle\int_0^1 y\sqrt{1+4y^2}\, dy = \frac{1}{12}(1+4y^2)^{\frac{3}{2}}\Big|_0^1 = \frac{5\sqrt{5}-1}{12}$

令 $x = \sin\theta$ 则 $dx = \cos\theta\, d\theta$

$\therefore \displaystyle\int_0^1 (1-x^2)\, dx = \int_0^{\frac{\pi}{2}} \cos^3\theta\, dx = \int_0^{\frac{\pi}{2}} (1-\sin^2\theta)\cos\theta\, dx = \left(\sin\theta - \frac{\sin^3\theta}{3}\right)\Big|_0^{\frac{\pi}{2}} = \frac{2}{3}$

$\therefore \displaystyle\iint_S f(x,y,z)\, dA = \int_0^1 y\sqrt{1+4y^2}\, dy \int_0^1 (1-x^2)\, dx = \frac{5\sqrt{5}-1}{12} \cdot \frac{2}{3} = \frac{5\sqrt{5}-1}{18}$

Example 10.

$$\iint_S y^2 + 2yz \, dA = ?, \quad S \text{为平面 } 2x + y + 2z = 4 \text{ 在第一卦限的所围平面}$$

【解】

令 $\vec{r}(x,y) = \left(x, y, 2 - x - \dfrac{y}{2}\right)$ 且 $R$ 为所围平面投影至 $xy$ 平面的封闭区域

则 $R = \{(x,y): 2x + y \le 4, x \ge 0, y \ge 0\}$

且 $\displaystyle\iint_S y^2 + 2yz \, dA = \iint_R \left(y^2 + 2y(2 - x - \dfrac{y}{2})\right)\left|\dfrac{\partial \vec{r}}{\partial x} \times \dfrac{\partial \vec{r}}{\partial y}\right| dxdy$

$\because \dfrac{\partial \vec{r}}{\partial x} = (1, 0, -1)$ 且 $\dfrac{\partial \vec{r}}{\partial y} = \left(0, 1, -\dfrac{1}{2}\right)$ $\quad \therefore \dfrac{\partial \vec{r}}{\partial x} \times \dfrac{\partial \vec{r}}{\partial y} = \left(1, \dfrac{1}{2}, 1\right) \Rightarrow \left|\dfrac{\partial \vec{r}}{\partial x} \times \dfrac{\partial \vec{r}}{\partial y}\right| = \dfrac{3}{2}$

$\therefore \displaystyle\iint_S y^2 + 2yz \, dA = \iint_R \left(y^2 + 2y(2 - x - \dfrac{y}{2})\right)\left|\dfrac{\partial \vec{r}}{\partial x} \times \dfrac{\partial \vec{r}}{\partial y}\right| dxdy$

$= \dfrac{3}{2} \displaystyle\int_0^2 \int_0^{4-2x} y^2 + 2y \cdot \left(2 - x - \dfrac{y}{2}\right) dydx$

$\because \displaystyle\int_0^{4-2x} y^2 + 2y \cdot \left(2 - x - \dfrac{y}{2}\right) dy = \int_0^{4-2x} 4y - 2xy \, dy = 2y^2 - xy^2 \big|_0^{4-2x} = 4(2-x)^3$

$\therefore \displaystyle\int_0^2 \int_0^{4-2x} y^2 + 2y \cdot \dfrac{1}{2} \cdot (4 - 2x - y) \, dydx = \int_0^2 4(2-x)^3 \, dx = -(2-x)^4 \big|_0^2 = 16$

$\therefore \displaystyle\iint_S y^2 + 2yz \, dA = 24$

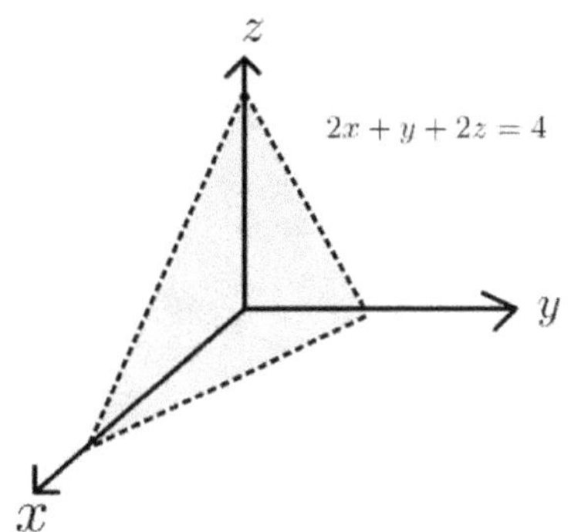


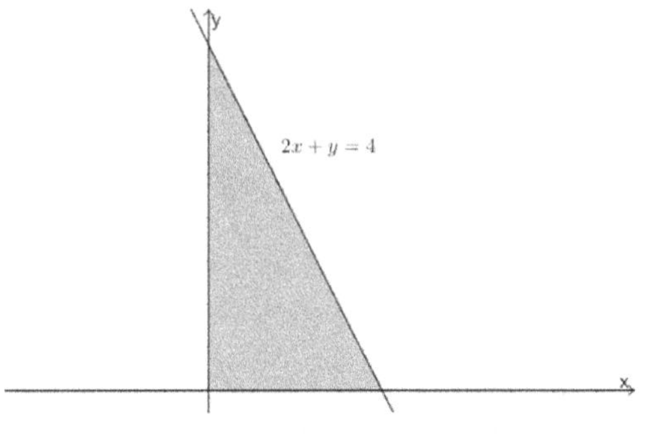

Example 11.

$$求 \iint_S z^2 \, dA = ?, \quad S: z = \sqrt{r^2 - x^2 - y^2}$$

【解】

令 $\vec{r}(\theta, \varphi) = (r\cos\theta\sin\varphi, r\sin\theta\sin\varphi, r\cos\varphi)$ 且 $R = \left\{(\theta, \varphi): 0 \le \theta \le 2\pi, 0 \le \varphi \le \dfrac{\pi}{2}\right\}$

則 $\iint_S z^2 \, dA = \iint_R r^2 \cos^2\varphi \left|\dfrac{\partial\vec{r}}{\partial\varphi} \times \dfrac{\partial\vec{r}}{\partial\theta}\right| d\theta d\varphi$

$\because \dfrac{\partial\vec{r}}{\partial\theta} = (-r\sin\theta\sin\varphi, r\cos\theta\sin\varphi, 0)$ 且 $\dfrac{\partial\vec{r}}{\partial\varphi} = (r\cos\theta\cos\varphi, r\sin\theta\cos\varphi, -r\sin\varphi)$

$\therefore \dfrac{\partial\vec{r}}{\partial\varphi} \times \dfrac{\partial\vec{r}}{\partial\theta} = (r^2\cos\theta\sin^2\varphi, r^2\sin\theta\sin^2\varphi, r^2\sin\varphi\cos\varphi) \Rightarrow \left|\dfrac{\partial\vec{r}}{\partial\varphi} \times \dfrac{\partial\vec{r}}{\partial\theta}\right| = r^2\sin\varphi$

$\therefore \iint_S z^2 \, dA = \iint_R r^2 \cos^2\varphi \left|\dfrac{\partial\vec{r}}{\partial\varphi} \times \dfrac{\partial\vec{r}}{\partial\theta}\right| d\theta d\varphi = r^4 \int_0^{2\pi} \int_0^{\frac{\pi}{2}} \cos^2\varphi \sin\varphi \, d\varphi d\theta = \dfrac{2\pi r^4}{3}$

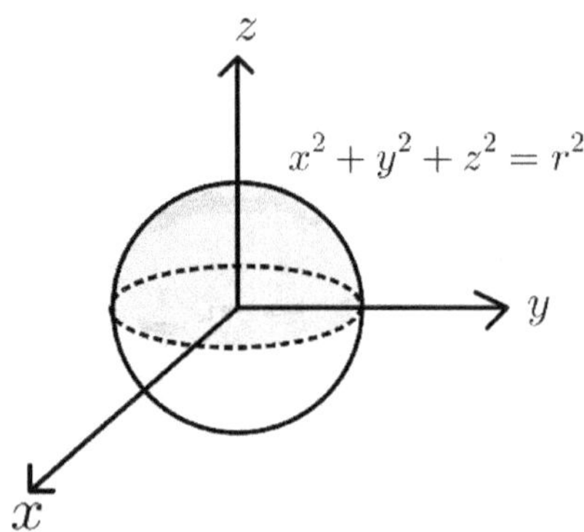


Example 12.

$$\text{Evaluate } \iint_S \sqrt{2z} \, dA = ?, \quad \text{over the surface } S: z = \dfrac{y^2}{2}, \quad 0 \le x \le 1, \quad 0 \le y \le 1$$

【解】

令 $\vec{r}(x,y) = \left(x, y, \dfrac{y^2}{2}\right)$ 且 $R$ 为 $S$ 曲面投影至 $xy$ 平面的封闭区域

則 $R = \{(x,y): 0 \le x \le 1, 0 \le y \le 1\}$ 且 $\displaystyle\iint_S \sqrt{2z}\, dA = \iint_R y\left|\dfrac{\partial \vec{r}}{\partial x} \times \dfrac{\partial \vec{r}}{\partial y}\right| dxdy$

$\because \dfrac{\partial \vec{r}}{\partial x} = (1,0,0)$ 且 $\dfrac{\partial \vec{r}}{\partial y} = (0,1,y)$ $\quad \therefore \dfrac{\partial \vec{r}}{\partial x} \times \dfrac{\partial \vec{r}}{\partial y} = (0,-y,1) \Rightarrow \left|\dfrac{\partial \vec{r}}{\partial x} \times \dfrac{\partial \vec{r}}{\partial y}\right| = (y^2+1)^{\frac{1}{2}}$

$$\iint_R y\left|\dfrac{\partial \vec{r}}{\partial x} \times \dfrac{\partial \vec{r}}{\partial y}\right| dxdy = \iint_R y(y^2+1)^{\frac{1}{2}} dxdy = \left.\dfrac{(y^2+1)^{\frac{3}{2}}}{3}\right|_0^1 = \dfrac{2\sqrt{2}-1}{3}$$

Example 13.

$\quad$ 求 $\displaystyle\iint_S xy\, dA = ?$, $S$ 为平面 $x + 2y + 3z = 6$ 在第一卦限的所围区域

【解】

令 $\vec{r}(x,y) = \left(x, y, 2 - \dfrac{x+2y}{3}\right)$ 且 $R$ 为所围平面投影至 $xy$ 平面的封闭区域

則 $R = \{(x,y): x + 2y \le 6, x \ge 0, y \ge 0\}$ 且 $\displaystyle\iint_S xy\, dA = \iint_R xy\left|\dfrac{\partial \vec{r}}{\partial x} \times \dfrac{\partial \vec{r}}{\partial y}\right| dxdy$

$\because \dfrac{\partial \vec{r}}{\partial x} = \left(1,0,-\dfrac{1}{3}\right)$ 且 $\dfrac{\partial \vec{r}}{\partial y} = \left(0,1,-\dfrac{2}{3}\right)$ $\quad \therefore \dfrac{\partial \vec{r}}{\partial x} \times \dfrac{\partial \vec{r}}{\partial y} = \left(\dfrac{1}{3}, \dfrac{2}{3}, 1\right) \Rightarrow \left|\dfrac{\partial \vec{r}}{\partial x} \times \dfrac{\partial \vec{r}}{\partial y}\right| = \dfrac{\sqrt{14}}{3}$

$\therefore \displaystyle\iint_S xy\, dA = \iint_R xy\left|\dfrac{\partial \vec{r}}{\partial x} \times \dfrac{\partial \vec{r}}{\partial y}\right| dxdy = \dfrac{\sqrt{14}}{3}\int_0^3 \int_0^{6-2y} xy\, dx\, dy = \dfrac{\sqrt{14}}{3}\int_0^3 y \cdot \left.\dfrac{x^2}{2}\right|_0^{6-2y} dxdy$

$= \dfrac{\sqrt{14}}{3}\int_0^3 18y - 12y^2 + 2y^3\, dy = \dfrac{\sqrt{14}}{3}\left.\left(9y^2 - 4y^3 + \dfrac{y^4}{2}\right)\right|_0^3 = \dfrac{\sqrt{14}}{3} \cdot \dfrac{27}{2} = \dfrac{9\sqrt{14}}{2}$

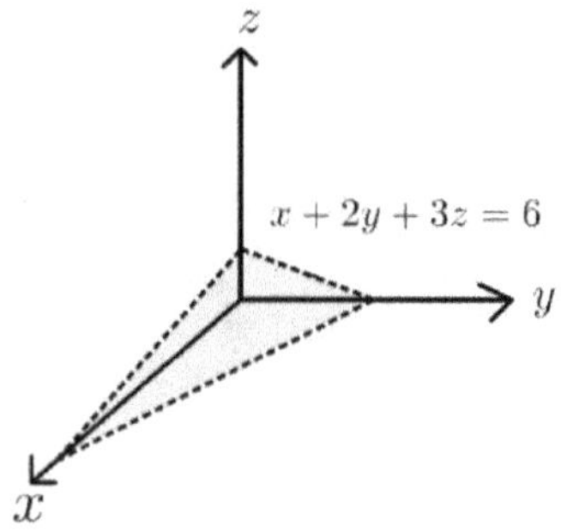


Example 14.

$$求 \iint_S x^2 + y^2 \, dA =?, \quad S: z = \sqrt{r^2 - x^2 - y^2}$$

【解】

令 $\vec{r}(\theta, \varphi) = (r\cos\theta\sin\varphi, r\sin\theta\sin\varphi, r\cos\varphi)$ 且 $R = \left\{ (\theta, \varphi) : 0 \leq \theta \leq 2\pi, 0 \leq \varphi \leq \dfrac{\pi}{2} \right\}$

則 $\displaystyle\iint_S x^2 + y^2 \, dA = \iint_R r^2 \sin^2\varphi \left| \dfrac{\partial \vec{r}}{\partial \varphi} \times \dfrac{\partial \vec{r}}{\partial \theta} \right| d\theta d\varphi$

$\because \dfrac{\partial \vec{r}}{\partial \theta} = (-r\sin\theta\sin\varphi, r\cos\theta\sin\varphi, 0)$ 且 $\dfrac{\partial \vec{r}}{\partial \varphi} = (r\cos\theta\cos\varphi, r\sin\theta\cos\varphi, -r\sin\varphi)$

$\therefore \dfrac{\partial \vec{r}}{\partial \varphi} \times \dfrac{\partial \vec{r}}{\partial \theta} = (r^2\cos\theta\sin^2\varphi, r^2\sin\theta\sin^2\varphi, r^2\sin\varphi\cos\varphi) \Rightarrow \left| \dfrac{\partial \vec{r}}{\partial \varphi} \times \dfrac{\partial \vec{r}}{\partial \theta} \right| = r^2\sin\varphi$

$\therefore \displaystyle\iint_S x^2 + y^2 \, dA = \iint_R r^2\sin^2\varphi \left| \dfrac{\partial \vec{r}}{\partial \varphi} \times \dfrac{\partial \vec{r}}{\partial \theta} \right| d\theta d\varphi = r^4 \int_0^{2\pi} \int_0^{\frac{\pi}{2}} \sin^3\varphi \, d\varphi d\theta$

$= r^4 \, 2\pi \left( -\cos\varphi + \dfrac{\cos\varphi}{3} \right) \Big|_0^{\frac{\pi}{2}} = \dfrac{4\pi r^4}{3}$

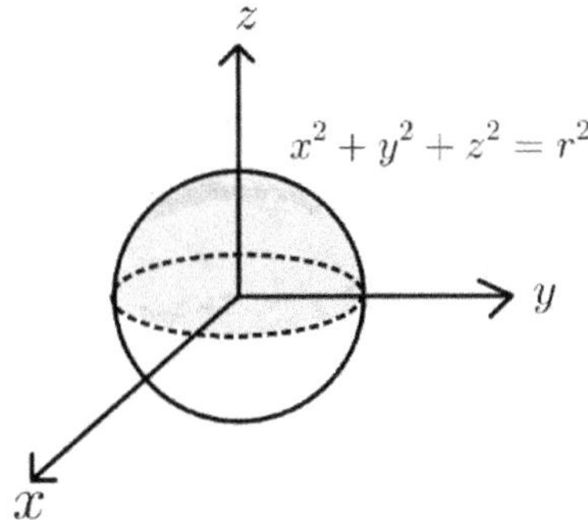



Example 15.

Evaluate $\displaystyle\iint_S z\, dA = ?$, over the conical surface $z = \sqrt{x^2 + y^2}$ between $z = 0$ and $z = 1$

【解】

令 $\vec{r}(x, y) = \left(x, y, \sqrt{x^2 + y^2}\right)$ 且 $R$ 為曲面 $S$ 投影至 $xy$ 平面的封閉區域

則 $R = \{(x, y): 0 \le x^2 + y^2 \le 1\}$ 且 $\displaystyle\iint_S z\, dA = \iint_R \sqrt{x^2 + y^2}\left|\frac{\partial \vec{r}}{\partial x} \times \frac{\partial \vec{r}}{\partial y}\right| dxdy$

$\because \dfrac{\partial \vec{r}}{\partial x} = \left(1, 0, \dfrac{x}{\sqrt{x^2 + y^2}}\right)$ 且 $\dfrac{\partial \vec{r}}{\partial y} = \left(0, 1, \dfrac{y}{\sqrt{x^2 + y^2}}\right)$

$\therefore \dfrac{\partial \vec{r}}{\partial x} \times \dfrac{\partial \vec{r}}{\partial y} = \left(\dfrac{x}{\sqrt{x^2 + y^2}}, -\dfrac{y}{\sqrt{x^2 + y^2}}, 1\right) \Rightarrow \left|\dfrac{\partial \vec{r}}{\partial x} \times \dfrac{\partial \vec{r}}{\partial y}\right| = \sqrt{1 + \dfrac{x^2}{x^2 + y^2} + \dfrac{y^2}{x^2 + y^2}} = \sqrt{2}$

$\therefore \displaystyle\iint_S z\, dA = \iint_{\sqrt{x^2+y^2} \le 1} \sqrt{x^2 + y^2}\left|\frac{\partial \vec{r}}{\partial x} \times \frac{\partial \vec{r}}{\partial y}\right| dxdy = \iint_{\sqrt{x^2+y^2} \le 1} \sqrt{x^2 + y^2}\,\sqrt{2}\,dxdy$

令 $x = r\cos\theta,\ y = r\sin\theta$ 且 $R = \{(r, \theta): 0 \le r \le 1, 0 \le \theta \le 2\pi\}$

$\therefore \displaystyle\iint_{\sqrt{x^2+y^2} \le 1} \sqrt{x^2 + y^2}\,\sqrt{2}\,dxdy = \sqrt{2}\int_0^{2\pi}\int_0^1 r\,\sqrt{2}\,r\,drd\theta = \dfrac{2\sqrt{2}\pi}{3}$

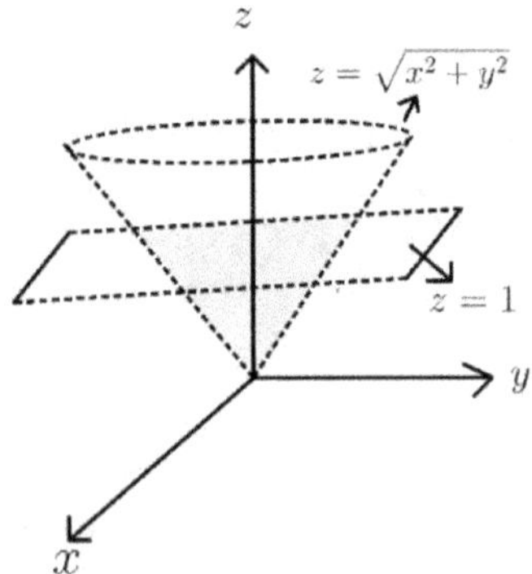



Example 16.

Evaluate $\displaystyle\iint_S z\,dA =?$, where $S$ is the hyperbolic $z^2 = 1 + x^2 + y^2$ between $z = 0$ and $z = \sqrt{5}$.

【解】

令 $\vec{r}(x,y) = \left(x, y, \sqrt{1 + x^2 + y^2}\right)$ 且 $R$ 为曲面$S$投影至$xy$平面的封闭区域

則 $R = \{(x,y): 0 \le x^2 + y^2 \le 4\}$ 且 $\displaystyle\iint_S z\,dA = \iint_R \sqrt{x^2 + y^2}\left|\frac{\partial \vec{r}}{\partial x} \times \frac{\partial \vec{r}}{\partial y}\right| dxdy$

$\because \dfrac{\partial \vec{r}}{\partial x} = \left(1, 0, \dfrac{x}{\sqrt{1 + x^2 + y^2}}\right)$ 且 $\dfrac{\partial \vec{r}}{\partial y} = \left(0, 1, \dfrac{y}{\sqrt{1 + x^2 + y^2}}\right)$

$\therefore \dfrac{\partial \vec{r}}{\partial x} \times \dfrac{\partial \vec{r}}{\partial y} = \left(\dfrac{x}{\sqrt{1 + x^2 + y^2}}, \dfrac{y}{\sqrt{1 + x^2 + y^2}}, 1\right) \Rightarrow \left|\dfrac{\partial \vec{r}}{\partial x} \times \dfrac{\partial \vec{r}}{\partial y}\right| = \sqrt{1 + \dfrac{x^2 + y^2}{1 + x^2 + y^2}}$

$\therefore \displaystyle\iint_S z\,dA = \iint_{\sqrt{x^2+y^2} \le 2} \sqrt{1 + x^2 + y^2}\left|\frac{\partial \vec{r}}{\partial x} \times \frac{\partial \vec{r}}{\partial y}\right| dxdy$

$= \displaystyle\iint_{\sqrt{x^2+y^2} \le 2} \sqrt{1 + x^2 + y^2}\sqrt{1 + \dfrac{x^2 + y^2}{1 + x^2 + y^2}}\,dxdy$

令 $x = r\cos\theta$, $y = r\sin\theta$ 且 $R = \{(r,\theta): 0 \le r \le 2, 0 \le \theta \le 2\pi\}$

$$\iint_{\sqrt{x^2+y^2}\le 2} \sqrt{1+x^2+y^2}\sqrt{1+\frac{x^2+y^2}{1+x^2+y^2}}\,dxdy = \int_0^{2\pi}\int_0^2 r\sqrt{1+2r^2}\,drd\theta$$

$$= 2\pi \times \left.\frac{(1+2r^2)^{\frac{3}{2}}}{6}\right|_0^2 = \frac{26\pi}{3}$$

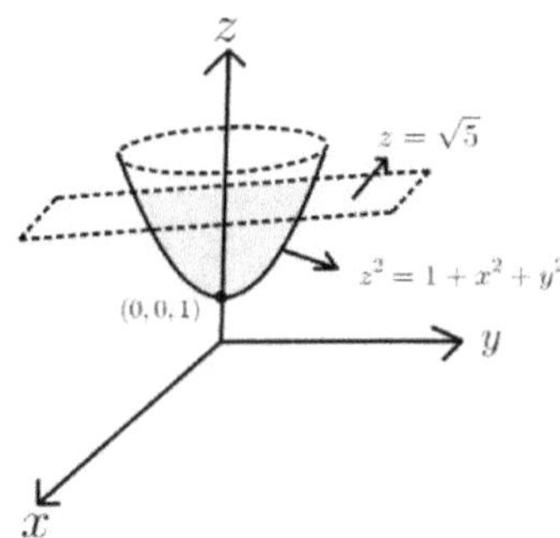


Example 17.

$$求 \iint_S x^2 + y^2\, dA =?, \quad S: z = 1 - (x^2 + y^2), \quad z > 0$$

【解】

令 $\vec{r}(x,y) = (x, y, 1-(x^2+y^2))$ 且 $R$ 為曲面$S$投影至$xy$平面的封閉区域

則 $R = \{(x,y): 0 \le x^2+y^2 \le 1\}$ 且 $\displaystyle\iint_S x^2+y^2\, dA = \iint_R (x^2+y^2)\left|\frac{\partial\vec{r}}{\partial x}\times\frac{\partial\vec{r}}{\partial y}\right|dxdy$

$\because \dfrac{\partial\vec{r}}{\partial x} = (1,0,-2x)$ 且 $\dfrac{\partial\vec{r}}{\partial y} = (0,1,-2y)$  $\therefore \dfrac{\partial\vec{r}}{\partial x}\times\dfrac{\partial\vec{r}}{\partial y} = (2x, 2y, 1)$

$\Rightarrow \left|\dfrac{\partial\vec{r}}{\partial x}\times\dfrac{\partial\vec{r}}{\partial y}\right| = \sqrt{1+4x^2+4y^2}$

令 $x = r\cos\theta$, $y = r\sin\theta$ 且 $R = \{(r,\theta): 0 \le r \le 1, 0 \le \theta \le 2\pi\}$

$$\therefore \iint_S x^2 + y^2 \, dA = \iint_R (x^2 + y^2) \left| \frac{\partial \vec{r}}{\partial x} \times \frac{\partial \vec{r}}{\partial y} \right| dxdy$$

$$= \iint_{0 \leq x^2 + y^2 \leq 1} (x^2 + y^2)\sqrt{1 + 4x^2 + 4y^2} \, dxdy = \int_0^{2\pi} \int_0^1 r^2 \sqrt{1 + 4r^2} \, r \, dr d\theta$$

$$= 2\pi \int_0^1 r^3 \sqrt{1 + 4r^2} \, dr$$

令 $t = \sqrt{1 + 4r^2}$ 則 $\int_0^1 r^3 \sqrt{1 + 4r^2} \, dr = \int_0^{\sqrt{5}} \frac{t^2 - 1}{4} \cdot t \cdot \frac{t}{4} dt = \frac{1}{16}\left( \frac{10\sqrt{5}}{3} + \frac{2}{15} \right)$

$$\therefore \iint_S x^2 + y^2 \, dA = \frac{\pi}{8}\left( \frac{10\sqrt{5}}{3} + \frac{2}{15} \right)$$

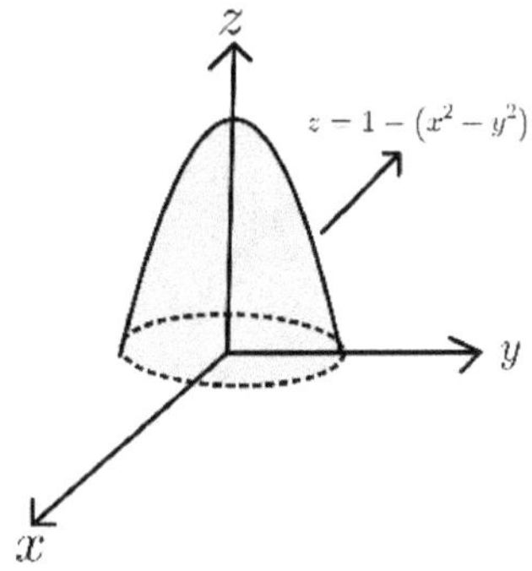


Example 18.

$$求 \iint_S x^3 \sin y \, dA =?, \quad S: z = x^3, 0 \leq x \leq 1, 0 \leq y \leq \frac{\pi}{2}$$

【解】

令 $\vec{r}(x, y) = (x, y, x^3)$ 且 $R$ 为曲面$S$投影至$xy$平面的封闭区域

則 $R = \left\{ (x, y): 0 \leq x \leq 1, 0 \leq y \leq \frac{\pi}{2} \right\}$ 且 $\iint_S x^3 \sin y \, dA = \iint_R x^3 \sin y \left| \frac{\partial \vec{r}}{\partial x} \times \frac{\partial \vec{r}}{\partial y} \right| dxdy$

$\because \dfrac{\partial \vec{r}}{\partial x} = (1, 0, 3x^2)$ 且 $\dfrac{\partial \vec{r}}{\partial y} = (0, 1, 0)$ $\therefore \dfrac{\partial \vec{r}}{\partial x} \times \dfrac{\partial \vec{r}}{\partial y} = (-3x^2, 0, 1) \Rightarrow \left| \dfrac{\partial \vec{r}}{\partial x} \times \dfrac{\partial \vec{r}}{\partial y} \right| = \sqrt{1 + 9x^4}$

$$\iint_S x^3 \sin y \, dA = \iint_R x^3 \sin y \left| \frac{\partial \vec{r}}{\partial x} \times \frac{\partial \vec{r}}{\partial y} \right| dxdy = \int_0^{\frac{\pi}{2}} \int_0^1 x^3 \sin y \sqrt{1 + 9x^4} \, dxdy$$

$$= \int_0^{\frac{\pi}{2}} \frac{1}{54} (1 + 9x^4)^{\frac{3}{2}} \Big|_0^1 \sin y \, dy = \frac{1}{54} (10\sqrt{10} - 1)$$

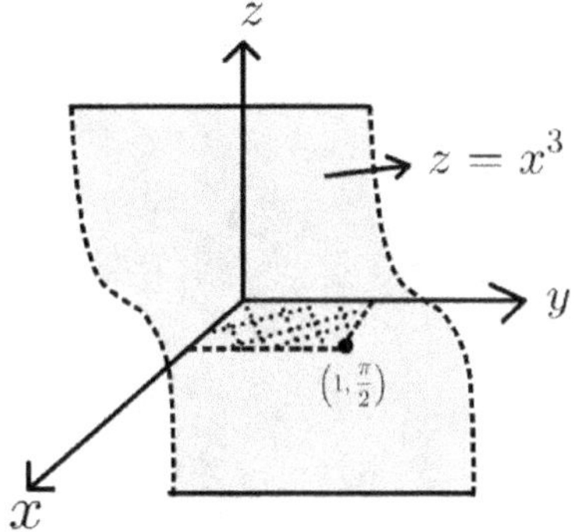


Example 19.

$$求 \iint_S \frac{xy}{z} dA = ?, \quad S: z = x^2 + y^2, \ 4 \le x^2 + y^2 \le 16, \ x, y \ge 0$$

【解】

令 $\vec{r}(x, y) = (x, y, x^2 + y^2)$ 且 $R = \{(x, y): 4 \le x^2 + y^2 \le 16, x, y \ge 0\}$

則 $\displaystyle \iint_S \frac{xy}{z} dA = \iint_R \frac{xy}{x^2 + y^2} \left| \frac{\partial \vec{r}}{\partial x} \times \frac{\partial \vec{r}}{\partial y} \right| dxdy$

$\because \dfrac{\partial \vec{r}}{\partial x} = (1, 0, 2x)$ 且 $\dfrac{\partial \vec{r}}{\partial y} = (0, 1, 2y)$ $\therefore \dfrac{\partial \vec{r}}{\partial x} \times \dfrac{\partial \vec{r}}{\partial y} = (-2x, -2y, 1)$

$\therefore \left| \dfrac{\partial \vec{r}}{\partial x} \times \dfrac{\partial \vec{r}}{\partial y} \right| = \sqrt{1 + 4x^2 + 4y^2}$

$\therefore \displaystyle \iint_S \frac{xy}{z} dA = \iint_R \frac{xy}{x^2 + y^2} \left| \frac{\partial \vec{r}}{\partial x} \times \frac{\partial \vec{r}}{\partial y} \right| dxdy = \iint_{4 \le x^2 + y^2 \le 16} \frac{xy\sqrt{1 + 4x^2 + 4y^2}}{x^2 + y^2} dxdy$

令 $x = r\cos\theta$, $y = r\sin\theta$ 且 $R = \left\{ (r, \theta): 2 \le r \le 4, 0 \le \theta \le \dfrac{\pi}{2} \right\}$ 則

$$\iint_{4 \le x^2 + y^2 \le 16} \frac{xy\sqrt{1 + 4x^2 + 4y^2}}{x^2 + y^2} dxdy = \int_0^{\frac{\pi}{2}} \int_2^4 \frac{r\cos\theta \, r\sin\theta \, \sqrt{1 + 4r^2}}{r^2} r \, dr d\theta$$

$$= \int_0^{\frac{\pi}{2}} \int_2^4 r\cos\theta \sin\theta \sqrt{1+4r^2}\, dr d\theta = \int_2^4 r\sqrt{1+4r^2}\, dr \int_0^{\frac{\pi}{2}} \cos\theta \sin\theta\, d\theta$$

$$= \left.\frac{(1+4r^2)^{\frac{3}{2}}}{12}\right|_2^4 \cdot \left.\left(\frac{-\cos 2\theta}{4}\right)\right|_0^{\frac{\pi}{2}} = \frac{65\sqrt{65} - 17\sqrt{17}}{24}$$

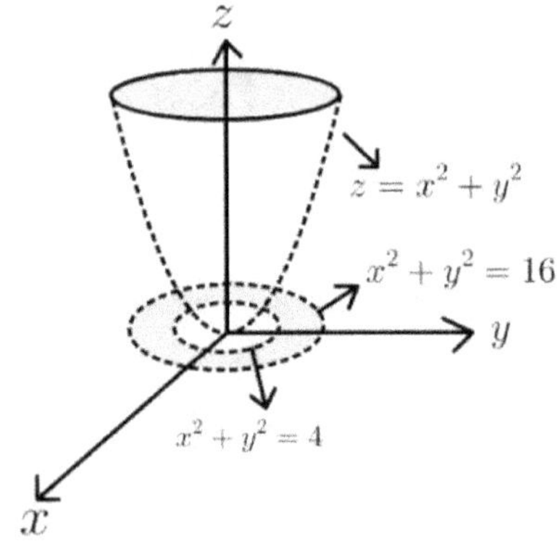


## 9.5.4    $\vec{F}$为向量函数, 给曲面S求通量

求曲面的通量主要都是考虑可定向曲面, 底下先介绍可定向曲面的定义
**【定义】**
如果曲面$S$上的任意点都存在一个连续的单位法向量$\vec{n}$则称曲面$S$为可定向曲面

假设$S$为一可定向的曲面
$$\vec{r}(s,t) = \big(x(s,t), y(s,t), z(s,t)\big), \quad \forall (s,t) \in D$$
具有单位法向量$\vec{n}$, 假设参数定义域$D$为一个矩形, 将其分割为数个小矩形 $D_{ij}$, 长为 $\Delta s$, 宽为 $\Delta t$, $D_{ij}$ 的中心点为 $\left(s_i^*, t_j^*\right)$, $D_{ij}$ 对应曲面 $S$ 上的区域为 $S_{ij}$ 并且 $\left(s_i^*, t_j^*\right)$ 对应曲面 $S_{ij}$ 上的点为 $E_{ij}^*$, 其位置向量为 $\vec{r}\left(s_i^*, t_j^*\right)$; 因此, 曲面$S$被分割成数个片段$S_{ij}$, 如下图

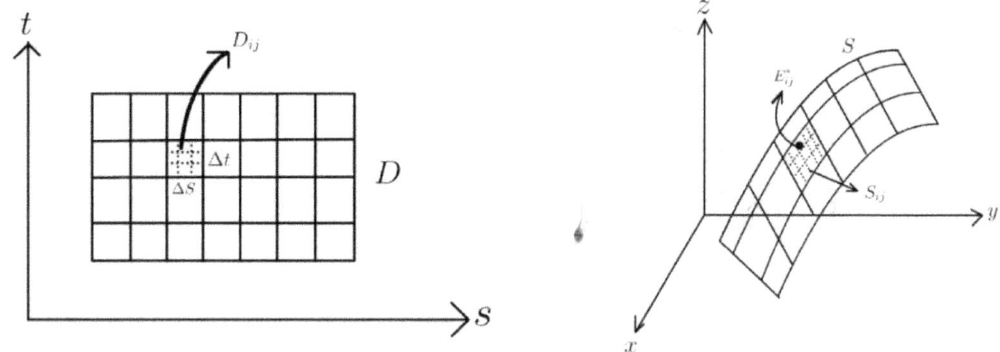

考虑一具有密度$\rho(x,y,z)$且速度场$v(x,y,z)$的流体通过曲面$S$,则每单位面积的流量以$\rho v$表示; 将$D$分割成无数个小矩形使得曲面$S$分割成无数个小曲面$S_{ij}$且$S_{ij}$几乎是个平面,则在法向量$\vec{n}$的方向上每单位时间通过$S_{ij}$的流体质量,近似估计相当于

$$\rho v \cdot \vec{n} \Delta S_{ij}$$

其中$\rho, v$以及$\vec{n}$在$S_{ij}$的某点取值; 将所有通过$S_{ij}$的流体质量相加,得到了曲面$S$的流体质量的近似估计

$$\sum_{i=1}^{m}\sum_{j=1}^{n} \rho v \cdot \vec{n} \Delta S_{ij}$$

当$S_{ij}$切割至任意小则$\displaystyle\sum_{i=1}^{m}\sum_{j=1}^{n} \rho v \cdot \vec{n} \Delta S_{ij}$将逼近曲面$S$的流体质量,取极限之后可得到函数$\rho v \cdot \vec{n}$ 在 $S$ 上的表面积分,即

$$\iint_S \rho v \cdot \vec{n}\, dS = \iint_S \rho(x,y,z)v(x,y,z) \cdot \vec{n}(x,y,z)\, dS = \lim_{m\to\infty, n\to\infty} \sum_{i=1}^{m}\sum_{j=1}^{n} \rho v \cdot \vec{n} \Delta S_{ij}$$

这在物理上代表流经 $S$ 的流速, 定义$\vec{F} = \rho v$, 其中$\vec{F}$是$\mathbb{R}^3$中的向量场。因此

$$\iint_S \rho(x,y,z)v(x,y,z) \cdot \vec{n}(x,y,z)\, dS = \iint_S \vec{F} \cdot \vec{n}\, dS$$

藉由上述黎曼和的极限定义向量函数$\vec{F}$在曲面$S$上的表面积

## 【定义】

假设连续向量场$\vec{F}$定义在可定向曲面$S$上, 并具有单位法向量$\vec{n}$,则$\vec{F}$在$S$上的表面积分为

$$\iint_S \vec{F} \cdot d\vec{S} = \iint_S \vec{F} \cdot \vec{n}\, dS$$

此积分称为$\vec{F}$在$S$上的通量

接下来的重点是给定可定向参数曲面 $\vec{r}(s,t)$, 如何找到单位法向量, 假设参数化曲面为

$$\vec{r}(s,t) = \big(x(s,t), y(s,t), z(s,t)\big), \quad \forall (s,t) \in D$$

则可以找到两个连续且方向相反的单位向量场, 分别是

$$\frac{\vec{r}_s(s,t) \times \vec{r}_t(s,t)}{|\vec{r}_s(s,t) \times \vec{r}_t(s,t)|} \quad 且 \quad -\frac{\vec{r}_s(s,t) \times \vec{r}_t(s,t)}{|\vec{r}_s(s,t) \times \vec{r}_t(s,t)|}$$

## 【定义】

假设 $S$ 为可定向曲面

(i) 若 $\dfrac{\vec{r}_s(s,t) \times \vec{r}_t(s,t)}{|\vec{r}_s(s,t) \times \vec{r}_t(s,t)|}$ 的 $z$ 坐标为正, $\forall (s,t) \in D$ 则 $S$ 为朝上定向(oriented upward)

(ii) 若 $\dfrac{\vec{r}_s(s,t) \times \vec{r}_t(s,t)}{|\vec{r}_s(s,t) \times \vec{r}_t(s,t)|}$ 的 $z$ 坐标为负, $\forall (s,t) \in D$ 则 $S$ 为朝下定向(oriented downward)

一般而言, 题目会告诉我们 $S$ 为朝上定向或者是朝下定向

同理, 如果 $\dfrac{\vec{r}_s(s,t) \times \vec{r}_t(s,t)}{|\vec{r}_s(s,t) \times \vec{r}_t(s,t)|}$ 的 $y$ 坐标为正则是朝着 $y$ 轴正向的方向定向, 依此类推

## 【定义】

假设 $S$ 为封闭曲面

(i) 如果单位法向量朝外指向非封闭的区域则 $S$ 为朝外定向(oriented outward)或正定向(oriented positive)

(ii) 如果单位法向量朝内指向封闭的区域则 $S$ 为朝内定向(oriented inward)或负定向(oriented negative )

一般而言, 如果给定封闭曲面 $S$ 计算通量, 且没有特别指明 $S$ 为朝外定向或朝内定向, 惯例上指的是朝外定向

为了计算上述定义中的表面积分, 考虑位于切平面内的一个近似平行四边形的面积用以逼近区域 $\Delta S_{ij}$, 之前在表面积的讨论中, 有一个 $\Delta S_{ij}$ 的近似估计

$$\Delta S_{ij} \approx \big|\vec{r}_s(s_i^*, t_j^*) \times \vec{r}_t(s_i^*, t_j^*)\big| \Delta s \Delta t$$

所以

$$\iint_S \vec{F} \cdot \vec{n}\, dS = \lim_{m \to \infty, n \to \infty} \sum_{i=1}^m \sum_{j=1}^n \vec{F} \cdot \vec{n} \Delta S_{ij} = \lim_{m \to \infty, n \to \infty} \sum_{i=1}^m \sum_{j=1}^n \vec{F} \cdot \vec{n} \big|\vec{r}_s(s_i^*, t_j^*) \times \vec{r}_t(s_i^*, t_j^*)\big| \Delta s \Delta t$$

$$= \lim_{m \to \infty, n \to \infty} \sum_{i=1}^m \sum_{j=1}^n \vec{F} \cdot \frac{\vec{r}_s(s_i^*, t_j^*) \times \vec{r}_t(s_i^*, t_j^*)}{|\vec{r}_s(s_i^*, t_j^*) \times \vec{r}_t(s_i^*, t_j^*)|} \cdot \big|\vec{r}_s(s_i^*, t_j^*) \times \vec{r}_t(s_i^*, t_j^*)\big| \Delta s \Delta t$$

$$= \lim_{m\to\infty,n\to\infty} \sum_{i=1}^{m}\sum_{j=1}^{n} \vec{F}(s_i^*,t_j^*)\cdot\left(\vec{r}_s(s_i^*,t_j^*)\times\vec{r}_t(s_i^*,t_j^*)\right)\Delta s\Delta t = \iint_D \vec{F}(\vec{r}(s,t))\cdot(\vec{r}_s\times\vec{r}_t)dsdt$$

藉由上述观察, 求向量函数的曲面积分相当于求双重积分, 特别的是双重积分的被积分函数为

$$\vec{F}(\vec{r}(s,t))\cdot(\vec{r}_s\times\vec{r}_t)$$

因此求向量函数曲面的积分相当于先求得被积分函数之后, 再求双重积分; 底下针对曲面的积分转换成双重积分后, 有哪些不同类型作初步的说明

考虑曲面$S$: $z = g(x,y)$, 则 $\vec{n} = \dfrac{\vec{r}_x(x,y)\times\vec{r}_y(x,y)}{\left|\vec{r}_x(x,y)\times\vec{r}_y(x,y)\right|}$

因为

$$\frac{\partial\vec{r}}{\partial x} = (1,0,g_x(x,y)), \quad \frac{\partial\vec{r}}{\partial y} = \left(0,1,g_y(x,y)\right) \quad \text{且} \quad \frac{\partial\vec{r}}{\partial x}\times\frac{\partial\vec{r}}{\partial y} = (-g_x(x,y),-g_y(x,y),1)$$

所以

$$\vec{n} = \frac{\vec{r}_x(x,y)\times\vec{r}_y(x,y)}{\left|\vec{r}_x(x,y)\times\vec{r}_y(x,y)\right|} = \frac{1}{\sqrt{1+\left(g_x(x,y)\right)^2+\left(g_1(x,y)\right)^2}}(-g_x(x,y),-g_y(x,y),1)$$

$\because \dfrac{\vec{r}_x(x,y)\times\vec{r}_y(x,y)}{\left|\vec{r}_x(x,y)\times\vec{r}_y(x,y)\right|}$ 的 z 坐标为正 $\quad\therefore S$为朝上定向

曲面是 oriented upward 有各式各样的样貌,底下列举常见 oriented upward 的曲面

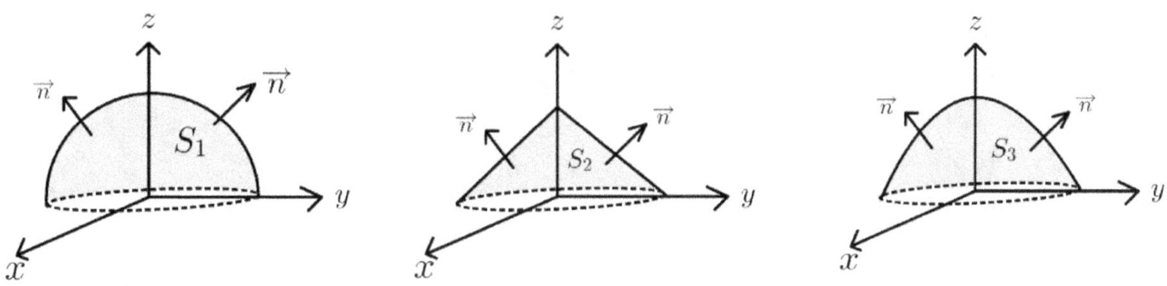

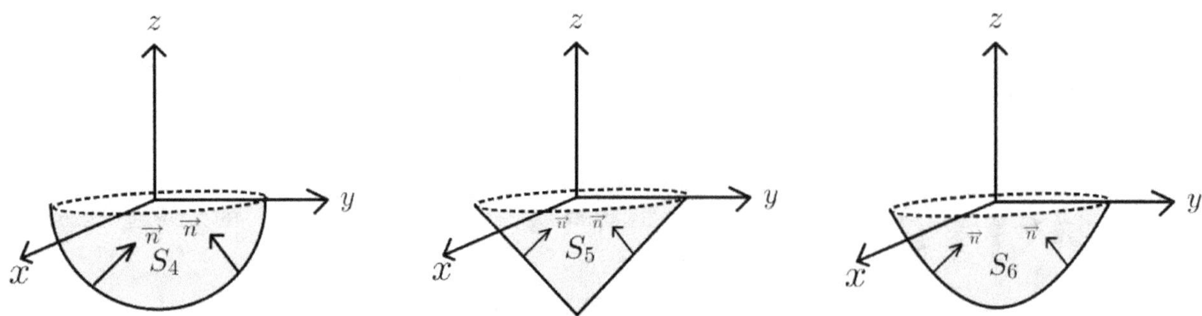

曲面是 oriented downward 有各式各样的样貌, 底下列举常见 oriented downward的曲面

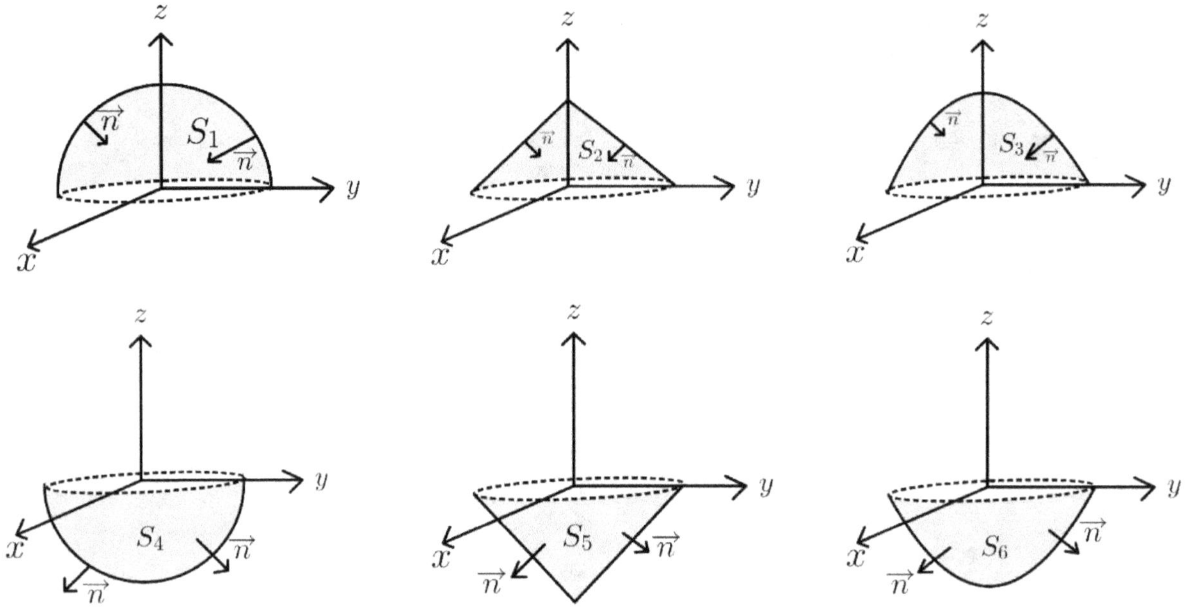

因此, 如果曲面$S: z = g(x,y)$为朝上定向, 定义域为 $a \leq x \leq b$ 且 $c \leq y \leq d$, 假设$\vec{F}(x,y,z)$为三维空间中的向量值函数, 令 $\vec{r}(x,y) = (x, y, g(x,y))$ 且 R为曲面$S$投影于$xy$平面上的封闭区域, 即

$$R = \{(x,y): a \leq x \leq b, c \leq y \leq d\}$$

则$\vec{F} \cdot \vec{n}$ 在 曲面$S$ 上的曲面积分表示为：

$$\iint_S \vec{F} \cdot \vec{n}\, dA = \iint_R \vec{F}(\vec{r}(x,y)) \cdot \frac{\partial \vec{r}}{\partial x} \times \frac{\partial \vec{r}}{\partial y}\, dxdy = \iint_R \vec{F}(x, y, g(x,y)) \cdot \frac{\partial \vec{r}}{\partial x} \times \frac{\partial \vec{r}}{\partial y}\, dxdy$$

由于 $\dfrac{\partial \vec{r}}{\partial x} = (1, 0, g_x(x,y))$, $\dfrac{\partial \vec{r}}{\partial y} = (0, 1, g_y(x,y))$ 且 $\dfrac{\partial \vec{r}}{\partial x} \times \dfrac{\partial \vec{r}}{\partial y} = (-g_x(x,y), -g_y(x,y), 1)$

所以公式可改写为：

$$\iint_S \vec{F} \cdot \vec{n}\, dA = \iint_R \vec{F}(x, y, g(x, y)) \cdot (-g_x(x, y), -g_y(x, y), 1)\, dxdy$$

因此, 如果$S$是朝下定向则

$$\iint_S \vec{F} \cdot \vec{n}\, dA = \iint_R \vec{F}(x, y, g(x, y)) \cdot (g_x(x, y), g_y(x, y), -1)\, dxdy$$

特别的是, 如果曲面$S$为$z = g(x, y)$, 投影至$xy$平面为$\{(x, y): x^2 + y^2 \leq a\}$, 藉由极坐标表示

$$x = r\cos\theta,\, y = r\sin\theta$$

如果$S$是朝上定向则

$$\iint_S \vec{F} \cdot \vec{n}\, dA = \int_0^{2\pi} \int_0^a \vec{F}(x, y, g(x, y)) \cdot (-g_x(x, y), -g_y(x, y), 1)\big|_{x=r\cos\theta, y=r\sin\theta}\, drd\theta$$

如果$S$是朝下定向则

$$\iint_S \vec{F} \cdot \vec{n}\, dA = \int_0^{2\pi} \int_0^a \vec{F}(x, y, g(x, y)) \cdot (g_x(x, y), g_y(x, y), -1)\big|_{x=r\cos\theta, y=r\sin\theta}\, drd\theta$$

假设曲面为一球面$S: x^2 + y^2 + z^2 = r^2$, 给定三维向量函数$\vec{F}(x, y, z)$, 目标是计算

通量$\iint_S \vec{F} \cdot \vec{n}\, dA$; 假设$\vec{r}(\theta, \varphi) = (r\cos\theta\sin\varphi, r\sin\theta\sin\varphi, r\cos\varphi)$ 则

$$\iint_S \vec{F} \cdot \vec{n}\, dA = \iint_R \vec{F}(\vec{r}(\theta, \varphi)) \cdot \frac{\partial \vec{r}}{\partial \varphi} \times \frac{\partial \vec{r}}{\partial \theta}\, d\theta d\varphi$$

两者的外积为

$$\frac{\partial \vec{r}}{\partial \varphi} \times \frac{\partial \vec{r}}{\partial \theta} = (r^2 \cos\theta \sin^2\varphi, r^2\sin\theta\sin^2\varphi, r^2\sin\varphi\cos\varphi)$$

将此外积代入积分表达式中得到:

$$\iint_S \vec{F} \cdot \vec{n}\, dA = \iint_R \vec{F}(\vec{r}(\theta, \varphi)) \cdot \frac{\partial \vec{r}}{\partial \varphi} \times \frac{\partial \vec{r}}{\partial \theta}\, d\theta d\varphi$$

$$= \iint_R \vec{F}(r\cos\theta\sin\varphi, r\sin\theta\sin\varphi, r\cos\varphi) \cdot$$

$$(r^2\cos\theta\sin^2\varphi, r^2\sin\theta\sin^2\varphi, r^2\sin\varphi\cos\varphi)d\theta d\varphi,$$

其中$R = \{(\theta, \varphi): 0 \leq \theta \leq 2\pi, 0 \leq \varphi \leq \pi\}$

上述给出了曲面在不同的情境之下计算向量函数曲面积分的公式, 与计算曲面通量相

关有两个重要定理, Stokes' Theorem 以及 Gauss's Theorem; 当曲面为非封闭曲面时, Stokes' Theorem 是计算非封闭曲面的通量以及计算封闭空间曲线线积分的重要桥梁; 当曲面为封闭曲面时, Gauss's Theorem 是计算封闭曲面的通量以及计算纯量函数三重积分的重要桥梁; 因此无论曲面是否封闭, 求曲面的通量是个重要的环节, 读者务必孰悉各种题型

考试类型:

Type 1.

給定曲面$S$: $\vec{r}(s,t)$, $\forall\, a \leq s \leq b, c \leq t \leq d$, $S$是朝上定向, $\vec{F}(x,y,z)$为三维空间的向量函数, 求通过曲面$S$的通量 $\iint_S \vec{F} \cdot \vec{n}\, dA =?$

解题流程:

Step1.

令 $R = \{(s,t): a \leq s \leq b, c \leq t \leq d\}$ 则 $\iint_S \vec{F} \cdot \vec{n}\, dA = \iint_R \vec{F} \cdot \dfrac{\partial \vec{r}}{\partial s} \times \dfrac{\partial \vec{r}}{\partial t}\, ds\, dt$

Step2.

求 $\dfrac{\partial \vec{r}}{\partial s} \times \dfrac{\partial \vec{r}}{\partial t} =?$, $\iint_R \vec{F} \cdot \dfrac{\partial \vec{r}}{\partial s} \times \dfrac{\partial \vec{r}}{\partial t}\, ds\, dt =?$

如果$S$是朝下定向则求 $-\iint_R \vec{F} \cdot \dfrac{\partial \vec{r}}{\partial s} \times \dfrac{\partial \vec{r}}{\partial t}\, ds\, dt =?$

Type 2.

給定曲面$S$: $z = g(x,y)$, $\forall\, a \leq x \leq b, c \leq y \leq d$, $S$是朝上定向且$\vec{F}(x,y,z)$为三维空间的向量函数, 求通过曲面$S$的通量 $\iint_S \vec{F} \cdot \vec{n}\, dA =?$

解题流程:

Step1.

令 $\vec{r}(x,y) = (x, y, g(x,y))$ 且 $R = \{(x,y): a \leq x \leq b, c \leq y \leq d\}$ 则

$$\iint_S \vec{F} \cdot \vec{n}\, dA = \iint_R \vec{F} \cdot \dfrac{\partial \vec{r}}{\partial x} \times \dfrac{\partial \vec{r}}{\partial y}\, dx\, dy = \iint_R \vec{F}(x,y,g(x,y)) \cdot \dfrac{\partial \vec{r}}{\partial x} \times \dfrac{\partial \vec{r}}{\partial y}\, dx\, dy$$

Step2.

$\because \dfrac{\partial \vec{r}}{\partial x} = (1, 0, g_x(x,y))$ 且 $\dfrac{\partial \vec{r}}{\partial y} = (0, 1, g_y(x,y))$ $\therefore \dfrac{\partial \vec{r}}{\partial x} \times \dfrac{\partial \vec{r}}{\partial y} = (-g_x(x,y), -g_y(x,y), 1)$

Step3.

$$\therefore \iint_S \vec{F} \cdot \vec{n}\, dA = \iint_R \vec{F}(x,y,g(x,y)) \cdot (-g_x(x,y), -g_y(x,y), 1)dxdy$$

Step4.

藉由 Fubini's Theorem

$$\iint_R \vec{F}(x,y,g(x,y)) \cdot (-g_x(x,y), -g_y(x,y), 1)dxdy$$

$$= \int_a^b \int_c^d \vec{F}(x,y,g(x,y)) \cdot (-g_x(x,y), -g_y(x,y), 1)dxdy$$

如果$S$是朝下定向则

$$求 \iint_S \vec{F} \cdot \vec{n}\, dA = -\int_a^b \int_c^d \vec{F}(x,y,g(x,y)) \cdot (-g_x(x,y), -g_y(x,y), 1)dxdy$$

<u>范例说明:</u>

$$求 \iint_S \vec{F} \cdot \vec{n}\, dA = ?$$

(I)若$\vec{F}(x,y,z) = \left(6y^2, 0, 2x^2 + \sqrt{xy}\right)$, 曲面$S$是$x + y + z = 1$在第一卦限的朝上定向区域

令$\vec{r}(x,y,z) = (x, y, 1 - x - y)$ 且 $R$为曲面$S$投影至$xy$平面的封闭区域

则 $R = \{(x,y): x + y \leq 1, x \geq 0, y \geq 0\}$ 且 $\displaystyle\iint_S \vec{F} \cdot \vec{n}\, dA = \iint_R \vec{F} \cdot \frac{\partial \vec{r}}{\partial x} \times \frac{\partial \vec{r}}{\partial y} dxdy$

$$\because \frac{\partial \vec{r}}{\partial x} = (1, 0, -1) \text{ 且 } \frac{\partial \vec{r}}{\partial y} = (0, 1, -1) \quad \therefore \frac{\partial \vec{r}}{\partial x} \times \frac{\partial \vec{r}}{\partial y} = (1, 1, 1)$$

$$\because \vec{F} \cdot \frac{\partial \vec{r}}{\partial x} \times \frac{\partial \vec{r}}{\partial y} = (6y^2, 0, 2x^2) \cdot \frac{\partial \vec{r}}{\partial x} \times \frac{\partial \vec{r}}{\partial y} = 2(x^2 + 3y^2), \quad \forall (x,y,z) \in S$$

$$\therefore \iint_S \vec{F} \cdot \vec{n}\, dA = \iint_R \vec{F} \cdot \frac{\partial \vec{r}}{\partial x} \times \frac{\partial \vec{r}}{\partial y} dxdy = 2 \iint_R (x^2 + 3y^2)\, dxdy$$

$$= 2 \int_0^1 \int_0^{1-y} x^2 + 3y^2\, dxdy = 2 \int_0^1 \left(\frac{x^3}{3} + 3y^2 x\right)\Big|_{x=0}^{x=1-y} dy$$

$$= 2 \int_0^1 \left(\frac{(1-y)^3}{3} + 3y^2(1-y)\right) dy = \frac{2}{3}$$

(II)若$\vec{F}(x,y,z) = (e^y, e^x + 6y, 12y)$,　曲面$S$是$x + y + z = 6$在第一卦限的朝上定向区域

令$\vec{r}(x,y,z) = (x,y,6-x-y)$且$R$为曲面$S$投影至$xy$平面的封闭区域

则 $R = \{(x,y): x + y \leq 6, x \geq 0, y \geq 0\}$ 且 $\iint_S \vec{F} \cdot \vec{n}\, dA = \iint_R \vec{F} \cdot \frac{\partial \vec{r}}{\partial x} \times \frac{\partial \vec{r}}{\partial y}\, dxdy$

$\because \frac{\partial \vec{r}}{\partial x} = (1,0,-1)$ 且 $\frac{\partial \vec{r}}{\partial y} = (0,1,-1)$ 　$\therefore \frac{\partial \vec{r}}{\partial x} \times \frac{\partial \vec{r}}{\partial y} = (1,1,1)$

$\because \vec{F} \cdot \frac{\partial \vec{r}}{\partial x} \times \frac{\partial \vec{r}}{\partial y} = e^y + e^x + 18y,\ \ \forall (x,y,z) \in S$

$\therefore \iint_S \vec{F} \cdot \vec{n}\, dA = \iint_R \vec{F} \cdot \frac{\partial \vec{r}}{\partial x} \times \frac{\partial \vec{r}}{\partial y}\, dxdy = \iint_R e^y + e^x + 18y\, dxdy$

$= \int_0^6 \int_0^{6-x} e^y + e^x + 18y\, dydx = 634 + 2e^6$

Type 3.

給定曲面$S: z = g(x,y)$,　$S$是朝上定向且投影至$xy$平面 $= \{(x,y): x^2 + y^2 \leq a^2\}$, 假设向

量场 $\vec{F} = (f_1(x,y,z), f_2(x,y,z), f_3(x,y,z))$,　求 $\iint_S \vec{F} \cdot \vec{n}\, dA =?$

<u>补充说明</u>

曲面$S$通常为椭圆球、圆球、椭圆椎、圆锥、椭圆抛物面

解题流程:

Step1.

令$\vec{r}(x,y) = (x,y,g(x,y))$ 且 $R = \{(x,y): x^2 + y^2 \leq a^2\}$

则 $\iint_S \vec{F} \cdot \vec{n}\, dA = \iint_R \vec{F} \cdot \frac{\partial \vec{r}}{\partial x} \times \frac{\partial \vec{r}}{\partial y}\, dxdy$

Step2.

$\because \frac{\partial \vec{r}}{\partial x} = (1,0,g_x(x,y)),\ \ \frac{\partial \vec{r}}{\partial y} = \left(0,1,g_y(x,y)\right)$ 　$\therefore \frac{\partial \vec{r}}{\partial x} \times \frac{\partial \vec{r}}{\partial y} = (-g_x(x,y), -g_y(x,y), 1)$

$\therefore \iint_S \vec{F} \cdot \vec{n}\, dA = \iint_R \vec{F} \cdot \frac{\partial \vec{r}}{\partial x} \times \frac{\partial \vec{r}}{\partial y}\, dxdy = \iint_R \vec{F} \cdot (-g_x(x,y), -g_y(x,y), 1)\, dxdy$

$$= \iint_R \left(-f_1 \cdot g_x(x,y) - f_2 \cdot g_y(x,y) + f_3\right)\Big|_{z=g(x,y)} dxdy$$

Step3.

令$x = r\cos\theta$, $y = r\sin\theta$ 则$\{(x,y): x^2 + y^2 \le a^2\} = \{(r,\theta): 0 \le r \le a, 0 \le \theta \le 2\pi\}$

且 $dxdy = \left\|\begin{vmatrix} \dfrac{\partial x}{\partial r} & \dfrac{\partial x}{\partial \theta} \\[6pt] \dfrac{\partial y}{\partial r} & \dfrac{\partial y}{\partial \theta} \end{vmatrix}\right\| drd\theta = \left\|\begin{vmatrix} \cos\theta & -r\sin\theta \\ \sin\theta & r\cos\theta \end{vmatrix}\right\| drd\theta = r\,drd\theta$

Step4.

$$\iint_R \left(-f_1 \cdot g_x(x,y) - f_2 \cdot g_y(x,y) + f_3\right)\Big|_{z=g(x,y)} dxdy$$

$$= \int_0^{2\pi} \int_0^a -f_1(r\cos\theta, r\sin\theta, g(r\cos\theta, r\sin\theta)) \cdot g_x(r\cos\theta, r\sin\theta) \cdot r$$

$$-f_2(r\cos\theta, r\sin\theta, g(r\cos\theta, r\sin\theta)) \cdot g_y(r\cos\theta, r\sin\theta) \cdot r$$

$$+f_3(r\cos\theta, r\sin\theta, g(r\cos\theta, r\sin\theta)) \cdot r\,drd\theta$$

如果$S$是朝下定向则

$$\iint_S \vec{F} \cdot \vec{n}\, dA$$

$$= -\int_0^{2\pi} \int_0^a -f_1(r\cos\theta, r\sin\theta, g(r\cos\theta, r\sin\theta)) \cdot g_x(r\cos\theta, r\sin\theta) \cdot r$$

$$-f_2(r\cos\theta, r\sin\theta, g(r\cos\theta, r\sin\theta)) \cdot g_y(r\cos\theta, r\sin\theta) \cdot r$$

$$+f_3(r\cos\theta, r\sin\theta, g(r\cos\theta, r\sin\theta)) \cdot r\,drd\theta$$

<u>范例说明:</u>

求 $\displaystyle\iint_S \vec{F} \cdot \vec{n}\, dA =?$

(I)若$\vec{F}(x,y,z) = (yz, -1, 1)$,  曲面 $S: z = \sqrt{x^2 + y^2}$, $\forall x^2 + y^2 \le a^2$, $S$是朝上定向

令$R = \{(x,y): x^2 + y^2 \le a^2\}$ 且 $x = r\cos\theta$, $y = r\sin\theta$ 则

$$\iint_S \vec{F} \cdot \vec{n}\, dA = \iint_R \vec{F} \cdot \frac{\partial \vec{r}}{\partial x} \times \frac{\partial \vec{r}}{\partial y} dxdy = \iint_{x^2+y^2 \le a^2} -xy + y(x^2 + y^2)^{-\frac{1}{2}} + 1\, dxdy$$

$$= \int_0^{2\pi} \int_0^a (-r^2 \cos\theta \sin\theta + \sin\theta + 1)\, r\,drd\theta = \pi a^2$$

Type 4.

给定球面 $S: x^2 + y^2 + z^2 = r^2$, 假设 $\vec{F}(x, y, z)$ 为三维空间向量函数, 求 $\iint_S \vec{F} \cdot \vec{n} dA =?$

解题流程:

Step1.

令 $\vec{r}(\theta, \varphi) = (r\cos\theta\sin\varphi, r\sin\theta\sin\varphi, r\cos\varphi)$ 且 $R = \{(\theta, \varphi): 0 \le \theta \le 2\pi, 0 \le \varphi \le \pi\}$

则 $\iint_S \vec{F} \cdot \vec{n} dA = \iint_R \vec{F}(r\cos\theta\sin\varphi, r\sin\theta\sin\varphi, r\cos\varphi) \cdot \dfrac{\partial\vec{r}}{\partial\varphi} \times \dfrac{\partial\vec{r}}{\partial\theta} d\theta d\varphi$

Step2.

$\because \dfrac{\partial\vec{r}}{\partial\theta} = (-r\sin\theta\sin\varphi, r\cos\theta\sin\varphi, 0)$ 且 $\dfrac{\partial\vec{r}}{\partial\varphi} = (r\cos\theta\cos\varphi, r\sin\theta\cos\varphi, -r\sin\varphi)$

$\therefore \dfrac{\partial\vec{r}}{\partial\varphi} \times \dfrac{\partial\vec{r}}{\partial\theta} = (r^2\cos\theta\sin^2\varphi, r^2\sin\theta\sin^2\varphi, r^2\sin\varphi\cos\varphi)$

Step3.

将 $\dfrac{\partial\vec{r}}{\partial\varphi} \times \dfrac{\partial\vec{r}}{\partial\theta} = (r^2\cos\theta\sin^2\varphi, r^2\sin\theta\sin^2\varphi, r^2\sin\varphi\cos\varphi)$

代入 $\iint_R \vec{F}(r\cos\theta\sin\varphi, r\sin\theta\sin\varphi, r\cos\varphi) \cdot \dfrac{\partial\vec{r}}{\partial\varphi} \times \dfrac{\partial\vec{r}}{\partial\theta} d\theta d\varphi$

$\therefore \iint_S \vec{F} \cdot \vec{n} dA = \iint_R \vec{F}(\vec{r}(\theta, \varphi)) \cdot \dfrac{\partial\vec{r}}{\partial\varphi} \times \dfrac{\partial\vec{r}}{\partial\theta} d\theta d\varphi$

$= \iint_R \vec{F}(r\cos\theta\sin\varphi, r\sin\theta\sin\varphi, r\cos\varphi)$

$\qquad\qquad \cdot (r^2\cos\theta\sin^2\varphi, r^2\sin\theta\sin^2\varphi, r^2\sin\varphi\cos\varphi) d\theta d\varphi$

<u>范例说明:</u>

求 $\iint_S \vec{F} \cdot \vec{n} dA =?$, 其中 $S: x^2 + y^2 + z^2 = r^2$

(I) 若 $\vec{F}(x, y, z) = (z, y, x)$, 令 $R = \{(\theta, \varphi): 0 \le \theta \le 2\pi, 0 \le \varphi \le \pi\}$

则 $\iint_S \vec{F} \cdot \vec{n} dA = \iint_R (r\cos\varphi, r\sin\theta\sin\varphi, r\cos\theta\sin\varphi)$

$\qquad\qquad \cdot (r^2\cos\theta\sin^2\varphi, r^2\sin\theta\sin^2\varphi, r^2\sin\varphi\cos\varphi) d\theta d\varphi = \dfrac{4\pi r^3}{3}$

(II) 若 $\vec{F}(x, y, z) = (x, y, z)$, 令 $R = \{(\theta, \varphi): 0 \le \theta \le 2\pi, 0 \le \varphi \le \pi\}$

则 $\displaystyle\iint_S \vec{F} \cdot \vec{n}\, dA = \iint_R (r\cos\theta\sin\varphi, r\sin\theta\sin\varphi, r\cos\varphi)$
$$\cdot (r^2\cos\theta\sin^2\varphi, r^2\sin\theta\sin^2\varphi, r^2\sin\varphi\cos\varphi)d\theta d\varphi = 4\pi r^3$$

Example 1.

假设 $\vec{\mathbf{F}} = (y+2, 2+4x, xyz)$, 求通过朝上定向曲面$S: y = x^2, \ 0 \le x \le 2, 0 \le z \le 2$ 的通量 $\displaystyle\iint_S \vec{F} \cdot \vec{n}\, dA =$?

【解】

$\because S$是朝上定向曲面

令 $\vec{r}(x, y, z) = (x, x^2, z)$ 且 $R$ 为曲面$S$投影至$xz$平面的封闭区域

则 $R = \{(x, z): 0 \le x \le 2, 0 \le z \le 2\}$ 且 $\displaystyle\iint_S \vec{F} \cdot \vec{n}\, dA = \iint_R \vec{F} \cdot \frac{\partial \vec{r}}{\partial x} \times \frac{\partial \vec{r}}{\partial z}\, dxdz$

$\because \dfrac{\partial \vec{r}}{\partial x} = (1, 2x, 0)$ 且 $\dfrac{\partial \vec{r}}{\partial z} = (0, 0, 1)$ $\quad \therefore \dfrac{\partial \vec{r}}{\partial x} \times \dfrac{\partial \vec{r}}{\partial z} = (2x, -1, 0)$

$\Rightarrow \vec{F} \cdot \dfrac{\partial \vec{r}}{\partial x} \times \dfrac{\partial \vec{r}}{\partial z} = (y+2, 2+4x, xyz) \cdot (2x, -1, 0) = 2xy + 4x - 2 - 4x = 2x^3 - 2,$

$\forall (x, y, z) \in S$

$\therefore \displaystyle\iint_S \vec{F} \cdot \vec{n}\, dA = \iint_R \vec{F} \cdot \frac{\partial \vec{r}}{\partial x} \times \frac{\partial \vec{r}}{\partial z}\, dxdz = \int_0^2 \int_0^2 2x^3 - 2\, dxdz = 8$

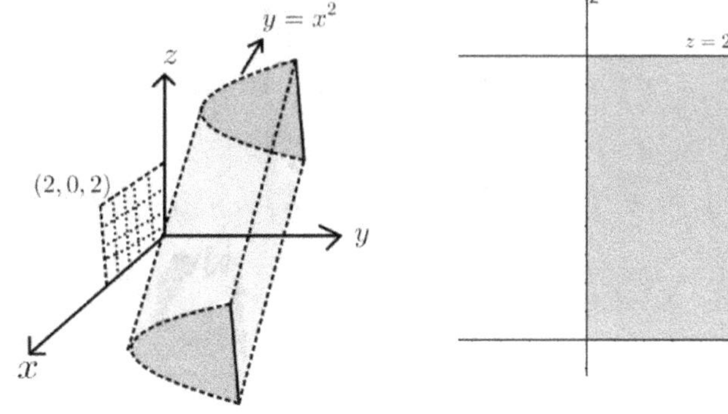


Example 2.

假设 $\vec{F} = (yz, -1, 1)$，求通过朝下定向曲面 $S\colon z = \sqrt{x^2 + y^2}$，$x^2 + y^2 \leq a^2$ 的通量

$$\iint_S \vec{F} \cdot \vec{n}\, dA = ?$$

【解】

$\because S$ 是朝下定向

令 $\vec{r}(x, y, z) = \left(x, y, (x^2 + y^2)^{\frac{1}{2}}\right)$ 且 $R$ 为曲面 $S$ 投影至 $xy$ 平面的封闭区域

则 $R = \{(x, y)\colon x^2 + y^2 \leq a^2\}$ 且 $\displaystyle\iint_S \vec{F} \cdot \vec{n}\, dA = -\iint_R \vec{F} \cdot \frac{\partial \vec{r}}{\partial x} \times \frac{\partial \vec{r}}{\partial y}\, dxdy$

$\because \dfrac{\partial \vec{r}}{\partial x} = \left(1, 0, x(x^2 + y^2)^{-\frac{1}{2}}\right)$ 且 $\dfrac{\partial \vec{r}}{\partial y} = \left(0, 1, y(x^2 + y^2)^{-\frac{1}{2}}\right)$

$\therefore \dfrac{\partial \vec{r}}{\partial x} \times \dfrac{\partial \vec{r}}{\partial y} = \left(-x(x^2 + y^2)^{-\frac{1}{2}}, -y(x^2 + y^2)^{-\frac{1}{2}}, 1\right)$

$\Rightarrow \vec{F} \cdot \dfrac{\partial \vec{r}}{\partial x} \times \dfrac{\partial \vec{r}}{\partial y} = (yz, -1, 1) \cdot \left(-x(x^2 + y^2)^{-\frac{1}{2}}, -y(x^2 + y^2)^{-\frac{1}{2}}, 1\right)$

$= -xy + y(x^2 + y^2)^{-\frac{1}{2}} + 1, \quad \forall (x, y, z) \in S$

$\therefore \displaystyle\iint_S \vec{F} \cdot \vec{n}\, dA = -\iint_R \vec{F} \cdot \frac{\partial \vec{r}}{\partial x} \times \frac{\partial \vec{r}}{\partial y}\, dxdy = \iint_{x^2 + y^2 \leq a^2} xy - y(x^2 + y^2)^{-\frac{1}{2}} - 1\, dxdy$

令 $x = r\cos\theta$，$y = r\sin\theta$ 则 $\{(x, y)\colon x^2 + y^2 \leq a^2\} = \{(r, \theta)\colon 0 \leq r \leq a, 0 \leq \theta \leq 2\pi\}$

$\therefore \displaystyle\iint_{x^2 + y^2 \leq a^2} xy - y(x^2 + y^2)^{-\frac{1}{2}} - 1\, dxdy = \int_0^{2\pi} \int_0^a (r^2 \cos\theta \sin\theta - \sin\theta - 1)\, r\, dr\, d\theta$

$= -\pi a^2$

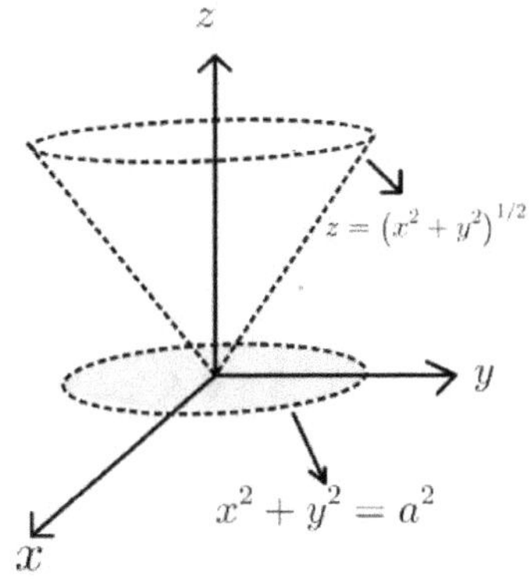



Example 3.

假设 $\vec{F} = \left(6y^2 - \sqrt{xy}, 0, 2x^2 + \sqrt{xy}\right)$, $S$是 $x + y + z = 1$ 在第一卦限的区域, 求$\vec{F}$ 通过朝上定向平面$S$上的通量 $\iint_S \vec{F} \cdot \vec{n}\, dA =?$

【解】

$\because S$是朝上定向

令 $\vec{r}(x, y, z) = (x, y, 1 - x - y)$ 且 $R$ 为所围平面投影至$xy$平面的封闭区域

则 $R = \{(x, y) : x + y \leq 1, x \geq 0, y \geq 0\}$ 且 $\displaystyle\iint_S \vec{F} \cdot \vec{n}\, dA = \iint_R \vec{F} \cdot \frac{\partial \vec{r}}{\partial x} \times \frac{\partial \vec{r}}{\partial y}\, dxdy$

$\because \dfrac{\partial \vec{r}}{\partial x} = (1, 0, -1)$ 且 $\dfrac{\partial \vec{r}}{\partial y} = (0, 1, -1)$ $\quad \therefore \dfrac{\partial \vec{r}}{\partial x} \times \dfrac{\partial \vec{r}}{\partial y} = (1, 1, 1)$

$\because \vec{F} \cdot \dfrac{\partial \vec{r}}{\partial x} \times \dfrac{\partial \vec{r}}{\partial y} = \left(6y^2 - \sqrt{xy}, 0, 2x^2 + \sqrt{xy}\right) \cdot \dfrac{\partial \vec{r}}{\partial x} \times \dfrac{\partial \vec{r}}{\partial y} = 2(x^2 + 3y^2), \quad \forall (x, y, z) \in S$

$\therefore \displaystyle\iint_S \vec{F} \cdot \vec{n}\, dA = \iint_R \vec{F} \cdot \frac{\partial \vec{r}}{\partial x} \times \frac{\partial \vec{r}}{\partial y}\, dxdy = 2\iint_R x^2 + 3y^2\, dxdy = 2 \int_0^1 \int_0^{1-y} x^2 + 3y^2\, dxdy$

$\displaystyle = 2 \int_0^1 \left(\frac{x^3}{3} + 3y^2 x\right)\Bigg|_{x=0}^{x=1-y} dy = 2 \int_0^1 \frac{(1-y)^3}{3} + 3y^2(1-y)\, dy = \frac{2}{3}$

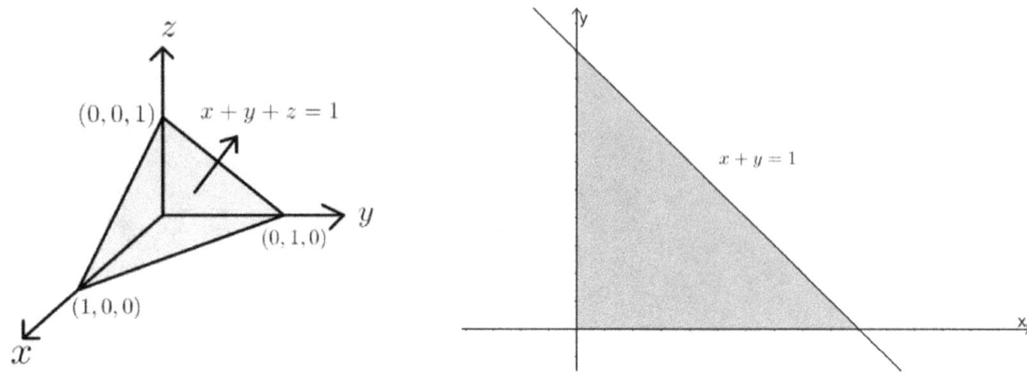

Example 4.

Assume $\vec{F} = (xy, yz, xz)$ and $S$ is the part of the paraboloid $z = 4 - x^2 - y^2$ that lies above the square $0 \le x \le 1, 0 \le y \le 1$ with upward orientation. Find $\displaystyle\iint_S \vec{F} \cdot \vec{n}\, dA$.

【解】

$\because S$ 是朝上定向

令 $\vec{r}(x, y, z) = (x, y, 4 - x^2 - y^2)$ 且 $R = \{(x, y): 0 \le x \le 1, 0 \le y \le 1\}$

则 $\displaystyle\iint_S \vec{F} \cdot \vec{n}\, dA = \iint_R \vec{F} \cdot \frac{\partial \vec{r}}{\partial x} \times \frac{\partial \vec{r}}{\partial y}\, dx\, dy$

$\because \dfrac{\partial \vec{r}}{\partial x} = (1, 0, -2x)$ 且 $\dfrac{\partial \vec{r}}{\partial y} = (0, 1, -2y)$   $\therefore \dfrac{\partial \vec{r}}{\partial x} \times \dfrac{\partial \vec{r}}{\partial y} = (2x, 2y, 1)$

$\because \vec{F} \cdot \dfrac{\partial \vec{r}}{\partial x} \times \dfrac{\partial \vec{r}}{\partial y} = (xy, yz, xz) \cdot (2x, 2y, 1) = 2x^2 y + 2y^2 z + xz = z(2y^2 + x) + 2x^2 y$

$= (4 - x^2 - y^2)(2y^2 + x) + + 2x^2 y = 8y^2 + 4x - 2x^2 y^2 - x^3 - 2y^4 - xy^2 + 2x^2 y,$

$\forall (x, y, z) \in S$

$\therefore \displaystyle\iint_S \vec{F} \cdot \vec{n}\, dA = \iint_R \vec{F} \cdot \frac{\partial \vec{r}}{\partial x} \times \frac{\partial \vec{r}}{\partial y}\, dx\, dy$

$= \displaystyle\int_0^1 \int_0^1 8y^2 + 4x - 2x^2 y^2 - x^3 - 2y^4 - xy^2 + 2x^2 y\, dy\, dx = \int_0^1 \frac{34}{15} + \frac{11x}{3} + \frac{x^2}{3} - x^3 dx$

$= \dfrac{34}{15} + \dfrac{11}{6} + \dfrac{1}{9} - \dfrac{1}{4} = \dfrac{713}{180}$

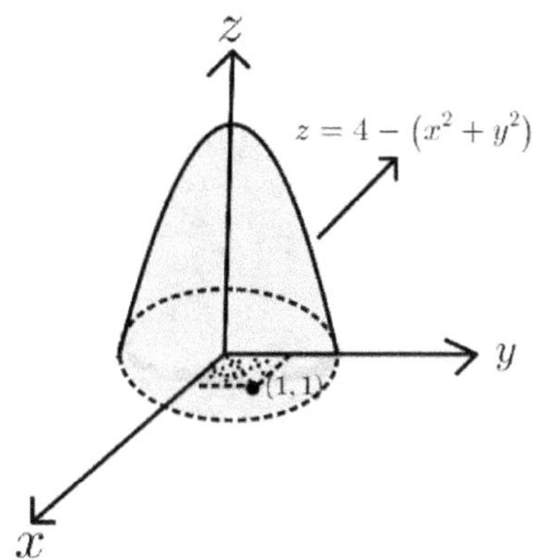


Example 5.

假设 $\vec{F} = (x, -z, y)$,　求通过曲面 $S: x^2 + y^2 + z^2 = r^2$ 于第一卦限朝内定向的通量

$$\iint_S \vec{F} \cdot \vec{n} \, dA = ?$$

【解】

$\because S$ 是朝内定向

令 $\vec{r}(\theta, \varphi) = (r\cos\theta\sin\varphi, r\sin\theta\sin\varphi, r\cos\varphi)$ 且 $R = \left\{ (\theta, \varphi): 0 \le \theta \le \dfrac{\pi}{2}, 0 \le \varphi \le \dfrac{\pi}{2} \right\}$

则 $\displaystyle\oiint_S \vec{F} \cdot \vec{n} \, dA = -\iint_R \vec{F} \cdot \frac{\partial\vec{r}}{\partial\varphi} \times \frac{\partial\vec{r}}{\partial\theta} \, d\theta d\varphi$

$\because \dfrac{\partial\vec{r}}{\partial\theta} = (-r\sin\theta\sin\varphi, r\cos\theta\sin\varphi, 0)$ 且 $\dfrac{\partial\vec{r}}{\partial\varphi} = (r\cos\theta\cos\varphi, r\sin\theta\cos\varphi, -r\sin\varphi)$

$\therefore \dfrac{\partial\vec{r}}{\partial\varphi} \times \dfrac{\partial\vec{r}}{\partial\theta} = (r^2\cos\theta\sin^2\varphi, r^2\sin\theta\sin^2\varphi, r^2\sin\varphi\cos\varphi)$

$\therefore \vec{F} \cdot \dfrac{\partial\vec{r}}{\partial\varphi} \times \dfrac{\partial\vec{r}}{\partial\theta}$

$= (r\cos\theta\sin\varphi, -r\cos\varphi, r\sin\theta\sin\varphi) \cdot (r^2\cos\theta\sin^2\varphi, r^2\sin\theta\sin^2\varphi, r^2\sin\varphi\cos\varphi)$

$= r^3(\cos^2\theta\sin^3\varphi - \sin\theta\sin^2\varphi\cos\varphi + \sin\theta\sin^2\varphi\cos\varphi) = r^3\cos^2\theta\sin^3\varphi$

$\therefore \displaystyle\oiint_S \vec{F} \cdot \vec{n} \, dA = -r^3 \int_0^{\frac{\pi}{2}} \int_0^{\frac{\pi}{2}} \cos^2\theta\sin^3\varphi \, d\varphi d\theta = -r^3 \int_0^{\frac{\pi}{2}} \cos^2\theta \, d\theta \int_0^{\frac{\pi}{2}} \sin^3\varphi \, d\varphi$

$$= -r^3 \int_0^{\frac{\pi}{2}} \frac{1 + \cos 2\theta}{2} \, d\theta \int_0^{\frac{\pi}{2}} \sin^3 \varphi \, d\varphi = -r^3 \cdot \frac{\pi}{4} \cdot \left( -\cos \varphi + \frac{\cos^3 \varphi}{3} \right) \Big|_0^{\frac{\pi}{2}} = \frac{\pi r^3}{6}$$

**Example 6.**

Let $\vec{\mathbf{F}} = (0, y, -z)$. Assume $S$ consists of the paraboloid $S_1: y = x^2 + z^2, 0 \leq y \leq 1$ and the disk $S_2: x^2 + z^2 \leq 1, y = 1$ with outward orientation. Find $\oiint_S \vec{\mathbf{F}} \cdot \vec{n} dA =?$

【解】

令 $S_1: y = x^2 + z^2, 0 \leq y \leq 1, \ S_2: x^2 + z^2 \leq 1, y = 1$

$\because S_1$ 是朝 $y$ 轴负向方向作定向

令 $\vec{r}(x, y, z) = (x, x^2 + z^2, z)$ 且 $R$ 为 $S_1$ 曲面投影至 $xz$ 平面的封闭区域

则 $R = \{(x, y): x^2 + z^2 \leq 1\}$ 且 $\iint_{S_1} \vec{F} \cdot \vec{n} \, dA = -\iint_R \vec{F} \cdot \frac{\partial \vec{r}}{\partial x} \times \frac{\partial \vec{r}}{\partial z} \, dxdz$

$\because \dfrac{\partial \vec{r}}{\partial x} = (1, 2x, 0)$ 且 $\dfrac{\partial \vec{r}}{\partial z} = (0, 2z, 1)$ $\quad \therefore \dfrac{\partial \vec{r}}{\partial x} \times \dfrac{\partial \vec{r}}{\partial z} = (2x, -1, 2z)$

$\therefore \vec{F} \cdot \dfrac{\partial \vec{r}}{\partial x} \times \dfrac{\partial \vec{r}}{\partial z} = (0, y, -z) \cdot (2x, -1, 2z) = -y - 2z^2 = -x^2 - 3z^2, \ \forall (x, y, z) \in S_1$

$\therefore \iint_{S_1} \vec{F} \cdot \vec{n} \, dA = -\iint_R \vec{F} \cdot \dfrac{\partial \vec{r}}{\partial x} \times \dfrac{\partial \vec{r}}{\partial z} \, dxdz = \iint_{x^2+z^2 \leq 1} x^2 + 3z^2 dxdz$

令 $x = r \cos \theta, z = r \sin \theta$ 则 $\{(x, y): x^2 + z^2 \leq 1\} = \{(r, \theta): 0 \leq r \leq 1, 0 \leq \theta \leq 2\pi\}$

且 $dxdz = \left\| \begin{vmatrix} \dfrac{\partial x}{\partial r} & \dfrac{\partial x}{\partial \theta} \\ \dfrac{\partial z}{\partial r} & \dfrac{\partial z}{\partial \theta} \end{vmatrix} \right\| drd\theta = \left\| \begin{vmatrix} \cos \theta & -r \sin \theta \\ \sin \theta & r \cos \theta \end{vmatrix} \right\| drd\theta = rdrd\theta$

$\therefore \iint_{x^2+z^2 \leq 1} x^2 + 3z^2 dxdz = \int_0^{2\pi} \int_0^1 r^3 \cos^2 \theta + 3r^3 \sin^2 \theta \, drd\theta$

$= \int_0^{2\pi} \dfrac{1}{4} \left( \dfrac{1 + \cos \theta}{2} \right) + \dfrac{3}{4} \left( \dfrac{1 - \cos \theta}{2} \right) d\theta = \pi$

令 $\vec{s}(x, y, z) = (x, 1, z)$

$\because \dfrac{\partial \vec{s}}{\partial x} = (1, 0, 0)$ 且 $\dfrac{\partial \vec{s}}{\partial z} = (0, 0, 1)$ $\quad \therefore \dfrac{\partial \vec{r}}{\partial x} \times \dfrac{\partial \vec{r}}{\partial z} = (0, -1, 0)$

$$\because \iint_{S_2} \vec{\mathbf{F}} \cdot \vec{n}\,dA = \iint_{S_2} dA = \iint_{x^2+z^2\leq 1} (0,y,-z) \cdot (0,-1,0)\,dxdz = -\pi$$

$$\therefore \oiint_S \vec{\mathbf{F}} \cdot \vec{n}\,dA = \iint_{S_1} \vec{\mathbf{F}} \cdot \vec{n}\,dA + \iint_{S_2} \vec{\mathbf{F}} \cdot \vec{n}\,dA = 0$$

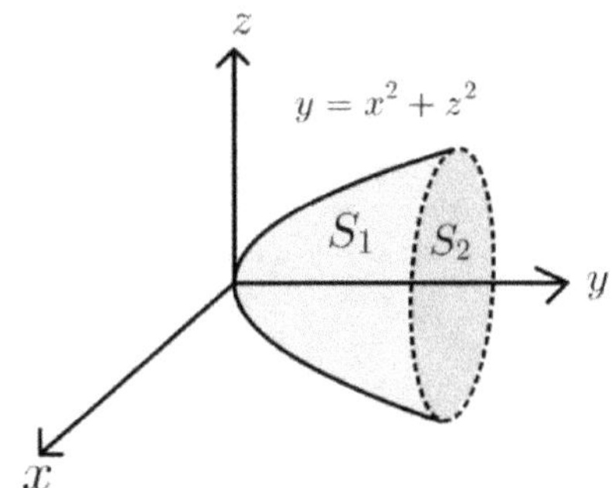

Example 7.

假设$\vec{\mathbf{F}} = (e^y, e^x + 6y, 12y)$, $S$是$x + y + z = 6$在第一卦限的区域, 求$\vec{\mathbf{F}}$通过朝上定向

平面$S$的通量 $\displaystyle\iint_S \vec{F} \cdot \vec{n}\,dA =?$

【解】

$\because S$是朝上定向

令 $\vec{r}(x,y,z) = (x,y,6-x-y)$ 且$R$ 为所围平面投影至$xy$平面的封闭区域

则 $R = \{(x,y): x+y \leq 6, x \geq 0, y \geq 0\}$ 且 $\displaystyle\iint_S \vec{F} \cdot \vec{n}\,dA = \iint_R \vec{F} \cdot \frac{\partial \vec{r}}{\partial x} \times \frac{\partial \vec{r}}{\partial y}\,dxdy$

$\because \dfrac{\partial \vec{r}}{\partial x} = (1,0,-1)$ 且 $\dfrac{\partial \vec{r}}{\partial y} = (0,1,-1)$ $\quad \therefore \dfrac{\partial \vec{r}}{\partial x} \times \dfrac{\partial \vec{r}}{\partial y} = (1,1,1)$

$\because \vec{F} \cdot \dfrac{\partial \vec{r}}{\partial x} \times \dfrac{\partial \vec{r}}{\partial y} = (e^y, e^x + 6y, 12y) \cdot (1,1,1) = e^y + e^x + 18y, \quad \forall (x,y,z) \in S$

$\therefore \displaystyle\iint_S \vec{F} \cdot \vec{n}\,dA = \iint_R \vec{F} \cdot \frac{\partial \vec{r}}{\partial x} \times \frac{\partial \vec{r}}{\partial y}\,dxdy = \iint_R e^y + e^x + 18y\,dxdy$

$= \displaystyle\int_0^6 \int_0^{6-x} e^y + e^x + 18y\,dydx = 634 + 2e^6$

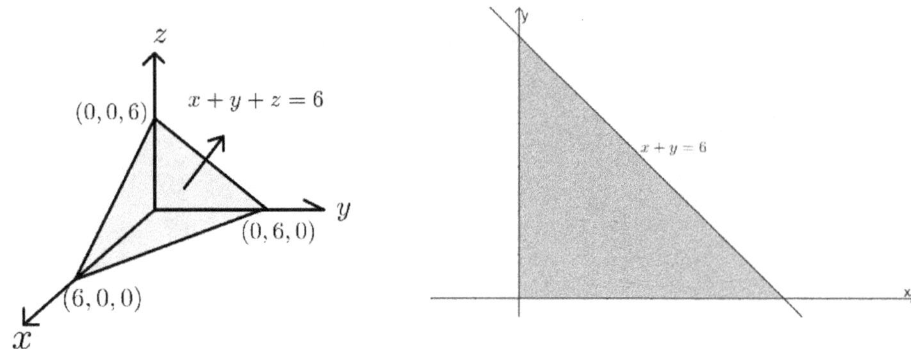


Example 8.

假设$\vec{\mathbf{F}} = (y, x, z)$, 曲面$S$为$z = 4 - x^2 - y^2$, $z = 0$ 所围封闭曲面, 求通过曲面$S$ 的通量, $\oiint_S \vec{F} \cdot \vec{n} dA =$?

【解】

令 $S_1: z = 4 - x^2 - y^2, z \geq 0$

$\because S_1$是朝上定向

令 $\vec{r}(x, y, z) = (x, y, 4 - x^2 - y^2)$ 且 $R$ 为曲面$S_1$投影至$xy$平面的封闭区域

则 $R = \{(x, y): \ x^2 + y^2 \leq 4\}$ 且 $\iint_{S_1} \vec{F} \cdot \vec{n} \, dA = \iint_R \vec{F} \cdot \frac{\partial \vec{r}}{\partial x} \times \frac{\partial \vec{r}}{\partial y} dxdy$

$\because \frac{\partial \vec{r}}{\partial x} = (1, 0, -2x)$ 且 $\frac{\partial \vec{r}}{\partial y} = (0, 1, -2y)$ $\quad \therefore \frac{\partial \vec{r}}{\partial x} \times \frac{\partial \vec{r}}{\partial y} = (2x, 2y, 1)$

$\because \vec{F} \cdot \frac{\partial \vec{r}}{\partial x} \times \frac{\partial \vec{r}}{\partial y} = (y, x, z) \cdot (2x, 2y, 1) = 4xy + z = 4xy + 4 - x^2 - y^2, \ \ \forall (x, y, z) \in S_1$

$\therefore \iint_{S_1} \vec{F} \cdot \vec{n} \, dA = \iint_{x^2 + y^2 \leq 4} 4xy + 4 - x^2 - y^2 \, dxdy$

令 $x = r \cos\theta, y = r \sin\theta$ 则 $\{(x, y): x^2 + y^2 \leq 4\} = \{(r, \theta): 0 \leq r \leq 2, 0 \leq \theta \leq 2\pi\}$

$\therefore \iint_{x^2 + y^2 \leq 4} 4xy + 4 - x^2 - y^2 \, dxdy = \int_0^{2\pi} \int_0^2 (4r^2 \cos\theta \sin\theta + 4 - r^2) \, rdrd\theta$

$$= 2\pi \left( 2r^2 - \frac{r^4}{4} \right)\Bigg|_{r=0}^{r=2} = 8\pi$$

令 $S_2: z = 0, x^2 + y^2 \leq 4,\ S_2$是朝下定向

令 $\vec{s}(x,y,z) = (x,y,0)$ 則 $\displaystyle\iint_{S_2} \vec{F} \cdot \vec{n}\, dA = -\iint_{S_2} \vec{F} \cdot \frac{\partial \vec{s}}{\partial x} \times \frac{\partial \vec{s}}{\partial y}\, dxdy$

$\because \dfrac{\partial \vec{s}}{\partial x} = (1,0,0)$ 且 $\dfrac{\partial \vec{s}}{\partial y} = (0,1,0)$ $\quad \therefore \dfrac{\partial \vec{s}}{\partial x} \times \dfrac{\partial \vec{s}}{\partial y} = (0,0,1)$

$\therefore \displaystyle\iint_{S_2} \vec{F} \cdot \vec{n}\, dA = \iint_{S_2} 0\, dA = 0 \quad \therefore \oiint_{S} \vec{F} \cdot \vec{n}\, dA = \iint_{S_1} \vec{F} \cdot \vec{n}\, dA + \iint_{S_2} \vec{F} \cdot \vec{n}\, dA = 8\pi$

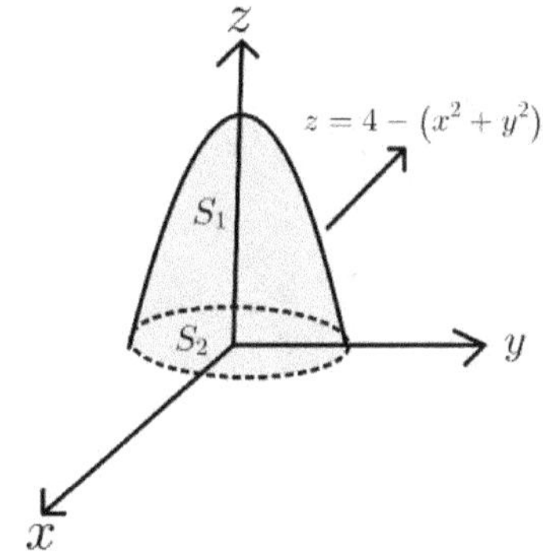


Example 9.

Let $\vec{F} = (x,y,z)$. Determine the flux across the triangle $S$: $(a,0,0), (0,a,0), (0,0,a)$, $a > 0$, with upward orientation.

【解】

$\because S$ 是朝上定向

令 $\vec{r}(x,y,z) = (x,y,a-x-y)$ 且 $R = \{(x,y): 0 \leq x \leq a, 0 \leq y \leq a - x\}$

則 $\displaystyle\iint_{S} \vec{F} \cdot \vec{n}\, dA = \iint_{R} \vec{F} \cdot \frac{\partial \vec{r}}{\partial x} \times \frac{\partial \vec{r}}{\partial y}\, dxdy$

$\because \dfrac{\partial \vec{r}}{\partial x} = (1,0,-1)$ 且 $\dfrac{\partial \vec{r}}{\partial y} = (0,1,-1)$ $\quad \therefore \dfrac{\partial \vec{r}}{\partial x} \times \dfrac{\partial \vec{r}}{\partial y} = (1,1,1)$

$$\because \vec{F} \cdot \frac{\partial \vec{r}}{\partial x} \times \frac{\partial \vec{r}}{\partial y} = (x,y,z) \cdot (1,1,1) = x+y+z = x+y+a-x-y = a, \quad \forall (x,y,z) \in S$$

$$\iint_R \vec{F} \cdot \frac{\partial \vec{r}}{\partial x} \times \frac{\partial \vec{r}}{\partial y} dxdy = \int_0^a \int_0^{a-x} a\, dydx = a\int_0^a a-xdx = a\left(ax - \frac{x^2}{2}\right)\Big|_0^a = \frac{a^3}{2}$$

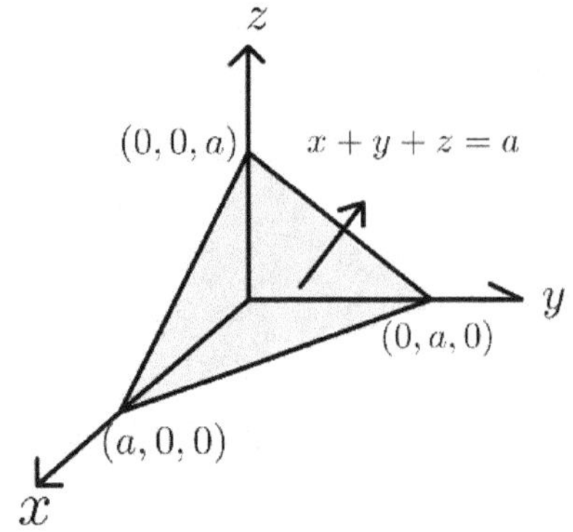

Example 10.

    Let $\vec{F} = (x^2, -y^2, 0)$. Determine the flux across the triangle $S$: $(a,0,0), (0,a,0), (0,0,a)$, $a > 0$, with upward orientation.

【解】

$\because S$ 是朝上定向

令 $\vec{r}(x,y,z) = (x,y,a-x-y)$ 且 $R = \{(x,y): 0 \le x \le a, 0 \le y \le a-x\}$

則 $\iint_S \vec{F} \cdot \vec{n}\, dA = \iint_R \vec{F} \cdot \frac{\partial \vec{r}}{\partial x} \times \frac{\partial \vec{r}}{\partial y} dxdy$

$\because \dfrac{\partial \vec{r}}{\partial x} = (1,0,-1)$ 且 $\dfrac{\partial \vec{r}}{\partial y} = (0,1,-1)$ $\quad \therefore \dfrac{\partial \vec{r}}{\partial x} \times \dfrac{\partial \vec{r}}{\partial y} = (1,1,1)$

$\because \vec{F} \cdot \dfrac{\partial \vec{r}}{\partial x} \times \dfrac{\partial \vec{r}}{\partial y} = (x^2, -y^2, 0) \cdot (1,1,1) = x^2 - y^2, \quad \forall (x,y,z) \in S$

$\therefore \iint_R \vec{F} \cdot \dfrac{\partial \vec{r}}{\partial x} \times \dfrac{\partial \vec{r}}{\partial y} dxdy = \int_0^a \int_0^{a-x} x^2 - y^2\, dydx = \int_0^a x^2(a-x) - \dfrac{(a-x)^3}{3} dx$

$= \int_0^a -\dfrac{a^3}{3} + a^2 x - \dfrac{2x^3}{3} dx = a^4\left(-\dfrac{1}{3} + \dfrac{1}{2} - \dfrac{1}{6}\right) = 0$

Example 11.

假設 $\vec{F} = (z, y, x)$, 求通过朝外定向曲面 $S: x^2 + y^2 + z^2 = r^2$ 的通量 $\displaystyle\oiint_S \vec{F} \cdot \vec{n} dA =?$

【解】

$\because S$ 是朝外定向

令 $\vec{r}(\theta, \varphi) = (r\cos\theta\sin\varphi, r\sin\theta\sin\varphi, r\cos\varphi)$ 且 $R = \{(\theta, \varphi): 0 \le \theta \le 2\pi, 0 \le \varphi \le \pi\}$

則 $\displaystyle\oiint_S \vec{F} \cdot \vec{n} dA = \iint_R \vec{F} \cdot \frac{\partial \vec{r}}{\partial \varphi} \times \frac{\partial \vec{r}}{\partial \theta} d\theta d\varphi$

$\because \dfrac{\partial \vec{r}}{\partial \theta} = (-r\sin\theta\sin\varphi, r\cos\theta\sin\varphi, 0)$ 且 $\dfrac{\partial \vec{r}}{\partial \varphi} = (r\cos\theta\cos\varphi, r\sin\theta\cos\varphi, -r\sin\varphi)$

$\therefore \dfrac{\partial \vec{r}}{\partial \varphi} \times \dfrac{\partial \vec{r}}{\partial \theta} = (r^2\cos\theta\sin^2\varphi, r^2\sin\theta\sin^2\varphi, r^2\sin\varphi\cos\varphi)$

$\therefore \vec{F} \cdot \dfrac{\partial \vec{r}}{\partial \varphi} \times \dfrac{\partial \vec{r}}{\partial \theta}$

$= (r\cos\varphi, r\sin\theta\sin\varphi, r\cos\theta\sin\varphi) \cdot (r^2\cos\theta\sin^2\varphi, r^2\sin\theta\sin^2\varphi, r^2\sin\varphi\cos\varphi)$

$= r^3(2\cos\theta\sin^2\varphi\cos\varphi + \sin^3\varphi\sin^2\theta)$

$\therefore \displaystyle\oiint_S \vec{F} \cdot \vec{n} dA = \iint_R \vec{F} \cdot \frac{\partial \vec{r}}{\partial \varphi} \times \frac{\partial \vec{r}}{\partial \theta} d\theta d\varphi$

$= r^3 \displaystyle\int_0^{2\pi} \int_0^{\pi} 2\cos\theta\sin^2\varphi\cos\varphi + \sin^3\varphi\sin^2\theta \, d\varphi d\theta$

$\because \displaystyle\int_0^{\pi} \sin^2\varphi\cos\varphi \, d\varphi = \frac{\sin^3\varphi}{3}\Big|_0^{\pi} = 0 \quad \therefore \int_0^{2\pi}\int_0^{\pi} 2\cos\theta\sin^2\varphi\cos\varphi \, d\varphi d\theta = 0$

$\therefore \displaystyle\int_0^{2\pi}\int_0^{\pi} (2\cos\theta\sin^2\varphi\cos\varphi + \sin^3\varphi\sin^2\theta)d\varphi d\theta = \int_0^{2\pi}\int_0^{\pi} \sin^3\varphi\sin^2\theta \, d\varphi d\theta$

$= \displaystyle\int_0^{2\pi} \sin^2\theta \, d\theta \int_0^{\pi} \sin^3\varphi \, d\varphi = \int_0^{2\pi} \frac{1-\cos 2\theta}{2} d\theta \int_0^{\pi} \sin\varphi(1-\cos^2\varphi)d\varphi$

$= \pi \displaystyle\int_0^{\pi} \sin\varphi(1-\cos^2\varphi)d\varphi$

$$\because \int_0^\pi \sin\varphi\,(1 - \cos^2\varphi)\,d\varphi = -\cos\varphi\,|_0^\pi + \frac{\cos^3\varphi}{3}\Big|_0^\pi = \frac{4}{3}$$

$$\therefore \oiint_S \vec{F}\cdot\vec{n}\,dA = r^3 \int_0^{2\pi}\int_0^\pi (2\cos\theta\sin^2\varphi\cos\varphi + \sin^3\varphi\sin^2\theta)\,d\varphi\,d\theta = \frac{4\pi r^3}{3}$$

Example 12.

假设$\vec{F} = (-xze^y, xze^y, 2z)$, 求通过朝上定向曲面 $S: x + y + z = 1$ 于第一卦限的通量

$$\iint_S \vec{F}\cdot\vec{n}\,dA = ?$$

【解】

$\because S$是朝上定向

令 $\vec{r}(x,y,z) = (x, y, 1 - x - y)$ 且 $R = \{(x,y):\ x + y \le 1, x \ge 0, y \ge 0\}$

則 $\displaystyle\iint_S \vec{F}\cdot\vec{n}\,dA = \iint_R \vec{F}\cdot\frac{\partial\vec{r}}{\partial x}\times\frac{\partial\vec{r}}{\partial y}\,dxdy$

$\because \dfrac{\partial\vec{r}}{\partial x} = (1,0,-1)$ 且 $\dfrac{\partial\vec{r}}{\partial y} = (0,1,-1)$  $\therefore \dfrac{\partial\vec{r}}{\partial x}\times\dfrac{\partial\vec{r}}{\partial y} = (1,1,1)$

$\because \vec{F}\cdot\dfrac{\partial\vec{r}}{\partial x}\times\dfrac{\partial\vec{r}}{\partial y} = (-xze^y, xze^y, 2z)\cdot(1,1,1) = 2z = 2(1 - x - y),\ \ \forall(x,y,z)\in S$

$\therefore \displaystyle\iint_S \vec{F}\cdot\vec{n}\,dA = \iint_R \vec{F}\cdot\frac{\partial\vec{r}}{\partial x}\times\frac{\partial\vec{r}}{\partial y}\,dxdy = 2\iint_R 1 - x - y\,dxdy$

$$= 2\int_0^1\int_0^{1-x} 1 - x - y\,dydx = 2\int_0^1 y - xy - \frac{y^2}{2}\Big|_{y=0}^{y=1-x}\,dx$$

$$= 2\int_0^1 (1-x) - x(1-x) - \frac{(1-x)^2}{2}\,dx = 2\int_0^1 \left(\frac{x^2}{2} - x + \frac{1}{2}\right)dx = \frac{1}{3}$$

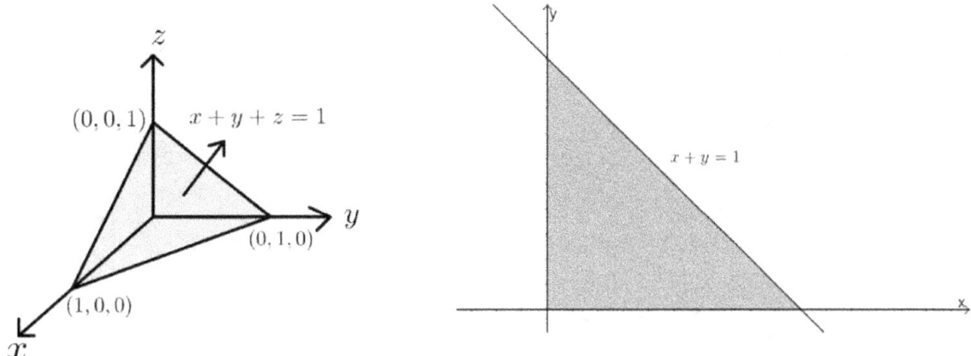

Example 13.

Let $\vec{\mathbf{F}} = \left(x, -y, \dfrac{3z}{2}\right)$. Determine the flux across $S: z = \dfrac{2}{3}\left(x^{\frac{3}{2}} + y^{\frac{3}{2}}\right)$ with $0 \le x \le 1$,

$0 \le y \le 1 - x$, with upward orientation

【解】

$\because S$ 是朝上定向

令 $\vec{r}(x, y, z) = \left(x, y, \dfrac{2}{3}\left(x^{\frac{3}{2}} + y^{\frac{3}{2}}\right)\right)$ 且 $R = \{(x, y): 0 \le x \le 1, 0 \le y \le 1 - x\}$

則 $\displaystyle\iint_S \vec{F} \cdot \vec{n}\, dA = \iint_R \vec{F} \cdot \dfrac{\partial \vec{r}}{\partial x} \times \dfrac{\partial \vec{r}}{\partial y}\, dxdy$

$\because \dfrac{\partial \vec{r}}{\partial x} = \left(1, 0, x^{\frac{1}{2}}\right)$ 且 $\dfrac{\partial \vec{r}}{\partial y} = \left(0, 1, y^{\frac{1}{2}}\right)$  $\therefore \dfrac{\partial \vec{r}}{\partial x} \times \dfrac{\partial \vec{r}}{\partial y} = \left(-x^{\frac{1}{2}}, -y^{\frac{1}{2}}, 1\right)$

$\because \vec{F} \cdot \dfrac{\partial \vec{r}}{\partial x} \times \dfrac{\partial \vec{r}}{\partial y} = \left(x, -y, \dfrac{3z}{2}\right) \cdot \left(-x^{\frac{1}{2}}, -y^{\frac{1}{2}}, 1\right) = \left(x, -y, x^{\frac{3}{2}} + y^{\frac{3}{2}}\right) \cdot \left(-x^{\frac{1}{2}}, -y^{\frac{1}{2}}, 1\right) = 2y^{\frac{3}{2}}$,

$\forall (x, y, z) \in S$

$\displaystyle\iint_R \vec{F} \cdot \dfrac{\partial \vec{r}}{\partial x} \times \dfrac{\partial \vec{r}}{\partial y}\, dxdy = \int_0^1 \int_0^{1-x} 2y^{\frac{3}{2}}\, dydx = \dfrac{4}{5} \int_0^1 (1-x)^{\frac{5}{2}} dx$

令 $1 - x = u$ 則 $\displaystyle\iint_S \vec{F} \cdot \vec{n}\, dA = \dfrac{4}{5} \int_0^1 (1-x)^{\frac{5}{2}} dx = -\dfrac{4}{5} \int_1^0 u^{\frac{5}{2}}\, du = \dfrac{8}{35}$

Example 14.

Let $\vec{F} = (0, y^2, 0)$. Determine the flux across $S: z = \dfrac{2}{3}\left(x^{\frac{3}{2}} + y^{\frac{3}{2}}\right)$ with $0 \leq x \leq 1$,

$0 \leq y \leq 1 - x$, with upward orientation

【解】

$\because S$ 是朝上定向

令 $\vec{r}(x, y, z) = \left(x, y, \dfrac{2}{3}\left(x^{\frac{3}{2}} + y^{\frac{3}{2}}\right)\right)$ 且 $R = \{(x, y): 0 \leq x \leq 1, 0 \leq y \leq 1 - x\}$

則 $\displaystyle\iint_S \vec{F} \cdot \vec{n}\, dA = \iint_R \vec{F} \cdot \dfrac{\partial \vec{r}}{\partial x} \times \dfrac{\partial \vec{r}}{\partial y}\, dxdy$

$\because \dfrac{\partial \vec{r}}{\partial x} = \left(1, 0, x^{\frac{1}{2}}\right)$ 且 $\dfrac{\partial \vec{r}}{\partial y} = \left(0, 1, y^{\frac{1}{2}}\right)$ $\therefore \dfrac{\partial \vec{r}}{\partial x} \times \dfrac{\partial \vec{r}}{\partial y} = \left(-x^{\frac{1}{2}}, -y^{\frac{1}{2}}, 1\right)$

$\because \vec{F} \cdot \dfrac{\partial \vec{r}}{\partial x} \times \dfrac{\partial \vec{r}}{\partial y} = (0, y^2, 0) \cdot \left(-x^{\frac{1}{2}}, -y^{\frac{1}{2}}, 1\right) = -y^{\frac{5}{2}}, \ \forall (x, y, z) \in S$

$\displaystyle\iint_R \vec{F} \cdot \dfrac{\partial \vec{r}}{\partial x} \times \dfrac{\partial \vec{r}}{\partial y}\, dxdy = \int_0^1 \int_0^{1-x} -y^{\frac{5}{2}}\, dydx = -\dfrac{2}{7}\int_0^1 y^{\frac{7}{2}}\Big|_0^{1-x} dx = -\dfrac{2}{7}\int_0^1 (1-x)^{\frac{7}{2}}dx$

令 $1 - x = u$ 則 $\displaystyle\iint_S \vec{F} \cdot \vec{n}\, dA = -\dfrac{2}{7}\int_0^1 (1-x)^{\frac{7}{2}}dx = \dfrac{2}{7}\int_1^0 u^{\frac{7}{2}}\, du = -\dfrac{4}{63}$

Example 15.

假设$\vec{\mathbf{F}} = (x, y, z)$，求通过朝外定向曲面 $S: x^2 + y^2 + z^2 = r^2$ 的通量 $\displaystyle\oiint_S \vec{F} \cdot \vec{n}dA =?$

【解】

$\because S$ 是朝外定向

令 $\vec{r}(r, \theta, \varphi) = (r\cos\theta \sin\varphi, r\sin\theta \sin\varphi, r\cos\varphi)$ 且 $R$ 为球面$S$投影至$\theta\varphi$平面的封闭区域

則 $R = \{(\theta, \varphi): 0 \leq \theta \leq 2\pi, 0 \leq \varphi \leq \pi\}$ 且 $\displaystyle\oiint_S \vec{F} \cdot \vec{n}dA = \iint_R \vec{F} \cdot \dfrac{\partial \vec{r}}{\partial \varphi} \times \dfrac{\partial \vec{r}}{\partial \theta}\, d\theta d\varphi$

$\because \dfrac{\partial \vec{r}}{\partial \theta} = (-r\sin\theta \sin\varphi, r\cos\theta \sin\varphi, 0)$ 且 $\dfrac{\partial \vec{r}}{\partial \varphi} = (r\cos\theta \cos\varphi, r\sin\theta \cos\varphi, -r\sin\varphi)$

$$\therefore \frac{\partial \vec{r}}{\partial \varphi} \times \frac{\partial \vec{r}}{\partial \theta} = (r^2 \cos\theta \sin^2\varphi, r^2 \sin\theta \sin^2\varphi, r^2 \sin\varphi \cos\varphi)$$

$$\therefore \vec{F} \cdot \frac{\partial \vec{r}}{\partial \varphi} \times \frac{\partial \vec{r}}{\partial \theta}$$

$$= (r\cos\theta \sin\varphi, r\sin\theta \sin\varphi, r\cos\varphi) \cdot (r^2 \cos\theta \sin^2\varphi, r^2 \sin\theta \sin^2\varphi, r^2 \sin\varphi \cos\varphi)$$

$$= r^3(\cos^2\theta \sin^3\varphi + \sin^3\varphi \sin^2\theta + \sin\varphi \cos^2\varphi) = r^3(\sin^3\varphi + \sin\varphi \cos^2\varphi) = r^3 \sin\varphi$$

$$\therefore \oiint_S \vec{F} \cdot \vec{n}\, dA = \iint_R \vec{F} \cdot \frac{\partial \vec{r}}{\partial \varphi} \times \frac{\partial \vec{r}}{\partial \theta}\, d\theta d\varphi = r^3 \int_0^{2\pi} \int_0^{\pi} \sin\varphi\, d\varphi d\theta = r^3 \cdot 2\pi \cdot 2 = 4\pi r^3$$

Example 16.

$$求 \oiint_S \vec{\mathbf{F}} \cdot \vec{n}\, dA = ?, \quad \vec{\mathbf{F}} = (0, 4yz, 0), \quad S: x^2 + y^2 + z^2 = r^2$$

【解】

$\because S$ 是朝外定向

令 $\vec{r}(\theta, \varphi) = (r\cos\theta \sin\varphi, r\sin\theta \sin\varphi, r\cos\varphi)$ 且 $R = \{(\theta, \varphi): 0 \leq \theta \leq 2\pi, 0 \leq \varphi \leq \pi\}$

則 $\oiint_S \vec{\mathbf{F}} \cdot \vec{n}\, dA = \iint_R \vec{F} \cdot \dfrac{\partial \vec{r}}{\partial \varphi} \times \dfrac{\partial \vec{r}}{\partial \theta}\, d\theta d\varphi$

$$\because \frac{\partial \vec{r}}{\partial \theta} = (-r\sin\theta \sin\varphi, r\cos\theta \sin\varphi, 0) \text{ 且 } \frac{\partial \vec{r}}{\partial \varphi} = (r\cos\theta \cos\varphi, r\sin\theta \cos\varphi, -r\sin\varphi)$$

$$\therefore \frac{\partial \vec{r}}{\partial \varphi} \times \frac{\partial \vec{r}}{\partial \theta} = (r^2 \cos\theta \sin^2\varphi, r^2 \sin\theta \sin^2\varphi, r^2 \sin\varphi \cos\varphi)$$

$$\therefore \vec{F} \cdot \frac{\partial \vec{r}}{\partial \varphi} \times \frac{\partial \vec{r}}{\partial \theta} = (0, 4r^2 \sin\theta \sin\varphi \cos\varphi, 0) \cdot (r^2 \cos\theta \sin^2\varphi, r^2 \sin\theta \sin^2\varphi, r^2 \sin\varphi \cos\varphi)$$

$$= 4r^4 \sin^3\varphi \sin^2\theta \cos\varphi$$

$$\therefore \oiint_S \vec{\mathbf{F}} \cdot \vec{n}\, dA = \int_0^{2\pi} \int_0^{\pi} 4r^4 \sin^3\varphi \sin^2\theta \cos\varphi\, d\varphi dr d\theta$$

$$\because \int_0^\pi \sin^3 \varphi \cos \varphi \, d\varphi = \left.\frac{\sin^4 \varphi}{4}\right|_0^\pi = 0 \qquad \therefore \oiint_S \vec{\mathbf{F}} \cdot \vec{n} \, dA = 0$$

Example 17.

$$\text{求} \quad \oiint_S \vec{\mathbf{F}} \cdot \vec{n} \, dA =?, \quad \vec{\mathbf{F}} = (x - y + z, 2x, 1), \quad S = S_1 \cup S_1 \text{ 为朝外定向曲面, 其中}$$

$$S_1 : z = x^2 + y^2, z \leq 1, \quad S_2 : x^2 + y^2 \leq 1, z = 1$$

【解】

$\because S_1$是朝下定向

令 $\vec{r}(x, y, z) = (x, y, x^2 + y^2)$ 且 $R$ 为$S_1$曲面投影至$xy$平面的封闭区域

则 $R = \{(x, y): x^2 + y^2 \leq 1\}$ 且 $\iint_{S_1} \vec{\mathbf{F}} \cdot \vec{n} \, dA = -\iint_R \vec{\mathbf{F}} \cdot \dfrac{\partial \vec{r}}{\partial x} \times \dfrac{\partial \vec{r}}{\partial y} \, dxdy$

$\because \dfrac{\partial \vec{r}}{\partial x} = (1, 0, 2x)$ 且 $\dfrac{\partial \vec{r}}{\partial y} = (0, 1, 2y)$ $\quad \therefore \dfrac{\partial \vec{r}}{\partial x} \times \dfrac{\partial \vec{r}}{\partial y} = (-2x, -2y, 1)$

$\therefore \vec{F} \cdot \dfrac{\partial \vec{r}}{\partial x} \times \dfrac{\partial \vec{r}}{\partial y} = (x - y + z, 2x, 1) \cdot (-2x, -2y, 1) = -2x(x - y + x^2 + y^2) - 4xy + 1,$

$\forall (x, y, z) \in S_1$

$\therefore \iint_{S_1} \vec{F} \cdot \vec{n} \, dA = -\iint_R \vec{F} \cdot \dfrac{\partial \vec{r}}{\partial x} \times \dfrac{\partial \vec{r}}{\partial y} \, dxdy = \iint_{x^2 + y^2 \leq 1} 2x^2 + 2xy + 2x(x^2 + y^2) - 1 \, dxdy$

令 $x = r\cos\theta, y = r\sin\theta$ 则 $\{(x, y): x^2 + y^2 \leq 1\} = \{(r, \theta): 0 \leq r \leq 1, 0 \leq \theta \leq 2\pi\}$

且 $dxdy = \left\| \begin{vmatrix} \dfrac{\partial x}{\partial r} & \dfrac{\partial x}{\partial \theta} \\ \dfrac{\partial y}{\partial r} & \dfrac{\partial y}{\partial \theta} \end{vmatrix} \right\| drd\theta = \left\| \begin{vmatrix} \cos\theta & -r\sin\theta \\ \sin\theta & r\cos\theta \end{vmatrix} \right\| drd\theta = r \, drd\theta$

$\therefore \iint_{x^2 + y^2 \leq 1} 2x^2 + 2xy + 2x(x^2 + y^2) - 1 \, dxdy$

$$= \int_0^{2\pi} \int_0^1 (2r^2 \cos^2\theta + 2r^2 \cos\theta \sin\theta + 2r^3 \cos\theta - 1) r \, drd\theta = -\frac{\pi}{2}$$

$\because \iint_{S_2} \vec{\mathbf{F}} \cdot \vec{n} \, dA = \iint_{S_2} dA = \pi \qquad \therefore \iint_{S_1} \vec{\mathbf{F}} \cdot \vec{n} \, dA + \iint_{S_2} \vec{\mathbf{F}} \cdot \vec{n} \, dA = \frac{\pi}{2}$

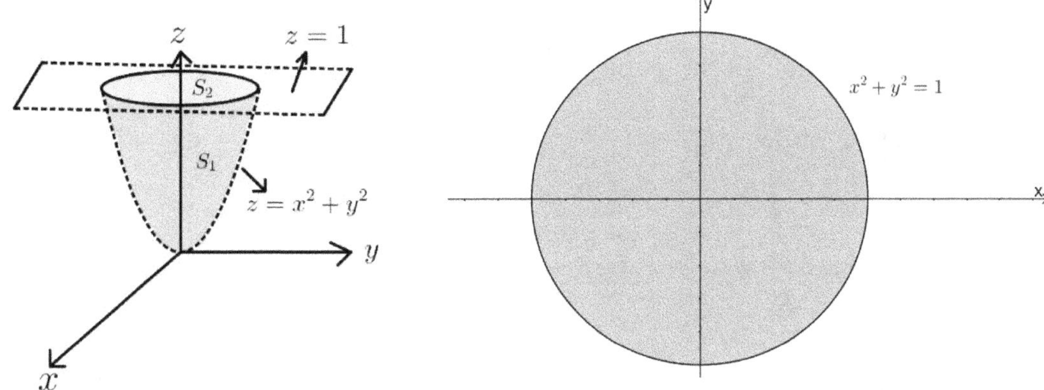

Example 18.

假設曲面 $S: \vec{r}(s,t) = (s+t, s-t, 1+2s+t), \ 0 \le s \le 2, 0 \le t \le 1$ 且

$\vec{\mathbf{F}} = (ze^{xy}, -3ze^{xy}, xy)$，求通过朝上定向曲面 $S$的通量 $\iint_S \vec{F} \cdot \vec{n}\, dA = ?$

【解】

$\because S$ 朝上定向

令 $\vec{r}(s,t) = (s+t, s-t, 1+2s+t)$ 且 $R = \{(s,t): 0 \le s \le 2, 0 \le t \le 1\}$

則 $\iint_S \vec{F} \cdot \vec{n}\, dA = \iint_R \vec{F} \cdot \dfrac{\partial \vec{r}}{\partial s} \times \dfrac{\partial \vec{r}}{\partial t}\, ds\, dt$

$\because \dfrac{\partial \vec{r}}{\partial s} = (1,1,2)$ 且 $\dfrac{\partial \vec{r}}{\partial t} = (1,-1,1)$ $\quad \therefore \dfrac{\partial \vec{r}}{\partial s} \times \dfrac{\partial \vec{r}}{\partial t} = (3,1,-2)$

$\because \vec{F} \cdot \dfrac{\partial \vec{r}}{\partial s} \times \dfrac{\partial \vec{r}}{\partial t} = (ze^{xy}, -3ze^{xy}, xy) \cdot (3,1,-2) = -2xy = -2(s^2 - t^2), \ \forall (x,y,z) \in S$

$\therefore \iint_S \vec{F} \cdot \vec{n}\, dA = \iint_R \vec{F} \cdot \dfrac{\partial \vec{r}}{\partial s} \times \dfrac{\partial \vec{r}}{\partial t}\, ds\, dt = -2 \iint_R s^2 - t^2\, ds\, dt = -2 \int_0^2 \int_0^1 s^2 - t^2\, dt\, ds$

$= -2 \int_0^2 s^2 - \dfrac{1}{3}\, ds = -2 \left( \dfrac{s^3 - s}{3} \right) \Big|_0^2 = -4$

Example 19.

Let $\vec{F} = (x, 2y, 3z)$. Assume $S$ is the cube with the vertices $(\pm 1, \pm 1, \pm 1)$ with outward orientation. Find $\displaystyle\oiint_S \vec{F} \cdot \vec{n}\, dA =?$

【解】

令 $S_1: x = 1,\ \ S_2: x = -1,\ \ S_3: y = 1,\ \ S_4: y = -1,\ \ S_5: z = 1,\ \ S_6: z = -1,$
則 $S = S_1 \cup S_2 \cup S_3 \cup S_4 \cup S_5 \cup S_6$

$As\ x = 1, \vec{n} = (1,0,0),\ \ \displaystyle\iint_{S_1} \vec{F} \cdot \vec{n}\, dA = \int_{-1}^{1}\int_{-1}^{1} x\, dydz = 4x = 4$

$As\ x = -1, \vec{n} = (-1,0,0),\ \ \displaystyle\iint_{S_2} \vec{F} \cdot \vec{n}\, dA = \int_{-1}^{1}\int_{-1}^{1} -x\, dydz = -4x = 4$

$As\ y = 1, \vec{n} = (0,1,0),\ \ \displaystyle\iint_{S_3} \vec{F} \cdot \vec{n}\, dA = \int_{-1}^{1}\int_{-1}^{1} 2y\, dxdz = 8y = 8$

$As\ y = -1, \vec{n} = (0,-1,0),\ \ \displaystyle\iint_{S_4} \vec{F} \cdot \vec{n}\, dA = \int_{-1}^{1}\int_{-1}^{1} -2y\, dxdz = -8y = 8$

$As\ z = 1, \vec{n} = (0,0,1),\ \ \displaystyle\iint_{S_5} \vec{F} \cdot \vec{n}\, dA = \int_{-1}^{1}\int_{-1}^{1} 3z\, dxdy = 12z = 12$

$As\ z = -1, \vec{n} = (0,0,-1),\ \ \displaystyle\iint_{S_6} \vec{F} \cdot \vec{n}\, dA = \int_{-1}^{1}\int_{-1}^{1} -3z\, dxdy = -12z = 12$

$\therefore \displaystyle\oiint_S \vec{F} \cdot \vec{n}\, dA = 48$

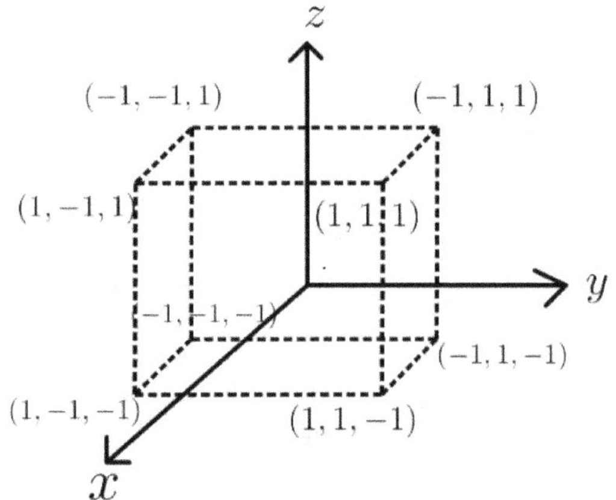


Example 20.

Let $\vec{F} = (x^2, y^2, z^2)$. Assume $S$ is the boundary of the solid half- cylinder

$0 \leq z \leq \sqrt{1-y^2}, 0 \leq x \leq 2$ with outward orientation. Find $\iint_S \vec{F} \cdot \vec{n}\, dA = ?$

【解】

令$S_1: z = \sqrt{1-y^2}, 0 \leq x \leq 2, 0 \leq y \leq 1$, $\quad S_2: y^2 + z^2 \leq 1, x = 0$, $\quad S_3: y^2 + z^2 \leq 1, x = 2$

則 $S = S_1 \cup S_2 \cup S_3$
$\because S_1$ 是朝上定向

令 $\vec{r}(x,y,z) = \left(x, y, \sqrt{1-y^2}\right)$ 且 $R = \{(x,y): 0 \leq x \leq 2, 0 \leq y \leq 1\}$

則 $\iint_{S_1} \vec{F} \cdot \vec{n}\, dA = \iint_R \vec{F} \cdot \dfrac{\partial \vec{r}}{\partial x} \times \dfrac{\partial \vec{r}}{\partial y}\, dxdy$

$\because \dfrac{\partial \vec{r}}{\partial x} = (1,0,0)$ 且 $\dfrac{\partial \vec{r}}{\partial y} = \left(0, 1, -y(1-y^2)^{-\frac{1}{2}}\right)$ $\quad \therefore \dfrac{\partial \vec{r}}{\partial x} \times \dfrac{\partial \vec{r}}{\partial y} = \left(0, -y(1-y^2)^{-\frac{1}{2}}, 1\right)$

$\because \vec{F} \cdot \dfrac{\partial \vec{r}}{\partial x} \times \dfrac{\partial \vec{r}}{\partial y} = (x^2, y^2, 1-y^2) \cdot \left(0, -y(1-y^2)^{-\frac{1}{2}}, 1\right) = -y^3(1-y^2)^{-\frac{1}{2}} + 1 - y^2,$

$\forall (x,y,z) \in S_1$
令 $y = \sin\theta$ 則 $dy = \cos\theta\, d\theta$

$\iint_R \vec{F} \cdot \dfrac{\partial \vec{r}}{\partial x} \times \dfrac{\partial \vec{r}}{\partial y}\, dxdy = \int_0^2 \int_0^1 -y^3(1-y^2)^{-\frac{1}{2}} + 1 - y^2\, dydx$

$= 2\int_0^1 -y^3(1-y^2)^{-\frac{1}{2}} + 1 - y^2\, dy = 2\int_0^{\frac{\pi}{2}} -\sin^3\theta + \cos^3\theta\, d\theta$

$= 2 \cdot \left( \left(-\cos\theta + \dfrac{\cos^3\theta}{3}\right)\Big|_0^{\frac{\pi}{2}} + \left(\sin\theta + \dfrac{\sin^3\theta}{3}\right)\Big|_0^{\frac{\pi}{2}} \right) = \dfrac{8}{3}$

$\because \iint_{S_2} \vec{F} \cdot \vec{n}\, dA = \iint_{S_2} dA = 0$ 且 $\iint_{S_3} \vec{F} \cdot \vec{n}\, dA = \iint_{S_3} dA = 2\pi$

$\therefore \iint_S \vec{F} \cdot \vec{n}\, dA = \dfrac{8}{3} + 2\pi$

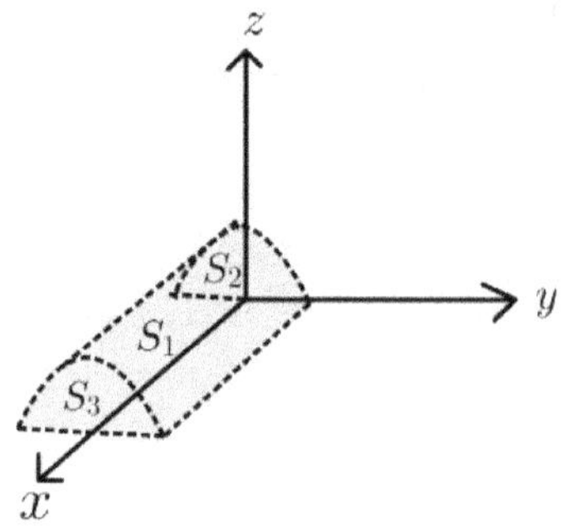

Example 21.

Let $\vec{F} = (0, xz, -xy)$. Determine the flux across $S: z = xy$ with $0 \le x \le 1, 0 \le y \le 2$, with upward orientation.

【解】

$\because S$ 是朝上定向

令 $\vec{r}(x, y, z) = (x, y, xy)$ 且 $R = \{(x, y): 0 \le x \le 1, 0 \le y \le 2\}$

則 $\displaystyle\iint_S \vec{F} \cdot \vec{n}\, dA = \iint_R \vec{F} \cdot \frac{\partial \vec{r}}{\partial x} \times \frac{\partial \vec{r}}{\partial y}\, dxdy$

$\because \dfrac{\partial \vec{r}}{\partial x} = (1, 0, y)$ 且 $\dfrac{\partial \vec{r}}{\partial y} = (0, 1, x)$ $\quad \therefore \dfrac{\partial \vec{r}}{\partial x} \times \dfrac{\partial \vec{r}}{\partial y} = (-y, -x, 1)$

$\because \vec{F} \cdot \dfrac{\partial \vec{r}}{\partial x} \times \dfrac{\partial \vec{r}}{\partial y} = (0, xz, -xy) \cdot (-y, -x, 1) = -x^2 z - xy = -x^3 y - xy, \quad \forall (x, y, z) \in S$

$$\iint_R \vec{F} \cdot \frac{\partial \vec{r}}{\partial x} \times \frac{\partial \vec{r}}{\partial y}\, dxdy = \int_0^1 \int_0^2 -x^3 y - xy\, dydx = \int_0^a -2x^3 - 2x\, dx = \left( -\frac{x^4}{2} - x^2 \right) \Big|_0^1 = -\frac{3}{2}$$

Example 22.

Calculate the flux of $\vec{F} = (x, y, z)$ out of the cylindrical surface with parameterization $S: \vec{r}(s, t) = (a \cos s, a \sin s, t), 0 \le s \le 2\pi, 0 \le t \le k$.

【解】

令 $\vec{r}(s, t) = (a\cos s, a \sin s, t)$ 且 $R = \{(s, t): 0 \le s \le 2\pi, 0 \le t \le k\}$

則 $\displaystyle\iint_S \vec{F} \cdot \vec{n} \, dA = \iint_R \vec{F} \cdot \frac{\partial \vec{r}}{\partial s} \times \frac{\partial \vec{r}}{\partial t} \, dsdt$

$\because \dfrac{\partial \vec{r}}{\partial s} = (-a\sin s, a\cos s, 0)$ 且 $\dfrac{\partial \vec{r}}{\partial t} = (0,0,1)$ $\quad \therefore \dfrac{\partial \vec{r}}{\partial s} \times \dfrac{\partial \vec{r}}{\partial t} = (a\cos s, a\sin s, 0)$

$\because \vec{F} \cdot \dfrac{\partial \vec{r}}{\partial s} \times \dfrac{\partial \vec{r}}{\partial t} = (a\cos s, a\sin s, t) \cdot (a\cos s, a\sin s, 0) = a^2, \ \forall(x,y,z) \in S$

$\displaystyle\iint_R \vec{F} \cdot \frac{\partial \vec{r}}{\partial s} \times \frac{\partial \vec{r}}{\partial t} \, dsdt = \int_0^{2\pi} \int_0^k a^2 \, dtds = 2\pi a^2 k$

Example 23.

$\quad \vec{\mathbf{F}} = (y, z, x), \ S$是$(a,0,0), (0,a,0), (0,0,a)$于第一象限所围朝上定向平面,

$\quad$ 求 $\displaystyle\iint_S \vec{\mathbf{F}} \cdot \vec{n} \, dA = ?$

【解】

$\because S$ 是朝上定向

令 $\vec{r}(x,y,z) = (x, y, a - x - y)$ 且 $R = \{(x,y): 0 \le x \le a, 0 \le y \le a - x\}$

則 $\displaystyle\iint_S \vec{\mathbf{F}} \cdot \vec{n} \, dA = \iint_R \vec{\mathbf{F}} \cdot \frac{\partial \vec{r}}{\partial x} \times \frac{\partial \vec{r}}{\partial y} \, dxdy$

$\because \dfrac{\partial \vec{r}}{\partial x} = (1, 0, -1)$ 且 $\dfrac{\partial \vec{r}}{\partial y} = (0, 1, -1)$ $\quad \therefore \dfrac{\partial \vec{r}}{\partial x} \times \dfrac{\partial \vec{r}}{\partial y} = (1,1,1)$

$\because \vec{\mathbf{F}} \cdot \dfrac{\partial \vec{r}}{\partial x} \times \dfrac{\partial \vec{r}}{\partial y} = (y, z, x) \cdot (1,1,1) = x + y + z = x + y + a - x - y = a, \ \forall(x,y,z) \in S$

$\displaystyle\iint_R \vec{\mathbf{F}} \cdot \frac{\partial \vec{r}}{\partial x} \times \frac{\partial \vec{r}}{\partial y} \, dxdy = \int_0^a \int_0^{a-x} a \, dydx = a\int_0^a a - x\,dx = a\left(ax - \frac{x^2}{2}\right)\Big|_0^a = \frac{a^3}{2}$

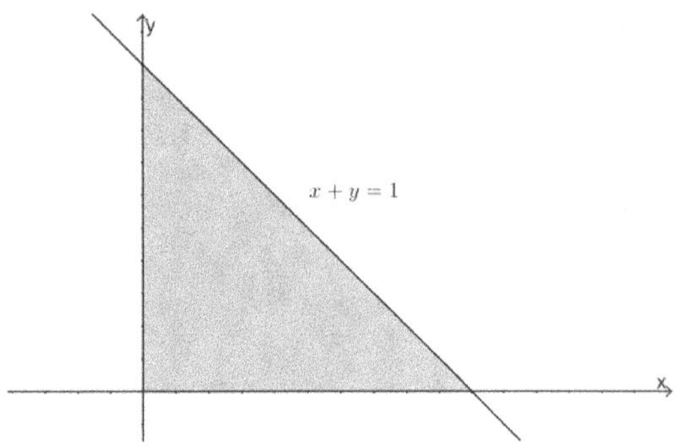


## 9.6  Stokes' Theorem

如之前所述，Green's Theorem 的旋度形式的表示为

$$\oint_C \vec{F} \cdot d\vec{r} = \iint_R \left(\nabla \times \vec{F}\right) \cdot \boldsymbol{k}\, dA$$

如果扩展至三维空间，直观的推论为

$$\oint_C \vec{F} \cdot d\vec{r} = \iint_S \left(\nabla \times \vec{F}\right) \cdot \vec{n}\, dA$$

其中 $S$ 为空间中非封闭的曲面且 $C$ 为 $S$ 的边界(空间中的封闭曲线)

  Stokes' Theorem 不仅是微积分基本定理在更高维度的推广，同时也是 Green's Theorem 的高维推广；Green's Theorem 是将平面区域的双重积分与平面上二维封闭曲线之线积分做转换，Stokes' Theorem 则是将三维空间封闭曲线的线积分与曲面 $S$ 上之曲面积分做转换；假设非封闭曲面 $S$ 的边界为封闭空间曲线 $C$，Stokes' Theorem 说明可以将 $\vec{F}$ 沿着封闭空间曲线 $C$ 的线积分转换为 $\vec{F}$ 旋度在非封闭曲面 $S$ 上的曲面积分，相反地，也可藉由求 $\vec{F}$ 沿着封闭空间曲线 $C$ 的线积分得到 $\vec{F}$ 旋度在非封闭曲面 $S$ 的曲面积分(如下图所示)

  Stokes' Theorem 重要的成立条件为曲面 $S$ 是非封闭曲面且其边界是空间封闭曲线；再者，曲面 $S$ 的定向与边界曲线 $C$ 是否为逆时针绕行的正向曲线有重要的关联；如果曲面 $S$ 的定向已经给定则法向量的方向确定，藉由右手法则，大拇指先指向法向量的方向，其余四根手指头顺势弯曲的方向则为曲线 $C$ 的绕行方向；另一方面，如果是给定封闭曲线 $C$ 所绕行的方向，也可藉由右手法则找出对应法向量的方向，藉此判断曲面为朝上定向或者是朝下定向(如下图)

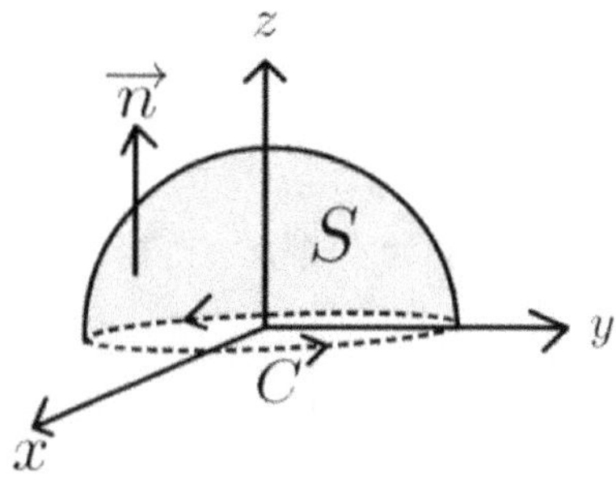

原本求通量的做法是投影到 $xy$ 平面再求双重积分，而 Stokes' Theorem 是将问题转成求三维空间封闭曲线的线积分，反之，求三维空间封闭曲线的线积分也可藉由 Stokes' Theorem 转成求空间中非封闭曲面的通量问题（如下图）

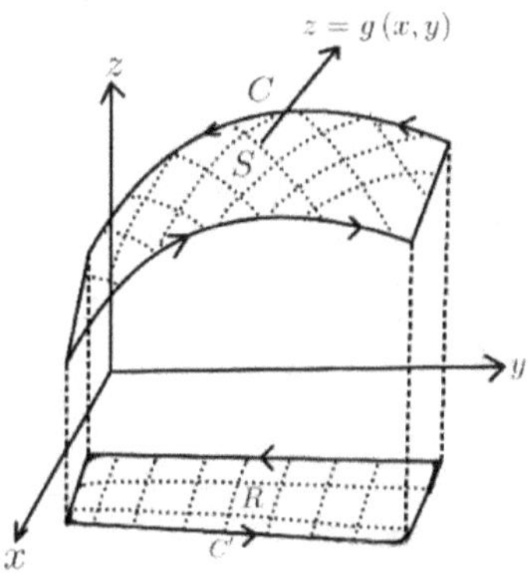

为了说明 Stokes' Theorem, 先回顾曲面积分的公式, 给定朝上定向曲面 $S$: $z = g(x, y)$, $\forall\, a \le x \le b, c \le y \le d$, 假设 $\vec{F}(x, y, z)$ 为三维空间的向量函数, 则

$$\iint_S \vec{F} \cdot \vec{n}\, dA = \int_a^b \int_c^d \vec{F}(x, y, g(x, y))\left(-g_x(x, y), -g_y(x, y), 1\right) dxdy$$

如果 $\vec{F} = (F_1, F_2, F_3)$ 则 $\displaystyle \iint_S \vec{F} \cdot \vec{n}\, dA = \iint_D -F_1 \frac{\partial g}{\partial x} - F_2 \frac{\partial g}{\partial y} + F_3 \, dA$

$$\because \operatorname{curl} \vec{F} = \nabla \times \vec{F} = \begin{Vmatrix} \boldsymbol{i} & \boldsymbol{j} & \boldsymbol{k} \\ \dfrac{\partial}{\partial x} & \dfrac{\partial}{\partial y} & \dfrac{\partial}{\partial z} \\ F_1 & F_2 & F_3 \end{Vmatrix} = \left(\frac{\partial F_3}{\partial y} - \frac{\partial F_2}{\partial z}\right)\boldsymbol{i} + \left(\frac{\partial F_1}{\partial z} - \frac{\partial F_3}{\partial x}\right)\boldsymbol{j} + \left(\frac{\partial F_2}{\partial x} - \frac{\partial F_1}{\partial y}\right)\boldsymbol{k}$$

若将上式的 $\vec{F}$ 用 $\operatorname{curl}\vec{F}$ 取代则

$$\iint_S (\nabla \times \vec{F}) \cdot \vec{n}\, dA = \iint_D -\left(\frac{\partial F_3}{\partial y} - \frac{\partial F_2}{\partial z}\right)\frac{\partial z}{\partial x} - \left(\frac{\partial F_1}{\partial z} - \frac{\partial F_3}{\partial x}\right)\frac{\partial z}{\partial y} + \left(\frac{\partial F_2}{\partial x} - \frac{\partial F_1}{\partial y}\right) dA$$

**【定理】** Stokes' Theorem

假设

1. 空间曲面$S$是朝上定向的非封闭光滑曲面,边界曲线$C$是简单封闭且分段光滑的正向曲线

2. $\vec{F}$ 为一个向量场,各分量在一个包含$S$ 的开集合上具有连续的偏导数则

$$\oint_C \vec{F} \cdot d\vec{r} = \iint_S (\nabla \times \vec{F}) \cdot \vec{n}\, dA$$

Proof:

假设曲面$S$：ㅤ$z = g(x,y), (x,y) \in D$，ㅤ$R$为曲面$S$投影于$xy$平面的平面区域，ㅤ其封闭边界曲线为$C'$，藉由向量函数的曲面公式

$$\iint_S \vec{F} \cdot \vec{n}\, dA = \iint_R \vec{F}(x,y,g(x,y)) \cdot (-g_x(x,y), -g_y(x,y), 1)\, dxdy$$

将上述的$\vec{F}$替换为$\nabla \times \vec{F}$

$$\because \operatorname{curl} \vec{F} = \nabla \times \vec{F} = \left\| \begin{matrix} \boldsymbol{i} & \boldsymbol{j} & \boldsymbol{k} \\ \dfrac{\partial}{\partial x} & \dfrac{\partial}{\partial y} & \dfrac{\partial}{\partial z} \\ F_1 & F_2 & F_3 \end{matrix} \right\| = \left(\frac{\partial F_3}{\partial y} - \frac{\partial F_2}{\partial z}\right)\boldsymbol{i} + \left(\frac{\partial F_1}{\partial z} - \frac{\partial F_3}{\partial x}\right)\boldsymbol{j} + \left(\frac{\partial F_2}{\partial x} - \frac{\partial F_1}{\partial y}\right)\boldsymbol{k}$$

$$\therefore \iint_S (\nabla \times \vec{F}) \cdot \vec{n}\, dA = \iint_R (\nabla \times \vec{F})(x,y,g(x,y)) \cdot (-g_x(x,y), -g_y(x,y), 1)\, dxdy$$

$$= \iint_R \left(\frac{\partial F_3}{\partial y} - \frac{\partial F_2}{\partial z}, \frac{\partial F_1}{\partial z} - \frac{\partial F_3}{\partial x}, \frac{\partial F_2}{\partial x} - \frac{\partial F_1}{\partial y}\right) \cdot (-g_x(x,y), -g_y(x,y), 1)\, dxdy$$

$$= \iint_R -\left(\frac{\partial F_3}{\partial y} - \frac{\partial F_2}{\partial z}\right)\frac{\partial z}{\partial x} - \left(\frac{\partial F_1}{\partial z} - \frac{\partial F_3}{\partial x}\right)\frac{\partial z}{\partial y} + \left(\frac{\partial F_2}{\partial x} - \frac{\partial F_1}{\partial y}\right) dA$$

其中 $F_1, F_2, F_3$ 于 $(x,y,g(x,y))$ 取值

令曲线 $C'$的参数式：$x = x(t), y = y(t), \ a \leq t \leq b$ 则

曲线$C$：$x = x(t), y = y(t), z = g(x(t),y(t)), \ a \leq t \leq b$

By Chain Rule,

$$\oint_C \vec{F} \cdot d\vec{r} = \int_a^b F_1 \frac{dx}{dt} + F_2 \frac{dy}{dt} + F_3 \frac{dz}{dt}\, dt = \int_a^b F_1 \frac{dx}{dt} + F_2 \frac{dy}{dt} + F_3 \left(\frac{\partial z}{\partial x}\frac{dx}{dt} + \frac{\partial z}{\partial y}\frac{dy}{dt}\right) dt$$

$$= \int_a^b \frac{dx}{dt}\left(F_1 + F_3 \frac{\partial z}{\partial x}\right) + \frac{dy}{dt}\left(F_2 + F_3 \frac{\partial z}{\partial x}\right) dt = \int_{C'} \left(F_1 + F_3 \frac{\partial z}{\partial x}\right) dx + \left(F_2 + F_3 \frac{\partial z}{\partial x}\right) dy$$

By Green's Theorem

$$= \int_{C'} \left(F_1 + F_3 \frac{\partial z}{\partial x}\right) dx + \left(F_2 + F_3 \frac{\partial z}{\partial x}\right) dy = \iint_R \frac{\partial}{\partial x}\left(F_2 + F_3 \frac{\partial z}{\partial x}\right) - \frac{\partial}{\partial y}\left(F_1 + F_3 \frac{\partial z}{\partial x}\right) dA$$

$\because F_1, F_2, F_3$ are functions of $x, y,$ and $z$ and $z$ is a function of $x$ and $y$

Applying the Chain Rule to right-hand side integral, we have

$$\iint_R \frac{\partial}{\partial x}\left(F_2 + F_3 \frac{\partial z}{\partial x}\right) - \frac{\partial}{\partial y}\left(F_1 + F_3 \frac{\partial z}{\partial x}\right) dA$$

$$= \iint_R \left(\frac{\partial F_2}{\partial x} + \frac{\partial F_2}{\partial z}\frac{\partial z}{\partial x} + \frac{\partial F_3}{\partial x}\frac{\partial z}{\partial y} + \frac{\partial F_3}{\partial z}\frac{\partial z}{\partial x}\frac{\partial z}{\partial y} + F_3 \frac{\partial^2 z}{\partial x \partial y}\right)$$

$$- \left(\frac{\partial F_1}{\partial y} + \frac{\partial F_1}{\partial z}\frac{\partial z}{\partial y} + \frac{\partial F_3}{\partial y}\frac{\partial z}{\partial x} + \frac{\partial F_3}{\partial z}\frac{\partial z}{\partial y}\frac{\partial z}{\partial x} + F_3 \frac{\partial^2 z}{\partial y \partial x}\right) dA$$

$$= \iint_R -\left(\frac{\partial F_3}{\partial y} - \frac{\partial F_2}{\partial z}\right)\frac{\partial z}{\partial x} - \left(\frac{\partial F_1}{\partial z} - \frac{\partial F_3}{\partial x}\right)\frac{\partial z}{\partial y} + \left(\frac{\partial F_2}{\partial x} - \frac{\partial F_1}{\partial y}\right) dA = \iint_S (\nabla \times \vec{F}) \cdot \vec{n} dA$$

$$\therefore \oint_C \vec{F} \cdot d\vec{r} = \iint_S (\nabla \times \vec{F}) \cdot \vec{n} dA$$

Stokes' Theorem 有个值得讨论的对应关系，就是曲面的法向量与封闭曲线绕行方向的关系；如果曲面$S$是朝上定向则封闭曲线$C$绕行方向为逆时针方向（正定向）；可以藉由右手法则，大拇指指向曲面朝上的法向量，其余四根手指头顺势弯曲的方向则为曲线的绕行方向，因此，如果曲面为朝上定向则封闭曲线$C$（四根手指头顺势弯曲的方向）为逆时针方（正定向）且

$$\oint_C \vec{F} \cdot d\vec{r} = \iint_S (\nabla \times \vec{F}) \cdot \vec{n} dA$$

如果曲面是朝下定向则封闭曲线 $C'$（四根手指头顺势弯曲的方向）绕行方向为负定向且

$$\oint_C \vec{F} \cdot d\vec{r} = -\iint_S (\nabla \times \vec{F}) \cdot \vec{n} dA = -\oint_{C'} \vec{F} \cdot d\vec{r}$$

上述的对应关系必须清楚的理解，否则计算出来的答案可能因为差个负号就错了；此外，另一个重要值得讨论的是不同曲面可能有相同的边界，藉由 Stokes' Theorem 可以推论这些不同曲面有相同边界会有相同的通量，为了让读者能更清楚掌握这些细微的规则以及 Stokes' Theorem 的应用情境，底下利用图示来说明；假设 $S_1, S_2, S_3$ 分别为上半球，上半圆锥，上半抛物曲面且这三者有共同边界$C$，藉由右手法则$C$所绕行的方向为逆时针方（正定向），藉由 Stokes' Theorem

$$\oint_C \vec{F} \cdot d\vec{r} = \iint_{S_1} (\nabla \times \vec{F}) \cdot \vec{n} dA = \iint_{S_2} (\nabla \times \vec{F}) \cdot \vec{n} dA = \iint_{S_3} (\nabla \times \vec{F}) \cdot \vec{n} dA$$

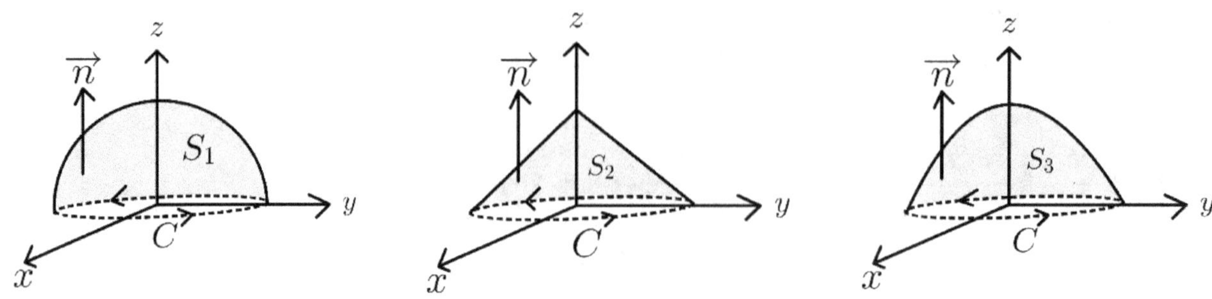

如下图所示, 假设 $S_4, S_5, S_6$ 分别为下半球, 下半圆锥, 下半抛物曲面且这三者有共同边界 $C'$, 藉由右手法则 $C'$ 所绕行的方向为顺时钟方向 (负定向), 且 $C$ 与 $C'$ 两者为相同曲线只是方向相反, 藉由 Stokes' Theorem

$$\oint_{C'} \vec{F} \cdot d\vec{r} = \iint_{S_4} (\nabla \times \vec{F}) \cdot \vec{n}\,dA = \iint_{S_5} (\nabla \times \vec{F}) \cdot \vec{n}\,dA = \iint_{S_6} (\nabla \times \vec{F}) \cdot \vec{n}\,dA = -\oint_{C} \vec{F} \cdot d\vec{r}$$

藉由上述观察, 可以得到一个 Stokes' Theorem 直观的推论, 一条空间封闭曲线可以引导出各式各样的非封闭曲面, 然而这些曲面的通量值会一样或者只是差个负号

Stokes' Theorem 的考试类型包含: 使用 Stokes' Theorem 说明线积分与积分路径无关、使用 Stokes' Theorem 把封闭线积分转成非封闭曲面的曲面积分、使用 Stokes' Theorem 把非封闭曲面的曲面积分转成封闭线积分

## 9.6.1 $\quad \nabla \times \vec{F} = 0$, 线积分与积分路径无关

给定向量值函数 $\vec{F}$, 试证线积分 $\displaystyle\int_{C} \vec{F} \cdot d\vec{r}$ 与路径无关

考试类型:

Type 1.

給定 $(x_0, y_0, z_0), (x_1, y_1, z_1)$, 试证 $\int_{(x_0,y_0,z_0)}^{(x_1,y_1,z_1)} \vec{F} \cdot d\vec{r}$ 与路径无关并求此积分

解题流程:

Step1.

令 $C_1$、$C_2$ 为從 $(x_0, y_0, z_0)$ 至 $(x_1, y_1, z_1)$ 的任意两条曲线

令 $C$ 为 $C_1$、$C_2$ 所围成的逆时针封闭曲线且 $S$ 为所围封闭区域

Step2.

Claim: $\nabla \times \vec{F} = 0$

藉由 Stokes' Theorem, $\oint_C \vec{F} \cdot d\vec{r} = \iint_S (\nabla \times \vec{F}) \cdot \vec{n} dA = 0$

$\therefore \int_{C_1} \vec{F} \cdot d\vec{r} - \int_{C_2} \vec{F} \cdot d\vec{r} = 0 \Rightarrow \int_{C_1} \vec{F} \cdot d\vec{r} = \int_{C_2} \vec{F} \cdot d\vec{r}$ $\therefore \int_C \vec{F} \cdot d\vec{r}$ 与积分路径无关

Step3.

$\because \int_C \vec{F} \cdot d\vec{r}$ 与积分路径无关 $\quad \therefore \exists G(x, y, z)$ 使得 $\nabla G(x, y, z) = \vec{F}(x, y, z)$

且 $\int_{(x_0,y_0,z_0)}^{(x_1,y_1,z_1)} \vec{F} \cdot d\vec{r} = \int_{(x_0,y_0,z_0)}^{(x_1,y_1,z_1)} \nabla G \cdot d\vec{r} = \int_{(x_0,y_0,z_0)}^{(x_1,y_1,z_1)} dG = G(x_1, y_1, z_1) - G(x_0, y_0, z_0)$

Step4.

$\because \nabla G(x, y, z) = \vec{F}(x, y, z) = (f_1, f_2, f_3)$

$\therefore \begin{cases} \dfrac{\partial G}{\partial x} = f_1(x, y, z) \\ \dfrac{\partial G}{\partial y} = f_2(x, y, z) \\ \dfrac{\partial G}{\partial z} = f_3(x, y, z) \end{cases} \quad \therefore \begin{cases} G(x, y, z) = \int f_1(x, y, z) dx + c \\ G(x, y, z) = \int f_2(x, y, z) dy + c \\ G(x, y, z) = \int f_3(x, y, z) dz + c \end{cases}$

找出明确 $G(x, y, z)$ 同时满足上述三个等式接着计算 $G(x_1, y_1, z_1) - G(x_0, y_0, z_0)$

<u>范例说明:</u>

給定 $(x_0, y_0, z_0), (x_1, y_1, z_1)$, 试证 $\int_{(x_0,y_0,z_0)}^{(x_1,y_1,z_1)} \vec{F} \cdot d\vec{r}$ 与路径无关

令 $C_1$、$C_2$ 为從 $(x_0, y_0, z_0)$ 至 $(x_1, y_1, z_1)$ 的任意两条曲线

令 $C$ 为 $C_1$、$C_2$ 所围成的逆时针封闭曲线且 $S$ 为所围封闭区域

(I)若 $\vec{F} = (2xy - y^4 + 3, x^2 - 4xy^3, 0)$ 则 $\nabla \times \vec{F} = (0,0,0)$

(II)若 $\vec{F} = (4x^3 y, x^4, 0)$ 则 $\nabla \times \vec{F} = (0,0,0)$

(III)若$\vec{\mathrm{F}} = (10x^4 - 2xy^3, -3x^2y^2, 0)$ 则 $\nabla \times \vec{\mathrm{F}} = (0,0,0)$

(IV)若$\vec{\mathrm{F}} = \left(yz^2, xz^2 + ze^{yz}, 2xyz + ye^{yz} + \dfrac{1}{1+z}\right)$ 则 $\nabla \times \vec{\mathrm{F}} = (0,0,0)$

(V)若$\vec{\mathrm{F}} = (2xyz, x^2z, x^2y)$ 则 $\nabla \times \vec{\mathrm{F}} = (0,0,0)$

藉由 Stokes' Theorem, $\displaystyle\oint_C \vec{\mathrm{F}} \cdot d\vec{r} = \iint_S (\nabla \times \vec{\mathrm{F}}) \cdot \vec{n}\, dA = 0$

$\therefore \displaystyle\int_{C_1} \vec{\mathrm{F}} \cdot d\vec{r} - \int_{C_2} \vec{\mathrm{F}} \cdot d\vec{r} = 0 \Rightarrow \int_{C_1} \vec{\mathrm{F}} \cdot d\vec{r} = \int_{C_2} \vec{\mathrm{F}} \cdot d\vec{r}$ $\therefore \displaystyle\int_C \vec{F} \cdot d\vec{r}$ 与积分路径无关

Example 1.

$\displaystyle\text{求} \oint_C (x^2y\cos x + 2xy\sin x - y^2e^x)dx + (x^2\sin x - 2ye^x)dy =?$, $C: x^{\frac{2}{3}} + y^{\frac{2}{3}} = a^{\frac{2}{3}}$

为逆时针封闭曲线

【解】

令$\vec{\mathrm{F}} = (x^2y\cos x + 2xy\sin x - y^2e^x, x^2\sin x - 2ye^x, 0)$

Claim: $\nabla \times \vec{\mathrm{F}} = 0$

$$\nabla \times \vec{\mathrm{F}} = \begin{vmatrix} \vec{i} & \vec{j} & \vec{k} \\ \dfrac{\partial}{\partial x} & \dfrac{\partial}{\partial y} & \dfrac{\partial}{\partial z} \\ x^2y\cos x + 2xy\sin x - y^2e^x & x^2\sin x - 2ye^x & 0 \end{vmatrix}$$

$= (0, 0, 2x\sin x + x^2\cos x - 2ye^x - (x^2\cos x + 2x\sin x - 2ye^x)) = (0,0,0)$

$\because \vec{F}$的各分量一阶偏导数存在且连续

藉由 Stokes' Theorem 则 $\displaystyle\oint_C \vec{\mathrm{F}} \cdot d\vec{r} = \iint_S (\nabla \times \vec{\mathrm{F}}) \cdot \vec{n}\, dA = 0$

Example 2.

$\displaystyle\text{试证} \int_{(0,0)}^{(3,1)} (2xy - y^4 + 3)dx + (x^2 - 4xy^3)dy$ 与 $(0,0),(3,1)$ 的路径无关并求此积分

【解】

令$C_1$、$C_2$为從$(0,0)$至$(3,1)$的任意两条曲线

令$C$为$C_1$、$C_2$所围成的逆时针封闭曲线且$S$为所围封闭区域

令$\vec{\mathrm{F}} = (2xy - y^4 + 3, x^2 - 4xy^3, 0)$

Claim: $\nabla \times \vec{\mathrm{F}} = 0$

$$\nabla\times\vec{F} = \begin{vmatrix} \vec{\imath} & \vec{\jmath} & \vec{k} \\ \dfrac{\partial}{\partial x} & \dfrac{\partial}{\partial y} & \dfrac{\partial}{\partial z} \\ 2xy-y^4+3 & x^2-4xy^3 & 0 \end{vmatrix} = \vec{\imath}\cdot 0 + \vec{\jmath}\cdot 0 + \vec{k}\cdot 0 = (0,0,0)$$

$\because \vec{F}$ 的各分量一阶偏导数存在且连续

藉由 Stokes' Theorem 则 $\displaystyle\oint_C \vec{F}\cdot d\vec{r} = \iint_S \left(\nabla\times\vec{F}\right)\cdot\vec{n}\,dA = 0$

$\therefore \displaystyle\int_{C_1} \vec{F}\cdot d\vec{r} - \int_{C_2}\vec{F}\cdot d\vec{r} = 0 \Rightarrow \int_{C_1}\vec{F}\cdot d\vec{r} = \int_{C_2}\vec{F}\cdot d\vec{r}$

$\therefore \displaystyle\int_{(0,0)}^{(3,1)}(2xy-y^4+3)dx + (x^2-4xy^3)dy$ 与积分路径无关

$\therefore \exists\, G(x,y,z)$ 使得 $\nabla G(x,y,z) = \vec{F}$

且 $\displaystyle\int_{(0,0)}^{(3,1)}\vec{F}\cdot d\vec{r} = \int_{(0,0)}^{(3,1)}\nabla G\cdot d\vec{r} = \int_{(0,0)}^{(3,1)} dG = G(3,1) - G(0,0)$

$$\therefore \begin{cases} \dfrac{\partial G}{\partial x} = 2xy - y^4 + 3 \\[2mm] \dfrac{\partial G}{\partial y} = x^2 - 4xy^3 \qquad \therefore G(x,y,z) = x^2y - xy^4 + 3x + c \\[2mm] \dfrac{\partial G}{\partial z} = 0 \end{cases}$$

$\therefore \displaystyle\int_{(0,0)}^{(3,1)}(2xy-y^4+3)dx + (x^2-4xy^3)dy = x^2y - xy^4 + 3x\,\Big|_{(0,0)}^{(3,1)} = 15$

Example 3.

$$\text{试证} \int_{(1,2)}^{(3,4)}(6xy^2-y^3)dx + (6x^2y-3xy^2)dy \text{ 与其路径无关并求此积分}$$

【解】

令 $C_1$、$C_2$ 为从 $(3,4)$ 至 $(1,2)$ 的任意两条曲线

令 $C$ 为 $C_1$、$C_2$ 所围成的逆时针封闭曲线且 $S$ 为所围封闭区域

令 $\vec{F} = (6xy^2 - y^3, 6x^2y - 3xy^2, 0)$

Claim: $\nabla\times\vec{F} = 0$

$$\nabla\times\vec{F} = \begin{vmatrix} \vec{\imath} & \vec{\jmath} & \vec{k} \\ \dfrac{\partial}{\partial x} & \dfrac{\partial}{\partial y} & \dfrac{\partial}{\partial z} \\ 6xy^2-y^3 & 6x^2y-3xy^2 & 0 \end{vmatrix} = \left(0,0,12x-3y^2-(12xy-3y^2)\right) = (0,0,0)$$

$\because \vec{F}$ 的各分量一阶偏导数存在且连续

藉由 Stokes' Theorem 则 $\oint_C \vec{F} \cdot d\vec{r} = \iint_S (\nabla \times \vec{F}) \cdot \vec{n} dA = 0$

$\therefore \int_{C_1} \vec{F} \cdot d\vec{r} - \int_{C_2} \vec{F} \cdot d\vec{r} = 0 \Rightarrow \int_{C_1} \vec{F} \cdot d\vec{r} = \int_{C_2} \vec{F} \cdot d\vec{r}$

$\therefore \int_{(1,2)}^{(3,4)} (6xy^2 - y^3)dx + (6x^2y - 3xy^2)dy$ 与积分路径无关

$\therefore \exists G(x, y, z)$ 使得 $\nabla G(x, y, z) = \vec{F}$

且 $\int_{(1,2)}^{(3,4)} \vec{F} \cdot d\vec{r} = \int_{(1,2)}^{(3,4)} \nabla G \cdot d\vec{r} = \int_{(1,2)}^{(3,4)} dG = G(3,4) - G(1,2)$

$$\therefore \begin{cases} \dfrac{\partial G}{\partial x} = 6xy^2 - y^3 \\ \dfrac{\partial G}{\partial y} = 6x^2y - 3xy^2 \\ \dfrac{\partial G}{\partial z} = 0 \end{cases} \quad \therefore G(x, y, z) = 3x^2y^2 - xy^3 + c$$

$\therefore \int_C \vec{F} \cdot d\vec{r} = 3x^2y^2 - xy^3 \Big|_{(1,2)}^{(3,4)} = 236$

Example 4.

$$試証 \int_{(0,0)}^{(a,b)} 4x^3y dx + x^4 dy \ 与 \ (0,0), (a,b) 的路径无关并求此积分$$

【解】

令 $C_1$、$C_2$ 为從 $(0,0)$ 至 $(a,b)$ 的任意两条曲线

令 $C$ 为 $C_1$、$C_2$ 所围成的逆时针封闭曲线且 $S$ 为所围封闭区域

令 $\vec{F} = (4x^3y, x^4, 0)$

Claim: $\nabla \times \vec{F} = 0$

$$\nabla \times \vec{F} = \begin{vmatrix} \vec{i} & \vec{j} & \vec{k} \\ \dfrac{\partial}{\partial x} & \dfrac{\partial}{\partial y} & \dfrac{\partial}{\partial z} \\ 4x^3y & x^4 & 0 \end{vmatrix} = (0, 0, 4x^3 - 4x^3) = (0, 0, 0)$$

$\because \vec{F}$ 的各分量一阶偏导数存在且连续

藉由 Stokes' Theorem 则 $\oint_C \vec{F} \cdot d\vec{r} = \iint_S (\nabla \times \vec{F}) \cdot \vec{n} dA = 0$

$$\therefore \int_{C_1} \vec{\mathbf{F}} \cdot d\vec{r} - \int_{C_2} \vec{\mathbf{F}} \cdot d\vec{r} = 0 \Rightarrow \int_{C_1} \vec{\mathbf{F}} \cdot d\vec{r} = \int_{C_2} \vec{\mathbf{F}} \cdot d\vec{r}$$

$$\therefore \int_{(0,0)}^{(a,b)} 4x^3 y\,dx + x^4 dy \text{ 与积分路径无关}$$

$$\therefore \exists G(x,y,z) \text{ 使得 } \nabla G(x,y,z) = \vec{\mathbf{F}}$$

$$\text{且} \int_{(0,0)}^{(a,b)} \vec{\mathbf{F}} \cdot d\vec{r} = \int_{(0,0)}^{(a,b)} \nabla G \cdot d\vec{r} = \int_{(0,0)}^{(a,b)} dG = G(a,b) - G(0,0)$$

$$\therefore \begin{cases} \dfrac{\partial G}{\partial x} = 4x^3 y \\[2mm] \dfrac{\partial G}{\partial y} = x^4 \\[2mm] \dfrac{\partial G}{\partial z} = 0 \end{cases} \qquad \therefore G(x,y,z) = x^4 y + c$$

$$\therefore \int_{(0,0)}^{(a,b)} 4x^3 y\,dx + x^4 dy = x^4 y + c \Big|_{(0,0)}^{(a,b)} = a^4 b$$

## Example 5.

$$求 \int_{(0,0)}^{(-2,-1)} (10x^4 - 2xy^3)dx - 3x^2 y^2 dy = ?, \quad \text{假设积分路径为 } x^4 - 6xy^3 = 4y^2$$

【解】

令 $C_1$、$C_2$ 为從 $(0,0)$ 至 $(-2,-1)$ 的任意两条曲线

令 $C$ 为 $C_1$、$C_2$ 所围成的逆时针封闭曲线且 $S$ 为所围封闭区域

令 $\vec{\mathbf{F}} = (10x^4 - 2xy^3, -3x^2 y^2, 0)$

Claim: $\nabla \times \vec{\mathbf{F}} = 0$

$$\nabla \times \vec{\mathbf{F}} = \begin{vmatrix} \vec{i} & \vec{j} & \vec{k} \\[1mm] \dfrac{\partial}{\partial x} & \dfrac{\partial}{\partial y} & \dfrac{\partial}{\partial z} \\[1mm] 10x^4 - 2xy^3 & -3x^2 y^2 & 0 \end{vmatrix} = (0,0,-6xy^2 + 6xy^2) = (0,0,0)$$

$\because \vec{F}$ 的各分量一阶偏导数存在且连续

藉由 Stokes' Theorem 则 $\displaystyle\oint_C \vec{\mathbf{F}} \cdot d\vec{r} = \iint_S (\nabla \times \vec{\mathbf{F}}) \cdot \vec{n}\,dA = 0$

$$\therefore \int_{C_1} \vec{\mathbf{F}} \cdot d\vec{r} - \int_{C_2} \vec{\mathbf{F}} \cdot d\vec{r} = 0 \Rightarrow \int_{C_1} \vec{\mathbf{F}} \cdot d\vec{r} = \int_{C_2} \vec{\mathbf{F}} \cdot d\vec{r}$$

$$\therefore \int_{(0,0)}^{(-2,-1)} (10x^4 - 2xy^3)dx - 3x^2y^2dy \text{ 与积分路径无关}$$

$$\therefore \exists G(x,y,z) \text{ 使得 } \nabla G(x,y,z) = \vec{F}$$

$$\text{且} \int_{(0,0)}^{(-2,-1)} \vec{F} \cdot d\vec{r} = \int_{(0,0)}^{(-2,-1)} \nabla G \cdot d\vec{r} = \int_{(0,0)}^{(-2,-1)} dG = G(-2,-1) - G(0,0)$$

$$\therefore \begin{cases} \dfrac{\partial G}{\partial x} = 10x^4 - 2xy^3 \\[2mm] \dfrac{\partial G}{\partial y} = -3x^2y^2 \qquad \therefore G(x,y,z) = 2x^5 - x^2y^3 + c \\[2mm] \dfrac{\partial G}{\partial z} = 0 \end{cases}$$

$$\therefore \int_{(0,0)}^{(-2,-1)} (10x^4 - 2xy^3)dx - 3x^2y^2dy = 2x^5 - x^2y^3 \Big|_{(0,0)}^{(-2,-1)} = -60$$

Example 6.

$$\text{求} \int_{(0,0)}^{\left(\frac{\pi}{2},1\right)} (x^2 + 6xy - 2y^2)dx + (3x^2 - 4xy + 2y)dy =?, \text{ 假设积分路径为 } y = \sin x$$

【解】

令 $C_1$、$C_2$ 为從 $(0,0)$ 至 $\left(\dfrac{\pi}{2},1\right)$ 的任意两条曲线

令 $C$ 为 $C_1$、$C_2$ 所围成的逆时针封闭曲线且 $S$ 为所围封闭区域

令 $\vec{F} = (x^2 + 6xy - 2y^2, 3x^2 - 4xy + 2y, 0)$

Claim: $\nabla \times \vec{F} = 0$

$$\nabla \times \vec{F} = \begin{vmatrix} \vec{i} & \vec{j} & \vec{k} \\[2mm] \dfrac{\partial}{\partial x} & \dfrac{\partial}{\partial y} & \dfrac{\partial}{\partial z} \\[2mm] x^2 + 6xy - 2y^2 & 3x^2 - 4xy + 2y & 0 \end{vmatrix} = (0,0,6x - 4y - (6x - 4y)) = (0,0,0)$$

$\because \vec{F}$ 的各分量一阶偏导数存在且连续

藉由 Stokes' Theorem 则 $\oint_C \vec{F} \cdot d\vec{r} = \iint_S (\nabla \times \vec{F}) \cdot \vec{n}dA = 0$

$$\therefore \int_{C_1} \vec{F} \cdot d\vec{r} - \int_{C_2} \vec{F} \cdot d\vec{r} = 0 \Rightarrow \int_{C_1} \vec{F} \cdot d\vec{r} = \int_{C_2} \vec{F} \cdot d\vec{r}$$

$$\therefore \int_{(0,0)}^{(\frac{\pi}{2},1)} (x^2 + 6xy - 2y^2)dx + (3x^2 - 4xy + 2y)dy \text{ 与积分路径无关}$$

$$\therefore \exists G(x,y,z) \text{ 使得 } \nabla G(x,y,z) = \vec{F}$$

$$\text{且} \int_{(0,0)}^{(\frac{\pi}{2},1)} \vec{F} \cdot d\vec{r} = \int_{(0,0)}^{(\frac{\pi}{2},1)} \nabla G \cdot d\vec{r} = \int_{(0,0)}^{(\frac{\pi}{2},1)} dG = G(\frac{\pi}{2},1) - G(0,0)$$

$$\therefore \begin{cases} \dfrac{\partial G}{\partial x} = x^2 + 6xy - 2y^2 \\ \dfrac{\partial G}{\partial y} = 3x^2 - 4xy + 2y \\ \dfrac{\partial G}{\partial z} = 0 \end{cases} \qquad \therefore G(x,y,z) = 3x^2y - 2xy^2 + y^2 + \frac{x^3}{3} + c$$

$$\therefore \int_{(0,0)}^{(\frac{\pi}{2},1)} (x^2 + 6xy - 2y^2)dx + (3x^2 - 4xy + 2y)dy = 3x^2y - 2xy^2 + y^2 + \frac{x^3}{3} \Big|_{(0,0)}^{(\frac{\pi}{2},1)}$$

$$= 3\left(\frac{\pi}{2}\right)^2 - 2\left(\frac{\pi}{2}\right) + 1 + \frac{\left(\frac{\pi}{2}\right)^3}{3}$$

Example 7.

$$求 \int_C \vec{F} \cdot d\vec{r} = ?, \quad \vec{F} = (2xy + z^3, x^2, 3xz^2), \quad C \text{为}(1,-2,1)\text{至}(3,1,4)\text{的任意曲线}$$

【解】

令$C_1$、$C_2$为從$(1,-2,1)$至$(3,1,4)$的任意两条曲线

令$C$为$C_1$、$C_2$所围成的逆时针封闭曲线且$S$为所围封闭区域

Claim: $\nabla \times \vec{F} = 0$

$$\nabla \times \vec{F} = \begin{vmatrix} \vec{i} & \vec{j} & \vec{k} \\ \dfrac{\partial}{\partial x} & \dfrac{\partial}{\partial y} & \dfrac{\partial}{\partial z} \\ 2xy + z^3 & x^2 & 3xz^2 \end{vmatrix} = (0, 3z^2 - 3z^2, 2x - 2x) = (0,0,0)$$

$\because \vec{F}$的各分量一阶偏导数存在且连续

藉由 Stokes' Theorem 则 $\oint_C \vec{F} \cdot d\vec{r} = \iint_S (\nabla \times \vec{F}) \cdot \vec{n} dA = 0$

$$\therefore \int_{C_1} \vec{F}\cdot d\vec{r} - \int_{C_2} \vec{F}\cdot d\vec{r} = 0 \Rightarrow \int_{C_1} \vec{F}\cdot d\vec{r} = \int_{C_2} \vec{F}\cdot d\vec{r} \quad \therefore \int_C \vec{F}\cdot d\vec{r}\ \text{与积分路径无关}$$

$\therefore \exists G(x,y,z)$ 使得 $\nabla G(x,y,z) = \vec{F}$

且 $\displaystyle\int_{(1,-2,1)}^{(3,1,4)} \vec{F}\cdot d\vec{r} = \int_{(1,-2,1)}^{(3,1,4)} \nabla G \cdot d\vec{r} = \int_{(1,-2,1)}^{(3,1,4)} dG = G(3,1,4) - G(1,-2,1)$

$$\therefore \begin{cases} \dfrac{\partial G}{\partial x} = 2xy + z^3 \\[2mm] \dfrac{\partial G}{\partial y} = x^2 \\[2mm] \dfrac{\partial G}{\partial z} = 3xz^2 \end{cases} \qquad \therefore G(x,y,z) = x^2 y + xz^3 + c$$

$$\therefore \int_C \vec{F}\cdot d\vec{r} = x^2 y + xz^3 \Big|_{(1,-2,1)}^{(3,1,4)} = 202$$

Example 8.

$$\text{求} \int_C yz^2\,dx + (xz^2 + ze^{yz})\,dy + \left(2xyz + ye^{yz} + \frac{1}{1+z}\right)dz\,,\ \text{其中曲线}$$

$$C = \{(t, t^2, t^3): 0 \le t \le 1\}$$

【解】

令 $C_1$、$C_2$ 为從 $(0,0,0)$ 至 $(1,1,1)$ 的任意两条曲线

令 $C$ 为 $C_1$、$C_2$ 所围成的逆时针封闭曲线且 $S$ 为所围封闭区域

令 $\vec{F} = \left(yz^2, xz^2 + ze^{yz}, 2xyz + ye^{yz} + \dfrac{1}{1+z}\right)$

Claim: $\nabla \times \vec{F} = 0$

$$\nabla \times \vec{F} = \begin{vmatrix} \vec{i} & \vec{j} & \vec{k} \\[2mm] \dfrac{\partial}{\partial x} & \dfrac{\partial}{\partial y} & \dfrac{\partial}{\partial z} \\[2mm] yz^2 & xz^2 + ze^{yz} & 2xyz + ye^{yz} + \dfrac{1}{1+z} \end{vmatrix}$$

$$= (2xz + e^{yz} + zye^{yz} - (2xz + e^{yz} + yze^{yz}), 2yz - 2yz, z^2 - z^2) = (0,0,0)$$

$\because \vec{F}$ 的各分量一阶偏导数存在且连续

藉由 Stokes' Theorem 则 $\displaystyle\oint_C \vec{F}\cdot d\vec{r} = \iint_S (\nabla \times \vec{F})\cdot \vec{n}\,dA = 0$

$$\therefore \int_{C_1} \vec{F} \cdot d\vec{r} - \int_{C_2} \vec{F} \cdot d\vec{r} = 0 \Rightarrow \int_{C_1} \vec{F} \cdot d\vec{r} = \int_{C_2} \vec{F} \cdot d\vec{r}$$

$$\therefore \int_C yz^2 dx + (xz^2 + ze^{yz})dy + \left(2xyz + ye^{yz} + \frac{1}{1+z}\right) dz \text{ 与积分路径无关}$$

$\therefore \exists G(x,y,z)$ 使得 $\nabla G(x,y,z) = \vec{F}$

且 $\displaystyle\int_{(0,0,0)}^{(1,1,1)} \vec{F} \cdot d\vec{r} = \int_{(0,0,0)}^{(1,1,1)} \nabla G \cdot d\vec{r} = \int_{(0,0,0)}^{(1,1,1)} dG = G(1,1,1) - G(0,0,0)$

$$\therefore \begin{cases} \dfrac{\partial G}{\partial x} = yz^2 \\[2mm] \dfrac{\partial G}{\partial y} = xz^2 + ze^{yz} \\[2mm] \dfrac{\partial G}{\partial z} = 2xyz + ye^{yz} + \dfrac{1}{1+z} \end{cases} \qquad \therefore G(x,y,z) = xyz^2 + e^{yz} + \ln(1+z) + c$$

$$\therefore \int_C yz^2 dx + (xz^2 + ze^{yz})dy + \left(2xyz + ye^{yz} + \frac{1}{1+z}\right) dz$$

$$= xyz^2 + e^{yz} + \ln(1+z)\Big|_{(0,0,0)}^{(1,1,1)} = e + \ln 2$$

Example 9.

$$求 \int_C \vec{F} \cdot d\vec{r} =?, \quad \vec{F} = (2xyz, x^2z, x^2y), \quad C 为 (0,0,0) 至 (1,1,1) 的任意曲线$$

【解】

令 $C_1$、$C_2$ 为從 $(0,0,0)$ 至 $(1,1,1)$ 的任意两条曲线

令 $C$ 为 $C_1$、$C_2$ 所围成的逆时针封闭曲线且 $S$ 为所围封闭区域

令 $\vec{F} = (2xyz, x^2z, x^2y)$

Claim: $\nabla \times \vec{F} = 0$

$$\nabla \times \vec{F} = \begin{vmatrix} \vec{i} & \vec{j} & \vec{k} \\ \dfrac{\partial}{\partial x} & \dfrac{\partial}{\partial y} & \dfrac{\partial}{\partial z} \\ 2xyz & x^2z & x^2y \end{vmatrix} = (x^2 - x^2, 2xy - 2xy, 2xz - 2xz) = (0,0,0)$$

$\because \vec{F}$ 的各分量一阶偏导数存在且连续

藉由 Stokes' Theorem 则 $\displaystyle\oint_C \vec{F} \cdot d\vec{r} = \iint_S (\nabla \times \vec{F}) \cdot \vec{n}\, dA = 0$

$$\therefore \int_{C_1} \vec{F} \cdot d\vec{r} - \int_{C_2} \vec{F} \cdot d\vec{r} = 0 \Rightarrow \int_{C_1} \vec{F} \cdot d\vec{r} = \int_{C_2} \vec{F} \cdot d\vec{r} \quad \therefore \int_C \vec{F} \cdot d\vec{r} \text{ 与积分路径无关}$$

$$\therefore \exists G(x,y,z) \text{ 使得 } \nabla G(x,y,z) = \vec{F}$$

$$\text{且} \int_{(0,0,0)}^{(1,1,1)} \vec{F} \cdot d\vec{r} = \int_{(0,0,0)}^{(1,1,1)} \nabla G \cdot d\vec{r} = \int_{(0,0,0)}^{(1,1,1)} dG = G(1,1,1) - G(0,0,0)$$

$$\therefore \begin{cases} \dfrac{\partial G}{\partial x} = 2xyz \\[2mm] \dfrac{\partial G}{\partial y} = x^2 z \\[2mm] \dfrac{\partial G}{\partial z} = x^2 y \end{cases} \quad \therefore G(x,y,z) = x^2 yz + c \quad \therefore \int_C \vec{F} \cdot d\vec{r} = x^2 yz \Big|_{(0,0,0)}^{(1,1,1)} = 1$$

## 9.6.2 $\quad \nabla \times \vec{F} \neq 0$,把封闭线积分转成非封闭曲面积分

应用情境为如果线积分的被积分函数较为复杂而使用 Stokes' Theorem 转成面积分之后,曲面积分的被积分函数较为干净

考试类型:

Type 1.

$C$ 为空间中逆时针封闭曲线,试证 $\displaystyle\oint_C \vec{F} \cdot d\vec{r}$ 只与 $C$ 所围面积有关,与 $C$ 的位置形状无关

解题流程:

Claim: $(\nabla \times \vec{F}) \cdot \vec{n} = \text{constant}$

假设 $\vec{F} = (f_1, f_2, f_3)$,令 $C$ 所围封闭区域为 R

藉由 Stokes' Theorem, $\displaystyle\oint_C \vec{F} \cdot d\vec{r} = \iint_R (\nabla \times \vec{F}) \cdot \vec{n}\, dA$

其中 $\vec{n}$ 为单位法向量 且 $\nabla \times \vec{F} = \begin{vmatrix} \vec{i} & \vec{j} & \vec{k} \\[1mm] \dfrac{\partial}{\partial x} & \dfrac{\partial}{\partial y} & \dfrac{\partial}{\partial z} \\[1mm] f_1 & f_2 & f_3 \end{vmatrix}$

若 $(\nabla \times \vec{F}) \cdot \vec{n} = c$ $(c \ is \ \text{constant})$ 则 $\displaystyle\oint_C \vec{F} \cdot d\vec{r} = \iint_R (\nabla \times \vec{F}) \cdot \vec{n}\, dA = \iint_R c\, dA$

$\therefore \oint_C \vec{\mathbf{F}} \cdot d\vec{r}$ 只与 $C$ 所围面积有关，与 $C$ 的形状与位置无关

<u>范例说明：</u>

$C$ 为空间中逆时针封闭曲线，试证 $\oint_C \vec{\mathbf{F}} \cdot d\vec{r}$ 只与 $C$ 所围面积有关，与 $C$ 的形状位置无关

令 $C$ 所围封闭区域为 R

藉由 Stokes' Theorem，$\oint_C \vec{\mathbf{F}} \cdot d\vec{r} = \iint_R (\nabla \times \vec{\mathbf{F}}) \cdot \vec{n} dA$

其中 $\vec{n}$ 为单位法向量 且 $\nabla \times \vec{\mathbf{F}} = \begin{vmatrix} \vec{\imath} & \vec{\jmath} & \vec{k} \\ \dfrac{\partial}{\partial x} & \dfrac{\partial}{\partial y} & \dfrac{\partial}{\partial z} \\ f_1 & f_2 & f_3 \end{vmatrix}$

(I) 若 $\vec{\mathbf{F}} = (3y, -5z, x)$ 且 $C$ 是 $x + 2y + 2z = 5$ 平面上任意逆时针封闭曲线

则 $(\nabla \times \vec{\mathbf{F}}) \cdot \vec{n} = (5, -1, -3) \cdot \left(\dfrac{1}{3}, \dfrac{2}{3}, \dfrac{2}{3}\right) = -1$

藉由 Stokes' Theorem，

$\oint_C \vec{\mathbf{F}} \cdot d\vec{r} = \iint_R (\nabla \times \vec{\mathbf{F}}) \cdot \vec{n} dA = \iint_R -1 \, dA$

$\therefore \oint_C \vec{\mathbf{F}} \cdot d\vec{r}$ 只与 $C$ 所围面积有关，与 $C$ 的形状与位置无关

(II) 若 $\vec{\mathbf{F}} = (z, -2x, 4y)$ 且 $C$ 是 $x + y + z = 2$ 平面上任意逆时针封闭曲线

则 $(\nabla \times \vec{\mathbf{F}}) \cdot \vec{n} = (4, 1, -2) \cdot \left(\dfrac{1}{\sqrt{3}}, \dfrac{1}{\sqrt{3}}, \dfrac{1}{\sqrt{3}}\right) = \sqrt{3}$

藉由 Stokes' Theorem，

$\oint_C \vec{\mathbf{F}} \cdot d\vec{r} = \iint_R (\nabla \times \vec{\mathbf{F}}) \cdot \vec{n} dA = \iint_R \sqrt{3} \, dA$

$\therefore \oint_C \vec{\mathbf{F}} \cdot d\vec{r}$ 只与 $C$ 所围面积有关，与 $C$ 的形状与位置无关

Type 2.

给定空间中非封闭朝上定向曲面 $S: z = g(x, y), \forall a \leq x \leq b, c \leq y \leq d, C$ 为 $S$ 的边界

(逆时针封闭曲线),$\vec{F}$各分量的一阶偏导数存在且连续，求 $\oint_C \vec{F} \cdot d\vec{r} =?$

解题流程：

Step1.

藉由 Stokes' Theorem，$\oint_C \vec{F} \cdot d\vec{r} = \iint_S (\nabla \times \vec{F}) \cdot \vec{n} dA$

Step2.

$\because S$为朝上定向

令 $\vec{r}(x,y) = (x, y, g(x,y))$ 且 $R$ 为曲面$S$投影至$xy$平面的封闭区域

则 $R = \{(x,y) : a \le x \le b, c \le y \le d\}$ 且 $\iint_S (\nabla \times \vec{F}) \cdot \vec{n}\, dA = \iint_R (\nabla \times \vec{F}) \cdot \dfrac{\partial \vec{r}}{\partial x} \times \dfrac{\partial \vec{r}}{\partial y} dxdy$

Step3.

$\because \dfrac{\partial \vec{r}}{\partial x} = (1, 0, g_x(x,y))$, $\dfrac{\partial \vec{r}}{\partial y} = (0, 1, g_y(x,y))$ $\therefore \dfrac{\partial \vec{r}}{\partial x} \times \dfrac{\partial \vec{r}}{\partial y} = (-g_x(x,y), -g_y(x,y), 1)$

$\therefore \iint_R (\nabla \times \vec{F}) \cdot \dfrac{\partial \vec{r}}{\partial x} \times \dfrac{\partial \vec{r}}{\partial y} dxdy = \iint_R (\nabla \times \vec{F}) \cdot (-g_x(x,y), -g_y(x,y), 1)dxdy$

Step4.

藉由 Fubini's Theorem

$\iint_R (\nabla \times \vec{F}) \cdot (-g_x(x,y), -g_y(x,y), 1)dxdy$

$= \int_c^d \int_a^b (\nabla \times \vec{F}) \cdot (-g_x(x,y), -g_y(x,y), 1)\, dxdy$

$= \int_c^d \int_a^b -f_1 \cdot g_x(x,y) - f_2 \cdot g_y(x,y) + f_3\, dxdy$

其中 $\nabla \times \vec{F} = (f_1(x,y,g(x,y)), f_2(x,y,g(x,y)), f_3(x,y,g(x,y)))$

Step5.

求 $\int_c^d \int_a^b -f_1 \cdot g_x(x,y) - f_2 \cdot g_y(x,y) + f_3\, dxdy =?$

如果$S$是朝下定向

则 $\oint_C \vec{F} \cdot d\vec{r} = \int_c^d \int_a^b f_1 \cdot g_x(x,y) + f_2 \cdot g_y(x,y) - f_3\, dxdy$

<u>范例说明：</u>

求 $\oint_C \vec{\mathbf{F}}\cdot d\vec{r} =?$

(I)若 $\vec{\mathbf{F}} = (z^2, y^2, x)$，$C$ 是 $(a,0,0),(0,a,0),(0,0,a)$ 所围逆时针封闭边界

令 $\vec{r}(x,y) = (x,y,a-x-y)$ 则 $\dfrac{\partial \vec{r}}{\partial x} = (1,0,-1)$，$\dfrac{\partial \vec{r}}{\partial y} = (0,1,-1)$ $\therefore \dfrac{\partial \vec{r}}{\partial x} \times \dfrac{\partial \vec{r}}{\partial y} = (1,1,1)$

$$\because \nabla \times \vec{\mathbf{F}} = \begin{vmatrix} \vec{\imath} & \vec{\jmath} & \vec{k} \\ \dfrac{\partial}{\partial x} & \dfrac{\partial}{\partial y} & \dfrac{\partial}{\partial z} \\ z^2 & y^2 & x \end{vmatrix} = (0, 2z-1, 0)$$

藉由 Stokes' Theorem,

$$\oint_C \vec{\mathbf{F}}\cdot d\vec{r} = \iint_S (\nabla\times\vec{\mathbf{F}})\cdot\vec{n}\,dA = \iint_{0\leq x+y\leq a} (0,2z-1,0)\cdot(1,1,1)\,dxdy$$

$$= \iint_{0\leq x+y\leq a} 2z-1\,dxdy = \iint_{0\leq x+y\leq a} 2(a-x-y)-1\,dxdy = \frac{a^3}{3} - \frac{a^2}{2}$$

Type 3.

给定空间中非封闭朝上定向曲面 $S: z = g(x,y)$，$S$ 投影至 $xy$ 平面 $= \{(x,y): x^2+y^2 \leq a^2\}$，

$C$ 为 $S$ 的边界（逆时针封闭曲线），$\vec{\mathbf{F}}$ 各分量的一阶偏导数存在且连续，求 $\oint_C \vec{\mathbf{F}}\cdot d\vec{r} =?$

解题流程:

Step1.

藉由 Stokes' Theorem, $\quad \oint_C \vec{\mathbf{F}}\cdot d\vec{r} = \iint_S (\nabla\times\vec{\mathbf{F}})\cdot\vec{n}\,dA$

Step2.

$\because S$ 为朝上定向

令 $\vec{r}(x,y) = (x,y,g(x,y))$ 且 $R$ 为曲面 $S$ 投影至 $xy$ 平面的封闭区域

则 $R = \{(x,y): x^2+y^2 \leq a^2\}$ 且 $\iint_S (\nabla\times\vec{\mathbf{F}})\cdot\vec{n}\,dA = \iint_R (\nabla\times\vec{\mathbf{F}})\cdot\dfrac{\partial \vec{r}}{\partial x}\times\dfrac{\partial \vec{r}}{\partial y}\,dxdy$

Step3.

$\because \dfrac{\partial \vec{r}}{\partial x} = (1,0,g_x(x,y))$，$\dfrac{\partial \vec{r}}{\partial y} = (0,1,g_y(x,y))$ $\therefore \dfrac{\partial \vec{r}}{\partial x}\times\dfrac{\partial \vec{r}}{\partial y} = (-g_x(x,y), -g_y(x,y), 1)$

$\therefore \iint_R (\nabla\times\vec{\mathbf{F}})\cdot\dfrac{\partial \vec{r}}{\partial x}\times\dfrac{\partial \vec{r}}{\partial y}\,dxdy = \iint_R (\nabla\times\vec{\mathbf{F}})\cdot(-g_x(x,y), -g_y(x,y), 1)\,dxdy$

$$= \iint_R -f_1 \cdot g_x(x,y) - f_2 \cdot g_y(x,y) + f_3 \, dxdy$$

其中 $\nabla \times \vec{F} = (f_1(x,y,g(x,y)), f_2(x,y,g(x,y)), f_3(x,y,g(x,y)))$

Step4.

令 $x = r\cos\theta, y = r\sin\theta$ 则 $\{(x,y): x^2 + y^2 \leq a^2\} = \{(r,\theta): 0 \leq r \leq a, 0 \leq \theta \leq 2\pi\}$

且 $dxdy = \left\| \begin{vmatrix} \dfrac{\partial x}{\partial r} & \dfrac{\partial x}{\partial \theta} \\ \dfrac{\partial y}{\partial r} & \dfrac{\partial y}{\partial \theta} \end{vmatrix} \right\| dr d\theta = \left\| \begin{vmatrix} \cos\theta & -r\sin\theta \\ \sin\theta & r\cos\theta \end{vmatrix} \right\| dr d\theta = r \, dr d\theta$

Step5.

$$\iint_R -f_1 \cdot g_x(x,y) - f_2 \cdot g_y(x,y) + f_3 \, dxdy$$

$$= \int_0^{2\pi} \int_0^a \left( -f_1 \cdot g_x(r\cos\theta, r\sin\theta) - f_2 \cdot g_y(r\cos\theta, r\sin\theta) + f_3 \right) \cdot r \, dr d\theta$$

如果 $S$ 是朝下定向，

则 $\displaystyle\oint_C \vec{F} \cdot d\vec{r} = \int_0^{2\pi} \int_0^a \left( f_1 \cdot g_x(r\cos\theta, r\sin\theta) + f_2 \cdot g_y(r\cos\theta, r\sin\theta) - f_3 \right) \cdot r \, dr d\theta$

<u>范例说明：</u>

(I) $\vec{F} = (y^3, -x^3, 0)$ 且 $C: x^2 + y^2 = 4, \; z = 0$ 为逆时针封闭曲线，求 $\displaystyle\oint_C \vec{F} \cdot d\vec{r} = ?$

令 $S = \{(x,y,z): x^2 + y^2 \leq 4, z = 0\}$

$\because \vec{F}$ 各分量的一阶偏导数存在且连续

藉由 Stokes' Theorem， $\displaystyle\oint_C \vec{F} \cdot d\vec{r} = \iint_S (\nabla \times \vec{F}) \cdot \vec{n} dA$

$$\iint_S (\nabla \times \vec{F}) \cdot \vec{n} dA = \iint_S (0,0,-3x^2 - 3y^2) \cdot (0,0,1) dxdy = -3 \iint_{x^2+y^2 \leq 4} x^2 + y^2 \, dxdy$$

$$= -24\pi$$

(II) $\vec{F} = (e^x y - y^3, e^x + x^3, 0)$ 且 $C: x^2 + y^2 = 6, z = 0$ 为逆时针封闭曲线，求 $\displaystyle\oint_C \vec{F} \cdot d\vec{r} = ?$

令 $S = \{(x,y,z): x^2 + y^2 \leq 6, z = 0\}$

$\because \vec{F}$ 各分量的一阶偏导数存在且连续

藉由 Stokes' Theorem， $\displaystyle\oint_C \vec{F} \cdot d\vec{r} = \iint_S (\nabla \times \vec{F}) \cdot \vec{n} dA$

$$\iint_S (\nabla \times \vec{F}) \cdot \vec{n} \, dA = \iint_S (0,0,3x^2+3y^2) \cdot (0,0,1) \, dxdy = 3 \iint_{x^2+y^2 \leq 36} x^2 + y^2 \, dxdy$$
$$= 1944\pi$$

Example 1.

藉由 Stokes' Theorem 求 $\oint_C \vec{F} \cdot d\vec{r} = ?$，$C: x^2 + y^2 = 4$，$z = 1$，$\vec{F} = (x, 2z - x, y^2)$，

$C$ 为逆时针封闭曲线

【解】

令 $S = \{(x,y,z): x^2 + y^2 \leq 4, z = 1\}$ 为朝上定向，$\because \vec{F}$ 的各分量一阶偏导数存在且连续

藉由 Stokes' Theorem， $\oint_C \vec{F} \cdot d\vec{r} = \iint_S (\nabla \times \vec{F}) \cdot \vec{n} \, dA$

$\because S$ 为朝上定向

令 $\vec{r}(x,y) = (x,y,1)$，$R = \{(x,y): x^2 + y^2 \leq 4\}$ 则

$$\iint_S (\nabla \times \vec{F}) \cdot \vec{n} \, dA = \iint_R (\nabla \times \vec{F}) \cdot \frac{\partial \vec{r}}{\partial x} \times \frac{\partial \vec{r}}{\partial y} \, dxdy$$

$\because \dfrac{\partial \vec{r}}{\partial x} = (1,0,0)$，$\dfrac{\partial \vec{r}}{\partial y} = (0,1,0)$ $\therefore \dfrac{\partial \vec{r}}{\partial x} \times \dfrac{\partial \vec{r}}{\partial y} = (0,0,1)$

$$\because \nabla \times \vec{F} = \begin{vmatrix} \vec{i} & \vec{j} & \vec{k} \\ \dfrac{\partial}{\partial x} & \dfrac{\partial}{\partial y} & \dfrac{\partial}{\partial z} \\ x & 2z - x & y^2 \end{vmatrix} = (2y + 2, 0, -1)$$

$$\therefore \iint_R (\nabla \times \vec{F}) \cdot \frac{\partial \vec{r}}{\partial x} \times \frac{\partial \vec{r}}{\partial y} \, dxdy = \iint_R (2y+2, 0, -1) \cdot (0,0,1) \, dxdy = - \iint_{x^2+y^2 \leq 4} dxdy$$
$$= -4\pi$$

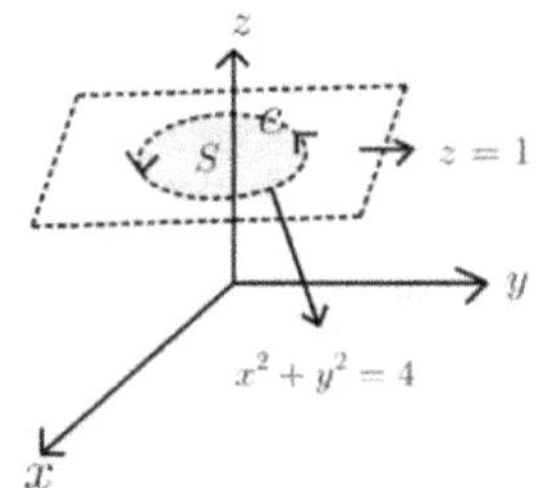

Example 2.

假设 $C$ 是 $x + 2y + 2z = 5$ 平面上任意正向封闭曲线，试证 $\oint_C 3ydx - 5zdy + xdz$

只与 $C$ 所围面积有关 与 $C$ 的形狀与位置无关

【解】

令 $C$ 在平面 $x + 2y + 2z = 5$ 所围封闭区域为 R

令 $\vec{\mathbf{F}} = (3y, -5z, x)$，$\because \vec{F}$ 的各分量一阶偏导数存在且连续

藉由 Stokes' Theorem 则 $\oint_C 3ydx - 5zdy + xdz = \oint_C \vec{\mathbf{F}} \cdot d\vec{r} = \iint_R (\nabla \times \vec{\mathbf{F}}) \cdot \vec{n} dA$

令 $\vec{n}$ 为 $x + 2y + 2z = 5$ 单位法向量 则 $\vec{n} = \left(\dfrac{1}{3}, \dfrac{2}{3}, \dfrac{2}{3}\right)$

$$\because \nabla \times \vec{\mathbf{F}} = \begin{vmatrix} \vec{i} & \vec{j} & \vec{k} \\ \dfrac{\partial}{\partial x} & \dfrac{\partial}{\partial y} & \dfrac{\partial}{\partial z} \\ 3y & -5z & x \end{vmatrix} = (5, -1, -3)$$

$$\therefore \iint_R (\nabla \times \vec{\mathbf{F}}) \cdot \vec{n} dA = \iint_R (5, -1, -3) \cdot \left(\dfrac{1}{3}, \dfrac{2}{3}, \dfrac{2}{3}\right) dA = -\iint_R dA$$

$$\therefore \oint_C 3ydx - 5zdy + xdz = -\iint_R dA$$

$$\therefore \oint_C 3ydx - 5zdy + xdz \text{ 只与 } C \text{ 所围面积有关 与 } C \text{ 的形狀与位置无关}$$

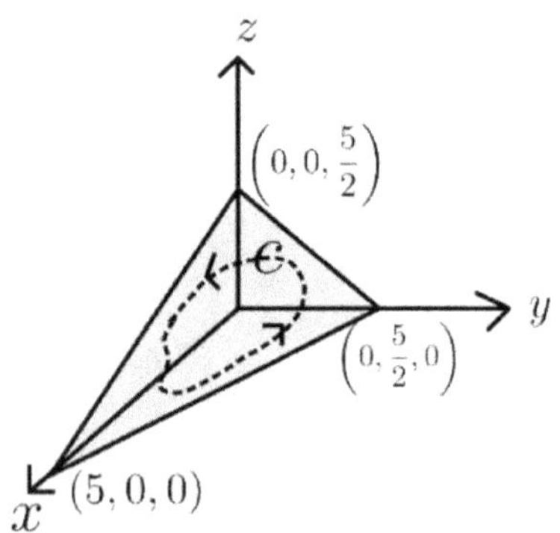


Example 3.

假设 $\vec{\mathbf{F}} = (-y^2, x, z^2)$，$C$ 为 $y + z = 2, x^2 + y^2 = 1$ 交集的逆时针封闭曲线，

藉由 Stokes' Theorem 求 $\oint_C \vec{F} \cdot d\vec{r} = ?$

【解】

令 $S = \{(x, y, z): y + z = 2, x^2 + y^2 \leq 1\}$ 为朝上定向

$\because \vec{F}$ 的各分量一阶偏导数存在且连续

藉由 Stokes' Theorem $\oint_C \vec{F} \cdot d\vec{r} = \iint_S (\nabla \times \vec{F}) \cdot \vec{n} dA$

$\because S$ 为朝上定向

令 $\vec{r}(x, y) = (x, y, 2 - y)$ 且 $R = \{(x, y): x^2 + y^2 \leq 1\}$

则 $\iint_S (\nabla \times \vec{F}) \cdot \vec{n} dA = \iint_R (\nabla \times \vec{F}) \cdot \dfrac{\partial \vec{r}}{\partial x} \times \dfrac{\partial \vec{r}}{\partial y} dxdy$

$\because \dfrac{\partial \vec{r}}{\partial x} = (1, 0, 0), \quad \dfrac{\partial \vec{r}}{\partial y} = (0, 1, -1) \quad \therefore \dfrac{\partial \vec{r}}{\partial x} \times \dfrac{\partial \vec{r}}{\partial y} = (0, 1, 1)$

$\because \nabla \times \vec{F} = \begin{vmatrix} \vec{i} & \vec{j} & \vec{k} \\ \dfrac{\partial}{\partial x} & \dfrac{\partial}{\partial y} & \dfrac{\partial}{\partial z} \\ -y^2 & x & z^2 \end{vmatrix} = (0, 0, 1 - 2y)$

$\therefore \iint_R (\nabla \times \vec{F}) \cdot \dfrac{\partial \vec{r}}{\partial x} \times \dfrac{\partial \vec{r}}{\partial y} dxdy = \iint_R (0, 0, 1 - 2y) \cdot (0, 1, 1) dxdy = \iint_{x^2 + y^2 \leq 1} 1 - 2ydxdy$

令 $x = r\cos\theta, y = r\sin\theta$ 则 $\{(x, y): x^2 + y^2 \leq 1\} = \{(r, \theta): 0 \leq r \leq 1, 0 \leq \theta \leq 2\pi\}$

且 $dxdy = \begin{Vmatrix} \begin{vmatrix} \dfrac{\partial x}{\partial r} & \dfrac{\partial x}{\partial \theta} \\ \dfrac{\partial y}{\partial r} & \dfrac{\partial y}{\partial \theta} \end{vmatrix} \end{Vmatrix} drd\theta = \begin{Vmatrix} \begin{vmatrix} \cos\theta & -r\sin\theta \\ \sin\theta & r\cos\theta \end{vmatrix} \end{Vmatrix} drd\theta = rdrd\theta$

$\therefore \iint_{x^2 + y^2 \leq 1} 1 - 2ydxdy = \int_0^{2\pi} \int_0^1 (1 - 2r\sin\theta)r \, drd\theta = \int_0^{2\pi} \dfrac{1}{2} - \dfrac{2\sin\theta}{3} d\theta = \pi$

$\therefore \oint_C \vec{F} \cdot d\vec{r} = \pi$

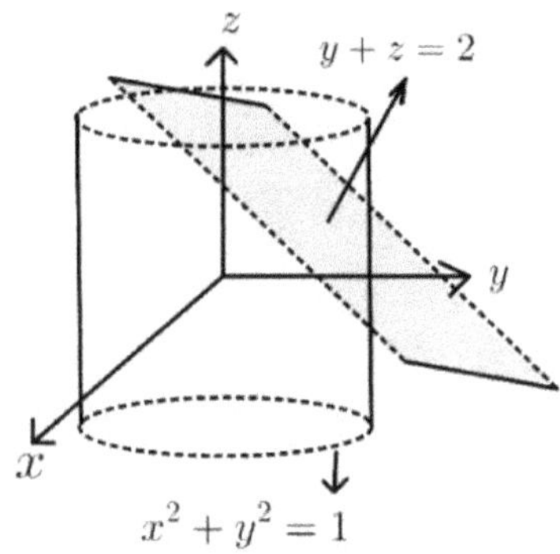


Example 4.

$$\vec{F} = \left(-z, x, \frac{y^2 z}{2}\right), \quad 假设曲线C为 z = a \ 与 \ z = \sqrt{x^2 + y^2} 的逆时针封闭交线,求$$

$$\oint_C \vec{F} \cdot d\vec{r} = ? \quad (1)直接计算 \quad (2)使用 \ \text{Stoke's Theorem}$$

【解】

(1)

令曲线 $C = \{(x, y, z): \sqrt{x^2 + y^2} = a, z = a\}$

令 $\vec{r}(\theta) = (a\cos\theta, a\sin\theta, a), \ 0 \le \theta \le 2\pi$ 则 $\vec{r}'(\theta) = (-a\sin\theta, a\cos\theta, 0)$

$$\oint_C \vec{F} \cdot d\vec{r} = \int_0^{2\pi} \vec{F}(\vec{r}(\theta)) \cdot \vec{r}'(\theta)\, d\theta = \int_0^{2\pi} \left(-a, a\cos\theta, \frac{a^3\sin^2\theta}{2}\right) \cdot (-a\sin\theta, a\cos\theta, 0)\, d\theta$$

$$= \int_0^{2\pi} a^2\sin\theta + a^2\cos^2\theta\, d\theta = \int_0^{2\pi} a^2\cos^2\theta\, d\theta = a^2 \int_0^{2\pi} \frac{1+\cos\theta}{2}\, d\theta = \pi a^2$$

(2)

$$\nabla \times \vec{F} = \begin{Vmatrix} \boldsymbol{i} & \boldsymbol{j} & \boldsymbol{k} \\ \dfrac{\partial}{\partial x} & \dfrac{\partial}{\partial y} & \dfrac{\partial}{\partial z} \\ F_1 & F_2 & F_3 \end{Vmatrix} = \begin{Vmatrix} \boldsymbol{i} & \boldsymbol{j} & \boldsymbol{k} \\ \dfrac{\partial}{\partial x} & \dfrac{\partial}{\partial y} & \dfrac{\partial}{\partial z} \\ -z & x & \dfrac{y^2 z}{2} \end{Vmatrix} = (yz, -1, 1)$$

令曲面 $S$ 为朝上定向: $z = \sqrt{x^2 + y^2}, \ z \le a$

藉由 Stokes' Theorem, $\quad \oint_C \vec{F} \cdot d\vec{r} = \iint_S (\nabla \times \vec{F}) \cdot \vec{n}\, dA$

$\because S$ 为朝上定向

令 $\vec{r}(x, y, z) = \left(x, y, (x^2 + y^2)^{\frac{1}{2}}\right)$ 且 $R$ 为 $S$ 曲面投影至 $xy$ 平面的封闭区域

则 $R = \{(x, y): x^2 + y^2 \leq a^2\}$ 且 $\displaystyle\iint_S (\nabla \times \vec{\mathrm{F}}) \cdot \vec{n}\, dA = \iint_R (\nabla \times \vec{\mathrm{F}}) \cdot \frac{\partial \vec{r}}{\partial x} \times \frac{\partial \vec{r}}{\partial y}\, dxdy$

$\because \dfrac{\partial \vec{r}}{\partial x} = \left(1, 0, x(x^2 + y^2)^{-\frac{1}{2}}\right)$ 且 $\dfrac{\partial \vec{r}}{\partial y} = \left(0, 1, y(x^2 + y^2)^{-\frac{1}{2}}\right)$

$\therefore \dfrac{\partial \vec{r}}{\partial x} \times \dfrac{\partial \vec{r}}{\partial y} = \left(-x(x^2 + y^2)^{-\frac{1}{2}}, -y(x^2 + y^2)^{-\frac{1}{2}}, 1\right)$

$\Rightarrow (\nabla \times \vec{\mathrm{F}}) \cdot \dfrac{\partial \vec{r}}{\partial x} \times \dfrac{\partial \vec{r}}{\partial y} = (yz, -1, 1) \cdot \left(-x(x^2 + y^2)^{-\frac{1}{2}}, -y(x^2 + y^2)^{-\frac{1}{2}}, 1\right)$

$= -xy + y(x^2 + y^2)^{-\frac{1}{2}} + 1, \quad \forall (x, y, z) \in S$

$\therefore \displaystyle\iint_S (\nabla \times \vec{\mathrm{F}}) \cdot \vec{n}\, dA = \iint_R (\nabla \times \vec{\mathrm{F}}) \cdot \frac{\partial \vec{r}}{\partial x} \times \frac{\partial \vec{r}}{\partial y}\, dxdy$

$= \displaystyle\iint_{x^2+y^2 \leq a^2} -xy + y(x^2 + y^2)^{-\frac{1}{2}} + 1\, dxdy$

令 $x = r\cos\theta, y = r\sin\theta$ 则 $\{(x, y): x^2 + y^2 \leq a^2\} = \{(r, \theta): 0 \leq r \leq a, 0 \leq \theta \leq 2\pi\}$

$\therefore \displaystyle\iint_{x^2+y^2 \leq a^2} -xy + y(x^2 + y^2)^{-\frac{1}{2}} + 1\, dxdy = \int_0^{2\pi} \int_0^a (-r^2 \cos\theta \sin\theta + \sin\theta + 1)\, rdrd\theta$

$= \pi a^2$

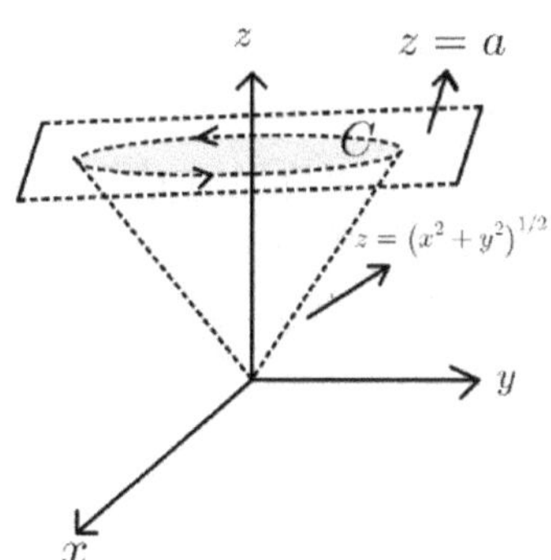


Example 5.

$$\vec{\mathrm{F}} = \left(0, \frac{2x^3}{3}, 2y^3\right), \quad \text{曲线 } C \text{ 为 } (1,0,0), (0,1,0), (0,0,1) \text{ 所围逆时针封闭曲线, 求} \oint_C \vec{\mathrm{F}} \cdot d\vec{r}$$

(1)直接计算 (2)使用 Stoke's Theorem

【解】

(1)

令 $C_1: (-t+1, t, 0)$, $C_2: (0, -t+1, t)$, $C_3: (t, 0, -t+1)$ 则 $C = C_1 \cup C_2 \cup C_3$

$$\therefore \oint_C \vec{F} \cdot d\vec{r} = \int_{C_1} \vec{F} \cdot d\vec{r} + \int_{C_2} \vec{F} \cdot d\vec{r} + \int_{C_3} \vec{F} \cdot d\vec{r}$$

$$\therefore \int_{C_1} \vec{F} \cdot d\vec{r} = \int_0^1 \frac{2(1-t)^3}{3} dt = -\int_1^0 \frac{2u^3}{3} du = \frac{1}{6}$$

$$\int_{C_2} \vec{F} \cdot d\vec{r} = \int_0^1 2(1-t)^3 dt = -\int_1^0 2u^3 du = \frac{1}{2} \quad \text{且} \quad \int_{C_3} \vec{F} \cdot d\vec{r} = 0$$

$$\therefore \oint_C \vec{F} \cdot d\vec{r} = \frac{2}{3}$$

(2)

$$\nabla \times \vec{F} = \begin{Vmatrix} \boldsymbol{i} & \boldsymbol{j} & \boldsymbol{k} \\ \dfrac{\partial}{\partial x} & \dfrac{\partial}{\partial y} & \dfrac{\partial}{\partial z} \\ F_1 & F_2 & F_3 \end{Vmatrix} = \begin{Vmatrix} \boldsymbol{i} & \boldsymbol{j} & \boldsymbol{k} \\ \dfrac{\partial}{\partial x} & \dfrac{\partial}{\partial y} & \dfrac{\partial}{\partial z} \\ 0 & \dfrac{2x^3}{3} & 2y^3 \end{Vmatrix} = (6y^2, 0, 2x^2)$$

令 $S$ 为 $(1,0,0), (0,1,0), (0,0,1)$ 在第一象限所围朝上定向平面

藉由 Stokes' Theorem, $\oint_C \vec{F} \cdot d\vec{r} = \iint_S (\nabla \times \vec{F}) \cdot \vec{n} dA$

$\because S$ 为朝上定向平面

令 $\vec{r}(x, y, z) = (x, y, 1-x-y)$ 且 $R$ 为 $S$ 曲面投影至 $xy$ 平面的封闭区域

则 $R = \{(x, y): x + y \leq 1, x \geq 0, y \geq 0\}$ 且 $\iint_S (\nabla \times \vec{F}) \cdot \vec{n} \, dA = \iint_R (\nabla \times \vec{F}) \cdot \dfrac{\partial \vec{r}}{\partial x} \times \dfrac{\partial \vec{r}}{\partial y} dxdy$

$\because \dfrac{\partial \vec{r}}{\partial x} = (1, 0, -1)$ 且 $\dfrac{\partial \vec{r}}{\partial y} = (0, 1, -1)$ $\quad \therefore \dfrac{\partial \vec{r}}{\partial x} \times \dfrac{\partial \vec{r}}{\partial y} = (1, 1, 1)$

$\because (\nabla \times \vec{F}) \cdot \dfrac{\partial \vec{r}}{\partial x} \times \dfrac{\partial \vec{r}}{\partial y} = (6y^2, 0, 2x^2) \cdot \dfrac{\partial \vec{r}}{\partial x} \times \dfrac{\partial \vec{r}}{\partial y} = 2(x^2 + 3y^2), \quad \forall (x, y, z) \in S$

$\therefore \iint_S (\nabla \times \vec{F}) \cdot \vec{n} \, dA = \iint_R (\nabla \times \vec{F}) \cdot \dfrac{\partial \vec{r}}{\partial x} \times \dfrac{\partial \vec{r}}{\partial y} dxdy = 2\iint_R x^2 + 3y^2 \, dxdy$

$$= 2\int_0^1 \int_0^{1-y} x^2 + 3y^2 \, dxdy = 2\int_0^1 \left( \frac{x^3}{3} + 3y^2 x \right)\Big|_{x=0}^{x=1-y} dy = 2\int_0^1 \frac{(1-y)^3}{3} + 3y^2(1-y) \, dy$$

$$= \frac{2}{3}$$

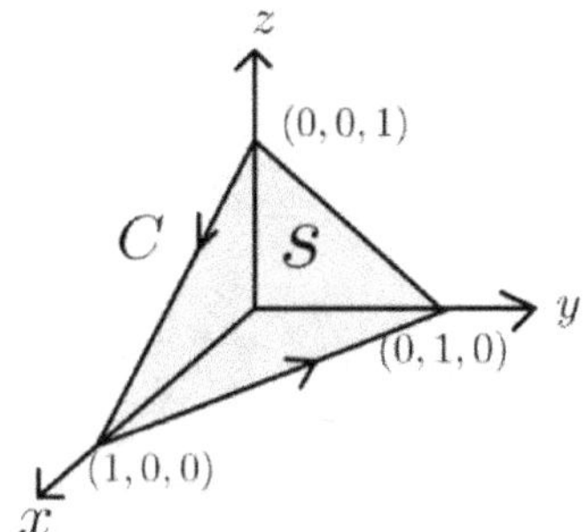

Example 6.

$$\vec{\mathrm{F}} = (yz, 0, 0), \quad C\text{为逆时针封闭圆}: x^2 + z^2 = 1, y = 1, \quad \text{求} \oint_C \vec{\mathrm{F}} \cdot d\vec{r} = ? \text{ (1)直接计算}$$

(2)使用 Stoke's Theorem

【解】

(1)

令$C = \{(x, y, z): x^2 + z^2 = 1, y = 1\}$，令 $\vec{r}(\theta) = (\cos\theta, 1, \sin\theta)$ 则 $\vec{r}'(\theta) = (-\sin\theta, 0, \cos\theta)$

$$\oint_C \vec{\mathrm{F}} \cdot d\vec{r} = \int_0^{2\pi} \vec{\mathrm{F}}(\vec{r}(\theta)) \cdot \vec{r}'(\theta)\, d\theta = \int_0^{2\pi} (\sin\theta, 0, 0) \cdot (-\sin\theta, 0, \cos\theta)\, d\theta$$

$$= -\int_0^{2\pi} \sin^2\theta\, d\theta = \int_0^{2\pi} \frac{1 - \cos\theta}{2}\, d\theta = -\pi$$

(2)

$$\nabla \times \vec{\mathrm{F}} = \begin{Vmatrix} \boldsymbol{i} & \boldsymbol{j} & \boldsymbol{k} \\ \dfrac{\partial}{\partial x} & \dfrac{\partial}{\partial y} & \dfrac{\partial}{\partial z} \\ F_1 & F_2 & F_3 \end{Vmatrix} = \begin{Vmatrix} \boldsymbol{i} & \boldsymbol{j} & \boldsymbol{k} \\ \dfrac{\partial}{\partial x} & \dfrac{\partial}{\partial y} & \dfrac{\partial}{\partial z} \\ yz & 0 & 0 \end{Vmatrix} = (0, y, -z)$$

令$S = \{(x, y, z): y = x^2 + z^2, 0 \le y \le 1\}$为朝$y$轴正向方向作定向

藉由 Stokes' Theorem, $\displaystyle \oint_C \vec{\mathrm{F}} \cdot d\vec{r} = \iint_S (\nabla \times \vec{\mathrm{F}}) \cdot \vec{n}\, dA$

令 $\vec{r}(x, y, z) = (x, x^2 + z^2, z)$ 且 $R$为$S$曲面投影至$xz$平面的封闭区域

则$R = \{(x, y): x^2 + z^2 \le 1\}$ 且 $\displaystyle \iint_S (\nabla \times \vec{\mathrm{F}}) \cdot \vec{n}\, dA = \iint_R (\nabla \times \vec{\mathrm{F}}) \cdot \frac{\partial \vec{r}}{\partial x} \times \frac{\partial \vec{r}}{\partial z}\, dx dz$

$$\because \frac{\partial \vec{r}}{\partial x} = (1, 2x, 0) \text{ 且 } \frac{\partial \vec{r}}{\partial z} = (0, 2z, 1) \quad \therefore \frac{\partial \vec{r}}{\partial x} \times \frac{\partial \vec{r}}{\partial z} = (2x, -1, 2z)$$

$$\therefore \left(\nabla \times \vec{F}\right) \cdot \frac{\partial \vec{r}}{\partial x} \times \frac{\partial \vec{r}}{\partial z} = (0, y, -z) \cdot (2x, -1, 2z) = -y - 2z^2 = -x^2 - 3z^2, \quad \forall (x, y, z) \in S$$

$$\therefore \iint_S \left(\nabla \times \vec{F}\right) \cdot \vec{n} \, dA = \iint_R \left(\nabla \times \vec{F}\right) \cdot \frac{\partial \vec{r}}{\partial x} \times \frac{\partial \vec{r}}{\partial z} \, dx \, dz = \iint_{x^2 + z^2 \leq 1} -x^2 - 3z^2 \, dx \, dz$$

令 $x = r \cos\theta, z = r \sin\theta$ 则 $\{(x, y): x^2 + z^2 \leq 1\} = \{(r, \theta): 0 \leq r \leq 1, 0 \leq \theta \leq 2\pi\}$

$$且 \; dx\,dz = \left\| \begin{vmatrix} \dfrac{\partial x}{\partial r} & \dfrac{\partial x}{\partial \theta} \\ \dfrac{\partial z}{\partial r} & \dfrac{\partial z}{\partial \theta} \end{vmatrix} \right\| dr\,d\theta = \begin{vmatrix} \cos\theta & -r\sin\theta \\ \sin\theta & r\cos\theta \end{vmatrix} dr\,d\theta = r\,dr\,d\theta$$

$$\therefore \iint_{x^2 + z^2 \leq 1} -x^2 - 3z^2 \, dx\,dz = \int_0^{2\pi} \int_0^1 -r^3 \cos^2\theta - 3r^3 \sin^2\theta \, dr\,d\theta$$

$$= \int_0^{2\pi} -\frac{1}{4}\left(\frac{1 + \cos\theta}{2}\right) - \frac{3}{4}\left(\frac{1 - \cos\theta}{2}\right) d\theta = -\pi$$

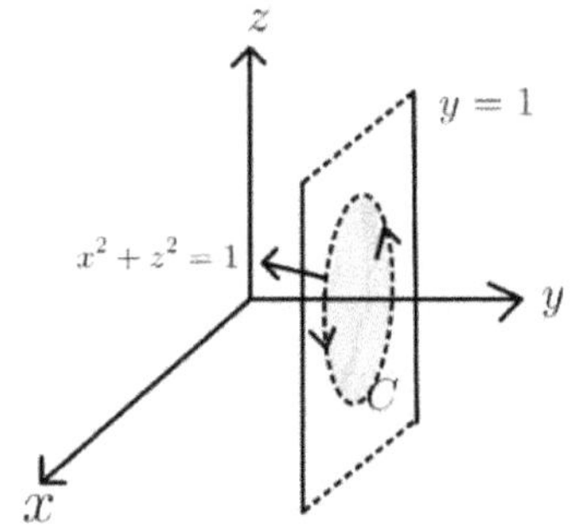


Example 7.

假设 $C$ 是 $x + y + z = 2$ 平面上任意正向封闭曲线，试证 $\oint_C z\,dx - 2x\,dy + 4y\,dz$

只与 $C$ 所围面积有关，与 $C$ 的形狀与位置无关

【解】

令 $C$ 在平面 $x + y + z = 2$ 所围封闭区域为 R

令 $\vec{F} = (z, -2x, 4y)$，$\because \vec{F}$ 的各分量一阶偏导数存在且连续

藉由 Stokes' Theorem, $\oint_C z\,dx - 2x\,dy + 4y\,dz = \oint_C \vec{F} \cdot d\vec{r} = \iint_R \left(\nabla \times \vec{F}\right) \cdot \vec{n}\,dA$

令 $\vec{n}$ 为 $x + y + z = 2$ 单位法向量 则 $\vec{n} = \left(\dfrac{1}{\sqrt{3}}, \dfrac{1}{\sqrt{3}}, \dfrac{1}{\sqrt{3}}\right)$

$$\because \nabla \times \vec{F} = \begin{vmatrix} \vec{\imath} & \vec{\jmath} & \vec{k} \\ \dfrac{\partial}{\partial x} & \dfrac{\partial}{\partial y} & \dfrac{\partial}{\partial z} \\ z & -2x & 4y \end{vmatrix} = (4,1,-2)$$

$$\therefore \iint_R (\nabla \times \vec{F}) \cdot \vec{n}\, dA = \iint_R (4,1,-2) \cdot \left(\frac{1}{\sqrt{3}}, \frac{1}{\sqrt{3}}, \frac{1}{\sqrt{3}}\right) dA = \frac{3}{\sqrt{3}} \iint_R dA$$

$$\therefore \oint_C z\,dx - 2x\,dy + 4y\,dz = \sqrt{3} \iint_R dA$$

$$\therefore \oint_C z\,dx - 2x\,dy + 4y\,dz \text{ 只与} C \text{所围面积有关，与} C \text{的形狀与位置无关}$$

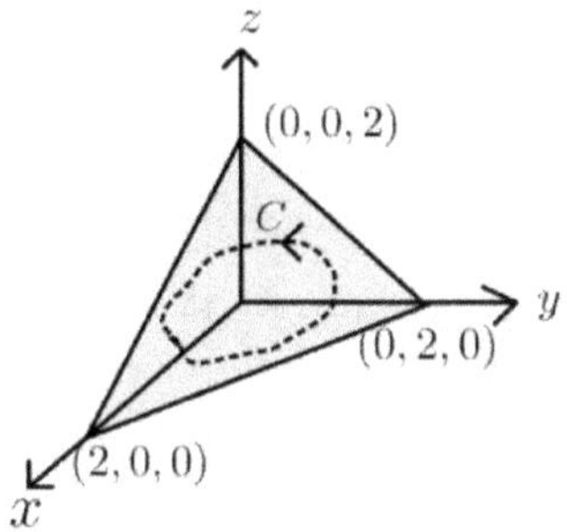


Example 8.

藉由 Stokes' Theorem 求 $\oint_C \vec{F} \cdot d\vec{r} = ?$, $\vec{F} = (z^2, y^2, x)$, $C$ 是 $(a,0,0), (0,a,0),$ $(0,0,a)$ 所围成的逆时针封闭曲线

【解】

令 $S = \{(x,y,z): x,y,z \geq 0, x+y+z = a\}$ 为朝上定向平面

$\because \vec{F}$ 的各分量一阶偏导数存在且连续

藉由 Stokes' Theorem, $\oint_C z^2\,dx + y^2\,dy + x\,dz = \oint_C \vec{F} \cdot d\vec{r} = \iint_S (\nabla \times \vec{F}) \cdot \vec{n}\, dA$

$\because S$ 为朝上定向

令 $\vec{r}(x,y) = (x, y, a-x-y)$ 且 $R = \{(x,y): 0 \leq x+y \leq a\}$

则 $\displaystyle\iint_S (\nabla \times \vec{F}) \cdot \vec{n}\, dA = \iint_R (\nabla \times \vec{F}) \cdot \frac{\partial \vec{r}}{\partial x} \times \frac{\partial \vec{r}}{\partial y}\, dx\, dy$

$\because \dfrac{\partial \vec{r}}{\partial x} = (1,0,-1)$, $\dfrac{\partial \vec{r}}{\partial y} = (0,1,-1)$ $\quad \therefore \dfrac{\partial \vec{r}}{\partial x} \times \dfrac{\partial \vec{r}}{\partial y} = (1,1,1)$

$$\because \nabla\times\vec{\mathbf{F}} = \begin{vmatrix} \vec{i} & \vec{j} & \vec{k} \\ \frac{\partial}{\partial x} & \frac{\partial}{\partial y} & \frac{\partial}{\partial z} \\ z^2 & y^2 & x \end{vmatrix} = (0, 2z-1, 0) = (0, 2(a-x-y)-1, 0), \quad \forall (x,y,z)\in S$$

$$\therefore \iint_R (\nabla\times\vec{\mathbf{F}})\cdot\frac{\partial\vec{r}}{\partial x}\times\frac{\partial\vec{r}}{\partial y}\,dxdy = \iint_{0\leq x+y\leq a} 2(a-x-y)-1\,dxdy$$

藉由 Fubini's Theorem

$$\iint_{0\leq x+y\leq a} 2(a-x-y)-1\,dxdy = \int_0^a\int_0^{a-y}(2a-1)-2x-2y\,dxdy$$

$$= \int_0^a a(a-1)-y(2a-1)+y^2\,dy = \frac{a^3}{3}-\frac{a^2}{2}$$

Example 9.

$$\vec{\mathbf{F}} = \left(x^2z, 0, -\frac{y^3}{3}\right), C\text{为 } (a,0,0), (0,a,0), (0,0,a)\text{所围逆时针封闭曲线}(a>0),$$

$$\text{求}\oint_C \vec{F}\cdot d\vec{r} =? \text{ (1)直接计算 (2)使用 Stoke's Theorem}$$

【解】

(1)

令 $C_1: (-at+a, at, 0)$, $C_2: (0, -at+a, at)$, $C_3: (at, 0, -at+a)$ 则 $C = C_1\cup C_2\cup C_3$

$$\therefore \oint_C \vec{F}\cdot d\vec{r} = \int_{C_1}\vec{F}\cdot d\vec{r} + \int_{C_2}\vec{F}\cdot d\vec{r} + \int_{C_3}\vec{F}\cdot d\vec{r}$$

$$\because \int_{C_1}\vec{F}\cdot d\vec{r} = -\int_0^1 \frac{(at)^3}{3}\,dt = 0$$

$$\int_{C_2}\vec{F}\cdot d\vec{r} = -a^3\int_0^1\frac{(1-t)^3}{3}\,dt = -\frac{a^3}{12} \quad \text{且} \quad \int_{C_3}\vec{F}\cdot d\vec{r} = a^3\int_0^1 t^2(1-t)\,dt = \frac{a^3}{12}$$

$$\therefore \oint_C \vec{F}\cdot d\vec{r} = \int_{C_1}\vec{F}\cdot d\vec{r} + \int_{C_2}\vec{F}\cdot d\vec{r} + \int_{C_3}\vec{F}\cdot d\vec{r} = 0$$

(2)

$$\nabla\times\vec{\mathbf{F}} = \begin{Vmatrix} i & j & k \\ \frac{\partial}{\partial x} & \frac{\partial}{\partial y} & \frac{\partial}{\partial z} \\ F_1 & F_2 & F_3 \end{Vmatrix} = \begin{Vmatrix} i & j & k \\ \frac{\partial}{\partial x} & \frac{\partial}{\partial y} & \frac{\partial}{\partial z} \\ x^2z & 0 & -\frac{y^3}{3} \end{Vmatrix} = (-y^2, x^2, 0)$$

令$S$为$(a,0,0),(0,a,0),(0,0,a)$在第一象限所围朝上定向平面

藉由 Stokes' Theorem，$\displaystyle\oint_C \vec{\mathbf{F}} \cdot d\vec{r} = \iint_S (\nabla \times \vec{\mathbf{F}}) \cdot \vec{n}\,dA$

$\because S$为朝上定向平面

令$\vec{r}(x,y,z) = (x,y,a-x-y)$且$R = \{(x,y): 0 \leq x \leq a, 0 \leq y \leq a-x\}$

则$\displaystyle\iint_S (\nabla \times \vec{\mathbf{F}}) \cdot \vec{n}\,dA = \iint_R (\nabla \times \vec{\mathbf{F}}) \cdot \frac{\partial \vec{r}}{\partial x} \times \frac{\partial \vec{r}}{\partial y}\,dxdy$

$\because \dfrac{\partial \vec{r}}{\partial x} = (1,0,-1)$且$\dfrac{\partial \vec{r}}{\partial y} = (0,1,-1)$   $\therefore \dfrac{\partial \vec{r}}{\partial x} \times \dfrac{\partial \vec{r}}{\partial y} = (1,1,1)$

$\because (\nabla \times \vec{\mathbf{F}}) \cdot \dfrac{\partial \vec{r}}{\partial x} \times \dfrac{\partial \vec{r}}{\partial y} = (-y^2, x^2, 0) \cdot (1,1,1) = x^2 - y^2, \quad \forall (x,y,z) \in S$

$\therefore \displaystyle\iint_R (\nabla \times \vec{\mathbf{F}}) \cdot \frac{\partial \vec{r}}{\partial x} \times \frac{\partial \vec{r}}{\partial y}\,dxdy = \int_0^a \int_0^{a-x} x^2 - y^2\,dydx = \int_0^a x^2(a-x) - \frac{(a-x)^3}{3}\,dx$

$\displaystyle = \int_0^a -\frac{a^3}{3} + a^2 x - \frac{2x^3}{3}\,dx = a^4 \left( -\frac{1}{3} + \frac{1}{2} - \frac{1}{6} \right) = 0$

Example 10.

$$\vec{\mathbf{F}} = \left( 0, -\frac{x^2 y}{2}, -\frac{x^2 z}{2} \right), \quad S: z = xy, 0 \leq x \leq 1, 0 \leq y \leq 2 \text{ 且 } C \text{ 为曲面} S \text{ 的正向边界}$$

曲线，使用 Stoke's Theorem 求 $\displaystyle\oint_C \vec{\mathbf{F}} \cdot d\vec{r} =?$

【解】

$$\nabla \times \vec{\mathbf{F}} = \begin{Vmatrix} \vec{i} & \vec{j} & \vec{k} \\ \dfrac{\partial}{\partial x} & \dfrac{\partial}{\partial y} & \dfrac{\partial}{\partial z} \\ F_1 & F_2 & F_3 \end{Vmatrix} = \begin{Vmatrix} \vec{i} & \vec{j} & \vec{k} \\ \dfrac{\partial}{\partial x} & \dfrac{\partial}{\partial y} & \dfrac{\partial}{\partial z} \\ 0 & -\dfrac{x^2 y}{2} & -\dfrac{x^2 z}{2} \end{Vmatrix} = (0, xz, -xy)$$

藉由 Stokes' Theorem，$\displaystyle\oint_C \vec{\mathbf{F}} \cdot d\vec{r} = \iint_S (\nabla \times \vec{\mathbf{F}}) \cdot \vec{n}\,dA$

令$\vec{r}(x,y,z) = (x,y,xy)$且$R = \{(x,y): 0 \leq x \leq 1, 0 \leq y \leq 2\}$

则 $\displaystyle\iint_S (\nabla \times \vec{F}) \cdot \vec{n}\, dA = \iint_R (\nabla \times \vec{F}) \cdot \frac{\partial \vec{r}}{\partial x} \times \frac{\partial \vec{r}}{\partial y} dxdy$

$\because \dfrac{\partial \vec{r}}{\partial x} = (1,0,y)$ 且 $\dfrac{\partial \vec{r}}{\partial y} = (0,1,x)$  $\therefore \dfrac{\partial \vec{r}}{\partial x} \times \dfrac{\partial \vec{r}}{\partial y} = (-y,-x,1)$

$\because (\nabla \times \vec{F}) \cdot \dfrac{\partial \vec{r}}{\partial x} \times \dfrac{\partial \vec{r}}{\partial y} = (0,xz,-xy) \cdot (-y,-x,1) = -x^2 z - xy = -x^3 y - xy, \quad \forall (x,y,z) \in S$

$\displaystyle\iint_R (\nabla \times \vec{F}) \cdot \frac{\partial \vec{r}}{\partial x} \times \frac{\partial \vec{r}}{\partial y} dxdy = \int_0^1 \int_0^2 -x^3 y - xy\, dydx = \int_0^a -2x^3 - 2x\, dx = \left( -\frac{x^4}{2} - x^2 \right)\Big|_0^1$

$= -\dfrac{3}{2}$

Example 11.

藉由 Stokes' Theorem 求 $\displaystyle\oint_C (e^x y - y^3)dx + (e^x + x^3)\, dy =?$，其中 $C$ 半径为 6 的

逆时针封闭圆：$x = 6\cos\theta, y = 6\sin\theta, 0 \le \theta \le 2\pi$

【解】

令 $S = \{(x,y,z): x^2 + y^2 \le 36, z = 0\}$ 为朝上定向

令 $\vec{F} = (e^x y - y^3, e^x + x^3, 0)$，$\because \vec{F}$ 的各分量一阶偏导数存在且连续

藉由 Stokes' Theorem，$\displaystyle\oint_C \vec{F} \cdot d\vec{r} = \iint_S (\nabla \times \vec{F}) \cdot \vec{n} dA$

$\because S$ 为朝上定向

令 $\vec{r}(x,y) = (x,y,0)$ 且 $R = \{(x,y): x^2 + y^2 \le 36\}$

则 $\displaystyle\iint_S (\nabla \times \vec{F}) \cdot \vec{n} dA = \iint_R (\nabla \times \vec{F}) \cdot \frac{\partial \vec{r}}{\partial x} \times \frac{\partial \vec{r}}{\partial y} dxdy$

$\because \dfrac{\partial \vec{r}}{\partial x} = (1,0,0), \dfrac{\partial \vec{r}}{\partial y} = (0,1,0)$  $\therefore \dfrac{\partial \vec{r}}{\partial x} \times \dfrac{\partial \vec{r}}{\partial y} = (0,0,1)$

$$\because \nabla \times \vec{F} = \begin{vmatrix} \vec{i} & \vec{j} & \vec{k} \\ \dfrac{\partial}{\partial x} & \dfrac{\partial}{\partial y} & \dfrac{\partial}{\partial z} \\ e^x y - y^3 & e^x + x^3 & 0 \end{vmatrix} = (0,0,3x^2 + 3y^2)$$

$$\therefore \iint_R (\nabla \times \vec{F}) \cdot \frac{\partial \vec{r}}{\partial x} \times \frac{\partial \vec{r}}{\partial y} dxdy = \iint_R (0,0,3x^2 + 3y^2) \cdot (0,0,1) dxdy$$

$$= 3 \iint_{x^2+y^2 \leq 36} x^2 + y^2 \, dxdy$$

令 $x = r\cos\theta$，$y = r\sin\theta$ 则 $\{(x,y): x^2 + y^2 \leq 36\} = \{(r,\theta): 0 \leq r \leq 6, 0 \leq \theta \leq 2\pi\}$

$$\text{且 } dxdy = \begin{Vmatrix} \dfrac{\partial x}{\partial r} & \dfrac{\partial x}{\partial \theta} \\ \dfrac{\partial y}{\partial r} & \dfrac{\partial y}{\partial \theta} \end{Vmatrix} drd\theta = \begin{Vmatrix} \cos\theta & -r\sin\theta \\ \sin\theta & r\cos\theta \end{Vmatrix} drd\theta = r\,drd\theta$$

$$\therefore 3 \iint_{x^2+y^2 \leq 36} x^2 + y^2 \, dxdy = 3 \int_0^{2\pi} \int_0^6 r^3 \, drd\theta = 1944\pi$$

Example 12.

藉由 Stokes' Theorem 求 $\oint_C y^3 dx - x^3 \, dy =?$，$C$ 半径为 2 的逆时针封闭圆：

$$x^2 + y^2 = 4$$

【解】

令 $S = \{(x,y,z): x^2 + y^2 \leq 4, z = 0\}$ 为朝上定向

令 $\vec{F} = (y^3, -x^3, 0)$，$\because \vec{F}$ 的各分量一阶偏导数存在且连续

藉由 Stokes' Theorem，$\oint_C \vec{F} \cdot d\vec{r} = \iint_S (\nabla \times \vec{F}) \cdot \vec{n} dA$

$\because S$ 为朝上定向

令 $\vec{r}(x,y) = (x,y,0)$ 且 $R = \{(x,y): x^2 + y^2 \leq 4\}$

则 $\iint_S (\nabla \times \vec{F}) \cdot \vec{n} dA = \iint_R (\nabla \times \vec{F}) \cdot \dfrac{\partial \vec{r}}{\partial x} \times \dfrac{\partial \vec{r}}{\partial y} dxdy$

$\because \dfrac{\partial \vec{r}}{\partial x} = (1,0,0)$，$\dfrac{\partial \vec{r}}{\partial y} = (0,1,0)$ $\therefore \dfrac{\partial \vec{r}}{\partial x} \times \dfrac{\partial \vec{r}}{\partial y} = (0,0,1)$

$$\because \nabla\times\vec{F} = \begin{vmatrix} \vec{\imath} & \vec{\jmath} & \vec{k} \\ \dfrac{\partial}{\partial x} & \dfrac{\partial}{\partial y} & \dfrac{\partial}{\partial z} \\ y^3 & -x^3 & 0 \end{vmatrix} = (0,0,-3x^2-3y^2)$$

$$\therefore \iint_R (\nabla\times\vec{F})\cdot\frac{\partial\vec{r}}{\partial x}\times\frac{\partial\vec{r}}{\partial y}dxdy = \iint_R (0,0,-3x^2-3y^2)\cdot(0,0,1)dxdy$$

$$= -3\iint_{x^2+y^2\leq 4} x^2+y^2 dxdy$$

令 $x = r\cos\theta, y = r\sin\theta$ 则 $\{(x,y):x^2+y^2\leq 4\} = \{(r,\theta):0\leq r\leq 2, 0\leq\theta\leq 2\pi\}$

$$且\ dxdy = \left\|\begin{vmatrix} \dfrac{\partial x}{\partial r} & \dfrac{\partial x}{\partial\theta} \\ \dfrac{\partial y}{\partial r} & \dfrac{\partial y}{\partial\theta} \end{vmatrix}\right\| drd\theta = \left\|\begin{vmatrix} \cos\theta & -r\sin\theta \\ \sin\theta & r\cos\theta \end{vmatrix}\right\| drd\theta = rdrd\theta$$

$$\therefore -3\iint_{x^2+y^2\leq 4} x^2+y^2 dxdy = -3\int_0^{2\pi}\int_0^2 r^3\,drd\theta = -24\pi$$

### 9.6.3    $\nabla\times\vec{F}\neq 0$,把非封闭曲面积分转成封闭线积分

同理,也可使用 Stokes' Theorem 把非封闭曲面积分的问题转成封闭曲线积分的问题

考试类型:

Type 1.

使用 Stokes' Theorem 求 $\iint_S (\nabla\times\vec{F})\cdot\vec{n}dA =?$,其中 $S$ 为朝上定向非封闭的光滑曲面,

$S$ 的边界 $= \{(x,y,z):x^2+y^2=a^2, z=c\}$,$\vec{F}$ 各分量的一阶偏导数存在且连续

解题流程:

Step1.

令 $C = \{(x,y,z):x^2+y^2=a^2, z=c\}$

Step2.

令 $x = a\cos\theta$, $y = a\sin\theta$ 则 $\vec{r}(\theta) = (a\cos\theta, a\sin\theta, c)$ $\therefore \vec{r}'(\theta) = (-a\sin\theta, a\cos\theta, 0)$

Step3.

藉由 Stokes' Theorem

$$\iint_S (\nabla \times \vec{\mathbf{F}}) \cdot \vec{n} \, dA = \oint_C \vec{\mathbf{F}} \cdot d\vec{r} = \int_0^{2\pi} \vec{\mathbf{F}}(\vec{r}(\theta)) \cdot \vec{r}'(\theta) d\theta$$

$$= \int_0^{2\pi} \big(f_1(a\cos\theta, a\sin\theta, c), f_2(a\cos\theta, a\sin\theta, c)\big) \cdot (-a\sin\theta, a\cos\theta) d\theta$$

其中 $\vec{\mathbf{F}} = \big(f_1(x,y,z), f_2(x,y,z), f_3(x,y,z)\big)$

Step4.

求 $\displaystyle\int_0^{2\pi} \big(f_1(a\cos\theta, a\sin\theta, c), f_2(a\cos\theta, a\sin\theta, c)\big) \cdot (-a\sin\theta, a\cos\theta) d\theta =?$

如果 $S$ 为朝下定向非封闭的光滑曲面则求

$$-\int_0^{2\pi} \big(f_1(a\cos\theta, a\sin\theta, c), f_2(a\cos\theta, a\sin\theta, c)\big) \cdot (-a\sin\theta, a\cos\theta) d\theta$$

<u>范例说明:</u>

(I) $\vec{\mathbf{F}} = (-y^3\cos z, x^3 e^z, -e^z)$ 且 $S$ 为朝上定向上半球球面: $x^2 + y^2 + (z-a)^2 = 2a^2, z > 0$

求 $\displaystyle\iint_S (\nabla \times \vec{\mathbf{F}}) \cdot \vec{n} \, dA =?$

令 $C = \{(x,y,z): x^2 + y^2 = a^2, z = 0\}$

令 $x = a\cos\theta$, $y = a\sin\theta$ 则 $\vec{r}(\theta) = (a\cos\theta, a\sin\theta, 0)$ $\therefore \vec{r}'(\theta) = (-a\sin\theta, a\cos\theta, 0)$

藉由 Stokes' Theorem

$$\iint_S (\nabla \times \vec{\mathbf{F}}) \cdot \vec{n} \, dA = \oint_C \vec{\mathbf{F}} \cdot d\vec{r} = \int_0^{2\pi} \vec{\mathbf{F}}(\vec{r}(\theta)) \cdot \vec{r}'(\theta) d\theta$$

$$= \int_0^{2\pi} (-(a\sin\theta)^3, (a\cos\theta)^3, 1) \cdot (-a\sin\theta, a\cos\theta, 0) d\theta = \frac{3a^4\pi}{2}$$

(II) $\vec{\mathbf{F}} = (2x - y, 2y + z, xyz)$ 且 $S$ 为朝上定向上半球球面: $\dfrac{x^2}{a^2} + \dfrac{y^2}{b^2} + \dfrac{z^2}{c^2} = 1$, $z > 0$,

求 $\displaystyle\iint_S (\nabla \times \vec{\mathbf{F}}) \cdot \vec{n} \, dA =?$

令 $C = \{(x,y,z): \dfrac{x^2}{a^2} + \dfrac{y^2}{b^2} = 1, z = 0\}$

令 $x = a\cos\theta$, $y = b\sin\theta$ 则 $\vec{r}(\theta) = (a\cos\theta, b\sin\theta, 0)$ $\therefore \vec{r}'(\theta) = (-a\sin\theta, b\cos\theta, 0)$

藉由 Stokes' Theorem

$$\iint_S (\nabla \times \vec{\mathbf{F}}) \cdot \vec{n} \, dA = \oint_C \vec{\mathbf{F}} \cdot d\vec{r} = \int_0^{2\pi} \vec{\mathbf{F}}(\vec{r}(\theta)) \cdot \vec{r}'(\theta) d\theta$$

$$= \int_0^{2\pi} (2a\cos\theta - b\sin\theta, 2b\sin\theta, 0) \cdot (-a\sin\theta, b\cos\theta, 0) d\theta = ab\pi$$

Type 2.

使用 Stokes' Theorem 求 $\iint_S (\nabla \times \vec{\mathbf{F}}) \cdot \vec{n}\, dA =?$，$S$ 为朝上定向非封闭的光滑曲面，$S$ 的封闭边界 $C$ 为分段光滑曲线,即 $C = \displaystyle\bigcup_{i=1}^{n} C_i$ 且 $C_i : \vec{r}_i(t)$，$\vec{\mathbf{F}}$ 各分量的一阶偏导数存在且连续

解题流程:

Step1.

藉由 Stokes' Theorem,$\quad \displaystyle\iint_S (\nabla \times \vec{\mathbf{F}}) \cdot \vec{n}\, dA = \oint_C \vec{\mathbf{F}} \cdot d\vec{r} = \sum_{i=1}^{n} \int_{C_i} \vec{\mathbf{F}}\left(\vec{r}_i(t)\right) \cdot \vec{r}_i'(t)\, dt$

Step2.

求 $\displaystyle\sum_{i=1}^{n} \int_{C_i} \vec{\mathbf{F}}\left(\vec{r}_i(t)\right) \cdot \vec{r}_i'(t)\, dt$

Example 1.

求 $\displaystyle\iint_S (\nabla \times \vec{\mathbf{F}}) \cdot \vec{n}\, dA =?$，$S = \{(x,y,z) : x^2 + y^2 \leq 1, z = 0\}$ 且 $S$ 为朝上定向,

$\vec{\mathbf{F}} = (y^2, x^2, 0)$　(1)直接计算　(2)使用 Stoke's Theorem

【解】

(1)

$\because \nabla \times \vec{\mathbf{F}} = \begin{vmatrix} \vec{i} & \vec{j} & \vec{k} \\ \dfrac{\partial}{\partial x} & \dfrac{\partial}{\partial y} & \dfrac{\partial}{\partial z} \\ y^2 & x^2 & 0 \end{vmatrix} = (0, 0, 2x - 2y)$

$\because S$ 为朝上定向

令 $\vec{r}(x,y) = (x, y, 0)$，令 $R$ 为曲面 $S$ 投影至 $xy$ 平面的封闭区域

则 $R = \{(x,y) : x^2 + y^2 \leq 1\}$ 且 $\displaystyle\iint_S (\nabla \times \vec{\mathbf{F}}) \cdot \vec{n}\, dA = \iint_R (\nabla \times \vec{\mathbf{F}}) \cdot \dfrac{\partial \vec{r}}{\partial x} \times \dfrac{\partial \vec{r}}{\partial y}\, dx\, dy$

$$\because \frac{\partial \vec{r}}{\partial x} = (1,0,0), \quad \frac{\partial \vec{r}}{\partial y} = (0,1,0) \quad \therefore \frac{\partial \vec{r}}{\partial x} \times \frac{\partial \vec{r}}{\partial y} = (0,0,1)$$

$$\therefore \iint_S (\nabla \times \vec{\mathbf{F}}) \cdot \vec{n}\, dA = 2 \iint_{x^2+y^2 \leq 1} x - y\, dxdy$$

令 $x = r\cos\theta$, $y = r\sin\theta$ 则 $2 \iint_{x^2+y^2 \leq 1} x - y\, dxdy = 2 \int_0^{2\pi} \int_0^1 (r\cos\theta - r\sin\theta) r\, dr d\theta$

$$\because \int_0^{2\pi} \cos\theta - \sin\theta\, d\theta = 0 \quad \therefore \iint_S (\nabla \times \vec{\mathbf{F}}) \cdot \vec{n}\, dA = 2 \int_0^{2\pi} \int_0^1 (r\cos\theta - r\sin\theta) r\, dr d\theta = 0$$

(2)

令 $C$ 为曲面 $S$ 与 $z = 0$ 交集的逆时针封闭曲线则 $C = \{(x,y,z): x^2 + y^2 = 1, z = 0\}$

令 $\vec{r}(\theta) = (\cos\theta, \sin\theta, 0)$, $0 \leq \theta \leq 2\pi$ $\therefore \vec{r}'(\theta) = (-\sin\theta, \cos\theta, 0)$

$\because \vec{F}$ 的各分量一阶偏导数存在且连续，藉由 Stokes' Theorem

则 $$\iint_S (\nabla \times \vec{\mathbf{F}}) \cdot \vec{n}\, dA = \oint_C \vec{\mathbf{F}} \cdot d\vec{r} = \int_0^{2\pi} \vec{\mathbf{F}}(\vec{r}(\theta)) \cdot \vec{r}'(\theta)\, d\theta$$

$$= \int_0^{2\pi} (\sin^2, \cos^2, 0) \cdot (-\sin\theta, \cos\theta, 0)\, d\theta = \int_0^{2\pi} -\sin^3\theta + \cos^3\theta\, d\theta$$

$$\because -\sin^3\theta + \cos^3\theta = (\cos\theta - \sin\theta)(\cos^2\theta + \sin\theta\cos\theta + \sin^2\theta)$$

$$= (\cos\theta - \sin\theta)(1 + \sin\theta\cos\theta) = (\cos\theta - \sin\theta) - \sin^2\theta\cos\theta + \sin\theta\cos^2\theta$$

$$\therefore \int_0^{2\pi} -\sin^3\theta + \cos^3\theta\, d\theta = \int_0^{2\pi} (\cos\theta - \sin\theta) - \sin^2\theta\cos\theta + \sin\theta\cos^2\theta\, d\theta$$

$$= -\frac{\sin^3\theta}{3} + \frac{\cos^3\theta}{3} \bigg|_0^{2\pi} = 0$$

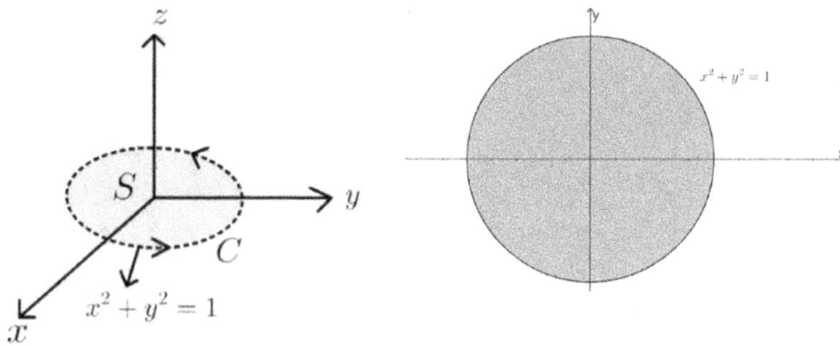


Example 2.

$\vec{\mathbf{F}} = (0, -xz, -xy)$, $S$ 为非封闭朝 $y$ 轴负向作定向的抛物面 $y = x^2 + z^2$, $0 \leq y \leq 1$,

$$\text{求} \iint_S \left(\nabla \times \vec{F}\right) \cdot \vec{n} dA = ? \quad (1)\text{直接计算} \quad (2) \text{使用 Stokes' Theorem}$$

【解】

(1)

$$\because \text{curl } \vec{F} = \nabla \times \vec{F} = \left\| \begin{matrix} \boldsymbol{i} & \boldsymbol{j} & \boldsymbol{k} \\ \frac{\partial}{\partial x} & \frac{\partial}{\partial y} & \frac{\partial}{\partial z} \\ 0 & -xz & -xy \end{matrix} \right\| = (0, y, -z)$$

$\because S$ 朝 $y$ 轴负向作定向

令 $\vec{r}(x, y, z) = (x, x^2 + z^2, z)$ 且 $R$ 为曲面 $S$ 投影至 $xz$ 平面的封闭区域

则 $R = \{(x, y): x^2 + z^2 \leq 1\}$ 且 $\iint_S \left(\nabla \times \vec{F}\right) \cdot \vec{n} \, dA = -\iint_R \left(\nabla \times \vec{F}\right) \cdot \frac{\partial \vec{r}}{\partial x} \times \frac{\partial \vec{r}}{\partial z} dxdz$

$\because \frac{\partial \vec{r}}{\partial x} = (1, 2x, 0)$ 且 $\frac{\partial \vec{r}}{\partial y} = (0, 2z, 1)$ $\quad \therefore \frac{\partial \vec{r}}{\partial x} \times \frac{\partial \vec{r}}{\partial y} = (2x, -1, 2z)$

$\therefore \left(\nabla \times \vec{F}\right) \cdot \frac{\partial \vec{r}}{\partial x} \times \frac{\partial \vec{r}}{\partial z} = (0, y, -z) \cdot (2x, -1, 2z) = -y - 2z^2 = -x^2 - 3z^2, \quad \forall (x, y, z) \in S$

$\therefore \iint_S \left(\nabla \times \vec{F}\right) \cdot \vec{n} \, dA = -\iint_R \left(\nabla \times \vec{F}\right) \cdot \frac{\partial \vec{r}}{\partial x} \times \frac{\partial \vec{r}}{\partial z} dxdz = \iint_{x^2 + z^2 \leq 1} x^2 + 3z^2 dxdz$

令 $x = r\cos\theta, z = r\sin\theta$ 则 $\{(x, y): x^2 + z^2 \leq 1\} = \{(r, \theta): 0 \leq r \leq 1, 0 \leq \theta \leq 2\pi\}$

且 $dxdz = \left\| \begin{matrix} \frac{\partial x}{\partial r} & \frac{\partial x}{\partial \theta} \\ \frac{\partial z}{\partial r} & \frac{\partial z}{\partial \theta} \end{matrix} \right\| drd\theta = \left| \begin{matrix} \cos\theta & -r\sin\theta \\ \sin\theta & r\cos\theta \end{matrix} \right| drd\theta = rdrd\theta$

$\therefore \iint_{x^2+z^2 \leq 1} x^2 + 3z^2 dxdz = \int_0^{2\pi} \int_0^1 r^3 \cos^2\theta + 3r^3 \sin^2\theta \, drd\theta$

$= \int_0^{2\pi} \frac{1}{4}\left(\frac{1 + \cos\theta}{2}\right) + \frac{3}{4}\left(\frac{1 - \cos\theta}{2}\right) d\theta = \pi$

(2)

令 $C: x^2 + z^2 = 1, \; y = 1$ 为逆时针封闭曲线

$\because S$ 朝 $y$ 轴负向作定向且 $\vec{F}$ 的各分量一阶偏导数存在且连续

藉由 Stokes' Theorem, $\iint_S \left(\nabla \times \vec{F}\right) \cdot \vec{n} dA = -\oint_C \vec{F} \cdot d\vec{r}$

令 $\vec{r}(t) = (\cos t, 1, \sin t)$ 则 $\vec{r}'(t) = (-\sin t, 0, \cos t)$

$$\therefore -\oint_C \vec{\mathrm{F}} \cdot d\vec{r} = -\int_0^{2\pi} \vec{\mathrm{F}}\big(\vec{r}(t)\big) \cdot \vec{r}'(t)\, dt$$

$$= -\int_0^{2\pi} (0, -\cos t \sin t, -\cos t) \cdot (-\sin t, 0, \cos t)\, dt = \int_0^{2\pi} \cos^2 t\, dt = \pi$$

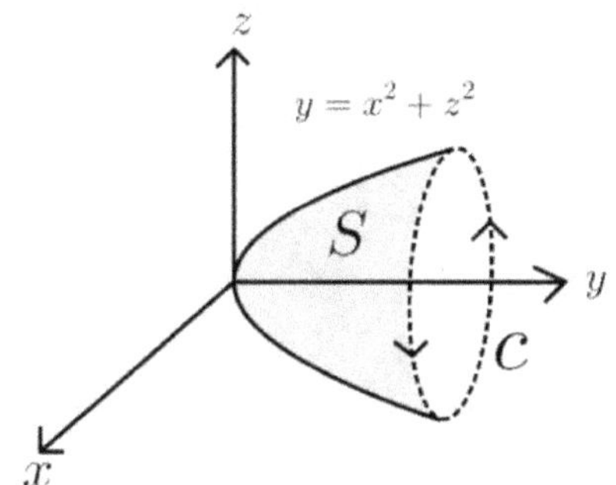


## Example 3.

使用 Stokes' Theorem 求 $\iint_S (\nabla \times \vec{\mathrm{F}}) \cdot \vec{n}\, dA =?$, $\vec{\mathrm{F}} = (-y^3 \cos z, x^3 e^z, -e^z)$,

$S: x^2 + y^2 + (z-a)^2 = 2a^2$, $z > 0$ 为朝上定向的上半球球面

【解】

令 $C$ 为曲面 $S$ 与 $z = 0$ 交集的逆时针封闭曲线则 $C = \{(x,y,z): x^2 + y^2 = a^2, z = 0\}$

令 $x = a\cos\theta, y = a\sin\theta$ 则 $\vec{r}(\theta) = (a\cos\theta, a\sin\theta, 0)$ $\therefore \vec{r}'(\theta) = (-a\sin\theta, a\cos\theta, 0)$

$\because S$ 为朝上定向

$\because \vec{F}$ 的各分量一阶偏导数存在且连续, 藉由 Stokes' Theorem

则 $\iint_S (\nabla \times \vec{\mathrm{F}}) \cdot \vec{n}\, dA = \oint_C \vec{\mathrm{F}} \cdot d\vec{r} = \int_0^{2\pi} \vec{\mathrm{F}}(\vec{r}(\theta)) \cdot \vec{r}'(\theta)\, d\theta$

$$= \int_0^{2\pi} (-(a\sin\theta)^3, (a\cos\theta)^3, 1) \cdot (-a\sin\theta, a\cos\theta, 0)\, d\theta = a^4 \int_0^{2\pi} \sin^4\theta + \cos^4\theta\, d\theta$$

$\because \sin^4\theta + \cos^4\theta = (\sin^2\theta + \cos^2\theta)^2 - 2\sin^2\theta\cos^2\theta = 1 - 2(\sin\theta\cos\theta)^2$

$$= 1 - 2\left(\frac{\sin 2\theta}{2}\right)^2 = 1 - \frac{\sin^2 2\theta}{2} = 1 - \frac{1 - \cos 4\theta}{4} = \frac{3 - \cos 4\theta}{4}$$

且 $\int_0^{2\pi} \sin^4\theta + \cos^4\theta\, d\theta = \int_0^{2\pi} \frac{3 - \cos 4\theta}{4}\, d\theta = \frac{3\pi}{2}$

$$\therefore \iint_S (\nabla \times \vec{\mathrm{F}}) \cdot \vec{n}\, dA = \frac{3a^4\pi}{2}$$

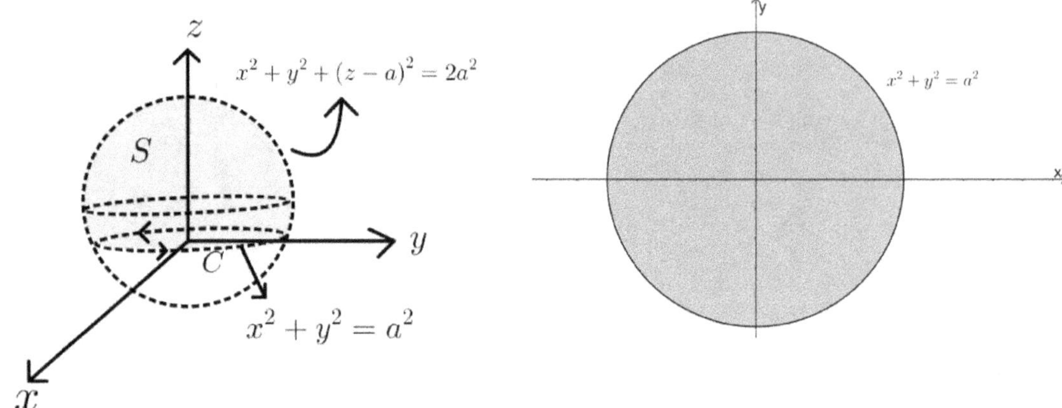



Example 4.

$$求 \iint_S (\nabla \times \vec{F}) \cdot \vec{n} \, dA = ?, \quad S = \{(x, y, z): x^2 + y^2 \leq 4, z = 1\} 为朝上定向平面,$$

$$\vec{F} = (x, 2z - x, y^2) \ (1)直接计算 \ (2)使用 \ Stoke's \ Theorem$$

【解】

(1)

$$\because \nabla \times \vec{F} = \begin{vmatrix} \vec{i} & \vec{j} & \vec{k} \\ \dfrac{\partial}{\partial x} & \dfrac{\partial}{\partial y} & \dfrac{\partial}{\partial z} \\ x & 2z - x & y^2 \end{vmatrix} = (2y - 2, 0, -1)$$

$\because S$ 为朝上定向

令 $\vec{r}(x, y) = (x, y, 1)$, 令 $R$ 为曲面 $S$ 投影至 $xy$ 平面的封闭区域

则 $R = \{(x, y): x^2 + y^2 \leq 4\}$ 且 $\displaystyle\iint_S (\nabla \times \vec{F}) \cdot \vec{n} \, dA = \iint_R (\nabla \times \vec{F}) \cdot \dfrac{\partial \vec{r}}{\partial x} \times \dfrac{\partial \vec{r}}{\partial y} \, dx \, dy$

$\because \dfrac{\partial \vec{r}}{\partial x} = (1, 0, 0), \ \dfrac{\partial \vec{r}}{\partial y} = (0, 1, 0) \quad \therefore \dfrac{\partial \vec{r}}{\partial x} \times \dfrac{\partial \vec{r}}{\partial y} = (0, 0, 1)$

$$\iint_S (\nabla \times \vec{F}) \cdot \vec{n} \, dA = \iint_R (2y - 2, 0, -1) \cdot (0, 0, 1) \, dx \, dy = \iint_{x^2 + y^2 \leq 4} -1 \, dA = -4\pi$$

(2)

令 $C$ 为逆时针封闭曲线 $= \{(x, y, z): x^2 + y^2 = 4, z = 1\}$

令 $x = 2\cos\theta$ , $y = 2\sin\theta$ 则 $\vec{r}(\theta) = (2\cos\theta, 2\sin\theta, 0)$ $\therefore \vec{r}'(\theta) = (-2\sin\theta, 2\cos\theta, 0)$

$\because \vec{F}$ 的各分量一阶偏导数存在且连续, 藉由 Stokes' Theorem

则 $\displaystyle\iint_S (\nabla \times \vec{F}) \cdot \vec{n}\,dA = \oint_C \vec{F} \cdot d\vec{r} = \int_0^{2\pi} \vec{F}(\vec{r}(\theta)) \cdot \vec{r}'(\theta)\,d\theta$

$= \displaystyle\int_0^{2\pi} (2\cos\theta, 2 - 2\cos\theta, 4\sin^2\theta) \cdot (-2\sin\theta, 2\cos\theta, 0)\,d\theta$

$= \displaystyle\int_0^{2\pi} -4\sin\theta\cos\theta + 4\cos\theta - 4\cos^2\theta\,d\theta = \int_0^{2\pi} -4\cos^2\theta\,d\theta = -4\int_0^{2\pi} \frac{1 + \cos 2\theta}{2}\,d\theta$

$= -4\pi$

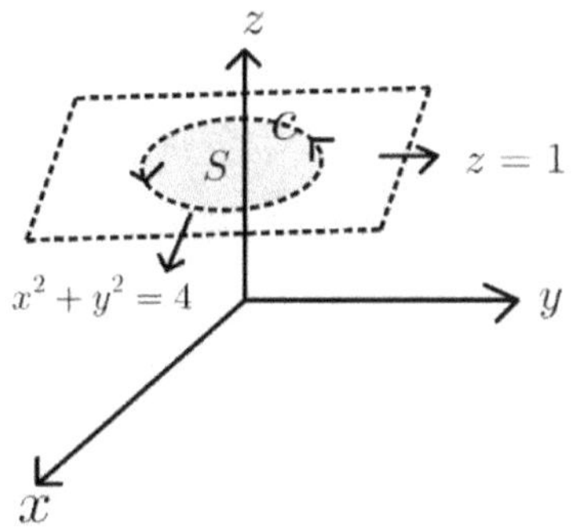


Example 5.

$\vec{F} = \left(0, \dfrac{2x^3}{3}, 2y^3\right)$, $S$ 为 $x + y + z = 1$ 在第一卦限所围朝上定向平面,

求 $\displaystyle\iint_S (\nabla \times \vec{F}) \cdot \vec{n}\,dA = ?$ (1) 直接计算 (2) 使用 Stokes' Theorem

【解】

(1)

$\because \operatorname{curl} \vec{F} = \nabla \times \vec{F} = \left\|\begin{array}{ccc} \boldsymbol{i} & \boldsymbol{j} & \boldsymbol{k} \\ \dfrac{\partial}{\partial x} & \dfrac{\partial}{\partial y} & \dfrac{\partial}{\partial z} \\ 0 & \dfrac{2x^3}{3} & 2y^3 \end{array}\right\| = (6y^2, 0, 2x^2)$

$\because S$ 为朝上定向

令 $\vec{r}(x, y, z) = (x, y, 1 - x - y)$ 且 $R$ 为平面 $S$ 投影至 $xy$ 平面的封闭区域

则 $R = \{(x, y): x + y \leq 1, x \geq 0, y \geq 0\}$ 且 $\displaystyle\iint_S (\nabla \times \vec{F}) \cdot \vec{n}\, dA = \iint_R (\nabla \times \vec{F}) \cdot \frac{\partial \vec{r}}{\partial x} \times \frac{\partial \vec{r}}{\partial y}\, dxdy$

$\because \dfrac{\partial \vec{r}}{\partial x} = (1, 0, -1)$ 且 $\dfrac{\partial \vec{r}}{\partial y} = (0, 1, -1)$ $\quad \therefore \dfrac{\partial \vec{r}}{\partial x} \times \dfrac{\partial \vec{r}}{\partial y} = (1, 1, 1)$

$\because (\nabla \times \vec{F}) \cdot \dfrac{\partial \vec{r}}{\partial x} \times \dfrac{\partial \vec{r}}{\partial y} = (6y^2, 0, 2x^2) \cdot \dfrac{\partial \vec{r}}{\partial x} \times \dfrac{\partial \vec{r}}{\partial y} = 2(x^2 + 3y^2), \quad \forall (x, y, z) \in S$

$\therefore \displaystyle\iint_S (\nabla \times \vec{F}) \cdot \vec{n}\, dA = \iint_R (\nabla \times \vec{F}) \cdot \frac{\partial \vec{r}}{\partial x} \times \frac{\partial \vec{r}}{\partial y}\, dxdy = 2 \iint_R x^2 + 3y^2\, dxdy$

$= 2 \displaystyle\int_0^1 \int_0^{1-y} x^2 + 3y^2\, dxdy = 2 \int_0^1 \left( \frac{x^3}{3} + 3y^2 x \right) \Big|_{x=0}^{x=1-y} dy = 2 \int_0^1 \frac{(1-y)^3}{3} + 3y^2(1-y)\, dy$

$= \dfrac{2}{3}$

(2)

令 $C_1: \vec{r}_1(t) = (1 - t, t, 0), \quad C_2: \vec{r}_2(t) = (0, 1 - t, t), \quad C_3: \vec{r}_3(t) = (t, 0, 1 - t), 0 \leq t \leq 1$

则 $\vec{r}_1'(t) = (-1, 1, 0), \quad \vec{r}_2'(t) = (0, -1, 1), \quad \vec{r}_3'(t) = (1, 0, -1)$ 且 $C = C_1 \cup C_2 \cup C_3$

藉由 Stokes' Theorem, $\displaystyle\iint_S (\nabla \times \vec{F}) \cdot \vec{n}\, dA = \oint_C \vec{F} \cdot d\vec{r}$

$\therefore \displaystyle\oint_C \vec{F} \cdot d\vec{r} = \int_{C_1} \vec{F}(\vec{r}_1(t)) \cdot \vec{r}_1'(t)\, dt + \int_{C_2} \vec{F}(\vec{r}_2(t)) \cdot \vec{r}_2'(t)\, dt + \int_{C_3} \vec{F}(\vec{r}_3(t)) \cdot \vec{r}_3'(t)\, dt$

$= \dfrac{2}{3} \displaystyle\int_0^1 (1 - t)^3\, dt + 2 \int_0^1 (1 - t)^3\, dt = \dfrac{2}{3}$

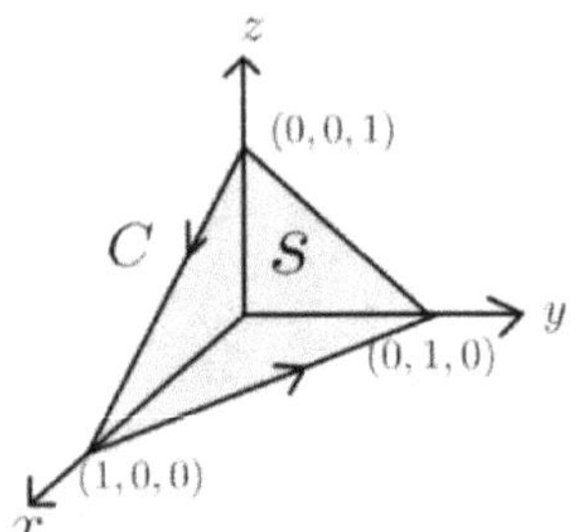

Example 6.

$$\vec{\mathbf{F}} = \left(\frac{z^2}{2}, \frac{x^2}{2}, \frac{y^2}{2}\right), \quad S是(a,0,0),(0,a,0),(0,0,a)于第一卦限朝上定向的平面,$$

$$求 \iint_S (\nabla \times \vec{\mathbf{F}}) \cdot \vec{n}\, dA =?\ (1)直接计算\ (2)\ 使用\ \text{Stokes' Theorem}$$

【解】

(1)

$$\text{curl}\ \vec{\mathbf{F}} = \nabla \times \vec{\mathbf{F}} = \begin{Vmatrix} \boldsymbol{i} & \boldsymbol{j} & \boldsymbol{k} \\ \dfrac{\partial}{\partial x} & \dfrac{\partial}{\partial y} & \dfrac{\partial}{\partial z} \\ \dfrac{z^2}{2} & \dfrac{x^2}{2} & \dfrac{y^2}{2} \end{Vmatrix} = (y, z, x)$$

$\because S$ 为朝上定向

令 $\vec{r}(x,y,z) = (x, y, a - x - y)$ 且 $R = \{(x,y): 0 \le x \le a, 0 \le y \le a - x\}$

$$则 \iint_S (\nabla \times \vec{\mathbf{F}}) \cdot \vec{n}\, dA = \iint_R (\nabla \times \vec{\mathbf{F}}) \cdot \frac{\partial \vec{r}}{\partial x} \times \frac{\partial \vec{r}}{\partial y}\, dx\, dy$$

$$\because \frac{\partial \vec{r}}{\partial x} = (1, 0, -1)\ 且\ \frac{\partial \vec{r}}{\partial y} = (0, 1, -1)\quad \therefore \frac{\partial \vec{r}}{\partial x} \times \frac{\partial \vec{r}}{\partial y} = (1,1,1)$$

$$\because (\nabla \times \vec{\mathbf{F}}) \cdot \frac{\partial \vec{r}}{\partial x} \times \frac{\partial \vec{r}}{\partial y} = (y,z,x) \cdot (1,1,1) = x + y + z = x + y + a - x - y = a, \quad \forall (x,y,z) \in S$$

$$\iint_R (\nabla \times \vec{\mathbf{F}}) \cdot \frac{\partial \vec{r}}{\partial x} \times \frac{\partial \vec{r}}{\partial y}\, dx\, dy = \int_0^a \int_0^{a-x} a\, dy\, dx = a \int_0^a a - x\, dx = a\left(ax - \frac{x^2}{2}\right)\Big|_0^a = \frac{a^3}{2}$$

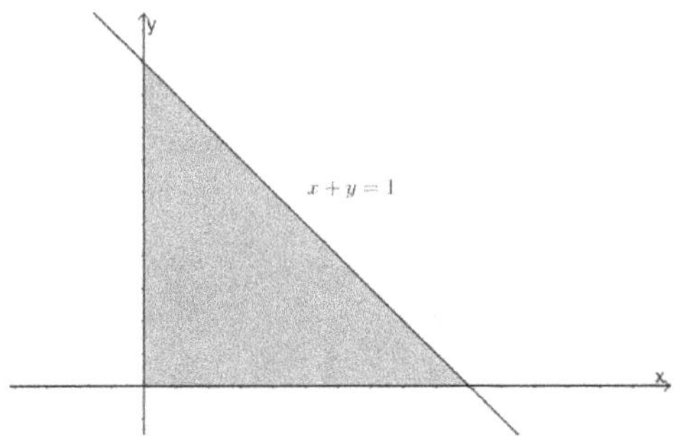

(2)

令 $C_1: \vec{r}_1(t) = (a - at, at, 0), C_2: \vec{r}_2(t) = (0, a - at, at), C_3: \vec{r}_3(t) = (at, 0, a - at), 0 \leq t \leq 1$

则 $\vec{r}_1'(t) = (-a, a, 0), \quad \vec{r}_2'(t) = (0, -a, a), \quad \vec{r}_3'(t) = (a, 0, -a)$ 且 $C = C_1 \cup C_2 \cup C_3$

藉由 Stokes' Theorem, $\quad \iint_S (\nabla \times \vec{\mathbf{F}}) \cdot \vec{n}\, dA = \oint_C \vec{\mathbf{F}} \cdot d\vec{r}$

$$\therefore \oint_C \vec{\mathbf{F}} \cdot d\vec{r} = \int_{C_1} \vec{\mathbf{F}}(\vec{r}_1(t)) \cdot \vec{r}_1'(t)\, dt + \int_{C_2} \vec{\mathbf{F}}(\vec{r}_2(t)) \cdot \vec{r}_2'(t)\, dt + \int_{C_3} \vec{\mathbf{F}}(\vec{r}_3(t)) \cdot \vec{r}_3'(t)\, dt$$

$$= \frac{a}{2}\int_0^1 (a - at)^3\, dt \cdot 3 = \frac{a^3}{2}$$

Example 7.

使用 Stokes' Theorem 求 $\iint_S (\nabla \times \vec{\mathbf{F}}) \cdot \vec{n}\, dA = ?$, 其中 $\vec{\mathbf{F}} = (2x - y, 2y + z, xyz)$,

$S: \dfrac{x^2}{a^2} + \dfrac{y^2}{b^2} + \dfrac{z^2}{c^2} = 1, \ z > 0$ 为朝上定向的上半椭球球面

【解】

令 $C$ 为曲面 $S$ 与 $z = 0$ 交集的逆时针封闭曲线则 $C = \{(x, y, z): \dfrac{x^2}{a^2} + \dfrac{y^2}{b^2} = 1, z = 0\}$

令 $x = a\cos\theta, y = b\sin\theta$ 则 $\vec{r}(\theta) = (a\cos\theta, b\sin\theta, 0) \quad \therefore \vec{r}'(\theta) = (-a\sin\theta, b\cos\theta, 0)$

$\because S$ 为朝上定向

$\because \vec{F}$ 的各分量一阶偏导数存在且连续, 藉由 Stokes' Theorem

则 $\iint_S (\nabla \times \vec{\mathbf{F}}) \cdot \vec{n}\, dA = \oint_C \vec{\mathbf{F}} \cdot d\vec{r} = \int_0^{2\pi} \vec{\mathbf{F}}(\vec{r}(\theta)) \cdot \vec{r}'(\theta)\, d\theta$

$$= \int_0^{2\pi} (2a\cos\theta - b\sin\theta, 2b\sin\theta, 0) \cdot (-a\sin\theta, b\cos\theta, 0)\, d\theta$$

$$= \int_0^{2\pi} -2a^2 \sin\theta\cos\theta + ab\sin^2\theta + 2b^2\sin\theta\cos\theta\, d\theta$$

$$= \int_0^{2\pi} 2(b^2 - a^2)\sin\theta\cos\theta + ab\sin^2\theta\, d\theta = \int_0^{2\pi} (b^2 - a^2)\sin 2\theta + ab\left(\frac{1 - \cos 2\theta}{2}\right) d\theta$$

$$= ab\pi$$

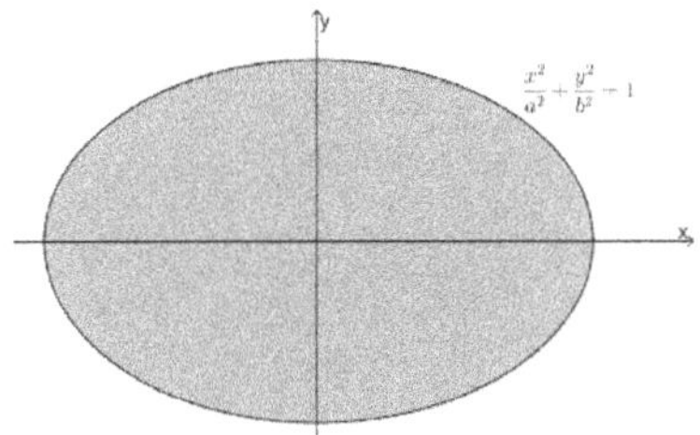

## Example 8.

使用 Stokes' Theorem 求 $\iint_S (\nabla \times \vec{F}) \cdot dS =?$, $\vec{F} = (xz, yz, xy)$, $S$ 为 $x^2 + y^2 + z^2 = 4$

在圆柱 $x^2 + y^2 = 1$ 內且 $z > 0$ 的朝上定向部分球面

【解】

令 $C$ 为曲面 $S$ 与 $x^2 + y^2 = 1$ 交集的逆时针封闭曲线则 $C = \{(x, y, z): x^2 + y^2 = 1, z = \sqrt{3}\}$

令 $x = \cos\theta, y = \sin\theta$ 则 $\vec{r}(\theta) = (\cos\theta, \sin\theta, \sqrt{3})$ $\therefore \vec{r}'(\theta) = (-\sin\theta, \cos\theta, 0)$

$\because S$ 为朝上定向

$\because \vec{F}$ 的各分量一阶偏导数存在且连续，藉由 Stokes' Theorem

则 $\displaystyle\iint_S (\nabla \times \vec{F}) \cdot \vec{n}\,dA = \oint_C \vec{F} \cdot d\vec{r} = \int_0^{2\pi} \vec{F}(\vec{r}(\theta)) \cdot \vec{r}'(\theta)\, d\theta$

$\displaystyle = \int_0^{2\pi} (\sqrt{3}\cos\theta, \sqrt{3}\sin\theta, \sin\theta\cos\theta) \cdot (-\sin\theta, \cos\theta, 0)\,d\theta = \sqrt{3} \int_0^{2\pi} 0\,d\theta = 0$

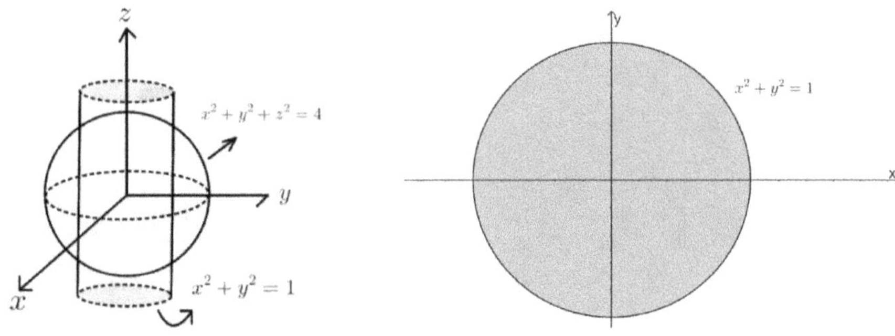

## Example 9.

假设 $\vec{F} = (0, e^x + 16xy, e^y + 6y^2)$, $S$ 是 $x + y + z = 6$ 在第一卦限朝上定向的平面，

求 $\displaystyle\iint_S (\nabla \times \vec{F}) \cdot \vec{n}\,dA =?$ (1) 直接计算 (2) 使用 Stokes' Theorem

【解】

(1)

$$\text{curl } \vec{F} = \nabla \times \vec{F} = \begin{Vmatrix} \boldsymbol{i} & \boldsymbol{j} & \boldsymbol{k} \\ \dfrac{\partial}{\partial x} & \dfrac{\partial}{\partial y} & \dfrac{\partial}{\partial z} \\ 0 & e^x + 16xy & e^y + 6y^2 \end{Vmatrix} = (e^y + 12y, 0, e^x + 6y)$$

$\because S$ 为朝上定向

令 $\vec{r}(x,y,z) = (x, y, 6 - x - y)$ 且 $R$ 为 $S$ 曲面投影至 $xy$ 平面的封闭区域

则 $R = \{(x,y): x + y \leq 6, x \geq 0, y \geq 0\}$ 且 $\displaystyle\iint_S (\nabla \times \vec{F}) \cdot \vec{n}\, dA = \iint_R (\nabla \times \vec{F}) \cdot \dfrac{\partial \vec{r}}{\partial x} \times \dfrac{\partial \vec{r}}{\partial y}\, dxdy$

$\because \dfrac{\partial \vec{r}}{\partial x} = (1, 0, -1)$ 且 $\dfrac{\partial \vec{r}}{\partial y} = (0, 1, -1)$ $\quad \therefore \dfrac{\partial \vec{r}}{\partial x} \times \dfrac{\partial \vec{r}}{\partial y} = (1, 1, 1)$

$\because (\nabla \times \vec{F}) \cdot \dfrac{\partial \vec{r}}{\partial x} \times \dfrac{\partial \vec{r}}{\partial y} = (e^y + 12y, 0, e^x + 6y) \cdot (1,1,1) = e^y + e^x + 18y, \quad \forall (x,y,z) \in S$

$\therefore \displaystyle\iint_S (\nabla \times \vec{F}) \cdot \vec{n}\, dA = \iint_R (\nabla \times \vec{F}) \cdot \dfrac{\partial \vec{r}}{\partial x} \times \dfrac{\partial \vec{r}}{\partial y}\, dxdy = \iint_R e^y + e^x + 18y\, dxdy$

$$= \int_0^6 \int_0^{6-x} e^y + e^x + 18y\, dydx = 634 + 2e^6$$

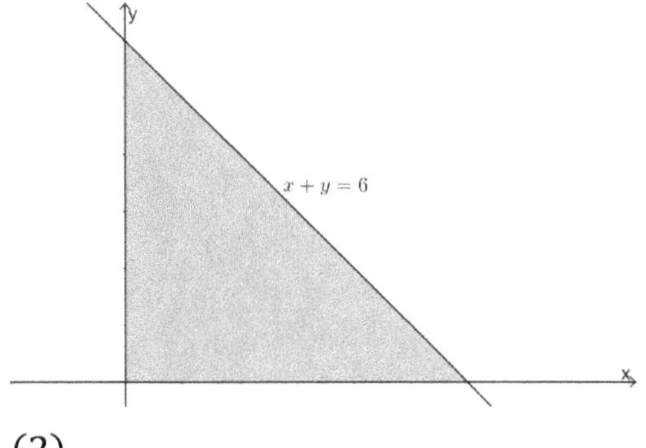


(2)

令 $C_1: \vec{r}_1(t) = (6 - 6t, 6t, 0), C_2: \vec{r}_2(t) = (0, 6 - 6t, 6t), C_3: \vec{r}_3(t) = (6t, 0, 6 - 6t)$

则 $\vec{r}_1'(t) = (-6, 6, 0), \quad \vec{r}_2'(t) = (0, -6, 6), \quad \vec{r}_3'(t) = (6, 0, -6)$ 且 $C = C_1 \cup C_2 \cup C_3$

藉由 Stokes' Theorem, $\displaystyle\iint_S (\nabla \times \vec{F}) \cdot \vec{n}\, dA = \oint_C \vec{F} \cdot d\vec{r}$

$$\therefore \oint_C \vec{F} \cdot d\vec{r} = \int_{C_1} \vec{F}(\vec{r}_1(t)) \cdot \vec{r}_1'(t)dt + \int_{C_2} \vec{F}(\vec{r}_2(t)) \cdot \vec{r}_2'(t)dt + \int_{C_3} \vec{F}(\vec{r}_3(t)) \cdot \vec{r}_3'(t)dt$$

$$\because \int_{C_1} \vec{F}(\vec{r}_1(t)) \cdot \vec{r}_1'(t)dt = 6\int_0^1 e^{6-6t} + 96t(6-6t)\, dt = e^6 - 1 + 576$$

$$\int_{C_2} \vec{F}(\vec{r}_2(t)) \cdot \vec{r}_2'(t)dt = -6\int_0^1 dt + 6\int_0^1 e^{6-6t} + 6(6-6t)^2\, dt = e^6 + 65$$

$$\int_{C_3} \vec{F}(\vec{r}_3(t)) \cdot \vec{r}_3'(t)dt = -6\int_0^1 dt = -6$$

$$\therefore \oint_C \vec{F} \cdot d\vec{r} = 634 + 2e^6$$

Example 10.

求 $\vec{F} = (2x - y, -yz^2, -zy^2)$ 的旋度在球面 $S: x^2 + y^2 + z^2 = 1$ 的上半表面上法线分量的面积分 (1) 直接计算　(2) 使用 Stoke's Theorem

【解】

(1)

$$\because \nabla \times \vec{F} = \begin{vmatrix} \vec{i} & \vec{j} & \vec{k} \\ \dfrac{\partial}{\partial x} & \dfrac{\partial}{\partial y} & \dfrac{\partial}{\partial z} \\ 2x - y & -yz^2 & -zy^2 \end{vmatrix} = (-2zy - (-2yz), 0, 1) = (0, 0, 1)$$

$\because S$ 为朝上定向

$$令 \vec{r}(x, y) = \left(x, y, \sqrt{1 - (x^2 + y^2)}\right)$$

令 $R$ 为球面投影至 $xy$ 平面的封闭区域 则 $R = \{(x, y): x^2 + y^2 \leq 1\}$

$$且 \iint_S (\nabla \times \vec{F}) \cdot \vec{n}\, dA = \iint_R (\nabla \times \vec{F}) \cdot \frac{\partial \vec{r}}{\partial x} \times \frac{\partial \vec{r}}{\partial y}\, dxdy$$

$$\because \frac{\partial \vec{r}}{\partial x} = \left(1, 0, -x(1-(x^2+y^2))^{-\frac{1}{2}}\right), \quad \frac{\partial \vec{r}}{\partial y} = \left(0, 1, -y(1-(x^2+y^2))^{-\frac{1}{2}}\right),$$

$$\therefore \frac{\partial \vec{r}}{\partial x} \times \frac{\partial \vec{r}}{\partial y} = \left( x(1-(x^2+y^2))^{-\frac{1}{2}}, y(1-(x^2+y^2))^{-\frac{1}{2}}, 1 \right)$$

$$\therefore (\nabla \times \vec{F}) \cdot \frac{\partial \vec{r}}{\partial x} \times \frac{\partial \vec{r}}{\partial y} = (0,0,1) \cdot \left( x(1-(x^2+y^2))^{-\frac{1}{2}}, y(1-(x^2+y^2))^{-\frac{1}{2}}, 1 \right) = 1$$

$$\therefore \iint_S (\nabla \times \vec{F}) \cdot \vec{n}\, dA = \iint_R (\nabla \times \vec{F}) \cdot \frac{\partial \vec{r}}{\partial x} \times \frac{\partial \vec{r}}{\partial y}\, dxdy = \iint_R dxdy = \pi$$

(2)

令 $C$ 为球面与 $z=0$ 交集的逆时针封闭曲线则 $C = \{(x,y,z): x^2+y^2=1, z=0\}$

令 $x=\cos\theta, y=\sin\theta$ 则 $\vec{r}(\theta) = (\cos\theta, \sin\theta, 0)$ $\quad \therefore \vec{r}'(\theta) = (-\sin\theta, \cos\theta, 0)$

$\because S$ 为朝上定向

$\because \vec{F}$ 的各分量一阶偏导数存在且连续,藉由 Stokes' Theorem

$$则 \iint_S (\nabla \times \vec{F}) \cdot \vec{n}\, dA = \oint_C \vec{F} \cdot d\vec{r} = \int_0^{2\pi} \vec{F}(\vec{r}(\theta)) \cdot \vec{r}'(\theta)\, d\theta$$

$$= \int_0^{2\pi} (2\cos\theta - \sin\theta, 0, 0) \cdot (-\sin\theta, \cos\theta, 0) d\theta = \int_0^{2\pi} -2\sin\theta\cos\theta + \sin^2\theta\, d\theta$$

$$= \int_0^{2\pi} -2\sin\theta\cos\theta + \frac{1-\cos 2\theta}{2}\, d\theta = \pi$$

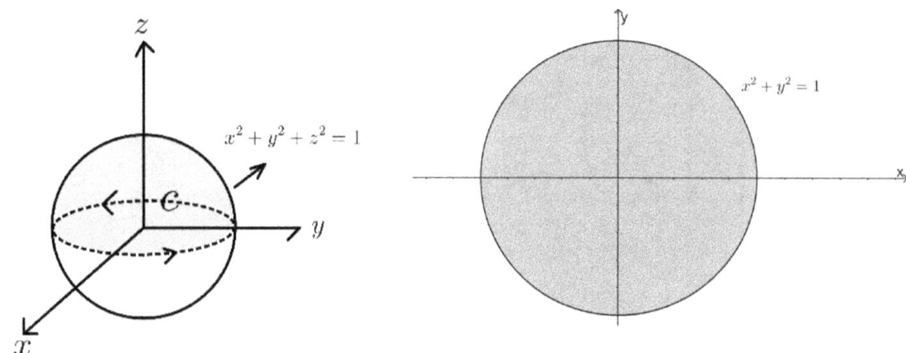


Example 11.

　　求 $\vec{F} = (2x-y, -yz^2, -zy^2)$ 的旋度在球面 $S: x^2+y^2+z^2=1$ 的下半表面上法线分量的面积分 (1) 直接计算 (2) 使用 Stoke's Theorem

【解】

(1)

$$\because \nabla\times\vec{F} = \begin{vmatrix} \vec{i} & \vec{j} & \vec{k} \\ \dfrac{\partial}{\partial x} & \dfrac{\partial}{\partial y} & \dfrac{\partial}{\partial z} \\ 2x-y & -yz^2 & -zy^2 \end{vmatrix} = (-2zy-(-2yz),0,1) = (0,0,1)$$

$\because S$ 为朝下定向

令 $\vec{r}(x,y) = \left(x,y,-\sqrt{1-(x^2+y^2)}\right)$，令 $R$ 为球面投影至 $xy$ 平面的封闭区域

则 $R = \{(x,y): x^2+y^2 \leq 1\}$ 且 $\displaystyle\iint_S (\nabla\times\vec{F})\cdot\vec{n}\,dA = -\iint_R (\nabla\times\vec{F})\cdot\dfrac{\partial\vec{r}}{\partial x}\times\dfrac{\partial\vec{r}}{\partial y}\,dxdy$

$\because \dfrac{\partial\vec{r}}{\partial x} = \left(1,0,x(1-(x^2+y^2))^{-\frac{1}{2}}\right),\quad \dfrac{\partial\vec{r}}{\partial y} = \left(0,1,y(1-(x^2+y^2))^{-\frac{1}{2}}\right)$

$\therefore \dfrac{\partial\vec{r}}{\partial x}\times\dfrac{\partial\vec{r}}{\partial y} = \left(-x(1-(x^2+y^2))^{-\frac{1}{2}},-y(1-(x^2+y^2))^{-\frac{1}{2}},1\right)$

$\therefore -(\nabla\times\vec{F})\cdot\dfrac{\partial\vec{r}}{\partial x}\times\dfrac{\partial\vec{r}}{\partial y} = -(0,0,1)\cdot\left(x(1-(x^2+y^2))^{-\frac{1}{2}},y(1-(x^2+y^2))^{-\frac{1}{2}},1\right) = -1$

$\therefore \displaystyle\iint_S (\nabla\times\vec{F})\cdot\vec{n}\,dA = -\iint_R (\nabla\times\vec{F})\cdot\dfrac{\partial\vec{r}}{\partial x}\times\dfrac{\partial\vec{r}}{\partial y}\,dxdy = -\iint_R dxdy = -\pi$

(2)

令 $C$ 为球面与 $z=0$ 交集的逆时针封闭曲线则 $C = \{(x,y,z): x^2+y^2=1, z=0\}$

令 $x=\cos\theta, y=\sin\theta$ 则 $\vec{r}(\theta) = (\cos\theta,\sin\theta,0)$ $\therefore \vec{r}'(\theta) = (-\sin\theta,\cos\theta,0)$

$\because S$ is oriented downward

$\because \vec{F}$ 的各分量一阶偏导数存在且连续，藉由 Stokes' Theorem

则 $\displaystyle\iint_S (\nabla\times\vec{F})\cdot\vec{n}\,dA = -\oint_C \vec{F}\cdot d\vec{r} = -\int_0^{2\pi} \vec{F}(\vec{r}(\theta))\cdot\vec{r}'(\theta)\,d\theta$

$\displaystyle = -\int_0^{2\pi} (2\cos\theta-\sin\theta,0,0)\cdot(-\sin\theta,\cos\theta,0)\,d\theta = -\int_0^{2\pi} -2\sin\theta\cos\theta+\sin^2\theta\,d\theta$

$\displaystyle = -\int_0^{2\pi} -2\sin\theta\cos\theta+\frac{1-\cos 2\theta}{2}\,d\theta = -\pi$

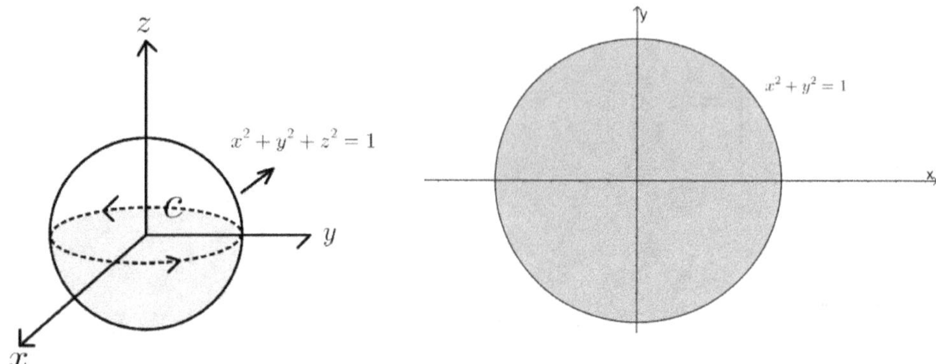



Example 12.

$$求 \iint_S (\nabla \times \vec{F}) \cdot \vec{n} \, dA = ?, \quad S = \{(x, y, z): 4z = x^2 + y^2, z \leq 1\} \text{为朝下定向曲面,}$$

$$\vec{F} = (x, 2z - x, y^2) \quad (1)\text{直接计算} \quad (2)\text{使用 Stoke's Theorem}$$

【解】

(1)

$$\because \nabla \times \vec{F} = \begin{vmatrix} \vec{i} & \vec{j} & \vec{k} \\ \dfrac{\partial}{\partial x} & \dfrac{\partial}{\partial y} & \dfrac{\partial}{\partial z} \\ x & 2z - x & y^2 \end{vmatrix} = (2y - 2, 0, -1)$$

$\because S$ 为朝下定向

$$令 \vec{r}(x, y) = \left(x, y, \frac{x^2 + y^2}{4}\right) \text{ 且 } R \text{为曲面} S \text{投影至} xy \text{平面的封闭区域}$$

$$则 R = \{(x, y): x^2 + y^2 \leq 4\} \text{ 且 } \iint_S (\nabla \times \vec{F}) \cdot \vec{n} \, dA = -\iint_R (\nabla \times \vec{F}) \cdot \frac{\partial \vec{r}}{\partial x} \times \frac{\partial \vec{r}}{\partial y} \, dx \, dy$$

$$\because \frac{\partial \vec{r}}{\partial x} = \left(1, 0, \frac{x}{2}\right) \text{ 且 } \frac{\partial \vec{r}}{\partial y} = \left(0, 1, \frac{y}{2}\right) \qquad \therefore \frac{\partial \vec{r}}{\partial x} \times \frac{\partial \vec{r}}{\partial y} = \left(-\frac{x}{2}, -\frac{y}{2}, 1\right)$$

$$\iint_S (\nabla \times \vec{F}) \cdot \vec{n} \, dA = -\iint_R (\nabla \times \vec{F}) \cdot \frac{\partial \vec{r}}{\partial x} \times \frac{\partial \vec{r}}{\partial y} \, dx \, dy$$

$$= -\iint_R (2y - 2, 0, -1) \cdot \left(-\frac{x}{2}, -\frac{y}{2}, 1\right) dx \, dy = \iint_{x^2 + y^2 \leq 4} xy - x + 1 \, dx \, dy$$

$$令 x = r\cos\theta, y = r\sin\theta \text{ 则}$$

$$\iint_{x^2+y^2\le 4} xy - x + 1\, dxdy = \int_0^{2\pi}\int_0^2 (r^2\cos\theta\sin\theta - r\cos\theta + 1)\, rdrd\theta = 4\pi$$

(2)

令 $C = \{(x,y,z): x^2+y^2=4, z=1\}$ 为逆时针封闭曲线

令 $x = 2\cos\theta$ , $y = 2\sin\theta$ 则 $\vec{r}(\theta) = (2\cos\theta, 2\sin\theta, 1)$ ∴ $\vec{r}'(\theta) = (-2\sin\theta, 2\cos\theta, 0)$

∵ $S$ is oriented downward

∵ $\vec{F}$ 的各分量一阶偏导数存在且连续,藉由 Stokes' Theorem

则 $$\iint_S (\nabla\times\vec{\mathbf{F}})\cdot\vec{n}\,dA = -\oint_C \vec{\mathbf{F}}\cdot d\vec{r} = -\int_0^{2\pi}\vec{\mathbf{F}}(\vec{r}(\theta))\cdot\vec{r}'(\theta)\,d\theta$$

$$= -\int_0^{2\pi}(2\cos\theta, 2-2\cos\theta, 4\sin^2\theta)\cdot(-2\sin\theta, 2\cos\theta, 0)\,d\theta$$

$$= \int_0^{2\pi} 4\sin\theta\cos\theta - 4\cos\theta + 4\cos^2\theta\,d\theta = \int_0^{2\pi} 4\cos^2\theta\,d\theta = 4\int_0^{2\pi}\frac{1+\cos 2\theta}{2}\,d\theta$$

$$= 4\pi$$

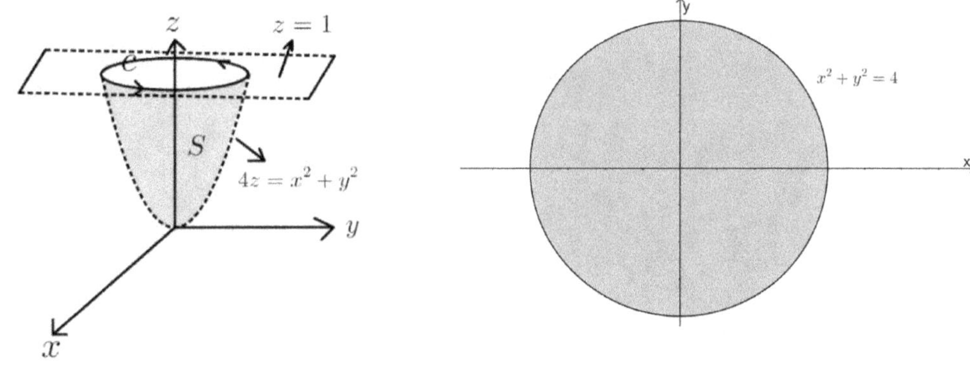


Example 13.

$$\vec{\mathbf{F}} = \left(-z, x, \frac{y^2 z}{2}\right),\ S: z = \sqrt{x^2+y^2},\ z\le a\ \text{为朝下定向曲面,} \ \text{求} \iint_S (\nabla\times\vec{\mathbf{F}})\cdot\vec{n}\,dA =?$$

(1)直接计算 (2) 使用 Stokes' Theorem

【解】

(1)

$$\because \operatorname{curl}\vec{F} = \nabla\times\vec{F} = \begin{Vmatrix} \boldsymbol{i} & \boldsymbol{j} & \boldsymbol{k} \\ \dfrac{\partial}{\partial x} & \dfrac{\partial}{\partial y} & \dfrac{\partial}{\partial z} \\ -z & x & \dfrac{y^2 z}{2} \end{Vmatrix} = (yz, -1, 1)$$

$\because S$ 为朝下定向曲面

令 $\vec{r}(x,y,z) = \left(x, y, (x^2+y^2)^{\frac{1}{2}}\right)$ 且 $R$ 为 $S$ 曲面投影至 $xy$ 平面的封闭区域

则 $R = \{(x,y): x^2+y^2 \leq a^2\}$ 且 $\iint_S (\nabla\times\vec{F})\cdot\vec{n}\,dA = -\iint_R (\nabla\times\vec{F})\cdot\dfrac{\partial\vec{r}}{\partial x}\times\dfrac{\partial\vec{r}}{\partial y}\,dxdy$

$\because \dfrac{\partial\vec{r}}{\partial x} = \left(1, 0, x(x^2+y^2)^{-\frac{1}{2}}\right)$ 且 $\dfrac{\partial\vec{r}}{\partial y} = \left(0, 1, y(x^2+y^2)^{-\frac{1}{2}}\right)$

$\therefore \dfrac{\partial\vec{r}}{\partial x}\times\dfrac{\partial\vec{r}}{\partial y} = \left(-x(x^2+y^2)^{-\frac{1}{2}}, -y(x^2+y^2)^{-\frac{1}{2}}, 1\right)$

$\Rightarrow (\nabla\times\vec{F})\cdot\dfrac{\partial\vec{r}}{\partial x}\times\dfrac{\partial\vec{r}}{\partial y} = (yz, -1, 1)\cdot\left(-x(x^2+y^2)^{-\frac{1}{2}}, -y(x^2+y^2)^{-\frac{1}{2}}, 1\right)$

$= -xy + y(x^2+y^2)^{-\frac{1}{2}} + 1, \quad \forall (x,y,z)\in S$

$\therefore \iint_S (\nabla\times\vec{F})\cdot\vec{n}\,dA = -\iint_R (\nabla\times\vec{F})\cdot\dfrac{\partial\vec{r}}{\partial x}\times\dfrac{\partial\vec{r}}{\partial y}\,dxdy = \iint_R xy - y(x^2+y^2)^{-\frac{1}{2}} - 1\,dxdy$

令 $x = r\cos\theta$, $y = r\sin\theta$ 则 $\{(x,y): x^2+y^2 \leq a^2\} = \{(r,\theta): 0\leq r\leq a, 0\leq\theta\leq 2\pi\}$

$\therefore \iint_R xy - y(x^2+y^2)^{-\frac{1}{2}} - 1\,dxdy = \int_0^{2\pi}\int_0^a (r^2\cos\theta\sin\theta - \sin\theta - 1)\,rdrd\theta = -\pi a^2$

(2)

令 $C: z = a, x^2+y^2 = a^2$ 为逆时针封闭曲线

$\because S$ 为朝下定向且 $\vec{F}$ 的各分量一阶偏导数存在且连续

藉由 Stokes' Theorem, $\displaystyle\iint_S (\nabla \times \vec{F}) \cdot \vec{n}\,dA = -\oint_C \vec{F} \cdot d\vec{r}$

令 $\vec{r}(t) = (a\cos t, a\sin t, a)$, $\ 0 \le t \le 2\pi$ 则 $\vec{r}'(t) = (-a\sin t, a\cos t, 0)$

$$\therefore \oint_C \vec{F} \cdot d\vec{r} = -\int_0^{2\pi} \vec{F}(\vec{r}(t)) \cdot \vec{r}'(t)\,dt$$

$$= -\int_0^{2\pi} \left(-a, a\cos t, \frac{a^3 \sin^2 t}{2}\right) \cdot (-a\sin t, a\cos t, 0)\,dt = -\int_0^{2\pi} a^2 \sin t + a^2 \cos^2 t\,dt$$

$$= -2\pi \cdot \frac{a^2}{2} = -\pi a^2$$

## 9.7  Gauss's Theorem

之前在说明 Green's Theorem 时有提到, Green's Theorem 的散度形式可改写为

$$\oint_C \vec{F} \cdot \vec{n}\,ds = \iint_R \nabla \cdot \vec{F}(x, y)\,dA$$

其中 $C$ 是平面区域 $R$ 的正向（逆时针）边界曲线, 将其扩展至三维空间的直观猜测为

$$\oiint_S \vec{F} \cdot \vec{n}\,dA = \iiint_V \nabla \cdot \vec{F}(x, y, z)\,dV$$

其中 $S$ 是封闭体积 $V$ 的边界表面

若 $\vec{F}$ 的一阶偏导数存在且连续则上述等式成立且称为 The Divergence Theorem(散度定理, 或 Gauss's Theorem), 值得注意的是 The Divergence Theorem、Stokes' Theorem 与 Green's Theorem 都是微积分基本定理的扩展, 此外, The Divergence Theorem 可以将向量场 $\vec{F}$ 穿越封闭曲面 $S$ 的通量（曲面积分）转换为 $\vec{F}$ 的散度在封闭体积 $V$ 的三重积分, 例如, 假设一个封闭曲面由 $S_1, S_2, S_3$ 组成（如下图）, 原本求通量的方法是求各自的通量再相加, 而藉由 The Divergence Theorem 则将问题转成纯量函数的体积分, 相反地, 也可藉由计算 $\vec{F}$ 在封闭曲

面$S$的通量(曲面积分)求得$\vec{F}$的散度在封闭体积 V 的三重积分

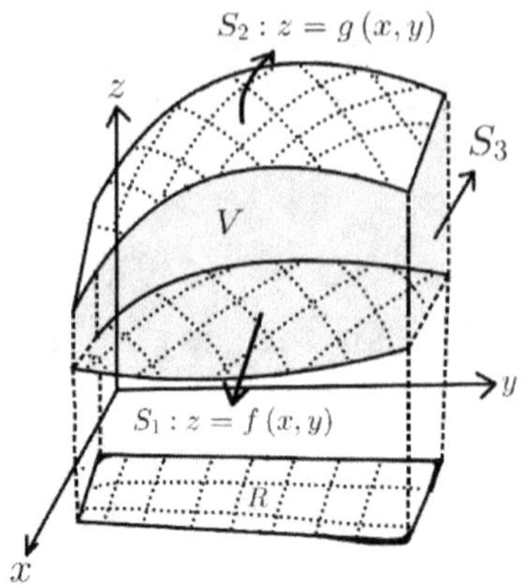

    The Divergence Theorem 的应用情境是曲面为封闭曲面，而 Stokes' Theorem 是非封闭曲面，   Green's Theorem 是出现二维封闭曲线，底下展示常见的封闭曲面

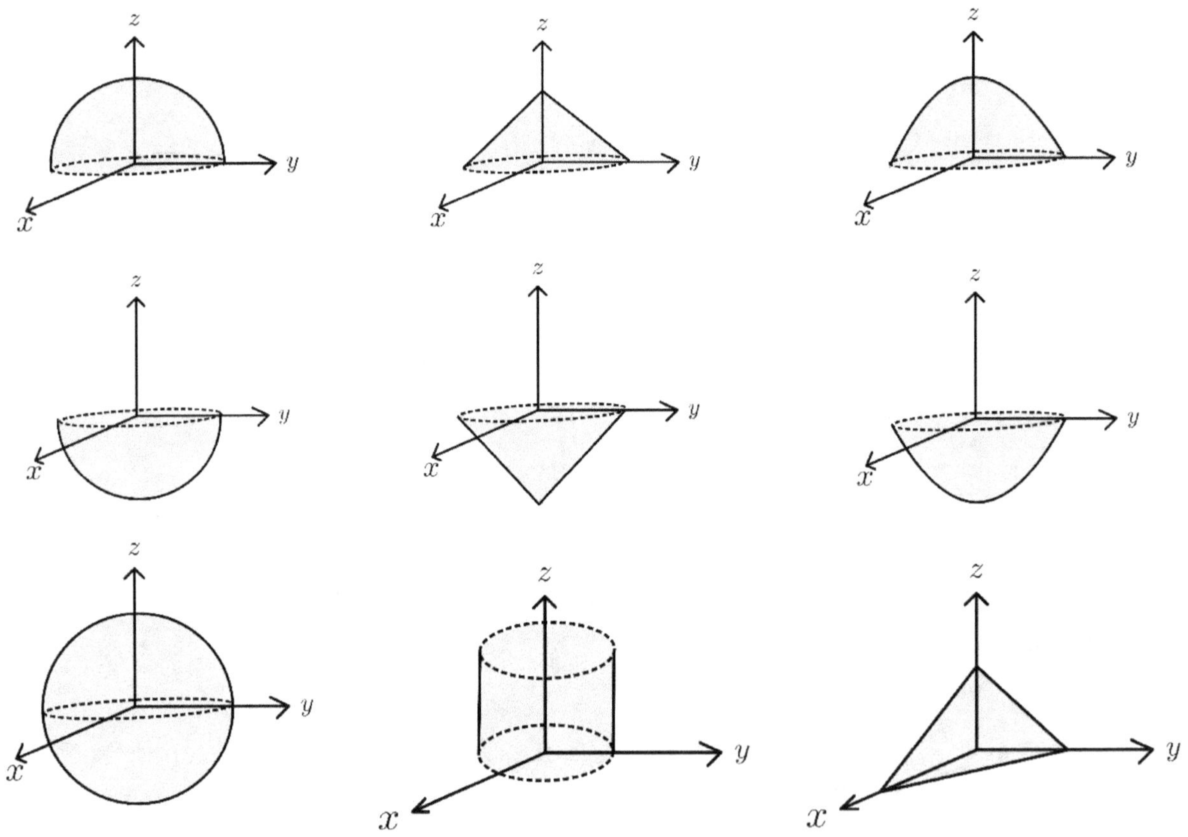

    以直观的角度而言，$\vec{F}$于封闭曲面的通量转成三重积分后,被积分函数为何应该是$\vec{F}$散度的形式，假设体积V切成无限多个小长方体，取一个顶点为$(x, y, z)$且边长为$\Delta x$，$\Delta y$,

$\Delta z$的小长方体为例，通过这个小长方体的通量是通过六个面的通量之和，考虑与$yz$平面平行的边，穿过底面的通量近似为

$$F_1(x, y, z)\Delta y\Delta z$$

穿过顶面的通量近似为

$$F_1(x + \Delta x, y, z)\Delta y\Delta z$$

顶面与底面通量的差异为

$$\big(F_1(x + \Delta x, y, z) - F_1(x, y, z)\big)\Delta y\Delta z$$

根据线性逼近

$$F_1(x + \Delta x, y, z) \approx F_1(x, y, z) + \frac{\partial F_1}{\partial x}\Delta x$$

顶面与底面通量的差异近似估计为

$$\big(F_1(x + \Delta x, y, z) - F_1(x, y, z)\big)\Delta y\Delta z = \left(F_1(x, y, z) + \frac{\partial F_1}{\partial x}\Delta x - F_1(x, y, z)\right)\Delta y\Delta z$$

$$= \frac{\partial F_1}{\partial x}\Delta x\Delta y\Delta z$$

同理，另外两组的近似差异分别为

$$\frac{\partial F_2}{\partial y}\Delta x\Delta y\Delta z \quad 与 \quad \frac{\partial F_3}{\partial z}\Delta x\Delta y\Delta z$$

将这些加起来，我们得到

$$\left(\frac{\partial F_1}{\partial x} + \frac{\partial F_2}{\partial y} + \frac{\partial F_3}{\partial z}\right)\Delta x\Delta y\Delta z$$

因此，得到了$\vec{F}$散度的形式

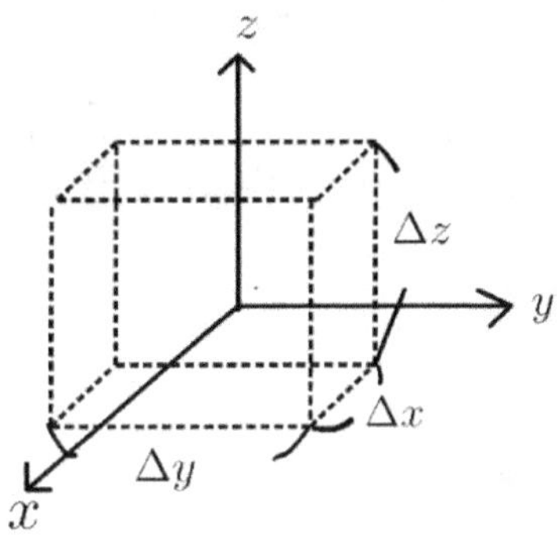

The Divergence Theorem 的考试类型包含把封闭面积分转成能直接计算的体积分、把封闭面积分转成的体积分再搭配坐标转换、把体积分转成封闭的面积分

【**定义**】Divergence of $\vec{\mathrm{F}}$

假设 $\vec{\mathrm{F}}(x,y,z) = F_1\mathbf{i} + F_2\mathbf{j} + F_3\mathbf{k}$ 定义于 V 的一个连续向量场，则 $\vec{\mathrm{F}}$ 的散度定义为

$$\nabla \cdot \vec{\mathrm{F}} = \frac{\partial F_1}{\partial x} + \frac{\partial F_2}{\partial y} + \frac{\partial F_3}{\partial z}$$

【**定义**】Simple Solid Regions

*A* solid region V is said to be of

(i) type I if $\exists$ continuous functions $f(x,y), g(x,y)$ such that

$\quad$ $V = \{(x,y,z): (x,y) \in \mathrm{R}, f(x,y) \leq z \leq g(x,y)\}$，$R$为 $V$投影至$xy$平面的区域

(ii) type II if $\exists$ continuous functions $f(y,z), g(y,z)$ such that

$\quad$ $V = \{(x,y,z): (y,z) \in \mathrm{R}, f(y,z) \leq x \leq g(y,z)\}$，$R$为 $V$投影至$yz$平面的区域

(ii) type III if $\exists$ continuous functions $f(x,z), g(x,z)$ such that

$\quad$ $V = \{(x,y,z): (x,z) \in \mathrm{R}, f(x,z) \leq y \leq g(x,z)\}$，$R$为 $V$投影至$xz$平面的区域

(iv) simple solid region if it is concurrently of types I, II, and III.


以 simple solid region 为例，图形如下图所示，特别的是垂直于$xy$平面的平面$S_3$有时可能不存在

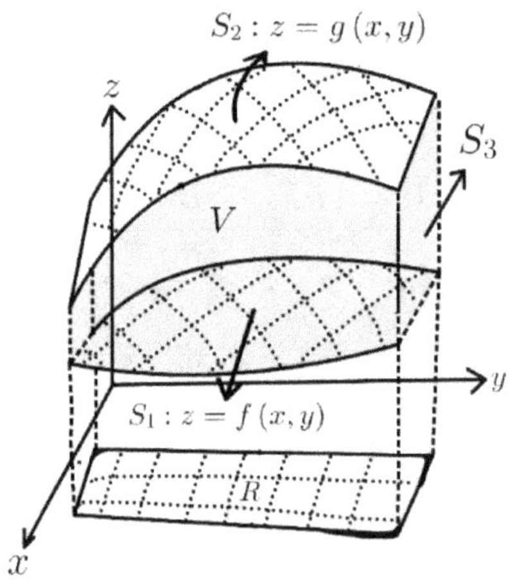


【**定理**】The Divergence Theorem

假设

1.$S$ 为一具有朝外定向的封闭曲面，所围区域为 V 且 V 为 simple solid region

2.向量场$F(x,y,z) = F_1\mathbf{i} + F_2\mathbf{j} + F_3\mathbf{k}$的各分量函数在包含 V 的开集合上有连续的偏导数则

$$\oiint_{S} \vec{\mathrm{F}} \cdot \vec{n}\, dA = \iiint_{V} \nabla \cdot \vec{\mathrm{F}}\, dV$$

Proof:

$$\because \oiint_S \vec{F} \cdot \vec{n}\, dA = \iint_S F_1\mathbf{i} \cdot \vec{n}\, dA + \iint_S F_2\mathbf{j} \cdot \vec{n}\, dA + \iint_S F_3\mathbf{k} \cdot \vec{n}\, dA$$

且

$$\iiint_V \nabla \cdot \vec{F}\, dV = \iiint_V \frac{\partial F_1}{\partial x}\, dV + \iiint_V \frac{\partial F_2}{\partial y}\, dV + \iiint_V \frac{\partial F_3}{\partial z}\, dV$$

所以

若 $\begin{cases} \displaystyle\iint_S F_1\mathbf{i} \cdot \vec{n}\, dA = \iiint_V \frac{\partial F_1}{\partial x}\, dV \\[2ex] \displaystyle\iint_S F_2\mathbf{j} \cdot \vec{n}\, dA = \iiint_V \frac{\partial F_2}{\partial y}\, dV \\[2ex] \displaystyle\iint_S F_3\mathbf{k} \cdot \vec{n}\, dA = \iiint_V \frac{\partial F_3}{\partial z}\, dV \end{cases}$ 则 $\oiint_S \vec{F} \cdot \vec{n}\, dA = \iiint_V \nabla \cdot \vec{F}\, dV$

Claim: $\displaystyle\iint_S F_3\mathbf{k} \cdot \vec{n}\, dA = \iiint_V \frac{\partial F_3}{\partial z}\, dV$

令 V 为 solid region of type I 且包含垂直于 $xy$ 平面的平面

令 $R$ 为封闭曲面 $S$ 投影至 $xy$ 平面的封闭区域且曲面 $S$ 为三块曲面所组成 $S_1, S_2$ 与 $S_3$，其中 $S_1$ 为下方曲面：$z = f(x, y)$，$S_2$ 为上方曲面：$z = g(x, y)$，$S_3$ 为垂直于 $xy$ 平面的平面

且 $V = \{(x, y, z): (x, y) \in R, f(x, y) \le z \le g(x, y)\}$,

藉由微积分基本定理

$$\iiint_V \frac{\partial F_3}{\partial z}\, dV = \iint_R \int_{f(x,y)}^{g(x,y)} \frac{\partial F_3}{\partial z}\, dz\, dx\, dy = \iint_R F_3(x, y, g) - F_3(x, y, f)\, dx\, dy$$

另一方面

$$\iint_S F_3\mathbf{k} \cdot \vec{n}\, dA = \iint_{S_1} F_3\mathbf{k} \cdot \vec{n}\, dA + \iint_{S_2} F_3\mathbf{k} \cdot \vec{n}\, dA + \iint_{S_3} F_3\mathbf{k} \cdot \vec{n}\, dA$$

$$\because \iint_{S_2} F_3\mathbf{k} \cdot \vec{n}\, dA = \iint_R F_3(x, y, g)\, dx\, dy \text{ and } \iint_{S_1} F_3\mathbf{k} \cdot \vec{n}\, dA = -\iint_R F_3(x, y, f)\, dx\, dy$$

$\because \mathbf{k}$ 是垂直方向且 $S_3$ 的 $\vec{n}$ 是水平方向 $\therefore \displaystyle\iint_{S_3} F_3\mathbf{k} \cdot \vec{n}\, dA = 0$

$$\therefore \iint_S F_3\mathbf{k} \cdot \vec{n}\, dA = \iint_R F_3(x, y, g)\, dx\, dy - \iint_R F_3(x, y, f)\, dx\, dy$$

$$= \iint_R F_3(x, y, g) - F_3(x, y, f)\, dx\, dy = \iiint_V \frac{\partial F_3}{\partial z}\, dV$$

$$\therefore \iint_S F_3 \mathbf{k} \cdot \vec{n}\, dA = \iiint_V \frac{\partial F_3}{\partial z}\, dV.$$

$\therefore$ V 若为 simple solid region of type I 且沒有包含垂直于$xy$平面的平面, 僅由$S_1, S_2$组成,
则上述等式仍成立,

Claim: $\displaystyle \iint_S F_1 \mathbf{i} \cdot \vec{n}\, dA = \iiint_V \frac{\partial F_1}{\partial x}\, dV$

令 V 为 solid region of type II

同理可证

$$\iint_S F_1 \mathbf{i} \cdot \vec{n}\, dA = \iiint_V \frac{\partial F_1}{\partial x}\, dV$$

Claim: $\displaystyle \iint_S F_2 \mathbf{j} \cdot \vec{n}\, dA = \iiint_V \frac{\partial F_2}{\partial y}\, dV$

令 V 为 solid region of type III

同理可证

$$\iint_S F_2 \mathbf{j} \cdot \vec{n}\, dA = \iiint_V \frac{\partial F_2}{\partial y}\, dV$$

$$\Rightarrow \oiint_S \vec{F} \cdot \vec{n}\, dA = \iiint_V \nabla \cdot \vec{F}\, dV$$

## 9.7.1　把封闭曲面积分转成能直接计算的体积分

使用 Gauss's Theorem 把封闭面积分转成能直接计算的体积分

考试类型:

Type 1.

假设向量场$\vec{F}$各分量的一阶偏导数存在且连续且 $\nabla \cdot \vec{F} = c\,(常数)$, $S: x^2 + y^2 + z^2 = r^2$,

则流出$S$的 flux(通量) $\displaystyle \oiint_S \vec{F} \cdot \vec{n}\, dA = \frac{4c\pi r^3}{3}$

解题流程:

Step1.

$\because \vec{F}$各分量的一阶偏导数存在且连续, 藉由 Gauss's Theorem, $\displaystyle \oiint_S \vec{F} \cdot \vec{n}\, dA = \iiint_V \nabla \cdot \vec{F}\, dV$

Step2.

$$\because \nabla \cdot \vec{F} = c \quad \therefore \iiint_V \nabla \cdot \vec{F}\, dV = \iiint_{x^2+y^2+z^2 \leq r^2} c\, dxdydz = c \cdot \frac{4\pi}{3} \cdot r^3$$

范例说明:

(I)若 $\vec{F} = (x^2 y, 2y + z^2, 2z - 2xyz)$, $S: x^2 + y^2 + z^2 = r^2$ 为朝外定向, 求 $\oiint_S \vec{F} \cdot \vec{n}\, dA = ?$

$\because \vec{F}$ 各分量的一阶偏导数存在且连续

藉由 Gauss's Theorem,

$$\oiint_S \vec{F} \cdot \vec{n}\, dA = \iiint_V \nabla \cdot \vec{F}\, dV = \iiint_{x^2+y^2+z^2 \leq r^2} 4\, dxdydz = 4 \cdot \frac{4\pi}{3} \cdot r^3 = \frac{16\pi r^3}{3}$$

(II)若 $\vec{F} = (x + y + \sin z^2, y + e^{x^2}, z + \ln(x^2 + y^2 + 1))$, $S: x^2 + y^2 + z^2 = r^2$ 为朝外定向,

求 $\oiint_S \vec{F} \cdot \vec{n}\, dA = ?$

$\because \vec{F}$ 各分量的一阶偏导数存在且连续

藉由 Gauss's Theorem,

$$\oiint_S \vec{F} \cdot \vec{n}\, dA = \iiint_V \nabla \cdot \vec{F}\, dV = \iiint_{x^2+y^2+z^2 \leq r^2} 3\, dxdydz = 3 \cdot \frac{4\pi}{3} \cdot r^3 = 4\pi r^3$$

Type 2.

假设 $\vec{F}$ 各分量的一阶偏导数存在且连续且 $\nabla \cdot \vec{F} = c$, $c \in \mathbb{R}$, $S: x^2 + y^2 = a^2, 0 \leq z \leq b$

为朝外定向封闭曲面, $a, b > 0$, 则 $\oiint_S \vec{F} \cdot \vec{n}\, dA = \pi a^2 bc$

解题流程:

Step1.

$\because \vec{F}$ 各分量的一阶偏导数存在且连续

藉由 Gauss's Theorem, $\oiint_S \vec{F} \cdot \vec{n}\, dA = \iiint_V \nabla \cdot \vec{F}\, dV$

Step2.

$\because \nabla \cdot \vec{F} = c \quad \therefore \iiint_V \nabla \cdot \vec{F}\, dV = \iiint_V c\, dxdydz = c \cdot \pi a^2 b$

范例说明:

(I)若$\vec{F} = (x, e^x + y, 2z),\ S: x^2 + y^2 = a^2,\ 0 \le z \le b,\ $ 求 $\oiint_S \vec{F} \cdot \vec{n}\, dA =?$

藉由 Gauss's Theorem,

则 $\oiint_S \vec{F} \cdot \vec{n}\, dA = \iiint_V \nabla \cdot \vec{F}\, dV = \iiint_V 4\, dxdydz = 4\pi a^2 b$

Example 1.

$$求 \oiint_{S_1} \vec{F} \cdot \vec{n}\, dA - \oiint_{S_2} \vec{F} \cdot \vec{n}\, dA =?,\ \ \vec{F} = (x, 2y + z, z + x^2),\ \ S_1: x^2 + y^2 + z^2 = 4,$$

$S_2: x^2 + y^2 + z^2 = 1$ 皆为朝外定向

【解】

令 $V = \{(x, y, z): 1 \le x^2 + y^2 + z^2 \le 4\}$

$\because \vec{F}$ 的各分量一阶偏导数存在且连续

藉由 Gauss's Theorem 则 $\oiint_{S_1} \vec{F} \cdot \vec{n}\, dA - \oiint_{S_2} \vec{F} \cdot \vec{n}\, dA = \iiint_V \nabla \cdot \vec{F}\, dV$

$\because \nabla \cdot \vec{F} = 1 + 2 + 1 = 4$

$\therefore \iiint_V \nabla \cdot \vec{F}\, dV = \iiint_{1 \le x^2 + y^2 + z^2 \le 4} 4\, dxdydz = 4\left(\dfrac{4\pi}{3} \cdot 2^3 - \dfrac{4\pi}{3} \cdot 1^3\right) = \dfrac{112\pi}{3}$

Example 2.

$$求向量场\vec{F} = (x^2 y, 2y + z^2, 2z - 2xyz) 流出 S 的\ \text{flux}(通量) =?,\ \ S: x^2 + y^2 + z^2 = r^2$$

【解】

$\because \vec{F}$ 的各分量一阶偏导数存在且连续

藉由 Gauss's Theorem 则 $\oiint_S \vec{F} \cdot \vec{n}\, dA = \iiint_V \nabla \cdot \vec{F}\, dV,\ \ V: x^2 + y^2 + z^2 \le r^2$

$\because \nabla \cdot \vec{F} = 2xy + 2 + 2 - 2xy = 4$

$\therefore \iiint_V \nabla \cdot \vec{F}\, dV = \iiint_{x^2 + y^2 + z^2 \le r^2} 4\, dxdydz = 4 \cdot \dfrac{4\pi}{3} \cdot r^3 = \dfrac{16\pi r^3}{3}$

Example 3.

$$求 \oiint_S \vec{F} \cdot \vec{n}\, dA =?,\ \ \vec{F} = (x, y, z),\ \ S: x^2 + y^2 + z^2 = 1\ 为朝外定向$$

【解】

$\because \vec{F}$ 的各分量一阶偏导数存在且连续

藉由 Gauss's Theorem 则 $\qquad \oiint_S \vec{F} \cdot \vec{n}\,dA = \iiint_V \nabla \cdot \vec{F}\,dV, \quad V: x^2 + y^2 + z^2 \leq 1$

$\because \nabla \cdot \vec{F} = 1 + 1 + 1 = 3$

$\therefore \iiint_V \nabla \cdot \vec{F}\,dV = \iiint_{x^2+y^2+z^2 \leq 1} 3\,dxdydz = 3\left(\frac{4\pi}{3} \cdot 1^3\right) = 4\pi$

Example 4.

$\quad$ 求向量场 $\vec{F} = \left(x + y + \sin z^2,\, y + e^{x^2},\, z + \ln(x^2 + y^2 + 1)\right)$ 流出 $S$ 的 flux(通量)$=?$,
$S: x^2 + y^2 + z^2 = r^2$

【解】

$\because \vec{F}$ 的各分量一阶偏导数存在且连续

藉由 Gauss's Theorem 则 $\qquad \oiint_S \vec{F} \cdot \vec{n}\,dA = \iiint_V \nabla \cdot \vec{F}\,dV, \quad V: x^2 + y^2 + z^2 \leq r^2$

$\because \nabla \cdot \vec{F} = 1 + 1 + 1 = 3$

$\therefore \iiint_V \nabla \cdot \vec{F}\,dV = \iiint_{x^2+y^2+z^2 \leq r^2} 3\,dxdydz = 3 \cdot \frac{4\pi}{3} \cdot r^3 = 4\pi r^3$

Example 5.

$\quad$ 求 $\oiint_S \vec{F} \cdot \vec{n}\,dA = ?, \quad \vec{F} = (2x + ye^z,\, e^x - ye^z,\, e^z + 3z - xy),\quad S: x^2 + y^2 + z^2 = 9$
$\quad$ 为朝外定向

【解】

$\because \vec{F}$ 的各分量一阶偏导数存在且连续

藉由 Gauss's Theorem 则 $\qquad \oiint_S \vec{F} \cdot \vec{n}\,dA = \iiint_V \nabla \cdot \vec{F}\,dV, \quad V: x^2 + y^2 + z^2 \leq 9$

$\because \nabla \cdot \vec{F} = 2 - e^z + e^z + 3 = 5$

$\therefore \iiint_V \nabla \cdot \vec{F}\,dV = \iiint_{x^2+y^2+z^2 \leq 9} 5\,dxdydz = 5\left(\frac{4\pi}{3} \cdot 3^3\right) = 180\pi$

Example 6.

$$\text{求} \oiint_S \vec{F} \cdot \vec{n}\, dA = ?, \quad \vec{F} = (x, e^x + y, 2z), \quad S: x^2 + y^2 = a^2, \quad 0 \le z \le b \text{为朝外定向}$$

【解】

令 $V = \{(x, y, z): x^2 + y^2 \le a^2, 0 \le z \le b\}$

$\because \vec{F}$ 的各分量一阶偏导数存在且连续

藉由 Gauss's Theorem 则 $\displaystyle \oiint_S \vec{F} \cdot \vec{n}\, dA = \iiint_V \nabla \cdot \vec{F}\, dV$

$\because \nabla \cdot \vec{F} = 1 + 1 + 2 = 4 \quad \therefore \iiint_V \nabla \cdot \vec{F}\, dV = \iiint_V 4\, dx\, dy\, dz = 4\pi a^2 b$

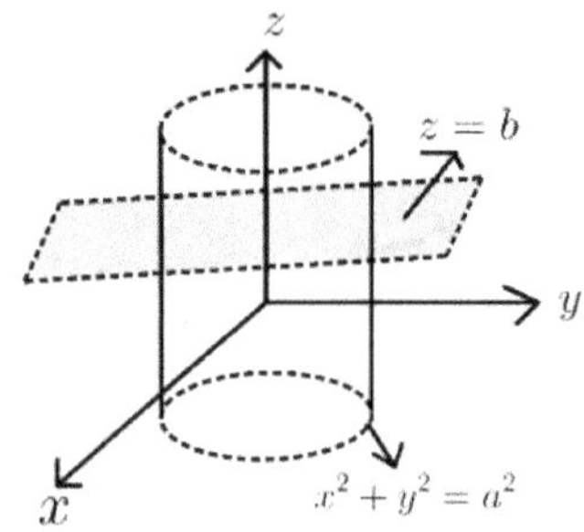


Example 7.

$$\text{求} \oiint_S (2x + 2y + z^2)\, dA = ?, \quad S: x^2 + y^2 + z^2 = r^2 \text{ 为朝外定向}$$

【解】

令 $\vec{F} = (2, 2, z)$, $\quad \because \vec{F}$ 的各分量一阶偏导数存在且连续

藉由 Gauss's Theorem 则 $\displaystyle \oiint_S \vec{F} \cdot \vec{n}\, dA = \iiint_V \nabla \cdot \vec{F}\, dV, \quad V: x^2 + y^2 + z^2 \le r^2$

$\because \nabla \cdot \vec{F} = 0 + 0 + 1 = 1 \quad \therefore \iiint_V \nabla \cdot \vec{F}\, dV = \iiint_V 1\, dx\, dy\, dz = \dfrac{4\pi}{3} \cdot r^3 = \dfrac{4\pi r^3}{3}$

Example 8.

Let $\vec{F} = (x, 2y, 3z)$. Assume $S$ is the cube with the vertices $(\pm 1, \pm 1, \pm 1)$ with outward orientation. Find $\displaystyle \oiint_S \vec{F} \cdot \vec{n}\, dA$ by using Gauss's Theorem

【解】

令 $V = \{(x, y, z): -1 \le x \le 1, -1 \le y \le 1, -1 \le z \le 1\}$

$\because \vec{F}$ 的各分量一阶偏导数存在且连续,

藉由 Gauss's Theorem 则 $\quad \oiint_S \vec{F} \cdot \vec{n}\, dA = \iiint_V \nabla \cdot \vec{F}\, dV$

$\because \nabla \cdot \vec{F} = 1 + 2 + 3 = 6 \quad \therefore \iiint_V \nabla \cdot \vec{F}\, dV = \iiint_V 6\, dx\,dy\,dz = 6 \cdot 2 \cdot 2 \cdot 2 = 48$

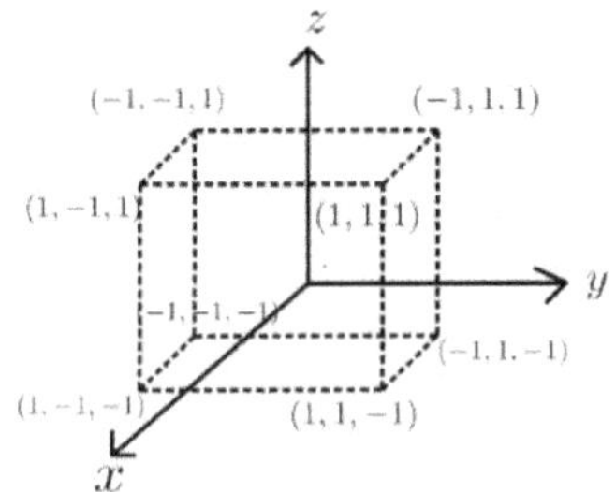


## 9.7.2 　把封闭曲面积分转成体积分

应用情境为封闭曲面积分的被积分向量值函数 $\vec{F}$ 本身较为复杂,其散度 $\nabla \cdot \vec{F}$ 较为干净

考试类型:

Type 1.

假设 $\vec{F}$ 的各分量一阶偏导数存在且连续,$S: x^2 + y^2 + z^2 = r^2$,$r > 0$ 且 $S$ 为朝外定向

求 $\oiint_S \vec{F} \cdot \vec{n}\, dA = ?$

解题流程:

Step1.

$\because \vec{F}$ 的各分量一阶偏导数存在且连续

藉由 Gauss's Theorem,$\quad \oiint_S \vec{F} \cdot \vec{n}\, dA = \iiint_V \nabla \cdot \vec{F}\, dV$

Step2.

$\iiint_V \nabla \cdot \vec{F}\, dV = \iiint_{x^2+y^2+z^2 \le r^2} f(x,y,z)\, dx\,dy\,dz$,　其中 $\nabla \cdot \vec{F} = f(x,y,z)$

Step3.

令 $x = \rho \sin\varphi \cos\theta\,,\, y = \rho \sin\varphi \sin\theta\,,\, z = \rho \cos\varphi$ 则

$$dxdydz = \begin{Vmatrix} \dfrac{\partial x}{\partial \rho} & \dfrac{\partial x}{\partial \varphi} & \dfrac{\partial x}{\partial \theta} \\[2mm] \dfrac{\partial y}{\partial \rho} & \dfrac{\partial y}{\partial \varphi} & \dfrac{\partial y}{\partial \theta} \\[2mm] \dfrac{\partial z}{\partial \rho} & \dfrac{\partial z}{\partial \varphi} & \dfrac{\partial z}{\partial \theta} \end{Vmatrix} d\rho d\varphi d\theta$$

$$= \begin{Vmatrix} \sin\varphi\cos\theta & \rho\cos\varphi\cos\theta & -\rho\sin\varphi\sin\theta \\ \sin\varphi\sin\theta & \rho\cos\varphi\sin\theta & -\rho\sin\varphi\cos\theta \\ \cos\varphi & -\rho\sin\varphi & 0 \end{Vmatrix} d\rho d\varphi d\theta = \rho^2 \sin\varphi \, d\rho d\varphi d\theta$$

Step4.

$$\therefore \iiint_{x^2+y^2+z^2 \leq r^2} f(x,y,z)\, dxdydz$$

$$= \int_0^\pi \int_0^{2\pi} \int_0^r f(\rho\sin\varphi\cos\theta, \rho\sin\varphi\sin\theta, \rho\cos\varphi) \cdot \rho^2 \sin\varphi \, d\rho d\theta d\varphi$$

$$求 \int_0^\pi \int_0^{2\pi} \int_0^r f(\rho\sin\varphi\cos\theta, \rho\sin\varphi\sin\theta, \rho\cos\varphi) \cdot \rho^2 \sin\varphi \, d\rho d\theta d\varphi = ?$$

<u>范例说明:</u>

(I) $\vec{F} = (2x + 3xy^2 + z, xy^2 + 3x^2y + e^z, z^3 - 2xyz)$, $S$ 为朝外定向: $x^2 + y^2 + z^2 = r^2$

求 $\oiint_S \vec{F} \cdot \vec{n} dA$

$\because \vec{F}$ 的各分量一阶偏导数存在且连续

藉由 Gauss's Theorem

$$\oiint_S \vec{F} \cdot \vec{n} dA = \iiint_V \nabla \cdot \vec{F} \, dV = \iiint_{x^2+y^2+z^2 \leq r^2} 2 + 3x^2 + 3y^2 + 3z^2 \, dxdydz$$

令 $x = \rho\sin\varphi\cos\theta$, $y = \rho\sin\varphi\sin\theta$, $z = \rho\cos\varphi$ 则 $dxdydz = \rho^2 \sin\varphi \, d\rho d\varphi d\theta$

$$\therefore \iiint_{x^2+y^2+z^2 \leq r^2} 2 + 3x^2 + 3y^2 + 3z^2 \, dxdydz = \frac{8\pi r^3}{3} + \int_0^\pi \int_0^{2\pi} \int_0^r 3\rho^2 \cdot \rho^2 \sin\varphi \, d\rho d\theta d\varphi$$

$$= \frac{8\pi r^3}{3} + \frac{12\pi r^5}{5}$$

(II) 若 $\vec{F} = (x^3 + z, y^3 + e^{-z}, z^3)$, $S: x^2 + y^2 + z^2 = a^2$ 且 $S$ 为朝外定向, 求 $\oiint_S \vec{F} \cdot \vec{n} dA$

$\because \vec{F}$ 的各分量一阶偏导数存在且连续

藉由 Gauss's Theorem

$$\oiint_S \vec{F} \cdot \vec{n}\, dA = \iiint_V \nabla \cdot \vec{F}\, dV = \iiint_{x^2+y^2+z^2 \leq a^2} 3x^2 + 3y^2 + 3z^2 \, dxdydz$$

令 $x = \rho \sin\varphi \cos\theta$ , $y = \rho \sin\varphi \sin\theta$ , $z = \rho \cos\varphi$ 则 $dxdydz = \rho^2 \sin\varphi \, d\rho d\varphi d\theta$

$$\therefore \iiint_{x^2+y^2+z^2 \leq a^2} 3x^2 + 3y^2 + 3z^2 \, dxdydz = 3\int_0^\pi \int_0^{2\pi} \int_0^a \rho^2 \cdot \rho^2 \sin\varphi \, d\rho d\theta d\varphi = \frac{12\pi a^5}{5}$$

Type 2.

假设向量场 $\vec{F}$ 的各分量一阶偏导数存在且连续, $S$ 为 $x^2 + y^2 = a^2$ , $0 \leq z \leq b$ 所围朝外

定向封闭曲面, 求 $\oiint_S \vec{F} \cdot \vec{n}\, dA =$ ?

解题流程:

Step1.

$\because \vec{F}$ 的各分量一阶偏导数存在且连续

藉由 Gauss's Theorem, $\oiint_S \vec{F} \cdot \vec{n}\, dA = \iiint_V \nabla \cdot \vec{F}\, dV$

Step2.

$$\iiint_V \nabla \cdot \vec{F}\, dV = \iiint_V f(x,y) \, dxdydz = \iint_R \int_0^b f(x,y) dz dxdy = b \iint_R f(x,y) \, dxdy$$

其中 $\nabla \cdot \vec{F} = f(x,y)$ , $R = \{(x,y): x^2 + y^2 \leq a^2\}$

Step3.

令 $x = r\cos\theta$ , $y = r\sin\theta$ 则 $\{(x,y): x^2 + y^2 \leq a^2\} = \{(r,\theta): 0 \leq r \leq a, 0 \leq \theta \leq 2\pi\}$

$$\text{且 } dxdy = \left\| \begin{matrix} \dfrac{\partial x}{\partial r} & \dfrac{\partial x}{\partial \theta} \\ \dfrac{\partial y}{\partial r} & \dfrac{\partial y}{\partial \theta} \end{matrix} \right\| drd\theta = \left\| \begin{matrix} \cos\theta & -r\sin\theta \\ \sin\theta & r\cos\theta \end{matrix} \right\| drd\theta = r drd\theta$$

$$\therefore b \iint_R f(x,y) \, dxdy = b \int_0^{2\pi} \int_0^a f(r\cos\theta, r\sin\theta) r drd\theta$$

Step4.

求 $\displaystyle\int_0^{2\pi} \int_0^a f(r\cos\theta, r\sin\theta) r drd\theta =$ ?

<u>范例说明:</u>

(I)假设 $\vec{F} = (2x^3, x^2y, x^2z)$, $S: x^2 + y^2 = a^2$, $0 \le z \le b$, 求 $\displaystyle\oiint_S \vec{F} \cdot \vec{n}\,dA = ?$

令 $R = \{(x,y): x^2 + y^2 \le a^2\}$ 且 $x = r\cos\theta, y = r\sin\theta$

则 $\{(x,y): x^2 + y^2 \le a^2\} = \{(r,\theta): 0 \le r \le a, 0 \le \theta \le 2\pi\}$

藉由 Gauss's Theorem

$$\oiint_S \vec{F} \cdot \vec{n}\,dA = \iiint_V \nabla \cdot \vec{F}\,dV = \iint_R \int_0^b 8x^2\,dz\,dx\,dy = b\iint_{x^2+y^2 \le a^2} 8x^2\,dx\,dy$$

$$= b\int_0^{2\pi}\int_0^a 8r^2\cos^2\theta\,r\,dr\,d\theta = 2\pi a^4 b$$

Type 3.

假设 $\vec{F}$ 的各分量一阶偏导数存在且连续, $S$ 为曲面 $z = g(x,y)$ 与 $z = 0$ 所围朝外定向封

闭曲面, 其中 $z = g(x,y)$ 投影至 $xy$ 平面 $= \{(x,y): x^2 + y^2 \le a^2\}$, 求 $\displaystyle\oiint_S \vec{F} \cdot \vec{n}\,dA = ?$

解题流程:

Step1.

令 $R = \{(x,y): x^2 + y^2 \le a^2\}$ 且 $x = r\cos\theta, y = r\sin\theta$

则 $\{(x,y): x^2 + y^2 \le a^2\} = \{(r,\theta): 0 \le r \le a, 0 \le \theta \le 2\pi\}$

Step2.

$\because \vec{F}$ 的各分量一阶偏导数存在且连续

藉由 Gauss's Theorem

$$\oiint_S \vec{F} \cdot \vec{n}\,dA = \iiint_V \nabla \cdot \vec{F}\,dV = \iint_R \int_0^{g(x,y)} f(x,y,z)\,dz\,dx\,dy, \text{ 其中 } \nabla \cdot \vec{F} = f(x,y,z)$$

Step3.

$$\iint_R \int_0^{g(x,y)} f(x,y,z)\,dz\,dx\,dy = \iint_R h(x,y)\,dx\,dy = \int_0^{2\pi}\int_0^a h(r\cos\theta, r\sin\theta)\,r\,dr\,d\theta$$

其中 $h(x,y) = \displaystyle\int_0^{g(x,y)} f(x,y,z)\,dz$

求 $\displaystyle\int_0^{2\pi}\int_0^a h(r\cos\theta, r\sin\theta)\,r\,dr\,d\theta$

<u>范例说明:</u>

(I)若 $\vec{F} = (x^2, xy, z)$, $S$ 为 $z = 4 - x^2 - y^2$ 与 $xy$ 平面所围封闭曲面, 求 $\oiint_S \vec{F} \cdot \vec{n} dA =$?

令 $R = \{(x, y): x^2 + y^2 \leq 4\}$ 且 $x = r\cos\theta, y = r\sin\theta$

则 $\{(x, y): x^2 + y^2 \leq 4\} = \{(r, \theta): 0 \leq r \leq 2, 0 \leq \theta \leq 2\pi\}$

藉由 Gauss's Theorem

$$\oiint_S \vec{F} \cdot \vec{n} dA = \iiint_V \nabla \cdot \vec{F} \, dV = \iint_R \int_0^{4-x^2-y^2} 3x + 1 \, dzdxdy$$

$$= \iint_R (3x + 1)(4 - x^2 - y^2) \, dxdy = \int_0^{2\pi} \int_0^2 (3\cos\theta + 1)(4 - r^2) r \, drd\theta = 8\pi$$


Type 4.

假设向量场 $\vec{F}$ 的各分量一阶偏导数存在且连续, 朝外定向封闭曲面 $S$ 所围体积为 V, 求

$\oiint_S \vec{F} \cdot \vec{n} dA =$?, 其中 $V = \{(x, y, z): x_0 \leq x \leq x_1, y_0 \leq y \leq y_1, z_0 \leq z \leq z_1\}$,

解题流程:

藉由 Gauss's Theorem, $\quad \oiint_S \vec{F} \cdot \vec{n} dA = \iiint_V \nabla \cdot \vec{F} \, dV = \int_{x_0}^{x_1} \int_{y_0}^{y_1} \int_{z_0}^{z_1} f(x, y, z) dzdydx$

其中 $\nabla \cdot \vec{F} = f(x, y, z)$

<u>范例说明:</u>

求 $\oiint_S \vec{F} \cdot \vec{n} dA =$?

(I) $\vec{F} = (xy + 3, y^2 + e^{zx}, \cos xy)$ 且 $S$ 为 $z = 1 - x^2$, $y = 0, z = 0$, $y + z = 2$ 所围封闭曲面

藉由 Gauss's Theorem, $\quad \oiint_S \vec{F} \cdot \vec{n} dA = \iiint_V \nabla \cdot \vec{F} \, dV = 3\int_{-1}^1 \int_0^{1-x} \int_0^{2-z} y \, dydzdx = \dfrac{184}{35}$

(II) $\vec{F} = (e^x \sin y, e^x \cos y, yz^2 + xy)$ 且 $S$ 为 $0 \leq x \leq 1, 0 \leq y \leq 1, 0 \leq z \leq 2$ 所围封闭曲面

藉由 Gauss's Theorem, $\quad \oiint_S \vec{F} \cdot \vec{n} dA = \iiint_V \nabla \cdot \vec{F} \, dV = \int_0^1 \int_0^1 \int_0^2 2yz \, dzdydx = 2$


Example 1.

$\quad \vec{F} = (2x + x^3 + ye^z, e^x - ye^z + y^3, e^z + 3z - xy + z^3)$, $S: x^2 + y^2 + z^2 = 4$,

$\quad$ 藉由 Gauss's Theorem 求 $\oiint_S \vec{F} \cdot \vec{n} dA =$?,

**【解】**

令 $V: x^2 + y^2 + z^2 \leq 4$

$\because \vec{F}$ 的各分量一阶偏导数存在且连续, 藉由 Gauss's Theorem 则 $\oiint_S \vec{F} \cdot \vec{n} \, dA = \iiint_V \nabla \cdot \vec{F} \, dV$

$\because \nabla \cdot \vec{F} = 5 + 3x^2 + 3y^2 + 3z^2$ $\therefore \iiint_V \nabla \cdot \vec{F} \, dV = \iiint_{x^2+y^2+z^2\leq 4} 5 + 3x^2 + 3y^2 + 3z^2 \, dxdydz$

令 $x = \rho \sin\varphi \cos\theta$, $y = \rho \sin\varphi \sin\theta$, $z = \rho \cos\varphi$ 则

$$dxdydz = \begin{Vmatrix} \dfrac{\partial x}{\partial \rho} & \dfrac{\partial x}{\partial \varphi} & \dfrac{\partial x}{\partial \theta} \\ \dfrac{\partial y}{\partial \rho} & \dfrac{\partial y}{\partial \varphi} & \dfrac{\partial y}{\partial \theta} \\ \dfrac{\partial z}{\partial \rho} & \dfrac{\partial z}{\partial \varphi} & \dfrac{\partial z}{\partial \theta} \end{Vmatrix} d\rho d\varphi d\theta$$

$$= \begin{Vmatrix} \sin\varphi\cos\theta & \rho\cos\varphi\cos\theta & -\rho\sin\varphi\sin\theta \\ \sin\varphi\sin\theta & \rho\cos\varphi\sin\theta & -\rho\sin\varphi\cos\theta \\ \cos\varphi & -\rho\sin\varphi & 0 \end{Vmatrix} d\rho d\varphi d\theta = \rho^2 \sin\varphi \, d\rho d\varphi d\theta$$

其中 $0 \leq \rho \leq 2$, $0 \leq \varphi \leq \pi$, $0 \leq \theta \leq 2\pi$

$$\therefore \iiint_{x^2+y^2+z^2\leq 4} 5 + 3x^2 + 3y^2 + 3z^2 \, dxdydz$$

$$= 5 \cdot \frac{4\pi}{3} \cdot 2^3 + \int_0^\pi \int_0^{2\pi} \int_0^3 3\rho^2 \cdot \rho^2 \sin\varphi \, d\rho d\theta d\varphi = \frac{160\pi}{3} + \frac{1152\pi}{5} = \frac{1952\pi}{15}$$

Example 2.

藉由 Gauss's Theorem 求向量场 $\vec{F} = (2x + 3xy^2 + z, xy^2 + 3x^2y + e^z, z^3 - 2xyz)$

流出 $S$ 的 flux(通量) $=?$, 其中 $S: x^2 + y^2 + z^2 = r^2$

**【解】**

令 $V: x^2 + y^2 + z^2 \leq r^2$

$\because \vec{F}$ 的各分量一阶偏导数存在且连续, 藉由 Gauss Theorem 则 $\oiint_S \vec{F} \cdot \vec{n} \, dA = \iiint_V \nabla \cdot \vec{F} \, dV$

$\because \nabla \cdot \vec{F} = 2 + 3y^2 + 2xy + 3x^2 + 3z^2 - 2xy = 2 + 3x^2 + 3y^2 + 3z^2$

$\therefore \iiint_V \nabla \cdot \vec{F} \, dV = \iiint_{x^2+y^2+z^2\leq r^2} 2 + 3x^2 + 3y^2 + 3z^2 \, dxdydz$

令 $x = \rho \sin\varphi \cos\theta$, $y = \rho \sin\varphi \sin\theta$, $z = \rho \cos\varphi$ 则

$$dxdydz = \begin{Vmatrix} \dfrac{\partial x}{\partial \rho} & \dfrac{\partial x}{\partial \varphi} & \dfrac{\partial x}{\partial \theta} \\ \dfrac{\partial y}{\partial \rho} & \dfrac{\partial y}{\partial \varphi} & \dfrac{\partial y}{\partial \theta} \\ \dfrac{\partial z}{\partial \rho} & \dfrac{\partial z}{\partial \varphi} & \dfrac{\partial z}{\partial \theta} \end{Vmatrix} d\rho d\varphi d\theta$$

$$= \begin{Vmatrix} \sin\varphi\cos\theta & \rho\cos\varphi\cos\theta & -\rho\sin\varphi\sin\theta \\ \sin\varphi\sin\theta & \rho\cos\varphi\sin\theta & -\rho\sin\varphi\cos\theta \\ \cos\varphi & -\rho\sin\varphi & 0 \end{Vmatrix} d\rho d\varphi d\theta = \rho^2\sin\varphi\, d\rho d\varphi d\theta$$

其中 $0 \le \rho \le r,\ 0 \le \varphi \le \pi,\ 0 \le \theta \le 2\pi$

$$\therefore \iiint_{x^2+y^2+z^2 \le r^2} 2 + 3x^2 + 3y^2 + 3z^2\, dxdydz$$

$$= 2 \cdot \frac{4\pi \cdot r^3}{3} + \int_0^\pi \int_0^{2\pi} \int_0^r 3\rho^2 \cdot \rho^2 \sin\varphi\, d\rho d\theta d\varphi = \frac{8\pi r^3}{3} + \frac{12\pi r^5}{5}$$

Example 3.

$$\text{求} \oiint_S \vec{F} \cdot \vec{n}\, dA = ?, \quad \text{其中} \vec{F} = (x^2 y + x^3, 2y + z^2 + y^3, 4z - 2xyz + z^3),$$

$$S: x^2 + y^2 + z^2 = 1 \text{ 為朝外定向}$$

【解】

令 $V: x^2 + y^2 + z^2 \le 1$

$\because \vec{F}$ 的各分量一階偏導數存在且連續，藉由 Gauss Theorem 則 $\quad \oiint_S \vec{F} \cdot \vec{n}\, dA = \iiint_V \nabla \cdot \vec{F}\, dV$

$\because \nabla \cdot \vec{F} = 6 + 3x^2 + 3y^2 + 3z^2 \quad \therefore \iiint_V \nabla \cdot \vec{F}\, dV = \iiint_{x^2+y^2+z^2 \le 1} 6 + 3x^2 + 3y^2 + 3z^2\, dxdydz$

令 $x = \rho \sin\varphi \cos\theta,\, y = \rho \sin\varphi \sin\theta,\, z = \rho \cos\varphi$ 則

$$dxdydz = \begin{Vmatrix} \dfrac{\partial x}{\partial \rho} & \dfrac{\partial x}{\partial \varphi} & \dfrac{\partial x}{\partial \theta} \\ \dfrac{\partial y}{\partial \rho} & \dfrac{\partial y}{\partial \varphi} & \dfrac{\partial y}{\partial \theta} \\ \dfrac{\partial z}{\partial \rho} & \dfrac{\partial z}{\partial \varphi} & \dfrac{\partial z}{\partial \theta} \end{Vmatrix} d\rho d\varphi d\theta$$

$$= \begin{vmatrix} \sin\varphi\cos\theta & \rho\cos\varphi\cos\theta & -\rho\sin\varphi\sin\theta \\ \sin\varphi\sin\theta & \rho\cos\varphi\sin\theta & -\rho\sin\varphi\cos\theta \\ \cos\varphi & -\rho\sin\varphi & 0 \end{vmatrix} d\rho d\varphi d\theta = \rho^2\sin\varphi\, d\rho d\varphi d\theta$$

其中 $0 \leq \rho \leq 1,\ 0 \leq \varphi \leq \pi,\ 0 \leq \theta \leq 2\pi$

$$\therefore \iiint_{x^2+y^2+z^2\leq 1} 6+3x^2+3y^2+3z^2\, dxdydz = 6 \cdot \frac{4\pi}{3} + \int_0^\pi \int_0^{2\pi} \int_0^1 3\rho^2 \cdot \rho^2 \sin\varphi\, d\rho d\theta d\varphi$$

$$= 8\pi + \frac{12\pi}{5} = \frac{52\pi}{15}$$

Example 4.

藉由 Gauss's Theorem 求 $\vec{F} = (0,4yz,0)$ 通过曲面 $S: x^2 + y^2 + z^2 = a^2$ 的通量，

$$\oiint_S \vec{F}\cdot\vec{n}dA =?$$

【解】

令 $V: x^2 + y^2 + z^2 \leq a^2$

$\because \vec{F}$ 的各分量一阶偏导数存在且连续，藉由 Gauss's Theorem 则 $\displaystyle\oiint_S \vec{F}\cdot\vec{n}dA = \iiint_V \nabla\cdot\vec{F}\, dV$

$\because \nabla\cdot\vec{F} = 4z \quad \therefore \displaystyle\iiint_V \nabla\cdot\vec{F}\, dV = \iiint_{x^2+y^2+z^2\leq a^2} 4z\, dxdydz$

令 $x = \rho\sin\varphi\cos\theta, y = \rho\sin\varphi\sin\theta, z = \rho\cos\varphi$ 则

$$dxdydz = \begin{vmatrix} \dfrac{\partial x}{\partial \rho} & \dfrac{\partial x}{\partial \varphi} & \dfrac{\partial x}{\partial \theta} \\[2mm] \dfrac{\partial y}{\partial \rho} & \dfrac{\partial y}{\partial \varphi} & \dfrac{\partial y}{\partial \theta} \\[2mm] \dfrac{\partial z}{\partial \rho} & \dfrac{\partial z}{\partial \varphi} & \dfrac{\partial z}{\partial \theta} \end{vmatrix} d\rho d\varphi d\theta$$

$$= \begin{vmatrix} \sin\varphi\cos\theta & \rho\cos\varphi\cos\theta & -\rho\sin\varphi\sin\theta \\ \sin\varphi\sin\theta & \rho\cos\varphi\sin\theta & -\rho\sin\varphi\cos\theta \\ \cos\varphi & -\rho\sin\varphi & 0 \end{vmatrix} d\rho d\varphi d\theta = \rho^2\sin\varphi\, d\rho d\varphi d\theta$$

其中 $0 \leq \rho \leq a,\ 0 \leq \varphi \leq \pi,\ 0 \leq \theta \leq 2\pi$

$$\therefore \iiint_{x^2+y^2+z^2\leq a^2} 4z\, dxdydz = \int_0^\pi \int_0^{2\pi} \int_0^a 4\rho\cos\varphi \cdot \rho^2 \sin\varphi\, d\rho d\theta d\varphi = 0$$

Example 5.

藉由 Gauss's Theorem 求 $\oiint_S \vec{F} \cdot \vec{n} dA =?$,  $\vec{F} = (x - y + z, 2x, 1), S$ 为 $z = x^2 + y^2$, $z = 1$ 所围朝外定向封闭曲面

【解】

令 $S_1: z = x^2 + y^2, z < 1$,  且 $S_2: x^2 + y^2 \le 1, z = 1$ 则 $S = S_1 \cup S_2$

令 $V = \{(x, y, z): x^2 + y^2 \le z, 0 \le z \le 1\}$

$\because \vec{F}$ 的各分量一阶偏导数存在且连续, 藉由 Gauss's Theorem 则 $\oiint_S \vec{F} \cdot \vec{n} dA = \iiint_V \nabla \cdot \vec{F} \, dV$

$\because \nabla \cdot \vec{F} = 1$   $\therefore \iiint_V \nabla \cdot \vec{F} \, dV = \iiint_V 1 \, dxdydz$

令 $R = \{(x, y): x^2 + y^2 \le 1\}$

则 $\iiint_V 1 \, dxdydz = \iint_R \int_0^{1-x^2-y^2} dzdxdy = \iint_R 1 - x^2 - y^2 dxdy$

令 $x = r\cos\theta, y = r\sin\theta$ 则 $\{(x, y): x^2 + y^2 \le 1\} = \{(r, \theta): 0 \le r \le 1, 0 \le \theta \le 2\pi\}$

且 $dxdy = \left\| \begin{vmatrix} \dfrac{\partial x}{\partial r} & \dfrac{\partial x}{\partial \theta} \\ \dfrac{\partial y}{\partial r} & \dfrac{\partial y}{\partial \theta} \end{vmatrix} \right\| drd\theta = \left\| \begin{vmatrix} \cos\theta & -r\sin\theta \\ \sin\theta & r\cos\theta \end{vmatrix} \right\| drd\theta = rdrd\theta$

$\therefore \iint_R 1 - x^2 - y^2 dxdy = \int_0^{2\pi} \int_0^1 (1-r^2)rdrd\theta = 2\pi \cdot \left( \dfrac{r^2}{2} - \dfrac{r^4}{4} \right)\Big|_0^1 = \dfrac{\pi}{2}$

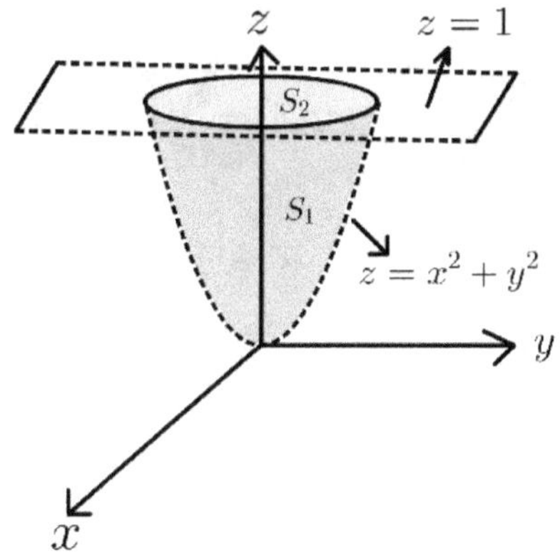


Example 6.

Let $\vec{F} = (x, 2y, 3z)$. Assume $V$ is the volume of cube with the vertices $(\pm 1, \pm 1, \pm 1)$ and

$S$ is the boundary of V.  Find $\oiint_S \vec{F} \cdot \vec{n}\, dA$  by using Gauss's Theorem.

【解】

令 $V = \{(x, y, z): -1 \le x \le 1, -1 \le y \le 1, -1 \le z \le 1\}$

$\because \vec{F}$ 的各分量一阶偏导数存在且连续，藉由 Gauss's Theorem 则 $\oiint_S \vec{F} \cdot \vec{n}\, dA = \iiint_V \nabla \cdot \vec{F}\, dV$

$\because \nabla \cdot \vec{F} = 6 \quad \therefore \iiint_V \nabla \cdot \vec{F}\, dV = \iiint_V 6\, dx\, dy\, dz = 6 \cdot 2 \cdot 2 \cdot 2 = 48$

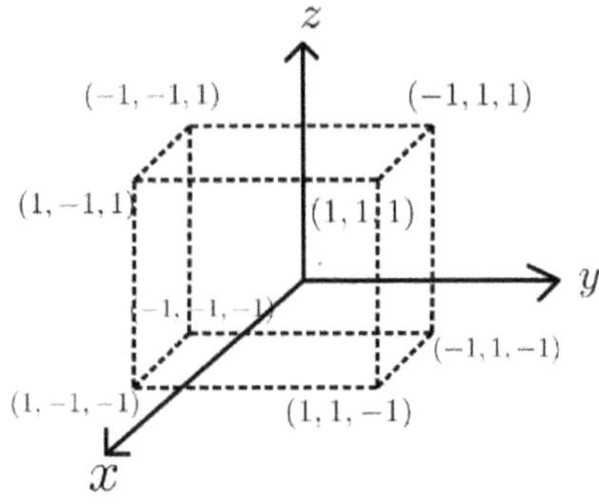


Example 7.

藉由 Gauss's Theorem 求 $\oiint_S \vec{F} \cdot \vec{n}\, dA = ?$,  其中 $\vec{F} = (x^3 + z, y^3 + e^{-z}, z^3)$,

$S: x^2 + y^2 + z^2 = a^2$

【解】

令 $V: x^2 + y^2 + z^2 \le a^2$

$\because \vec{F}$ 的各分量一阶偏导数存在且连续，藉由 Gauss's Theorem 则 $\oiint_S \vec{F} \cdot \vec{n}\, dA = \iiint_V \nabla \cdot \vec{F}\, dV$

$\because \nabla \cdot \vec{F} = 3x^2 + 3y^2 + 3z^2 \quad \therefore \iiint_V \nabla \cdot \vec{F}\, dV = \iiint_{x^2+y^2+z^2 \le a^2} 3x^2 + 3y^2 + 3z^2\, dx\, dy\, dz$

令 $x = \rho \sin\varphi \cos\theta\,, y = \rho \sin\varphi \sin\theta\,, z = \rho \cos\varphi$ 则

$$dx\, dy\, dz = \begin{Vmatrix} \dfrac{\partial x}{\partial \rho} & \dfrac{\partial x}{\partial \varphi} & \dfrac{\partial x}{\partial \theta} \\[2mm] \dfrac{\partial y}{\partial \rho} & \dfrac{\partial y}{\partial \varphi} & \dfrac{\partial y}{\partial \theta} \\[2mm] \dfrac{\partial z}{\partial \rho} & \dfrac{\partial z}{\partial \varphi} & \dfrac{\partial z}{\partial \theta} \end{Vmatrix} d\rho\, d\varphi\, d\theta$$

$$= \begin{Vmatrix} \sin\varphi\cos\theta & \rho\cos\varphi\cos\theta & -\rho\sin\varphi\sin\theta \\ \sin\varphi\sin\theta & \rho\cos\varphi\sin\theta & -\rho\sin\varphi\cos\theta \\ \cos\varphi & -\rho\sin\varphi & 0 \end{Vmatrix} d\rho d\varphi d\theta = \rho^2\sin\varphi\, d\rho d\varphi d\theta$$

其中 $0 \le \rho \le a,\ 0 \le \varphi \le \pi,\ 0 \le \theta \le 2\pi$

$$\therefore \iiint_{x^2+y^2+z^2\le a^2} 3x^2+3y^2+3z^2\, dxdydz = 3\int_0^\pi\int_0^{2\pi}\int_0^a \rho^2\cdot\rho^2\sin\varphi\, d\rho d\theta d\varphi = \frac{12\pi a^5}{5}$$

Example 8.

 藉由 Gauss's Theorem 求 $\oiint_S \vec{F}\cdot\vec{n}dA =?,\ \vec{F} = (xy+3, y^2+e^{zx}, \cos xy),$

 $S$ 为 $z = 1-x^2,\ y = 0,\ z = 0,\ y+z = 2$ 所围朝外定向封闭曲面

【解】

令 $V = \{(x,y,z): -1 \le x \le 1, 0 \le z \le 1-x, 0 \le y \le 2-z\}$

$\because \vec{F}$ 的各分量一阶偏导数存在且连续，藉由 Gauss's Theorem 则 $\oiint_S \vec{F}\cdot\vec{n}dA = \iiint_V \nabla\cdot\vec{F}\, dV$

$\because \nabla\cdot\vec{F} = y + 2y + 0 = 3y \quad \therefore \iiint_V \nabla\cdot\vec{F}\, dV = 3\int_{-1}^1\int_0^{1-x}\int_0^{2-z} ydydzdx = \frac{184}{35}$

Example 9.

 藉由 Gauss's Theorem 求 $\oiint_S \vec{F}\cdot\vec{n}dA =?,\ \vec{F} = (2x^3, x^2y, x^2z),\ S$ 为 $x^2+y^2 = a^2,$

 $z = 0, z = b$ 所围封闭朝外定向曲面

【解】

令 $S_1: x^2+y^2 = a^2,\ 0 \le z \le b, S_2: x^2+y^2 \le a^2,\ z = 0,\ S_3: x^2+y^2 \le a^2,\ z = b$
则 $S = S_1 \cup S_2 \cup S_3$
令 V 为 $S$ 所围体积则 $V = \{(x,y,z): x^2+y^2 \le a^2, 0 \le z \le b\}$

$\because \vec{F}$ 的各分量一阶偏导数存在且连续，藉由 Gauss's Theorem 则 $\oiint_S \vec{F}\cdot\vec{n}dA = \iiint_V \nabla\cdot\vec{F}\, dV$

$\because \nabla\cdot\vec{F} = 6x^2 + x^2 + x^2 = 8x^2 \quad \therefore \iiint_V \nabla\cdot\vec{F}\, dV = \iiint_V 8x^2\, dxdydz$

令 $R = \{(x,y): x^2+y^2 \le a^2\}$

则 $\iiint_V 8x^2\, dxdydz = 8\iint_R\int_0^b x^2 dzdxdy = 8b\iint_{x^2+y^2\le a^2} x^2 dxdy$

令 $x = r\cos\theta\,, y = r\sin\theta$ 则 $\{(x,y): x^2 + y^2 \le a^2\} = \{(r,\theta): 0 \le r \le a, 0 \le \theta \le 2\pi\}$

且 $dxdy = \left\|\begin{vmatrix} \dfrac{\partial x}{\partial r} & \dfrac{\partial x}{\partial \theta} \\[2mm] \dfrac{\partial y}{\partial r} & \dfrac{\partial y}{\partial \theta} \end{vmatrix}\right\| drd\theta = \left\|\begin{vmatrix} \cos\theta & -r\sin\theta \\ \sin\theta & r\cos\theta \end{vmatrix}\right\| drd\theta = rdrd\theta$

$\therefore 8b \iint_{x^2+y^2 \le a^2} x^2 dxdy = 8b \int_0^{2\pi} \int_0^a r^2 \cos^2\theta\, rdrd\theta = 8b \int_0^{2\pi} \cos^2\theta\, d\theta \int_0^a r^3 dr$

$\because \int_0^{2\pi} \cos^2\theta\, d\theta = \int_0^{2\pi} \dfrac{1 + \cos 2\theta}{2} d\theta = \pi$ 且 $\int_0^a r^3 dr = \dfrac{a^4}{4}$ $\therefore \oiint_S \vec{F} \cdot \vec{n} dA = 2\pi a^4 b$

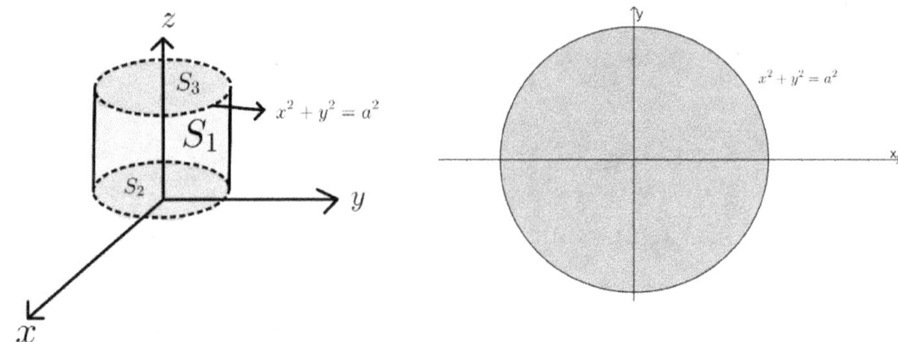


## Example 10.

$\vec{F} = (xye^z, xy^2z^3, -ye^z)$, $S$ 为坐标平面与 $x = 3, y = 2, z = 1$ 所围朝外定向封闭曲面，

藉由 Gauss's Theorem 求 $\oiint_S \vec{F} \cdot \vec{n} dA$

【解】

令 $V$ 为 $S$ 所围体积则 $V = \{(x,y,z): 0 \le x \le 3, 0 \le y \le 2, 0 \le z \le 1\}$

$\because \vec{F}$ 的各分量一阶偏导数存在且连续，藉由 Gauss's Theorem 则 $\oiint_S \vec{F} \cdot \vec{n} dA = \iiint_V \nabla \cdot \vec{F}\, dV$

$\iiint_V \nabla \cdot \vec{F}\, dV = \iiint_V ye^z + 2xyz^3 - ye^z\, dV = 2 \iiint_V xyz^3\, dV = 2 \int_0^3 \int_0^2 \int_0^1 xyz^3\, dzdydx$

$= 2 \cdot \dfrac{9}{2} \cdot \dfrac{4}{2} \cdot \dfrac{1}{4} = \dfrac{9}{2}$

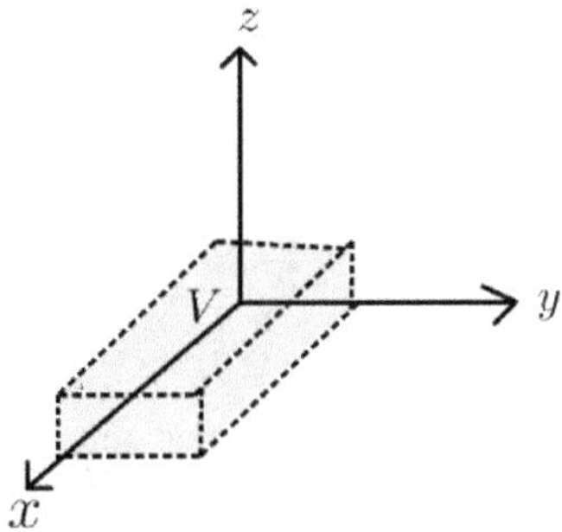

**Example 11.**

$\vec{F} = (3xy^2, xe^z, z^3)$, $S$ 为 $y^2 + z^2 = 1$, $x = -1, x = 2$ 所围封闭朝外定向曲面,

藉由 Gauss's Theorem 求 $\oiint_S \vec{F} \cdot \vec{n}\,dA$

【解】

令 $V$ 为 $S$ 所围体积则 $V = \{(x, y, z) : -1 \le x \le 2, \ y^2 + z^2 \le 1\}$

$\because \vec{F}$ 的各分量一阶偏导数存在且连续, 藉由 Gauss's Theorem 则 $\oiint_S \vec{F} \cdot \vec{n}\,dA = \iiint_V \nabla \cdot \vec{F}\,dV$

$$\iiint_V \nabla \cdot \vec{F}\,dV = \iiint_V 3y^2 + 3z^2\,dV = \iint_R \int_{-1}^{2} 3y^2 + 3z^2\,dxdydz = 9 \iint_R y^2 + z^2\,dydx$$

令 $y = r\cos\theta$, $z = r\sin\theta$ 则 $R = \{(y, z) : y^2 + z^2 \le 1\} = \{(r, \theta) : 0 \le r \le 1, 0 \le \theta \le 2\pi\}$

$$\therefore 9 \iint_R y^2 + z^2\,dydx = 9 \int_0^{2\pi} \int_0^1 r^3\,drd\theta = 9 \cdot 2\pi \cdot \frac{1}{4} = \frac{9\pi}{2}$$

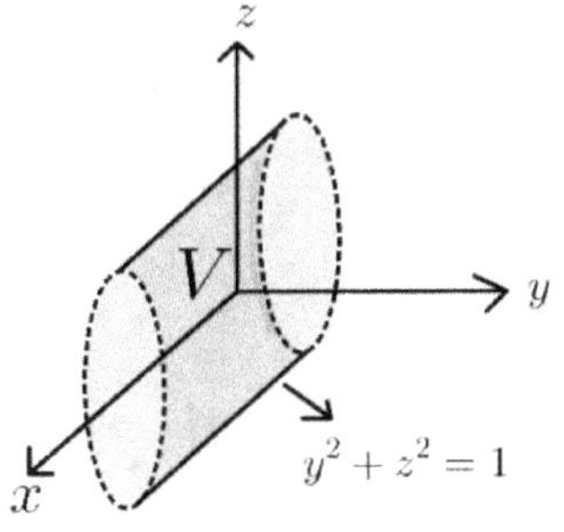


**Example 12.**

藉由 Gauss's Theorem 求 $\oiint_S x^4 + y^4 + z^4\, dA =?$, $\ S: x^2 + y^2 + z^2 = a^2$

【解】

∵ 球面的单位法向量 $\vec{n} = (x, y, z)$,  令 $\vec{F} = (x^3, y^3, z^3)$ 则 $\vec{F} \cdot \vec{n} = x^4 + y^4 + z^4$

令 $V: x^2 + y^2 + z^2 \leq a^2$

∵ $\vec{F}$ 的各分量一阶偏导数存在且连续,  藉由 Gauss's Theorem 则 $\oiint_S \vec{F} \cdot \vec{n}\, dA = \iiint_V \nabla \cdot \vec{F}\, dV$

∵ $\nabla \cdot \vec{F} = 3x^2 + 3y^2 + 3z^2$  ∴ $\iiint_V \nabla \cdot \vec{F}\, dV = \iiint_{x^2+y^2+z^2 \leq a^2} 3x^2 + 3y^2 + 3z^2\, dxdydz$

令 $x = \rho \sin\varphi \cos\theta$, $y = \rho \sin\varphi \sin\theta$, $z = \rho \cos\varphi$  则

$$dxdydz = \begin{Vmatrix} \dfrac{\partial x}{\partial \rho} & \dfrac{\partial x}{\partial \varphi} & \dfrac{\partial x}{\partial \theta} \\[2mm] \dfrac{\partial y}{\partial \rho} & \dfrac{\partial y}{\partial \varphi} & \dfrac{\partial y}{\partial \theta} \\[2mm] \dfrac{\partial z}{\partial \rho} & \dfrac{\partial z}{\partial \varphi} & \dfrac{\partial z}{\partial \theta} \end{Vmatrix} d\rho d\varphi d\theta$$

$$= \begin{Vmatrix} \sin\varphi\cos\theta & \rho\cos\varphi\cos\theta & -\rho\sin\varphi\sin\theta \\ \sin\varphi\sin\theta & \rho\cos\varphi\sin\theta & -\rho\sin\varphi\cos\theta \\ \cos\varphi & -\rho\sin\varphi & 0 \end{Vmatrix} d\rho d\varphi d\theta = \rho^2 \sin\varphi\, d\rho d\varphi d\theta$$

其中 $0 \leq \rho \leq a, 0 \leq \varphi \leq \pi, 0 \leq \theta \leq 2\pi$

∴ $\iiint_{x^2+y^2+z^2 \leq a^2} 3x^2 + 3y^2 + 3z^2\, dxdydz = 3 \int_0^\pi \int_0^{2\pi} \int_0^a \rho^2 \cdot \rho^2 \sin\varphi\, d\rho d\theta d\varphi = \dfrac{12\pi a^5}{5}$

Example 13.

藉由 Gauss's Theorem 求 $\oiint_S \vec{F} \cdot \vec{n}\, dA =?$,  $\vec{F} = (xy^2, yz^2, zx^2)$,  $S: x^2 + y^2 + z^2 = r^2$

【解】

令 $V: x^2 + y^2 + z^2 \leq r^2$

∵ $\vec{F}$ 的各分量一阶偏导数存在且连续,  藉由 Gauss's Theorem 则 $\oiint_S \vec{F} \cdot \vec{n}\, dA = \iiint_V \nabla \cdot \vec{F}\, dV$

∵ $\nabla \cdot \vec{F} = x^2 + y^2 + z^2$  ∴ $\iiint_V \nabla \cdot \vec{F}\, dV = \iiint_{x^2+y^2+z^2 \leq r^2} x^2 + y^2 + z^2\, dxdydz$

令 $x = \rho \sin\varphi \cos\theta$, $y = \rho \sin\varphi \sin\theta$, $z = \rho \cos\varphi$  则

$$dxdydz = \begin{Vmatrix} \dfrac{\partial x}{\partial \rho} & \dfrac{\partial x}{\partial \varphi} & \dfrac{\partial x}{\partial \theta} \\[2mm] \dfrac{\partial y}{\partial \rho} & \dfrac{\partial y}{\partial \varphi} & \dfrac{\partial y}{\partial \theta} \\[2mm] \dfrac{\partial z}{\partial \rho} & \dfrac{\partial z}{\partial \varphi} & \dfrac{\partial z}{\partial \theta} \end{Vmatrix} d\rho d\varphi d\theta$$

$$= \begin{Vmatrix} \sin\varphi\cos\theta & \rho\cos\varphi\cos\theta & -\rho\sin\varphi\sin\theta \\ \sin\varphi\sin\theta & \rho\cos\varphi\sin\theta & -\rho\sin\varphi\cos\theta \\ \cos\varphi & -\rho\sin\varphi & 0 \end{Vmatrix} d\rho d\varphi d\theta = \rho^2 \sin\varphi\, d\rho d\varphi d\theta$$

其中 $0 \le \rho \le r, 0 \le \varphi \le \pi, 0 \le \theta \le 2\pi$

$$\therefore \iiint_{x^2+y^2+z^2 \le r^2} x^2 + y^2 + z^2 \, dxdydz = \int_0^\pi \int_0^{2\pi} \int_0^r \rho^2 \cdot \rho^2 \sin\varphi\, d\rho d\theta d\varphi = \frac{4\pi r^5}{5}$$

**Example 14.**

藉由 Gauss's Theorem 求 $\displaystyle\oiint_S \vec{F} \cdot \vec{n} dA =?$, $\vec{F} = (e^x \sin y, e^x \cos y, yz^2 + xy)$,

$S: 0 \le x \le 1, \ 0 \le y \le 1, \ 0 \le z \le 2$

【解】

令 $V = \{(x, y, z): 0 \le x \le 1, \ 0 \le y \le 1, \ 0 \le z \le 2\}$

$\because \vec{F}$ 的各分量一阶偏导数存在且连续, 藉由 Gauss's Theorem 则 $\displaystyle\oiint_S \vec{F} \cdot \vec{n} dA = \iiint_V \nabla \cdot \vec{F}\, dV$

$\because \nabla \cdot \vec{F} = e^x \sin y - e^x \sin y + 2yz = 2yz$ $\quad \therefore \iiint_V \nabla \cdot \vec{F}\, dV = \int_0^1 \int_0^1 \int_0^2 2yz\, dz dy dx = 2$

**Example 15.**

$\vec{F} = (x^2 z^3, 2xyz^3, xz^4)$, $S$ 是以 $(\pm 1, \pm 2, \pm 3)$ 为顶点所形成长方体的表面, 藉由

Gauss's Theorem 求 $\displaystyle\oiint_S \vec{F} \cdot \vec{n} dA =?$,

【解】

令 $V$ 为 $S$ 所围体积则 $V = \{(x, y, z): -1 \le x \le 1, \ -2 \le y \le 3, \ -3 \le z \le 3\}$

$\because \vec{F}$ 的各分量一阶偏导数存在且连续, 藉由 Gauss's Theorem 则 $\displaystyle\oiint_S \vec{F} \cdot \vec{n} dA = \iiint_V \nabla \cdot \vec{F}\, dV$

$$\iiint_V \nabla \cdot \vec{F}\, dV = \iiint_V 2xz^3 + 2xz^3 + 4xz^3\, dV = 8\iiint_V xz^3\, dV = 8\int_{-1}^{1}\int_{-1}^{2}\int_{-3}^{3} xz^3\, dzdydx$$

$$= 0$$

## Example 16.

$\vec{F} = (x^4, -x^3 z^2, 4xy^2 z)$, 假设 $S$ 是 $x^2 + y^2 = 1$，$z = x + 2$ 以及 $z = 0$ 所围封闭

朝外定向曲面，藉由 Gauss's Theorem 求 $\oiint_S \vec{F} \cdot \vec{n}\, dA =?$

【解】

令 $V$ 为 $S$ 所围体积则 $V = \{(x,y,z): x^2 + y^2 \le 1, 0 \le z \le x + 2\}$

$\because \vec{F}$ 的各分量一阶偏导数存在且连续，藉由 Gauss's Theorem 则 $\oiint_S \vec{F} \cdot \vec{n}\, dA = \iiint_V \nabla \cdot \vec{F}\, dV$

$\therefore \iiint_V \nabla \cdot \vec{F}\, dV = \iiint_V 4x^3 + 4xy^2\, dV = \iint_R \int_0^{x+2} 4x^3 + 4xy^2 dzdxdy$

$= 4\iint_R x(x^2 + y^2)(x + 2)\, dydx$

令 $x = r\cos\theta$, $y = r\sin\theta$ 则 $R = \{(x,y): x^2 + y^2 \le 1\} = \{(r,\theta): 0 \le r \le 1, 0 \le \theta \le 2\pi\}$

$\therefore 4\iint_R x(x^2 + y^2)(x + 2)\, dydx = 4\int_0^{2\pi}\int_0^1 r^5 \cos^2\theta\, drd\theta = 4 \cdot 2\pi \cdot \frac{1}{6}\cdot\frac{1}{2} = \frac{2\pi}{3}$

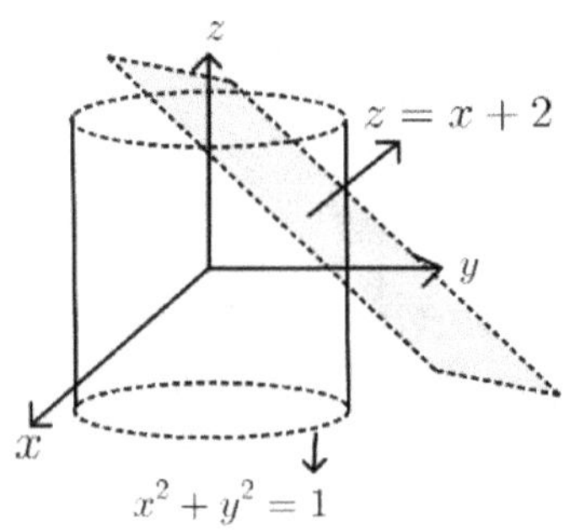


## Example 17.

藉由 Gauss's Theorem 求向量场 $\vec{F} = (2x, xz, z^2)$ 流出 $S$ 的 flux(通量) $=?$，$S$ 为抛物面

$z = 9 - x^2 - y^2$ 与 $xy$ 平面所围区域

【解】

令$V$为 $S$所围体积则 $V = \{(x, y, z): x^2 + y^2 \leq 9, 0 \leq z \leq 9 - x^2 - y^2\}$

$\because \vec{F}$的各分量一阶偏导数存在且连续, 藉由 Gauss's Theorem 则 $\oiint_S \vec{F} \cdot \vec{n}dA = \iiint_V \nabla \cdot \vec{F} \, dV$

$\because \nabla \cdot \vec{F} = 2 + 0 + 2z = 2 + 2z \quad\quad \therefore \iiint_V \nabla \cdot \vec{F} \, dV = \iiint_V 2 + 2z \, dxdydz$

令$R = \{(x, y): x^2 + y^2 \leq 9\}$

则 $\iiint_V 2 + 2z \, dxdydz = \iint_R \int_0^{9-x^2-y^2} 2 + 2zdzdxdy = \iint_R 2z + z^2|_0^{9-x^2-y^2} dxdy$

$= \iint_R 2(9 - x^2 - y^2) + (9 - x^2 - y^2)^2 \, dxdy$

令$x = r \cos \theta, y = r \sin \theta$ 则 $\{(x, y): x^2 + y^2 \leq 9\} = \{(r, \theta): 0 \leq r \leq 3, 0 \leq \theta \leq 2\pi\}$

且 $dxdy = \left\| \begin{array}{cc} \dfrac{\partial x}{\partial r} & \dfrac{\partial x}{\partial \theta} \\ \dfrac{\partial y}{\partial r} & \dfrac{\partial y}{\partial \theta} \end{array} \right\| drd\theta = \left\| \begin{array}{cc} \cos \theta & -r \sin \theta \\ \sin \theta & r \cos \theta \end{array} \right\| drd\theta = rdrd\theta$

$\therefore \iint_R 2(9 - x^2 - y^2) + (9 - x^2 - y^2)^2 \, dxdy = \int_0^{2\pi} d\theta \int_0^3 r^5 - 20r^3 + 99rdr = 2\pi \cdot 162$

$= 324\pi$

$\therefore \oiint_S \vec{F} \cdot \vec{n}dA = 324\pi$

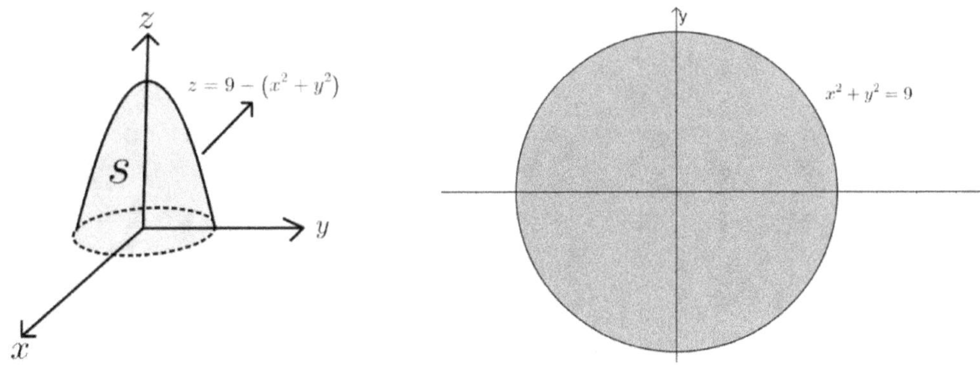


Example 18.

藉由 Gauss's Theorem 求向量场 $\vec{F} = (x^2, xy, z)$流出$S$的 flux(通量) =?, $S$为抛物面 $z = 4 - x^2 - y^2$与 $xy$平面所围区域

【解】

令 $V$ 为 $S$ 所围体积则 $V = \{(x,y,z): x^2 + y^2 \leq 4, 0 \leq z \leq 4 - x^2 - y^2\}$

藉由 Gauss's Theorem 则 $\oiint_S \vec{F} \cdot \vec{n}\,dA = \iiint_V \nabla \cdot \vec{F}\,dV$

$\because \nabla \cdot \vec{F} = 2x + x + 1 = 3x + 1 \quad \therefore \iiint_V \nabla \cdot \vec{F}\,dV = \iiint_V 3x + 1\,dxdydz$

令 $R = \{(x,y): x^2 + y^2 \leq 4 \}$

则 $\iiint_V 3x + 1\,dxdydz = \iint_R \int_0^{4-x^2-y^2} 3x + 1\,dzdxdy = \iint_R (3x+1)(4 - x^2 - y^2)\,dxdy$

令 $x = r\cos\theta, y = r\sin\theta$

则 $\{(x,y): x^2 + y^2 \leq 4\} = \{(r,\theta): 0 \leq r \leq 2, 0 \leq \theta \leq 2\pi\}$

且 $dxdy = \left\| \begin{vmatrix} \dfrac{\partial x}{\partial r} & \dfrac{\partial x}{\partial \theta} \\ \dfrac{\partial y}{\partial r} & \dfrac{\partial y}{\partial \theta} \end{vmatrix} \right\| drd\theta = \left\| \begin{vmatrix} \cos\theta & -r\sin\theta \\ \sin\theta & r\cos\theta \end{vmatrix} \right\| drd\theta = rdrd\theta$

$\therefore \iint_R (3x+1)(4 - x^2 - y^2)\,dxdy = \int_0^{2\pi} \int_0^2 (3\cos\theta + 1)(4 - r^2)rdrd\theta$

$\because \int_0^{2\pi} \int_0^2 3\cos\theta\,(4 - r^2)rdrd\theta = 0$

$\therefore \int_0^{2\pi} \int_0^2 (3\cos\theta + 1)(4 - r^2)rdrd\theta = \int_0^{2\pi} \int_0^2 (4 - r^2)rdrd\theta = 8\pi$

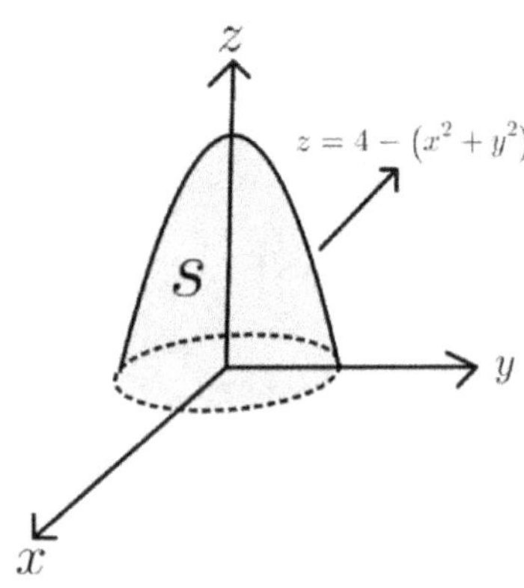


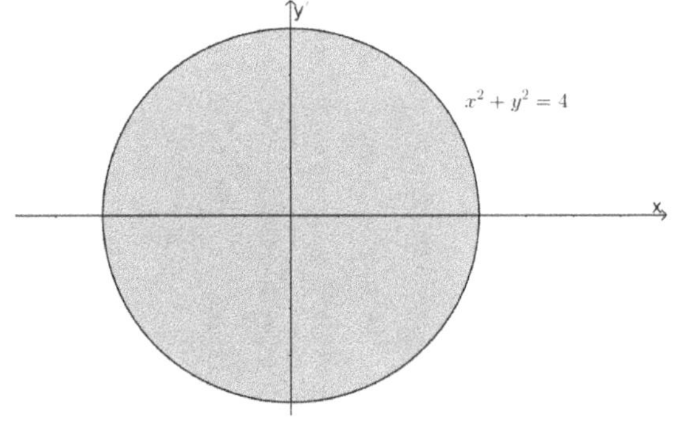


Example 19.

藉由 Gauss's Theorem 求 $\vec{F} = (y, x, z)$ 通过 $z = 4 - x^2 - y^2,\ z = 0$ 所围封闭曲面朝外

定向的通量, $\displaystyle\oiint_S \vec{F}\cdot\vec{n}\,dA =?$

【解】

令 $V$ 为 $S$ 所围体积则 $V = \{(x,y,z): x^2+y^2 \le 4, 0 \le z \le 4-x^2-y^2\}$

$\because \vec{F}$ 的各分量一阶偏导数存在且连续, 藉由 Gauss's Theorem 则 $\displaystyle\oiint_S \vec{F}\cdot\vec{n}\,dA = \iiint_V \nabla\cdot\vec{F}\,dV$

$\because \nabla\cdot\vec{F} = 1 \quad \therefore \displaystyle\iiint_V \nabla\cdot\vec{F}\,dV = \iiint_V 1\,dxdydz$

令 $R = \{(x,y): x^2+y^2 \le 4\}$

则 $\displaystyle\iiint_V 1\,dxdydz = \iint_R \int_0^{4-x^2-y^2} dzdxdy = \iint_R 4-x^2-y^2\,dxdy$

令 $x = r\cos\theta, y = r\sin\theta$ 则 $\{(x,y): x^2+y^2 \le 4\} = \{(r,\theta): 0 \le r \le 2, 0 \le \theta \le 2\pi\}$

且 $dxdy = \left\|\begin{array}{cc} \dfrac{\partial x}{\partial r} & \dfrac{\partial x}{\partial \theta} \\ \dfrac{\partial y}{\partial r} & \dfrac{\partial y}{\partial \theta} \end{array}\right\| drd\theta = \left\|\begin{array}{cc} \cos\theta & -r\sin\theta \\ \sin\theta & r\cos\theta \end{array}\right\| drd\theta = r\,drd\theta$

$\therefore \displaystyle\iint_R 4-x^2-y^2\,dxdy = \int_0^{2\pi}\int_0^2 (4-r^2)r\,drd\theta = 2\pi\cdot 4 = 8\pi$

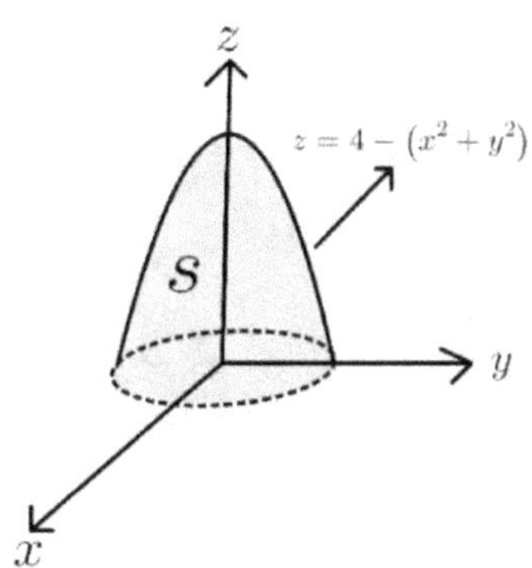


Example 20.

　　藉由 Gauss's Theorem 求 $\displaystyle\oiint_S \vec{F}\cdot\vec{n}\,dA =?$, $\vec{F} = (x+x^3, y+y^3, z+z^3)$,

　　$S: x^2+y^2+z^2 = r^2$

【解】

令 $V: x^2+y^2+z^2 \le r^2$

$\because \vec{F}$ 的各分量一阶偏导数存在且连续, 藉由 Gauss's Theorem 则 $\oiint_S \vec{F} \cdot \vec{n} dA = \iiint_V \nabla \cdot \vec{F} \, dV$

$\because \nabla \cdot \vec{F} = 3 + 3x^2 + 3y^2 + 3z^2$

$\therefore \iiint_V \nabla \cdot \vec{F} \, dV = \iiint_{x^2+y^2+z^2 \leq r^2} 3 + 3x^2 + 3y^2 + 3z^2 \, dxdydz$

令 $x = \rho \sin\varphi \cos\theta \, , y = \rho \sin\varphi \sin\theta \, , z = \rho \cos\varphi$ 则

$$dxdydz = \begin{Vmatrix} \dfrac{\partial x}{\partial \rho} & \dfrac{\partial x}{\partial \varphi} & \dfrac{\partial x}{\partial \theta} \\[2mm] \dfrac{\partial y}{\partial \rho} & \dfrac{\partial y}{\partial \varphi} & \dfrac{\partial y}{\partial \theta} \\[2mm] \dfrac{\partial z}{\partial \rho} & \dfrac{\partial z}{\partial \varphi} & \dfrac{\partial z}{\partial \theta} \end{Vmatrix} d\rho d\varphi d\theta$$

$$= \begin{Vmatrix} \sin\varphi \cos\theta & \rho\cos\varphi \cos\theta & -\rho\sin\varphi \sin\theta \\ \sin\varphi \sin\theta & \rho\cos\varphi \sin\theta & -\rho\sin\varphi \cos\theta \\ \cos\varphi & -\rho\sin\varphi & 0 \end{Vmatrix} d\rho d\varphi d\theta = \rho^2 \sin\varphi \, d\rho d\varphi d\theta$$

where $0 \leq \rho \leq r, 0 \leq \varphi \leq \pi, 0 \leq \theta \leq 2\pi$

$\therefore \iiint_{x^2+y^2+z^2 \leq r^2} 3 + 3x^2 + 3y^2 + 3z^2 \, dxdydz$

$= 3 \cdot \dfrac{4\pi \cdot r^3}{3} + \int_0^\pi \int_0^{2\pi} \int_0^r 3\rho^2 \cdot \rho^2 \sin\varphi \, d\rho d\theta d\varphi = 4\pi r^3 + \dfrac{12\pi r^5}{5}$

Example 21.

$\quad S: x^2 + y^2 + z^2 = 1, \ \vec{F} = \left( x + x^3 + y + \sin z^2 \, , y + y^3 + e^{x^2} , z + z^3 + \ln x^2 y^2 \right),$

$\quad$ 藉由 Gauss's Theorem 求 $\oiint_S \vec{F} \cdot \vec{n} dA =?$

【解】

令 $V: x^2 + y^2 + z^2 \leq 1$

$\because \vec{F}$ 的各分量一阶偏导数存在且连续, 藉由 Gauss's Theorem 则 $\oiint_S \vec{F} \cdot \vec{n} dA = \iiint_V \nabla \cdot \vec{F} \, dV$

$\because \nabla \cdot \vec{F} = 3 + 3x^2 + 3y^2 + 3z^2 \ \therefore \iiint_V \nabla \cdot \vec{F} \, dV = \iiint_{x^2+y^2+z^2 \leq 1} 3 + 3x^2 + 3y^2 + 3z^2 \, dxdydz$

令 $x = \rho \sin\varphi \cos\theta \, , y = \rho \sin\varphi \sin\theta \, , z = \rho \cos\varphi$ 则

$$dxdydz = \begin{Vmatrix} \dfrac{\partial x}{\partial \rho} & \dfrac{\partial x}{\partial \varphi} & \dfrac{\partial x}{\partial \theta} \\[2mm] \dfrac{\partial y}{\partial \rho} & \dfrac{\partial y}{\partial \varphi} & \dfrac{\partial y}{\partial \theta} \\[2mm] \dfrac{\partial z}{\partial \rho} & \dfrac{\partial z}{\partial \varphi} & \dfrac{\partial z}{\partial \theta} \end{Vmatrix} d\rho d\varphi d\theta$$

$$= \begin{Vmatrix} \sin\varphi\cos\theta & \rho\cos\varphi\cos\theta & -\rho\sin\varphi\sin\theta \\ \sin\varphi\sin\theta & \rho\cos\varphi\sin\theta & -\rho\sin\varphi\cos\theta \\ \cos\varphi & -\rho\sin\varphi & 0 \end{Vmatrix} d\rho d\varphi d\theta = \rho^2\sin\varphi\, d\rho d\varphi d\theta$$

where $0 \le \rho \le 1, 0 \le \varphi \le \pi, 0 \le \theta \le 2\pi$

$$\iiint_{x^2+y^2+z^2 \le 1} 3 + 3x^2 + 3y^2 + 3z^2\, dxdydz = 3 \cdot \frac{4\pi}{3} + \int_0^\pi \int_0^{2\pi} \int_0^1 3\rho^2 \cdot \rho^2 \sin\varphi\, d\rho d\theta d\varphi$$

$$= 4\pi + \frac{12\pi}{5} = \frac{32\pi}{5}$$

Example 22.

$\vec{F} = (x^2\sin y, x\cos y, -xz\sin y)$, $S: x^2 + y^2 + z^2 = 8$, 藉由 Gauss's Theorem

求 $\oiint_S \vec{F} \cdot \vec{n}\, dA$

【解】

令 $V$ 为 $S$ 所围体积则 $V: x^2 + y^2 + z^2 \le 8$

$\because \vec{F}$ 的各分量一阶偏导数存在且连续，藉由 Gauss's Theorem 则 $\oiint_S \vec{F} \cdot \vec{n}\, dA = \iiint_V \nabla \cdot \vec{F}\, dV$

$$\iiint_V \nabla \cdot \vec{F}\, dV = \iiint_V 2x\sin y - x\sin y - x\sin y\, dV = 0$$

Example 23.

$\vec{F} = (x^3 - 3y, 2yz + 1, xyz)$, 假设 $S$ 为平面 $x = \pm1, y = \pm1, z = \pm1$ 所围朝外定向

封闭曲面，藉由 Gauss's Theorem 求 $\oiint_S \vec{F} \cdot \vec{n}\, dA = ?$

【解】

令 $V$ 为 $S$ 所围体积则 $V = \{(x, y, z): -1 \le x \le 1, -1 \le y \le 1, -1 \le z \le 1\}$

$\because \vec{F}$ 的各分量一阶偏导数存在且连续, 藉由 Gauss's Theorem 则 $\oiint_S \vec{F} \cdot \vec{n}\,dA = \iiint_V \nabla \cdot \vec{F}\,dV$

$$\iiint_V \nabla \cdot \vec{F}\,dV = \iiint_V 3x^2 + 2z + xy\,dV$$

$$= \int_{-1}^1 \int_{-1}^1 \int_{-1}^1 3x^2\,dxdydz + \int_{-1}^1 \int_{-1}^1 \int_{-1}^1 2z\,dzdydx + \int_{-1}^1 \int_{-1}^1 \int_{-1}^1 xy\,dxdydz$$

$$\because \int_{-1}^1 \int_{-1}^1 \int_{-1}^1 3x^2\,dxdydz = 8, \quad \int_{-1}^1 \int_{-1}^1 \int_{-1}^1 2z\,dzdydx = 0, \quad \int_{-1}^1 \int_{-1}^1 \int_{-1}^1 xy\,dxdydz = 0$$

$$\therefore \oiint_S \vec{F} \cdot \vec{n}\,dA = \iiint_V \nabla \cdot \vec{F}\,dV = \iiint_V 3x^2 + 2z + xy\,dV = 8$$

Example 24.

藉由 Gauss's Theorem 求 $\oiint_S \vec{F} \cdot \vec{n}\,dA = ?$, $\vec{F} = (x^4, -x^3z^2, 4xy^2z)$, $S$ 为 $x^2 + y^2 = 1$,

$z = x + 2$, $z = 0$ 所围封闭朝外定向曲面

【解】

令 $V$ 为 $S$ 所围体积则 $V = \{(x, y, z): x^2 + y^2 \leq 1, 0 \leq z \leq x + 2\}$

$\because \vec{F}$ 的各分量一阶偏导数存在且连续, 藉由 Gauss's Theorem 则 $\oiint_S \vec{F} \cdot \vec{n}\,dA = \iiint_V \nabla \cdot \vec{F}\,dV$

$\because \nabla \cdot \vec{F} = 4x^3 + 0 + 4xy^2 = 4x(x^2 + y^2)$ $\therefore \iiint_V \nabla \cdot \vec{F}\,dV = \iiint_V 4x(x^2 + y^2)\,dxdydz$

令 $R = \{(x, y): x^2 + y^2 \leq 1\}$

则 $\iiint_V 4x(x^2 + y^2)\,dxdydz = \iint_R \int_0^{x+2} 4x(x^2 + y^2)\,dzdxdy$

$$= \iint_R 4x(x^2 + y^2)(x + 2)\,dxdy$$

令 $x = r\cos\theta, y = r\sin\theta$ 则 $\{(x, y): x^2 + y^2 \leq 1\} = \{(r, \theta): 0 \leq r \leq 1, 0 \leq \theta \leq 2\pi\}$

且 $dxdy = \left\| \begin{matrix} \dfrac{\partial x}{\partial r} & \dfrac{\partial x}{\partial \theta} \\ \dfrac{\partial y}{\partial r} & \dfrac{\partial y}{\partial \theta} \end{matrix} \right\| drd\theta = \left\| \begin{matrix} \cos\theta & -r\sin\theta \\ \sin\theta & r\cos\theta \end{matrix} \right\| drd\theta = r\,drd\theta$

$\therefore \iint_R 4x(x^2 + y^2)(x + 2))\,dxdy = \int_0^{2\pi} \int_0^1 4r^3 \cos\theta\,(r\cos\theta + 2)r\,drd\theta$

$$\because \int_0^{2\pi} \int_0^1 4r^3 \cos\theta \cdot 2r\,dr\,d\theta = 0$$

$$\therefore \int_0^{2\pi} \int_0^1 4r^3 \cos\theta\,(r\cos\theta + 2)r\,dr\,d\theta = \int_0^{2\pi} \int_0^1 4r^3 \cos\theta\,(r\cos\theta)r\,dr\,d\theta$$

$$= 4 \int_0^{2\pi} \int_0^1 r^5 \frac{1+\cos 2\theta}{2}\,dr\,d\theta = 2 \int_0^{2\pi} \int_0^1 r^5\,dr\,d\theta = \frac{2\pi}{3}$$

$$\therefore \oiint_S \vec{F} \cdot \vec{n}\,dA = \frac{2\pi}{3}$$

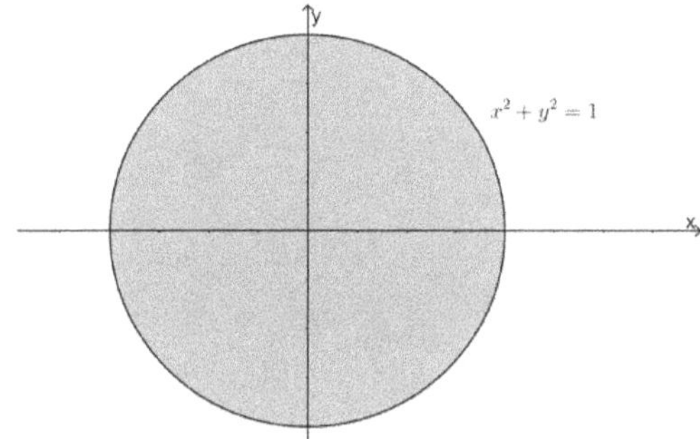


Example 25.

$$\vec{F} = (xz^2 + \sin z, x^2 y - z^3 + \cos x, 2xy + y^2 z), \quad S: x^2 + y^2 + z^2 = a^2, z \geq 0,$$

藉由 Gauss's Theorem 求 $\oiint_S \vec{F} \cdot \vec{n}\,dA = ?$

【解】

$\because$ 球面的单位法向量 $\vec{n} = (x, y, z)$, 令 $\vec{F} = (xz^2 + \sin z, x^2 y - z^3 + \cos x, 2xy + y^2 z)$

令 $V: x^2 + y^2 + z^2 \leq a^2, z \geq 0$

$\because \vec{F}$ 的各分量一阶偏导数存在且连续, 藉由 Gauss's Theorem 则 $\oiint_S \vec{F} \cdot \vec{n}\,dA = \iiint_V \nabla \cdot \vec{F}\,dV$

$\because \nabla \cdot \vec{F} = x^2 + y^2 + z^2$  $\therefore \iiint_V \nabla \cdot \vec{F}\,dV = \iiint_{x^2+y^2+z^2 \leq a^2} x^2 + y^2 + z^2\,dx\,dy\,dz$

令 $x = \rho\sin\varphi\cos\theta, y = \rho\sin\varphi\sin\theta, z = \rho\cos\varphi$ 则

$$dx\,dy\,dz = \left\| \begin{vmatrix} \dfrac{\partial x}{\partial \rho} & \dfrac{\partial x}{\partial \varphi} & \dfrac{\partial x}{\partial \theta} \\[2mm] \dfrac{\partial y}{\partial \rho} & \dfrac{\partial y}{\partial \varphi} & \dfrac{\partial y}{\partial \theta} \\[2mm] \dfrac{\partial z}{\partial \rho} & \dfrac{\partial z}{\partial \varphi} & \dfrac{\partial z}{\partial \theta} \end{vmatrix} \right\| d\rho\,d\varphi\,d\theta$$

$$= \begin{Vmatrix} \sin\varphi\cos\theta & \rho\cos\varphi\cos\theta & -\rho\sin\varphi\sin\theta \\ \sin\varphi\sin\theta & \rho\cos\varphi\sin\theta & -\rho\sin\varphi\cos\theta \\ \cos\varphi & -\rho\sin\varphi & 0 \end{Vmatrix} d\rho d\varphi d\theta = \rho^2 \sin\varphi\, d\rho d\varphi d\theta$$

其中 $0 \le \rho \le a, 0 \le \varphi \le \dfrac{\pi}{2}, 0 \le \theta \le 2\pi$

$$\therefore \iiint_{x^2+y^2+z^2 \le a^2} x^2 + y^2 + z^2\, dxdydz = \int_0^{\frac{\pi}{2}} \int_0^{2\pi} \int_0^a \rho^2 \cdot \rho^2 \sin\varphi\, d\rho d\theta d\varphi = \frac{2\pi a^5}{5}$$

## 9.7.3　把体积分转成封闭的曲面积分

考试类型:

Type 1.

假设向量场 $\vec{F}$ 的各分量一阶偏导数存在且连续，$S: x^2 + y^2 + z^2 = r^2$，$V$ 为 $S$ 所围体积，

藉由 Gauss's Theorem 求 $\iiint_V \nabla \cdot \vec{F}\, dV = ?$

解题流程:

Step1.

藉由 Gauss's Theorem, $\iiint_V \nabla \cdot \vec{F}\, dV = \oiint_S \vec{F} \cdot \vec{n} dA$

Step2.

令 $\vec{r}(\theta, \varphi) = (r\cos\theta\sin\varphi, r\sin\theta\sin\varphi, r\cos\varphi)$ 且 $R = \{(\theta, \varphi) : 0 \le \theta \le 2\pi, 0 \le \varphi \le \pi\}$

则 $\oiint_S \vec{F} \cdot \vec{n} dA = \iint_R \vec{F} \cdot \dfrac{\partial \vec{r}}{\partial \varphi} \times \dfrac{\partial \vec{r}}{\partial \theta} d\theta d\varphi$

Step3.

$\because \dfrac{\partial \vec{r}}{\partial \theta} = (-r\sin\theta\sin\varphi, r\cos\theta\sin\varphi, 0)$ 且 $\dfrac{\partial \vec{r}}{\partial \varphi} = (r\cos\theta\cos\varphi, r\sin\theta\cos\varphi, -r\sin\varphi)$

$\therefore \dfrac{\partial \vec{r}}{\partial \varphi} \times \dfrac{\partial \vec{r}}{\partial \theta} = (r^2\cos\theta\sin^2\varphi, r^2\sin\theta\sin^2\varphi, r^2\sin\varphi\cos\varphi)$

Step4.

$$\oiint_S \vec{F} \cdot \vec{n} dA = \int_0^{2\pi} \int_0^{\pi} (f_1(x,y,z) r^2\cos\theta\sin^2\varphi + f_2(x,y,z) r^2\sin\theta\sin^2\varphi$$
$$+ f_3(x,y,z) r^2\sin\varphi\cos\varphi)\big|_{x=r\cos\theta\sin\varphi, y=r\sin\theta\sin\varphi, z=r\cos\varphi} d\varphi d\theta$$

其中 $\vec{F} = \big(f_1(x,y,z), f_2(x,y,z), f_3(x,y,z)\big)$

<u>范例说明:</u>

(I) 求 $\iiint_V \nabla \cdot \vec{F}\, dV = ?$，其中 $\vec{F} = (0, 4yz, 0)$，$S: x^2 + y^2 + z^2 = r^2$，$V$ 为 $S$ 所围体积

$\because \vec{F}$ 的各分量一阶偏导数存在且连续

藉由 Gauss's Theorem 则 $\iiint_V \nabla \cdot \vec{F}\, dV = \oiint_S \vec{F} \cdot \vec{n}\, dA$

令 $\vec{r}(\theta, \varphi) = (r\cos\theta\sin\varphi, r\sin\theta\sin\varphi, r\cos\varphi)$ 且 $R = \{(\theta, \varphi): 0 \le \theta \le 2\pi, 0 \le \varphi \le \pi\}$

则 $\oiint_S \vec{F} \cdot \vec{n}\, dA = \iint_R \vec{F} \cdot \dfrac{\partial\vec{r}}{\partial\varphi} \times \dfrac{\partial\vec{r}}{\partial\theta}\, d\theta\, d\varphi$

$\because \dfrac{\partial\vec{r}}{\partial\theta} = (-r\sin\theta\sin\varphi, r\cos\theta\sin\varphi, 0)$ 且 $\dfrac{\partial\vec{r}}{\partial\varphi} = (r\cos\theta\cos\varphi, r\sin\theta\cos\varphi, -r\sin\varphi)$

$\therefore \dfrac{\partial\vec{r}}{\partial\varphi} \times \dfrac{\partial\vec{r}}{\partial\theta} = (r^2\cos\theta\sin^2\varphi, r^2\sin\theta\sin^2\varphi, r^2\sin\varphi\cos\varphi)$

$\vec{F} \cdot \dfrac{\partial\vec{r}}{\partial\varphi} \times \dfrac{\partial\vec{r}}{\partial\theta} = (0, 4r^2\sin\theta\sin\varphi\cos\varphi, 0) \cdot (r^2\cos\theta\sin^2\varphi, r^2\sin\theta\sin^2\varphi, r^2\sin\varphi\cos\varphi)$

$= 4r^4\sin^3\varphi\sin^2\theta\cos\varphi$

$\therefore \oiint_S \vec{F} \cdot \vec{n}\, dA = \int_0^{2\pi} \int_0^{\pi} 4r^4\sin^3\varphi\sin^2\theta\cos\varphi\, d\varphi\, d\theta = 0$

(II) 求 $\iiint_V \nabla \cdot \vec{F}\, dV = ?$，其中 $\vec{F} = (x, y, z)$，$S: x^2 + y^2 + z^2 = r^2$，$V$ 为 $S$ 所围体积

$\because \vec{F}$ 的各分量一阶偏导数存在且连续

藉由 Gauss's Theorem 则 $\iiint_V \nabla \cdot \vec{F}\, dV = \oiint_S \vec{F} \cdot \vec{n}\, dA$

令 $\vec{r}(\theta, \varphi) = (r\cos\theta\sin\varphi, r\sin\theta\sin\varphi, r\cos\varphi)$ 且 $R = \{(\theta, \varphi): 0 \le \theta \le 2\pi, 0 \le \varphi \le \pi\}$

则 $\oiint_S \vec{F} \cdot \vec{n}\, dA = \iint_R \vec{F} \cdot \dfrac{\partial\vec{r}}{\partial\varphi} \times \dfrac{\partial\vec{r}}{\partial\theta}\, d\theta\, d\varphi$

$$\because \frac{\partial \vec{r}}{\partial \theta} = (-r\sin\theta\sin\varphi, r\cos\theta\sin\varphi, 0) \quad \text{且} \quad \frac{\partial \vec{r}}{\partial \varphi} = (r\cos\theta\cos\varphi, r\sin\theta\cos\varphi, -r\sin\varphi)$$

$$\therefore \frac{\partial \vec{r}}{\partial \varphi} \times \frac{\partial \vec{r}}{\partial \theta} = (r^2\cos\theta\sin^2\varphi, r^2\sin\theta\sin^2\varphi, r^2\sin\varphi\cos\varphi)$$

$$\vec{F} \cdot \frac{\partial \vec{r}}{\partial \varphi} \times \frac{\partial \vec{r}}{\partial \theta}$$

$$= (r\cos\theta\sin\varphi, r\sin\theta\sin\varphi, r\cos\varphi) \cdot (r^2\cos\theta\sin^2\varphi, r^2\sin\theta\sin^2\varphi, r^2\sin\varphi\cos\varphi)$$

$$= r^3(\cos^2\theta\sin^3\varphi + \sin^3\varphi\sin^2\theta + \sin\varphi\cos^2\varphi) = r^3\sin\varphi$$

$$\therefore \oiint_S \vec{F} \cdot \vec{n}\,dA = r^3 \int_0^{2\pi} \int_0^{\pi} \sin\varphi\,d\varphi d\theta = 4\pi r^3$$

Type 2.

假设向量场$\vec{F}$的各分量一阶偏导数存在且连续, $S$为 $z = g(x,y), z = 0$ 所围封闭朝外定向

空间曲面,所围体积为 V, $S$投影至$xy$平面区域 $= \{(x,y): x^2 + y^2 \leq a^2\}$, 求 $\iiint_V \nabla \cdot \vec{F}\,dV$

补充说明:

曲面$S$通常为椭圆球、圆球、椭圆椎、圆锥、椭圆抛物面

解题流程:

Step1.

令$S_1: z = g(x,y)$, $S_2: z = 0$ 且$x^2 + y^2 \leq a^2$

藉由 Gauss's Theorem, $\iiint_V \nabla \cdot \vec{F}\,dV = \oiint_S \vec{F} \cdot \vec{n}\,dA = \iint_{S_1} \vec{F} \cdot \vec{n}\,dA + \iint_{S_2} \vec{F} \cdot \vec{n}\,dA$

Step2.

假设 $S_1$ 为朝上定向

令 $\vec{r}(x,y) = (x, y, g(x,y))$且 $R$ 为$S_1$投影至$xy$平面的封闭区域 $= \{(x,y): x^2 + y^2 \leq a^2\}$

则 $\iint_{S_1} \vec{F} \cdot \vec{n}\,dA = \iint_R \vec{F} \cdot \frac{\partial \vec{r}}{\partial x} \times \frac{\partial \vec{r}}{\partial y}\,dxdy$

Step3.

$$\because \frac{\partial \vec{r}}{\partial x} = (1, 0, g_x(x,y)) \quad \text{且} \quad \frac{\partial \vec{r}}{\partial y} = (0, 1, g_y(x,y)) \quad \therefore \frac{\partial \vec{r}}{\partial x} \times \frac{\partial \vec{r}}{\partial y} = (-g_x(x,y), -g_y(x,y), 1)$$

$$\iint_{S_1} \vec{F} \cdot \vec{n}\, dA = \iint_R \vec{F} \cdot \frac{\partial \vec{r}}{\partial x} \times \frac{\partial \vec{r}}{\partial y}\, dxdy = \iint_R \vec{F} \cdot (-g_x(x,y), -g_y(x,y), 1)\, dxdy$$

$$= \iint_R \left( -f_1 \cdot g_x(x,y) - f_2 \cdot g_y(x,y) + f_3 \right)\Big|_{z=g(x,y)}\, dxdy$$

其中 $\vec{F} = (f_1, f_2, f_3)$

Step4.

令 $x = r\cos\theta$, $y = r\sin\theta$ 则 $\{(x,y): x^2 + y^2 \leq a^2\} = \{(r,\theta): 0 \leq r \leq a, 0 \leq \theta \leq 2\pi\}$

且 $dxdy = \left\| \begin{vmatrix} \dfrac{\partial x}{\partial r} & \dfrac{\partial x}{\partial \theta} \\ \dfrac{\partial y}{\partial r} & \dfrac{\partial y}{\partial \theta} \end{vmatrix} \right\| drd\theta = \left\| \begin{vmatrix} \cos\theta & -r\sin\theta \\ \sin\theta & r\cos\theta \end{vmatrix} \right\| drd\theta = r\, drd\theta$

Step5.

$$\iint_R \left( -f_1 \cdot g_x(x,y) - f_2 \cdot g_y(x,y) + f_3 \right)\Big|_{z=g(x,y)}\, dxdy$$

$$= \int_0^{2\pi} \int_0^a -f_1(r\cos\theta, r\sin\theta, g(r\cos\theta, r\sin\theta)) \cdot g_x(r\cos\theta, r\sin\theta) \cdot r$$

$$-f_2(r\cos\theta, r\sin\theta, g(r\cos\theta, r\sin\theta)) \cdot g_y(r\cos\theta, r\sin\theta) \cdot r$$

$$+f_3(r\cos\theta, r\sin\theta, g(r\cos\theta, r\sin\theta)) \cdot r\, drd\theta$$

Step6.

求 $\displaystyle\iint_{S_2} \vec{F} \cdot \vec{n}\, dA = ?$

范例说明:

(I) 若 $\vec{F} = (y, x, z)$ 且 V 为 $z = 4 - x^2 - y^2$, $z = 0$ 所围封闭体积，藉由 Gauss's Theorem 求

$$\iiint_V \nabla \cdot \vec{F}\, dV = ?$$

令 $S_1: z = 4 - x^2 - y^2$, $S_2: z = 0$ 且 $x^2 + y^2 \leq 4$

藉由 Gauss's Theorem, $\displaystyle\iiint_V \nabla \cdot \vec{F}\, dV = \iint_{S_1} \vec{F} \cdot \vec{n}\, dA + \iint_{S_2} \vec{F} \cdot \vec{n}\, dA$

$$\iint_{S_1} \vec{F} \cdot \vec{n}\, dA = \iint_{x^2+y^2\leq 4} 4xy + z\, dxdy = \iint_{x^2+y^2\leq 4} 4xy + 4 - x^2 - y^2\, dxdy$$

令 $x = r\cos\theta$, $y = r\sin\theta$

$$\iint_{x^2+y^2\le 4} 4xy + 4 - x^2 - y^2 \, dxdy = \int_0^{2\pi}\int_0^2 (4r^2\cos\theta\sin\theta + 4 - r^2)\, rdrd\theta = 8\pi$$

$$\because \frac{\partial \vec{s}}{\partial x} = (1,0,0) \ \text{且} \ \frac{\partial \vec{s}}{\partial y} = (0,1,0) \ \therefore \frac{\partial \vec{s}}{\partial x}\times\frac{\partial \vec{s}}{\partial y} = (0,0,1) \ \therefore \iint_{S_2} \vec{F}\cdot\vec{n}\, dA = \iint_{S_2} 0\, dA = 0$$

$$\therefore \iint_{S_1} \vec{F}\cdot\vec{n}\, dA + \iint_{S_2} \vec{F}\cdot\vec{n}\, dA = 8\pi$$

Type 3.

假设$\vec{F}$的各分量一阶偏导数存在且连续，$S$由数个曲面组成的封闭曲面，$S = \bigcup_{j=1}^{n} S_j$,

$n \ge 2$，V 为$S$所围封闭体积，藉由 Gauss's Theorem 求 $\iiint_V \nabla\cdot\vec{F}\, dV = ?$

补充说明：

例如体积 V 为一个长方体

解题流程：

Step1.

藉由 Gauss's Theorem， $\iiint_V \nabla\cdot\vec{F}\, dV = \oiint_S \vec{F}\cdot\vec{n}dA = \sum_{j=1}^{n} \iint_{S_j} \vec{F}\cdot\vec{n}\, dA$

Step2.

如果$S_j$皆是平行坐标平面的平面，则可直接计算求$\iint_{S_j} \vec{F}\cdot\vec{n}\, dA$

此外，如果$S_j$为朝上定向曲面：$z = g(x,y)$，定义域为$a \le x \le b$ 且 $c \le y \le d$ 则
令$\vec{r}(x,y) = (x,y,g(x,y))$ 且 $R$ 为曲面$S$投影于$xy$平面上的封闭区域，即
$R = \{(x,y): a \le x \le b, c \le y \le d\}$则公式可改写为：

$$\iint_{S_j} \vec{F}\cdot\vec{n}\, dA = \iint_R \vec{F}(x,y,g(x,y))\cdot(-g_x(x,y), -g_y(x,y), 1)dxdy$$

Example 1.

藉由 Gauss's Theorem 求 $\iiint_V \nabla \cdot \vec{F}\, dV = ?$, $\vec{F} = (0, 4yz, 0)$, $V: x^2 + y^2 + z^2 \leq r^2$

【解】

令 $S: x^2 + y^2 + z^2 = r^2$, $\because \vec{F}$ 的各分量一阶偏导数存在且连续

藉由 Gauss's Theorem 则 $\iiint_V \nabla \cdot \vec{F}\, dV = \oiint_S \vec{F} \cdot \vec{n}\, dA$

令 $\vec{r}(\theta, \varphi) = (r\cos\theta \sin\varphi, r\sin\theta \sin\varphi, r\cos\varphi)$

令 $R = \{(\theta, \varphi): 0 \leq \theta \leq 2\pi, 0 \leq \varphi \leq \pi\}$ 则 $\oiint_S \vec{F} \cdot \vec{n}\, dA = \iint_R \vec{F} \cdot \dfrac{\partial \vec{r}}{\partial \varphi} \times \dfrac{\partial \vec{r}}{\partial \theta}\, d\theta d\varphi$

$\because \dfrac{\partial \vec{r}}{\partial \theta} = (-r\sin\theta \sin\varphi, r\cos\theta \sin\varphi, 0)$ 且 $\dfrac{\partial \vec{r}}{\partial \varphi} = (r\cos\theta \cos\varphi, r\sin\theta \cos\varphi, -r\sin\varphi)$

$\therefore \dfrac{\partial \vec{r}}{\partial \varphi} \times \dfrac{\partial \vec{r}}{\partial \theta} = (r^2 \cos\theta \sin^2\varphi, r^2 \sin\theta \sin^2\varphi, r^2 \sin\varphi \cos\varphi)$

$\therefore \vec{F} \cdot \dfrac{\partial \vec{r}}{\partial \varphi} \times \dfrac{\partial \vec{r}}{\partial \theta} = (0, 4r^2 \sin\theta \sin\varphi \cos\varphi, 0) \cdot (r^2 \cos\theta \sin^2\varphi, r^2 \sin\theta \sin^2\varphi, r^2 \sin\varphi \cos\varphi)$

$= 4r^4 \sin^3\varphi \sin^2\theta \cos\varphi$

$\therefore \oiint_S \vec{F} \cdot \vec{n}\, dA = \int_0^{2\pi} \int_0^{\pi} 4r^4 \sin^3\varphi \sin^2\theta \cos\varphi \, d\varphi d\theta$

$\because \int_0^{\pi} \sin^3\varphi \cos\varphi \, d\varphi = \dfrac{\sin^4\varphi}{4}\Big|_0^{\pi} = 0$ $\qquad \therefore \oiint_S \vec{F} \cdot \vec{n}\, dA = 0$


Example 2.

Let $\vec{F} = (x, 2y, 3z)$. Assume $V$ is the volume of cube with the vertices $(\pm 1, \pm 1, \pm 1)$.

Find $\iiint_V \nabla \cdot \vec{F}\, dV$ by using Gauss's Theorem.

【解】

令 $S_1: x = 1$, $S_2: x = -1$, $S_3: y = 1$, $S_4: y = -1$, $S_5: z = 1$, $S_6: z = -1$ 且 $S = \bigcup_{j=1}^{6} S_j$

$\because \vec{F}$ 的各分量一阶偏导数存在且连续

藉由 Gauss's Theorem 则 $\iiint_V \nabla \cdot \vec{F}\, dV = \oiint_S \vec{F} \cdot \vec{n}\, dA$

$As\ x = 1, \vec{n} = (1,0,0),\quad \iint_{S_1} \vec{F} \cdot \vec{n}\, dA = \int_{-1}^{1} \int_{-1}^{1} x\, dydz = 4x = 4$

$As\ x = -1, \vec{n} = (-1,0,0),\quad \iint_{S_2} \vec{F} \cdot \vec{n}\, dA = \int_{-1}^{1} \int_{-1}^{1} -x\, dydz = -4x = 4$

$As\ y = 1, \vec{n} = (0,1,0),\quad \iint_{S_3} \vec{F} \cdot \vec{n}\, dA = \int_{-1}^{1} \int_{-1}^{1} 2y\, dxdz = 8y = 8$

$As\ y = -1, \vec{n} = (0,-1,0),\quad \iint_{S_4} \vec{F} \cdot \vec{n}\, dA = \int_{-1}^{1} \int_{-1}^{1} -2y\, dxdz = -8y = 8$

$As\ z = 1, \vec{n} = (0,0,1),\quad \iint_{S_5} \vec{F} \cdot \vec{n}\, dA = \int_{-1}^{1} \int_{-1}^{1} 3z\, dxdy = 12z = 12$

$As\ z = -1, \vec{n} = (0,0,-1),\quad \iint_{S_6} \vec{F} \cdot \vec{n}\, dA = \int_{-1}^{1} \int_{-1}^{1} -3z\, dxdy = -12z = 12$

$\therefore \oiint_S \vec{F} \cdot \vec{n}\, dA = 48$

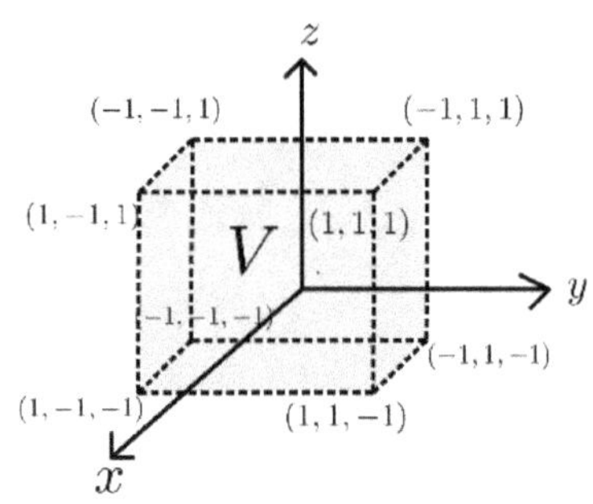


Example 3.

$\quad$ 求 $\iiint_V \nabla \cdot \vec{F}\, dV =?,\quad \vec{F} = (x,y,z),\quad V: x^2 + y^2 + z^2 \leq r^2$ (1)使用 Gauss's Theorem

$\quad$ (2) 直接计算

【解】

(1)

令 $S: x^2 + y^2 + z^2 = r^2$，$\because \vec{F}$ 的各分量一阶偏导数存在且连续

藉由 Gauss's Theorem 则 $\displaystyle\iiint_V \nabla \cdot \vec{F}\, dV = \oiint_S \vec{F} \cdot \vec{n}\, dA$

令 $\vec{r}(r, \theta, \varphi) = (r\cos\theta\sin\varphi, r\sin\theta\sin\varphi, r\cos\varphi)$

令 $R = \{(\theta, \varphi): 0 \le \theta \le 2\pi, 0 \le \varphi \le \pi\}$ 则 $\displaystyle\oiint_S \vec{F} \cdot \vec{n}\, dA = \iint_R \vec{F} \cdot \frac{\partial\vec{r}}{\partial\varphi} \times \frac{\partial\vec{r}}{\partial\theta}\, d\theta d\varphi$

$\because \dfrac{\partial\vec{r}}{\partial\theta} = (-r\sin\theta\sin\varphi, r\cos\theta\sin\varphi, 0)$ 且 $\dfrac{\partial\vec{r}}{\partial\varphi} = (r\cos\theta\cos\varphi, r\sin\theta\cos\varphi, -r\sin\varphi)$

$\therefore \dfrac{\partial\vec{r}}{\partial\varphi} \times \dfrac{\partial\vec{r}}{\partial\theta} = (r^2\cos\theta\sin^2\varphi, r^2\sin\theta\sin^2\varphi, r^2\sin\varphi\cos\varphi)$

$\therefore \vec{F} \cdot \dfrac{\partial\vec{r}}{\partial\varphi} \times \dfrac{\partial\vec{r}}{\partial\theta}$

$= (r\cos\theta\sin\varphi, r\sin\theta\sin\varphi, r\cos\varphi) \cdot (r^2\cos\theta\sin^2\varphi, r^2\sin\theta\sin^2\varphi, r^2\sin\varphi\cos\varphi)$

$= r^3(\cos^2\theta\sin^3\varphi + \sin^3\varphi\sin^2\theta + \sin\varphi\cos^2\varphi) = r^3\sin\varphi$

$\therefore \displaystyle\oiint_S \vec{F} \cdot \vec{n}\, dA = \iint_R \vec{F} \cdot \frac{\partial\vec{r}}{\partial\varphi} \times \frac{\partial\vec{r}}{\partial\theta}\, d\theta d\varphi = r^3 \int_0^{2\pi} \int_0^{\pi} \sin\varphi\, d\varphi d\theta = 4\pi r^3$

$\therefore \displaystyle\iiint_V \nabla \cdot \vec{F}\, dV = \oiint_S \vec{F} \cdot \vec{n}\, dA = 4\pi r^3$

(2)

$$\iiint_V \nabla \cdot \vec{F}\, dV = \iiint_{x^2+y^2+z^2 \le r^2} 3\, dxdydz = 4\pi r^3$$

Example 4.

$\vec{F} = (yz, -1, 1)$, V 为 $z = \sqrt{x^2 + y^2}$ 与 $z = a$ 所围体积, 藉由 Gauss's Theorem

求 $\displaystyle\iiint_V \nabla \cdot \vec{F}\, dV = ?$

【解】

令 $S_1: z = \sqrt{x^2 + y^2},\ z \leq a,\ S_2: z = a,\ x^2 + y^2 \leq a^2$

$\because \vec{F}$ 的各分量一阶偏导数存在且连续

藉由 Gauss's Theorem 则 $\iiint_V \nabla \cdot \vec{F}\, dV = \iint_{S_1} \vec{F} \cdot \vec{n}\, dA + \iint_{S_2} \vec{F} \cdot \vec{n}\, dA$

$\because S_1$ 为朝下定向

令 $\vec{r}(x,y,z) = \left(x, y, (x^2 + y^2)^{\frac{1}{2}}\right)$ 且 $R$ 为曲面 $S$ 投影至 $xy$ 平面的封闭区域

则 $R = \{(x,y): x^2 + y^2 \leq a^2\}$ 且 $\iint_{S_1} \vec{F} \cdot \vec{n}\, dA = -\iint_R \vec{F} \cdot \frac{\partial \vec{r}}{\partial x} \times \frac{\partial \vec{r}}{\partial y}\, dxdy$

$\because \dfrac{\partial \vec{r}}{\partial x} = \left(1, 0, x(x^2 + y^2)^{-\frac{1}{2}}\right)$ 且 $\dfrac{\partial \vec{r}}{\partial y} = \left(0, 1, y(x^2 + y^2)^{-\frac{1}{2}}\right)$

$\therefore \dfrac{\partial \vec{r}}{\partial x} \times \dfrac{\partial \vec{r}}{\partial y} = \left(-x(x^2 + y^2)^{-\frac{1}{2}}, -y(x^2 + y^2)^{-\frac{1}{2}}, 1\right)$

$\Rightarrow \vec{F} \cdot \dfrac{\partial \vec{r}}{\partial x} \times \dfrac{\partial \vec{r}}{\partial y} = (yz, -1, 1) \cdot \left(-x(x^2 + y^2)^{-\frac{1}{2}}, -y(x^2 + y^2)^{-\frac{1}{2}}, 1\right)$

$= -xy + y(x^2 + y^2)^{-\frac{1}{2}} + 1,\ \forall (x,y,z) \in S_1$

$\therefore \iint_{S_1} \vec{F} \cdot \vec{n}\, dA = -\iint_R \vec{F} \cdot \dfrac{\partial \vec{r}}{\partial x} \times \dfrac{\partial \vec{r}}{\partial y}\, dxdy = \iint_{x^2 + y^2 \leq a^2} xy - y(x^2 + y^2)^{-\frac{1}{2}} - 1\, dxdy$

令 $x = r\cos\theta,\ y = r\sin\theta$ 则 $\{(x,y): x^2 + y^2 \leq a^2\} = \{(r,\theta): 0 \leq r \leq a, 0 \leq \theta \leq 2\pi\}$

$\therefore \iint_{x^2 + y^2 \leq a^2} xy - y(x^2 + y^2)^{-\frac{1}{2}} - 1\, dxdy = \int_0^{2\pi} \int_0^a (r^2 \cos\theta \sin\theta - \sin\theta - 1)\, rdrd\theta$

$= -\pi a^2$

令 $S_2: z = a, x^2 + y^2 \leq a^2$

令 $\vec{s}(x,y,z) = (x, y, a)$ 则 $\iint_{S_2} \vec{F} \cdot \vec{n}\, dA = \iint_{S_2} \vec{F} \cdot \dfrac{\partial \vec{s}}{\partial x} \times \dfrac{\partial \vec{s}}{\partial y}\, dxdy$

$$\because \frac{\partial \vec{s}}{\partial x} = (1,0,0) \text{ 且 } \frac{\partial \vec{s}}{\partial y} = (0,1,0) \qquad \therefore \frac{\partial \vec{s}}{\partial x} \times \frac{\partial \vec{s}}{\partial y} = (0,0,1)$$

$$\therefore \iint_{S_2} \vec{F} \cdot \vec{n} \, dA = \iint_{S_2} 1 \, dA = \pi a^2 \quad \therefore \iint_{S} \vec{F} \cdot \vec{n} \, dA = 0$$

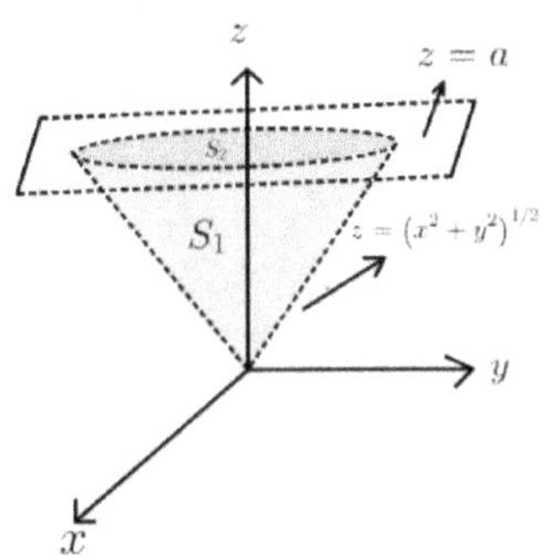


Example 5.

藉由 Gauss's Theorem 求 $\iiint_V \nabla \cdot \vec{F} \, dV =?$，$\vec{F} = (x - y + z, 2x, 1)$，V 为抛物面

$z = x^2 + y^2$，与 $z = 1$ 所围体积

【解】

令 $S_1: z = x^2 + y^2, \ z \leq 1, \ S_2: z = 1, \ x^2 + y^2 \leq 1$

$\because \vec{F}$ 的各分量一阶偏导数存在且连续

藉由 Gauss's Theorem 则 $\iiint_V \nabla \cdot \vec{F} \, dV = \iint_{S_1} \vec{F} \cdot \vec{n} \, dA + \iint_{S_2} \vec{F} \cdot \vec{n} \, dA$

$\because S_1$ 为朝下定向

令 $\vec{r}(x,y,z) = (x, y, x^2 + y^2)$ 且 $R$ 为 $S_1$ 曲面投影至 $xy$ 平面的封闭区域

则 $R = \{(x,y): x^2 + y^2 \leq 1\}$ 且 $\iint_{S_1} \vec{F} \cdot \vec{n} \, dA = -\iint_R \vec{F} \cdot \frac{\partial \vec{r}}{\partial x} \times \frac{\partial \vec{r}}{\partial y} \, dxdy$

$\because \frac{\partial \vec{r}}{\partial x} = (1,0,2x) \text{ 且 } \frac{\partial \vec{r}}{\partial y} = (0,1,2y) \quad \therefore \frac{\partial \vec{r}}{\partial x} \times \frac{\partial \vec{r}}{\partial y} = (-2x, -2y, 1)$

$$\therefore -\vec{F} \cdot \frac{\partial \vec{r}}{\partial x} \times \frac{\partial \vec{r}}{\partial y} = (x - y + z, 2x, 1) \cdot (-2x, -2y, 1) = 2x^2 + 2xy + 2x(x^2 + y^2) - 1,$$

$$\forall (x, y, z) \in S_1$$

$$\therefore \iint_{S_1} \vec{F} \cdot \vec{n}\, dA = -\iint_R \vec{F} \cdot \frac{\partial \vec{r}}{\partial x} \times \frac{\partial \vec{r}}{\partial y}\, dxdy = \iint_{x^2 + y^2 \leq 1} 2x^2 + 2xy + 2x(x^2 + y^2) - 1\, dxdy$$

令 $x = r\cos\theta, y = r\sin\theta$ 则 $\{(x,y): x^2 + y^2 \leq 1\} = \{(r, \theta): 0 \leq r \leq 1, 0 \leq \theta \leq 2\pi\}$

$$\text{且 } dxdy = \left\| \begin{vmatrix} \dfrac{\partial x}{\partial r} & \dfrac{\partial x}{\partial \theta} \\ \dfrac{\partial y}{\partial r} & \dfrac{\partial y}{\partial \theta} \end{vmatrix} \right\| drd\theta = \left\| \begin{vmatrix} \cos\theta & -r\sin\theta \\ \sin\theta & r\cos\theta \end{vmatrix} \right\| drd\theta = r\, drd\theta$$

$$\therefore \iint_{x^2 + y^2 \leq 1} 2x^2 + 2xy + 2x(x^2 + y^2) - 1\, dxdy$$

$$= \int_0^{2\pi} \int_0^1 (2r^2\cos^2\theta + 2r^2\cos\theta\sin\theta + 2r^3\cos\theta - 1) r\, drd\theta = -\frac{\pi}{2}$$

$$\because \iint_{S_2} \vec{F} \cdot \vec{n}\, dA = \iint_{S_2} dA = \pi \quad \therefore \iiint_V \nabla \cdot \vec{F}\, dV = \iint_{S_1} \vec{F} \cdot \vec{n}\, dA + \iint_{S_2} \vec{F} \cdot \vec{n}\, dA = \frac{\pi}{2}$$

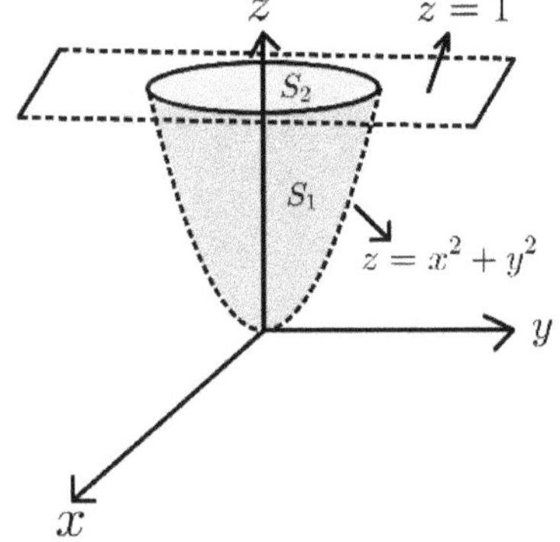


Example 6.

$$求 \iiint_V \nabla \cdot \vec{F}\, dV, \vec{F} = (y, x, z), \ V \text{ 为 } z = 4 - x^2 - y^2, z \geq 0 \text{ 所围封闭区域}$$

　　(1)使用 Gauss's Theorem　(2)直接计算

【解】

(1)

令 $S_1: z = 4 - x^2 - y^2, z \geq 0, \ S_2: x^2 + y^2 \leq 4, \ z = 0$

$\because \vec{F}$ 的各分量一阶偏导数存在且连续

藉由 Gauss's Theorem 则 $\iiint_V \nabla \cdot \vec{F}\, dV = \iint_{S_1} \vec{F} \cdot \vec{n}\, dA + \iint_{S_2} \vec{F} \cdot \vec{n}\, dA$

令 $S_1: z = 4 - x^2 - y^2$, 令 $\vec{r}(x,y,z) = (x, y, 4 - x^2 - y^2)$

$\because S_1$ 为朝上定向

令 $R$ 为 $S_1$ 曲面投影至 $xy$ 平面的封闭区域

则 $R = \{(x,y): x^2 + y^2 \le 4\}$ 且 $\iint_{S_1} \vec{F} \cdot \vec{n}\, dA = \iint_R \vec{F} \cdot \dfrac{\partial \vec{r}}{\partial x} \times \dfrac{\partial \vec{r}}{\partial y}\, dxdy$

$\because \dfrac{\partial \vec{r}}{\partial x} = (1, 0, -2x)$ 且 $\dfrac{\partial \vec{r}}{\partial y} = (0, 1, -2y)$ $\quad \therefore \dfrac{\partial \vec{r}}{\partial x} \times \dfrac{\partial \vec{r}}{\partial y} = (2x, 2y, 1)$

$\therefore \iint_{S_1} \vec{F} \cdot \vec{n}\, dA = \iint_{x^2 + y^2 \le 4} (4xy + z)\, dxdy = \iint_{x^2 + y^2 \le 4} 4xy + 4 - x^2 - y^2\, dxdy$

令 $x = r\cos\theta$, $y = r\sin\theta$ 则 $\{(x,y): x^2 + y^2 \le 4\} = \{(r,\theta): 0 \le r \le 2, 0 \le \theta \le 2\pi\}$

$\therefore \iint_{x^2 + y^2 \le 4} 4xy + 4 - x^2 - y^2\, dxdy$

$= \int_0^{2\pi} \int_0^2 (4r^2 \cos\theta \sin\theta + 4 - r^2)\, r\, dr\, d\theta = 2\pi \left( 2r^2 - \dfrac{r^4}{4} \right) \Big|_{r=0}^{r=2} = 8\pi$

令 $S_2: z = 0, x^2 + y^2 \le 4$, $\quad \because S_2$ 为朝下定向

令 $\vec{s}(x,y,z) = (x, y, 0)$ 则 $\iint_{S_2} \vec{F} \cdot \vec{n}\, dA = -\iint_{S_2} \vec{F} \cdot \dfrac{\partial \vec{s}}{\partial x} \times \dfrac{\partial \vec{s}}{\partial y}\, dxdy$

$\because \dfrac{\partial \vec{s}}{\partial x} = (1, 0, 0)$ 且 $\dfrac{\partial \vec{s}}{\partial y} = (0, 1, 0)$ $\quad \therefore \dfrac{\partial \vec{s}}{\partial x} \times \dfrac{\partial \vec{s}}{\partial y} = (0, 0, 1)$

$\therefore -\iint_{S_2} \vec{F} \cdot \dfrac{\partial \vec{s}}{\partial x} \times \dfrac{\partial \vec{s}}{\partial y}\, dxdy = \iint_{S_2} 0\, dA = 0$

$$\therefore \iiint_V \nabla \cdot \vec{F}\, dV = \iint_{S_1} \vec{F} \cdot \vec{n}\, dA + \iint_{S_2} \vec{F} \cdot \vec{n}\, dA = 8\pi$$

(2)

令 $R = \{(x, y):\ x^2 + y^2 \le 4\}$

则 $\displaystyle\iiint_V \nabla \cdot \vec{F}\, dV = \iint_R \int_0^{4-(x^2+y^2)} dz\, dx\, dy = \iint_R 4 - (x^2 + y^2)\, dx\, dy$

令 $x = r\cos\theta$ , $y = r\sin\theta$  则 $\{(x,y): x^2 + y^2 \le 4\} = \{(r,\theta): 0 \le r \le 2, 0 \le \theta \le 2\pi\}$

$$\therefore \iint_R 4 - (x^2 + y^2)\, dx\, dy = \int_0^{2\pi}\int_0^2 (4 - r^2)\, r\, dr\, d\theta = 2\pi \cdot \left.\frac{-(4-r^2)^2}{4}\right|_0^2 = 8\pi$$

$$\therefore \iiint_V \nabla \cdot \vec{F}\, dV = 8\pi$$

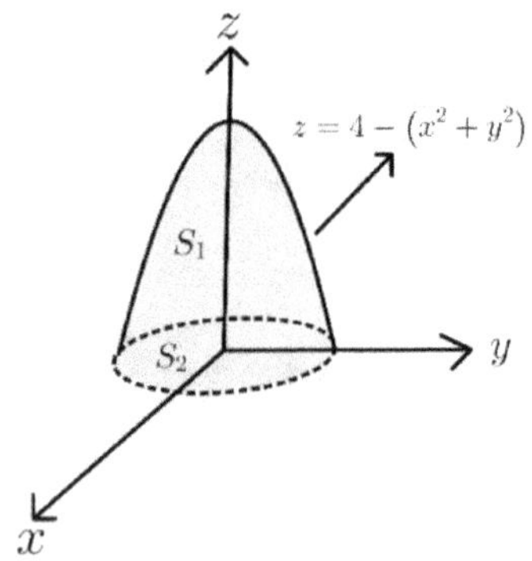


Example 7.

Let $\vec{F} = (x^3 - 3y, 2yz + 1, xyz)$. Assume $V$ is the cube bounded by planes $x = \pm 1$, $y = \pm 1, z = \pm 1$. Find $\displaystyle\iiint_V \nabla \cdot \vec{F}\, dV$ by using Gauss's Theorem.

【解】

令 $S_1: x = 1$, $S_2: x = -1$, $S_3: y = 1$, $S_4: y = -1$, $S_5: z = 1$, $S_6: z = -1$ 且 $S = \bigcup_{i=1}^{6} S_i$

$\because \vec{F}$ 的各分量一阶偏导数存在且连续

藉由 Gauss's Theorem 则 $\iiint_V \nabla \cdot \vec{F}\, dV = \oiint_S \vec{F} \cdot \vec{n}\, dA$

$As\ x = 1, \vec{n} = (1,0,0),\quad \iint_{S_1} \vec{F} \cdot \vec{n}\, dA = \int_{-1}^{1} \int_{-1}^{1} 1 - 3y\, dydz = 4$

$As\ x = -1, \vec{n} = (-1,0,0),\quad \iint_{S_2} \vec{F} \cdot \vec{n}\, dA = -\int_{-1}^{1} \int_{-1}^{1} (-1 - 3y)\, dydz = 4$

$As\ y = 1, \vec{n} = (0,1,0),\quad \iint_{S_3} \vec{F} \cdot \vec{n}\, dA = \int_{-1}^{1} \int_{-1}^{1} 2z + 1\, dxdz = 2$

$As\ y = -1, \vec{n} = (0,-1,0),\quad \iint_{S_4} \vec{F} \cdot \vec{n}\, dA = -\int_{-1}^{1} \int_{-1}^{1} -2z + 1\, dxdz = -2$

$As\ z = 1, \vec{n} = (0,0,1),\quad \iint_{S_5} \vec{F} \cdot \vec{n}\, dA = \int_{-1}^{1} \int_{-1}^{1} xy\, dxdy = 0$

$As\ z = -1, \vec{n} = (0,0,-1),\quad \iint_{S_6} \vec{F} \cdot \vec{n}\, dA = -\int_{-1}^{1} \int_{-1}^{1} xy\, dxdy = 0$

$\therefore \oiint_S \vec{F} \cdot \vec{n}\, dA = 8$

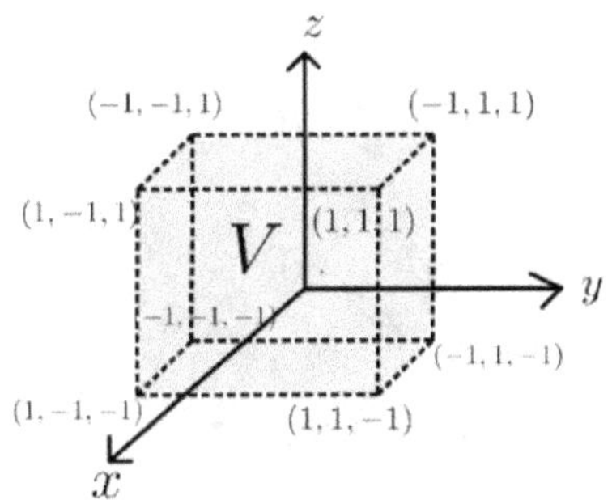


Example 8.

求 $\iiint_V \nabla \cdot \vec{F}\, dV = ?$, $\vec{F} = (z, y, x)$, $V: x^2 + y^2 + z^2 \leq r^2$  (1)使用 Gauss's Theorem

(2) 直接计算

【解】

(1)

令 $S: x^2 + y^2 + z^2 = r^2$

$\because \vec{F}$ 的各分量一阶偏导数存在且连续，藉由 Gauss's Theorem 则 $\iiint_V \nabla \cdot \vec{F}\, dV = \oiint_S \vec{F} \cdot \vec{n}\, dA$

令 $\vec{r}(r, \theta, \varphi) = (r\cos\theta\sin\varphi, r\sin\theta\sin\varphi, r\cos\varphi)$

且 $R = \{(\theta, \varphi): 0 \le \theta \le 2\pi, 0 \le \varphi \le \pi\}$ 则 $\oiint_S \vec{F} \cdot \vec{n}\, dA = \iint_R \vec{F} \cdot \dfrac{\partial \vec{r}}{\partial \varphi} \times \dfrac{\partial \vec{r}}{\partial \theta}\, d\theta d\varphi$

$\because \dfrac{\partial \vec{r}}{\partial \theta} = (-r\sin\theta\sin\varphi, r\cos\theta\sin\varphi, 0)$ 且 $\dfrac{\partial \vec{r}}{\partial \varphi} = (r\cos\theta\cos\varphi, r\sin\theta\cos\varphi, -r\sin\varphi)$

$\therefore \dfrac{\partial \vec{r}}{\partial \varphi} \times \dfrac{\partial \vec{r}}{\partial \theta} = (r^2\cos\theta\sin^2\varphi, r^2\sin\theta\sin^2\varphi, r^2\sin\varphi\cos\varphi)$

$\therefore \vec{F} \cdot \dfrac{\partial \vec{r}}{\partial \varphi} \times \dfrac{\partial \vec{r}}{\partial \theta}$

$= (r\cos\varphi, r\sin\theta\sin\varphi, r\cos\theta\sin\varphi) \cdot (r^2\cos\theta\sin^2\varphi, r^2\sin\theta\sin^2\varphi, r^2\sin\varphi\cos\varphi)$

$= r^3(2\cos\theta\sin^2\varphi\cos\varphi + \sin^3\varphi\sin^2\theta)$

$\therefore \iint_R \vec{F} \cdot \dfrac{\partial \vec{r}}{\partial \varphi} \times \dfrac{\partial \vec{r}}{\partial \theta}\, d\theta d\varphi = r^3 \int_0^{2\pi} \int_0^{\pi} 2\cos\theta\sin^2\varphi\cos\varphi + \sin^3\varphi\sin^2\theta\, d\varphi d\theta$

$\because \int_0^{\pi} \sin^2\varphi\cos\varphi\, d\varphi = \left.\dfrac{\sin^3\varphi}{3}\right|_0^{\pi} = 0 \quad \therefore \int_0^{2\pi} \int_0^{\pi} 2\cos\theta\sin^2\varphi\cos\varphi\, d\varphi d\theta = 0$

$\therefore \int_0^{2\pi} \int_0^{\pi} (2\cos\theta\sin^2\varphi\cos\varphi + \sin^3\varphi\sin^2\theta)\, d\varphi d\theta = \int_0^{2\pi} \int_0^{\pi} \sin^3\varphi\sin^2\theta\, d\varphi d\theta$

$= \int_0^{2\pi} \sin^2\theta\, d\theta \int_0^{\pi} \sin^3\varphi\, d\varphi = \int_0^{2\pi} \left(\dfrac{1 - \cos 2\theta}{2}\right) d\theta \int_0^{\pi} \sin\varphi\,(1 - \cos^2\varphi)\, d\varphi$

$= \pi \int_0^{\pi} \sin\varphi\,(1 - \cos^2\varphi)\, d\varphi$

$\because \int_0^{\pi} \sin\varphi\,(1 - \cos^2\varphi)\, d\varphi = -\cos\varphi\,\big|_0^{\pi} + \left.\dfrac{\cos^3\varphi}{3}\right|_0^{\pi} = \dfrac{4}{3}$

$$\therefore \oiint_S \vec{\mathrm{F}} \cdot \vec{n}\, dA = r^3 \int_0^{2\pi} \int_0^{\pi} (2\cos\theta\sin^2\varphi\cos\varphi + \sin^3\varphi\sin^2\theta)\, d\varphi d\theta = \frac{4\pi r^3}{3}$$

$$\therefore \iiint_V \nabla \cdot \vec{\mathrm{F}}\, dV = \frac{4\pi r^3}{3}$$

(2)

$$\iiint_V \nabla \cdot \vec{\mathrm{F}}\, dV = \iiint_{x^2+y^2+z^2 \leq r^2} 1\, dxdydz = \frac{4\pi r^3}{3}$$

Example 9.

$$\vec{\mathrm{F}} = (0, y, -z), \ \text{假设} \ V \ \text{为抛物面} \ y = x^2 + z^2 \ \text{与} \ y = 1 \ \text{所围体积,} \ \text{求} \iiint_V \nabla \cdot \vec{\mathrm{F}}\, dV$$

(1)使用 Gauss's Theorem　(2) 直接计算

【解】

(1)

令 $S_1: y = x^2 + z^2, 0 \leq y \leq 1,\ S_2:\ x^2 + z^2 \leq 1, y = 1$

$\because \vec{F}$ 的各分量一阶偏导数存在且连续

藉由 Gauss's Theorem 则 $\displaystyle \iiint_V \nabla \cdot \vec{\mathrm{F}}\, dV = \iint_{S_1} \vec{F} \cdot \vec{n}\, dA + \iint_{S_2} \vec{F} \cdot \vec{n}\, dA$

$\because S_1$ 为朝 $y$ 轴负向方向作定向

令 $\vec{r}(x,y,z) = (x, x^2 + z^2, z)$　且 $R$ 为 $S_1$ 曲面投影至 $xz$ 平面的封闭区域

则 $R = \{(x,y): x^2 + z^2 \leq 1\}$ 且 $\displaystyle \iint_{S_1} \vec{F} \cdot \vec{n}\, dA = -\iint_R \vec{F} \cdot \frac{\partial \vec{r}}{\partial x} \times \frac{\partial \vec{r}}{\partial z}\, dxdz$

$\because \dfrac{\partial \vec{r}}{\partial x} = (1, 2x, 0)$ 且 $\dfrac{\partial \vec{r}}{\partial z} = (0, 2z, 1)$　　$\therefore \dfrac{\partial \vec{r}}{\partial x} \times \dfrac{\partial \vec{r}}{\partial z} = (2x, -1, 2z)$

$\therefore \vec{F} \cdot \dfrac{\partial \vec{r}}{\partial x} \times \dfrac{\partial \vec{r}}{\partial z} = (0, y, -z) \cdot (2x, -1, 2z) = -y - 2z^2 = -x^2 - 3z^2,\ \forall (x,y,z) \in S_1$

$\therefore \displaystyle \iint_{S_1} \vec{F} \cdot \vec{n}\, dA = -\iint_R \vec{F} \cdot \frac{\partial \vec{r}}{\partial x} \times \frac{\partial \vec{r}}{\partial z}\, dxdz = \iint_{x^2+z^2 \leq 1} x^2 + 3z^2\, dxdz$

令 $x = r\cos\theta, z = r\sin\theta$ 则 $\{(x,y): x^2 + z^2 \leq 1\} = \{(r,\theta): 0 \leq r \leq 1, 0 \leq \theta \leq 2\pi\}$

$$\text{且 } dxdz = \left\| \begin{vmatrix} \dfrac{\partial x}{\partial r} & \dfrac{\partial x}{\partial \theta} \\[2mm] \dfrac{\partial z}{\partial r} & \dfrac{\partial z}{\partial \theta} \end{vmatrix} \right\| drd\theta = \left\| \begin{vmatrix} \cos\theta & -r\sin\theta \\ \sin\theta & r\cos\theta \end{vmatrix} \right\| drd\theta = rdrd\theta$$

$$\therefore \iint_{x^2+z^2\le 1} x^2 + 3z^2 \, dxdz = \int_0^{2\pi} \int_0^1 r^3\cos^2\theta + 3r^3\sin^2\theta \, drd\theta$$

$$= \int_0^{2\pi} \frac{1}{4}\left(\frac{1+\cos\theta}{2}\right) + \frac{3}{4}\left(\frac{1-\cos\theta}{2}\right) d\theta = \pi$$

$$\text{令 } \vec{s}(x,y,z) = (x,1,z)$$

$$\because \frac{\partial \vec{s}}{\partial x} = (1,0,0) \ \text{且} \ \frac{\partial \vec{s}}{\partial z} = (0,0,1) \qquad \therefore \frac{\partial \vec{r}}{\partial x} \times \frac{\partial \vec{r}}{\partial z} = (0,-1,0)$$

$$\because \iint_{S_2} \vec{F}\cdot\vec{n}\,dA = \iint_{S_2} dA = \iint_{x^2+z^2\le 1} (0,y,-z)\cdot(0,-1,0)dxdz = -\pi$$

$$\therefore \iint_{S_1} \vec{F}\cdot\vec{n}\,dA + \iint_{S_2} \vec{F}\cdot\vec{n}\,dA = 0$$

(2)

$$\oiint_S \vec{F}\cdot\vec{n}\,dA = \iiint_V \nabla\cdot\vec{F}\,dV = \iiint_V 0\,dV = 0$$

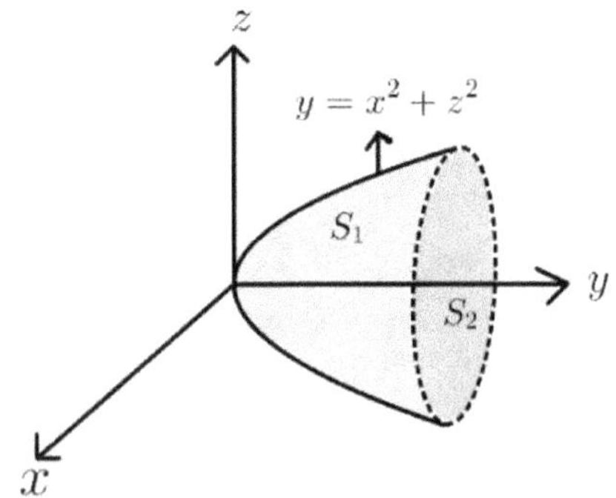


Example 10.

Let $\vec{F} = (x^2, y^2, z^2)$. Assume $V$ is bounded by the solid half-cylinder $0 \le z \le \sqrt{1-y^2}$, and the two disk planes: $x = 0, x = 2$. Find $\displaystyle\iiint_V \nabla\cdot\vec{F}\,dV$ by using Gauss's Theorem.

【解】

令 $S_1: 0 \le z \le \sqrt{1-y^2}, 0 \le x \le 2,$  $S_2:\ y^2 + z^2 \le 1, x = 0,$  $S_3:\ y^2 + z^2 \le 1, x = 2$

$$\text{且} S = \bigcup_{i=1}^{3} S_i$$

$\because \vec{F}$ 的各分量一阶偏导数存在且连续

藉由 Gauss's Theorem 则 $\displaystyle\iiint_V \nabla \cdot \vec{F}\, dV = \sum_{j=1}^{3} \iint_{S_j} \vec{F} \cdot \vec{n}\, dA$

$\because S_1$ 为朝上定向

令 $\vec{r}(x,y,z) = \left(x, y, \sqrt{1-y^2}\right)$ 且 $R = \{(x,y): 0 \le x \le 2, 0 \le y \le 1\}$

则 $\displaystyle\iint_{S_1} \vec{F} \cdot \vec{n}\, dA = \iint_R \vec{F} \cdot \frac{\partial \vec{r}}{\partial x} \times \frac{\partial \vec{r}}{\partial y}\, dxdy$

$\because \dfrac{\partial \vec{r}}{\partial x} = (1,0,0)$ 且 $\dfrac{\partial \vec{r}}{\partial y} = \left(0, 1, -y(1-y^2)^{-\frac{1}{2}}\right)$ $\quad \therefore \dfrac{\partial \vec{r}}{\partial x} \times \dfrac{\partial \vec{r}}{\partial y} = \left(0, -y(1-y^2)^{-\frac{1}{2}}, 1\right)$

$\because \vec{F} \cdot \dfrac{\partial \vec{r}}{\partial x} \times \dfrac{\partial \vec{r}}{\partial y} = (x^2, y^2, 1-y^2) \cdot \left(0, -y(1-y^2)^{-\frac{1}{2}}, 1\right) = -y^3(1-y^2)^{-\frac{1}{2}} + 1 - y^2,$

$\forall (x,y,z) \in S_1$

令 $y = \sin\theta$ 则 $dy = \cos\theta\, d\theta$

$$\iint_R \vec{F} \cdot \frac{\partial \vec{r}}{\partial x} \times \frac{\partial \vec{r}}{\partial y}\, dxdy = \int_0^2 \int_0^1 -y^3(1-y^2)^{-\frac{1}{2}} + 1 - y^2\, dydx$$

$$= 2\int_0^1 -y^3(1-y^2)^{-\frac{1}{2}} + 1 - y^2\, dy = 2\int_0^{\frac{\pi}{2}} -\sin^3\theta + \cos^3\theta\, d\theta$$

$$= 2 \cdot \left( \left(-\cos\theta + \frac{\cos^3\theta}{3}\right)\Big|_0^{\frac{\pi}{2}} + \left(\sin\theta + \frac{\sin^3\theta}{3}\right)\Big|_0^{\frac{\pi}{2}} \right) = \frac{8}{3}$$

$$\iint_{S_2} \vec{F} \cdot \vec{n}\, dA = \iint_{S_2} dA = 0 \quad \text{且} \quad \iint_{S_3} \vec{F} \cdot \vec{n}\, dA = \iint_{S_3} dA = 2\pi$$

$$\iiint_V \nabla \cdot \vec{F}\, dV = \frac{8}{3} + 2\pi$$

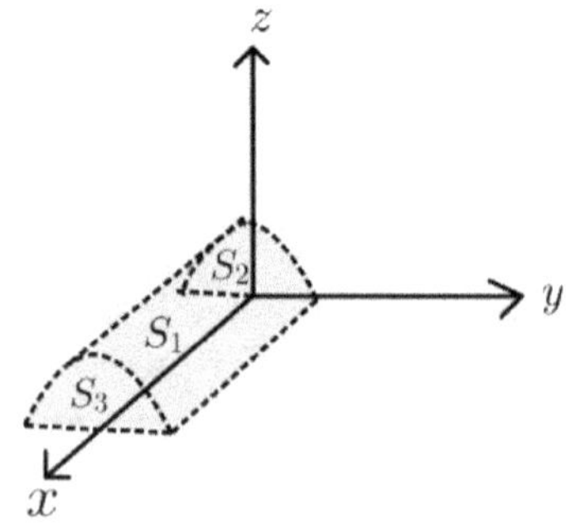

Example 11.

假设 $\vec{F} = (6y^2, 0, 2x^2)$, $V$ 是 $x + y + z = 1$ 在第一卦限的所围体积, 求 $\iiint_V \nabla \cdot \vec{F}\, dV$

(1)直接计算 (2)使用 Gauss's Theorem

【解】

(1)

$\because \nabla \cdot \vec{F} = 0 \quad \therefore \iiint_V \nabla \cdot \vec{F}\, dV = 0$

(2)

令 $S_1: x + y + z = 1, S_2: x = 0, S_3: y = 0, S_4: z = 0$ 且 $S = \bigcup_{j=1}^{4} S_j$

$\because \vec{F}$ 的各分量一阶偏导数存在且连续

藉由 Gauss's Theorem 则

$$\iiint_V \nabla \cdot \vec{F}\, dV = \sum_{j=1}^{4} \iint_{S_j} \vec{F} \cdot \vec{n}\, dA$$

$\because S_1$ 是朝上定向平面

令 $\vec{r}(x, y, z) = (x, y, 1 - x - y)$ 且 $R$ 为 $S_1$ 曲面投影至 $xy$ 平面的封闭区域

则 $R = \{(x, y): x + y \leq 1, x \geq 0, y \geq 0\}$ 且 $\iint_{S_1} \vec{F} \cdot \vec{n}\, dA = \iint_R \vec{F} \cdot \frac{\partial \vec{r}}{\partial x} \times \frac{\partial \vec{r}}{\partial y}\, dxdy$

$\because \frac{\partial \vec{r}}{\partial x} = (1, 0, -1)$ 且 $\frac{\partial \vec{r}}{\partial y} = (0, 1, -1) \quad \therefore \frac{\partial \vec{r}}{\partial x} \times \frac{\partial \vec{r}}{\partial y} = (1, 1, 1)$

$$\because \vec{F} \cdot \frac{\partial \vec{r}}{\partial x} \times \frac{\partial \vec{r}}{\partial y} = (6y^2, 0, 2x^2) \cdot \frac{\partial \vec{r}}{\partial x} \times \frac{\partial \vec{r}}{\partial y} = 2(x^2 + 3y^2), \quad \forall (x, y, z) \in S_1$$

$$\therefore \iint_{S_1} \vec{F} \cdot \vec{n} \, dA = \iint_R \vec{F} \cdot \frac{\partial \vec{r}}{\partial x} \times \frac{\partial \vec{r}}{\partial y} \, dxdy = 2 \iint_R x^2 + 3y^2 \, dxdy = 2 \int_0^1 \int_0^{1-y} x^2 + 3y^2 \, dxdy$$

$$= 2 \int_0^1 \left( \frac{x^3}{3} + 3y^2 x \right) \Bigg|_{x=0}^{x=1-y} dy = 2 \int_0^1 \frac{(1-y)^3}{3} + 3y^2(1-y) \, dy = \frac{2}{3}$$

As $x = 0, n = (-1, 0, 0)$,

$$\iint_{S_2} \vec{F} \cdot \vec{n} \, dA = -\int_0^1 \int_0^{1-z} 6y^2 \, dydz = -2 \int_0^1 (1-z)^3 \, dz = -\frac{1}{2}$$

As $y = 0, n = (0, 1, 0)$, $\quad \iint_{S_3} \vec{F} \cdot \vec{n} \, dA = 0$

As $z = 0, n = (0, 0, -1)$,

$$\iint_{S_4} \vec{F} \cdot \vec{n} \, dA = -\int_0^1 \int_0^{1-x} 2x^2 \, dydx = -\frac{1}{6}$$

$$\therefore \sum_{j=1}^4 \iint_{S_j} \vec{F} \cdot \vec{n} \, dA = 0$$

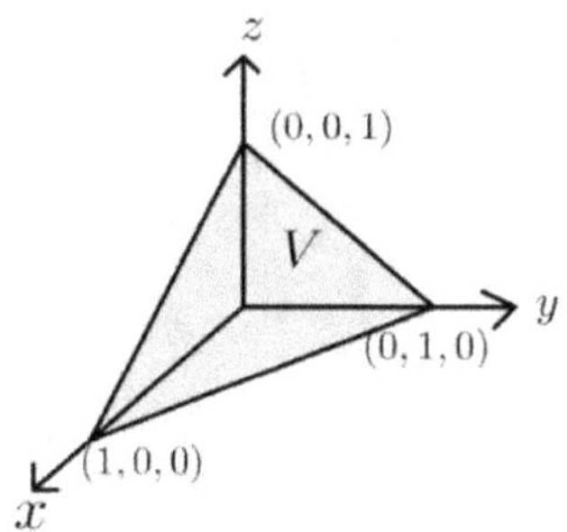


Example 12.

假设 $\vec{F} = (x, y, z)$, V 是 $(a, 0, 0), (0, a, 0), (0, 0, a)$ 与坐标平面所围体积, 求 $\iiint_V \nabla \cdot \vec{F} \, dV$

(1)直接计算 (2)使用 Gauss's Theorem

【解】

(1)

$$\iiint_V \nabla \cdot \vec{F}\, dV = \iiint_V 3\, dV = \frac{a^3}{2}$$

(2)

令 $S_1: x + y + z = a, S_2: x = 0, S_3: y = 0, S_4: z = 0$ 且 $S = \bigcup_{j=1}^{4} S_j$

$\because \vec{F}$ 的各分量一阶偏导数存在且连续

藉由 Gauss's Theorem 则 $\displaystyle \iiint_V \nabla \cdot \vec{F}\, dV = \sum_{j=1}^{4} \iint_{S_j} \vec{F} \cdot \vec{n}\, dA$

$\because S_1$ 为朝上定向

令 $\vec{r}(x, y, z) = (x, y, a - x - y)$ 且 $R = \{(x, y): 0 \leq x \leq a, 0 \leq y \leq a - x\}$

则 $\displaystyle \iint_{S_1} \vec{F} \cdot \vec{n}\, dA = \iint_R \vec{F} \cdot \frac{\partial \vec{r}}{\partial x} \times \frac{\partial \vec{r}}{\partial y}\, dxdy$

$\because \dfrac{\partial \vec{r}}{\partial x} = (1, 0, -1)$ 且 $\dfrac{\partial \vec{r}}{\partial y} = (0, 1, -1)$   $\therefore \dfrac{\partial \vec{r}}{\partial x} \times \dfrac{\partial \vec{r}}{\partial y} = (1, 1, 1)$

$\because \vec{F} \cdot \dfrac{\partial \vec{r}}{\partial x} \times \dfrac{\partial \vec{r}}{\partial y} = (x, y, z) \cdot (1, 1, 1) = x + y + z = x + y + a - x - y = a, \ \ \forall (x, y, z) \in S_1$

$$\iint_R \vec{F} \cdot \frac{\partial \vec{r}}{\partial x} \times \frac{\partial \vec{r}}{\partial y}\, dxdy = \int_0^a \int_0^{a-x} a\, dydx = a \int_0^a a - x\, dx = a \left( ax - \frac{x^2}{2} \right) \Big|_0^a = \frac{a^3}{2}$$

As $x = 0, n = (-1, 0, 0),$ $\displaystyle \iint_{S_2} \vec{F} \cdot \vec{n}\, dA = -\int_0^1 \int_0^{1-z} x\, dydz = 0$

As $y = 0, n = (0, 1, 0),$ $\displaystyle \iint_{S_3} \vec{F} \cdot \vec{n}\, dA = -\int_0^1 \int_0^{1-z} y\, dxdz = 0$

As $z = 0, n = (0, 0, -1),$ $\displaystyle \iint_{S_4} \vec{F} \cdot \vec{n}\, dA = -\int_0^1 \int_0^{1-x} z\, dxdy = 0$

$$\therefore \sum_{j=1}^{4} \iint_{S_j} \vec{F} \cdot \vec{n}\, dA = \frac{a^3}{2}$$

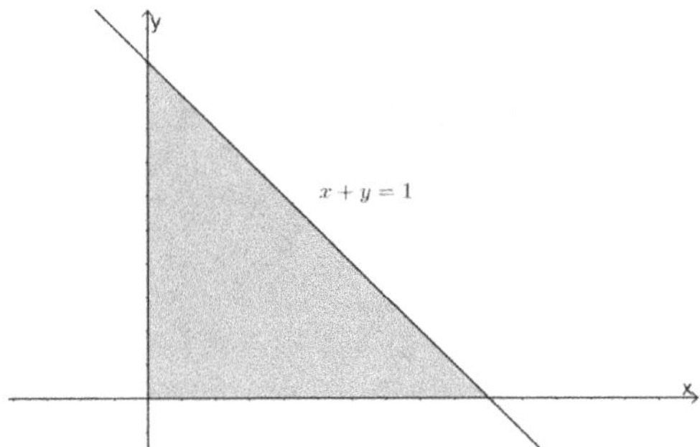

Example 13.

假设 $\vec{F} = (e^y, e^x + 6y, 12y)$, $V$ 是 $x + y + z = 6$ 在第一卦限所围体积, 求 $\iiint_V \nabla \cdot \vec{F}\, dV$

(1)直接计算 (2)使用 Gauss's Theorem

【解】

(1)

$$\iiint_V \nabla \cdot \vec{F}\, dV = \iiint_V 6\, dV = 6 \cdot \frac{1}{3} \cdot \frac{1}{2} \cdot 6 \cdot 6 \cdot 6 = 216$$

(2)

令 $S_1: x + y + z = 6, S_2: x = 0, S_3: y = 0, S_4: z = 0$ 且 $S = \bigcup_{j=1}^{4} S_j$

$\because \vec{F}$ 的各分量一阶偏导数存在且连续

藉由 Gauss's Theorem 则 $\displaystyle\iiint_V \nabla \cdot \vec{F}\, dV = \sum_{j=1}^{4} \iint_{S_j} \vec{F} \cdot \vec{n}\, dA$

$\because S_1$ 为朝上定向

令 $\vec{r}(x, y, z) = (x, y, 6 - x - y)$ 且 $R$ 为 $S_1$ 曲面投影至 $xy$ 平面的封闭区域

则 $R = \{(x, y): x + y \leq 6, x \geq 0, y \geq 0\}$ 且 $\displaystyle\iint_{S_1} \vec{F} \cdot \vec{n}\, dA = \iint_R \vec{F} \cdot \frac{\partial \vec{r}}{\partial x} \times \frac{\partial \vec{r}}{\partial y}\, dxdy$

$\because \dfrac{\partial \vec{r}}{\partial x} = (1, 0, -1)$ 且 $\dfrac{\partial \vec{r}}{\partial y} = (0, 1, -1)$  $\therefore \dfrac{\partial \vec{r}}{\partial x} \times \dfrac{\partial \vec{r}}{\partial y} = (1, 1, 1)$

$$\because \vec{F} \cdot \frac{\partial \vec{r}}{\partial x} \times \frac{\partial \vec{r}}{\partial y} = (e^y, e^x + 6y, 12y) \cdot (1,1,1) = e^y + e^x + 18y, \quad \forall (x, y, z) \in S_1$$

$$\therefore \iint_{S_1} \vec{F} \cdot \vec{n} \, dA = \iint_R \vec{F} \cdot \frac{\partial \vec{r}}{\partial x} \times \frac{\partial \vec{r}}{\partial y} \, dxdy = \iint_R e^y + e^x + 18y \, dxdy$$

$$= \int_0^6 \int_0^{6-x} e^y + e^x + 18y \, dydx = 634 + 2e^6$$

As $x = 0, n = (-1,0,0)$,

$$\iint_{S_2} \vec{F} \cdot \vec{n} \, dA = -\int_0^6 \int_0^{6-z} e^y \, dydz = -\int_0^6 e^{6-z} - 1 \, dz = 7 - e^6$$

As $y = 0, n = (0,-1,0)$,

$$\iint_{S_3} \vec{F} \cdot \vec{n} \, dA = -\int_0^6 \int_0^{6-z} e^x \, dxdz = -\int_0^6 e^{6-z} - 1 \, dz = 7 - e^6$$

As $z = 0, n = (0,0,-1)$,

$$\iint_{S_4} \vec{F} \cdot \vec{n} \, dA = -\int_0^6 \int_0^{6-x} 12y \, dydx = -432$$

$$\therefore \sum_{j=1}^4 \iint_{S_j} \vec{F} \cdot \vec{n} \, dA = 216$$

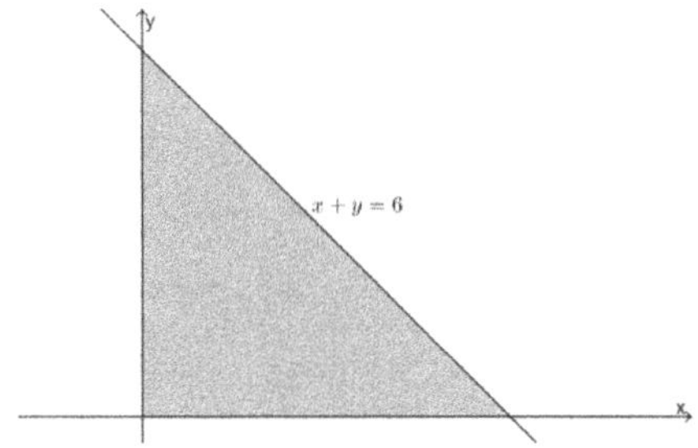


Example 14.

$$\vec{F} = (0, xz, -xy). \quad \text{V 为} z = xy, 0 \le x \le 1, 0 \le y \le 2 \text{ 所围体积,} \quad \text{求} \iiint_V \nabla \cdot \vec{F} \, dV$$

(1)直接计算 (2)使用 Gauss's Theorem

【解】

(1)

$$\because \nabla \cdot \vec{F} = 0 \quad \therefore \iiint_V \nabla \cdot \vec{F}\, dV = 0$$

(2)

$$令 S_1\colon z = xy, S_2\colon x = 1, S_3\colon y = 2, S_4\colon z = 0 \text{ 且 } S = \bigcup_{j=1}^{4} S_j$$

$\because \vec{F}$ 的各分量一阶偏导数存在且连续

藉由 Gauss's Theorem 则 
$$\iiint_V \nabla \cdot \vec{F}\, dV = \sum_{j=1}^{4} \iint_{S_j} \vec{F} \cdot \vec{n}\, dA$$

$\because S_1$ 为朝上定向

令 $\vec{r}(x, y, z) = (x, y, xy)$ 且 $R = \{(x, y)\colon 0 \le x \le 1, 0 \le y \le 2\}$

则 
$$\iint_{S_1} \vec{F} \cdot \vec{n}\, dA = \iint_R \vec{F} \cdot \frac{\partial \vec{r}}{\partial x} \times \frac{\partial \vec{r}}{\partial y}\, dxdy$$

$$\because \frac{\partial \vec{r}}{\partial x} = (1, 0, y) \quad 且 \quad \frac{\partial \vec{r}}{\partial y} = (0, 1, x) \quad \therefore \frac{\partial \vec{r}}{\partial x} \times \frac{\partial \vec{r}}{\partial y} = (-y, -x, 1)$$

$$\because \vec{F} \cdot \frac{\partial \vec{r}}{\partial x} \times \frac{\partial \vec{r}}{\partial y} = (0, xz, -xy) \cdot (-y, -x, 1) = -x^2 z - xy = -x^3 y - xy, \quad \forall (x, y, z) \in S_1$$

$$\iint_R \vec{F} \cdot \frac{\partial \vec{r}}{\partial x} \times \frac{\partial \vec{r}}{\partial y}\, dxdy = \int_0^1 \int_0^2 -x^3 y - xy\, dydx = \int_0^a -2x^3 - 2x\, dx = \left( -\frac{x^4}{2} - x^2 \right)\Big|_0^1 = -\frac{3}{2}$$

$$\text{As } x = 1, n = (1, 0, 0), \quad \iint_{S_2} \vec{F} \cdot \vec{n}\, dA = 0$$

$$\text{As } y = 2, n = (0, 1, 0), \quad \iint_{S_3} \vec{F} \cdot \vec{n}\, dA = \int_0^6 \int_0^{2x} xz\, dzdx = \frac{1}{2}$$

$$\text{As } z = 0, n = (0, 0, -1), \quad \iint_{S_4} \vec{F} \cdot \vec{n}\, dA = \int_0^2 \int_0^1 xy\, dxdy = 1$$

$$\therefore \iiint_V \nabla \cdot \vec{F}\, dV = \sum_{j=1}^{4} \iint_{S_j} \vec{F} \cdot \vec{n}\, dA = 0$$